Important Formulas

Quadratic Formula

If $ax^2 + bx + c = 0$, for $a \neq 0$, then

$$x = \frac{-b \pm \sqrt{b^2 - 4ac}}{2a}$$

Distance Formula

The distance d between $P_1 = (x_1, y_1)$ and $P_2 = (x_2, y_2)$ is

$$d = \sqrt{(x_2 - x_1)^2 + (y_2 - y_1)^2}$$

Slope Formula

The slope m of the line segment $\overline{AB}$, where $A = (x_1, y_1)$ and $B = (x_2, y_2)$ is

$$m = \frac{y_2 - y_1}{x_2 - x_1}$$

Geometry

Assume A = area, C = circumference, V = volume, S = surface area, r = radius, h = altitude, l = length, w = width, b (or a) = length of a base, and s = length of a side.

1	Square	$A = s^2$
2	Rectangle	$A = lw$
3	Parallelogram	$A = bh$
4	Triangle	$A = \frac{1}{2}bh$
5	Circle	$A = \pi r^2$; $C = 2\pi r$
6	Trapezoid	$A = \frac{1}{2}(a + b)h$
7	Cube	$S = 6s^2$; $V = s^3$
8	Rectangular Box	$S = 2(lw + wh + lh)$; $V = lwh$
9	Cylinder	$S = 2\pi rh$; $V = \pi r^2 h$
10	Sphere	$S = 4\pi r^2$; $V = \frac{4}{3}\pi r^3$
11	Cone	$S = \pi r\sqrt{r^2 + h^2}$; $V = \frac{1}{3}\pi r^2 h$

The Greek Alphabet

Alpha	Α	α	Nu	Ν	ν
Beta	Β	β	Xi	Ξ	ξ
Gamma	Γ	γ	Omicron	Ο	o
Delta	Δ	δ	Pi	Π	π
Epsilon	Ε	ϵ	Rho	Ρ	ρ
Zeta	Ζ	ζ	Sigma	Σ	σ
Eta	Η	η	Tau	Τ	τ
Theta	Θ	θ	Upsilon	ϒ	υ
Iota	Ι	ι	Phi	Φ	ϕ
Kappa	Κ	κ	Chi	Χ	χ
Lambda	Λ	λ	Psi	Ψ	ψ
Mu	Μ	μ	Omega	Ω	ω

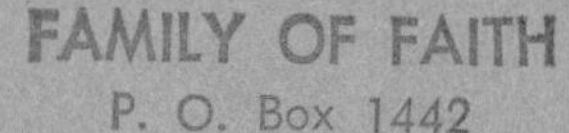

College Algebra

THIRD EDITION

College Algebra

WITH APPLICATIONS

M. A. Munem

MACOMB COLLEGE

D. J. Foulis

UNIVERSITY OF MASSACHUSETTS

Worth Publishers

College Algebra WITH APPLICATIONS, THIRD EDITION

Printed in the United States of America

Library of Congress Catalog Card Number: 90-71761

ISBN: 0-87901-499-7

Printing: 1 2 3 4 5–95 94 93 92 91

Printed on Recycled Paper. ♻

Design ▪ Malcolm Grear Designers

Art Director ▪ George Touloumes

Production Supervisor ▪ Stacey B. Alexander

Composition ▪ Syntax International

Printing and Binding ▪ R. R. Donnelley & Sons Company

Cover ▪ John Giannicchi/Science Source/Photo Researchers, Inc.

Worth Publishers

33 Irving Place

New York, New York 10003

Contents

Preface

The third edition of *College Algebra* is being published at the beginning of a new era in the teaching of mathematics in our colleges and universities. Educators agree that this decade will see major changes in the way mathematics is taught—critical thinking will be nurtured, rote learning will be de-emphasized, and mathematical comprehension will be fostered by increased classroom use of electronic computers and calculators. In revising our textbook, we have tried to provide a measured response to these trends while continuing to offer students a straightforward, readable book.

Prerequisites

In writing and revising this textbook, we had in mind a typical reader with the equivalent of two years of college-preparatory mathematics in algebra and plane geometry, or one who has taken a college-level course in introductory algebra. Determined students with less preparation should be able to use the textbook successfully, particularly if they supplement it with the accompanying *Student Guide with Solutions.*

Presentation

Topics are presented in brief sections that develop logically from basic to more advanced skills and concepts. Motivation for new ideas is provided by showing their application in real-world situations. Numerous illustrative examples are worked out in detail. When appropriate, specific problem-solving procedures are given.

Problems

Problems at the end of each section begin with simple drill-type exercises to build student confidence. A gradual progression to more advanced problems invites students to work to the best of their abilities.

Odd-numbered problems Many of the odd-numbered problems, particularly those at the beginning of each problem set, are similar in scope to the worked-out examples in the text. Answers to most of these problems, with appropriate graphs, are given in the back of the book. In general, *odd-numbered problems can be assigned for homework with an expectation of success by most students.*

Even-numbered problems Some of the even-numbered problems, particularly those toward the end of each problem set, are considerably more challenging than the odd-numbered ones. These problems are meant to encourage critical thinking and to confirm mastery of mathematical concepts and techniques. *Even-numbered problems toward the ends of the problem sets should be assigned with care to avoid demoralizing students who are not prepared to handle them.*

Colored problem numbers Problems with red numbers constitute a good representation of the main ideas of each section and thus could provide a suitable basic set for homework assignments. Instructors using these problems for homework may wish to augment the basic set with a few carefully selected even-numbered problems.

Calculator problems Problems for which the use of a calculator is appropriate are marked with the symbol [c]. *In many of these problems we have purposely avoided using "round numbers" so students will be prepared to work with numbers that actually arise in applied mathematics.* In particular, this affords an opportunity for students to practice using correct rounding-off techniques.

Review problem sets The review problems at the end of each chapter can be used in a variety of ways: Instructors may wish to use them for supplementary or extra-credit assignments or as a source of problems for quizzes and exams. Students may wish to use them to pinpoint areas where further study is needed. Often, these problems are not arranged by section so that students can practice recognizing types of problems without using the placement of the problem in the problem set as a clue.

Major Changes in this Edition

In keeping with the current trend toward de-emphasizing rote learning and fostering understanding and critical thinking, we have made a slight reduction in the number of drill problems and added a few more problems that test for comprehension. In the same spirit, we have rewritten and reorganized textual material to serve these ends. We are placing even more stress on the important idea of a mathematical model, and thus have moved the section on ratio, proportion, and variation up to Section 3 of Chapter 1. Other significant changes are as follows:

Calculators We have moved the introductory material on the use of calculators up to Section 2 of Chapter 1 and made extensive use of calculators in examples and problems throughout the book. We assume that *all* students using the book have access to a basic scientific calculator.

Graphing calculators Although we have indicated how graphing calculators can be used to facilitate the solution of problems, *the use of a graphing calculator is optional* in this edition. Examples and problems for which it would be reasonable to use a graphing calculator are marked with the symbol [gc]

Numerical methods The discussion of error involved in numerical calculations (pages 13 to 16) is now supplemented by a consideration of absolute, relative, and percent error (pages 129 to 130). More emphasis has been given to the bisection method for finding zeros (pages 263 to 265), and material on the use of function iteration and fixed points has been added (pages 130, 203 to 204, and 265 to 266).

Applications More applications of exponential functions to car and mortgage payments are now given in Section 5.1, and the idea of mathematical expectation is applied to games of chance in Section 7.6. (Consult the *Index of Applications*, pages 497 to 500, to locate these and other applied examples and problems.)

Chapter tests New to this edition are brief tests at the end of each chapter. Students can use these for practice and review and to determine readiness for class tests. All answers for chapter tests are provided at the back of the book.

Preparation for calculus More examples and problems have been added pertaining to algebraic ideas that are used in calculus.

Available Student Aids

Student Guide with Solutions The student guide has been thoroughly revised. It now includes worked-out solutions to every other odd-numbered problem in the textbook as well as supplementary problems in a tutorial format for each chapter. In addition, there are two practice tests for each chapter—a multiple-choice test and a problem-solving test. All answers and solutions to the tutorial problems and to the chapter tests are provided in the guide itself.

Graphics Discoveries Alice M. Kaseberg, Steven L. Myers, and Robert B. Thompson (Lane Community College) have prepared a guide that will assist students in exploring key topics in algebra using currently available graphing calculators.

Computer Software Robert J. Weaver of Mount Holyoke College has prepared instructional software on graphing polynomial functions, using the division algorithm, approximating zeros of polynomial functions using the bisection method, and solving general triangles. His disk, for use with IBM and compatible computers, is available to adopters on request.

Also, the *DERIVE*® computer software system is available at a special discount price to teachers and students in courses where this book has been adopted. For information regarding *DERIVE*, please call 1-800-255-2468.

Available Instructor Aids

Solutions Manual Step-by-step solutions to all problems in the textbook are available in this manual, prepared by Hyla Gold Foulis. A glance at the worked-out solutions will help the instructor to select homework problems at an appropriate level of difficulty for his or her class.

Test Bank For each of the seven chapters in *College Algebra*, the *Test Bank* provides instructors with 25 to 30 multi-version base, or template, questions. These questions are keyed to the learning objectives from the student's guide and cover the main ideas in each chapter. Approximately two-thirds of the 25 to 30 base

DERIVE is a registered trademark of Soft Warehouse, Inc.

questions are problem-solving questions, and the remainder are multiple-choice. Each of the base questions appears in six versions—three of them easy to medium and the other three medium to difficult—for a total of 150 to 180 questions per chapter.

Several versions of a computerized test-generation system make exam preparation quick and easy.

Test Manual Drawing on the questions in the *Test Bank*, the *Test Manual* provides six ready-made tests for each chapter. Four of these are short-answer tests—two easy to medium and two medium to difficult—and the remaining two are multiple-choice—one easy to medium and one medium to difficult. The tests are intended for use in a standard fifty-minute class.

The Video Tutor and The Video Tutor Learning Guides About twenty-five hours of videotaped lectures (in 15-minute segments) accompanied by learning guides for students have been prepared by A. E. T. Bentley (Capilano College). These offer detailed coverage of all topics in Chapters 1 to 5.

Overhead Transparencies A set of two-color acetate transparencies is available for use on an overhead projector. These include important figures in the textbook and statements of key definitions and theorems for use in lectures.

Acknowledgments

In revising our textbook, we have again drawn on our own experiences in teaching algebra and on feedback from our students. Especially valuable were the suggestions from other instructors who used the book and the thoughtful comments of our reviewers. We wish to thank all of these people and, in particular, to express our gratitude to the following:

James E. Bright
Clayton State College

Laura Cameron
University of New Mexico

Ray E. Collings
Tri-County Technical College

Johanna Danos
Windward Community College

Marcia Drost
Indiana University - Kokomo

Charles S. Johnson
Los Angeles Valley College

Karla V. Neal
Louisiana State University

Vera Preston
Austin Community College

Dr. Ralph M. Selig
Fairleigh Dickinson University

Henry Mark Smith
University of New Orleans

James M. Sullivan
Sierra College

Robert B. Thompson
Lane Community College

We are deeply indebted to Professor Wayne Hille of Wayne State University for assisting in the preparation of the *Student Guide with Solutions* and for proofreading the *Solutions Manual*. Special thanks are due to Professors Steve Fasbinder and

Victoria Ackerman of Macomb College who assisted in the proofreading of the book and the *Student Guide with Solutions*, and to Hyla Gold Foulis for reviewing each successive stage of the manuscript, proofreading pages, and preparing the *Solutions Manual* for the book. We also thank the people at Worth Publishers for their assistance.

M. A. Munem
D. J. Foulis

A Note to the Instructor on the Use of Calculators

Problems and examples for which the use of a calculator is recommended are marked with the symbol [c]. Answers and solutions were obtained using an HP-32S calculator—other calculators that use different internal routines may give slightly different results. In some cases we have carried out computations to the full number of significant digits available on our calculator. This will permit students to check their own calculator work and to see for themselves that different calculators may give answers that differ in the last decimal place or so. Instructors should alert their students to this possibility.

Problems for which the (optional) use of a graphing calculator is reasonable are marked with the symbol [gc]. Instructors who require the use of a graphing calculator may wish to have their students use the *Graphics Discoveries* by Alice M. Kaseberg, Steven L. Myers, and Robert B. Thompson as a supplement to this textbook.

Instructors should stress that it is undesirable to use a calculator (or graphing calculator) for problems that are *not* marked with the symbol [c] (or [gc]). The use of a calculator (or graphing calculator) for such problems can actually hinder the student's understanding.

The rule for rounding off numbers presented in Section 1.2 is consistent with the operation of most calculators with round-off capability. Some instructors may wish to mention the popular alternative round-off rule: If the first dropped digit is 5 and there are no nonzero digits to its right, round off so that the last retained digit is even.

Instructors should encourage their students to learn to use calculators efficiently—for instance, to do chain calculations using the memory features of the calculator. In some of our examples, we have shown the intermediate results of chain calculations so the students can check their calculator work; however, it should be emphasized that it is not necessary to write down these intermediate results when using the calculator.

Because there are so many different calculators available, we have not given detailed key-stroke instructions for calculator operation in this textbook. Students should be encouraged to consult the instruction manuals furnished with their calculators. This is particularly important in connection with the use of Horner's method for evaluating polynomials (page 248).

Finally, we have made no attempt to provide a systematic discussion of the inaccuracies inherent in computations with a calculator. An excellent account of calculator inaccuracy can be found in "Calculator Calculus and Roundoff Errors" by George Miel in *The American Mathematical Monthly*, 1980, vol. 87, pp. 243–252.

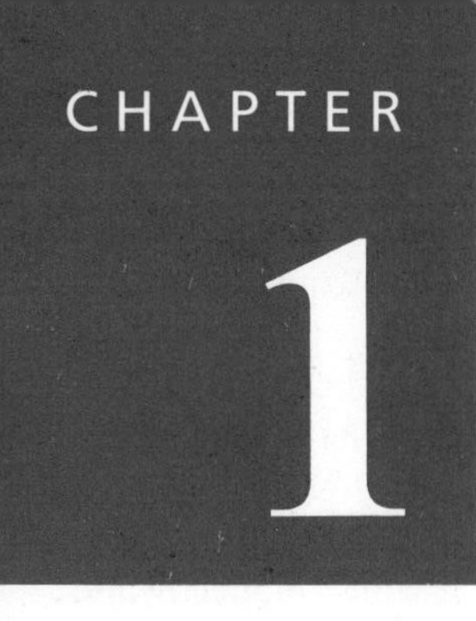

CHAPTER 1 Concepts of Algebra

The gravitational forces that govern the motions of the planets can be calculated algebraically by using Newton's law of universal gravitation.

This chapter is designed as a review of the basic concepts and methods of algebra. Its purpose is to help you attain the algebraic skills that are required throughout the textbook. Topics covered include the language and symbols of algebra, polynomials, fractions, exponents, radicals, and complex numbers.

1.1 The Algebra of Real Numbers

Algebra begins with a systematic study of the operations and rules of arithmetic. The operations of addition, subtraction, multiplication, and division serve as a basis for all arithmetic calculations. In order to achieve generality, letters of the alphabet are used in algebra to represent numbers. A letter such as x, y, a, or b can stand for a particular number (known or unknown), or it can stand for any number at all. A letter that represents an arbitrary number is called a **variable.**

The sum, difference, product, and quotient of two numbers, x and y, can be written as

$$x + y, \qquad x - y, \qquad x \times y, \qquad \text{and} \qquad x \div y.$$

In algebra, the notation $x \times y$ for the product of x and y is not often used because of the possible confusion of the letter x with the multiplication sign $\times$. The preferred notation is $x \cdot y$ or simply xy. Similarly, the notation $x \div y$ is usually avoided in favor of the fraction $\frac{x}{y}$ or x/y.

Algebraic notation—the "shorthand" of mathematics—is designed to clarify ideas and simplify calculations by permitting us to write expressions compactly and efficiently. For instance, $x + x + x + x + x$ can be written simply as $5x$. The use of exponents provides an economy of notation for products; for instance, $x \cdot x$ can be written simply as x^2 and $x \cdot x \cdot x$ as x^3. In general, if n is a positive integer,

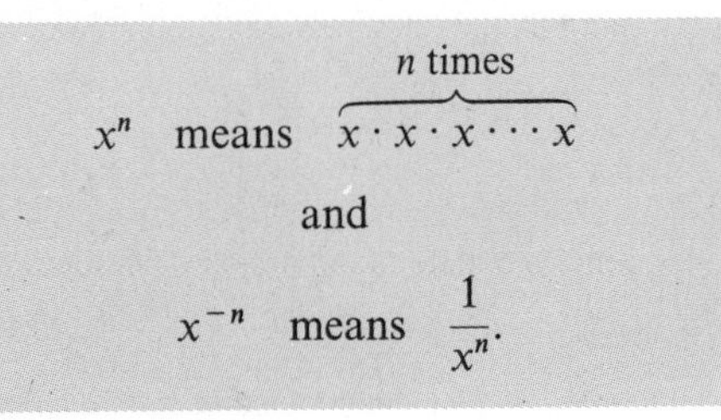

$$x^n \text{ means } \overbrace{x \cdot x \cdot x \cdots x}^{n \text{ times}}$$

and

$$x^{-n} \text{ means } \frac{1}{x^n}.$$

In using the **exponential notation x^n**, we refer to x as the **base** and n as the **exponent,** or the **power** to which the base is raised. When the exponent is negative, we must assume that the base is nonzero to avoid zero in the denominator.

By writing an **equals sign** (=) between two algebraic expressions, we obtain an **equation,** or **formula,** stating that the two expressions represent the same number. Using equations and formulas, we can express mathematical facts in compact, easily remembered forms. Formulas are used to express relationships among various quantities in such fields as geometry, physics, engineering, statistics, geology, business, medicine, economics, and the life sciences. Calculating the numerical value expressed by a formula when particular numbers are assigned to letters is known as **evaluation.**

Figure 1

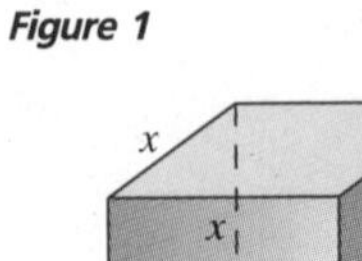

Example 1 **(a)** Write a formula for the volume V of a cube that has edges of length x units (Figure 1).

[C] **(b)*** Evaluate V when $x = 5.23$ centimeters.

Solution **(a)** $V = x \cdot x \cdot x = x^3$ cubic centimeters.

(b) When $x = 5.23$ centimeters, $V = (5.23)^3 = 143.055667$ cubic centimeters. ■

Example 2 A certain type of living cell divides every hour. Starting with one such cell in a culture, the number N of cells present at the end of t hours is given by the formula $N = 2^t$. Find the number of cells in the culture after 6 hours.

* Problems for which the use of a calculator is suggested are marked with the symbol [C] Recommendations for the use of a calculator are given in Section 1.2.

Solution Substituting $t = 6$ in the formula $N = 2^t$, we find that

$$N = 2^6 = 2 \cdot 2 \cdot 2 \cdot 2 \cdot 2 \cdot 2 = 64 \text{ cells.}$$

■

Basic Algebraic Properties of Real Numbers

The numbers used to measure real-world quantities such as length, area, volume, speed, electrical charge, efficiency, probability of rain, intensity of earthquakes, profit, body temperature, gross national product, growth rate, and so forth, are called **real numbers.** They include such numbers as

$$5, \quad -17, \quad \frac{17}{13}, \quad -\frac{2}{3}, \quad 0, \quad 2.71828, \quad \sqrt{2}, \quad -\frac{\sqrt{3}}{2}, \quad 3 \times 10^8, \qquad \text{and} \qquad \pi.$$

The basic algebraic properties of the real numbers can be expressed in terms of the two fundamental operations of addition and multiplication.

Basic Algebraic Properties of Real Numbers

Let a, b, and c denote real numbers.

1. *The Commutative Properties*

(i) $a + b = b + a$ **(ii)** $a \cdot b = b \cdot a$

The commutative properties say that the *order* in which we either add or multiply real numbers doesn't matter.

2. *The Associative Properties*

(i) $a + (b + c) = (a + b) + c$ **(ii)** $a \cdot (b \cdot c) = (a \cdot b) \cdot c$

The associative properties tell us that the way real numbers are *grouped* when they are either added or multiplied doesn't matter. Because of the associative properties, expressions such as $a + b + c$ and $a \cdot b \cdot c$ make sense without parentheses.

3. *The Distributive Properties*

(i) $a \cdot (b + c) = a \cdot b + a \cdot c$ **(ii)** $(b + c) \cdot a = b \cdot a + c \cdot a$

The distributive properties can be used to expand a product into a sum, such as

$$a(b + c + d) = ab + ac + ad,$$

or the other way around, to rewrite a sum as a product:

$$ax + bx + cx + dx + ex = (a + b + c + d + e)x.$$

4. *The Identity Properties*

(i) $a + 0 = 0 + a = a$ **(ii)** $a \cdot 1 = 1 \cdot a = a$

We call 0 the **additive identity** and 1 the **multiplicative identity** for the real numbers.

5. *The Inverse Properties*

(i) For each real number a, there is a real number $-a$, called the **additive inverse** of a, such that

$$a + (-a) = (-a) + a = 0.$$

(ii) For each real number $a \neq 0$, there is a real number $1/a$, called the **multiplicative inverse** of a, such that

$$a \cdot \frac{1}{a} = \frac{1}{a} \cdot a = 1.$$

Although the additive inverse of a, namely $-a$, is usually called the **negative** of a, you must be careful because *$-a$ isn't necessarily a negative number.* For instance, if $a = -2$, then $-a = -(-2) = 2$. Notice that *the multiplicative inverse $1/a$ is assumed to exist only if $a \neq 0$.* The real number $1/a$ is also called the **reciprocal** of a and is often written as a^{-1}.

Example 3 State one basic algebraic property of the real numbers to justify each statement.

(a) $7 + (-2) = (-2) + 7$

(b) $x + (3 + y) = (x + 3) + y$

(c) $a + (b + c)d = a + d(b + c)$

(d) $x[y + (z + w)] = xy + x(z + w)$

(e) $(x + y) + [-(x + y)] = 0$

(f) $(x + y) \cdot 1 = x + y$

(g) If $x + y \neq 0$, then $(x + y)[1/(x + y)] = 1$

Solution

(a) Commutative property for addition

(b) Associative property for addition

(c) Commutative property for multiplication

(d) Distributive property

(e) Additive inverse property

(f) Multiplicative identity property

(g) Multiplicative inverse property ■

Many of the important properties of the real numbers can be *derived* as results of the basic properties, although we shall not do so here. Among the more important **derived properties** are the following.

6. *The Cancellation Properties*

(i) If $a + x = a + y$, then $x = y$.

(ii) If $a \neq 0$ and $ax = ay$, then $x = y$.

7. *The Zero-Factor Properties*

(i) $a \cdot 0 = 0 \cdot a = 0$

(ii) If $a \cdot b = 0$, then $a = 0$ or $b = 0$ (or both).

8. *Properties of Negation*

(i) $-(-a) = a$ **(ii)** $(-a)b = a(-b) = -(ab)$

(iii) $(-a)(-b) = ab$ **(iv)** $-(a + b) = (-a) + (-b)$

The operations of subtraction and division are defined as follows.

Definition 1 Subtraction and Division

Let a and b be real numbers.

(i) The **difference** $a - b$ is defined by $a - b = a + (-b)$.

(ii) The **quotient** or **ratio** $a \div b$ or $\frac{a}{b}$ is defined only if $b \neq 0$. If $b \neq 0$, then by definition

$$\frac{a}{b} = a \cdot \frac{1}{b}.$$

For instance, $7 - 4$ means $7 + (-4)$ and $\frac{7}{4}$ means $7 \cdot \frac{1}{4}$. Because 0 has no multiplicative inverse, a/b is undefined when $b = 0$. Thus,

Division by zero is not allowed.

When $a \div b$ is written in the form a/b, it is called a **fraction** with **numerator *a*** and **denominator *b*.** Although the denominator can't be zero, there's nothing wrong with having a zero in the numerator. In fact, if $b \neq 0$,

$$\frac{0}{b} = 0 \cdot \frac{1}{b} = 0$$

by the zero-factor property 7(i).

We can now state the following additional derived property for fractions.

9. *The Negative of a Fraction*

If $b \neq 0$, then $\frac{-a}{b} = \frac{a}{-b} = -\frac{a}{b}$.

Because Properties 1 through 9 provide a foundation for the algebra of real numbers, it is essential for you to become familiar with them.

Sets of Real Numbers

Grouping or **classifying** is a familiar technique in the natural sciences for dealing with the immense diversity of things in the real world. For instance, in biology, plants and animals are divided into various phyla, and then into classes, orders, families, genera, and species. In much the same way, real numbers can be grouped or classified by singling out important features possessed by some numbers but not by others. By using the idea of a *set*, classification of real numbers can be accomplished with clarity and precision.

A **set** may be thought of as a collection of objects. Most sets considered in this textbook are sets of real numbers. Any one of the objects in a set is called an **element,** or **member,** of the set. Sets are denoted either by capital letters such as A, B, C or by braces $\{\ldots\}$ enclosing symbols for the elements in the set. Thus, if we write $\{1, 2, 3, 4, 5\}$, we mean the set whose elements are the numbers 1, 2, 3, 4, and 5. Two sets are said to be **equal** if they contain precisely the same elements.

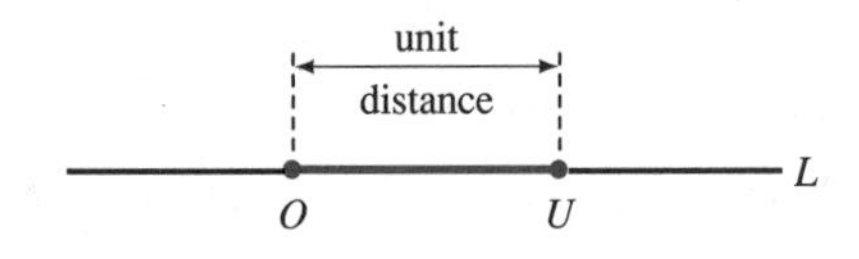

Sets of numbers and relations among such sets can often be visualized by the use of a **number line** or **coordinate axis.** A number line is constructed by fixing a point O called the **origin** and another point U called the **unit point** on a straight line L (Figure 2). The distance between O and U is called the **unit distance** and may be 1 inch, 1 centimeter, or 1 unit of whatever measure you choose. If the line L is horizontal, it is customary to place U to the right of O.

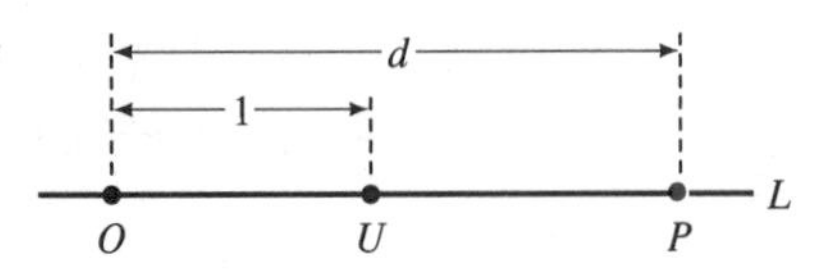

Each point P on the line L is now assigned a "numerical address" or **coordinate** x representing its signed distance from the origin, measured in terms of the given unit. Thus, $x = \pm d$, where d is the distance between O and P; the plus sign or minus sign is used to indicate whether P is to the right or left of O (Figure 3). Of course, the origin O is assigned the coordinate 0 (zero), and the unit point U is assigned the coordinate 1. On the resulting number scale (Figure 4), each point P has a corresponding numerical coordinate x and each real number x is the coordinate of a uniquely determined point P. It is convenient to use an arrowhead on the number line to indicate the direction in which the numerical coordinates are increasing (to the right in Figure 4).

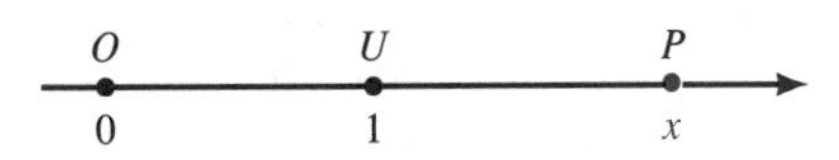

A set of numbers can be illustrated on a number line by shading or coloring the points whose coordinates are members of the set. For instance:

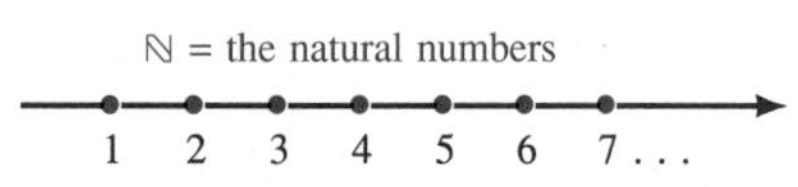

1. The **natural numbers,** also called **counting numbers** or **positive integers,** are the numbers, 1, 2, 3, 4, 5, and so on, obtained by adding 1 over and over again. The set $\{1, 2, 3, 4, 5, \ldots\}$ of all natural numbers, denoted by the symbol $\mathbb{N}$, is illustrated in Figure 5.

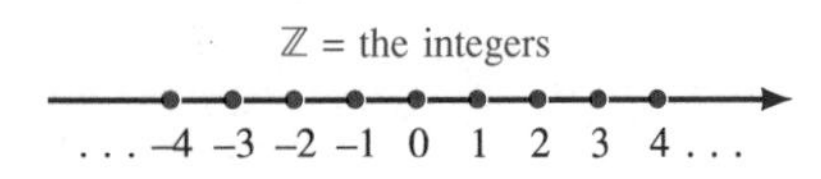

2. The **integers** consist of all the natural numbers, the negatives of the natural numbers, and zero. The set of all integers $\{\ldots, -4, -3, -2, -1, 0, 1, 2, 3, 4, \ldots\}$ denoted by the symbol $\mathbb{Z}$, is illustrated in Figure 6.

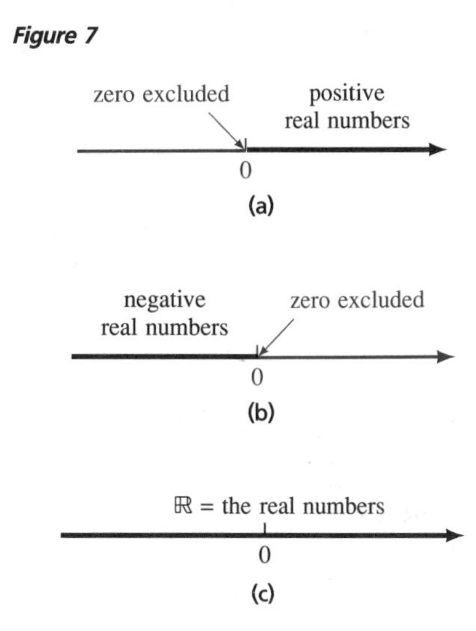

3. The **positive real numbers** correspond to points to the right of the origin (Figure 7a), and the **negative real numbers** correspond to points to the left of the origin (Figure 7b). The set of all real numbers is denoted by the symbol $\mathbb{R}$ (Figure 7c).

4. The **rational numbers** are those real numbers that can be written in the form a/b, where a and b are integers and $b \neq 0$. Since b may equal 1, every integer is a rational number. Other examples of rational numbers are $\frac{13}{2}$, $\frac{3}{4}$, and $-\frac{22}{7}$. The set of all rational numbers is denoted by the symbol $\mathbb{Q}$ (which reminds us that rational numbers are *quotients* of integers). Rational numbers in decimal form either **terminate** or begin to **repeat** the same pattern indefinitely.

5. The **irrational numbers** are the real numbers that are not rational. A real number is irrational if and only if its decimal representation is **nonterminating** and **nonrepeating.** Examples are $\sqrt{2} = 1.4142135\ldots$, $\sqrt{3} = 1.7320508\ldots$, and $\pi = 3.1415926\ldots$.

Rational Numbers and Decimals

By using long division, you can express a rational number as a decimal. For instance, if you divide 2 by 5, you will obtain $\frac{2}{5} = 0.4$, a terminating decimal. Similarly, if you divide 2 by 3, you will obtain $\frac{2}{3} = 0.66666\ldots$, a nonterminating, repeating decimal. A repeating decimal, such as $0.66666\ldots$, is often written as $0.\overline{6}$, where the overbar indicates the digit or digits that repeat; hence $\frac{2}{3} = 0.\overline{6}$.

Example 4 Express each rational number as a decimal.

(a) $-\frac{3}{5}$ **(b)** $\frac{3}{8}$ **(c)** $\frac{17}{6}$ **(d)** $\frac{3}{7}$

Solution **(a)** $-\frac{3}{5} = -0.6$ **(b)** $\frac{3}{8} = 0.375$ **(c)** $\frac{17}{6} = 2.83333\ldots = 2.8\overline{3}$
(d) $\frac{3}{7} = 0.428571428571428571\ldots = 0.\overline{428571}$ ■

Every terminating or repeating decimal represents a rational number. The following example illustrates how you can rewrite a terminating decimal as a quotient of integers.*

Example 5 Express each terminating decimal as a quotient of integers.

(a) 0.7 **(b)** -0.63 **(c)** 1.075

Solution **(a)** $0.7 = \frac{7}{10}$ **(b)** $-0.63 = -\frac{63}{100}$ **(c)** $1.075 = \frac{1075}{1000} = \frac{43}{40}$ ■

* For the technique for rewriting a repeating decimal as a quotient of integers, see Example 6 on page 82.

Fractions or decimals are often expressed as percents; for instance, 3% means $\frac{3}{100}$ or 0.03.

Example 6 Rewrite each percent as a decimal.

(a) 5.7% **(b)** 0.003%

Solution **(a)** $5.7\% = \frac{5.7}{100} = 0.057$ **(b)** $0.003\% = \frac{0.003}{100} = 0.00003$ ■

Example 7 Rewrite each rational number as a percent.

(a) $\frac{4}{5}$ **(b)** $\frac{1}{3}$

Solution **(a)** $\frac{4}{5} = 0.8 = 0.8 \times 100\% = 80\%$

(b) $\frac{1}{3} = 0.\bar{3} = 0.\bar{3} \times 100\% = 33.\bar{3}\%$ ■

Example 8 **(a)** What is 60% of 300?

(b) What percent of 2000 is 40?

Solution **(a)** $60\% \times 300 = 0.6 \times 300 = 180$

(b) $\frac{40}{2000} = 0.02 = 0.02 \times 100\% = 2\%$ ■

If a number increases, then the **percent of increase** is given by

$$\frac{\text{amount of increase}}{\text{original value}} \times 100\%.$$

If a number decreases, then the **percent of decrease** is given by

$$\frac{\text{amount of decrease}}{\text{original value}} \times 100\%.$$

C **Example 9** Juanita's weekly salary increases from \$305 to \$317.20. What is the percent of increase?

Solution The amount of increase is \$317.20 − \$305 = \$12.20. Since the original salary before the increase was \$305, the percent of increase is given by

$$\frac{12.20}{305} \times 100\% = 0.04 \times 100\% = 4\%.$$ ■

Many calculators have special keys to facilitate computations with percents. Check the instruction booklet furnished with your calculator to see if it has this feature.

Problem Set 1.1

In each problem set, problems with colored numbers constitute a good representation of the main ideas of the section. Note that some of the even-numbered problems may be considerably more challenging than the odd-numbered ones.

In Problems 1 and 2, rewrite each expression using exponential notation.

1. **(a)** $5 \cdot 5 \cdot 5 \cdot 5 \cdot 5 \cdot 5 \cdot 5 \cdot 5$
   (b) $(-7)(-7)(-7)(-7)$
   (c) $3 \cdot 3 \cdot 3 \cdot 3 \cdot 4 \cdot 4 \cdot 4$
   (d) $8 \cdot 8 + (-6)(-6)(-6)$

2. **(a)** $(x \cdot x \cdot x \cdot x)/(y \cdot y \cdot y)$
   (b) $y^2 y^2 y^2 / z^3$

In Problems 3 to 14, write a formula for the given quantity.

3. The number z that is twice the sum of x and y.

4. The perimeter P of a rectangle with length a units and width b units.

5. The area A of a circle of radius r units.

6. The surface area A of a cube with edges of length x units.

7. The number x that is 5% of a number n.

8. The volume V of a sphere of radius r units. [Consult Appendix II if you don't know the formula.]

9. The area A of a triangle with base b units and height h units.

10. The number N of living cells in a culture after t hours if there are N_0 cells when $t = 0$, if each cell divides into two cells at the end of each hour, and if no cells die.

11. The amount A dollars you owe after t years if you borrow p dollars at a simple interest rate r per year.

12. The number L of board feet of lumber in a tree d feet in diameter and h feet high. Assume for simplicity that the tree is a right circular cylinder. [*Note:* One board foot is the volume of a board with dimensions 1 foot by 1 foot by 1 inch.]

13. The volume V of a rectangular aquarium with length l units and with a square end having sides w units long (Figure 8).

Figure 8

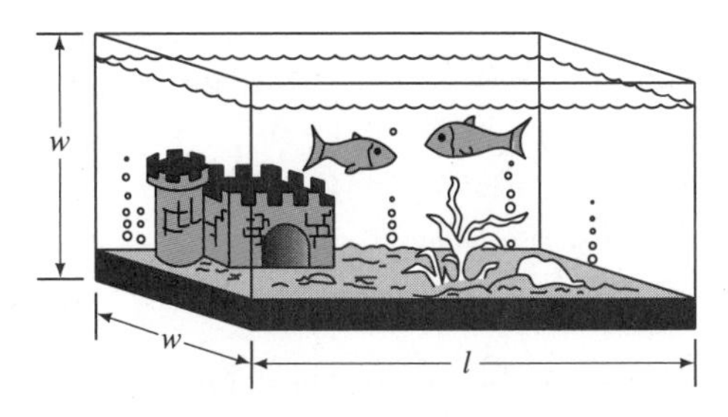

14. The surface area A of a spherical tank of radius r units. [*Hint:* See Appendix II.]

C 15. If p dollars is invested at a nominal annual interest rate r compounded n times per year, the investment will be worth $p\left(1 + \frac{r}{n}\right)^{nt}$ dollars at the end of t years. Suppose you invest \$5000 at 10% interest ($r = 0.10$) compounded semiannually ($n = 2$). What is your investment worth at the end of 2 years?

16. The *half-life* of a radioactive substance is the period of time T during which exactly half of the substance will undergo radioactive disintegration. If q represents the quantity of such a substance at time t, then $q = \frac{q_0}{2^{t/T}}$, where q_0 is the original amount of the substance when $t = 0$. Potassium 42, which is used as a biological tracer, has a half-life of $T = 12.5$ hours. If a certain quantity of potassium 42 is injected into an organism, and if none is lost by excretion, what percentage will remain after $t = 50$ hours?

In Problems 17 and 18, state one basic algebraic property of the real numbers to justify each statement.

17. **(a)** $5 + (-3) = (-3) + 5$
    (b) $5 \cdot (3 + 7) = (3 + 7) \cdot 5$
    (c) $(-13)(-12) = (-12)(-13)$
    (d) $5 \cdot (3 + 7) = 5 \cdot 3 + 5 \cdot 7$

(e) $1 \cdot (3 + 7) = 3 + 7$

(f) $[(3)(4)](5) = (3)[(4)(5)]$

(g) $4 + (x + z) = (4 + x) + z$

(h) $5 \cdot \frac{1}{5} = 1$

18. **(a)** $4 \cdot (x + 0) = 4x$

(b) $(x - y) + [-(x - y)] = 0$

(c) $x[(-y) + y] = x(-y) + xy$

(d) $(-4) \cdot \dfrac{1}{(-4)} = 1$

(e) $(4 + x) + [-(4 + x)] = 0$

(f) $x + [y + (z + w)] = (x + y) + (z + w)$

(g) $(ab)(cd) = [(ab)c]d$

(h) $(a + b)(c + d) = a(c + d) + b(c + d)$

In Problems 19 and 20, state one of the derived algebraic properties of the real numbers to justify each statement.

19. **(a)** $-(-5) = 5$ **(b)** $(-3)(-x) = 3x$

(c) $5(-6) = -(5)(6)$ **(d)** $(u + v) \cdot 0 = 0$

(e) If $7 + x = 7 + x^2$, then $x = x^2$.

20. **(a)** If $2y = 2y^{-1}$, then $y = y^{-1}$.

(b) If $(2x + 3)(x + 1) = 0$, then $2x + 3 = 0$ or $x + 1 = 0$.

(c) If $x^2 = 0$, then $x = 0$.

(d) $\dfrac{h+1}{-2} = -\dfrac{h+1}{2}$ **(e)** $\dfrac{-3}{2-p} = \dfrac{3}{-(2-p)}$

21. In each of the following calculations a *mistake* has been made. Find the mistake and make the correct calculation.

(a) $(3 + 5)^2 = 3^2 + 5^2 = 9 + 25 = 34$?

(b) $\dfrac{1}{3+5} = \dfrac{1}{3} + \dfrac{1}{5} = \dfrac{5}{15} + \dfrac{3}{15} = \dfrac{8}{15}$?

22. **(a)** Does the operation of subtraction have the associative property; that is, is it always true that $a - (b - c) = (a - b) - c$?

(b) Does the operation of division have the associative property?

In Problems 23 and 24, list all of the elements that belong to each set and illustrate the set on a number line.

23. A is the set of all even integers between 3 and 11.

24. B is the set of all natural numbers between $-\frac{1}{2}$ and $\frac{7}{2}$.

In Problems 25 and 26, indicate whether each statement is true or false.

25. **(a)** $\frac{1}{2}$ is an element of $\mathbb{N}$.

(b) $\frac{1}{2}$ is an element of $\mathbb{Q}$.

(c) $\sqrt{2}$ is an element of $\mathbb{R}$.

(d) $\sqrt{2}$ is an element of $\mathbb{Q}$.

(e) $\frac{1}{2} + \frac{2}{3}$ is an element of $\mathbb{Q}$.

26. **(a)** $\frac{1}{2} + \sqrt{2}$ is an element of $\mathbb{Q}$.

(b) Every natural number is an integer.

(c) Every natural number is a rational number.

(d) Every positive real number is a rational number.

(e) $\sqrt{\pi}$ is a rational number.

In Problems 27 and 28, express each rational number as a decimal.

27. **(a)** $\frac{4}{5}$ **(b)** $-\frac{3}{8}$ **(c)** $-\frac{13}{50}$ **(d)** $\frac{-17}{200}$

(e) $-\frac{5}{8}$ **(f)** $-\frac{5}{3}$ **(g)** $\frac{15}{7}$

28. **(a)** $-\frac{7}{25}$ **(b)** $\frac{3}{50}$ **(c)** $\frac{7}{100}$ **(d)** $\frac{200}{500}$

(e) $\frac{75}{30}$ **(f)** $\frac{11}{12}$ **(g)** $\frac{-23}{7}$

In Problems 29 and 30, express each terminating decimal as a quotient of integers.

29. **(a)** 0.41 **(b)** -0.032 **(c)** -0.581

(d) 91.2 **(e)** 1.0451

30. **(a)** -0.54 **(b)** 22.61 **(c)** 2.691

(d) 0.00012 **(e)** -2.00002

In Problems 31 and 32, rewrite each percent as a decimal.

31. **(a)** 11% **(b)** 1.03%

(c) 432% **(d)** 0.0006%

32. **(a)** 99.44% **(b)** 315%

(c) 0.001% **(d)** 1.00001%

In Problems 33 and 34, rewrite each rational number as a percent.

33. **(a)** $\frac{1}{2}$ **(b)** $\frac{3}{125}$ **(c)** $\frac{2}{3}$ **(d)** $\frac{2}{7}$

34. **(a)** $\frac{7}{50}$ **(b)** 0.042 **(c)** $\frac{5}{8}$ **(d)** 1.245

35. **(a)** What is 7% of 30?

(b) What percent of 20 is 3?

36. In the United States, 5 people out of every 1000 are in the army and 8 people out of every 10,000 are army officers. What percent of army personnel are officers?

37. The cost of a book increases from $20 to $24. What is the percent of increase?

38. In April, Alfie's salary is $380 per week. In May, his April salary is increased by 5%. In June, his May salary is decreased by 5%. What is Alfie's salary in June?

39. The cost of a calculator decreases from $24 to $20. What is the percent of decrease?

[c] **40.** The price of a new car is $11,342.99 plus 6% sales tax. Find:

(a) The amount of sales tax in dollars and

(b) The total cost of the car including the sales tax.

41. A small town decreases its annual budget from $800,000 to $600,000. What is the percent of decrease?

1.2 Calculators, Scientific Notation, and Rounding Off

A scientific calculator with keys for exponential, logarithmic, and trigonometric functions will expedite some of the calculations required in this textbook. Problems or groups of problems for which the use of a calculator is suggested are marked with the symbol [c]. A calculator with graphing features will quickly generate many of the graphs in the homework problems. Examples and problems for which a graphing calculator may be used are marked with the symbol [gc]

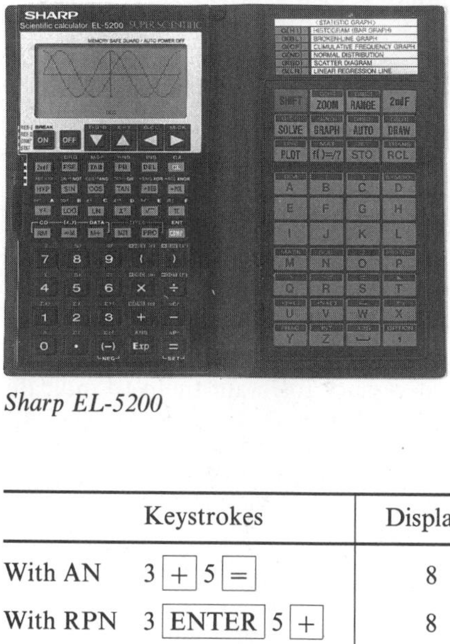

There are two types of calculators available, those using algebraic notation (AN) and those using reverse Polish notation (RPN). Advocates of AN claim that it is more "natural," while supporters of RPN say that it is just as "natural" and avoids the parentheses required when sequential calculations are made in AN. Before purchasing a scientific calculator, you should familiarize yourself with both AN and RPN so that you can make an intelligent decision based on your own preferences.

Sharp EL-5200

The distinction between AN and RPN is shown at the left by the computations of $3 + 5$ and 2^4. Thus, in using RPN, both numbers are entered *first*, and *then* the key for the desired operation is pressed.

	Keystrokes	Display
With AN	3 [+] 5 [=]	8
With RPN	3 [ENTER] 5 [+]	8
With AN	2 [y^x] 4 [=]	16
With RPN	2 [ENTER] 4 [y^x]	16

After acquiring any calculator, learn to use it properly by studying the instruction booklet furnished with it. In particular, practice performing chain calculations so you can do them as efficiently as possible using whatever "memory" features your calculator may possess to store intermediate results. After you learn *how* to use a calculator, it is imperative that you learn *when* to use it, and especially when *not* to use it.

Attempts to use a calculator for problems that are *not* marked with the symbol [c] can lead to bad habits, which not only waste time but actually hinder understanding.

Scientific Notation

In applied mathematics, very large and very small numbers are written in compact form by using integer powers of 10. For instance, the speed of light in vacuum,

$$c = 300{,}000{,}000 \text{ meters per second (approximately),}$$

can be written more compactly as

$$c = 3 \times 10^8 \text{ meters per second.}$$

More generally, a real number x is said to be expressed in **scientific notation** if it is written in the form

$$x = \pm p \times 10^n,$$

where n is an integer and p is a number greater than or equal to 1, but less than 10. The integer n is called the **characteristic** of x, the number p is called the **mantissa** of x, and the factor 10^n is called the **order of magnitude** of x.

Scientific calculators can be set to display numbers either in ordinary decimal form or in scientific notation. Furthermore, such a calculator will automatically switch to scientific notation whenever the number is too large or too small to be displayed in ordinary decimal form. When a number such as 2.579×10^{-13} is displayed, the multiplication sign and the base 10 usually do not appear, and the display shows simply

$$2.579 \quad -13$$

or perhaps

$$2.579 \quad \text{E} \; -13,$$

the E serving as a reminder that -13 is an exponent.

To change a number from ordinary decimal form to scientific notation, move the decimal point to obtain a number between 1 and 10 (one digit in front of the decimal point) and multiply by 10^n or by 10^{-n}, where n is the number of places the decimal point was moved to the left or to the right, respectively. Final zeros after the decimal point can be dropped unless it is necessary to retain them to indicate the accuracy of an approximation.

Example 1 Rewrite each statement so that all numbers are expressed in scientific notation.

(a) The volume of the earth is approximately

$$1{,}087{,}000{,}000{,}000{,}000{,}000{,}000 \text{ cubic meters.}$$

(b) The earth rotates about its axis with an angular speed of approximately 0.00417 degree per second.

Solution **(a)** We move the decimal point 21 places to the left

$$1{,}087{,}000{,}000{,}000{,}000{,}000{,}000$$

to obtain a number between 1 and 10 and multiply by 10^{21}, so that

$$1{,}087{,}000{,}000{,}000{,}000{,}000{,}000 = 1.087 \times 10^{21}.$$

Thus, the volume of the earth is approximately 1.087×10^{21} cubic meters.

(b) We move the decimal point three places to the right

$$0004.17$$

to obtain a number between 1 and 10 and multiply by 10^{-3}, so that

$$0.00417 = 4.17 \times 10^{-3}.$$

Thus, the earth rotates about its axis with an angular speed of approximately 4.17×10^{-3} degree per second. ■

The procedure above can be reversed whenever a number is given in scientific notation and we wish to rewrite it in ordinary decimal form.

Example 2 Rewrite the following numbers in ordinary decimal form.

(a) 7.71×10^5 **(b)** 6.32×10^{-8}

Solution **(a)** $7.71 \times 10^5 = 771000. = 771{,}000$

(b) $6.32 \times 10^{-8} = 0.00000000632 = 0.000{,}000{,}0632$. (Very small numbers, written in ordinary decimal form, are easier to read if commas are used to set off zeros in groups of three.) ■

Names and scientific prefixes for some integer powers of 10 are listed in Table 1. For instance, a kilometer is 1000 meters, and a micrometer is 1 millionth of a meter.

Table 1

Power of 10	Name	Prefix
10	Ten	deka
10^2	Hundred	hecto
10^3	Thousand	kilo
10^6	Million	mega
10^9	Billion	giga
10^{12}	Trillion	tera
10^{15}	Quadrillion	peta
10^{-1}	Tenth	deci
10^{-2}	Hundredth	centi
10^{-3}	Thousandth	milli
10^{-6}	Millionth	micro
10^{-9}	Billionth	nano
10^{-12}	Trillionth	pico
10^{-15}	Quadrillionth	femto

Approximations

Numbers produced by a calculator are often inexact, because the calculator can work only with a finite number of decimal places. For instance, a 10-digit calculator gives

$$2 \div 3 = 6.666666667 \times 10^{-1}$$

and

$$\sqrt{2} = 1.414213562,$$

both of which are **approximations** of the true values. Therefore:

> Unless we explicitly ask for numerical approximations or indicate that a calculater is recommended, it's usually best to leave answers in fractional form or as radical expressions.

Don't be too quick to pick up your calculator—answers such as $2/3$, $\sqrt{2}$, $(\sqrt{2} + \sqrt{3})/7$, *and* $\pi/4$ *are often preferred to much more lengthy decimal expressions that are only approximations.*

Most numbers obtained from measurements of real-world quantities are subject to error* and also have to be regarded as approximations. If the result of a measurement (or any calculation involving approximations) is expressed in scientific notation, $p \times 10^n$, it is usually understood that p should contain only **significant digits,** that is, digits that, except possibly for the last, are known to be correct or reliable. (The last digit may be off by one unit because the number was rounded off.) For instance, if we read in a physics textbook that

$$\text{one electron volt} = 1.60 \times 10^{-19} \text{ joule},$$

we understand that the digits 1, 6, and 0 are significant and we say that, *to an accuracy of three significant digits*, one electron volt is 1.60×10^{-19} joule.

To emphasize that a numerical value is only an approximation, we often use a wave-shaped equal sign, $\approx$. For instance,

$$\sqrt{2} \approx 1.414.$$

However, we sometimes use ordinary equal signs when dealing with inexact quantities, simply because it becomes tiresome to indicate repeatedly that approximations are involved.

Rounding Off

Some scientific calculators can be set to round off all displayed numbers to a particular number of decimal places or significant digits. However, it's easy enough to round off numbers without a calculator: Simply drop all unwanted digits to the right of the digits that are to be retained and increase the last retained digit by 1 if the first dropped digit is 5 or greater. It may be necessary to replace dropped digits by zeros in order to hold the decimal point; for instance, we round off 5157.3 to the nearest hundred as 5200.

Rounding off should be done in one step, rather than digit by digit. Digit-by-digit rounding off may produce an incorrect result. For instance, if 8.2347 is rounded off to four significant digits as 8.235, which in turn is rounded off to three significant digits, the result would be 8.24. However, 8.2347 is *correctly* rounded off in one step to three significant digits as 8.23.

Example 3 Round off the given number as indicated.

(a) 1.327 to the nearest tenth

(b) -19.8735 to the nearest thousandth

(c) 4671 to the nearest hundred

(d) 9.22345×10^7 to four significant digits

* For further details, see the material in Section 2.6 on absolute, relative, and percent error.

Solution **(a)** To the nearest tenth,

hundredths place

$$1.327 \approx 1.3$$

tenths place thousandths place

(b) To the nearest thousandth,

$$-19.8735 \approx -19.874$$

thousandths place

(c) To the nearest hundred,

$$4671 \approx 4700$$

hundreds place

(d) To four significant digits, $9.22345 \times 10^7 \approx 9.223 \times 10^7$ ■

If approximate numbers expressed in ordinary decimal form are added or subtracted, the result should be considered accurate only to as many decimal places as the least accurate of the numbers, and it should be rounded off accordingly. To add or subtract approximate numbers expressed in scientific notation, first convert the numbers to ordinary decimal form, then round off the result as above, and finally rewrite the answer in scientific notation to obtain the appropriate number of significant digits.

C Example 4 Suppose that each of the quantities $a = 1.7 \times 10^{-2}$, $b = 2.711 \times 10^{-2}$, and $c = 6.213455 \times 10^2$ is accurate only to the number of displayed digits. Find $a - b + c$ and express the result in scientific notation rounded off to an appropriate number of significant digits.

Solution Since we are adding and subtracting, we begin by rewriting

$$a = 0.017, \qquad b = 0.02711, \qquad \text{and} \qquad c = 621.3455$$

in ordinary decimal form. The least accurate of these numbers is a, which is accurate only to the nearest thousandth. Using a calculator, we find that

$$a - b + c = 621.33539,$$

but we must round off this result to the nearest thousandth and write

$$a - b + c = 621.335.$$

Finally, we rewrite the answer in scientific notation, so that

$$a - b + c = 621.335 = 6.21335 \times 10^2.$$ ■

If approximate numbers are multiplied or divided, the result should be considered accurate only to as many significant digits as the least accurate of the numbers, and it should be rounded off accordingly.

C Example 5 Suppose that each of the quantities $a = 2.15 \times 10^{-3}$ and $b = 2.874 \times 10^2$ is accurate only to the number of displayed digits. Calculate the indicated quantity and express it in scientific notation rounded off to an appropriate number of significant digits.

(a) ab **(b)** b/a **(c)** b^2

Solution **(a)** Using a calculator, we find that $ab = 6.1791 \times 10^{-1}$. Since a, the least accurate of the two factors, is accurate only to three significant digits, we must round off our answer to three significant digits and write $ab = 6.18 \times 10^{-1}$.

(b) Here we have $b/a = 1.336744186 \times 10^5$, but again we must round off our answer to three significant digits and write $b/a = 1.34 \times 10^5$.

(c) Here we have $b^2 = 8.259876 \times 10^4$, but, since b is accurate only to four significant digits, we must round off our answer to four significant digits and write $b^2 = 8.260 \times 10^4$. ■

Problem Set 1.2

In Problems 1 to 8, rewrite each number in scientific notation.

1. 15.500 **2.** 0.0043

3. 58,761,000 **4.** 77 million

5. 186,000,000,000 **6.** 420 trillion

7. 0.000,000,901 **8.** $(0.025)^{-5}$

In Problems 9 to 14, rewrite each number in ordinary decimal form.

9. 3.33×10^4 **10.** 1.732×10^{10}

11. 4.102×10^{-5} **12.** -8.255×10^{-11}

13. 1.001×10^7 **14.** -2.00×10^9

In Problems 15 to 20, rewrite each statement so that all numbers are expressed in scientific notation.

15. The image of one frame in a motion-picture film stays on the screen approximately 0.062 second.

16. One liter is defined to be 0.001,000,028 cubic meter.

17. An *astronomical unit* is defined to be the average distance between the earth and the sun, 92,900,000 miles, and a *parsec* is the distance at which one astronomical unit would subtend one second of arc, about 19,200,000,000,000 miles.

18. A *light year* is the distance that light, traveling at approximately 186,200 miles per second, traverses in one year. Thus, a light year is approximately 5,872,000,000,000 miles.

19. In physics, the average lifetime of a lambda particle is estimated to be 0.000,000,000,251 second.

20. In thermodynamics, the Boltzmann constant is 0.000,000,000,000,000,000,000,0138 joule per kelvin.

C In Problems 21 to 24, calculate the indicated quantity and write the answer in scientific notation, but do not round off.

21. (8,000)(2,000,000,000)(0.000,03)

22. $(0.000,006)^3(500,000,000)^{-4}$

23. $\dfrac{(7,000,000,000)^3}{0.0049}$

24. $\dfrac{(0.000{,}000{,}039)^2(591{,}000)^3}{(197{,}000)^2}$

C **25.** In electronics, $P = I^2R$ is the formula for the power P in watts dissipated by a resistance of R ohms through which a current of I amperes flows. Calculate P if $I = 1.43 \times 10^{-4}$ ampere and $R = 3.21 \times 10^4$ ohms.

C **26.** The mass of the sun is approximately 1.97×10^{29} kilograms, and our galaxy (the Milky Way) is estimated to have a total mass of 1.5×10^{11} suns. The mass of the known universe is at least 10^{11} times the mass of our galaxy. Calculate the approximate mass of the known universe.

In Problems 27 to 30, specify the accuracy of the indicated value in terms of significant digits.

27. A drop of water contains 1.7×10^{21} molecules.

28. The binding energy of the earth to the sun is 2.5×10^{33} joules.

29. One mile $= 6.3360 \times 10^4$ inches.

30. One atmosphere $= 1.01 \times 10^5$ newtons per square meter.

In Problems 31 to 38, round off the given number as indicated.

31. 5280 to the nearest hundred

32. 9.29×10^7 to the nearest million

33. 0.0145 to the nearest thousandth

34. 999 to the nearest ten

35. 111,111.11 to the nearest ten thousand

36. 5.872×10^{12} to three significant digits

37. 2.1448×10^{-13} to three significant digits

38. π to four significant digits

C In Problems 39 to 48, find the numerical value of the indicated quantity in scientific notation rounded off to an appropriate number of significant digits. Assume that the given values are accurate only to the number of displayed digits.

39. $a + b$ if $a = 2.0371 \times 10^2$ and $b = 2.7312 \times 10^1$

40. $a + b - c$ if $a = 1.450 \times 10^5$, $b = 7.63 \times 10^2$, and $c = 2.251 \times 10^3$

41. $a - b + c$ if $a = 4.900 \times 10^{-4}$, $b = 3.512 \times 10^{-6}$, and $c = 2.27 \times 10^{-7}$

42. $a + b - c + d$ if $a = 8.1370$, $b = 2.2 \times 10^1$, $c = 1 \times 10^{-3}$, and $d = 5.23 \times 10^{-4}$

43. ab if $a = 3.19 \times 10^2$ and $b = 4.732 \times 10^{-3}$

44. ab^2 if $a = 2.11 \times 10^4$ and $b = 1.009 \times 10^{-2}$

45. a^3 if $a = 1.02 \times 10^9$

46. $\dfrac{ab}{c}$ if $a = 7.71 \times 10^3$, $b = 3.250 \times 10^{-4}$, and $c = 1.09 \times 10^5$

47. $\dfrac{a^2}{b}$ if $a = 3.32 \times 10^2$ and $b = 3.18 \times 10^{-1}$

48. $\dfrac{a+b}{c}$ if $a = 4.163 \times 10^2$, $b = 2.142 \times 10^1$, and $c = 1.555 \times 10^3$

49. According to the U.S. Bureau of Economic Analysis, the Gross National Product (GNP) of the United States for the fiscal year ending in 1988 was $\$4.70237 \times 10^{12}$. According to the Office of Management and Budget, the outstanding gross debt for the same time period was $\$2.5816 \times 10^{12}$. Round off your answers to the following questions in an appropriate manner:

(a) How much more was the GNP than the debt in the fiscal year ending in 1988?

(b) If the population of the United States in 1988 is estimated to be 2.446×10^8, find the per-capita GNP (that is, GNP $\div$ population) in 1988.

50. In the BASIC language, often used for microcomputers, the product of X and Y is denoted by `X*Y`, 10 raised to the power N is denoted by `10^N`, and `INT(X)` is notation for the largest integer that is less than or equal to X. If N is a positive integer, show that the formula `INT(X*10^N+.5)/10^N` gives the value of X rounded off to N decimal places.

51. Let $x = p \times 10^n$ in scientific notation, and suppose that p is rounded off to a certain number of decimal places. Explain why x is thereby rounded off to one more significant digit than the number of decimal places to which p was rounded off.

1.3 Mathematical Models and the Idea of Variation

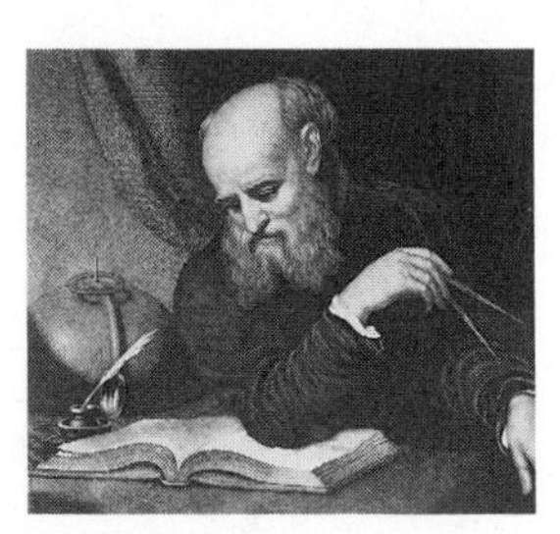

Galileo Galilei

In the later years of his life, the Italian scientist Galileo Galilei (1564–1642) wrote about his experiments with motion in a treatise called *Dialogues Concerning Two New Sciences.* Here he described his wonderful discovery that distances covered in consecutive equal time intervals by balls rolling down inclined planes are proportional to the successive odd positive integers. Thus, if d denotes the distance covered during the nth time interval, then

$$d = k(2n - 1),$$

where k is a suitable constant. Galileo determined that the constant k depends only on the incline and not on the mass of the ball or the material of which it is composed. He reasoned that the same result should hold for freely falling bodies if air resistance is neglected.

Galileo's discovery of a mathematical model for uniformly accelerated motion is considered to have been the beginning of the science of dynamics. A **mathematical model** is an equation or set of equations in which letters represent real-world quantities—for instance, distance and time intervals in the equation $d = k(2n - 1)$. A mathematical model is often an *idealization* of the real-world situation it supposedly describes. For instance, Galileo's mathematical model assumes a perfectly smooth inclined plane, no friction, and no air resistance, and it neglects the rotational moment of inertia of the balls as they roll down the inclined plane.

Direct and Inverse Variation

In this section, we study mathematical models related to the concept of **variation.** Thus, the letters y and x in the following definition could represent variable quantities in a mathematical model.

Definition 1 **Direct Variation**

If there is a constant k such that

$$y = kx$$

holds for all values of x, we say that **y is directly proportional to x** or that **y varies directly as x** (or *with* x). The constant k is called the **constant of proportionality** or the **constant of variation.**

For instance, if an automobile is moving at a constant speed of 55 miles per hour, then the distance d it travels in t hours is given by the formula

$$d = 55t.$$

Here d and t are variables in the mathematical model $d = 55t$ describing the motion of the automobile; the distance d is directly proportional to the time t, and the constant of proportionality is 55 miles per hour.

Suppose that x and y are variables and that

$$y = kx,$$

so that y is directly proportional to x with k as the constant of proportionality. Note that $y = 0$ when $x = 0$. Also, if $x \neq 0$, we have

$$\frac{y}{x} = k,$$

so the fraction, or ratio, y/x maintains the constant value k as x varies through nonzero values. Thus, if a nonzero value of x and the corresponding value of y are known, the value of k can be determined.

Example 1 Suppose that the rate r at which impulses are transmitted along a nerve fiber is directly proportional to the diameter d of the fiber. Given that $r = 20$ meters per second when $d = 6$ micrometers:

(a) Write a formula that expresses r in terms of d.

(b) Find r if $d = 4$ micrometers.

Solution Since r is directly proportional to d, we have

$$r = kd,$$

where k is the constant of proportionality.

(a) For $d \neq 0$, we have

$$k = \frac{r}{d}.$$

Substituting $r = 20$ meters per second and $d = 6$ micrometers into this equation, we find that $k = 20/6 = 10/3$. Therefore, for all values of d, we have

$$r = \tfrac{10}{3}d.$$

(b) When $d = 4$ micrometers, we have

$$r = \tfrac{10}{3}(4) = \tfrac{40}{3} \text{ meters per second.}$$ ■

If we say that **y varies directly as the square of x,** we mean that x and y are related by an equation of the form

$$y = kx^2$$

for some constant of variation k. For instance, the formula $A = \pi r^2$ for the area of a circle states that A is directly proportional to the square of the radius r, and that π is the constant of variation. More generally:

Definition 2 **Variation with the nth Power**

If there are constants n and k such that

$$y = kx^n$$

holds for all values of x, we say that **y is directly proportional to the nth power of x** or that **y varies directly as** (or *with*) **the nth power of x.**

Figure 1

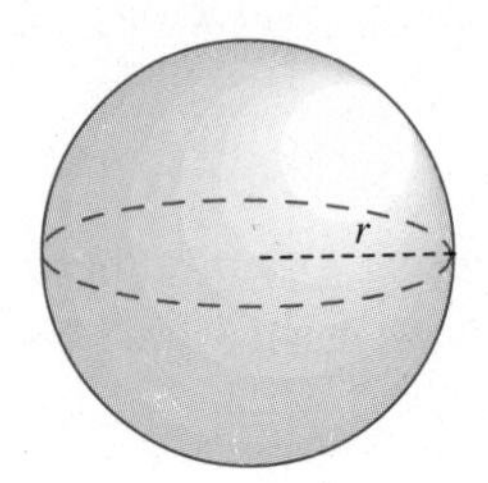

Example 2 The formula

$$V = \tfrac{4}{3}\pi r^3$$

for the volume of a sphere states that V is directly proportional to the cube of the radius r and that the constant of variation is $\frac{4}{3}\pi$ (Figure 1). Find the volume of a sphere of radius 2.71 centimeters and round off your answer appropriately.

Solution

$$V = \tfrac{4}{3}\pi r^3 = \tfrac{4}{3}\pi(2.71)^3 \approx 83.4 \text{ cubic centimeters}$$

Assuming that the radius $r = 2.71$ centimeters is correct to three significant digits, we have rounded off our answer to three significant digits. ■

The situation in which y is proportional to the reciprocal of x arises so frequently that it is given a special name, **inverse variation.**

Definition 3 **Inverse Variation**

If there is a constant k such that

$$y = \frac{k}{x}$$

for all nonzero values of x, then we say that **y is inversely proportional to x** or that **y varies inversely as x** (or *with* x).

Naturally, if $y = k/x^n$ for some constant k and some positive rational number n, we say that **y is inversely proportional to the nth power of x.**

Example 3 A company is informed by its bookkeeper that the number S of units of a certain item sold per month seems to be inversely proportional to the quantity $p + 20$, where p is the selling price per unit in dollars. Suppose that 800 units per month are sold when the price is \$10 per unit. How many units per month would be sold if the price were dropped to \$5 per unit?

Solution

Since S is inversely proportional to $p + 20$, there is a constant k such that

$$S = \frac{k}{p + 20}.$$

We know that $S = 800$ when $p = 10$, so we have

$$800 = \frac{k}{10 + 20} = \frac{k}{30}$$

or

$$k = 30(800) = 24{,}000.$$

Therefore, the formula

$$S = \frac{24{,}000}{p + 20}$$

gives S in terms of p. For $p = 5$ dollars, we have

$$S = \frac{24{,}000}{5 + 20} = \frac{24{,}000}{25} = 960,$$

so 960 units per month would be sold at a price of \$5 per unit. ■

Joint and Combined Variation

Often the value of a variable quantity depends on the values of several other quantities; for instance, the amount of simple interest on an investment depends on the interest rate, the amount invested, and the period of time involved. For compound interest, the amount of interest depends on an additional variable: how often the compounding takes place. A situation in which one variable depends on several others is called **combined variation.** An important type of combined variation is defined as follows.

Definition 4

Joint Variation

If a variable quantity y is proportional to the product of two or more variable quantities, we say that **y is jointly proportional** to these quantities, or that **y varies jointly as** (or *with*) these quantities.

For instance, if y is jointly proportional to u and v, then y is related to u and v by the formula

$$y = kuv,$$

where k is a constant. Sometimes the word "jointly" is omitted and we simply say that y is proportional to u and v.

Example 4 In chemistry, the absolute temperature T of a perfect gas varies jointly as its pressure P and its volume V. Given that $T = 500$ kelvins when $P = 50$ pounds per square inch and $V = 100$ cubic inches, find a formula for T in terms of P and V and find T when $P = 100$ pounds per square inch and $V = 75$ cubic inches.

Solution Since T varies jointly as P and V, there is a constant k such that $T = kPV$. Putting $T = 500$, $P = 50$, and $V = 100$, we find that

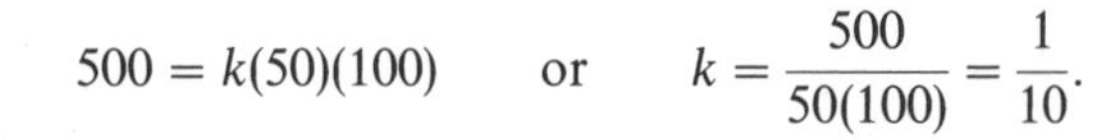

$$500 = k(50)(100) \qquad \text{or} \qquad k = \frac{500}{50(100)} = \frac{1}{10}.$$

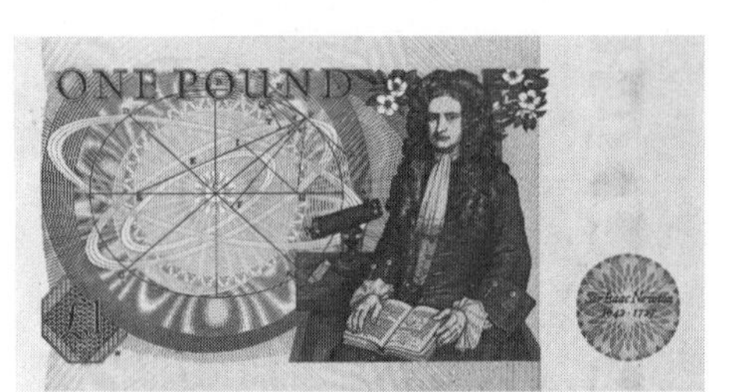

Sir Isaac Newton on a British one pound note. The apple blossoms above him remind us that an apple falling from a tree at his farm sparked his interest in the principles of gravitation. On the left is an illustration from his book the Principia, *which he holds in his hands, and in which he presented his universal law of gravitation.*

Thus, the desired formula is

$$T = \tfrac{1}{10}PV.$$

When $P = 100$ and $V = 75$, we have

$$T = \tfrac{1}{10}(100)(75) = 750 \text{ kelvins.}$$

■

We sometimes have joint variation together with inverse variation. Perhaps the most important historical discovery of this type of combined variation is *Newton's law of universal gravitation*, which states that the gravitational force of attraction F between two particles is jointly proportional to their masses m and M, and inversely proportional to the square of the distance d between them. In other words, F is related to m, M, and d by the formula

$$F = G\frac{mM}{d^2},$$

where G is the constant of proportionality.

C **Example 5** Careful measurements show that two 1-kilogram masses 1 meter apart exert a mutual gravitational attraction of 6.67×10^{-11} newton. (One pound of force is approximately 4.45 newtons.) The earth has a mass of 5.98×10^{24} kilograms. Find the earth's gravitational force on a space capsule that has a mass of 1000 kilograms and that is 10^8 meters from the center of the earth.

Solution Putting $F = 6.67 \times 10^{-11}$ newton, $m = 1$ kilogram, $M = 1$ kilogram, and $d = 1$ meter in the formula $F = G(mM/d^2)$, we find that

$$6.67 \times 10^{-11} = G\frac{1(1)}{1^2} = G,$$

so that, for arbitrary values of m, M, and d,

$$F = (6.67 \times 10^{-11})\frac{mM}{d^2}.$$

Now we substitute $M = 5.98 \times 10^{24}$, $m = 10^3$, and $d = 10^8$ to obtain

$$F = (6.67 \times 10^{-11}) \frac{10^3(5.98 \times 10^{24})}{(10^8)^2} \approx 39.9 \text{ newtons}$$

(or less than 9 pounds of gravitational force). ■

In solving Example 5 we used the fact—first proved by Newton himself using integral calculus—that in calculating the gravitational attraction of a homogeneous sphere, one can assume that all of the mass is concentrated at the center.

Problem Set 1.3

In Problems 1 to 14, relate the quantities by writing a formula involving a constant of proportionality k.

1. The electric charge Q stored on a parallel-plate capacitor is proportional to the potential difference V across the plates.
2. The energy E stored in a compressed spring varies as the square of the distance x by which it is compressed.
3. The distance d through which a body falls is directly proportional to the square of the amount of time t during which it falls.
4. According to a mathematical model formulated by Josef Stefan in 1879, the rate I at which a body radiates thermal energy is proportional to the fourth power of its absolute temperature T.
5. The power P provided by a wind generator is directly proportional to the cube of the wind speed v.
6. The kinetic energy K of a moving body is jointly proportional to its mass m and the square of its speed v.
7. The heat energy E conducted per hour through the wall of a house is jointly proportional to the area A of the wall and the difference T degrees between the inside and outside temperatures, and inversely proportional to the thickness d of the wall.
8. The maximum safe load W on a horizontal beam supported at both ends varies directly as the width w of the beam and the square of its depth d, and inversely as the length l of the beam.
9. The force F of repulsion between two electrically charged particles is directly proportional to the charges Q_1 and Q_2 on the particles, and inversely proportional to the square of the distance d between them (Coulomb's law).
10. The intensity I of a sound wave is jointly proportional to the square of its amplitude A, the square of its frequency n, the speed of sound v, and the density d of the air.
11. The rate r at which an infectious disease spreads is jointly proportional to the fraction p of the population that has the disease and the fraction $1 - p$ of the population that does not have the disease.
12. The frequency n of a guitar string is directly proportional to the square root of the string tension F, and inversely proportional to the length l of the string and the square root of its linear density (mass per unit length) d.
13. The rate of change R of the temperature of an object varies directly as the difference between the temperature T of the object and the temperature S of the environment in which it is placed (Newton's law of cooling).
14. In physiology, a useful mathematical model for the volume V of a human limb states that V is proportional to the product of the length L of the limb and the square of its circumference C.

In Problems 15 to 22, **(a)** find a formula relating the given quantities and **(b)** find the value of the indicated quantity under the given conditions.

15. y is directly proportional to x, and $y = 3$ when $x = 6$. Find y when $x = 12$.
16. y varies directly as x^3, and $y = 4$ when $x = 2$. Find y when $x = 4$.

17. w varies inversely as x, and $w = 3$ when $x = 4$. Find w when $x = 8$.

18. z is directly proportional to x and inversely proportional to y, and $z = 8$ when $x = 4$ and $y = 2$. Find z when $x = 8$ and $y = 4$.

19. u varies inversely as v^2, and $u = 2$ when $v = 10$. Find u when $v = 3$.

20. z is directly proportional to x^3 and inversely proportional to y^2, and $z = 2$ when $x = 2$ and $y = 4$. Find z when $x = -2$ and $y = 2$.

c **21.** V varies directly as T and inversely as P, and $V = 41.02$ when $T = 298.3$ and $P = 36.72$. Find V when $T = 324.9$ and $P = 22.04$.

c **22.** y varies jointly as x^2 and z and inversely as w, and $y = 12.3$ when $x = 3.98$, $z = 3.25$, and $w = 2.07$. Find y when $x = -4.41$, $z = 5.00$, and $w = -1.00$.

23. At a constant speed v on level ground, the number N of gallons of gasoline used by an automobile is directly proportional to the amount of time T during which it travels. If the automobile uses 6 gallons in 1.5 hours, how many gallons will it use in 10 hours?

c **24.** The distance d in which an automobile can stop when its brakes are applied varies directly as the square of its speed v. Skid marks from a car involved in an accident measured 173 feet. In a test, the same car going 40 miles per hour was braked to a panic stop in 88 feet. Was the driver exceeding the speed limit (55 miles per hour) when the accident occurred?

25. The time t days required to pick up litter along a stretch of highway is inversely proportional to the number n of people on the work team. If 5 people can do the job in 3 days, how many days would be required for 8 people to do it?

c **26.** The amount H of heat conducted per second through a cube is jointly proportional to the temperature difference T degrees between the opposite faces and the area A of one face, and inversely proportional to the length l of an edge of the cube. The constant of proportionality for this combined variation is called the *coefficient of conductivity* for the material of the cube. Find the coefficient of conductivity for a cube with an edge that is 45 centimeters long if 1.26 joules of heat per second are conducted through the cube when the temperature difference between opposite faces is 31.5° Celsius.

27. In physics, Hooke's law for a perfectly elastic spring states that the distance d that the spring is stretched is directly proportional to the stretching force F. Suppose that a force $F = 3$ newtons stretches the spring a distance $d = 0.2$ meter (Figure 2).

Figure 2

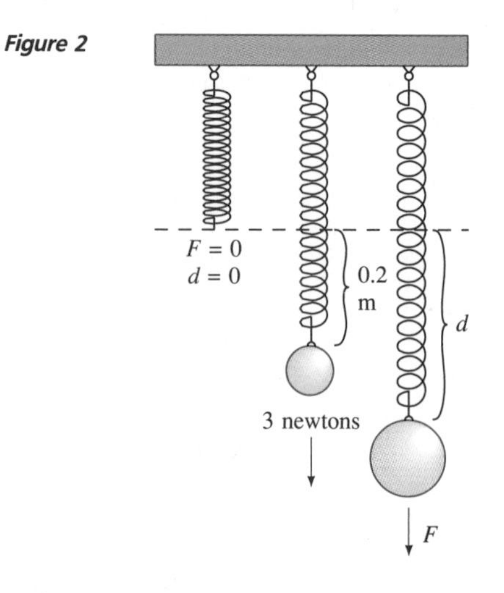

(a) Write a formula relating d to F.

(b) How far will a force of 9 newtons stretch the spring?

c **28.** Suppose that the weight w of men of a given body type varies directly as the cube of their height h. One man of this body type is 1.85 meters tall and weighs 80 kilograms. If a basketball player of this same body type is 2 meters tall, what is the basketball player's weight?

c **29.** The distance from the earth to the moon is approximately 3.8×10^8 meters, and the mass of the moon is approximately 7.3×10^{22} kilograms. Find the gravitational attraction of the moon on a person on earth whose mass is 70 kilograms. (See Example 5.)

C 30. Using the data in Problem 29, find the ratio of a person's weight on the moon to the same person's weight on earth. Take the radius of the moon to be 1.7×10^6 meters and the radius of the earth to be 6.4×10^6 meters.

C 31. Romeo's mass is 67 kilograms and Juliet's is 48 kilograms. Find the gravitational force of attraction between Romeo and Juliet when they are 0.5 meter apart.

C 32. A one-kilogram mass at the surface of the earth weighs 9.8 newtons and the radius of the earth is 6.4×10^6 meters. Find the mass of the earth.

1.4 Polynomials

An **algebraic expression** is an expression formed from any combination of numbers and variables by using the operations of addition, subtraction, multiplication, division, exponentiation (raising to powers), or extraction of roots. For instance,

$$7, \quad x, \quad 2x - 3y + 1, \quad \frac{5x^3 - 1}{4xy + 1}, \quad \pi r^2, \quad \text{and} \quad \pi r\sqrt{r^2 + h^2}$$

are algebraic expressions. By an algebraic expression *in* certain variables, we mean an expression that contains only those variables, and by a **constant,** we mean an algebraic expression that contains no variables at all. If numbers are substituted for the variables in an algebraic expression, the resulting number is called the **value** of the expression for these values of the variables.

Example 1 Find the value of $\dfrac{2x - 3y + 1}{xy^2}$ when $x = 2$ and $y = -1$.

Solution Substituting $x = 2$ and $y = -1$, we obtain

$$\frac{2(2) - 3(-1) + 1}{2(-1)^2} = \frac{4 + 3 + 1}{2(1)} = \frac{8}{2} = 4.$$

■

If an algebraic expression consists of parts connected by plus or minus signs, it is called an **algebraic sum,** and each of the parts, together with the sign preceding it, is called a **term.** For instance, in the algebraic sum

$$3x^2y - \frac{4xz^2}{y} + \pi x^{-1}y,$$

the terms are $3x^2y$, $-4xz^2/y$, and $\pi x^{-1}y$.

Any part of a term that is multiplied by the remaining part is called a **coefficient** of the remaining part. For instance, in the term $-4xz^2/y$, the coefficient of z^2/y is $-4x$, whereas the coefficient of xz^2/y is -4. A coefficient such as -4, which involves no variables, is called a **numerical coefficient.** Terms such as $5x^2y$ and $-12x^2y$, which differ only in their numerical coefficients, are called **like terms** or **similar terms.**

An algebraic expression such as $4\pi r^2$ can be considered an algebraic sum consisting of just one term. Such a one-termed expression is called a **monomial.** An algebraic sum with two terms is called a **binomial**, and an algebraic sum with three terms is called a **trinomial.** For instance, the expression $3x^2 + 2xy$ is a binomial, whereas $-2xy^{-1} + 3\sqrt{x} - 4$ is a trinomial. An algebraic sum with two or more terms is called a **multinomial.**

A **polynomial** is an algebraic sum in which no variables appear in denominators or under radical signs, and all variables that do appear are raised only to positive-integer powers. For instance, the trinomial

$$-2xy^{-1} + 3\sqrt{x} - 4$$

is *not* a polynomial; however, the trinomial

$$3x^2y^4 + \sqrt{2}xy - \tfrac{1}{2}$$

is a polynomial in the variables x and y. A term such as $-\frac{1}{2}$, which contains no variables, is called a **constant term** of the polynomial. The numerical coefficients of the terms in a polynomial are called the **coefficients** of the polynomial. The coefficients of the polynomial above are

$$3, \qquad \sqrt{2}, \qquad \text{and} \qquad -\tfrac{1}{2}.$$

The **degree of a term** in a polynomial is the sum of all the exponents of the variables in the term. In adding exponents, you should regard a variable with no exponent as being a first power. For instance, in the polynomial

$$9xy^7 - 12x^3yz^2 + 3x - 2,$$

the term $9xy^7$ has degree $1 + 7 = 8$, the term $-12x^3yz^2$ has degree $3 + 1 + 2 = 6$, and the term $3x$ has degree 1. The constant term, if it is nonzero, is always regarded as having degree 0.

The highest degree of all terms that appear with nonzero coefficients in a polynomial is called the **degree of the polynomial.** For instance, the polynomial considered above has degree 8. Although the constant monomial 0 is regarded as a polynomial, this particular polynomial is not assigned a degree.

Example 2 In each case, identify the algebraic expression as a monomial, binomial, trinomial, multinomial, and/or polynomial and specify the variables involved. For any polynomials, give the degree and the coefficients.

(a) $\dfrac{4x}{y}$ **(b)** $4x + 3y^{-1}$ **(c)** $-\frac{5}{3}x^2y + 8xy - 11$

Solution

(a) Monomial in x and y (*not* a polynomial because of the variable y in the denominator)

(b) Binomial in x and y, multinomial (*not* a polynomial because of the negative exponent on y)

(c) Trinomial, multinomial, polynomial in x and y of degree 3 with coefficients $-\frac{5}{3}$, 8, and -11 ■

A polynomial of degree n in a single variable x can be written in the **general form**

$$a_nx^n + a_{n-1}x^{n-1} + \cdots + a_2x^2 + a_1x + a_0,$$

in which $a_n, a_{n-1}, \ldots, a_2, a_1, a_0$ are the numerical coefficients, $a_n \neq 0$ (although any of the other coefficients can be zero), and a_0 is the constant term.

Addition and Subtraction of Polynomials

To find the sum of two or more polynomials, we use the associative and commutative properties of addition to group like terms together, and then we combine the like terms by using the distributive property.

Example 3 Find the following sum: $(2x^2 + 7x - 5) + (3x^2 - 11x + 8)$.

Solution

$$\begin{aligned}(2x^2 + 7x - 5) + (3x^2 - 11x + 8) &= (2x^2 + 3x^2) + (7x - 11x) + (-5 + 8)\\ &= 5x^2 - 4x + 3\end{aligned}$$ ■

To find the difference of two polynomials, we change the signs of all the terms in the polynomial being subtracted and then add.

Example 4 Find the following difference: $(3x^3 - 5x^2 + 8x - 3) - (5x^3 - 7x + 11)$.

Solution

$$\begin{aligned}&(3x^3 - 5x^2 + 8x - 3) - (5x^3 - 7x + 11)\\ &\quad = (3x^3 - 5x^2 + 8x - 3) + (-5x^3 + 7x - 11)\\ &\quad = (3x^3 - 5x^3) - 5x^2 + (8x + 7x) + (-3 - 11)\\ &\quad = -2x^3 - 5x^2 + 15x - 14\end{aligned}$$ ■

In adding or subtracting polynomials, you may prefer to use a vertical arrangement with like terms in the same columns.

Example 5 Perform the indicated operations:

$$(4x^3 + 7x^2y + 2xy^2 - 2y^3) + (2x^3 + xy^2 + 4y^3) + (4x^2y - 8xy^2 - 9y^3) - (y^3 - 7x^2y).$$

Solution First we use a vertical arrangement to perform the addition.

$$\begin{array}{rrrrr} & 4x^3 + & 7x^2y + & 2xy^2 - & 2y^3\\ (+) & 2x^3 & + & xy^2 + & 4y^3\\ (+) & & 4x^2y - & 8xy^2 - & 9y^3\\ \hline & 6x^3 + & 11x^2y - & 5xy^2 - & 7y^3\end{array}$$

Then we perform the subtraction vertically.

$$\begin{array}{rrrrr} & 6x^3 + & 11x^2y - & 5xy^2 - & 7y^3\\ (-) & - & 7x^2y & + & y^3\\ \hline & 6x^3 + & 18x^2y - & 5xy^2 - & 8y^3\end{array}$$ ■

Multiplication of Polynomials

To multiply two or more monomials, we use the commutative and associative properties of multiplication along with the following properties of exponents.

Properties of Exponents

Let a and b denote real numbers. Then, if m and n are positive integers,

(i) $a^m a^n = a^{m+n}$ **(ii)** $(a^m)^n = a^{mn}$ **(iii)** $(ab)^n = a^n b^n$

We verify (i) as follows and leave (ii) and (iii) as exercises (Problems 41 and 42).

$$a^m a^n = \overbrace{(a \cdot a \cdots a)}^{m \text{ factors}}\overbrace{(a \cdot a \cdots a)}^{n \text{ factors}} = \overbrace{a \cdot a \cdots a \cdot a \cdot a \cdots a}^{m+n \text{ factors}} = a^{m+n}.$$

Properties (i), (ii), and (iii) are useful for simplifying algebraic expressions containing exponents. In general, when the properties of real numbers are used to rewrite an algebraic expression as compactly as possible, or in a form so that further calculations are made easier, we say that the expression has been **simplified.** Thus, although the word "simplify" has no precise mathematical definition, its meaning is usually clear from the context in which it is used.

Example 6 Use the properties of exponents to simplify each expression.

(a) x^5x^4 **(b)** $2y^4y^6y^2z^2$
(c) $(5x^3y)(-3x^2y^4)$ **(d)** $(3x^2y^4)^2$
(e) $(-2x^4y^2)(3x^2y^3)(5xy^4)$ **(f)** $(x+3)^2(x+3)^4$

Solution

(a) $x^5x^4 = x^{5+4} = x^9$ [Exponent Property (i)]

(b) $2y^4y^6y^2z^2 = 2y^{4+6+2}z^2 = 2y^{12}z^2$ [Exponent Property (i)]

(c) $(5x^3y)(-3x^2y^4) = (5)(-3)(x^3x^2)(y^1y^4)$
$= -15x^5y^5$ [Exponent Property (i)]

(d) $(3x^2y^4)^2 = 3^2(x^2)^2(y^4)^2$ [Exponent Property (iii)]
$= 9x^{(2)(2)}y^{(4)(2)}$ [Exponent Property (ii)]
$= 9x^4y^8$

(e) $(-2x^4y^2)(3x^2y^3)(5xy^4) = (-2)(3)(5)(x^4x^2x)(y^2y^3y^4) = -30x^7y^9$

(f) $(x+3)^2(x+3)^4 = (x+3)^{2+4} = (x+3)^6$ ■

To multiply a polynomial by a monomial, we use the distributive property. Thus, we multiply each term of the polynomial by the monomial and then simplify the resulting products by using the properties of exponents.

Example 7 Find the product $(-3x^2y^3 + 5xy + 7)(4x^3y)$.

Solution

$$\begin{aligned}(-3x^2y^3 + 5xy + 7)(4x^3y) &= (-3x^2y^3)(4x^3y) + (5xy)(4x^3y) + 7(4x^3y)\\ &= -12x^5y^4 + 20x^4y^2 + 28x^3y\end{aligned}$$ ■

To multiply two polynomials, we again employ the distributive property. Thus, we multiply each term of the first polynomial by each term of the second and combine like terms. This procedure is sometimes called **expanding** the product.

Example 8 Expand the product $(x^4 - 5x^2 + 7)(3x^2 + 2)$.

Solution

$$\begin{aligned}(x^4 - 5x^2 + 7)(3x^2 + 2) &= (x^4 - 5x^2 + 7)(3x^2) + (x^4 - 5x^2 + 7)(2)\\ &= (3x^6 - 15x^4 + 21x^2) + (2x^4 - 10x^2 + 14)\\ &= 3x^6 - 13x^4 + 11x^2 + 14\end{aligned}$$

The same work is arranged vertically as follows:

$$\begin{array}{lrl} & x^4 - 5x^2 + 7 & \\ (\times) & 3x^2 + 2 & \\ \hline & 3x^6 - 15x^4 + 21x^2 & \longleftarrow (x^4 - 5x^2 + 7)(3x^2) \\ (+) & 2x^4 - 10x^2 + 14 & \longleftarrow (x^4 - 5x^2 + 7)(2) \\ \hline & 3x^6 - 13x^4 + 11x^2 + 14 & \longleftarrow \text{the sum of the above} \end{array}$$ ■

Certain products of polynomials occur so often that it is useful to know the expanded forms by heart. The following list contains some of these **special products.**

Special Products

If a, b, c, d, x, and y are real numbers, then:

1. $(a - b)(a + b) = a^2 - b^2$
2. $(a + b)^2 = a^2 + 2ab + b^2$
3. $(a - b)^2 = a^2 - 2ab + b^2$
4. $(a + b)(a^2 - ab + b^2) = a^3 + b^3$
5. $(a - b)(a^2 + ab + b^2) = a^3 - b^3$
6. $(a + b)^3 = a^3 + 3a^2b + 3ab^2 + b^3$
7. $(a - b)^3 = a^3 - 3a^2b + 3ab^2 - b^3$
8. $(ax + by)(cx + dy) = acx^2 + (ad + bc)xy + bdy^2$

You should verify the expansions in the list above by actually doing the multiplication. They should become so familiar that you use them automatically in your calculations.

Example 9 Use Special Products 1 to 8 to perform each multiplication.

(a) $(3x - 2y)(3x + 2y)$ **(b)** $(7x^2 + 3y^3)^2$ **(c)** $(3x + 4)(9x^2 - 12x + 16)$

(d) $[4(p + q) - 5]^2$ **(e)** $(5r + 6s)^3$ **(f)** $(2p - 3q + 4r)^2$

Solution

(a) $(3x - 2y)(3x + 2y) = (3x)^2 - (2y)^2 = 9x^2 - 4y^2$ (Special Product 1)

(b) $(7x^2 + 3y^3)^2 = (7x^2)^2 + 2(7x^2)(3y^3) + (3y^3)^2$ (Special Product 2)
$= 49x^4 + 42x^2y^3 + 9y^6$

(c) $(3x + 4)(9x^2 - 12x + 16)$
$= (3x + 4)[(3x)^2 - (3x)(4) + 4^2]$
$= (3x)^3 + 4^3$ (Special Product 4)
$= 27x^3 + 64$

(d) $[4(p + q) - 5]^2 = 16(p + q)^2 - 2[4(p + q)(5)] + 5^2$ (Special Product 3)
$= 16(p^2 + 2pq + q^2) - 40(p + q) + 25$ (Special Product 2)
$= 16p^2 + 32pq + 16q^2 - 40p - 40q + 25$

(e) $(5r + 6s)^3 = (5r)^3 + 3(5r)^2(6s) + 3(5r)(6s)^2 + (6s)^3$ (Special Product 6)
$= 125r^3 + 450r^2s + 540rs^2 + 216s^3$

(f) $(2p - 3q + 4r)^2$
$= [(2p - 3q) + 4r]^2$
$= (2p - 3q)^2 + 2(2p - 3q)(4r) + (4r)^2$ (Special Product 2)
$= 4p^2 - 12pq + 9q^2 + 16pr - 24qr + 16r^2$ ■

A general formula for expanding $(a + b)^n$ for any positive integer n is provided by the **binomial theorem:**

$$(a + b)^n = a^n + na^{n-1}b + \frac{n(n-1)}{1 \cdot 2}a^{n-2}b^2 + \frac{n(n-1)(n-2)}{1 \cdot 2 \cdot 3}a^{n-3}b^3 + \cdots + \frac{n(n-1)(n-2)\cdots(n-k+1)}{1 \cdot 2 \cdot 3 \cdot 4 \cdots k}a^{n-k}b^k + \cdots + nab^{n-1} + b^n$$

A proof and a more detailed discussion of this formula are given in Section 2 of the last chapter.

Example 10 Use the binomial theorem to find the expansion of $(2x + y)^4$.

Solution We substitute $a = 2x$, $b = y$, and $n = 4$ into the formula above to obtain

$$\begin{aligned}(2x + y)^4 &= (2x)^4 + 4(2x)^3y + \frac{4 \cdot 3}{1 \cdot 2}(2x)^2y^2 + \frac{4 \cdot 3 \cdot 2}{1 \cdot 2 \cdot 3}(2x)y^3 + y^4 \\ &= 16x^4 + 4(8x^3)y + 6(4x^2)y^2 + 4(2x)y^3 + y^4 \\ &= 16x^4 + 32x^3y + 24x^2y^2 + 8xy^3 + y^4.\end{aligned}$$ ■

Problem Set 1.4

In Problems 1 and 2, find the value of the algebraic expression for the given values of the variables.

1. (a) $-3xy + 5x + 2$ when $x = 2$ and $y = -1$
   (b) $\frac{1}{2}gt^2$ when $g = 32$ and $t = 5$

C 2. (a) prt when $p = 1254$, $r = 0.0820$, and $t = 3.750$
   (b) $\frac{1}{2}(-b + \sqrt{b^2 - 4ac})$ when $a = 3.35$, $b = 1.05$, and $c = -2.77$

In Problems 3 and 4, identify the algebraic expression as a monomial, binomial, trinomial, multinomial, and/or polynomial and specify the variables involved. For the polynomials, give the degree and the coefficients.

3. (a) $-4x^2$
   (b) $5x^3y^{-1} + 8x^2 + 1$
   (c) $3xz^2 - \frac{6}{11}x^2z - x - z + 2$
   (d) $\sqrt{2}x^5 + \pi x^4 - \sqrt{\pi}\,x^3 + \frac{12}{13}x^2 - 5$
   (e) $x^3 + y^3 + z^3 + w^3$

4. (a) $2\dfrac{x^4}{y} - 3\dfrac{xy^2}{z} + 2xyz - 17$
   (b) $\sqrt{x} + \sqrt{y} + \sqrt{x^2 + w^2}$
   (c) $-\dfrac{b}{2a} + \dfrac{\sqrt{b^2 - 4ac}}{2a}$
   (d) $pq - \pi$
   (e) $x^{-1}y^{-1}z^{-1}$

In Problems 5 to 8, perform the indicated operations.

5. (a) $(-5x + 1) + (7x + 11)$
   (b) $(4z^2 + 2) + (-3z^2 + z)$
   (c) $(7x^2 - x + 9) + (11x^2 + 2x - 4)$
   (d) $5s^2 + (\pi r^2 + 2s^2) - 2s^2$

6. (a) $3\pi r^2 + (2\pi^2 h - \pi r^2)$
   (b) $(n^3 - 5n^2 + 3n + 4) - (-2n^3 + 4n - 3)$
   (c) $(4t^3 - 3t^2 + 2t + 1) + (-2t^3 + 5t^2 + 3t + 4)$
   (d) $(5x^2y - 3xy^2 + 7xy - 11) + (-3x^2y + 7xy^2 + 4xy + 8)$
   (e) $(-4pq^2 + 5p^2q + 11pq + 7) - (-7pq^2 + 4pq - 5p + 4)$

7. (a) $(3x + 2) - (4x - 7)$
   (b) $(3z^2 + 7z + 5) - (-z^2 - 3z + 2)$
   (c) $(5t^2 + 4t - 3) + (-9t^2 + 2t + 1) - (3t^2 - 8t + 7)$
   (d) $x - \{[xy + (1 + x)] - (-xy + x^2)\}$

8. (a) $(3x^2 - xy + y^2 - 2x + y - 5) + (7x^2 + 2y^2 - x + 2) - (8x^2 + 4xy - 3y^2 - x)$
   (b) $(\frac{2}{3}x^2 - \frac{1}{4}x + \frac{1}{2}) - (\frac{1}{5}x^2 - \frac{1}{3}x + \frac{3}{4})$

In Problems 9 to 12, use the properties of exponents to simplify each expression.

9. (a) $(3x^4)(2x^3)$ (b) $2^{3n}2^{7n}$
   (c) $(3x + y)^2(3x + y)^4$ (d) $(t^4)^{12}$

10. (a) $(-y)^4(-y)^5$ (b) $(4t^{2n})(3t)^n$
    (c) $(5r - 3t)^5(3t - 5r)^2(3t - 5r)$ (d) $(-x^4)^{11}$

11. (a) $(u^n)^{5n}$ (b) $(3v)^4(2v)^2$
    (c) $(5a^2b)^2(3ab^2)^3$ (d) $[2(x + 3y)^2]^5[3(x + 3y)^4]^3$

12. (a) $[-(3x + y)^4]^3$ (b) $(-5r^4)^3$
    (c) $(rs)^n(r^2s)^{2n}(rs^2t)^{3n}$ (d) $(n^n)^n$

In Problems 13 to 24, expand each product.

13. $5x^3y(3x - 4xy + 4z)$
14. $-a^2b(-3a^2 + 4b + 2b^2)$
15. $(3x + 2y)(-4x + 5y)$
16. $(2p^3 + 5q)(3p^2 - q)$
17. $(x + 2y)(x^2 - 2xy + 4y^2)$
18. $(r^2 + 3r + s^2)(r^2 + 3r - s^2)$
19. $(5c + d)(5c - d)(25c^2 + d^2)$
20. $(6x^3 + 2x^2 - 3x + 1)(2x^2 - 4x + 3)$
21. $(x^2 + 7xy - y^2)(3x^2 - xy + y^2 - 2x - y + 2)$
22. $(2x^{2n} - x^n - 1)(3x^{2n} + 2x^n + 2)$
23. $(4p^2q - 3pq^2 + 5pq)(p^2q - 2pq^2 - 3pq)$
24. $(3x^3 - 2x^2 + x - 1)(x^4 - x^3 + 5x^2 - 2x + 1)$

In Problems 25 to 40, use the appropriate special products to perform each multiplication.

25. $(4 + 3x)^2$
26. $(5p^2 - 2q^2)^2$

27. $[(4t^2 + 1) - s]^2$

28. $(5u + 2v + w)^2$

29. $(3r - 2s)(3r + 2s)$

30. $(4a^n - b^n)(4a^n + b^n)$

31. $[(2a - 3b) - 4c][(2a - 3b) + 4c]$

32. $[(x + 3y) - 2z][(x + 3y) + 2z]$

33. $(2 + t)(4 - 2t + t^2)$

34. $(3 - u)(9 + 3u + u^2)$

35. $(2x^2 + 3y)^3$

36. $(p^n - 2q^n)^3$

37. $(2x - y + z)^2$

38. $(1 + y + z)[(1 + y)^2 - (1 + y)z + z^2]$

39. $(t^3 - 2t^2 + 5t + 2)^2$

40. $[(x + y)^2]^3$

41. If a is a real number and n and m are positive integers, show that $(a^m)^n = a^{mn}$.

42. If a and b are real numbers and n is a positive integer, show that $(ab)^n = a^n b^n$.

43. If a and b are real numbers, $x = a^2 - b^2$, $y = 2ab$, and $z = a^2 + b^2$, show that $x^2 + y^2 = z^2$.

44. Using the result of Problem 43, find several triples of positive integers x, y, z such that $x^2 + y^2 = z^2$. [*Hint:* Substitute positive integers for a and b.]

45. The length of a rectangle is $5x - 9$ units and the width is $2x + 3$ units (Figure 1).

Figure 1

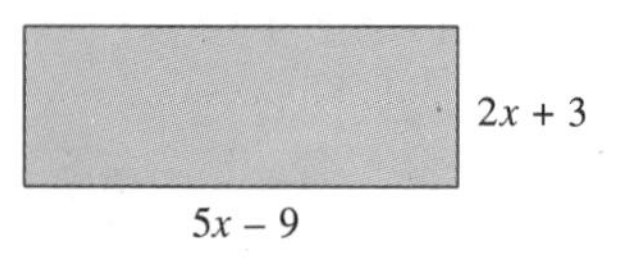

(a) Write a polynomial expression for the perimeter of the rectangle in terms of x and simplify the result.

(b) Write a polynomial expression for the area of the rectangle in terms of x and expand the result.

46. The radius of the base of a solid right circular cylinder is $3x - 2$ units and the height is $5x + 3$ units (Figure 2). Find an expression for the total surface area of the cylinder, including the top and bottom, and simplify the result. [*Hint:* Use $A = 2\pi r^2 + 2\pi rh$.]

Figure 2

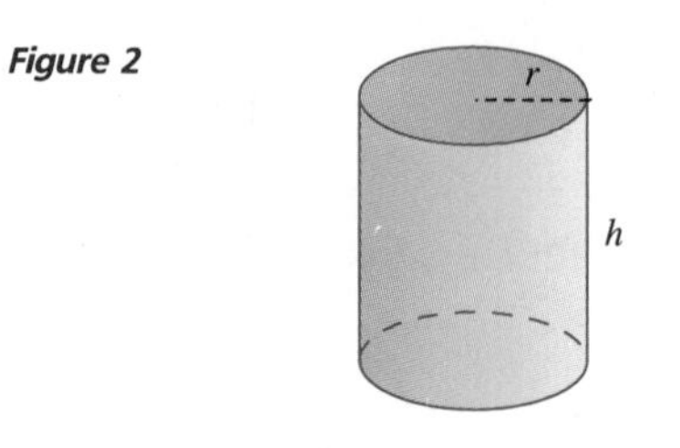

47. The edge of a cube is $4x + 3$ units (Figure 3). Write an expression for the volume of the cube in terms of x and then expand the result.

Figure 3

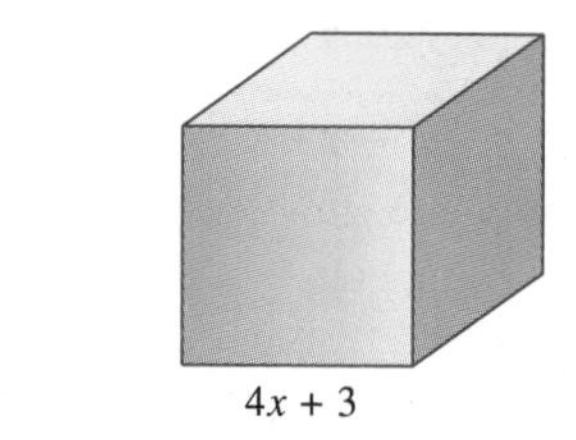

48. The radius of a sphere is $3x^2 + 2$ units (Figure 4). Write an expression for the volume of the sphere in terms of x and then expand the result.

Figure 4

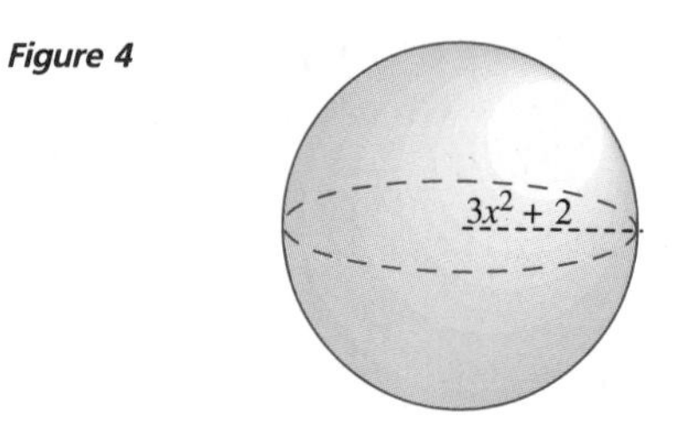

In Problems 49 to 52, use the binomial theorem to expand each expression.

49. $(x + 2y)^4$

50. $(x^2 + 3y^2)^4$

51. $(2x - y^2)^5$

52. $(x^2 - 2y)^6$

1.5 Factoring Polynomials

When two or more algebraic expressions are multiplied, each expression is called a **factor** of the product. For instance, in the product

$$(x - y)(x + y)(2x^2 - y)x,$$

the factors are $x - y$, $x + y$, $2x^2 - y$, and x. Often we are given a product in its expanded form and we need to find the original factors. The process of finding these factors is called **factoring.**

In this section, we confine our study of factoring to polynomials with integer coefficients. Thus, we shall not yet consider such possibilities as $5x^2 - y^2 = (\sqrt{5}x - y)(\sqrt{5}x + y)$, because $\sqrt{5}$ isn't an integer.

Of course, we can factor any polynomial "trivially" by writing it as 1 times itself or as -1 times its negative. A polynomial with integer coefficients that cannot be factored (except trivially) into two or more polynomials with integer coefficients is said to be **prime.** When a polynomial is written as a product of prime factors, we say that it is **factored completely.**

Removing a Common Factor

The distributive property can be used to factor a polynomial in which all the terms contain a common factor. The following example illustrates how to "*remove the common factor.*"

Example 1 Factor each polynomial by removing the common factor.

(a) $20x^2y + 8xy$ **(b)** $u(v + w) + 7v(v + w)$

Solution **(a)** Here $4xy$ is a common factor of the two terms, since

$$20x^2y = (4xy)(5x) \quad \text{and} \quad 8xy = (4xy)(2).$$

Therefore,

$$20x^2y + 8xy = (4xy)(5x) + (4xy)(2) = 4xy(5x + 2).$$

(b) Here the common factor is $v + w$, and we have

$$u(v + w) + 7v(v + w) = (u + 7v)(v + w).$$

■

Factoring by Recognizing Special Products

Success in factoring depends on your ability to recognize patterns in the polynomials to be factored—an ability that grows with practice. Special Products 1 to 8 in Section 1.4, read from right to left, suggest useful patterns; for instance, Special Product 1, read as $a^2 - b^2 = (a - b)(a + b)$, shows that a **difference of two squares** can always be factored.

The formula for factoring a difference of two squares has an interesting geometric interpretation that was known to the ancient Greeks. In Figure 1a the quantity $a^2 - b^2$ is represented by the shaded area A—literally a difference of two squares. If this area is cut apart as in Figure 1b and reassembled as in Figure 1c, it becomes evident that

$$a^2 - b^2 = A = A_1 + A_2 = (a - b)(a + b).$$

Figure 1

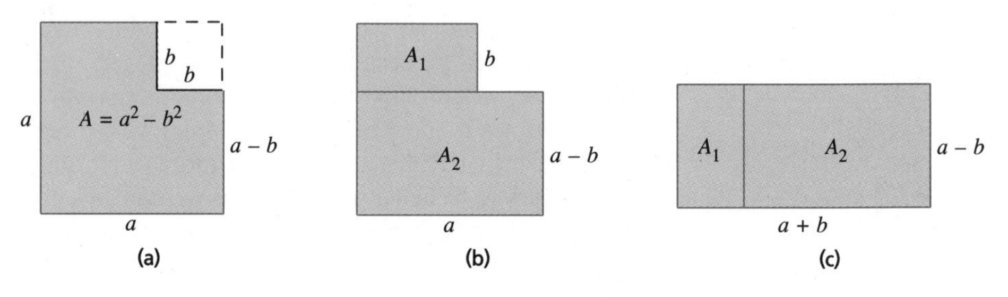

Special Product 5, read as

$$a^3 - b^3 = (a - b)(a^2 + ab + b^2),$$

shows that a **difference of two cubes** can also be factored. This can also be interpreted geometrically by representing $a^3 - b^3$ as the difference in the volumes of two cubes (Problem 66).

Example 2 Use the special products on page 29 to factor each expression.

(a) $25t^2 - 16s^2$ **(b)** $x^3 + 64y^3$ **(c)** $27r^3 - 8c^3$
(d) $16x^6 - (3y + 2z)^2$ **(e)** $81b^4 - 1$ **(f)** $(x + y)^3 - (z - w)^3$

Solution **(a)** Notice that $25t^2 = (5t)^2$ and $16s^2 = (4s)^2$, so the given expression is a difference of two squares. Thus, we have

$$25t^2 - 16s^2 = (5t)^2 - (4s)^2 = (5t - 4s)(5t + 4s).$$

(b) Since $64 = 4^3$, the expression $x^3 + 64y^3$ is a **sum of two cubes.** Using Special Product 4 in the form $a^3 + b^3 = (a + b)(a^2 - ab + b^2)$, with $a = x$ and $b = 4y$, we have

$$x^3 + 64y^3 = x^3 + (4y)^3 = (x + 4y)(x^2 - 4xy + 16y^2).$$

(c) Here we have a difference of two cubes, so

$$\begin{aligned} 27r^3 - 8c^3 = (3r)^3 - (2c)^3 &= (3r - 2c)[(3r)^2 + (3r)(2c) + (2c)^2] \\ &= (3r - 2c)(9r^2 + 6rc + 4c^2). \end{aligned}$$

(d) The expression is a difference of two squares, so

$$\begin{aligned}16x^6 - (3y + 2z)^2 &= (4x^3)^2 - (3y + 2z)^2\\ &= [4x^3 + (3y + 2z)][4x^3 - (3y + 2z)]\\ &= (4x^3 + 3y + 2z)(4x^3 - 3y - 2z).\end{aligned}$$

(e) Using Special Product 1 *twice*, we have

$$\begin{aligned}81b^4 - 1 &= (9b^2)^2 - 1\\ &= (9b^2 - 1)(9b^2 + 1)\\ &= (3b - 1)(3b + 1)(9b^2 + 1).\end{aligned}$$

(f) $(x + y)^3 - (z - w)^3$

$$\begin{aligned}&= [(x + y) - (z - w)][(x + y)^2 + (x + y)(z - w) + (z - w)^2]\\ &= (x + y - z + w)(x^2 + 2xy + y^2 + xz - xw + yz - yw + z^2 - 2zw + w^2)\end{aligned}$$ ■

Factoring Trinomials of the Type $ax^2 + bx + c$

If a, b, and c are integers and $a \neq 0$, it may be possible to factor a trinomial of the type $ax^2 + bx + c$ into a product of two binomials. For instance,

$$8x^2 + 22x + 15 = (2x + 3)(4x + 5).$$

Notice the relationship between the coefficients of the trinomial and the coefficients of the factors. The coefficients 8 and 15 are obtained as follows:

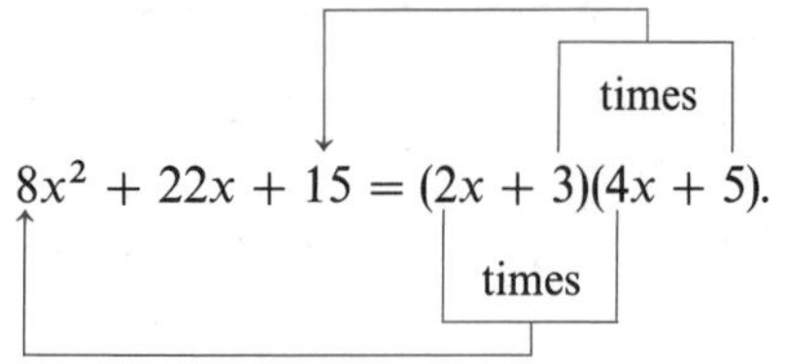

$$8x^2 + 22x + 15 = (2x + 3)(4x + 5).$$

times

times

The coefficient of the middle term in the trinomial, namely 22, is obtained as follows:

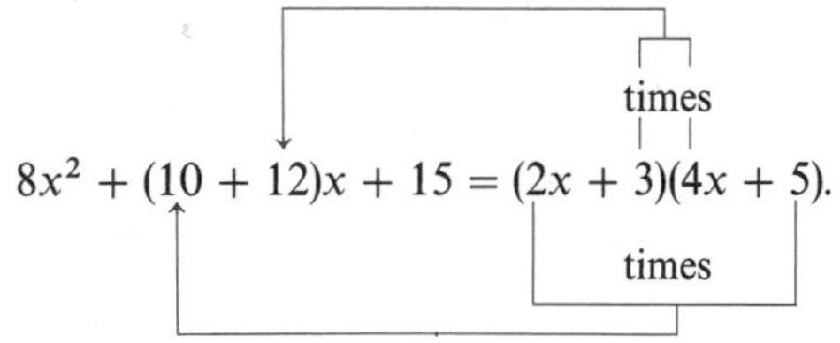

$$8x^2 + (10 + 12)x + 15 = (2x + 3)(4x + 5).$$

times

times

Thus, to factor a trinomial $ax^2 + bx + c$, begin by writing

$$ax^2 + bx + c = (\underset{?}{\uparrow}\ x + \underset{?}{\uparrow}\ )(\underset{?}{\uparrow}\ x + \underset{?}{\uparrow}\ ),$$

where the blanks are to be filled in with integers. The product of the unknown coefficients of x must be the integer a, and the product of the unknown constant terms must be the integer c. Just try all possible choices of such integers until you find a combination that gives the desired middle coefficient b. If all of the coefficients of the trinomial are positive, you need only try combinations of positive integers.

Example 3 Factor the trinomial $2x^2 + 9x + 4$.

Solution Since the only two positive integers whose product is 2 are 2 and 1, we can begin by writing

$$2x^2 + 9x + 4 = (2x + \underset{?}{\uparrow}\ )(x + \underset{?}{\uparrow}\ ).$$

Because 4 can be factored as $4 = 4 \cdot 1$ or as $4 = 2 \cdot 2$ or as $4 = 1 \cdot 4$, there are just three possible ways to fill the remaining blanks:

$$(2x + 4)(x + 1) \quad \text{or} \quad (2x + 2)(x + 2) \quad \text{or} \quad (2x + 1)(x + 4).$$

Now, $(2x + 4)(x + 1) = 2x^2 + 6x + 4$, and that isn't what we want. Also, $(2x + 2)(x + 2) = 2x^2 + 6x + 4$, and that isn't what we want either. But, $(2x + 1)(x + 4) = 2x^2 + 9x + 4$, and that *is* what we want! Therefore,

$$2x^2 + 9x + 4 = (2x + 1)(x + 4).$$

■

If some of the coefficients of the trinomial $ax^2 + bx + c$ are negative, you will have to try combinations with negative integers.

Example 4 Factor the trinomial $6x^2 + 13x - 5$.

Solution Because -5 can be factored as $-5 = 5 \cdot (-1)$ or as $-5 = (-5) \cdot 1$, we can begin by writing

$$6x^2 + 13x - 5 = (\underset{?}{\uparrow}\ x \pm 5)(\underset{?}{\uparrow}\ x \mp 1),$$

where the two blanks must still be filled in, and we must choose the correct algebraic signs. The possibilities for the blanks are given by $6 = 6 \cdot 1 = 3 \cdot 2 = 2 \cdot 3 = 1 \cdot 6$. Hence, the possible combinations are

$$\begin{array}{llll} & (6x + 5)(x - 1) & \text{or} & (6x - 5)(x + 1) \\ \text{or} & (3x + 5)(2x - 1) & \text{or} & (3x - 5)(2x + 1) \\ \text{or} & (2x + 5)(3x - 1) & \text{or} & (2x - 5)(3x + 1) \\ \text{or} & (x + 5)(6x - 1) & \text{or} & (x - 5)(6x + 1). \end{array}$$

We try each of these, one at a time, until we find that $(2x + 5)(3x - 1)$ works. Therefore,

$$6x^2 + 13x - 5 = (2x + 5)(3x - 1).$$

■

The method illustrated above can also be used to factor trinomials of the form $ax^2 + bxy + cy^2$.

Example 5 Factor the trinomial $6x^2 - 19xy + 3y^2$.

Solution We begin by writing

$$6x^2 - 19xy + 3y^2 = (\underset{?}{\uparrow}\, x - 3y)(\underset{?}{\uparrow}\, x - y).$$

The two minus signs provide for the positive coefficient $+3$ of y^2 and for the negative coefficient -19 of the middle term. Since $6 = 6 \cdot 1 = 3 \cdot 2 = 2 \cdot 3 = 1 \cdot 6$, there are four possible ways to fill in the two blanks. We try them one at a time until we find that $(x - 3y)(6x - y)$ works. Therefore,

$$6x^2 - 19xy + 3y^2 = (x - 3y)(6x - y).$$

■

If, in attempting to factor a trinomial $ax^2 + bx + c$ or $ax^2 + bxy + cy^2$, you find that none of the possible combinations works, you can conclude that the trinomial is prime. However, you can test to see if it is prime without bothering to try all these combinations just by evaluating the expression

$$b^2 - 4ac.$$

If $b^2 - 4ac$ is the square of an integer, then the trinomial can be factored; otherwise, it is prime. (You will see why this test works in Section 2.3.)

Example 6 Test each trinomial to see whether it is factorable or prime. If it can be factored, do so.

(a) $4x^2 - 8x + 3$ **(b)** $3x^2 - 5xy + y^2$

Solution **(a)** Here $a = 4$, $b = -8$, $c = 3$, and

$$b^2 - 4ac = (-8)^2 - 4(4)(3) = 64 - 48 = 16 = 4^2.$$

Therefore, the trinomial can be factored. Indeed,

$$4x^2 - 8x + 3 = (2x - 1)(2x - 3).$$

(b) Here $a = 3$, $b = -5$, $c = 1$, and

$$b^2 - 4ac = (-5)^2 - 4(3)(1) = 25 - 12 = 13.$$

Because 13 is not the square of an integer, $3x^2 - 5xy + y^2$ is prime. ■

Factoring by Grouping

Sometimes we have to group the terms of a polynomial in a certain way in order to see how it can be factored.

Example 7 Factor each expression by grouping the terms in a suitable way.

(a) $ac + d - c - ad$ **(b)** $x^2 + 4xy - 9c^2 + 4y^2$

(c) $4x^3 - 8x^2 - x + 2$ **(d)** $x^4 + 6x^2y^2 + 25y^4$

Solution

(a) $ac + d - c - ad = (ac - c) + (d - ad) = c(a - 1) + d(1 - a)$
$= c(a - 1) - d(a - 1) = (c - d)(a - 1)$

(b) $x^2 + 4xy - 9c^2 + 4y^2 = (x^2 + 4xy + 4y^2) - 9c^2 = (x + 2y)^2 - (3c)^2$
$= [(x + 2y) - 3c][(x + 2y) + 3c]$
$= (x + 2y - 3c)(x + 2y + 3c)$

(c) $4x^3 - 8x^2 - x + 2 = 4x^2(x - 2) - x + 2 = 4x^2(x - 2) - (x - 2)$
$= (4x^2 - 1)(x - 2) = (2x - 1)(2x + 1)(x - 2)$

(d) If the middle term were $10x^2y^2$ rather than $6x^2y^2$, we could factor the expression as $x^4 + 10x^2y^2 + 25y^4 = (x^2 + 5y^2)^2$. But the middle term can be changed to $10x^2y^2$ by adding $4x^2y^2$ to the $6x^2y^2$ already there and then subtracting $4x^2y^2$ at the end of the expression. Thus,

$$\begin{aligned} x^4 + 6x^2y^2 + 25y^4 &= (x^4 + 10x^2y^2 + 25y^4) - 4x^2y^2 \\ &= (x^2 + 5y^2)^2 - (2xy)^2 \\ &= [(x^2 + 5y^2) - 2xy][(x^2 + 5y^2) + 2xy] \\ &= (x^2 - 2xy + 5y^2)(x^2 + 2xy + 5y^2). \end{aligned}$$

■

Problem Set 1.5

In Problems 1 to 4, factor each polynomial by removing the common factor.

1. $a^3b + 2a^2b + a^2b^2$
2. $14a^2b - 35ab - 63ab^2$
3. $2(x - y)r^2 + 2(x - y)rh$
4. $a^2(s + 2t)^2 + a(-s - 2t)$

In Problems 5 to 16, use the special products (page 29) to factor each expression.

5. $64x^2 - 36y^2$
6. $4u^2 - 25v^2$
7. $81y^4 - z^4$
8. $x^4 - 16y^4$
9. $(x + y)^2 - 49z^2$
10. $100a^4 - (3a + 2b)^2$
11. $8x^3 + 1000$
12. $64a^3 + 27b^3$
13. $64x^3 - y^6$
14. $8a^3 + (b + c)^3$
15. $x^6 + 512y^6$
16. $m^9 + n^9$

In Problems 17 to 24, factor each trinomial.

17. $8x^2 + 10x - 7$
18. $30 - 49w + 6w^2$
19. $2v^2 + 3v - 20$
20. $8t^4 - 14t^2 - 15$
21. $4r^2 - 12rs + 9s^2$
22. $6(u + v)^2 - 5(u + v) - 6$
23. $6x^4 + 13x^2 + 6$
24. $8x^2 + 22x(y + 2z) + 5(y + 2z)^2$

In Problems 25 to 30, test the trinomial to see whether it is factorable or prime. If it can be factored, do so.

25. $x^2 + x + 1$
26. $16x^2 - 12xy + y^2$
27. $12r^2 - 11r + 2$
28. [C] $52a^2 - 37ab - 35b^2$
29. $2a^2 - 6a + 5$
30. [C] $33(x + y)^2 - 22(x + y) + 3$

In Problems 31 to 36, factor each expression by grouping the terms in a suitable way.

31. $3x - 2y - 6 + xy$
32. $3rs - 2t - 3rt + 2s$
33. $a^2x + a^2d - x - d$
34. $x^2 + 3x - y^2 + 3y$
35. $4x^2 + 12x - 9y^2 + 9$
36. $9a^4 + 8a^2b^2 + 4b^4$

In Problems 37 to 64, factor each polynomial completely.

37. $75x^2 - 48y^2$
38. $16(u + v)^2 - 81$
39. $7x(y - z) + 14x^2(y - z)$
40. $(x + y)^2 - (2z - 3)^2$
41. $33w^2 + 14wz - 40z^2$
42. $9t^2 + s^2 - r^4 - 6ts$
43. $64x^6 + (p - q)^3$
44. $x^8 - 16$
45. $x^5 - 4x^3 + x^2 - 4$
46. $2m^3np^2 + 7m^2n^2p^2 - 15mn^3p^2$
47. $12u^2v + 9 - 4u^2 - 27v$
48. $(x - 2y)^4 - (2x - y)^4$
49. $p^4 + q^4 - pq^3 - p^3q$
50. $2(3 - 4x)^2 + (x^2 - 8x + 5)(4x - 3)$
51. $x^4 + x^2y^2 + y^4$
52. $9r^4 + 2r^2s^2 + s^4$
53. $t^4 - 8t^2 + 16$
54. $u^4 + 5u^2v^2 + 9v^4$
55. $(2s + 1)a^3 - (2s + 1)b^3$
56. $s^8 - 82s^4 + 81$
57. $x^8 - 256y^8$
58. $4(x + y)^2z - 25w - 25z + 4(x + y)^2w$
59. $10x^2y^3z - 13xy^4z - 3y^5z$
60. $x^6 - 4 - x^4 + 4x^2$
61. $x^2 + x - 9y^2 - 3y$
62. $a^3b^3 + a^3 - b^3 - 1$
63. $3(3a - 2b)^2 + 3a - 2b - 14$
64. $3s^2t^{2n} - st^n - 10$

65. The following expressions were obtained as solutions to problems in a calculus textbook. Factor each expression and then simplify the factors.

(a) $2(3x^2 + 7)(6x)(5 - 3x)^3 + 3(3x^2 + 7)^2(-3)(5 - 3x)^2$

(b) $2(5t^2 + 1)(10t)(3t^4 + 2)^4 + 4(5t^2 + 1)(12t^3)(3t^4 + 2)^3$

66. Give a geometric interpretation for the formula $a^3 - b^3 = (a - b)(a^2 + ab + b^2)$ by considering the difference in the volumes of two cubes to be $a^3 - b^3$ (Figure 2).

Figure 2

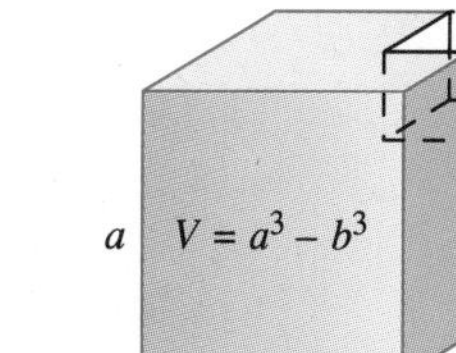

67. For purposes of shielding from stray electromagnetic radiation, an electronic device is housed in a metal cannister in the shape of a right circular cylinder with a total surface area A given by $A = 2\pi r^2 + 2\pi rh$ (Figure 3). Factor the right side of this formula.

Figure 3

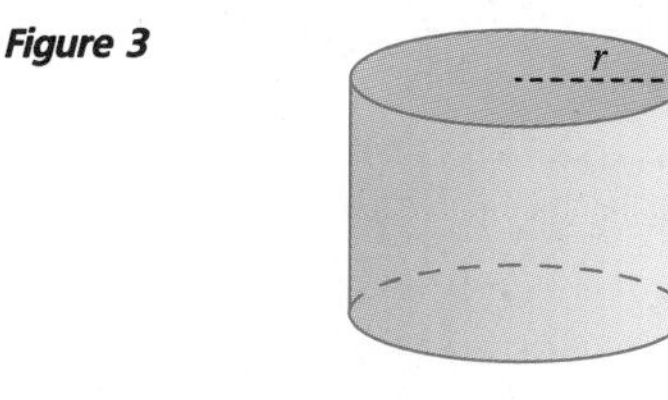

1.6 Fractions

If p and q are algebraic expressions, the quotient, or ratio, p/q is called a **fractional expression** (or simply a **fraction**) with **numerator** p and **denominator** q. Always remember that the denominator of a fraction cannot be zero. If $q = 0$, the expression p/q simply has no meaning. Therefore, *whenever we use a fractional expression, we shall automatically assume that the variables involved are restricted to numerical values that will give a nonzero denominator.*

A fractional expression in which both the numerator and the denominator are polynomials is called a **rational expression.** Examples of rational expressions are

$$\frac{2x}{1}, \quad \frac{3}{y}, \quad \frac{5x+3}{17}, \quad \frac{2st-t^3}{3s^4-t^4}, \quad \text{and} \quad \frac{x^2-1}{y^2-1}.$$

Notice that a fraction such as $3x/\sqrt{y}$ isn't a rational expression because its denominator isn't a polynomial. We say that a rational expression is **reduced to lowest terms** or **simplified** if its numerator and denominator have no common factors (other than 1 or -1). Thus, to simplify a rational expression, we factor both the numerator and the denominator into prime factors and then **cancel** common factors by using the following property.

The Cancellation Property for Fractions

If $q \neq 0$ and $k \neq 0$, then $\dfrac{pk}{qk} = \dfrac{p}{q}$.

Cancellation is usually indicated by slanted lines drawn through the canceled factors; for instance,

$$\frac{x^2-1}{x^2-3x+2} = \frac{\cancel{(x-1)}(x+1)}{\cancel{(x-1)}(x-2)} = \frac{x+1}{x-2}.$$

If one fraction can be obtained from another by canceling common factors or by multiplying numerator and denominator by the same nonzero expression, then the two fractions are said to be **equivalent.** Thus, the calculation above shows that

$$\frac{x^2-1}{x^2-3x+2} \quad \text{and} \quad \frac{x+1}{x-2}$$

are equivalent fractions.

Example 1 Reduce each fraction to lowest terms.

(a) $\dfrac{14x^7y}{56x^3y^2}$ **(b)** $\dfrac{cd-c^2}{c^2-d^2}$

(c) $\dfrac{5x^2-14x-3}{2x^2+x-21}$ **(d)** $\dfrac{16x-32}{8y(x-2)-4(x-2)}$

Solution **(a)** $\dfrac{14x^7y}{56x^3y^2} = \dfrac{\cancel{14x^3y}\cdot x^4}{\cancel{14x^3y}\cdot 4y} = \dfrac{x^4}{4y}$

(b) First, we factor the numerator and denominator, and then we use the fact that $d - c = -(c - d)$:

$$\frac{cd-c^2}{c^2-d^2} = \frac{c(d-c)}{(c-d)(c+d)} = \frac{-c\cancel{(c-d)}}{\cancel{(c-d)}(c+d)} = \frac{-c}{c+d}.$$

(c) $\dfrac{5x^2 - 14x - 3}{2x^2 + x - 21} = \dfrac{(5x+1)(x-3)}{(2x+7)(x-3)} = \dfrac{5x+1}{2x+7}$

(d) $\dfrac{16x - 32}{8y(x-2) - 4(x-2)} = \dfrac{16(x-2)}{(x-2)(8y-4)} = \dfrac{(4)(4)(x-2)}{4(x-2)(2y-1)} = \dfrac{4}{2y-1}$ ■

Multiplication and Division of Fractions

The following rules for multiplication and division of fractions can be derived from the basic algebraic properties of the real numbers and the definition of a quotient.

1. *Rule for Multiplication of Fractions*

If p, q, r, and s are real numbers, $q \neq 0$, and $s \neq 0$, then

$$\frac{p}{q} \cdot \frac{r}{s} = \frac{pr}{qs}.$$

For instance,

$$\frac{2}{3} \cdot \frac{5}{7} = \frac{2 \cdot 5}{3 \cdot 7} = \frac{10}{21}.$$

2. *Rule for Division of Fractions*

If p, q, r, and s are real numbers, $q \neq 0$, $r \neq 0$, and $s \neq 0$, then

$$\frac{p}{q} \div \frac{r}{s} = \frac{p/q}{r/s} = \frac{p}{q} \cdot \frac{s}{r} = \frac{ps}{qr}.$$

For instance,

$$\frac{2}{3} \div \frac{5}{7} = \frac{2}{3} \cdot \frac{7}{5} = \frac{14}{15} \quad \text{and} \quad \frac{3/4}{4/5} = \frac{3}{4} \cdot \frac{5}{4} = \frac{15}{16}.$$

Of course, the same rules apply to multiplying or dividing fractional expressions in general. Before multiplying fractions, it's a good idea to factor the numerators and denominators completely to reveal any factors that can be canceled.

Example 2 Perform the indicated operation and simplify the result.

(a) $\dfrac{x+4}{y} \cdot \dfrac{y^3}{5x+20}$

(b) $\dfrac{x^3 - 1}{5x^2 - 26x + 5} \cdot \dfrac{5x^2 + 9x - 2}{x^2 + x - 2}$

(c) $\dfrac{t^2 - 49}{t^2 - 5t - 14} \div \dfrac{2t^2 + 15t + 7}{2t^2 - 13t - 7}$

Solution (a) $\dfrac{x+4}{y} \cdot \dfrac{y^3}{5x+20} = \dfrac{x+4}{y} \cdot \dfrac{y \cdot y^2}{5(x+4)} = \dfrac{(x+4)y \cdot y^2}{y(5)(x+4)} = \dfrac{y^2}{5}$

(b)
$$\begin{aligned}\frac{x^3-1}{5x^2-26x+5}\cdot\frac{5x^2+9x-2}{x^2+x-2} &= \frac{(x-1)(x^2+x+1)}{(5x-1)(x-5)}\cdot\frac{(5x-1)(x+2)}{(x+2)(x-1)}\\ &= \frac{\cancel{(x-1)}(x^2+x+1)\cancel{(5x-1)}\cancel{(x+2)}}{\cancel{(5x-1)}(x-5)\cancel{(x+2)}\cancel{(x-1)}}\\ &= \frac{x^2+x+1}{x-5}\end{aligned}$$

(c)
$$\begin{aligned}\frac{t^2-49}{t^2-5t-14}\div\frac{2t^2+15t+7}{2t^2-13t-7} &= \frac{t^2-49}{t^2-5t-14}\cdot\frac{2t^2-13t-7}{2t^2+15t+7}\\ &= \frac{(t-7)(t+7)}{(t+2)(t-7)}\cdot\frac{(2t+1)(t-7)}{(2t+1)(t+7)}\\ &= \frac{\cancel{(t-7)}\cancel{(t+7)}\cancel{(2t+1)}(t-7)}{(t+2)\cancel{(t-7)}\cancel{(2t+1)}\cancel{(t+7)}} = \frac{t-7}{t+2}\end{aligned}$$
■

Addition and Subtraction of Fractions

Two or more fractions with the same denominator are said to have a **common denominator.** The following rules for adding and subtracting fractions with a common denominator can be derived from the basic algebraic properties of the real numbers and the definition of a quotient.

Rules for Addition and Subtraction of Fractions with a Common Denominator

If p, q, and r are real numbers and $q \neq 0$, then

$$\frac{p}{q}+\frac{r}{q}=\frac{p+r}{q} \quad \text{and} \quad \frac{p}{q}-\frac{r}{q}=\frac{p-r}{q}.$$

For instance,

$$\frac{3}{7}+\frac{2}{7}=\frac{3+2}{7}=\frac{5}{7} \quad \text{and} \quad \frac{3}{7}-\frac{2}{7}=\frac{3-2}{7}=\frac{1}{7}.$$

Again, the same rules apply to adding or subtracting fractional expressions.

Example 3 Perform each operation.

(a) $\dfrac{5x}{2x-1}+\dfrac{3x}{2x-1}$ **(b)** $\dfrac{5x}{(3x-2)^2}-\dfrac{3x}{(3x-2)^2}$

Solution

(a) $$\frac{5x}{2x-1}+\frac{3x}{2x-1}=\frac{5x+3x}{2x-1}=\frac{8x}{2x-1}$$

(b) $$\frac{5x}{(3x-2)^2}-\frac{3x}{(3x-2)^2}=\frac{5x-3x}{(3x-2)^2}=\frac{2x}{(3x-2)^2}$$
■

To add or subtract fractions that do not have a common denominator, you must rewrite the fractions so they do have the same denominator. To do this, multiply the numerator and denominator of each fraction by an appropriate quantity. For instance,

$$\frac{2}{3}+\frac{4}{5}=\frac{2\cdot 5}{3\cdot 5}+\frac{3\cdot 4}{3\cdot 5}=\frac{10}{15}+\frac{12}{15}=\frac{10+12}{15}=\frac{22}{15}.$$

More generally, if $q \neq s$, you can always add p/q and r/s as follows:

$$\frac{p}{q}+\frac{r}{s}=\frac{p\cdot s}{q\cdot s}+\frac{q\cdot r}{q\cdot s}=\frac{ps+qr}{qs}.$$

Example 4 Add the expressions $\frac{3x}{4x-1}$ and $\frac{2x}{3x-5}$.

Solution

$$\frac{3x}{4x-1}+\frac{2x}{3x-5}=\frac{3x(3x-5)}{(4x-1)(3x-5)}+\frac{(4x-1)(2x)}{(4x-1)(3x-5)}$$

$$=\frac{3x(3x-5)+(4x-1)(2x)}{(4x-1)(3x-5)}=\frac{9x^2-15x+8x^2-2x}{(4x-1)(3x-5)}$$

$$=\frac{17x^2-17x}{(4x-1)(3x-5)}=\frac{17x(x-1)}{(4x-1)(3x-5)}$$

■

When adding or subtracting fractions, it's usually best to use the **least common denominator,** abbreviated LCD. The LCD of two or more fractions is found as follows:

Step 1. Factor each denominator completely.

Step 2. Form the product of all the different prime factors in the denominators of the fractions, taking each factor the greatest number of times it occurs in any of the denominators.

Example 5 Find the LCD of the fractions, perform the indicated operations, and simplify the result.

(a) $\frac{8x}{x^2-9}+\frac{4}{5x-15}$ **(b)** $\frac{x}{x^3+x^2+x+1}-\frac{1}{x^3+2x^2+x}-\frac{1}{x^2+2x+1}$

Solution

(a) We factor the denominators to obtain

$$\frac{8x}{x^2-9}+\frac{4}{5x-15}=\frac{8x}{(x-3)(x+3)}+\frac{4}{5(x-3)}.$$

Here, the LCD is $5(x-3)(x+3)$. Now we rewrite each fraction as an equivalent fraction with denominator $5(x-3)(x+3)$. We do this by multiplying the numerator and denominator of the first fraction by 5, and the numerator and denominator of the second fraction by $x + 3$. Thus,

$$\begin{aligned}\frac{8x}{x^2-9}+\frac{4}{5x-15} &= \frac{8x}{(x-3)(x+3)}+\frac{4}{5(x-3)}\\ &= \frac{5(8x)}{5(x-3)(x+3)}+\frac{4(x+3)}{5(x-3)(x+3)}\\ &= \frac{5(8x)+4(x+3)}{5(x-3)(x+3)} = \frac{44x+12}{5(x-3)(x+3)} = \frac{4(11x+3)}{5(x-3)(x+3)}.\end{aligned}$$

(b) Factoring the denominators, we have

$$\begin{aligned}&\frac{x}{x^3+x^2+x+1}-\frac{1}{x^3+2x^2+x}-\frac{1}{x^2+2x+1}\\ &\quad= \frac{x}{x^2(x+1)+(x+1)}-\frac{1}{x(x^2+2x+1)}-\frac{1}{x^2+2x+1}\\ &\quad= \frac{x}{(x^2+1)(x+1)}-\frac{1}{x(x+1)^2}-\frac{1}{(x+1)^2},\end{aligned}$$

so the LCD is $x(x^2 + 1)(x + 1)^2$. Therefore,

$$\begin{aligned}&\frac{x}{(x^2+1)(x+1)}-\frac{1}{x(x+1)^2}-\frac{1}{(x+1)^2}\\ &\quad= \frac{x\cdot x(x+1)}{x(x^2+1)(x+1)^2}-\frac{x^2+1}{x(x^2+1)(x+1)^2}-\frac{x(x^2+1)}{x(x^2+1)(x+1)^2}\\ &\quad= \frac{x^2(x+1)-(x^2+1)-x(x^2+1)}{x(x^2+1)(x+1)^2} = \frac{x^3+x^2-x^2-1-x^3-x}{x(x^2+1)(x+1)^2}\\ &\quad= \frac{-1-x}{x(x^2+1)(x+1)^2} = -\frac{\cancel{(x+1)}}{x(x^2+1)(x+1)^{\cancel{2}}}\\ &\quad= -\frac{1}{x(x^2+1)(x+1)}.\end{aligned}$$

■

Complex Fractions

A fraction that contains one or more fractions in its numerator or denominator is called a **complex fraction.*** Examples are

$$\frac{\dfrac{3}{xy^2}}{\dfrac{2}{x^2y}}, \quad \frac{\dfrac{x}{y}-\dfrac{y}{x}}{\dfrac{1}{x}+\dfrac{1}{y}}, \quad \text{and} \quad \frac{x^2-\dfrac{1}{x}}{x+\dfrac{1}{x}+1}.$$

A complex fraction may be simplified by reducing its numerator and denominator (separately) to simple fractions and then dividing.

* There is no particular relationship between complex fractions and the complex numbers introduced later (in Section 1.9).

Example 6 Simplify the complex fraction $\dfrac{1+\dfrac{1}{x}}{x-\dfrac{1}{x}}$.

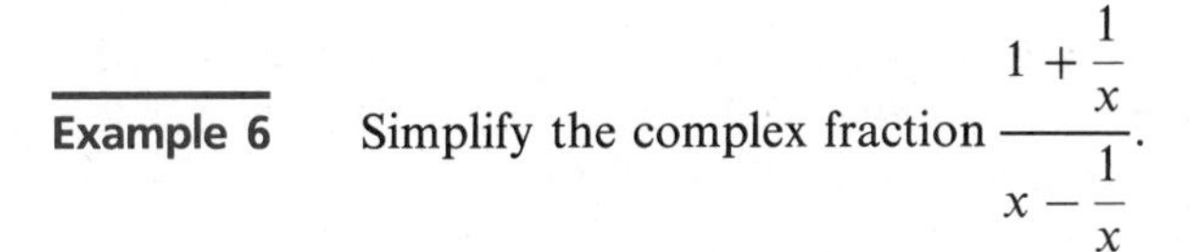

Solution

$$\frac{1+\dfrac{1}{x}}{x-\dfrac{1}{x}}=\frac{\dfrac{x+1}{x}}{\dfrac{x^2-1}{x}}=\frac{x+1}{x}\div\frac{x^2-1}{x}=\frac{x+1}{x}\cdot\frac{x}{x^2-1}$$

$$=\frac{(x+1)x}{(x^2-1)x}=\frac{x+1}{(x-1)(x+1)}=\frac{1}{x-1}$$ ■

An alternative method for simplifying a complex fraction is to multiply its numerator and denominator by the LCD of all fractions occurring in its numerator *and* denominator. The resulting fraction may then be simplified by cancellation.

Example 7 Simplify the complex fraction $\dfrac{\dfrac{1}{x^3}+\dfrac{2}{x^2y}+\dfrac{1}{xy^2}}{\dfrac{y}{x^2}-\dfrac{1}{y}}$.

Solution The LCD of x^3, x^2y, xy^2, x^2, and y is x^3y^2. Therefore,

$$\frac{\dfrac{1}{x^3}+\dfrac{2}{x^2y}+\dfrac{1}{xy^2}}{\dfrac{y}{x^2}-\dfrac{1}{y}}=\frac{x^3y^2\left(\dfrac{1}{x^3}+\dfrac{2}{x^2y}+\dfrac{1}{xy^2}\right)}{x^3y^2\left(\dfrac{y}{x^2}-\dfrac{1}{y}\right)}=\frac{y^2+2xy+x^2}{xy^3-x^3y}=\frac{(y+x)^2}{xy(y^2-x^2)}$$

$$=\frac{(y+x)^2}{xy(y-x)(y+x)}=\frac{y+x}{xy(y-x)}.$$ ■

Ratio and Proportion

Two quantities a and b are often compared by considering their ratio a/b. This has the advantage that it is independent of the *units** in which the two quantities are measured, just so they are both measured in the same units. For instance, we might compare the fuel efficiency of two automobiles by forming the ratio of the gas mileage of one to that of the other, and we could use either miles per gallon or kilometers per liter as the units of measurement for both of the automobiles.

* Here we are assuming that a multiplicative conversion factor is used to change the units of measurement—this would not be true, for instance, for a change from degrees Fahrenheit to degrees Celsius.

Example 8 The weight W newtons of an object is inversely proportional to the square of $(R + h)$, where R kilometers is the radius of the earth and h kilometers is the height of the object above the surface of the earth. Find the ratio W/W_0 of the weight W of an object at height h and its weight W_0 at the surface of the earth.

Solution We have

$$W = \frac{k}{(R + h)^2},$$

where k is the constant of variation. Putting $h = 0$, we find that

$$W_0 = \frac{k}{R^2}.$$

Therefore,

$$\frac{W}{W_0} = \frac{k/(R + h)^2}{k/R^2} = \frac{R^2}{(R + h)^2} = \left(\frac{R}{R + h}\right)^2.$$

■

In Example 8, the ratio W/W_0 will have the same value whether W and W_0 are measured in newtons or in, say, pounds. Likewise, the ratio $R/(R + h)$ will have the same value whether R and h are measured in kilometers or in, say, miles.

An equation such as

$$\frac{a}{b} = \frac{c}{d}$$

which expresses the equality of two ratios, is called a **proportion.** The proportion $a/b = c/d$ is often read, "a is to b as c is to d." If both sides of the equation

$$\frac{a}{b} = \frac{c}{d}$$

are multiplied by bd, we have

$$\frac{a}{\not b}\not b d = \frac{c}{\not d}b\not d \qquad \text{or} \qquad ad = bc.$$

Conversely, if we divide both sides of the equation $ad = bc$ by bd, we obtain

$$\frac{a\not d}{b\not d} = \frac{\not b c}{\not b d},$$

which is just the original proportion. Therefore, $a/b = c/d$ is equivalent to the equation $ad = bc$. You may want to use the memory device

$$\frac{a}{b} \times \frac{c}{d},$$

indicating that a is multiplied by d and that b is multiplied by c when

$$\frac{a}{b} = \frac{c}{d} \qquad \text{is rewritten as} \qquad ad = bc.$$

This procedure is called **cross-multiplication.**

Ecologists releasing banded duck

Example 9 In wildlife management, the population P of animals of a certain type that inhabit a particular area is often estimated by capturing M of these animals, banding or otherwise marking them, and releasing the marked animals. Later, after the marked animals have mixed with the rest of the animals in the area, a random sample consisting of S animals is captured and the number m of marked animals in the sample is counted. It is assumed that the ratio of marked animals in the population is the same as the ratio of marked animals in the sample, so that $M/P = m/S$. Solve for P in terms of M, S, and m.

Solution Cross-multiplying in the proportion

$$\frac{M}{P} = \frac{m}{S},$$

we find that

$$MS = Pm.$$

The last equation can be solved for P by dividing both sides by m. Thus,

$$P = \frac{MS}{m}. \qquad \blacksquare$$

Problem Set 1.6

In Problems 1 to 14, reduce each fraction to lowest terms.

1. $\dfrac{2a^2bxy^2}{6a^2xy}$
2. $\dfrac{18x^4(-y)(-z)^5}{30x(-y)^2(-z)}$
3. $\dfrac{cy + cz}{2y + 2z}$
4. $\dfrac{(2t + 6)^2}{4t^2 - 36}$
5. $\dfrac{t - 5}{25 - t^2}$
6. $\dfrac{c^2 - cd}{bc - bd}$
7. $\dfrac{6y^2 - y - 1}{2y^2 + 9y - 5}$
8. $\dfrac{r^2 - s^2}{s^4 - r^4}$
9. $\dfrac{(a + b)^2 - 4ab}{(a - b)^2}$
10. $\dfrac{(x - y)^2 - z^2}{(x + z)^2 - y^2}$
11. $\dfrac{6c^2 - 7c - 3}{4c^2 - 8c + 3}$
12. $\dfrac{6m(m - 1) - 12}{m^3 - 8 - (m - 2)^2}$
13. $\dfrac{3(r + t)^2 + 17(r + t) + 10}{2(r + t)^2 + 7(r + t) - 15}$
14. $\dfrac{c^2 - d^2 + c - d}{c^2 + 2cd + d^2 - 1}$

In Problems 15 to 28, perform the indicated operations and simplify the result.

15. $\dfrac{c^2 - 8c}{c - 8} \cdot \dfrac{c + 2}{c}$
16. $\dfrac{r^4 - 16}{(r - 2)^2} \cdot \dfrac{r - 2}{4 - r^2}$
17. $\dfrac{3x^2 + 15}{x^2 + 6x + 15} \cdot \dfrac{x^2 + 2x + 1}{x^2 - 1}$
18. $\dfrac{x(2x - 9) - 5}{20 + x(1 - x)} \cdot \dfrac{x(3x + 14) + 8}{x(2x - 11) - 6}$
19. $\dfrac{c^2}{a^2 - 1} \div \dfrac{c^2}{a - 1}$
20. $\dfrac{z^2 - 49}{z^2 - 5z - 14} \div \dfrac{z + 7}{2z^2 - 13z - 7}$
21. $\dfrac{x^3 - y^3}{x - y} \div \dfrac{x^2 + xy + y^2}{x^2 - 2xy + y^2}$

22. $\dfrac{xy^3 - 4x^2y^2}{y - x} \div \dfrac{16x^2y^2 - y^4}{4x^2 - 3xy - y^2}$

23. $\dfrac{14u^2 + 23u + 3}{2u^2 + u - 3} \div \dfrac{7u^2 + 15u + 2}{2u^2 - 3u + 1}$

24. $\dfrac{(t+1)^2 - 9}{(t + \frac{1}{2})^2 - \frac{49}{4}} \div \dfrac{(t - \frac{1}{2})^2 - \frac{9}{4}}{t^2 - 3t}$

25. $\dfrac{x(3x + 22) - 16}{x(x + 11) + 24} \div \dfrac{x(3x + 13) - 10}{x(2x + 13) + 15}$

26. $\dfrac{w(3w - 17) + 10}{3w(3w - 4) + 4} \div \dfrac{20 + w(1 - w)}{w(6w - 7) + 2}$

27. $\left(\dfrac{2t^2 + 9t - 5}{t^2 + 10t + 21} \div \dfrac{2t^2 - 13t + 6}{3t^2 + 11t + 6}\right) \cdot \dfrac{t^2 + 3t - 28}{3t^2 - 10t - 8}$

28. $\left(\dfrac{a(3a + 17) + 10}{a(a + 10) + 25} \cdot \dfrac{a^2 - 25}{a(3a + 11) + 6}\right) \div \dfrac{a(5a - 24) - 5}{a(5a + 16) + 3}$

In Problems 29 to 32, perform the indicated operations and simplify the result.

29. $\dfrac{x^2}{x + 1} - \dfrac{1}{x + 1}$

30. $\dfrac{x^2}{x - 2} + \dfrac{3x}{x - 2} - \dfrac{10}{x - 2}$

31. $\dfrac{4u}{2u + 1} - \dfrac{3}{2u - 1} + \dfrac{1}{4}$

32. $\dfrac{3 + p}{2 - p} + \dfrac{1 + 2p}{1 + 3p + 2p^2}$

In Problems 33 to 40, find the LCD of the fractions, perform the indicated operations, and simplify the result.

33. $\dfrac{y^2 - 2}{y^2 - y - 2} + \dfrac{y + 1}{y - 2}$

34. $\dfrac{t + 3}{t^2 - t - 2} + \dfrac{2t - 1}{t^2 + 2t - 8}$

35. $\dfrac{2}{t - 2} + \dfrac{3}{t + 1} - \dfrac{t - 8}{t^2 - t - 2}$

36. $\dfrac{y}{y - z} - \dfrac{2y}{y + z} + \dfrac{3yz}{z^2 - y^2}$

37. $\dfrac{u}{(u^2 + 3)(u - 1)} - \dfrac{3u^2}{(u - 1)^2(u + 2)}$

38. $\dfrac{1}{(a - b)(a - c)} + \dfrac{1}{(b - a)(c - b)} - \dfrac{1}{(a - c)(b - c)}$

39. $\dfrac{2x}{4x^3 - 4x^2 + x} - \dfrac{x}{2x^3 - x^2} + \dfrac{1}{x^3}$

40. $\dfrac{3}{t - 3} - \dfrac{2}{t^2 + 3t} + \dfrac{10}{t^3 - 9t}$

In Problems 41 to 48, simplify each complex fraction.

41. $\dfrac{\frac{1}{d} - \frac{1}{c}}{c^2 - d^2}$

42. $\dfrac{\frac{x}{y} + 1}{(x + y)^2}$

43. $\dfrac{(a + b)^2}{\frac{1}{a} + \frac{1}{b}}$

44. $\dfrac{\frac{1}{a^2} + \frac{2}{ab} + \frac{1}{b^2}}{\frac{1}{a^2} - \frac{1}{b^2}}$

45. $\dfrac{\frac{1}{p^2} - \frac{1}{q^2}}{\frac{1}{p^3} - \frac{1}{q^3}}$

46. $\dfrac{\frac{a^3}{b^2} + b}{1 - \frac{a}{b} + \frac{a^2}{b^2}}$

47. $\dfrac{\frac{1}{x + y} - \frac{1}{x - y}}{\frac{2y}{x^2 - y^2}}$

48. $\dfrac{\frac{x^2 - y^2}{x^2 + y^2} + \frac{x^2 + y^2}{x^2 - y^2}}{\frac{x - y}{x + y} - \frac{x + y}{x - y}}$

The expressions in Problems 49 to 52 were obtained as solutions to problems in a calculus textbook. Simplify each expression.

49. $\dfrac{\frac{1}{(x + h)^2} - \frac{1}{x^2}}{h}$

50. $3\left(\dfrac{x + 1}{x^2 + 1}\right)^2 \dfrac{x^2 + 1 - (x + 1)(2x)}{(x^2 + 1)^2}$

51. $\dfrac{(7x + 2)^3(27x^2 + 2) - (9x^3 + 2x)(3)(7x + 2)^2(7)}{(7x + 2)^6}$

52. $\dfrac{(5x + 1)^8(3)(x^2 - 1)^2(2x) - (x^2 - 1)^3(8)(5x + 1)^7(5)}{(5x + 1)^{16}}$

In Problems 53 and 54, *an error in calculation has been made* in each part. Find the errors.

53. **(a)** $\dfrac{4 - x}{4} = 1 - x$?

(b) $\dfrac{3 + 4x}{5 - 3x} = \dfrac{7}{2}$?

(c) $\dfrac{(a + b)c}{d + c} = \dfrac{a + b}{d}$?

54. **(a)** $\dfrac{a}{x + y} = \dfrac{a}{x} + \dfrac{a}{y}$?

(b) $(a + b)\left(\dfrac{1}{a} + \dfrac{1}{b}\right) = a\left(\dfrac{1}{a}\right) + b\left(\dfrac{1}{b}\right) = 2$?

C 55. Suppose that $y = \dfrac{3x^2 - 4x - 4}{x - 2}$ and notice that the fraction on the right is undefined when $x = 2$. It is interesting (and important in calculus) to study the values of y when x is very close to 2. Use a calculator to complete the following tables:

x	1	1.25	1.50	1.75	1.9	1.99
y						

x	3	2.75	2.50	2.25	2.1	2.01
y						

What seems to be happening to y as x comes closer and closer to 2?

56. In electronics, if two resistors of resistance R_1 and R_2 ohms are connected in parallel (Figure 1), then the resistance R of the combination is given by

$$R = \frac{1}{(1/R_1) + (1/R_2)}.$$

Simplify this formula for R.

Figure 1

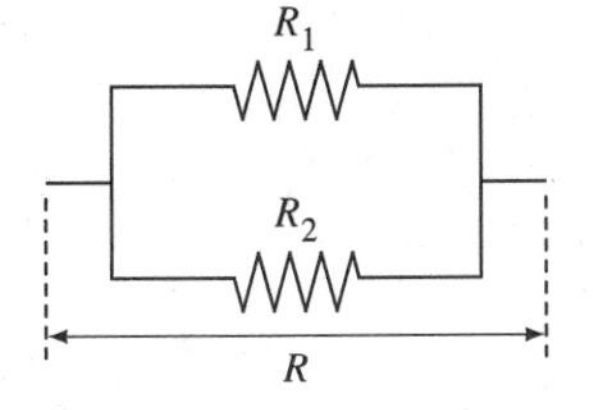

57. In optics, the distance x of an object from the focus of a thin converging lens of focal length f and the distance y of the corresponding image from the focus of the lens satisfy the relation $xy = f^2$ (Figure 2). If $p = x + f$ and $q = y + f$ are the distances from the object to the lens and from the image to the lens, show that

$$\frac{1}{p} + \frac{1}{q} = \frac{1}{f}.$$

Figure 2

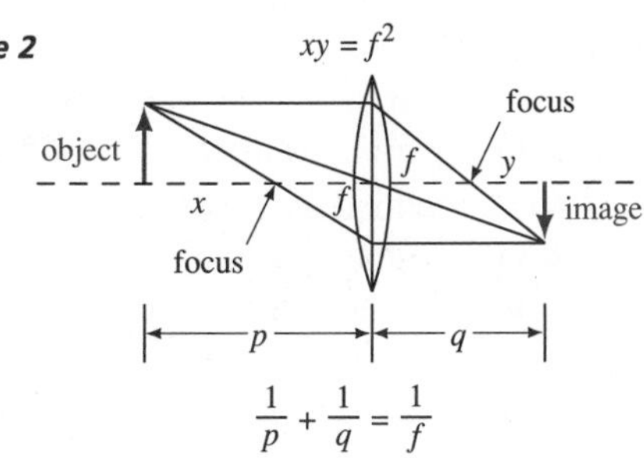

58. Two electric charges are located on a number line (Figure 3): a charge of -4 coulombs at $x = -3$ and a charge of -2 coulombs at $x = 3$. If a charge of $+1$ coulomb is placed at a position x between -3 and 3, Coulomb's law in physics implies that the net force exerted on the charge at x by the original two charges is proportional to the quantity $\dfrac{2}{(3 - x)^2} + \dfrac{-4}{(x + 3)^2}$. Simplify this quantity.

Figure 3

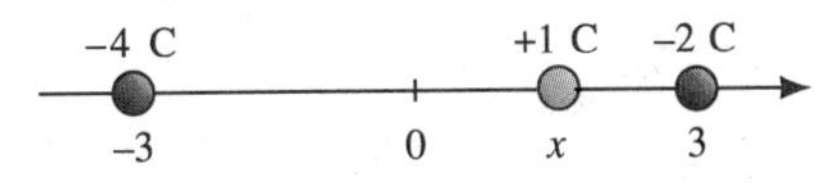

59. In each case, find the ratio of the first quantity to the second and write your answer in the form of a reduced fraction. If the quantities are measured in different units, you must first express both in the same units.

(a) 51 pounds to 68 pounds

(b) 700 grams to 21 kilograms (1 kilogram = 1000 grams.)

(c) $\frac{3}{4}$ yard to 15 inches

(d) 1500 cubic centimeters to $\frac{1}{10}$ cubic meter (1 meter = 100 centimeters.)

(e) 88 kilometers per hour to 11 miles per hour (Take 1 kilometer to be $\frac{5}{8}$ mile.)

60. Find the ratio of each pair of expressions if the ratio of a to b is $\frac{4}{5}$.

(a) $6a - 7b$ to $3a + 4b$ **(b)** $-7a$ to $3a + 13b$

61. The schematic diagram for a hydraulic lift is shown in Figure 4. If the pistons have cross-sectional areas A_1 and A_2, then the forces F_1 and F_2 satisfy the proportion $\dfrac{F_1}{F_2} = \dfrac{A_1}{A_2}$. Such a device, used to raise a dentist's chair, has pistons with cross-sectional areas of $A_1 = 1500\text{ cm}^2$ and $A_2 = 75\text{ cm}^2$. Find the force F_1 necessary to lift the chair if $F_2 = 1500$ newtons.

Figure 4

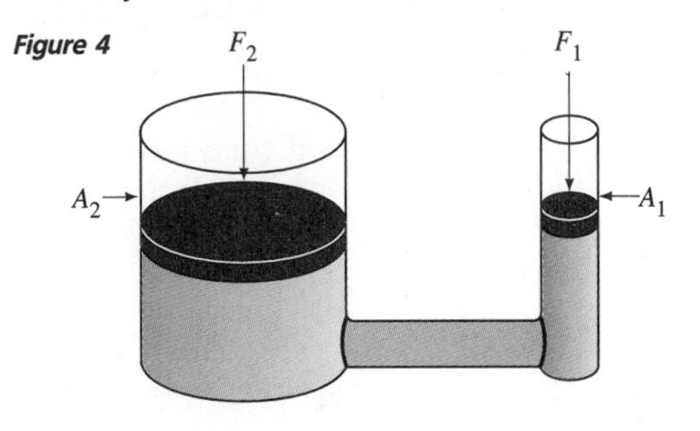

62. The ratio of sodium in 1 ounce of corn flakes to sodium in 1 ounce of an English muffin is $\frac{29}{9}$. If 1 ounce of corn flakes contains 300 milligrams of sodium, how many milligrams of sodium are in a 2-ounce English muffin?

C **63.** Gordo, who has been trying to lose weight, is flying in a jetliner at an altitude of 5 miles. He remembers from his college physics course that his weight W at this altitude will be less than his weight $W_0 = 255$ pounds on the ground. Find the resulting weight loss $W_0 - W$ if the radius of the earth is 3960 miles. (See Example 8.)

64. In some European countries, the dimensions of a standard sheet of typing paper are chosen so that if the paper is cut in half, the ratio of length to width of the half sheet is the same as the ratio of length to width of the original sheet. Find this ratio.

65. In order to estimate the number of fish in a lake, ecologists catch 100 fish, band them, and release the banded fish to mix with the rest of the fish. Later, 50 fish are caught from the lake, and 2 of them are found to be banded. Estimate the number of fish in the lake. (See Example 9.)

1.7 Radical Expressions

If n is a positive integer and a is a real number, then any real number x such that $x^n = a$ is called an **nth root of a.** For $n = 2$ an nth root is called a **square root,** and for $n = 3$ it is called a **cube root.** For instance, 4 is a cube root of 64, since $4^3 = 64$, and -3 is a cube root of -27, since $(-3)^3 = -27$. Because $5^2 = 25$ and $(-5)^2 = 25$, *both 5 and -5 are square roots of 25.*

If n is an *odd* positive integer, each real number a has exactly one (real) nth root x, and x has the same algebraic sign as a. If n is an *even* positive integer, each positive number has *two* (real) nth roots—one positive and one negative. However, if n is even, negative numbers have no real nth roots. (Do you see why?) Thus, we make the following definition.

Definition 1 **The Principal nth Root of a Number**

Let n be a positive integer and suppose that a is a real number.

(i) If a is a positive number, the **principal nth root of a,** $\sqrt[n]{a}$, is the positive nth root of a.

(ii) The **principal nth root of zero** is zero, that is, $\sqrt[n]{0} = 0$.

(iii) If a is a negative number, the **principal nth root of a,** $\sqrt[n]{a}$, is defined as a real number only when n is odd, in which case it is the negative number whose nth power is a.

Notice that $\sqrt[n]{a}$ is defined except when a is negative and n is even. Furthermore if $\sqrt[n]{a}$ is defined, it represents just *one* real number. For instance, $\sqrt{4} = 2$,* $\sqrt[3]{27} = 3$, $\sqrt[5]{32} = 2$, and $\sqrt[5]{-32} = -2$, whereas $\sqrt{-4}$ is undefined as a real number. Although it is correct to say that 2 and -2 are square roots of 4, it is *incorrect* to write $\sqrt{4} = \pm 2$ because the symbol $\sqrt{\ }$ denotes the **principal square root.**

* We usually write $\sqrt{\ }$ rather than $\sqrt[2]{\ }$.

In the expression $\sqrt[n]{a}$, the symbol $\sqrt[n]{}$ is called the **radical sign** (or simply the **radical**); the positive integer n is called the **index;** and the real number a under the radical is called the **radicand:**

$$\underset{\text{radical}}{\overset{\text{index}}{\sqrt[4]{625}}} = 5. \quad \text{(625: radicand; 5: principal fourth root of 625)}$$

Algebraic expressions containing radicals are called **radical expressions.** Examples are

$$\sqrt[4]{625}, \qquad 5 + \sqrt{x}, \qquad \text{and} \qquad \sqrt{t} - \sqrt[7]{t^4 - 1}.$$

A positive integer is called a **perfect *n*th power** if it is the nth power of an integer. Obviously, the nth root of a perfect nth power is an integer. For instance,

$$25 = 5^2 \text{ is a perfect square and } \sqrt{25} = 5$$
$$27 = 3^3 \text{ is a perfect cube and } \sqrt[3]{27} = 3$$
$$625 = 5^4 \text{ is a perfect fourth power and } \sqrt[4]{625} = 5.$$

Example 1 Find each principal root (if it is defined).

(a) $\sqrt{36}$ **(b)** $\sqrt[3]{64}$
(c) $\sqrt[3]{-64}$ **(d)** $\sqrt[6]{-64}$

Solution **(a)** $\sqrt{36} = 6$ **(b)** $\sqrt[3]{64} = 4$
(c) $\sqrt[3]{-64} = -4$ **(d)** $\sqrt[6]{-64}$ is undefined. ■

If both of the positive integers h and k are perfect nth powers, then $\sqrt[n]{h/k} = \sqrt[n]{h}/\sqrt[n]{k}$, and it follows that $\sqrt[n]{h/k}$ is a rational number. However, if the fraction h/k is reduced to lowest terms and if either h or k (or both) fails to be a perfect nth power, it can be shown that $\sqrt[n]{h/k}$ is an irrational number.

Example 2 Determine which of the indicated principal roots are rational numbers and evaluate those that are.

(a) $\sqrt[3]{\frac{8}{125}}$ **(b)** $\sqrt{\frac{2}{50}}$ **(c)** $\sqrt[4]{75}$

Solution **(a)** Both the numerator $8 = 2^3$ and the denominator $125 = 5^3$ are perfect cubes, so $\sqrt[3]{\frac{8}{125}} = \frac{2}{5}$.

(b) The fraction $\frac{2}{50}$ isn't reduced to lowest terms. But $\frac{2}{50} = \frac{1}{25}$ and both $1 = 1^2$ and $25 = 5^2$ are perfect squares, so $\sqrt{\frac{2}{50}} = \sqrt{\frac{1}{25}} = \frac{1}{5}$.

(c) The number 75 isn't a perfect 4th power, so $\sqrt[4]{75}$ isn't a rational number. ■

Radical expressions can often be simplified by using the following properties.

Properties of Radicals

Let a and b be real numbers and suppose that m and n are positive integers. Then, provided that all expressions are defined:

(i) $\sqrt[n]{ab} = \sqrt[n]{a}\sqrt[n]{b}$ **(ii)** $\sqrt[n]{\dfrac{a}{b}} = \dfrac{\sqrt[n]{a}}{\sqrt[n]{b}}$ **(iii)** $\sqrt[n]{\sqrt[m]{a}} = \sqrt[nm]{a}$

(iv) $\sqrt[n]{a^m} = (\sqrt[n]{a})^m$ **(v)** $\sqrt[n]{a^n} = (\sqrt[n]{a})^n = a$

Furthermore, if n is *odd,*

$$\sqrt[n]{-a} = -\sqrt[n]{a}.$$

In using these properties to simplify radical expressions, it's a good idea to factor the radicand in order to reveal any exponents that are multiples of the index.

Example 3 Use the properties of radicals to simplify each expression. Assume that variables are restricted to values for which all expressions are defined.

(a) $\sqrt{9x^3}$ **(b)** $\sqrt[3]{-16x^7y^4}$

(c) $\sqrt{\dfrac{8x^7}{9y^4}}$ **(d)** $\sqrt[3]{(-x^5)^2}$

(e) $\sqrt[7]{\sqrt[5]{x^{35}}}$ **(f)** $\sqrt[5]{a^4}\sqrt[5]{a^3}$

Solution **(a)** If $\sqrt{9x^3}$ is defined, then x cannot be negative and it follows that

$$\sqrt{9x^3} = \sqrt{9x^2 \cdot x} = \sqrt{9x^2}\sqrt{x} = 3x\sqrt{x}.$$

(b) $\sqrt[3]{-16x^7y^4} = -\sqrt[3]{16x^7y^4} = -\sqrt[3]{8x^6y^3 \cdot 2xy} = -\sqrt[3]{8x^6y^3}\sqrt[3]{2xy} = -2x^2y\sqrt[3]{2xy}$

[Here we used $\sqrt[3]{x^6} = \sqrt[3]{(x^2)^3} = x^2$.]

(c) $\sqrt{\dfrac{8x^7}{9y^4}} = \sqrt{\dfrac{4x^6}{9y^4}(2x)} = \sqrt{\dfrac{4x^6}{9y^4}}\sqrt{2x} = \dfrac{\sqrt{4x^6}}{\sqrt{9y^4}}\sqrt{2x} = \dfrac{2x^3}{3y^2}\sqrt{2x}$

(d) $\sqrt[3]{(-x^5)^2} = \sqrt[3]{x^{10}} = \sqrt[3]{x^9x} = \sqrt[3]{x^9}\sqrt[3]{x} = \sqrt[3]{(x^3)^3}\sqrt[3]{x} = x^3\sqrt[3]{x}$

(e) $\sqrt[7]{\sqrt[5]{x^{35}}} = \sqrt[35]{x^{35}} = x$

(f) $\sqrt[5]{a^4}\sqrt[5]{a^3} = \sqrt[5]{a^4a^3} = \sqrt[5]{a^7} = \sqrt[5]{a^5a^2} = \sqrt[5]{a^5}\sqrt[5]{a^2} = a\sqrt[5]{a^2}$ ■

Two or more radical expressions are said to be **like** or **similar** if, after being simplified, they contain the same index and radicand. For instance, $3\sqrt{5}$, $x\sqrt{5}$, and $\sqrt{5}/2$ are similar, but $3\sqrt{5}$ and $5\sqrt[3]{5}$ are not similar. To simplify a sum whose terms are radical expressions, we simplify each term and then combine similar terms.

Example 4 Simplify each sum. Assume that variables are restricted to values for which all expressions are defined.

(a) $7\sqrt{12} + \sqrt{75} - 5\sqrt{27}$ **(b)** $\sqrt{2x^2y} + x\sqrt{18y} - 3\sqrt{x^3y}$

(c) $\sqrt[3]{4a^5} + a\sqrt[3]{32a^2}$

Solution

(a)
$$\begin{aligned} 7\sqrt{12} + \sqrt{75} - 5\sqrt{27} &= 7\sqrt{4 \cdot 3} + \sqrt{25 \cdot 3} - 5\sqrt{9 \cdot 3} \\ &= 7\sqrt{4}\sqrt{3} + \sqrt{25}\sqrt{3} - 5\sqrt{9}\sqrt{3} \\ &= 7 \cdot 2\sqrt{3} + 5\sqrt{3} - 5 \cdot 3\sqrt{3} \\ &= 14\sqrt{3} + 5\sqrt{3} - 15\sqrt{3} \\ &= (14 + 5 - 15)\sqrt{3} = 4\sqrt{3} \end{aligned}$$

(b)
$$\begin{aligned} \sqrt{2x^2y} + x\sqrt{18y} - 3\sqrt{x^3y} &= \sqrt{x^2}\sqrt{2}\sqrt{y} + x\sqrt{9}\sqrt{2}\sqrt{y} - 3\sqrt{x^2}\sqrt{x}\sqrt{y} \\ &= x\sqrt{2}\sqrt{y} + 3x\sqrt{2}\sqrt{y} - 3x\sqrt{x}\sqrt{y} \\ &= 4x\sqrt{2}\sqrt{y} - 3x\sqrt{x}\sqrt{y} \\ &= x\sqrt{y}(4\sqrt{2} - 3\sqrt{x}) \end{aligned}$$

(c)
$$\begin{aligned} \sqrt[3]{4a^5} + a\sqrt[3]{32a^2} &= \sqrt[3]{a^3 \cdot 4a^2} + a\sqrt[3]{8 \cdot 4a^2} = \sqrt[3]{a^3}\sqrt[3]{4a^2} + a\sqrt[3]{8}\sqrt[3]{4a^2} \\ &= a\sqrt[3]{4a^2} + 2a\sqrt[3]{4a^2} = 3a\sqrt[3]{4a^2} \end{aligned}$$
■

Sums involving radicals are multiplied the same way polynomials are.

Example 5 Expand each product and simplify the result. Assume that x and y are nonnegative.

(a) $\sqrt{3}(\sqrt{27} + \sqrt{15})$ **(b)** $(3\sqrt{x} - 7\sqrt{y})(5\sqrt{x} + 2\sqrt{y})$

(c) $(\sqrt[3]{s} - \sqrt[3]{2})(\sqrt[3]{s^2} + \sqrt[3]{2s} + \sqrt[3]{4})$

Solution

(a)
$$\begin{aligned} \sqrt{3}(\sqrt{27} + \sqrt{15}) &= \sqrt{3}\sqrt{27} + \sqrt{3}\sqrt{15} \\ &= \sqrt{3}\sqrt{3}\sqrt{9} + \sqrt{3}\sqrt{3}\sqrt{5} = 3 \cdot 3 + 3\sqrt{5} = 9 + 3\sqrt{5} \end{aligned}$$

(b)
$$\begin{aligned} (3\sqrt{x} - 7\sqrt{y})(5\sqrt{x} + 2\sqrt{y}) &= 15(\sqrt{x})^2 + (3 \cdot 2 - 7 \cdot 5)\sqrt{x}\sqrt{y} - 14(\sqrt{y})^2 \\ &= 15x - 29\sqrt{xy} - 14y \end{aligned}$$

(c)
$$\begin{aligned} (\sqrt[3]{s} - \sqrt[3]{2})(\sqrt[3]{s^2} + \sqrt[3]{2s} + \sqrt[3]{4}) &= (\sqrt[3]{s} - \sqrt[3]{2})[(\sqrt[3]{s})^2 + \sqrt[3]{s}\sqrt[3]{2} + (\sqrt[3]{2})^2] \\ &= (\sqrt[3]{s})^3 - (\sqrt[3]{2})^3 = s - 2 \end{aligned}$$

[Here we used Special Product 5 on page 29, with $a = \sqrt[3]{s}$ and $b = \sqrt[3]{2}$.] ■

In part (c) of Example 5, the product of two radical expressions contains no radicals. Whenever the product of two radical expressions is free of radicals, we say that the two expressions are **rationalizing factors** for each other. For instance,

$$(2\sqrt{3} - 3\sqrt{x})(2\sqrt{3} + 3\sqrt{x}) = (2\sqrt{3})^2 - (3\sqrt{x})^2 = 12 - 9x,$$

so $2\sqrt{3} - 3\sqrt{x}$ is a rationalizing factor for $2\sqrt{3} + 3\sqrt{x}$.

Fractions containing radicals are sometimes easier to deal with if their denominators are free of radicals. To rewrite a fraction so that there are no radicals in the denominator, we multiply the numerator and denominator by a rationalizing factor for the denominator. This is called **rationalizing the denominator.**

Example 6 Rationalize the denominator of $\dfrac{3}{\sqrt{5}}$.

Solution

$$\frac{3}{\sqrt{5}} = \frac{3\sqrt{5}}{\sqrt{5}\sqrt{5}} = \frac{3\sqrt{5}}{5}$$ ■

Rationalizing factors can often be found by using the special products that yield the difference of two squares or the sum or difference of two cubes.

Example 7 Rationalize the denominator of each fraction and simplify the result. Assume that x is restricted to values for which all expressions are defined.

(a) $\dfrac{\sqrt{3}-1}{\sqrt{2}+1}$ **(b)** $\dfrac{3}{7\sqrt[3]{x}}$ **(c)** $\dfrac{\sqrt{x}-3}{\sqrt{x}+3}$ **(d)** $\dfrac{2}{\sqrt[3]{x}+2}$

Solution

(a) Since $(\sqrt{2}+1)(\sqrt{2}-1) = (\sqrt{2})^2 - 1^2 = 2 - 1 = 1$, it follows that $\sqrt{2}-1$ is a rationalizing factor for the denominator. Thus,

$$\frac{\sqrt{3}-1}{\sqrt{2}+1} = \frac{(\sqrt{3}-1)(\sqrt{2}-1)}{(\sqrt{2}+1)(\sqrt{2}-1)} = \frac{\sqrt{6}-\sqrt{3}-\sqrt{2}+1}{1} = \sqrt{6}-\sqrt{3}-\sqrt{2}+1.$$

(b) Since $\sqrt[3]{x}(\sqrt[3]{x^2}) = \sqrt[3]{x^3} = x$, we have

$$\frac{3}{7\sqrt[3]{x}} = \frac{3\sqrt[3]{x^2}}{7\sqrt[3]{x}\sqrt[3]{x^2}} = \frac{3\sqrt[3]{x^2}}{7x}.$$

(c)

$$\frac{\sqrt{x}-3}{\sqrt{x}+3} = \frac{(\sqrt{x}-3)(\sqrt{x}-3)}{(\sqrt{x}+3)(\sqrt{x}-3)} = \frac{(\sqrt{x})^2 - 6\sqrt{x} + 9}{(\sqrt{x})^2 - 9} = \frac{x - 6\sqrt{x} + 9}{x - 9}$$

(d) Using Special Product 4 on page 29, with $a = \sqrt[3]{x}$ and $b = 2$, we see that a rationalizing factor for the denominator is provided by

$$a^2 - ab + b^2 = (\sqrt[3]{x})^2 - (\sqrt[3]{x})(2) + 2^2 = \sqrt[3]{x^2} - 2\sqrt[3]{x} + 4.$$

Thus,

$$(\sqrt[3]{x}+2)(\sqrt[3]{x^2} - 2\sqrt[3]{x} + 4) = (a+b)(a^2 - ab + b^2) = a^3 + b^3$$
$$= (\sqrt[3]{x})^3 + 2^3 = x + 8,$$

and we have

$$\frac{2}{\sqrt[3]{x}+2} = \frac{2(\sqrt[3]{x^2} - 2\sqrt[3]{x} + 4)}{(\sqrt[3]{x}+2)(\sqrt[3]{x^2} - 2\sqrt[3]{x} + 4)} = \frac{2\sqrt[3]{x^2} - 4\sqrt[3]{x} + 8}{x+8}.$$ ■

In calculus, it is sometimes necessary to rationalize the *numerator* of a fraction.

Example 8 Rationalize the numerator of $\frac{\sqrt{x+h+1} - \sqrt{x+1}}{h}$.

Solution

$$\begin{aligned}\frac{\sqrt{x+h+1} - \sqrt{x+1}}{h} &= \frac{(\sqrt{x+h+1} - \sqrt{x+1})(\sqrt{x+h+1} + \sqrt{x+1})}{h(\sqrt{x+h+1} + \sqrt{x+1})} \\ &= \frac{(\sqrt{x+h+1})^2 - (\sqrt{x+1})^2}{h(\sqrt{x+h+1} + \sqrt{x+1})} \\ &= \frac{(x+h+1) - (x+1)}{h(\sqrt{x+h+1} + \sqrt{x+1})} \\ &= \frac{\cancel{h}}{\cancel{h}(\sqrt{x+h+1} + \sqrt{x+1})} \\ &= \frac{1}{\sqrt{x+h+1} + \sqrt{x+1}}\end{aligned}$$ ■

Problem Set 1.7

Throughout this problem set, assume that variables are restricted to values for which all radical expressions are defined.

In Problems 1 to 10, determine which of the indicated principal roots are rational numbers and evaluate those that are.

1. $\sqrt[3]{\frac{27}{64}}$
2. $\sqrt[5]{-32}$
3. $\sqrt{24}$
4. $\sqrt{0.04}$
5. $\sqrt[3]{\frac{16}{250}}$
6. $\sqrt[3]{-\frac{18}{54}}$
7. $\sqrt[7]{\frac{1}{128}}$
8. $\sqrt[4]{\frac{-1}{81}}$
9. $\sqrt[4]{\frac{81}{256}}$
10. $\sqrt[4]{(81)(256)}$

In Problems 11 to 36, simplify each radical expression.

11. $\sqrt{27x^3}$
12. $\sqrt[3]{4a^5}$
13. $\sqrt{75x^4y^9}$
14. $\sqrt[4]{81x^5y^{16}}$
15. $\sqrt[3]{24a^4b^5}$
16. $\sqrt[5]{-u^6v^7}$
17. $(\sqrt{x}-1)^4$
18. $(\sqrt[3]{2a}+b)^6$
19. $\frac{\sqrt{16x^4y^3}}{\sqrt{64x^{12}y}}$
20. $\sqrt[5]{\frac{1}{243b^5}}$
21. $\sqrt{\frac{6y}{7}}\sqrt{\frac{35}{72y^2}}$
22. $\sqrt[3]{\frac{u^3v^6}{(3uv^2)^3}}$
23. $\sqrt[3]{\frac{-27x^4}{y^{21}}}$
24. $\frac{\sqrt[6]{u^{15}v^{21}}}{\sqrt[6]{u^3v^6}}$
25. $(2\sqrt[3]{11})^3(-\tfrac{1}{2}\sqrt[5]{24})^5$
26. $\frac{(\frac{1}{2}\sqrt[3]{9}\sqrt{3})^4}{(3\sqrt[5]{3}\sqrt{3})^5}$

27. $\sqrt[4]{125t^2\sqrt{25t^4}}$

28. $\sqrt[5]{\sqrt[3]{x^{30}}}$

29. $\sqrt[4]{x^2y^{10}}\sqrt[4]{x^6y^9}$

30. $\sqrt[5]{s^3t^5}\sqrt[5]{st^5}\sqrt[5]{st}$

31. $\sqrt{\dfrac{u-v}{u+v}}\sqrt{\dfrac{u^2+2uv+v^2}{u^2-v^2}}$

32. $\dfrac{\sqrt{3xy^3z}\sqrt{2x^2yz^4}}{\sqrt{6x^3y^4z^3}}$

33. $\dfrac{\sqrt[3]{54x^2yz^4}}{\sqrt[3]{16xy^4z^3}}$

34. $\sqrt[5]{\dfrac{(a+b)^4(c+d)^3}{8}}\sqrt[5]{\dfrac{(a+b)^6}{4(c+d)^8}}$

35. $\dfrac{\sqrt[3]{x^2y^3}\sqrt[3]{125x^3y^2}}{\sqrt[3]{8x^3y^4}}$

36. $\sqrt[4]{u^3\sqrt{u\sqrt[3]{u^3}}}$

In Problems 37 to 44, simplify each algebraic sum.

37. $6\sqrt{2}-3\sqrt{8}+98$

38. $2\sqrt{54}-\sqrt{216}-7\sqrt{24}$

39. $5\sqrt[3]{81}-3\sqrt[3]{24}+\sqrt[3]{192}$

40. $8xy\sqrt{x^2y}-3\sqrt{x^4y^3}+5x^2\sqrt{y^3}$

41. $4t^3\sqrt{180t}-2t^2\sqrt{20t^3}-\sqrt{5t^7}$

42. $\sqrt[3]{375}-\sqrt[6]{576}-4\sqrt[9]{27}$

43. $\sqrt[4]{\dfrac{625}{216}}-\sqrt[4]{\dfrac{32}{27}}$

44. $\sqrt{\dfrac{a^3}{3b^3}}+ab\sqrt{\dfrac{a}{3b^5}}-\dfrac{b^2}{3}\sqrt{\dfrac{3a^3}{b^7}}$

In Problems 45 to 54, expand each product and simplify the result.

45. $\sqrt{6}(2\sqrt{3}-3\sqrt{2})$

46. $\sqrt{2x}(x\sqrt{2}-2\sqrt{x})$

47. $(2+\sqrt{3})(1-\sqrt{3})$

48. $(\sqrt{x}+\sqrt{3})^2$

49. $(\sqrt{5}-3\sqrt{2})(\sqrt{5}+3\sqrt{2})$

50. $(3\sqrt{x}-2\sqrt{y})(3\sqrt{x}+2\sqrt{y})$

51. $\sqrt{\sqrt{21}+\sqrt{5}}\sqrt{\sqrt{21}-\sqrt{5}}$

52. $\sqrt[3]{\sqrt{33}-\sqrt{6}}\sqrt[3]{\sqrt{33}+\sqrt{6}}$

53. $(\sqrt{3}+\sqrt{2}+1)(\sqrt{3}+\sqrt{2}-1)$

54. $(\sqrt{a}+\sqrt{b}-\sqrt{c})(\sqrt{a}+\sqrt{b}+\sqrt{c})$

In Problems 55 to 72, rationalize the denominator and simplify the result.

55. $\dfrac{8}{\sqrt{11}}$

56. $\dfrac{\sqrt{t}}{\sqrt{7tu}}$

57. $\dfrac{5\sqrt{2}}{3\sqrt{5}}$

58. $\dfrac{3x}{\sqrt{21x}}$

59. $\dfrac{\sqrt{2}}{3\sqrt{x+2y}}$

60. $\dfrac{10t}{3\sqrt{5(3t+7)}}$

61. $\dfrac{20}{\sqrt[3]{5}}$

62. $\dfrac{a^2b^3}{3\sqrt[4]{ab}}$

63. $\dfrac{\sqrt{3}}{\sqrt{3}+5}$

64. $\dfrac{5}{1-\sqrt{p}}$

65. $\dfrac{\sqrt{5}+\sqrt{2}}{\sqrt{5}-\sqrt{2}}$

66. $\dfrac{3\sqrt{p}+\sqrt{q}}{4\sqrt{p}-3\sqrt{q}}$

67. $\dfrac{\sqrt{a+5}+\sqrt{a}}{\sqrt{a+5}-\sqrt{a}}$

68. $\dfrac{2\sqrt{x^2-1}+\sqrt{x^2+1}}{3\sqrt{x^2-1}+\sqrt{x^2+1}}$

69. $\dfrac{5}{2-\sqrt[3]{x}}$

70. $\dfrac{2}{1+\sqrt{x}-\sqrt{y}}$

71. $\dfrac{\sqrt[3]{a}-\sqrt[3]{b}}{\sqrt[3]{a}+\sqrt[3]{b}}$

72. $\dfrac{1}{\sqrt{x}+\sqrt{y}+\sqrt{z}}$

In Problems 73 to 76, rationalize the *numerator* of each fraction and simplify the result.

73. $\dfrac{\sqrt{x+h}-\sqrt{x}}{h}$

74. $\dfrac{\sqrt[3]{x}-\sqrt[3]{a}}{x-a}$

75. $\dfrac{\sqrt{(x+h)^2+1}-\sqrt{x^2+1}}{h}$

76. $\dfrac{\dfrac{1}{\sqrt{x+h}}-\dfrac{1}{\sqrt{x}}}{h}$

77. Show that the expression

$$\frac{\sqrt{x}}{\sqrt{x+a}} - \frac{\sqrt{x+a}}{\sqrt{x}}$$

can be rewritten in the form

$$\frac{-a}{\sqrt{x(x+a)}}.$$

78. Let $r_1 = \dfrac{-b + \sqrt{b^2 - 4ac}}{2a}$ and $r_2 = \dfrac{-b - \sqrt{b^2 - 4ac}}{2a}$.

(a) Find $\dfrac{1}{r_1}$ and rationalize the denominator of the resulting expression.

(b) Find $r_1^2 - r_2^2$ and simplify the resulting expression.

C **79.** In electronics, the resonant frequency f hertz for a tuned circuit is given by the formula $f = 1/(2\pi\sqrt{LC})$, where L henrys is the inductance and C farads is the capacitance. Find f if $L = 3.57 \times 10^{-9}$ henry and $C = 1.21 \times 10^{-12}$ farad.

80. If V_0 is the velocity of sound in air at 0° Celsius, it is shown in physics that the velocity of sound in air at T degrees Celsius is given by $V = V_0/\sqrt{1 - kT}$, where k is a constant approximately equal to 3.66×10^{-3}. A physics textbook argues that if T is small, so that kT is very small, then $V \approx V_0\sqrt{1 + kT}$. Justify this approximation by explaining why if x is very small, then $1/\sqrt{1 - x} \approx \sqrt{1 + x}$. [*Hint:* Use the fact that if x is small, x^2 is much smaller.]

81. In astronomy, the relativistic Doppler shift formula

$$\nu = \nu_0 \frac{\sqrt{1 - (v^2/c^2)}}{1 + (v/c)}$$

is used to determine the frequency ν of light from a distant galaxy as measured at an observatory on earth. (ν is the Greek letter *nu*.) In this formula, c is the speed of light, v is the speed with which the galaxy is receding from the earth, and ν_0 is the frequency of the light emitted by the galaxy. The fact that the observed frequency ν is smaller than the original frequency is popularly known as the "red shift." If $b = v/c$, show that

$$\frac{\nu}{\nu_0} = \sqrt{\frac{1 - b}{1 + b}}.$$

Spiral galaxy in Ursa Major

82. In Problem 80, the physics textbook goes on to argue that if T is small, then $V_0\sqrt{1 + kT} \approx V_0[1 + (kT/2)]$. Justify this approximation by explaining why if x is very small, then $\sqrt{1 + x} \approx 1 + (x/2)$. [*Hint:* Begin by expanding the square of $1 + (x/2)$.]

83. A tract of land is subdivided into three square lots as shown in Figure 1. Find the perimeter of this tract in simplified radical form.

Figure 1

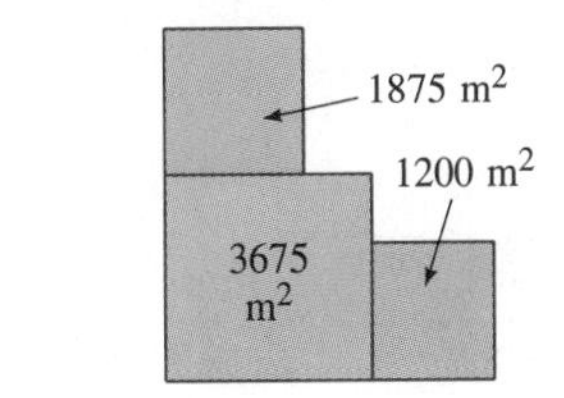

C **84.** The speed of sound in air at 0° Celsius is 331.4 meters per second.

(a) Use the formula $V = V_0/\sqrt{1 - kT}$ from Problem 80 to calculate the speed of sound in air at 20° Celsius.

(b) What is the error if the approximation $V \approx V_0[1 + (kT/2)]$ from Problem 82 is used to find the speed of sound in air at 20° Celsius?

C **85.** A biologist uses the mathematical model

$$P = 10{,}000\sqrt[4]{t^5 + 10t^2 + 9}$$

to estimate the population P of bacteria in a culture, where t is the time in hours since the culture was started. Find the approximate population P when $t = 2.5$ hours.

1.8 Rational Exponents

In Section 1.4 we established the following properties of positive integer exponents:

$$\textbf{(i)}\ a^m a^n = a^{m+n} \qquad \textbf{(ii)}\ (a^m)^n = a^{mn} \qquad \textbf{(iii)}\ (ab)^n = a^n b^n.$$

These properties follow from the definition

$$a^n = \overbrace{aa \cdots a}^{n \text{ factors}},$$

which makes sense only if n is a positive integer. In the present section we extend this definition so that arbitrary rational numbers can be used as exponents. The key idea is to extend the definition in such a way that Properties (i), (ii), and (iii) continue to hold.

We begin with the question of how to define a^0. If zero as an exponent is to obey Property (i), we must have

$$a^0 a^n = a^{0+n} = a^n.$$

If $a \neq 0$, this equation can hold only if $a^0 = 1$, which leads us to the following definition.

Definition 1 **Zero as an Exponent**

If a is any nonzero real number, we define $a^0 = 1$.

Notice that 0^0 *is not defined.*

According to Section 1.1, if $a \neq 0$ and n is a positive integer, then a^{-n} should be interpreted to mean $1/a^n$. To see why this is reasonable, notice that if Property (i) is to hold, we must have

$$a^{-n} a^n = a^{-n+n} = a^0 = 1.$$

If $a \neq 0$, this equation can hold only if $a^{-n} = 1/a^n$. For completeness, we now repeat our earlier definition.

Definition 2 **Negative Integer Exponents**

If a is any nonzero real number and n is a positive integer,

$$a^{-n} = \frac{1}{a^n}.$$

In other words, a^{-n} is the reciprocal of a^n. In particular,

$$a^{-1} = \frac{1}{a^1} = \frac{1}{a}.$$

Definitions 1 and 2 enable us to use *any integers*—positive, negative, or zero—as exponents, with the exception that 0^n is defined only when n is positive. You can verify that Properties (i), (ii), and (iii) hold for all integer exponents if $a \neq 0$ and $b \neq 0$. The following properties also hold for all integers m and n and all nonzero real numbers a and b (Problem 64).

(iv) $\left(\frac{a}{b}\right)^n = \frac{a^n}{b^n}$ **(v)** $\frac{a^m}{a^n} = a^{m-n}$ **(vi)** $\frac{a^{-m}}{b^{-n}} = \frac{b^n}{a^m}$ **(vii)** $\left(\frac{a}{b}\right)^{-1} = \frac{b}{a}$

Property (vi) permits us to move a factor in a numerator to the denominator of a fraction, or vice versa, simply by changing the sign of the exponent of the factor. By Property (vii), the reciprocal of a nonzero fraction is obtained by inverting the fraction.

Example 1 Rewrite each expression so it contains only positive exponents and simplify the result.

(a) 2^{-3} **(b)** $(7x^0)^{-2}$ **(c)** $(x^4)^{-2}$

(d) $\left(\frac{3}{x}\right)^{-4}$ **(e)** $\left(\frac{3}{2^{-1}}\right)^{-1}$ **(f)** $(x-y)^{-4}(x-y)^{13}$

(g) $(x^{-2}y^{-3})^{-4}$ **(h)** $\frac{5x^{-3}(a+b)^2}{15x^4(a+b)^{-5}}$ **(i)** $\frac{x^{-1}-y^{-1}}{x-y}$

Solution

(a) $2^{-3} = \frac{1}{2^3} = \frac{1}{8}$ (Definition 2)

(b) $(7x^0)^{-2} = (7 \cdot 1)^{-2}$ (Definition 1)

$= 7^{-2} = \frac{1}{7^2} = \frac{1}{49}$ (Definition 2)

(c) $(x^4)^{-2} = x^{-8} = \frac{1}{x^8}$ [Property (ii) and Definition 2]

(d) $\left(\frac{3}{x}\right)^{-4} = \frac{3^{-4}}{x^{-4}} = \frac{x^4}{3^4} = \frac{x^4}{81}$ [Properties (iv) and (vi)]

(e) $\left(\frac{3}{2^{-1}}\right)^{-1} = \frac{2^{-1}}{3} = \frac{1}{2 \cdot 3} = \frac{1}{6}$ [Properties (vii) and (vi)]

(f) $(x-y)^{-4}(x-y)^{13} = (x-y)^{-4+13} = (x-y)^9$ [Property (i)]

(g) $(x^{-2}y^{-3})^{-4} = (x^{-2})^{-4}(y^{-3})^{-4}$ [Property (iii)]

$= x^{(-2)(-4)}y^{(-3)(-4)} = x^8y^{12}$ [Property (ii)]

(h) $\frac{5x^{-3}(a+b)^2}{15x^4(a+b)^{-5}} = \frac{(a+b)^2(a+b)^5}{3x^4x^3} = \frac{(a+b)^7}{3x^7}$ [Properties (vi) and (i)]

(i) $\frac{x^{-1}-y^{-1}}{x-y} = \frac{\frac{1}{x}-\frac{1}{y}}{x-y} = \frac{\frac{y-x}{xy}}{x-y} = \frac{y-x}{xy} \cdot \frac{1}{x-y} = \frac{y-x}{xy(x-y)}$

$= \frac{-\cancel{(x-y)}}{xy\cancel{(x-y)}} = \frac{-1}{xy}$ ■

If m and n are integers with $n \neq 0$, how shall we define $a^{m/n}$? If Property (ii) is to hold, we must have

$$a^{m/n} = a^{(1/n)m} = (a^{1/n})^m,$$

so the basic question is how to define $a^{1/n}$. But, again, if Property (ii) is to hold, we must have

$$(a^{1/n})^n = a^{(1/n)n} = a^1 = a;$$

in other words, $a^{1/n}$ must be an nth root of a. Thus, we have the following definition.

Definition 3 **Rational Exponents**

Let a be a nonzero real number. Suppose that m and n are integers, that n is positive, and that the fraction m/n is reduced to lowest terms. Then, if $\sqrt[n]{a}$ exists,

$$a^{1/n} = \sqrt[n]{a} \qquad \text{and} \qquad a^{m/n} = (\sqrt[n]{a})^m = (a^{1/n})^m.$$

Also, if m/n is a positive rational number,

$$0^{m/n} = 0.$$

Notice that $a^{1/n}$ is just an alternative notation for $\sqrt[n]{a}$, the principal nth root of a. For instance,

$$25^{1/2} = \sqrt{25} = 5, \qquad 0^{1/7} = \sqrt[7]{0} = 0, \qquad (-27)^{1/3} = \sqrt[3]{-27} = -3,$$

and so forth. It is important to keep in mind that Definition 3 is to be used *only when $n > 0$ and m/n is reduced to lowest terms.* For instance, it is *incorrect* to write $(-8)^{2/6} = (\sqrt[6]{-8})^2$ because 2/6 isn't reduced to lowest terms. Instead, we must first reduce 2/6 to lowest terms and write

$$(-8)^{2/6} = (-8)^{1/3} = \sqrt[3]{-8} = -2.$$

Example 2 Find the value of each expression (if it is defined).

(a) $8^{4/3}$ **(b)** $81^{-3/4}$ **(c)** $(-7)^{3/2}$ **(d)** $(-64)^{8/12}$

Solution Using Definition 3, we have

(a) $8^{4/3} = (\sqrt[3]{8})^4 = 2^4 = 16$

(b) $81^{-3/4} = 81^{(-3)/4} = (\sqrt[4]{81})^{-3} = 3^{-3} = \dfrac{1}{3^3} = \dfrac{1}{27}$

(c) $(-7)^{3/2}$ is undefined since $\sqrt{-7}$ does not exist (as a real number).

(d) $(-64)^{8/12} = (-64)^{2/3} = (\sqrt[3]{-64})^2 = (-4)^2 = 16$ ■

A calculator with a y^x key (or the equivalent) may be used to find the approximate value of a number raised to a rational power. Consult the instruction booklet furnished with your calculator to learn the appropriate keystrokes.

Example 3 Find the approximate value of $(-3.22)^{5/3}$.

Solution Most calculators will return an error message if you attempt to raise a negative number to a fractional power. Thus, we begin by observing that since 3 and 5 are both odd,

$$(-3.22)^{5/3} = -(3.22^{5/3}).$$

Using a 10-digit calculator,* we find that

$$3.22^{5/3} \approx 7.021444265.$$

Therefore,

$$(-3.22)^{5/3} \approx -7.021444265.$$

■

For convenience, we now summarize the properties of rational exponents.

Properties of Rational Exponents

Let a and b be real numbers, suppose that p and q are rational numbers, and let n be a positive integer. Then, provided that all expressions are defined (as real numbers):

(i) $a^p a^q = a^{p+q}$ (ii) $(a^p)^q = a^{pq}, \quad a > 0$ (iii) $(ab)^p = a^p b^p$

(iv) $\left(\dfrac{a}{b}\right)^p = \dfrac{a^p}{b^p}$ (v) $\dfrac{a^p}{a^q} = a^{p-q}$ (vi) $\dfrac{a^{-p}}{b^{-q}} = \dfrac{b^q}{a^p}$

(vii) $\left(\dfrac{a}{b}\right)^{-1} = \dfrac{b}{a}$ (viii) $\sqrt[n]{a^p} = (\sqrt[n]{a})^p = a^{p/n}$ (ix) $a^{-p} = \dfrac{1}{a^p}$

As illustrated by the following example, these properties are especially useful for simplifying algebraic expressions containing rational exponents.

Example 4 Simplify each expression and write the answer so that it contains only positive exponents. You may assume that variables are restricted to values for which the properties of rational exponents hold.

(a) $\left(\dfrac{3}{2}\right)^{-2}$ (b) $\dfrac{x^3 y^{-2}}{x^{-1} y}$

(c) $(64x^{-3})^{-2/3} + 25^{-0.5}$ (d) $\left[\dfrac{(27x^2y^3)^{1/3}}{9x^{-2}y^4}\right]^{-1}$

(e) $\dfrac{(3x+2)^{1/2}(3x+2)^{-1/4}}{(3x+2)^{-3/4}}$ (f) $(a^{-1/2} - b^{-1/2})(a^{-1/2} + b^{-1/2})$

* Your calculator may give a slightly different result in the last digit or so—various types of calculators use different internal routines.

Solution

(a) $\left(\frac{3}{2}\right)^{-2} = \left(\frac{2}{3}\right)^{2} = \frac{4}{9}$

(b) $\frac{x^3y^{-2}}{x^{-1}y} = \frac{x^3x}{y^2y} = \frac{x^4}{y^3}$

(c) $(64x^{-3})^{-2/3} + 25^{-0.5} = 64^{-2/3}(x^{-3})^{-2/3} + \frac{1}{25^{1/2}}$

$$= (\sqrt[3]{64})^{-2}x^{(-3)(-2/3)} + \frac{1}{5}$$

$$= 4^{-2}x^2 + \frac{1}{5} = \frac{x^2}{4^2} + \frac{1}{5} = \frac{x^2}{16} + \frac{1}{5} = \frac{5x^2 + 16}{80}$$

(d) $\left[\frac{(27x^2y^3)^{1/3}}{9x^{-2}y^4}\right]^{-1} = \frac{9x^{-2}y^4}{(27x^2y^3)^{1/3}} = \frac{9x^{-2}y^4}{27^{1/3}x^{2/3}y^1}$

$$= \frac{9y^4y^{-1}}{3x^{2/3}x^2} = \frac{3y^{4-1}}{x^{2/3+2}} = \frac{3y^3}{x^{8/3}}$$

(e) $\frac{(3x+2)^{1/2}(3x+2)^{-1/4}}{(3x+2)^{-3/4}} = (3x+2)^{1/2}(3x+2)^{-1/4}(3x+2)^{3/4}$

$$= (3x+2)^{1/2-1/4+3/4} = (3x+2)^1 = 3x+2$$

(f) $(a^{-1/2} - b^{-1/2})(a^{-1/2} + b^{-1/2}) = (a^{-1/2})^2 - (b^{-1/2})^2$

$$= a^{-1} - b^{-1}$$

$$= \frac{1}{a} - \frac{1}{b} = \frac{b-a}{ab}$$ ■

For certain problems, particularly in calculus, it is necessary to simplify algebraic sums in which each term contains a rational power of the same expression. To do this, begin by factoring out the expression with the *smallest* rational power.

Example 5 Simplify the expression

$$\frac{-6x(2x+5)^{1/2} - (1-3x^2)(\tfrac{1}{2})(2x+5)^{-1/2}(2)}{[(2x+5)^{1/2}]^2}.$$

Solution

In the numerator, we begin by factoring out $(2x+5)^{-1/2}$ because $-\frac{1}{2}$ is smaller than $\frac{1}{2}$. Thus, the given expression is equivalent to

$$\frac{(2x+5)^{-1/2}[-6x(2x+5)^{(1/2)-(-1/2)} - (1-3x^2)]}{2x+5}$$

$$= \frac{(2x+5)^{-1/2}[-6x(2x+5) - (1-3x^2)]}{2x+5}$$

$$= \frac{-12x^2 - 30x - 1 + 3x^2}{(2x+5)^{1/2}(2x+5)} = -\frac{9x^2+30x+1}{(2x+5)^{3/2}}.$$ ■

Problem Set 1.8

In Problems 1 to 24, rewrite each expression so it contains only positive exponents, and simplify the result. Assume that n is a positive integer.

1. $\left(\frac{1}{3}\right)^{-3}$
2. $\frac{1}{7^{-2}}$
3. $(8x^0)^{-2}$
4. $(2^0y^{-2})^{-5}$
5. $x^{-2}y^4z^{-1}$
6. $[(3^0)/(4^{-2})]^{-1}$
7. $(-1)^{-1}$
8. $(x^{-3})^6(x^0)^{-2}$
9. $(4c^{-4})(-7c^6)$
10. $(a+b)^{-4}(a+b)^9$
11. $\frac{3x^{-5}}{6y^{-2}}$
12. $\frac{c^{-1}+d^{-1}}{cd}$
13. $(t^{-1}+3^{-2})^{-1}$
14. $\frac{a^{-1}-b^{-1}}{(a+b)^{-2}}$
15. $x^{2n-3}x^{3-3n}$
16. $t^nt^{1-n}t^{2n-4}$
17. $1-(p-1)^{-1}+(p+1)^{-1}$
18. $\frac{x^{-2}-y^{-2}}{x^{-1}+y^{-1}}$
19. $(t+2)^{-1}(t+2)^{-2}(t^2-4)^2$
20. $(x^{-2}y^{-2}z^{-3})^{-2n}$
21. $\left(\frac{a^{-3}}{b^{-3}}\right)^{-n}$
22. $\left[\frac{(cd)^{-2n}}{c^{-2n}d^{-2n}}\right]^{5n}$
23. $\frac{(p+q)^{-1}(p-q)^{-1}}{(p+q)^{-1}-(p-q)^{-1}}$
24. $\left[\frac{(5x)^{-1}(3x)^2y^{-3}}{15x^{-2}(25y^{-4})}\right]^{-4}$

In Problems 25 and 26, find the value of each expression (if it is defined). Do not use a calculator.

25. (a) $9^{3/2}$ (b) $(-8)^{5/3}$ (c) $32^{0.6}$ (d) $(-8)^{0.3}$ (e) $\left(-\frac{1}{8}\right)^{-6/9}$ (f) $6^{1/2}15^{1/2}10^{1/2}$

26. (a) $16^{-5/4}$ (b) $(-4)^{7/8}$ (c) $(-1)^{-10/6}$ (d) $\left(\frac{-8}{27}\right)^{4/6}$ (e) $\left(\frac{-1}{32}\right)^{1.8}$ (f) $(-0.125)^{-2/6}$

[C] In Problems 27 and 28, use a calculator to find approximate numerical values for each quantity.

27. (a) $7^{4/9}$ (b) $7^{9/4}$ (c) $(-7)^{1/3}$ (d) $(-7)^{-1/3}$

28. (a) $(67.2)^{2/3}$ (b) $(67.2)^{2.3}$ (c) $(-67.2)^{-2/3}$ (d) $(-67.2)^{-3/7}$ (e) $\pi^{-1.75}$

In Problems 29 to 46, simplify each expression and write the answer so it contains only positive exponents. (You may assume that variables are restricted to values for which the properties of rational exponents hold.)

29. $a^{1/2}a^{3/2}$
30. $m^{-1.4}m^{2.4}m^{-2}$
31. $(8p^9)^{4/3}$
32. $(81m^{12})^{-3/4}$
33. $(32u^{-5})^{-3/5}$
34. $[(3t+5)^{-7/5}]^{-10/7}$
35. $(16x^{-4})^{-3/4}$
36. $\left(\frac{2a^{3/2}b^{7/2}}{4a^2b^{-1}}\right)^{-4}$
37. $\left(\frac{x^{-1/3}}{x^{3/2}}\right)^6$
38. $\left(\frac{x^{1/2}}{y^2}\right)^4\left(\frac{y^{-1/3}}{x^{2/3}}\right)^3$
39. $\left(\frac{81r^{-12}}{16s^8}\right)^{-1/4}$
40. $\left(\frac{25x^{-16}y^{-8}}{4x^4y^{-2}}\right)^{-3/2}$
41. $\left(\frac{x^{m/3}y^{-3m/2}}{x^{-2m/3}y^{m/2}}\right)^{-2/m}$, $m>0$
42. $(x^{3/2}-y^{3/2})^2$
43. $\left[\frac{(3x+2y)^{1/2}(4r+3t)^{1/3}}{(4r+3t)^{1/2}(3x+2y)^{1/3}}\right]^6$
44. $[(2x+7)^{-3/4}]^{4/3}-[(2x+7)^{4/3}]^{-3/4}$
45. $(m+n)^{1/3}(m-n)^{1/3}(m^2-n^2)^{-2/3}$
46. $[(x^2+1)^{1/3}-1][(x^2+1)^{2/3}+(x^2+1)^{1/3}+1]$

In Problems 47 to 52, factor each expression and simplify the result.

47. $x^{3/2}+2x^{1/2}y+x^{-1/2}y^2$
48. $(2x-1)^{-1/2}(6x-3)+(2x-1)^{1/2}$
49. $(p-1)^{-2}-2(p-1)^{-3}(p+1)$
[*Hint:* -3 is smaller than -2.]
50. $-2(1-5a)^{-3}+3(3a-4)^{-1}(1-5a)^{-4}$

51. $2(4t-1)^{-1}(2t+1)^{-2}+4(4t-1)^{-2}(2t+1)^{-1}$
[*Hint:* $-2 < -1$.]

52. $2(2t+3)(4t-3)^{-1/4}+4(4t-3)^{3/4}$

The expressions in Problems 53 to 56 were obtained as answers to problems in a calculus textbook. Factor each expression and simplify the result.

53. $2(3x+x^{-1})(3-x^{-2})(6x-1)^5 + 30(3x+x^{-1})^2(6x-1)^4$

54. $-3(3t-1)^{-2}(2t+5)^{-3}-6(3t-1)^{-1}(2t+5)^{-4}$

55. $-14(7y+3)^{-3}(2y-1)^4+8(7y+3)^{-2}(2y-1)^3$

56. $-5(6u+u^{-1})^{-6}(6-u^{-2})(2u-2)^7 + 14(6u+u^{-1})(2u-2)^6$

57. In each of the calculations shown, *an error has been made.* In each case, find the error.

(a) $(-1)^{2/4}=[(-1)^2]^{1/4}=1^{1/4}=1$?
(b) $\sqrt[4]{(-4)^2}=(\sqrt[4]{-4})^2$ is undefined?
(c) $[(-2)(-8)]^{3/2}=(-2)^{3/2}(-8)^{3/2}$ is undefined?
(d) $(-32)^{0.2}=(-32)^{2/10}=(\sqrt[10]{-32})^2$ is undefined?
(e) $[(-1)^2]^{1/2}=(-1)^{2(1/2)}=(-1)^1=-1$?

[C] **58.** A mathematical model for the efficiency E (in percent) of an internal combustion gasoline engine in terms of the compression ratio R of the engine is given by the formula $E=(1-R^{-2/5})\cdot 100\%$. Find E, rounded off to two significant digits, if $R=\frac{11}{2}$.

[C] **59.** The shoe size of a normal man varies approximately as the 3/2 power of his height. If the average 6-foot man wears a size 11 shoe, what would you predict as the shoe size of a 7-foot basketball player?

60. The amount A of pollution entering the atmosphere in a certain region is found to be directly proportional to the 2/3 power of the number N of people in that region. If a population of $N=8000$ people produces $A=900$ tons of pollution per year, find a formula for A in terms of N.

[C] **61.** In astronomy, Kepler's third law states that the time T required for a planet to make one revolution about the sun is directly proportional to the 3/2 power of the maximum radius of its orbit. If the maximum radius of the earth's orbit is 93 million miles and the maximum radius of Mars' orbit is 142 million miles, how many days are required for Mars to make one revolution about the sun?

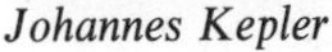

Johannes Kepler

[C] **62.** The mathematical model $P=15(2.72)^{-0.0004h}$ can be used to find the approximate atmospheric pressure in pounds per square inch at an altitude of h feet above sea level. Find the air pressure at the summit of Pikes Peak, the altitude of which is 14,100 feet.

[C] **63.** The mathematical model $N=9.60\times 10^{18}\,T^{3/2}$ is used to estimate the number N of electrons in a semiconductor at a temperature of T kelvins. Find N if $T=288$ kelvins.

64. Suppose that a and b are nonzero real numbers and that m and n are integers. By considering all possible cases in which m or n is positive, negative, or zero, verify Properties (iv), (v), (vi), and (vii) on page 59.

1.9 Complex Numbers

Because the square of a real number is nonnegative, there is no real number whose square is -1. Pondering this fact, the Italian physician and mathematician Geronimo Cardano (1501–1576) declared that numbers such as $\sqrt{-1}$ are "of hidden nature." Nevertheless, he calculated formally with these "hidden numbers" in a

Geronimo Cardano

remarkable and influential book, *Ars Magna* (*The Great Art*), published in 1545. Gradually, other mathematicians began to accept the idea of calculating with square roots of negative numbers, although they regarded such numbers as being fictitious or imaginary. During the nineteenth century these so-called imaginary numbers were linked with "real" objects in various ways, and it was shown that they are as legitimate as numbers of any other kind.

To launch our study of square roots of negative numbers we introduce the **imaginary unit**

$$i = \sqrt{-1}.$$

Leaving aside for now the question of just what i is, let's work with it a bit and see what happens. For the time being, we simply regard i or $\sqrt{-1}$ as an "invented number" with the property that

$$i^2 = -1.$$

Using i, we can define the **principal square root** of a negative number as follows:

If $c > 0$, then $\sqrt{-c} = \sqrt{c(-1)} = \sqrt{c}\sqrt{-1} = \sqrt{c}\,i = i\sqrt{c}$.

Example 1 Find each principal square root.

(a) $\sqrt{-3}$ **(b)** $\sqrt{-4}$

Solution

(a) $\sqrt{-3} = \sqrt{3(-1)} = \sqrt{3}\sqrt{-1} = \sqrt{3}i = i\sqrt{3}$

(b) $\sqrt{-4} = \sqrt{4(-1)} = \sqrt{4}\sqrt{-1} = 2i$ ■

Notice that if we multiply i by a real number such as $\sqrt{3}$ or 2, we simply have to write the result as $\sqrt{3}i$ or $2i$. More generally, we write the product of i and a real number b as bi or as ib. By further such multiplication, we get nothing new. For instance, if c is a real number and we multiply bi by c, we just get $(cb)i$, which again has the same form—a real number times i. On the other hand, if we multiply bi by ci, we get

$$(bi)(ci) = (bc)i^2 = (bc)(-1) = -bc,$$

which is a real number!

Adding two numbers of the form bi and ci, we obtain

$$bi + ci = (b + c)i,$$

which is another number of the same form—a real number times i. However, if we add a real number a to bi, we simply have to write the result as $a + bi$, which *is* something new! A number of the form

$$a + bi,$$

where a and b are real numbers, is called a **complex number.** Examples of complex numbers are

$$5 + 3i, \qquad -4 + 7i, \qquad \text{and} \qquad 2 + \sqrt{5}i.$$

Let's agree to write $a + (-b)i$ in the simpler form $a - bi$; for instance,

$$3 + (-6)i = 3 - 6i.$$

If we suppose that the basic algebraic properties of the real numbers continue to operate for complex numbers, then, for any real numbers a, b, c, and d, we have:

1. $(a + bi) + (c + di) = a + c + bi + di = (a + c) + (b + d)i$

2. $(a + bi) - (c + di) = a - c + bi - di = (a - c) + (b - d)i$

3. $(a + bi)(c + di) = ac + adi + bic + bidi = ac + bdi^2 + (ad + bc)i$
$= ac + bd(-1) + (ad + bc)i = (ac - bd) + (ad + bc)i$

We take 1, 2, and 3 above as definitions of **addition, subtraction,** and **multiplication** of complex numbers. Notice that the sum, difference, and product of complex numbers is again a complex number. In dealing with complex numbers, we understand that the real number 0 has its usual additive and multiplicative properties, so that

$$0 + bi = bi \qquad \text{and} \qquad 0i = 0.$$

Finally, let's agree that

$$a + bi = c + di \qquad \text{means that} \qquad a = c \qquad \text{and} \qquad b = d.$$

The real numbers a and b are called the **real part** and the **imaginary part** of the complex number $a + bi$. Thus, to say that two complex numbers are equal is to say that their real parts are equal and their imaginary parts are equal. In other words, a single equation involving complex numbers represents *two* equations involving real numbers!

Example 2 Express each complex number in the form $a + bi$, where a and b are real numbers.

(a) $(3 + 5i) + (6 + 5i)$ **(b)** $(2 - 3i) - (6 + 4i)$ **(c)** $(4 + 3i)(2 + 4i)$

Solution

(a) $(3 + 5i) + (6 + 5i) = 3 + 6 + 5i + 5i = 9 + 10i$

(b) $(2 - 3i) - (6 + 4i) = 2 - 6 - 3i - 4i = -4 - 7i$

(c) $(4 + 3i)(2 + 4i) = 8 + 16i + 6i + 12i^2 = 8 + 22i + 12(-1) = -4 + 22i$ ■

The set of all complex numbers, equipped with the algebraic operations of addition, subtraction, and multiplication, is called the **complex number system** and is denoted by the symbol $\mathbb{C}$. Note that a real number a can be regarded as a complex number whose imaginary part is zero: $a = a + 0i$; therefore, the real number system $\mathbb{R}$ forms part of the complex number system $\mathbb{C}$. It isn't difficult to show that the complex numbers have the same basic algebraic properties—commutative, associative, distributive, identity, and inverse—as the real numbers (Section 1.1, pages

3 and 4). Of these, the most intriguing is certainly the multiplicative inverse property—the fact that every nonzero complex number has a multiplicative inverse or reciprocal.

If $a + bi \neq 0$, you can obtain the **reciprocal** $1/(a + bi)$ by a technique similar to that for rationalizing the denominator of a fraction. Just *multiply numerator and denominator by* $a - bi$. Thus,

$$\frac{1}{a+bi} = \frac{a-bi}{(a+bi)(a-bi)} = \frac{a-bi}{a^2-(bi)^2} = \frac{a-bi}{a^2-b^2i^2} = \frac{a-bi}{a^2-b^2(-1)}$$
$$= \frac{a-bi}{a^2+b^2} = \left(\frac{a}{a^2+b^2}\right) + \left(\frac{-b}{a^2+b^2}\right)i.$$

The complex number $a - bi$ used in the calculation above is called the **complex conjugate** of $a + bi$. Complex conjugates are also useful in dealing with quotients of complex numbers. If $a + bi \neq 0$, the **quotient** $(c + di)/(a + bi)$ is defined to be the product of $c + di$ and $1/(a + bi)$.

To find the real and imaginary parts of a quotient of complex numbers, multiply the numerator and denominator by the complex conjugate of the denominator.

Example 3 Express each complex number in the form $a + bi$, where a and b are real numbers.

(a) $\dfrac{1}{4+3i}$ **(b)** $\dfrac{2-3i}{1-4i}$

Solution **(a)** Multiplying numerator and denominator by $4 - 3i$, the complex conjugate of $4 + 3i$, we obtain

$$\frac{1}{4+3i} = \frac{4-3i}{(4+3i)(4-3i)} = \frac{4-3i}{16-9i^2} = \frac{4-3i}{25} = \frac{4}{25} - \frac{3}{25}i.$$

(b) Multiplying numerator and denominator by $1 + 4i$, the complex conjugate of $1 - 4i$, we obtain

$$\frac{2-3i}{1-4i} = \frac{(2-3i)(1+4i)}{(1-4i)(1+4i)} = \frac{2+8i-3i-12i^2}{1-16i^2} = \frac{2+5i-12(-1)}{1-16(-1)}$$
$$= \frac{14+5i}{17} = \frac{14}{17} + \frac{5}{17}i.$$ ■

Complex numbers, like real numbers, can be denoted by letters of the alphabet and treated as variables or unknowns. The letters z and w are special favorites for this purpose. If $z = a + bi$, where a and b are real numbers, the complex conjugate of z is often written as $\bar{z} = a - bi$. Notice that

$$z\bar{z} = (a+bi)(a-bi) = a^2 - (bi)^2 = a^2 - b^2i^2 = a^2 - b^2(-1) = a^2 + b^2.$$

Thus $z\bar{z}$ is the sum of the squares of the real and imaginary parts of z, and so *it is always a nonnegative real number.*

Example 4 If $z = 3 + 4i$, find:

(a) $\bar{z}$ **(b)** $z + \bar{z}$ **(c)** $z - \bar{z}$ **(d)** $z\bar{z}$

Solution **(a)** $\bar{z} = 3 - 4i$ **(b)** $z + \bar{z} = (3 + 4i) + (3 - 4i) = 6$

(c) $z - \bar{z} = (3 + 4i) - (3 - 4i) = 8i$ **(d)** $z\bar{z} = (3 + 4i)(3 - 4i) = 3^2 + 4^2 = 25$ ■

As we have mentioned, a real number a can be regarded as a complex number $a + 0i$ with imaginary part zero. Thus, $\bar{a} = a - 0i = a$, so *each real number is its own complex conjugate.* Complex conjugation "preserves" all of the algebraic operations; that is, if z and w are complex numbers, then:

(i) $\overline{z + w} = \bar{z} + \bar{w}$ **(ii)** $\overline{z - w} = \bar{z} - \bar{w}$

(iii) $\overline{zw} = \bar{z}\bar{w}$ **(iv)** $\overline{\left(\dfrac{z}{w}\right)} = \dfrac{\bar{z}}{\bar{w}}$, $w \neq 0$.

We leave the verification of (i) to (iv) as an exercise (Problem 57). Properties (i) and (iii) can be extended to more than two complex numbers. For instance, if u, v, and w are three complex numbers, we can apply (i) twice to obtain

$$\overline{u + v + w} = \overline{(u + v) + w} = \overline{(u + v)} + \bar{w} = (\bar{u} + \bar{v}) + \bar{w} = \bar{u} + \bar{v} + \bar{w}.$$

Thus, for complex numbers, *the conjugate of a sum is the sum of the conjugates*, and a similar result holds for products.

Of course, integer powers of complex numbers are defined just as they are for real numbers: $z^2 = zz$, $z^3 = zzz$, and so on. And if $z \neq 0$, $z^0 = 1$, $z^{-1} = 1/z$, $z^{-2} = 1/z^2$, and so on. Notice that

$$\begin{array}{lll} i^1 = i & i^5 = i & i^9 = i \\ i^2 = -1 & i^6 = -1 & i^{10} = -1 \\ i^3 = -i & i^7 = -i & i^{11} = -i \\ i^4 = 1 & i^8 = 1 & i^{12} = 1 \end{array}$$

and so on. Thus, the positive integer powers of i endlessly repeat the pattern of the first four. Hence, if n is a positive integer and r is the remainder when n is divided by 4, then $i^n = i^r$.

Example 5 Find i^{59}.

Solution $i^{59} = i^{56+3} = i^{56}i^3 = (i^4)^{14}i^3 = 1^{14}(-i) = -i$ ■

Although complex numbers were originally introduced by Cardano and others to provide solutions for certain algebraic equations (such as $x^2 + 1 = 0$), they now have a wide variety of important applications in engineering and physics. For example, in 1893, Charles P. Steinmetz (1865–1923), an American electrical engineer born in Germany, developed a theory of alternating currents based on the complex numbers.

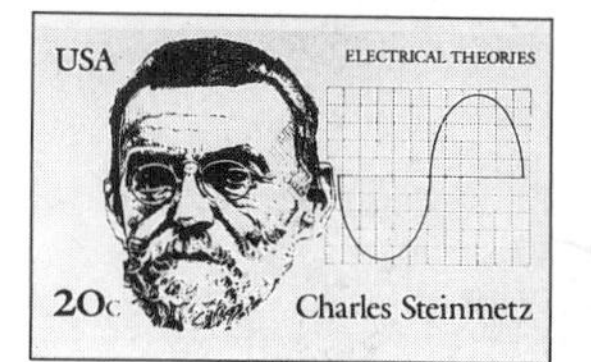

U.S. postage stamp honoring Charles P. Steinmetz

In direct-current theory, **Ohm's law**

$$E = IR$$

relates the electromotive force (voltage) E, the current I, and the resistance R. Because of inductive and capacitative effects, voltage and current may be out of phase in alternating-current circuits, and the equation $E = IR$ may no longer hold. Steinmetz saw that by representing voltage and current with *complex numbers* E and I, he could deal algebraically with phase differences. Furthermore, he combined the resistance R, the inductive effect X_L (called **inductive reactance**), and the capacitative effect X_C (called **capacitative reactance**) in a single complex number

$$Z = R + (X_L - X_C)i,$$

called **complex impedance.** Using the complex numbers E, I, and Z, Steinmetz showed that Ohm's law for alternating currents takes the form

$$E = IZ.$$

Today, these ideas of Steinmetz are used routinely by electrical engineers all over the world. It has been said that Steinmetz "generated electricity with the square root of minus one."

Problem Set 1.9

In Problems 1 and 2, find each principal square root.

1. **(a)** $\sqrt{-2}$ **(b)** $\sqrt{-27}$

2. **(a)** $\sqrt{-9}$ **(b)** $\sqrt{\sqrt[3]{-64}}$

In Problems 3 to 48, express each complex number in the form $a + bi$, where a and b are real numbers.

3. $3 + \sqrt{-16}$

4. $-2\sqrt{-8}$

5. $5\sqrt{-72}$

6. $-7 - \sqrt{-64}$

7. $(2 + 3i) + (7 - 2i)$

8. $(-1 + 2i) + (3 + 4i)$

9. $(4 + i) + 2(3 - i)$

10. $(4 + 2i) + 3(2 - 5i)$

11. $(5 - 4i) + 14$

12. $(\frac{1}{2} - \frac{2}{3}i) + (\frac{3}{4} + \frac{1}{6}i)$

13. $(3 + 2i) - (5 + 4i)$

14. $(\frac{1}{2} - \frac{4}{3}i) - \frac{1}{6}(5 + 7i)$

15. $2(5 + 4i) - 3(7 + 4i)$

16. $i - \frac{1}{2}(1 + 5i)$

17. $-(-3 + 5i) - (4 + 9i)$

18. $(\sqrt{2} + \sqrt{3}i) - \left(\frac{\sqrt{2}}{2} - \frac{\sqrt{3}}{3}i\right)$

19. $(2 + i)(1 + 5i)$

20. $i(3 + 7i)$

21. $(-7 + 3i)(-3 + 2i)$

22. $(\frac{2}{3} + \frac{3}{5}i)(\frac{1}{2} - \frac{1}{3}i)$

23. $-8i(5 + 8i)$

24. $(\sqrt{2} + \sqrt{3}i)(\sqrt{2} - \sqrt{3}i)$

25. $(-4i)(-5i)$

26. $(-2i)(3i)(-4i)$

27. $(\frac{1}{2} + \frac{1}{3}i)(\frac{1}{2} - \frac{1}{3}i)$

28. $(\sqrt{2} + \sqrt{3}i)^2$

29. i^{21}

30. i^{41}

31. i^{201}

32. $(1 + i)^3$

33. $(4 + 2i)^2$

34. $(1 + i)^4$

35. $\left(\frac{1}{2} + \frac{\sqrt{3}}{2}i\right)^2$

36. $(1 - i)^4$

37. $\frac{1}{2 + 3i}$

38. $\frac{1}{3 - 4i}$

39. $\frac{3}{7 + 2i}$

40. $\frac{-4i}{6 - i}$

41. $\frac{3 - 4i}{4 + 2i}$

42. $\frac{\pi + 4i}{2 - i}$

43. $\frac{7 + 2i}{3 - 5i}$

44. $\frac{1}{i}$

45. $\frac{2 - 6i}{3 + i} - \frac{4 + i}{3 + i}$

46. $\frac{(1 - i)(2 + i)}{(2 - 3i)(3 - 4i)}$

47. $\dfrac{3-2i}{3+i}+\dfrac{4i}{3-7i}$

48. $\left(\dfrac{3-i^7}{i^9-3}\right)^2$

In Problems 49 to 56, calculate **(a)** $\bar{z}$, **(b)** $z+\bar{z}$, **(c)** $z-\bar{z}$, and **(d)** $z\bar{z}$.

49. $z=2+i$

50. $z=i$

51. $z=-i$

52. $z=(1+i)^2$

53. $z=\dfrac{1+i}{1-i}$

54. $z=-3i^5$

55. $z=\dfrac{4-3i}{2+4i}$

56. $z=1+i+i^2+i^3$

57. Show that:

(a) $\overline{z+w}=\bar{z}+\bar{w}$

(b) $\overline{z-w}=\bar{z}-\bar{w}$

(c) $\overline{zw}=\bar{z}\bar{w}$

(d) If $w\neq 0$, $\overline{z/w}=\bar{z}/\bar{w}$.

58. Assume that $a+bi\neq 0$. By multiplying, show that

$$(a+bi)\left[\left(\frac{a}{a^2+b^2}\right)+\left(\frac{-b}{a^2+b^2}\right)i\right]=1.$$

59. If z is a nonzero complex number, show that $\overline{z^{-1}}=(\bar{z})^{-1}$.

60. If z is a complex number, show that:

(a) $\frac{1}{2}(z+\bar{z})$ is the real part of z

(b) $\dfrac{1}{2i}(z-\bar{z})$ is the imaginary part of z.

61. If $z=\dfrac{-3+\sqrt{7}i}{4}$, find and simplify the expression $2z^2+3z+2$.

62. Simplify the expression $\left(\dfrac{1-\sqrt{34}i}{5}\right)^2+\left(\dfrac{1+\sqrt{34}i}{5}\right)^2$.

63. In an alternating-current circuit, the impedance Z is given by the formula $Z=R+(X_L-X_C)i$. Find Z when $R=14.1$ ohms, $X_L=10.1$ ohms, and $X_C=12.2$ ohms.

C **64.** Using Ohm's law $E=IZ$ for the alternating-current circuit in Problem 63, find I if $E=25+3i$ volts. Write the answer in the form $a+bi$.

CHAPTER 1 Review Problem Set

In Problems 1 to 6, write a formula for the given quantity.

1. w is 3 times the product of x and y, divided by z.

2. x is 7% less than the number n.

3. s is one-half of the perimeter of a triangle with sides a, b, and c.

4. The surface area A of a rectangular box, open at the top and closed at the bottom, with length l, width w, and height h.

5. The population P of a town n years from now, if the current population is 1000 and the population triples every year. [*Hint:* After $n=1$ year, the population is 3000, after $n=2$ years, it is 9000, and so forth.]

6. The number N of board feet of lumber in n "two-by-fours," each of which is l feet long. [*Note:* The dimensions of a cross section of a "two-by-four" are actually 1.5 inches by 3.5 inches; one board foot is the volume of a board with dimensions 1 foot by 1 foot by 1 inch.]

7. The formula $K=\frac{1}{2}mv^2$ gives the kinetic energy K, in joules, of an object of mass m kilograms moving with a speed of v meters per second. A jogger with a mass of 70 kilograms is running at a speed of 3 meters per second. Find the kinetic energy of the jogger.

C **8.** If P dollars is borrowed and paid back in n equal periodic installments of R dollars each, including interest at the rate r per period on the unpaid balance, then

$$R=\frac{Pr}{1-(1+r)^{-n}}.$$

Find R if \$20,000 is borrowed and paid back in 5 equal yearly installments, including an interest of 10% per year ($r=0.1$) on the unpaid balance.

9. State one basic algebraic property of the real numbers to justify each statement.

(a) $3\cdot(-7)=(-7)\cdot 3$

(b) $3(x+2)=3x+3\cdot 2$

(c) $(-3) + (5 + \pi) = [(-3) + 5] + \pi$
(d) $0 + y^2 = y^2$
(e) $15 \cdot (x + y) = (x + y) \cdot 15$
(f) $3(\pi + 0) = 3\pi$
(g) $1 \cdot (a - b) = a - b$
(h) $6 \cdot (4 \cdot 3) = (6 \cdot 4) \cdot 3$
(i) $(-3) \cdot \frac{1}{(-3)} = 1$ **(j)** $\pi + (-\pi) = 0$

10. State one of the derived algebraic properties of the real numbers to justify each statement.

(a) $-[-(x + y)] = x + y$
(b) $(-4)(-\pi) = 4\pi$ **(c)** $(x^2 - y^2) \cdot 0 = 0$
(d) If $(3x^2 - 5)(2x - 1) = 0$, then $3x^2 - 5 = 0$ or $2x - 1 = 0$.

11. Find the mistake and correct each calculation.

(a) $\frac{1}{2} + \frac{1}{3} = \frac{2}{5}$?
(b) $\sqrt{9 + 16} = 3 + 4 = 7$?

12. If A is the set of all integers between $-\frac{3}{2}$ and $\frac{5}{2}$, list all of the elements that belong to A and sketch A on a number line.

13. Express each rational number as a decimal and as a percent.

(a) $\frac{11}{50}$ **(b)** $\frac{-3}{1000}$ **(c)** $-\frac{17}{200}$
(d) $-\frac{7}{8}$ **(e)** $\frac{130}{40}$ **(f)** $-\frac{7}{3}$

14. Express each percent as a decimal and as a quotient of integers.

(a) 49.5% **(b)** 0.007%
(c) 0.43% **(d)** 140%

15. The price of an automobile increases from \$8000 to \$8500. What is the percent of increase?

16. Employment at a factory decreases from 400 workers to 375 workers. What is the percent of decrease?

In Problems 17 and 18, rewrite each number in scientific notation.

17. (a) 57,120,000,000 **(b)** 0.000,000,714

18. (a) 731 billion **(b)** 33 millionths

In Problems 19 and 20, rewrite each number in ordinary decimal form.

19. (a) 1.732×10^7 **(b)** 3.12×10^{-8}

20. (a) -1.066×10^4 **(b)** -3.05×10^{-11}

In Problems 21 and 22, rewrite each statement so that all numbers are expressed in scientific notation.

21. (a) An amoeba weighs about 5 millionths of a gram.
(b) The weight of a tobacco mosaic virus is approximately 0.000,000,000,000,000,066 gram.

22. (a) The diameter of the star Betelgeuse is approximately 358,400,000 kilometers.
(b) In a game of bridge there is one chance in approximately 158,800,000,000 that a player will be dealt a hand containing all cards of the same suit, and there is one chance in approximately 2,235,000,000,000,000,000,000,000,000 that all four players will be dealt such a hand.

[C] **23.** Convert the given numbers to scientific notation and calculate the indicated quantity. Do not round off your answer.

(a) $\dfrac{(40{,}320{,}000{,}000)(0.000{,}007{,}703)}{21{,}000}$

(b) $\dfrac{(97{,}400{,}000)(705{,}000)(1{,}410{,}000)^2}{0.000{,}000{,}2209}$

24. In adding approximate numbers, does it ever matter whether you round the numbers off (to the number of decimal places in the least accurate of them) before adding them, instead of adding them and then rounding off the sum?

In Problems 25 and 26, specify the accuracy of the indicated value in terms of significant digits.

25. (a) The chances of being dealt a "full house" in five-card poker are about one in 6.94×10^2.
(b) Five thousand miles is about 3×10^8 inches.

26. (a) One British thermal unit is about 6.6×10^{21} electron volts.
(b) The standard value of the acceleration of gravity is $g = 9.80665$ meters per second squared.

In Problems 27 and 28, round off the given number as indicated.

27. (a) 17,450 to the nearest thousand
(b) 0.00251 to the nearest thousandth

28. (a) 7.2283×10^5 to three significant digits
(b) 2.71828 to four significant digits

C In Problems 29 to 34, find the numerical value in scientific notation of the indicated quantity rounded off to an appropriate number of significant digits. Assume that the given values are accurate only to the number of displayed digits.

29. $R_1 + R_2 + R_3$ if $R_1 = 2.7 \times 10^4$, $R_2 = 1.5 \times 10^3$, and $R_3 = 7 \times 10^3$

30. $(R_1^{-1} + R_2^{-1} + R_3^{-1})^{-1}$ if $R_1 = 1.7 \times 10^3$, $R_2 = 3.1 \times 10^4$, and $R_3 = 5 \times 10^3$

31. $\frac{1}{2}mv^2$ if $m = 5.98 \times 10^{24}$ and $v = 2.9770 \times 10^4$

32. mc^2 if $c = 3.00 \times 10^8$ and $m = 9.11 \times 10^{-31}$

33. $\dfrac{IB}{nex}$ if $I = 2.05 \times 10^2$, $B = 1.5$, $n = 8.4 \times 10^{28}$, $e = 1.6 \times 10^{-19}$, and $x = 1.3 \times 10^{-3}$

34. $\frac{4}{3}\pi r^3$ if $r = 6.4 \times 10^6$

In Problems 35 and 36, relate the quantities by writing an equation involving a constant of variation.

35. **(a)** The force F of the wind on a blade of a wind-powered generator varies jointly with the area A of the blade and the square of the wind speed v.

(b) The inductance L of a coil of wire is jointly proportional to the cross-sectional area A of the coil and the square of the number N of turns, and inversely proportional to the length l of the coil.

(c) The rate r at which a rumor is spreading in a population of size P is jointly proportional to the number N of people who have heard the rumor and the number $P - N$ of people who have not.

(d) The collector current I_C of a transistor is jointly proportional to its current ratio β and its base current I_B. [β is the small Greek letter *beta.*]

36. **(a)** The power P provided by a jet of water is jointly proportional to the cross-sectional area A of the jet and the cube of the speed v of the water in the jet.

(b) In a *voltage-controlled* electronic device, the change in output current, written ΔI, is directly proportional to the change in input voltage, written ΔV. [Δ is the capital Greek letter *delta.* It is often used to stand for "a change in."]

(c) In geology, it is found that the erosive force E of a swiftly flowing stream is directly proportional to the sixth power of the speed v of flow of the water.

In Problems 37 to 42, **(a)** find a formula relating the given quantities, and **(b)** find the value of the indicated quantity under the specified conditions.

37. y is directly proportional to x^2 and inversely proportional to $z + 3$, and $y = 4$ when $x = 2$ and $z = 1$. Find y when $x = 3$ and $z = 6$.

38. y is jointly proportional to $\sqrt[4]{x}$ and z^3, and $y = 32$ when $x = 16$ and $z = 2$. Find y when $x = 81$ and $z = \frac{1}{3}$.

39. w is directly proportional to x and inversely proportional to y, and $w = 7$ when $x/y = 3$. Find w when $x = 24$ and $y = 6$.

C **40.** y is directly proportional to x and inversely proportional to z^2, and $y = 1.422$ when $x = 0.4181$ and $z = 0.7135$. Find y when $x = 2.133$ and $z = 5.357$.

C **41.** The volume V of a sphere is directly proportional to the cube of its diameter. If a sphere of diameter 2 meters has a volume of approximately 4.19 cubic meters, find the approximate volume of a sphere of diameter 10 meters.

42. A company's sales volume S per month varies directly as the number A of dollars per month spent for advertising, and inversely as the product of the selling price x dollars per unit and the inflation index I. If S is 20,000 units when $A = \$10,000$, $x = \$400$, and $I = 5\%$, find S when $A = \$15,000$, $x = \$500$, and $I = 10\%$.

In Problems 43 and 44, specify the type of each algebraic expression (monomial, binomial, trinomial, multinomial, polynomial, constant, fraction, rational expression, or radical expression). For the polynomials, give the degree and the coefficients.

43. **(a)** $-2x^2$ **(b)** $\frac{x+y}{x-y}$

(c) $3x^2 - 5x - 1$ **(d)** $xy + \sqrt{x}$

(e) $\sqrt{2}x^3 - \sqrt[5]{7}$ **(f)** $\sqrt{\pi} + \frac{x}{y}$

44. **(a)** xy **(b)** $xy + x^{-1} - 1$

(c) $\frac{x^2 - y^2}{x^2 + 1}$ **(d)** $\sqrt{7}$

(e) $3x^3 - 2x^2 + 6x - y$ **(f)** $x^2 + x + \sqrt{x^2 + 1}$

In Problems 45 and 46, use the properties of exponents to simplify each expression.

45. **(a)** $5y^2 \cdot 4y^3$ **(b)** $(-x^2y)(x^4y^3)$ **(c)** $(-p^2)^4$
(d) $(-x^2y)^7$ **(e)** $(ab^n)(ab)^n$

46. **(a)** $(-6t^4)(-5t^6)t^{10}$ **(b)** $t^{3n}t^{2n}t^n$ **(c)** $(-q^3)^5$
(d) $[-(x+y)^2]^3$ **(e)** $(2x^n)^4$

In Problems 47 to 60, carry out the indicated operation and simplify the result.

47. $(x^3 - 2x^2 + 7x - 5) + (2x^3 - x^2 - 5x + 11)$

48. $(3x^3 - 3x^2 - 8x - 17) - (4x^3 + 5x - 7)$

49. $(xy^2 + 3)(2xy^2 + 1)$ **50.** $(s^3 + t^2)(s^3 - 2t^2)$

51. $(2p + 3)(p^2 - 4p + 1)$

52. $(x^2 + x - 2)(x^2 + 3x + 1)$

53. $(2x + 3)(x - 1)(x - 2)$

54. $(2x - y)(x + 3y)(3x + y)(3x - y)$

55. $(2x^2 - 5yz)^2$ **56.** $(3a^n - b^n)^2$

57. $(2x - y + 3z)^2$ **58.** $(p - q)^2(p + q)^2$

59. $(3x^3 - 2xy)^3$ **60.** $(p^n - 3)(p^{2n} + 3p^n + 9)$

In Problems 61 to 80, factor each polynomial completely.

61. $9x^2y^2 - 12xy^4$ **62.** $18r^2s + 12r^3s^4$

63. $(a + b)^2c^2 - (a + b)c^4$

64. $(3p + q)^3 - (3p + q)^2u$

65. $(x - y)^2 - z^2$ **66.** $x^8y^8 - 1$

67. $25x^2 - 49x^2y^2$ **68.** $(c - 2d)^4 - (3c - d)^4$

69. $8p^3 + 27q^3$ **70.** $(a + 2b)^3 - (a - 2b)^3$

71. $x^2 + 2x - 24$ **72.** $a^6 + 8a^3 + 16$

73. $6x^2 + 5xy - 6y^2$ **74.** $x^{2n} + x^n - 6$

75. $4u^2v^2 - 7uv - 2$

76. $2a^2 + 4ab + 2b^2 - a - b - 10$

77. $20 + 7x - 6x^2$

78. $x^2 + 2xy - 4x - 4y + y^2 + 4$

79. $p^2 + 9q^2 - 4 + 6pq$

80. $(a + b)^4 - 7(a + b)^2 + 1$

In Problems 81 to 84, reduce each fraction to lowest terms.

81. $\frac{t^2 + 5t + 6}{t^2 + 4t + 4}$ **82.** $\frac{2a^2 - 3a - 2}{2a^2 + 3a + 1}$

83. $\frac{c(c-2) - 3}{(c-2)(c+1)}$ **84.** $\frac{(b-2)(b+1) - 4}{(b+2)(b-2)}$

In Problems 85 to 96, perform the indicated operations and simplify the result.

85. $\frac{x^2 + 6x + 5}{2x^2 - 2x - 12} \cdot \frac{4x^2 - 36}{x^2 + 8x + 15}$

86. $\frac{2y^2 - 7y - 15}{5y^2 - 24y - 5} \cdot \frac{20y^2 + 14y + 2}{2y^2 + 11y + 12} \cdot \frac{3y^2 + y - 2}{10y^2 + 35y + 15}$

87. $\frac{x^2y - xy^2}{3x^2 - 9xy + 6y^2} \div \frac{x^3 + x^2y}{6x^3 - 6x^2y - 12xy^2}$

88. $\frac{14a^2 + 23a + 3}{2a^2 + a - 3} \div \frac{7a^2 + 15a + 2}{2a^2 - 3a + 1}$

89. $\frac{t^2 - 2t + 1}{t^2 + t} - \frac{t - 3}{t + 1}$

90. $\frac{a - 2b}{2ab - 6b^2} - \frac{b}{a^2 - 4ab + 3b^2}$

91. $\frac{2}{c-2}+\frac{1}{c+3}-\frac{10}{c^2+c-6}$

92. $\frac{3}{p+1}-\frac{3}{p^2+p}+\frac{6}{p^2-1}$

93. $\frac{1+\frac{6}{a-3}}{a+3}$

94. $\frac{2-\frac{3}{y+2}}{\frac{x}{y-1}+\frac{x}{y+2}}$

95. $\frac{\frac{6}{a^2+3a-10}-\frac{1}{a-2}}{\frac{1}{a-2}+1}$

96. $\frac{\frac{1}{x+y}-\frac{1}{x-y}}{\frac{2y}{x^2-y^2}}$

97. In each case, find the ratio of the first quantity to the second. Write your answer as a reduced fraction. If the quantities are measured in different units you must first express both in the same units.

(a) Twelve dollars to sixty-five cents.

(b) 100 milliwatts to 25 microwatts.

(c) 250 turns in the primary winding of a transformer to 10,000 turns in the secondary winding (Figure 1).

Figure 1

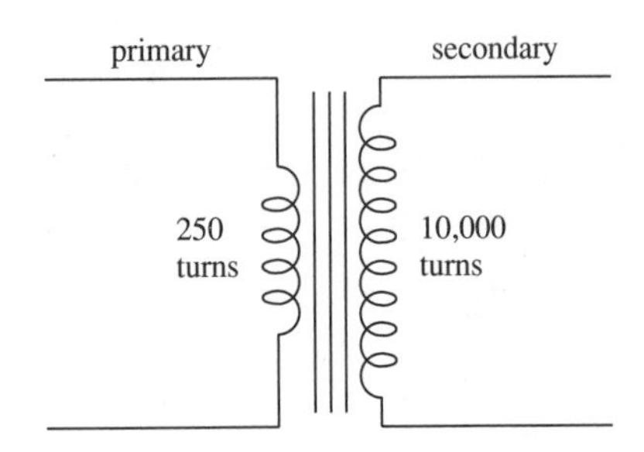

(d) 10 miles to 1000 feet.

(e) 50 ounces to 2 pounds.

(f) 1 kilometer to 1 mile.

98. Find the ratio of each pair of expressions if the ratio of a to b is $\frac{2}{3}$.

(a) $3a+2b$ to $4a-7b$ **(b)** $5a-3b$ to $3b+7a$

In Problems 99 and 100, solve each proportion for x.

99. $2/x=\frac{4}{5}$

100. $2x-3$ is to 3 as $1-x$ is to 5.

101. If $a/x=x/b$, x is called the *mean proportional* between a and b. Find the mean proportional between 2 and 8.

102. The ratio of the surface areas of two spheres is $\frac{9}{4}$. Find the ratio of their volumes. [*Hint:* The volume V and surface area A of a sphere of radius r are given by $V=\frac{4}{3}\pi r^3$ and $A=4\pi r^2$.]

103. The *voltage gain* of an electronic amplifier is defined to be the ratio of the output voltage to the input voltage. Find the voltage gain of an amplifier if the input voltage is 0.005 volt and the output voltage is 25 volts.

104. One of the indicators of the quality of instruction at an educational institution is the student-to-teacher ratio. Is it desirable that this ratio be large or small?

In Problems 105 to 108, determine which of the indicated principal roots are rational numbers and evaluate those that are. Do not use a calculator.

105. $\sqrt[5]{0.6}$ **106.** $\sqrt[3]{-\frac{108}{32}}$

107. $\sqrt[4]{-81}$ **108.** $\sqrt[3]{0.216}$

In Problems 109 to 114, use the properties of radicals to simplify each expression. Assume that variables are restricted to values for which all expressions are defined.

109. $\sqrt[3]{4x^2}\,\sqrt[3]{2x^4}$ **110.** $\frac{\sqrt[4]{64a^3b^2c}}{\sqrt[8]{16a^2b^{12}c^{10}}}$

111. $\sqrt[3]{\frac{(a+b)^9}{27a^3}}$

112. $\sqrt[n]{(a+b)^{4n}c^{2n}}\,\sqrt[m]{(a+b)^{2m}c^m}$

113. $\sqrt[3]{\frac{a^{14}\sqrt{a^6}}{a^7}}\,\sqrt[3]{\frac{5a^{12}}{a^{15}}}$ **114.** $\frac{\sqrt[4]{x^6y^3z^2}\,\sqrt[4]{x^3yz^6}}{\sqrt[4]{xy^2}}$

In Problems 115 to 130, perform the indicated operations and simplify the result. Rationalize all denominators (whenever possible). Assume that variables are restricted to values for which all expressions are defined.

115. $5\sqrt[3]{2p}+4\sqrt[3]{16p}$ **116.** $\sqrt[3]{250x^2}-6\sqrt[3]{16x^2}$

117. $\sqrt{6}(5-\sqrt{6})+\sqrt[3]{216}$ **118.** $(\sqrt{y}+1)(\sqrt{y}-2)$

119. $(\sqrt{2a}-\sqrt{3})(\sqrt{2a}+\sqrt{3})$

120. $(\sqrt{y+z} - 3\sqrt{x})(\sqrt{y+z} + 3\sqrt{x})$

121. $(\sqrt{a+b} - \sqrt{a})^2$

122. $(A - B)(A^2 + AB + B^2)$ where $A = \sqrt[3]{x+1}$ and $B = \sqrt[3]{x-1}$

123. $\dfrac{6}{\sqrt{2x}}$

124. $\dfrac{5}{\sqrt[3]{3p}}$

125. $\dfrac{\sqrt{a}}{\sqrt{a} - \sqrt{b}}$

126. $\dfrac{\sqrt{c}}{\sqrt{c} + \sqrt{d}}$

127. $\dfrac{\sqrt{a-1}}{1 + \sqrt{a-1}}$

128. $\dfrac{(x+1)^2}{\dfrac{x\sqrt{x}+1}{2x\sqrt{x}} - \dfrac{\sqrt{x}-1}{2\sqrt{x}}}$

129. $\dfrac{5}{\sqrt[3]{2} - 1}$

130. $\dfrac{6}{\sqrt[3]{x+y} - \sqrt[3]{x}}$

In Problems 131 and 132, rationalize the numerator.

131. $\dfrac{3\sqrt{x} + \sqrt{y}}{5}$

132. $\dfrac{\sqrt{x+h+2} - \sqrt{x+2}}{h}$

In Problems 133 to 142, simplify each expression and write it in a form containing only positive exponents.

133. $\left(\dfrac{x}{y^{-2}}\right)^{-1} + \left(\dfrac{y}{x^{-2}}\right)^{-1}$

134. $\dfrac{(a+2)^{-1} - (a-2)^{-1}}{(a+2)^{-1} + (a-2)^{-1}}$

135. $x^{-3}(x - x^{-1})$

136. $\dfrac{c^{-1} + c^{-2}}{c^{-3}}$

137. $(-3a^{-3})(-a^{-1})^3$

138. $(a^2b^{-4})^{-1}(a^{-3}b^2)^{-2}$

139. $\left(\dfrac{x^{-2}}{y^3}\right)^{-2}\left(\dfrac{x^{-3}}{y^{-4}}\right)^{-3}$

140. $\dfrac{(xy^{-1})^{-2}}{x} \cdot \left(\dfrac{x}{y^{-1}}\right)^{-3}$

141. $\dfrac{(5p^2)^{-2}(5p^5)^{-2}}{(5^{-1}p^{-2})^2}$

142. $\dfrac{(a^3b^2c^4)^{-2}(a^4b^2c)^{-1}}{(abc)^{-1}(a^2bc^3)^2}$

In Problems 143 and 144, find the value of each expression without using a calculator.

143. **(a)** $\left(\dfrac{8}{27}\right)^{2/3}$ **(b)** $32^{-1.8}$

144. **(a)** $243^{0.6}$ **(b)** $(64^{1/6} + 4096^{1/12})^{-2}$

In Problems 145 to 152, use the properties of rational exponents to simplify each expression and write it in a form containing only positive exponents. Assume that variables are restricted to values for which the properties of rational exponents hold.

145. $y^{-3/4}y^{2/3}y^{4/3}y^{-1/4}$

146. $x^{-1/2}(x^{3/2} + x^{1/2})$

147. $(a^{-1/4})^8(a^{-1/15})^{-45}a^2$

148. $a^{1/3}b^{1/3}\left[\left(\dfrac{a+b}{2}\right)^2 - \left(\dfrac{a-b}{2}\right)^2\right]^{-2/3}$

149. $(x^2y^{-1})^{-1/2}(x^{-3})^{-1/3}(y^{-2})^{-1/2}$

150. $(a^{1/m}b^{-m})^{-m}(a^{-m}b^{1/m})^m, \quad m > 0$

151. $\left(\dfrac{-64a^3}{b^6c^4}\right)^{-2/3}\left(\dfrac{8a^{1/3}b^{3/2}}{c^{1/3}}\right)^6$

152. $\left(\dfrac{a^{-3/5}b^{-1/3}c^{2/5}}{a^{-1/5}b^{-2/3}c^{1/5}}\right)^{15}$

In Problems 153 to 156, factor each expression and simplify the result.

153. $y^{-12}(x-y)(x+y)^{-3} + y^{-10}(x+y)^{-4}$

154. $a^{7/5}b^{-2/3} - a^{2/5}b^{1/3}$

155. $(y+2)^{-2/3}(y+1)^{2/3} + 2(y+2)^{1/3}(y+1)^{-1/3}$

156. $2(x+1)^{5/3}(x-2)^{-1/3} + (x^2 - x - 2)^{2/3}$

In Problems 157 to 172, express each complex number in the form $a + bi$, where a and b are real numbers.

157. $(3 + 2i) + (7 + 3i)$

158. $(5 - 7i) - (4 + 2i)$

159. $(7 - 4i) - (-6 + 4i)$

160. $(-\frac{5}{2} + 6i) + (-\frac{7}{2} - 3i)$

161. $(5 - 11i)(5 + 2i)$

162. $(5 + 2i)(7 + 3i)$

163. $(2 + 5i)(-2 + 4i)$

164. $(7 - 2i)(2 + 3i)$

165. $\dfrac{2 + 5i}{3 + 2i}$

166. $\dfrac{1 + 4i}{\sqrt{3} + 2i}$

167. i^{403}

168. i^{-21}

169. $\left(\dfrac{1}{3i}\right)^3$

170. $\dfrac{1}{(3 + 2i)^2}$

171. $(2 - 3i)\overline{(3 - 2i)}$

172. $\overline{(3 + 5i)}(3 + 5i)$

CHAPTER 1 Test

1. State one basic algebraic property of the real numbers to justify each statement.

   (a) $5 \cdot [(3+2)+7] = 5 \cdot (3+2) + 5 \cdot 7$
   (b) $(8 \cdot 3) \cdot \sqrt{2} = 8 \cdot (3 \cdot \sqrt{2})$
   (c) $-7 \cdot 1 = -7$
   (d) $(3a+b) + [-(3a+b)] = 0$
   (e) $5 \cdot (a+2b) = (a+2b) \cdot 5$
   (f) $-5 \cdot \dfrac{1}{(-5)} = 1$

2. Answer true or false.

   (a) $\dfrac{y^7 \cdot (-y)^2}{y^9} = y$ (b) $(-8)^{-1/3} = 2$
   (c) $(27 + x^3)^{1/3} = 3 + x$ (d) $\left(\dfrac{7}{3+x}\right)^{-1} = \dfrac{3}{7} + \dfrac{x}{7}$
   (e) $\dfrac{8}{\sqrt[3]{2}} = 4\sqrt[3]{4}$ (f) 8 is 40% of 200.
   (g) $4 + \sqrt{25} = 4 \pm 5$ (h) $(a+b)^{3/7} = \sqrt[7]{(a+b)^3}$
   (i) $0.\overline{7} = \frac{7}{9}$

3. Simplify each expression.

   (a) $\dfrac{3y^2 - 5y - 2}{y^2 - 4}$
   (b) $(a^2 - 7a + 1) - (4 + 7a + a^2)$
   (c) $(\sqrt{a} + \sqrt{b})^2(\sqrt{a} - \sqrt{b})^2$
   (d) $\dfrac{1}{a}[(a-2b)^3 - 11ab^2 + 6a^2b + 8b^3]$
   (e) $\left(\dfrac{-x^{3/2}}{x^{-1/3}}\right)^3$
   (f) $\sqrt[5]{\dfrac{27x^{11}y^6}{3x^2}}$
   (g) $(3^{-1} + 2^{-1})^{-1}$
   (h) $(3x - 2y)(9x^2 + 6xy + 4y^2)$

4. Factor completely.

   (a) $x^6 + 7x^3 - 8$
   (b) $(3x+5)^{1/3}(2)(4x+1)(4) + (4x+1)^2(\frac{1}{3})(3x+5)^{-2/3}(13)$

5. (a) Rationalize the denominator: $\dfrac{\sqrt{x}+2}{\sqrt{x}-2}$

   (b) Rationalize the numerator: $\dfrac{\sqrt{1-x}-1}{x}$

6. Perform each operation and simplify.

   (a) $\dfrac{x}{x^2-9} + \dfrac{1}{x-3}$ (b) $\dfrac{x+1}{2x-3} \div \dfrac{x^4-1}{2x^2-3x}$
   (c) $\dfrac{x - \dfrac{4}{x}}{\dfrac{2}{x} - 1}$

7. A manufacturer determines that the volume V of sales of a certain item is inversely proportional to the price p.

   (a) Find a formula relating V to p if $V = 2000$ sales per week when $p = \$5$.
   (b) What is the value of V when $p = \$4$?

8. A particular virus is in the shape of a sphere of radius 8.75×10^{-8} meter. What volume does the virus occupy?

9. The formula $d = rt$ relates distance, rate of speed, and time. To estimate the age of the universe based on the "big bang" theory, we can use the fact that the farthest known galaxies are approximately $d = 3 \times 10^{21}$ kilometers from us and are receding at a rate of 1.5×10^{11} kilometers per year. Solve for the (approximate) age of the universe t.

10. Write each complex number in the form $a + bi$.

    (a) $(3 - 4i) + (\frac{1}{2} + i)$ (b) $(2 - i)(3 + 2i)$
    (c) $\dfrac{7}{3 - i}$ (d) i^{33}

CHAPTER 2

Equations and Inequalities

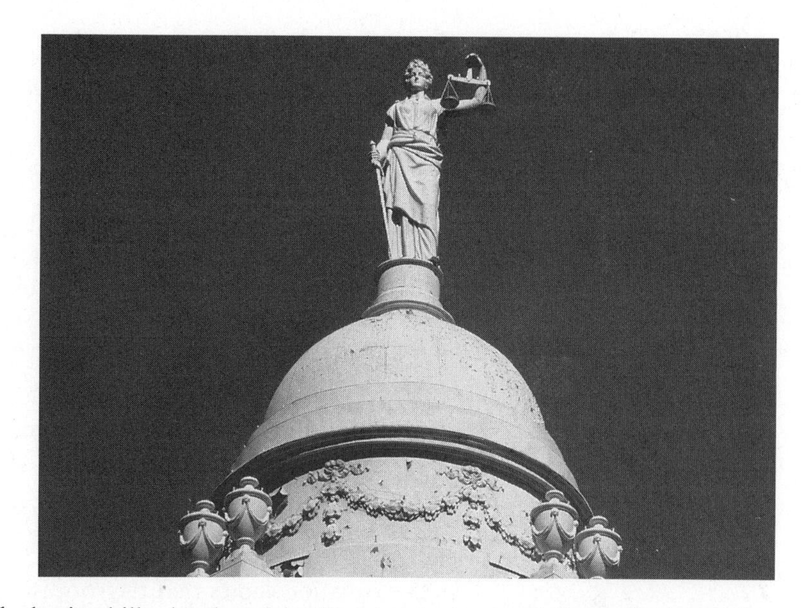

Equations and inequalities are illustrated by a balance, like the one shown here held by the Statue of Justice in Washington, D.C. When the balance is in equilibrium, the weights on either side are equal. When one of the weights is greater than the other, the balance responds to the inequality.

The basic algebraic skills developed in Chapter 1 are especially useful in solving the equations and inequalities that arise in practical applications of mathematics. In this chapter we discuss methods for solving equations and inequalities that contain just one variable. Equations and inequalities containing more than one variable are considered later in the book.

2.1 Equations

An equation containing a variable is neither true nor false until a particular number is substituted for the variable. If a true statement results from such a substitution, we say that the substitution **satisfies** the equation. For instance, the substitution $x = 3$ satisfies the equation $x^2 = 9$, but the substitution $x = 4$ does not.

An equation that is satisfied by every substitution for which both sides are defined is called an **identity.** For instance, $(x + 1)^2 = x^2 + 2x + 1$ is an identity, as is $(\sqrt{x})^2 = x$. An equation that is not an identity is called a **conditional equation.** For instance, $2x = 6$ is a conditional equation because there is at least one substitution (say, $x = 4$) that produces a false statement.

If the substitution $x = a$ satisfies an equation, we say that the number a is a **solution** or a **root** of the equation. Thus, 3 is a root of the equation $2x = 6$, but 4 is not. Two equations are said to be **equivalent** if they have exactly the same roots. Thus, the equation $2x - 6 = 0$ is equivalent to the equation $2x = 6$ because both equations have one and the same root, namely, $x = 3$.

You can change an equation into an equivalent one by performing any of the following operations:

1. *Add or subtract the same quantity on both sides of the equation.*

2. *Multiply or divide both sides of the equation by the same nonzero quantity.*

3. *Simplify one or both sides of the equation by using the methods described in Chapter 1.*

4. *Interchange the two sides of the equation.*

To **solve** an equation means to find all of its roots. The usual method for solving an equation is to write a sequence of equations, starting with the given one, in which each equation is equivalent to the previous one, but "simpler" in some sense. The last equation should either express the solution directly, or be so simple that its solution is obvious. For example, to solve the equation

$$2x - 6 = 0,$$

we begin by adding 6 to both sides to get the equivalent equation

$$2x = 6,$$

then we divide both sides by 2 to produce the equivalent equation

$$x = 3.$$

The last equation shows that the root is 3.

Variables representing quantities whose value or values we wish to find by solving equations are called **unknowns.** A common practice is to use letters toward the end of the alphabet for unknowns, and letters toward the beginning of the alphabet for **constants** whose values we can assign at will. In particular, the letter x is often used for an "unknown quantity," and the letters a, b, and c are used for constants. A **literal** or **general equation** is an equation containing, in addition to one or more unknowns, at least one letter that stands for a constant. For instance,

$$ax + b = 0$$

is a literal equation in which x is the unknown and the constant coefficients a and b can be assigned whatever values we please. If we let $a = 2$ and $b = -6$, we obtain

$$2x - 6 = 0,$$

whose solution is $x = 3$.

In applied mathematics, we cannot always follow the convention that unknowns are represented by letters toward the end of the alphabet, because certain symbols are reserved for special quantities. For instance, in physics, c is used for the speed of light, m is used for mass, v is used for velocity, and so on. We shall specify which letters represent unknowns to be solved for whenever it isn't clear from the context.

An equation such as $7x^3 + 3x^2 + x + 1 = 2x - 5$, in which both sides are polynomials in the unknown, is called a **polynomial equation.** By subtracting the polynomial on the right from both sides of the polynomial equation, we obtain an equivalent polynomial equation in **standard form** with zero on the right side:

$$7x^3 + 3x^2 + x + 1 = 2x - 5$$

$$7x^3 + 3x^2 + x + 1 - (2x - 5) = 0 \qquad \text{(We subtracted } 2x - 5 \text{ from both sides.)}$$

$$7x^3 + 3x^2 - x + 6 = 0. \qquad \text{(We combined like terms.)}$$

The last equation is in standard form. The **degree** of a polynomial equation is defined as the degree of the polynomial on the left side when the equation is in standard form. For instance, $7x^3 + 3x^2 + x + 1 = 2x - 5$ is a third-degree polynomial equation because, after it is written in standard form, $7x^3 + 3x^2 - x + 6 = 0$, the polynomial on the left side has degree 3.

First-Degree or Linear Equations

A **first-degree** or **linear equation** in x is written in standard form as

$$ax + b = 0 \qquad \text{with } a \neq 0.$$

This equation is solved as follows:

$$ax + b = 0$$

$$ax = -b \qquad \text{(We subtracted } b \text{ from both sides.)}$$

$$x = \frac{-b}{a}. \qquad \text{(We divided both sides by } a.)$$

In many cases, simple first-degree equations can be solved mentally. For example,

the solution of $5x = 10$ is $x = 2$,

and the solution of $2x + 3 = 0$ is $x = -\frac{3}{2}$.

In Examples 1 to 3, solve each equation.

Example 1 $29 - 2x = 15x - 5$

Solution

$$29 - 2x = 15x - 5$$

$$29 - 2x - 15x = -5 \qquad \text{(We subtracted } 15x \text{ from both sides.)}$$

$$29 - 17x = -5 \qquad \text{(We combined like terms.)}$$

$$-17x = -34 \qquad \text{(We subtracted 29 from both sides.)}$$

$$17x = 34 \qquad \text{(We multiplied both sides by } -1.)$$

$$x = \tfrac{34}{17} \qquad \text{(We divided both sides by 17.)}$$

$$x = 2$$

To guard against errors in arithmetic or algebra, it's a good idea to *check* the solution by substituting it back into the original equation. Thus, if we substitute $x = 2$ in the equation $29 - 2x = 15x - 5$, we obtain

$$29 - 2(2) = 15(2) - 5$$

or

$$25 = 25,$$

which shows that 2 is indeed the solution. ■

Example 2 $(2n + 3)(6n - 1) - 9 = 15n^2 - (3n - 2)(n - 2)$

Solution We begin by expanding the products on both sides of the equation:

$$(12n^2 + 16n - 3) - 9 = 15n^2 - (3n^2 - 8n + 4)$$

$$12n^2 + 16n - 12 = 12n^2 + 8n - 4 \qquad \text{(We collected like terms.)}$$

$$16n - 12 = 8n - 4 \qquad \text{(We subtracted } 12n^2 \text{ from both sides.)}$$

$$8n - 12 = -4 \qquad \text{(We subtracted } 8n \text{ from both sides.)}$$

$$8n = 8 \qquad \text{(We added 12 to both sides.)}$$

$$n = 1. \qquad \text{(We divided both sides by 8.)}$$

A check by substituting $n = 1$ into the original equation yields the equation $16 = 16$ and shows that our solution is correct. ■

Example 3 $\frac{1}{7}(3x - 1) - \frac{1}{5}(2x - 4) = 1$

Solution In order to clear the equation of fractions, we begin by multiplying both sides by 35, the LCD of the two fractions:

$$5(3x - 1) - 7(2x - 4) = 35$$

$$15x - 5 - 14x + 28 = 35 \qquad \text{(We expanded the products.)}$$

$$x + 23 = 35 \qquad \text{(We collected like terms.)}$$

$$x = 12. \qquad \text{(We subtracted 23 from both sides.)}$$

Again, a check will confirm the correctness of our work. ■

If, in solving an equation, you multiply both sides by an expression *containing the unknown*, you must always check the solution. The following example shows why.

Example 4 Solve the equation $\dfrac{1}{y(y-1)} - \dfrac{1}{y} = \dfrac{1}{y-1}$.

Solution Multiplying both sides of the equation by the LCD $y(y - 1)$ and simplifying, we have

$$\cancel{y}(\cancel{y-1})\frac{1}{\cancel{y}(\cancel{y-1})} - \cancel{y}(y-1)\frac{1}{\cancel{y}} = y(\cancel{y-1})\frac{1}{\cancel{y-1}};$$

that is, $$1 - (y - 1) = y \quad \text{or} \quad 2 - y = y.$$

Adding y to both sides of the last equation, we obtain

$$2 = 2y; \quad \text{that is,} \quad 2y = 2,$$

from which it follows that $y = 1$. We now check by substituting $y = 1$ in the original equation to obtain

$$\frac{1}{1(1-1)} - \frac{1}{1} = \frac{1}{1-1},$$

an equation in which neither side is defined because of the zeros in the denominators. In other words, the substitution $y = 1$ doesn't make the equation true—it makes the equation meaningless. Here, the correct conclusion is that the original equation *has no root.* ■

In Example 4, a solution $y = 1$ was found for the *final* equation, but it was not a solution of the *original* equation. What happened? Well, we multiplied both sides of the original equation by $y(y - 1)$, a quantity that equals zero when $y = 1$. But multiplication of both sides of an equation by zero does not produce an equivalent equation!

A fake "root" that doesn't satisfy the original equation (as in Example 4) is called an **extraneous root.*** Extraneous roots can be introduced when both sides of an equation are multiplied by a quantity containing the unknown or when both sides are raised to an even power (for instance, when both sides are squared). It's always a good idea to check your solution by substituting into the original equation, but when extraneous roots could be involved, you *must* make such a check.

C Example 5 Solve the equation $\frac{3}{7.35} - \frac{2}{p} = \frac{7}{0.463}$ and round off the answer to three significant digits.

Solution

$$\frac{2}{p} = \frac{3}{7.35} - \frac{7}{0.463}$$

$$p = \frac{2}{\frac{3}{7.35} - \frac{7}{0.463}}$$

Using a calculator to evaluate the complex fraction on the right, we find that

$$p \approx -0.135956134.$$

Substituting this value of p into the original equation, and again using the calculator to evaluate both sides, we obtain

$$15.118790497 = 15.118790497,$$

* Although this terminology is widely used, it is somewhat misleading. An extraneous root really *isn't* a root at all!

and so our answer is correct to within the limits of accuracy of the calculator. Thus, rounding off to three significant digits, we have

$$p \approx -0.136.$$

(Notice that we have rounded off only at the very end of our calculation—successively rounding off intermediate steps of a calculation can have an undesirable cumulative effect on the final result.) ■

The following example illustrates an interesting application of linear equations.

Example 6 Express the repeating decimal $0.32\overline{57}$ as a quotient of integers.

Solution Let $x = 0.32\overline{57}$. Then $100x = 32.57\overline{57}$. If we subtract $0.32\overline{57}$ from $32.57\overline{57}$, the repeating portion of the decimals cancels out:

$$\begin{array}{r} 32.57\overline{57} \\ -\ 0.32\overline{57} \\ \hline 32.25 \end{array}.$$

Therefore, $\quad 100x - x = 32.57\overline{57} - 0.32\overline{57} = 32.25;$

that is, $\quad 99x = 32.25 \quad \text{or} \quad 9900x = 3225.$

It follows that

$$x = \tfrac{3225}{9900} = \tfrac{43}{132}.$$

In general, the power of 10 used as a multiplier in this type of problem equals the number of digits in the repeating block. ■

Solution of Literal Equations

Literal equations containing one unknown can often be solved by using the methods illustrated above. It's usually a good idea to begin by trying to bring all terms containing the unknown to one side of the equation, and all terms not containing the unknown to the opposite side. As always, you must be careful not to divide by zero.

Example 7 Solve the equation $ax + 4c = b - 2x$ for x.

Solution

$$ax + 4c = b - 2x$$

$$ax + 2x + 4c = b$$ (We added $2x$ to both sides so that all terms containing x are on the left side.)

$$ax + 2x = b - 4c$$ (We subtracted $4c$ from both sides so that all terms not containing x are on the right side.)

$$(a + 2)x = b - 4c$$ (We used the distributive property.)

Now, provided that $a + 2 \neq 0$, we can divide both sides of the last equation by $a + 2$ to obtain the solution

$$x = \frac{b - 4c}{a + 2} \qquad \text{for } a \neq -2.$$

You can check this solution by substituting it into the original equation (Problem 60). ■

Example 8 The formula $S = 2\pi r^2 + 2\pi rh$ gives the total surface area S of a closed right circular cylinder of radius r and height h (Figure 1). Solve for h in terms of S and r.

Figure 1

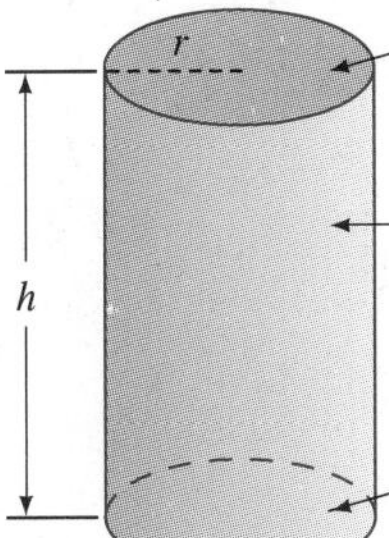

r

h

area of top πr^2

lateral surface area $2\pi rh$

area of bottom πr^2

Solution

$$S = 2\pi r^2 + 2\pi rh$$

$2\pi r^2 + 2\pi rh = S$ (We interchanged the two sides.)

$2\pi rh = S - 2\pi r^2$ (We subtracted $2\pi r^2$ from both sides to isolate the term containing h on the left side.)

$h = \dfrac{S - 2\pi r^2}{2\pi r}$ (We divided both sides by $2\pi r$.)

Because the radius of a cylinder must be positive, the denominator $2\pi r$ is nonzero. Again, we leave it as an exercise for you to check this solution (Problem 60). ■

Problem Set 2.1

In each problem set, problems with colored numbers constitute a good representation of the main ideas of the section. Note that some of the even-numbered problems may be considerably more challenging than the odd-numbered ones.

In Problems 1 and 2, solve each equation mentally for x.

1. **(a)** $3x + 6 = 0$ **(b)** $6x - 8 = 0$
(c) $\frac{2}{3}x - 3 = 0$

2. **(a)** $5x = -4$ **(b)** $\frac{1}{2}x + 2 = 0$
(c) $cx = d, c \neq 0$

C In Problems 3 and 4, solve each equation with the aid of a calculator. Round off all answers to the correct number of significant digits.

3. $31.02x + 47.71 = 0$

4. $(3.442 \times 10^{-14})x + (2.193 \times 10^9) = 0$

In Problems 5 to 18, solve each equation.

5. $3x - 2 = 7 + 2x$

6. $3x + 8 = 9 - 2x$

7. $9 - 2y = 12 - 3y$

8. $5p + 6 = 3p + 5$

9. $3(y + 6) = y - 1$

10. $10x - 1 - 7x + 3 = 7x - 10$

11. $14 - (3x - 30) = 15x - 10$

12. $7(2n + 5) - 6(n + 8) = 7$

13. $\frac{1}{2}y - \frac{2}{3}y = 7 - \frac{3}{4}y$

14. $\dfrac{n - 1}{3} + 3 = \dfrac{n + 14}{9}$

15. $\dfrac{5 + x}{6} - \dfrac{10 - x}{3} = 1$

16. $\frac{2(4x-5)}{3}+9=\frac{3(x+2)}{4}-\frac{13}{6}$

17. $(2x+3)^2=(2x-1)(2x+1)$

18. $(t-1)(2t+3)+(t+1)(t-4)=3t^2$

In Problems 19 to 28, solve each equation. Be sure to check for extraneous roots.

19. $\frac{10}{x}-2=\frac{5-x}{4x}$

20. $\frac{3-y}{3y}+\frac{1}{4}=\frac{1}{2y}$

21. $\frac{t}{t+4}=\frac{1}{2}$

22. $\frac{u-5}{u+5}+\frac{u+15}{u-5}=\frac{25}{25-u^2}+2$

23. $\frac{2}{x-2}+\frac{1}{x+1}=\frac{1}{(x-2)(x+1)}$

24. $\frac{2n}{n+7}-1=\frac{n}{n+3}+\frac{1}{(n+7)(n+3)}$

25. $\frac{1}{y-3}-\frac{1}{3-y}=\frac{1}{y^2-9}$

26. $\frac{5}{y-1}+\frac{1}{4-3y}=\frac{3}{6y-8}$

27. $\frac{t}{t^2-1}+\frac{2}{t+1}=\frac{1}{t^2-1}$

28. $\frac{1}{y-2}=\frac{3}{y+2}-\frac{6y}{y^2-4}$

In Problems 29 and 30, round off your answers to three significant digits.

[c] 29. $\frac{0.5}{9.37}-\frac{2}{x}=\frac{5}{0.713}$

[c] 30. $\frac{16.3}{3.92}+\frac{8}{y+1}=\frac{11}{0.832}$

In Problems 31 and 32, rewrite each repeating decimal as a quotient of integers.

31. **(a)** $0.\overline{2}$ **(b)** $3.41\overline{21}$ **(c)** $0.\overline{039}$
(d) $-1.00\overline{17}$ **(e)** $0.00\overline{7}$

32. **(a)** $0.\overline{121}$ **(b)** $-3.\overline{321}$
(c) $0.1\overline{523}$ **(d)** $0.\overline{285714}$

In Problems 33 to 42, solve each literal equation for the indicated unknown. Be careful not to divide by zero.

33. $5(2x+a)=bx-c$ for x

34. $7(2t+5a)-6(t+b)=3a$ for t

35. $\frac{ax+b}{c}=d+\frac{x}{4c}$ for x, if $c\neq 0$

36. $\frac{y-3a}{b}=\frac{2a}{b}+y$ for y, if $b\neq 0$

37. $\frac{x}{m}-\frac{a-x}{m}=d$ for x, if $m\neq 0$

38. $\frac{3ap-2b}{3b}-\frac{ap-a}{2b}=\frac{ap}{b}-\frac{2}{3}$ for p, if $b\neq 0$

39. $\frac{mn}{x}-bc=d+\frac{1}{x}$ for x

40. $\frac{1}{a}+\frac{a}{a+x}=\frac{a+x}{ax}$ for x, if $a\neq 0$

41. $\frac{2x}{x-b}=3-\frac{x-b}{x}$ for x

42. $\frac{x-2r}{25+x}+\frac{x+2r}{25-x}=\frac{4rs}{625-x^2}$ for x

In Problems 43 to 52, a formula used in the specified field of applied mathematics is given. In each case, solve for the indicated unknown.

43. $V=\pi r^2h$ for h (geometry)

44. $F=\frac{mv^2}{r}$ for m (mechanics)

45. $A=P(1+rt)$ for t (finance)

46. $PV=nRT$ for T (physics)

47. $C=\frac{5}{9}(F-32)$ for F (meteorology)

48. $\frac{1}{p}+\frac{1}{q}=\frac{1}{f}$ for f (optics)

49. $S=\frac{rl-a}{r-1}$ for r (economics)

50. $\frac{P_1V_1}{T_1}=\frac{P_2V_2}{T_2}$ for T_2 (thermodynamics)

51. $I=\frac{nE}{nr+R}$ for n (electrical engineering)

52. $E_b = (ZM_H + Nm_n - M_A)c^2$ for Z (nuclear physics)

In Problems 53 and 54, determine whether each equation is a conditional equation or an identity.

53. **(a)** $(4x + 3)^2 = 16x^2 + 24x + 9$

(b) $\dfrac{1}{(x+1)^2} = \dfrac{x}{x^3 + 2x^2 + x}$ **(c)** $\sqrt{x^2} = x$

54. **(a)** $\dfrac{1}{x} + \dfrac{1}{2} = \dfrac{2}{x+2}$ **(b)** $\sqrt{1 + x^2} = 1 + x$

(c) $\dfrac{1-x}{x^2-1} + \dfrac{x}{x+1} = \dfrac{x-1}{x+1}$

In Problems 55 to 58, determine whether the given equations are equivalent. Give reasons for your answers.

55. $x = 6$ and $x^2 = 36$

56. $(x-1)(x+2) = x^2$ and $(x-1)(x+2)x = x^3$

57. $x = 3$ and $x^3 = 27$

58. $\dfrac{t^2-1}{t+1} = t$ and $t^2 - 1 = t(t+1)$

59. The formula $H = (A + B\sqrt{V} - CV)(S - T)$ gives the heat loss (wind chill) H in Btu's per square foot of skin per hour if the air temperature is T degrees Fahrenheit and the wind speed is V miles per hour. Here $S =$ 91.4°F represents neutral skin temperature, and A, B, and C are constants determined experimentally to be $A = 2.14$, $B = 1.37$, and $C = 0.0916$. The *equivalent temperature* (**wind chill index**) is defined to be the air temperature T_E degrees Fahrenheit at which the same heat loss would occur if the wind speed were 4 miles per hour (the speed of a brisk walk). Thus,

$$(A + B\sqrt{V} - CV)(S - T) = [A + B\sqrt{4} - C(4)](S - T_E).$$

(a) By solving the last equation, find a formula for T_E in terms of T, V, S, A, B, and C.

C **(b)** Find the equivalent temperature T_E if $T = 25$°F and $V = 20$ miles per hour.

60. Check the solutions to Examples 7 and 8.

2.2 Applications Involving First-Degree Equations

Questions that arise in the real world are usually expressed in words, rather than in mathematical symbols. For example: What will be the monthly payment on my mortgage? How much insulation must I use in my house? What course should I fly to Boston? How safe is this new product? In order to answer such questions, it is necessary to have certain pertinent information. For instance, to determine the monthly payment on a mortgage, you need to know the amount of the mortgage, the interest rate, and the time period involved.

Problems in which a question is asked and pertinent information is supplied in the form of words are called "word problems" or "story problems" by students and teachers alike. In this section, we study word problems that can be worked by setting up an equation containing the unknown and solving it by the methods illustrated in Section 2.1. For working these problems, we recommend the following systematic procedure:

Step 1. Begin by reading the problem carefully, several times if necessary, until you understand it well. Draw a diagram whenever possible. Look for the question or questions you are to answer.

Step 2. List all of the unknown numerical quantities involved in the problem. It may be useful to arrange these quantities in a table or chart along with related known quantities. Select one of the unknown quantities on your list, one that seems to play a prominent role in the problem, and call it x. (Of course, any other letter will do as well.)

Step 3. Using information given or implied in the wording of the problem, write algebraic relationships among the numerical quantities listed in step 2. Relationships that express some of these quantities in terms of x are especially useful. Reread the problem, sentence by sentence, to make sure you have rewritten all the given information in algebraic form.

Step 4. Combine the algebraic relationships written in step 3 into a single equation containing only x and known numerical constants.

Step 5. Solve the equation for x. Use this value of x to answer the question or questions in step 1.

Step 6. Check your answer to make certain that it agrees with the facts in the problem.

Of course, a calculator is often useful to expedite arithmetic.

Example 1 One number is 15 less than a second number. Three times the first number added to twice the second number is 80. Find the two numbers.

Solution We follow the procedure just outlined.

Step 1. Question: What are the two numbers?

Step 2. Unknown quantities: *The first number* and *the second number*. Let $x =$ *the first number*. (See the alternative solution below, where we choose x to represent the second number.)

Step 3. Information given:

(i) *The first number* = *the second number* − 15; that is,

$$x = \textit{the second number} - 15.$$

(ii) 3(*the first number*) + 2(*the second number*) = 80; that is,

$$3x + 2(\textit{the second number}) = 80.$$

Step 4. From relationship (i) in step 3, we have

$$\textit{the second number} = x + 15.$$

Therefore, relationship (ii) can be written as

$$3x + 2(x + 15) = 80.$$

Step 5. Solving the equation $3x + 2(x + 15) = 80$, we obtain

$$3x + 2x + 30 = 80$$

$$5x = 50$$

$$x = 10.$$

Therefore $\quad$ *the first number* $= x = 10$

and $\quad$ *the second number* $= x + 15 = 10 + 15 = 25.$

Step 6. *Check:* Indeed, 10 is 15 less than 25 and $3(10) + 2(25) = 80$.

Alternative Solution

In step 2 above we could have let $x =$ *the second number.* With this assignment of the variable, relationship (i) in step 3 becomes *the first number* $= x - 15$, and relationship (ii) becomes $3(x - 15) + 2x = 80$, or $5x = 125$. Hence, *the second number* $= x = 25$ and *the first number* $= x - 15 = 10$. ■

Example 2

A suit is on sale for \$195. What was the original price of the suit if it has been discounted 25%?

Solution

Step 1. Question: What was the original price of the suit?

Step 2. Unknown quantities: *The original price of the suit* and *the amount of the discount in dollars.* Let $x =$ *the original price.*

Step 3. (*Original price*) − *discount* = sale price = 195 dollars; that is,

$$x - \textit{discount} = 195.$$

$$\textit{Discount} = 25\% \text{ of } \textit{original price} = 0.25x.$$

Step 4. $x - 0.25x = 195$

Step 5. $0.75x = 195$

$$x = \frac{195}{0.75}$$

$$= 260$$

The original price of the suit was \$260.

Step 6. *Check:* If a \$260 suit is discounted by 25%, the discount is (0.25)(\$260) = \$65 and the sale price is \$260 − \$65 = \$195. ■

In solving the following problem, we use the **simple interest formula**

$$I = Prt$$

where I is the **simple interest,** P is the **principal** (the amount invested), r is the **rate** of interest per interest period, and t is the number of interest periods.

Example 3

A businesswoman has invested a total of \$30,000 in two certificates. The first certificate pays 10.5% annual simple interest, and the second pays 9% annual simple interest. At the end of one year, her combined interest on the two certificates is \$2970. How much did she originally invest in each certificate?

Solution

Step 1. Question: What was the principal for each of the two certificates?

Step 2. Let $x =$ *the principal for the first certificate* in dollars. Quantities involved in the problem appear in the following table. Here the interest period is $t = 1$ year, so that $I = Pr$.

Certificate	Principal	Rate	Time	Simple Interest
First	x dollars	0.105	1	$0.105x$ dollars
Second	$30{,}000 - x$ dollars	0.09	1	$0.09(30{,}000 - x)$ dollars

Step 3. Most of the information given in the problem appears in the table. The only remaining fact is that

the combined simple interest = 2970 dollars.

Step 4. Because the sum of the simple interest on the two certificates is the combined simple interest,

$$0.105x + 0.09(30{,}000 - x) = 2970.$$

Step 5.

$$0.105x + 0.09(30{,}000 - x) = 2970$$

$$0.105x + 2700 - 0.09x = 2970$$

$$0.105x - 0.09x = 2970 - 2700$$

$$0.015x = 270$$

$$15x = 270{,}000$$

$$x = \frac{270{,}000}{15} = 18{,}000.$$

Therefore, \$18,000 was invested in the first certificate and

$$\$30{,}000 - \$18{,}000 = \$12{,}000$$

was invested in the second.

Step 6. *Check:* The simple interest on \$18,000 for one year at 10.5% is $0.105(\$18{,}000) = \$1890$. The simple interest on \$12,000 for one year at 9% is $0.09(\$12{,}000) = \$1080$. The total amount invested is

$$\$18{,}000 + \$12{,}000 = \$30{,}000$$

and the combined interest is $\$1890 + \$1080 = \$2970$. ■

Many word problems involving mixtures of substances or items can be worked by solving first-degree equations. Examples 4 and 5 illustrate how to solve typical **mixture problems.**

Example 4 A chemist has one solution containing a 10% concentration of acid and a second solution containing a 15% concentration of acid. How many milliliters of each should be mixed in order to obtain 10 milliliters of a solution containing a 12% concentration of acid?

Solution Let $x =$ *the number of milliliters of the first solution,* so that $10 - x =$ *the number of milliliters of the second solution.* The sketch in Figure 1 illustrates the information summarized in the table below.

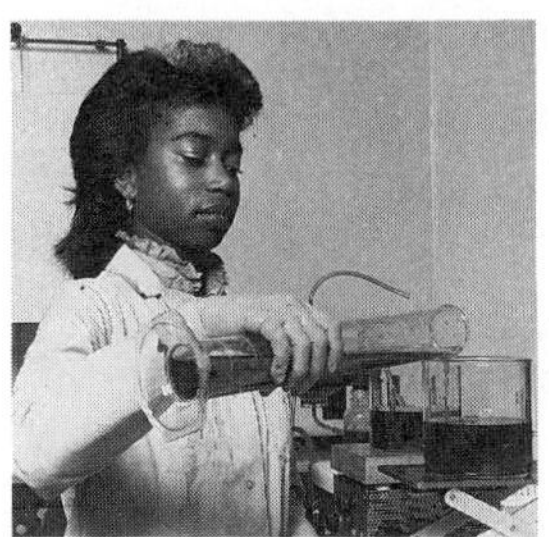

Figure 1

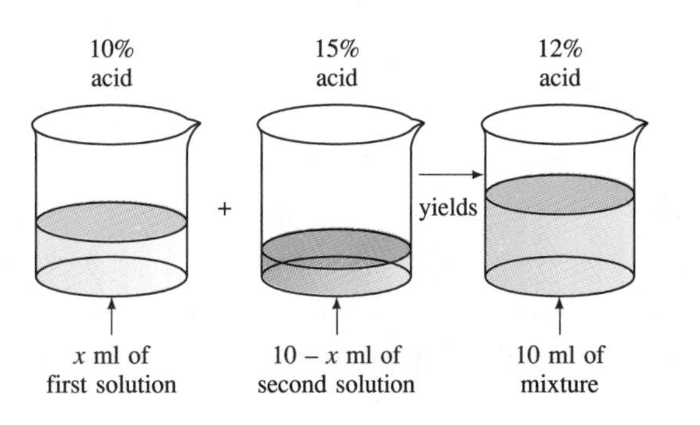

	Acid Concentration	Milliliters of Solution	Milliliters of Acid in Solution
First Solution	0.10	x	$0.10x$
Second Solution	0.15	$10 - x$	$0.15(10 - x)$
Mixture	0.12	10	$0.12(10) = 1.2$

Since the amount of acid in the mixture is the sum of the amounts in the two solutions,

$$0.10x + 0.15(10 - x) = 1.2$$

$$0.10x + 1.5 - 0.15x = 1.2$$

$$0.10x - 0.15x = 1.2 - 1.5$$

$$-0.05x = -0.3$$

$$5x = 30$$

$$x = 6.$$

Hence, 6 milliliters of the first solution and $10 - 6 = 4$ milliliters of the second solution should be mixed.

Check: In 6 milliliters of the first solution there is $(0.10)6 = 0.6$ milliliter of acid. In 4 milliliters of the second solution there is $(0.15)4 = 0.6$ milliliter of acid. Thus, there are $0.6 + 0.6 = 1.2$ milliliters of acid in the $6 + 4 = 10$ milliliters of the mixture. Therefore, the acid concentration of the mixture is $\frac{1.2}{10} = 0.12 = 12\%$. ■

Example 5 A vending machine for chewing gum accepts nickels, dimes, and quarters. When the coin box is emptied, the total value of the coins is found to be \$24.15. Find the number of coins of each kind in the box if there are twice as many nickels as quarters and five more dimes than nickels.

Solution Let n = *the number of nickels.* The following table summarizes the given information:

	Number of Coins	Individual Value	Total Value in Dollars
Nickels	n	\$0.05	$0.05n$
Dimes	$n + 5$	\$0.10	$0.10(n + 5)$
Quarters	$\frac{1}{2}n$	\$0.25	$0.25(\frac{1}{2}n)$

Since the coins have a total value of \$24.15,

$$0.05n + 0.10(n + 5) + 0.25(\tfrac{1}{2}n) = 24.15.$$

To solve this equation, we begin by multiplying both sides by 100 to remove the decimals:

$$\begin{aligned} 5n + 10(n + 5) + 25(\tfrac{1}{2}n) &= 2415 \\ 15n + \frac{25}{2}n &= 2415 - 50 = 2365 \\ 30n + 25n &= 4730 \\ 55n &= 4730 \\ n &= \frac{4730}{55} = 86. \end{aligned}$$

Therefore, there are 86 nickels, $86 + 5 = 91$ dimes, and $\frac{1}{2}(86) = 43$ quarters.

Check: $\$0.05(86) + \$0.10(91) + \$0.25(43) = \$24.15$. ■

Another type of applied problem involves objects that move a distance d at a constant rate r (also called speed) in t units of time. To solve these problems, use the formula

$$d = rt \qquad (\textit{distance} = \textit{rate} \times \textit{time}).$$

This formula can be rewritten as $r = d/t$ or as $t = d/r$.

Example 6 A jogger and a bicycle rider leave a field house at the same time and set out for a nearby town. The jogger runs at a constant speed of 16 kilometers per hour. At the end of 2 hours, the bicycle rider is 19.2 kilometers ahead of the jogger (Figure 2). How fast is the bicycle rider traveling, assuming that his speed is constant?

Figure 2

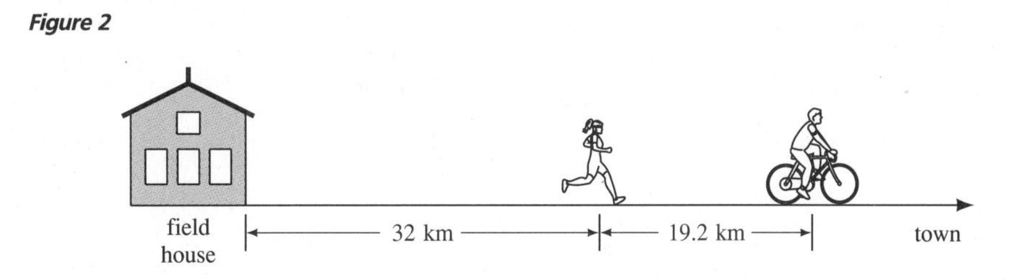

Solution Let $x =$ *the speed of the bicycle rider*. The following table summarizes the given information:

	Rate	Time	Rate × Time = Distance
Jogger	16 km/hr	2 hr	$2(16) = 32$ km
Bicyclist	x km/hr	2 hr	$2x$ km

Since the distance the bicyclist has traveled is 19.2 kilometers more than the jogger has run during the 2 hours,

$$2x = 32 + 19.2$$

$$2x = 51.2$$

$$x = 25.6.$$

Therefore, the speed of the bicyclist is 25.6 kilometers per hour.

Check: $2(25.6) = 51.2 = 32 + 19.2$. ■

Problems concerning a job that is done at a constant rate can often be solved by using the following principle:

If a job can be done in time t, then $1/t$ of the job can be done in one unit of time.

Example 7 At a factory, smokestack A pollutes the air 1.25 times as fast as smokestack B. How long would it take smokestack B, operating alone, to pollute the air by as much as both smokestacks do in 20 hours?

Solution

Here the job in question is polluting the air by as much as both smokestacks do in 20 hours. Let t be the time in hours required for smokestack B operating alone to do this job. Then $1/t$ of the job is done by smokestack B in one hour, and $1.25(1/t)$ of the job is done by smokestack A in one hour. The two smokestacks together accomplish

$$\frac{1}{t} + 1.25\left(\frac{1}{t}\right)$$

of the job in one hour. We also know that both smokestacks accomplish $\frac{1}{20}$ of the job in one hour. Therefore,

$$\frac{1}{t} + 1.25\left(\frac{1}{t}\right) = \frac{1}{20},$$

so that

$$\frac{2.25}{t} = \frac{1}{20}.$$

Solving this proportion by cross-multiplication, we find that

$$20(2.25) = t(1)$$
$$45 = t.$$

Therefore, it requires 45 hours for smokestack B to do the job alone.

Check: Is it true that $\frac{1}{45} + 1.25(\frac{1}{45}) = \frac{1}{20}$? Yes. ■

Problem Set 2.2

C In many of the following problems, a calculator may be useful to expedite the arithmetic.

1. The difference between two numbers is 12. If 2 is added to seven times the smaller number, the result is the same as if 2 is subtracted from three times the larger. Find the numbers.
2. Psychologists define the intelligence quotient (IQ) of a person to be 100 times the person's mental age divided by the chronological age. What is the chronological age of a person with an IQ of 150 and a mental age of 18?
3. A student took three examinations in a sociology course. If the average of the three grades was 87 and if the grades on the first and third examinations were 89 and 91, find the grade on the second examination.
4. In baseball, a player's batting average is found by dividing the number of safe hits by the number of times at bat. A certain player got 50 safe hits in his first 200 times at bat in the baseball season. He hits 0.300 for the rest of the season and finishes with a season average of 0.280. How many times was the player at bat during the entire season?
5. The Fahrenheit temperature corresponding to a particular Celsius temperature can be found by adding 32 to $\frac{9}{5}$ of the Celsius temperature.
   (a) Find the temperature at which the reading is the same on both the Fahrenheit and the Celsius scales.
   (b) When is the Celsius reading three times the Fahrenheit reading?

6. At a certain factory, twice as many men as women apply for work. If 5% of the people who apply are hired and 3% of the men who apply are hired, what percent of the women who apply are hired?

7. An antique car is 40 years older than a replica of the car. In 7 years the antique car will be five times as old as the replica will be then. What are the present ages of the two cars?

8. A retail outlet sells wood-burning stoves for $675. At this price, the profit on the stove is one-third of its cost to the retailer. What is the retailer's profit on the stove? [*Hint:* Cost + profit = selling price.]

9. At the end of the model year, a car dealer advertises that the list prices on all of last year's models have been discounted by 20%. What was the original list price of a car that has a discounted price of $9800?

10. A person invests part of $75,000 in a certificate that yields 8.5% simple annual interest, and the rest in a certificate that yields 9.2% simple annual interest. At the end of the year, the combined interest on the two certificates is $6606. How much was invested in each certificate?

11. To reduce their income tax, Jay and Joan invest a total of $140,000 in two municipal bonds, one that pays 6% tax-free simple annual interest and one that pays 6.5%. The total nontaxable income from both investments at the end of one year is $8775. How much was invested in each bond?

12. A factory pays time and a half for all hours worked above 40 hours per week. An employee who makes $7.30 per hour grossed $478.15 in one week. How many hours did the employee work that week?

13. The price of units in a condominium increases by 3% in April. The price increases again, by 2%, in July. What was the price of a unit before the April increase if its price after the July increase is $78,795?

14. The manager of a trust fund invests $210,000 in three enterprises. She invests three times as much at 8% as she does at 9%, and commits the rest at 10%. Her total annual income from the three investments will be $17,850. How much does she invest at each rate?

15. By installing a $120 thermostat that automatically reduces the temperature setting at night, a family hopes to cut its annual bill for heating oil by 8% and thereby recover the cost of the thermostat in fuel savings after 2 years. What was the family's annual fuel bill before installing the thermostat?

16. An archaeologist discovers a crown weighing 800 grams in the tomb of an ancient king. Evidence indicates that the crown is made of a mixture of gold and silver. Gold weighs 19.3 grams per cubic centimeter; silver weighs 10.5 grams per cubic centimeter; and it is found that the crown weighs 16.2 grams per cubic centimeter. How many grams of gold does the crown contain?

17. A metallurgist has 30 kilograms of an alloy that is 30% copper. How much of an alloy that is 65% copper must be added to make an alloy that is 55% copper?

18. A pharmacist is to prepare 15 milliliters of an ophthalmic solution that is 0.30% timolol maleate. The pharmacy has only a 0.25% and a 0.50% solution in stock. How much of each should be mixed to prepare the desired prescription?

19. A petroleum distributor has two gasohol storage tanks, the first containing 9% alcohol and the second containing 12% alcohol. An order is received for 300,000 gallons of gasohol containing 10% alcohol. How can this order be filled by mixing gasohol from the two storage tanks?

20. The cooling system of an automobile engine holds 16 liters of fluid. The system is filled with a mixture of 80% water and 20% antifreeze. It is necessary to increase the amount of antifreeze to 40%. This is to be done by draining some of the mixture and adding pure antifreeze to bring the total amount of fluid back up to 16 liters. How much of the mixture should be drained?

21. To generate hydrogen in a chemistry laboratory, a 40% solution of sulfuric acid is needed. How many milliliters of water must be mixed with 25 milliliters of an 88% solution of sulfuric acid to dilute it to the required 40% of acid?

22. It is desired to prepare V cubic units of a product containing $c\%$ of a certain ingredient. Supplies of this product containing $a\%$ and $b\%$ of the ingredient are available. Assuming that c is between a and b, find a formula for the number of cubic units x of the $a\%$ product that must be mixed with $V - x$ cubic units of the $b\%$ product to obtain the desired $c\%$ product.

23. A biologist wishes to determine the volume of blood in the circulatory system of an animal. She injects 6 milliliters of a 9% solution of a biologically inert chemical and waits until it is thoroughly mixed with the animal's blood. Then she withdraws a small sample of blood and determines that 0.03% of the sample consists of the biologically inert chemical. Find the original volume of blood in the circulatory system.

24. If x cubic units of a product containing $a\%$ of a certain ingredient is mixed with V cubic units of the product containing none of the ingredient, and if the resulting mixture contains $c\%$ of the ingredient, find a formula for V in terms of a, c, and x.

25. A bank teller has 73 bills in \$5, \$10, and \$20 denominations. The number of \$5 bills is 1 more than six times the number of \$10 bills. If the total value of the bills is \$945, how many bills of each type are there?

26. A bill of \$7.45 was paid with 32 coins: half dollars, quarters, dimes, and nickels. If there were seven more dimes than nickels and twice as many dimes as quarters, how many coins of each type were used? [*Hint:* Let x be the number of quarters.]

27. It takes Abner 3.5 minutes to run the same distance that Kim can run in 3 minutes. What is this distance if Kim runs 2 feet per second faster than Abner?

28. Two airplanes leave an airport at the same time and travel in opposite directions. One plane is traveling 64 kilometers per hour faster than the other. After 2 hours, they are 3200 kilometers apart. How fast is each plane traveling?

29. A cross-country skier starts from a certain point and travels at a constant speed of 4 kilometers per hour. A snowmobile starts from the same point 45 minutes later, follows the skier's tracks at a constant speed, and catches up to the skier in 10 minutes. What is the speed of the snowmobile in kilometers per hour?

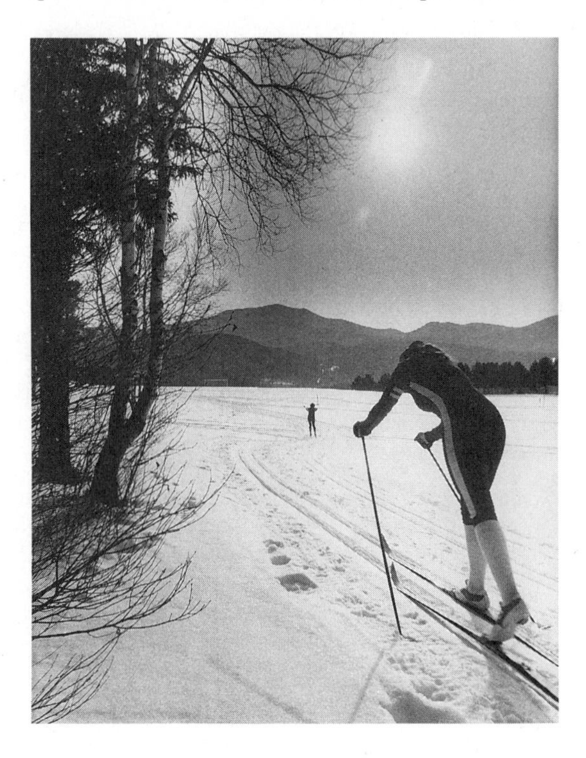

30. A driver plans to average 50 miles per hour on a trip from A to B. Her average speed for the first half of the distance from A to B is 45 miles per hour. How fast must she drive for the rest of the way?

31. A semitrailer 70 feet long is traveling 55 miles per hour. How fast must a car 18 feet long travel in order to pass the truck in 10 seconds?

32. A bus driver covers the distance from Boston to New York with an average speed of r miles per hour. On the return trip, the driver averages R miles per hour. Write a formula for the average speed of the driver for the round trip. [*Hint:* The answer is *not* $\frac{1}{2}(r + R)$.]

33. The primary (P) wave of an earthquake travels 1.7 times faster than the secondary (S) wave. Assuming that the S wave travels at 275 kilometers per minute and that the P wave is recorded at a seismic station 5.07 minutes before the S wave, how far from the station was the earthquake?

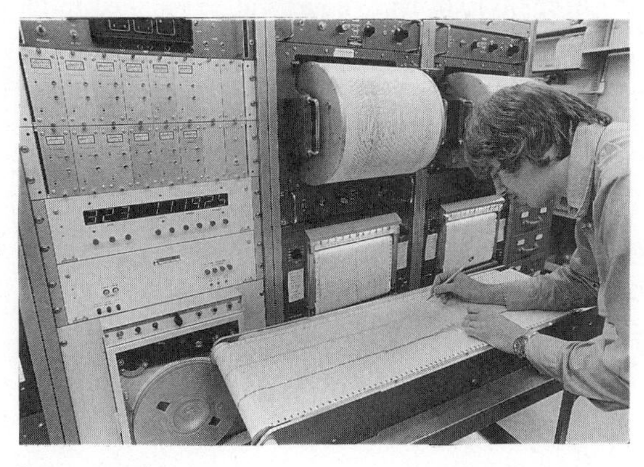

Seismograph

34. A commuter is picked up by her husband at the train station every afternoon. The husband leaves the house at the same time every day, always drives at the same speed, and regularly arrives at the station just as his wife's train pulls in. One day she takes a different train and arrives at the station one hour earlier than usual. She starts immediately to walk home at a constant speed. Her husband sees her along the road, picks her up, and drives her the rest of the way home. They arrive there 10 minutes earlier than usual.

    (a) How many minutes did she spend walking?

    (b) If the wife walks 4 miles an hour, how fast does the husband drive?

35. The perimeter of a rectangular basketball court is 80 meters, and its length is 2 meters shorter than twice its width. Find the length and width of the court.

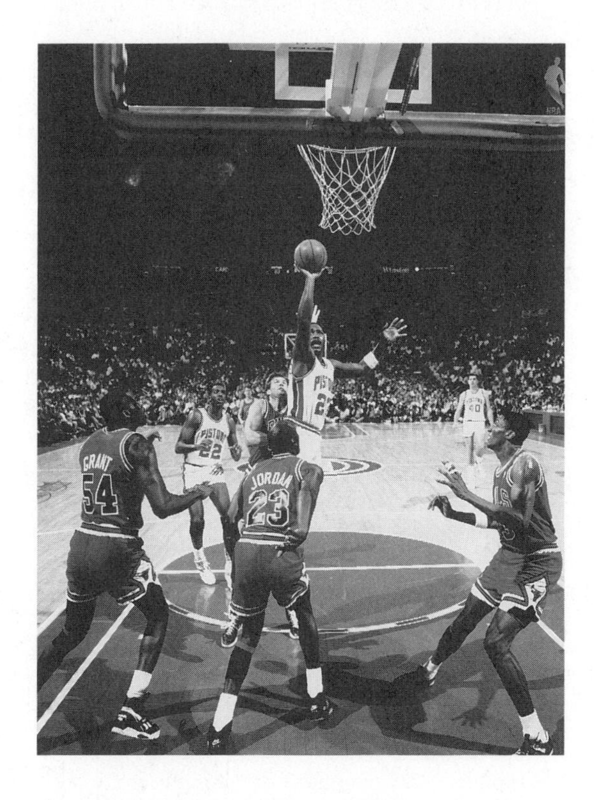

36. A certain species of Eubacteriales is a bacterium having the shape of a cylinder with hemispherical ends

(Figure 3). Find the length l meters of such a bacterium if its radius is $r = 5 \times 10^{-7}$ meter and its volume is $V = 1.9 \times 10^{-18}$ cubic meter.

Figure 3

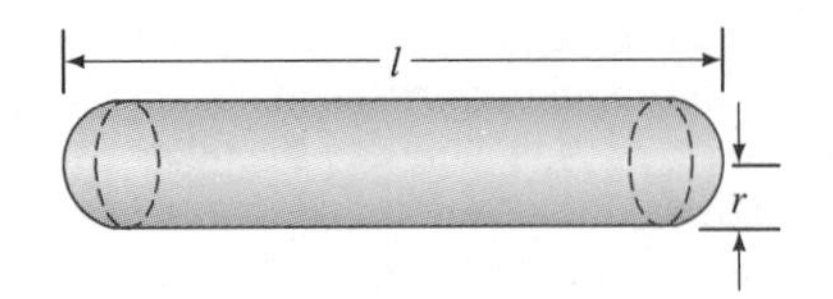

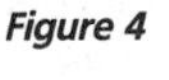 **37.** The Norman style of architecture is characterized by massive construction, carved decorations, and rectangular windows and doors surmounted by semicircular arches (Figure 4). If such a door has an area of 2.7 square meters and a width of 1 meter, find its height h meters to two significant digits.

Figure 4

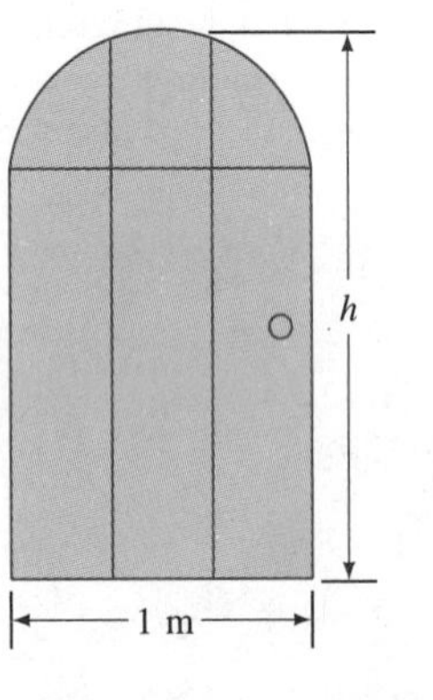

38. The circumference of the earth at the equator is 1.315×10^8 feet. Suppose that a steel belt is fitted tightly around the equator. If an additional 10 feet is added onto this belt and the slack is uniformly distributed around the earth, would you be able to crawl under the belt?

39. Factory A pollutes a lake twice as fast as factory B. The two factories operating together emit a certain amount of pollutant in 18 hours. How long would it take for factory A, operating alone, to produce the same amount of pollutant?

40. Student activists at Curmudgeon College have planned to distribute leaflets on the campus. The leaflets were to be run off on two machines, one electrically driven and the other operated by a hand crank. The electric machine produces copies four times as fast as the hand-cranked machine. The students figured that with both machines operating together they could run off the number of leaflets they needed in a total of 1.5 hours. However, because of a power blackout, they could use only the hand-cranked machine. How long did it take to run off the leaflets?

41. A computer can prepare a biweekly payroll in 10 hours. If a second computer is added, the two computers can do the job in only 3 hours. How long would it take the second computer alone to do the payroll?

42. A solar collector can generate 50 Btu's in 8 minutes. (One Btu is the amount of energy necessary to raise the temperature of one pound of water by one degree Fahrenheit.) A second solar collector can generate 50 Btu's in 5 minutes. The first collector is operated by itself for 1 minute; then the second collector is activated and both operate together. How long, after the second collector is activated, will it take to generate a total of 50 Btu's?

43. A large pump can fill a water tank in 6 hours, and a smaller pump can fill it in 10 hours (Figure 5). In order to fill an empty tank, the large pump is operated for 2 hours, and then the small pump is also turned on. In total, how many hours will be required to fill the tank in this way?

Figure 5

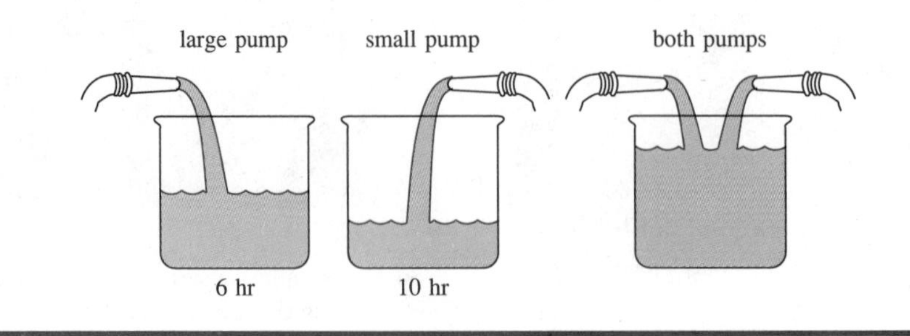

2.3 Second-Degree or Quadratic Equations

A **second-degree** or **quadratic equation** in x is written in standard form as

$$ax^2 + bx + c = 0, \qquad \text{with } a \neq 0.$$

We discuss three methods for solving quadratic equations: *factoring, completing the square,* and using the *quadratic formula.*

Solution by Factoring

When a quadratic equation is in standard form, it may be possible to factor its left side as a product of two first-degree polynomials. The equation can then be solved by setting each factor equal to zero and solving the resulting first-degree equations. This procedure is justified by the fact that a product of real numbers is zero if and only if at least one of the numbers is zero; that is, for real numbers p and q,

$$pq = 0 \qquad \text{if and only if} \qquad p = 0 \text{ or } q = 0.^*$$

In Examples 1 to 3, solve each equation by factoring.

Example 1 $15x^2 + 14x = 8$

Solution We begin by subtracting 8 from both sides of the equation to change it into standard form

$$15x^2 + 14x - 8 = 0.$$

Factoring the polynomial on the left, we obtain

$$(3x + 4)(5x - 2) = 0.$$

Now we set each factor equal to zero and solve the resulting first-degree equations:

$$3x + 4 = 0 \qquad\Big|\qquad 5x - 2 = 0$$
$$x = -\tfrac{4}{3} \qquad\Big|\qquad x = \tfrac{2}{5}.$$

Therefore, the roots are $-\frac{4}{3}$ and $\frac{2}{5}$. These roots may be checked by substitution into the original equation. ■

Example 2 $x^2 - 4x + 4 = 0$

Solution Factoring, and setting the factors equal to zero, we have

$$x^2 - 4x + 4 = 0$$
$$(x - 2)(x - 2) = 0$$
$$x - 2 = 0 \qquad\Big|\qquad x - 2 = 0$$
$$x = 2 \qquad\Big|\qquad x = 2.$$

* Note that complex numbers also have this property.

Therefore, there is only one solution, namely, $x = 2$. Again, this may be checked by substitution. ■

When the two first-degree polynomials obtained by factoring have the same root, as in Example 2 above, we call the result a **double root.**

Example 3 $\dfrac{x-6}{3x+4} - \dfrac{2x-3}{x+2} = 0$

Solution We begin by multiplying both sides of the equation by $(3x + 4)(x + 2)$, the LCD of the two fractions:

$$(\cancel{3x+4})(x+2)\frac{x-6}{\cancel{3x+4}} - (3x+4)(\cancel{x+2})\frac{2x-3}{\cancel{x+2}} = 0$$

$$(x+2)(x-6) - (3x+4)(2x-3) = 0$$

$$x^2 - 4x - 12 - (6x^2 - x - 12) = 0$$

$$x^2 - 4x - 12 - 6x^2 + x + 12 = 0$$

$$-5x^2 - 3x = 0$$

$$5x^2 + 3x = 0.$$

Factoring the left side of the last equation and setting the factors equal to zero, we obtain:

$$x(5x+3) = 0$$

$$x = 0 \quad \Big| \quad 5x + 3 = 0$$

$$x = -\tfrac{3}{5}$$

Because we multiplied both sides of the original equation by the expression $(3x + 4)(x + 2)$, which contains the unknown, we must check to be sure that $x = 0$ and $x = -\frac{3}{5}$ aren't extraneous roots. Substituting $x = 0$ in the left side of the original equation, we obtain

$$\frac{0-6}{3(0)+4} - \frac{2(0)-3}{0+2} = \frac{-6}{4} - \frac{-3}{2} = -\frac{3}{2} + \frac{3}{2} = 0,$$

so $x = 0$ is a solution. Substituting $x = -\frac{3}{5}$, we have

$$\frac{(-\frac{3}{5})-6}{3(-\frac{3}{5})+4} - \frac{2(-\frac{3}{5})-3}{(-\frac{3}{5})+2} = \frac{-33}{11} - \frac{-21}{7} = -3 + 3 = 0,$$

so $x = -\frac{3}{5}$ is also a solution. Hence, the roots are $-\frac{3}{5}$ and 0. ■

Solution by Completing the Square

A second method for solving quadratic equations is based on the idea of a **perfect square**—a polynomial that is the square of another polynomial. For example, $x^2 + 6x + 9$ is a perfect square because it is the square of $x + 3$. The polynomial $x^2 + 6x$ isn't a perfect square, but if we add 9 to it, we get the perfect square $x^2 + 6x + 9$.

More generally, by adding $(k/2)^2$ to $x^2 + kx$ we obtain a perfect square:

$$x^2 + kx + \left(\frac{k}{2}\right)^2 = \left(x + \frac{k}{2}\right)^2.$$

Thus, to create a perfect square from an expression of the form $x^2 + kx$:

> Add the square of half the coefficient of x.

This is called **completing the square.***

Example 4 Complete the square by adding a constant to each expression.

(a) $x^2 + 8x$ **(b)** $x^2 - 3x$

Solution **(a)** $x^2 + 8x + \left(\frac{8}{2}\right)^2 = x^2 + 8x + 16 = (x + 4)^2$

(b) $x^2 - 3x + \left(\frac{-3}{2}\right)^2 = x^2 - 3x + \frac{9}{4} = \left(x - \frac{3}{2}\right)^2$ ■

The following example illustrates how to solve a quadratic equation by completing the square.

Example 5 Solve the equation $2x^2 - 6x - 5 = 0$.

Solution We begin by adding 5 to both sides:

$$2x^2 - 6x = 5.$$

Next, we divide both sides of the equation by 2 so that the coefficient of x^2 will be 1:

$$x^2 - 3x = \tfrac{5}{2}.$$

* Note that this method of completing the square works *only when the coefficient of* x^2 *is* 1.

Now we can complete the square for the expression on the left by adding $\left(\frac{-3}{2}\right)^2 = \frac{9}{4}$ to both sides of the equation:

$$x^2 - 3x + \frac{9}{4} = \frac{5}{2} + \frac{9}{4}$$

$$\left(x - \frac{3}{2}\right)^2 = \frac{19}{4}.$$

It follows that $x - \frac{3}{2}$ is a square root of $\frac{19}{4}$. But there are two square roots of $\frac{19}{4}$: the principal square root $\sqrt{\frac{19}{4}} = \frac{\sqrt{19}}{2}$ and its negative $-\sqrt{\frac{19}{4}} = -\frac{\sqrt{19}}{2}$. Therefore,

$$x - \frac{3}{2} = \frac{\sqrt{19}}{2} \quad \text{or else} \quad x - \frac{3}{2} = -\frac{\sqrt{19}}{2};$$

that is,

$$x = \frac{3}{2} + \frac{\sqrt{19}}{2} = \frac{3 + \sqrt{19}}{2} \quad \text{or else} \quad x = \frac{3}{2} - \frac{\sqrt{19}}{2} = \frac{3 - \sqrt{19}}{2}.$$

These two solutions can be written in the compact form

$$x = \frac{3 \pm \sqrt{19}}{2}.$$

■

The Quadratic Formula

Let's apply the method of completing the square to solve for the unknown x in the literal equation

$$ax^2 + bx + c = 0, \qquad \text{with } a \neq 0.$$

First, we subtract c from both sides:

$$ax^2 + bx = -c.$$

Then, to prepare for completing the square, we divide both sides by a:

$$x^2 + \frac{b}{a}x = -\frac{c}{a}.$$

To complete the square, we add $\left[\frac{1}{2}\left(\frac{b}{a}\right)\right]^2 = \frac{b^2}{4a^2}$ to both sides:

$$x^2 + \frac{b}{a}x + \frac{b^2}{4a^2} = \frac{b^2}{4a^2} - \frac{c}{a}$$

$$\left(x + \frac{b}{2a}\right)^2 = \frac{b^2 - 4ac}{4a^2}.$$

It follows that

$$x + \frac{b}{2a} = \pm\sqrt{\frac{b^2 - 4ac}{4a^2}}$$

$$x = -\frac{b}{2a} \pm \frac{\sqrt{b^2 - 4ac}}{2a}.$$

Therefore, the roots of the quadratic equation $ax^2 + bx + c = 0$ can be found by using the **quadratic formula,**

$$x = \frac{-b \pm \sqrt{b^2 - 4ac}}{2a}.$$

Although the method of factoring is the quickest way to solve a quadratic equation when the factors are easily recognized, *we recommend using the quadratic formula in all other cases.* Quadratic equations are rarely solved by completing the square—this method was introduced primarily to show how the quadratic formula is derived.

Example 6 Use the quadratic formula to solve the equation $2x^2 - 5x + 1 = 0$.

Solution The equation is in the standard form $ax^2 + bx + c = 0$, with $a = 2$, $b = -5$, and $c = 1$. Substituting these values into the quadratic formula, we obtain

$$x = \frac{-b \pm \sqrt{b^2 - 4ac}}{2a} = \frac{-(-5) \pm \sqrt{(-5)^2 - 4(2)(1)}}{2(2)}$$

$$= \frac{5 \pm \sqrt{25 - 8}}{4} = \frac{5 \pm \sqrt{17}}{4}.$$

In other words, the two roots are $\frac{5 - \sqrt{17}}{4}$ and $\frac{5 + \sqrt{17}}{4}$. ■

C **Example 7** Using the quadratic formula and a calculator, find the roots of the quadratic equation $-1.32t^2 + 2.78t + 9.37 = 0$. Round off the answers to two decimal places.

Solution According to the quadratic formula, the roots are

$$t = \frac{-2.78 - \sqrt{(2.78)^2 - 4(-1.32)(9.37)}}{2(-1.32)} \approx 3.92$$

and

$$t = \frac{-2.78 + \sqrt{(2.78)^2 - 4(-1.32)(9.37)}}{2(-1.32)} \approx -1.81.$$ ■

The Discriminant and Complex Roots

The expression $b^2 - 4ac$, which appears under the radical sign in the quadratic formula

$$x = \frac{-b \pm \sqrt{b^2 - 4ac}}{2a},$$

is called the **discriminant** of the quadratic equation

$$ax^2 + bx + c = 0.$$

If a, b, and c are real numbers, you can use the algebraic sign of the discriminant to determine the number and the nature of the roots of the quadratic equation:

(i) If $b^2 - 4ac$ is *positive*, the equation has two real and unequal roots.

(ii) If $b^2 - 4ac = 0$, the equation has only one root—a double root.

(iii) If $b^2 - 4ac$ is *negative*, the equation has no real root—its roots are two complex numbers that are complex conjugates of each other.

Examples 6 and 7 illustrate (i), Example 2 illustrates (ii), and Example 9 below illustrates (iii).

Example 8 Use the discriminant to determine the nature of the roots of each quadratic equation without actually solving it.

(a) $5x^2 - x - 3 = 0$ **(b)** $9x^2 + 42x + 49 = 0$ **(c)** $x^2 - x + 1 = 0$

Solution **(a)** Here $a = 5, b = -1, c = -3$, and $b^2 - 4ac = (-1)^2 - 4(5)(-3) = 61$ is positive; hence, there are two unequal real roots.

(b) Here $a = 9$, $b = 42$, $c = 49$, and $b^2 - 4ac = 42^2 - 4(9)(49) = 0$; hence, the equation has just one root—a double root—and this root is a real number.

(c) Here $a = 1$, $b = -1$, $c = 1$, and $b^2 - 4ac = (-1)^2 - 4(1)(1) = -3$ is negative; hence, the equation has no real root. ■

Example 9 Use the quadratic formula to find the roots of the quadratic equation $x^2 - x + 1 = 0$.

Solution Using the quadratic formula, with $a = 1$, $b = -1$, and $c = 1$, we have

$$\begin{aligned} x &= \frac{-(-1) \pm \sqrt{(-1)^2 - 4(1)(1)}}{2(1)} \\ &= \frac{1 \pm \sqrt{-3}}{2} \\ &= \frac{1 \pm \sqrt{3}i}{2} = \frac{1}{2} \pm \frac{\sqrt{3}}{2}i. \end{aligned}$$

Thus, the roots are the complex conjugates

$$\frac{1}{2} - \frac{\sqrt{3}}{2}i \quad \text{and} \quad \frac{1}{2} + \frac{\sqrt{3}}{2}i.$$

■

A more detailed and more general discussion of complex roots of polynomial equations can be found in Section 4.6.

Applications Involving Quadratic Equations

Quadratic equations have many applications in the arts and sciences, business, economics, medicine, and engineering.

Example 10 An artist is planning an acrylic painting in which geometric shapes and masses form a rectangle surrounded by a stark white border of constant width. The rectangle will have dimensions 2 feet by 3 feet. To achieve vitality, the artist wants the area of the rectangle to be the same as the area of the border. Find the width of the border that will accomplish this.

Solution Let x feet be the width of the border. Subdividing the border into four rectangles as in Figure 1, we find that the area of the border in square feet is given by

$$2[x(2 + 2x)] + 2(3x) = 4x^2 + 10x.$$

Since the area of the rectangle is $2 \cdot 3 = 6$ square feet, we must solve the equation

$$4x^2 + 10x = 6.$$

This equation is equivalent to

$$2x^2 + 5x - 3 = 0.$$

Figure 1

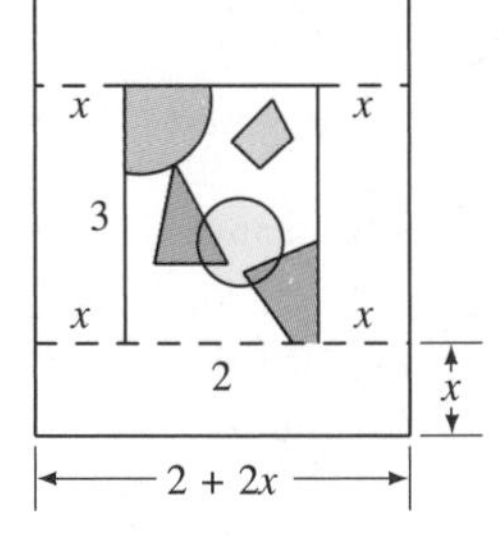

Factoring, and setting the factors equal to zero, we have

$$(2x - 1)(x + 3) = 0$$

$2x - 1 = 0$	$x + 3 = 0$
$x = \frac{1}{2}$	$x = -3.$

Since we cannot have a border with a negative width, only the first solution has meaning. Thus, the border of the painting should be $\frac{1}{2}$ foot wide. ■

Problem Set 2.3

In Problems 1 to 10, solve each equation by factoring.

1. $x^2 - 7x = 0$
2. $2x^2 - 5x = 0$
3. $x^2 + 2x = 3$
4. $z^2 - 2z = 35$
5. $x^2 - 4x = 21$
6. $2t^2 + 5t = -3$
7. $2z^2 - 7z - 15 = 0$
8. $10r^2 + 19r + 6 = 0$
9. $6y^2 - 13y = 5$
10. $54z^2 - 9z = 30$

In Problems 11 and 12, complete the square by adding a constant to each expression.

11. **(a)** $x^2 + 6x$ **(b)** $x^2 - 5x$ **(c)** $x^2 + \frac{3}{4}x$
12. **(a)** $x^2 - 6x$ **(b)** $x^2 + \frac{b}{a}x$ **(c)** $x^2 + \sqrt{2}x$

In Problems 13 and 14, complete the square to solve each equation.

13. **(a)** $x^2 + 4x - 15 = 0$ **(b)** $3r^2 + 6r = 4$
14. **(a)** $x^2 - 5x - 5 = 0$ **(b)** $5u^2 + 9u = 3$

In Problems 15 to 20, use the quadratic formula to solve each equation.

15. $5x^2 - 7x - 6 = 0$
16. $6x^2 - x - 2 = 0$
17. $3x^2 + 5x + 1 = 0$
18. $12y^2 + y - 1 = 0$
19. $2x^2 - x - 2 = 0$
20. $4u^2 - 11u + 3 = 0$

In Problems 21 to 32: **(a)** Use the discriminant to determine the nature of the roots of each quadratic equation. **(b)** Use the quadratic formula to find the roots of each equation.

21. $x^2 - 5x - 7 = 0$
22. $6t^2 - 2 = 7t$
23. $9x^2 - 6x + 1 = 0$
24. $4x^2 + 20x + 25 = 0$
25. $6y^2 + y + 3 = 0$
26. $10r^2 + 2r + 8 = 0$
27. $6x^2 + x - 1 = 0$
28. $10x^2 + 2x - 8 = 0$
29. $5x^2 - 4x + 2 = 0$
30. $4x^2 + 4x + 5 = 0$
31. $x^2 + x + 1 = 0$
32. $x^2 - 2x + \frac{3}{2} = 0$

In Problems 33 to 56, solve each equation by any method you wish.

33. $15x^2 + 4 = 23x$
34. $25y^2 - 20y + 4 = 0$
35. $4x(2x - 1) = 3$
36. $2x^2 + 7x + 6 = 0$
37. $x^2 + 6x = -9$
38. $3z^2 - 2z = 5$
39. $x^2 + 4x + 1 = 0$
40. $15 - 7u - 4u^2 = 0$
41. $9t^2 - 6t + 1 = 0$
42. $0.2x^2 - 1.2x + 1.7 = 0$
43. $z^2 + 2z = 8$
44. $8t^2 - 10t + 3 = 0$
45. $(8x + 19)x = 27$
46. $30(y^2 + 1) = 61y$
47. $25x - 40 + \frac{16}{x} = 0$
48. $36 + \frac{60}{s} = \frac{-25}{s^2}$
49. $x - 16 = \frac{105}{x}$
50. $\frac{10y + 19}{y} = \frac{15}{y^2}$
51. $\frac{5}{x + 4} - \frac{3}{x - 2} = 4$
52. $\frac{2y + 11}{2y + 8} = \frac{3y - 1}{y - 1}$
53. $\frac{2y - 5}{2y + 1} + \frac{6}{2y - 3} = \frac{7}{4}$
54. $\frac{t - 4}{t + 1} - \frac{15}{4} = \frac{t + 1}{t - 4}$
55. $\frac{x + 2}{x} + \frac{x}{x - 2} = 5$
56. $\frac{7u + 4}{u^2 - 6u + 8} - \frac{5}{2 - u} = \frac{u + 5}{u - 4}$

C In Problems 57 and 58, use the quadratic formula and a calculator to find the roots of each quadratic equation. Round off answers to two decimal places.

57. $44.04x^2 + 64.72x - 31.23 = 0$
58. $8.85x^2 - 71.23x + 94.73 = 0$

In Problems 59 and 60, a formula from applied mathematics is given. Solve each equation for the indicated unknown variable. State any necessary restrictions on the values of the remaining variables.

59. **(a)** $s = v_0 t + \frac{1}{2}gt^2$ for t (mechanics)
    (b) $LI^2 + RI + \frac{1}{C} = 0$ for I (electronics)
60. **(a)** $P = EI - RI^2$ for I (electrical engineering)
    (b) $A = 2\pi r^2 + 2\pi rh$ for r (geometry)

61. A swimming area in a small lake is to be treated periodically to control the growth of bacteria. The mathematical model $C = 2t^2 - 10t + 50$ is used to predict the bacteria count per milliliter t days after the treatment. If the bacteria count must not exceed 150, how often should the swimming area be treated?

62. In a medical laboratory, the quantity x milligrams of antigen present during an antigen–antibody reaction is related to the time t in minutes required for the precipitation of a fixed amount of antigen–antibody by the mathematical model

$$3t = 140 - 50x + 5x^2.$$

Find x if $t = 20$ minutes and x is at least 3 milligrams.

63. Working together, a carpenter and an apprentice can install drywall in a large living room in 4 hours. The apprentice, working alone, requires 6 hours more than the carpenter, working alone, to do this job. In how many hours can the carpenter do the job without the aid of the apprentice?

64. One machine can manufacture N items in x hours, and a second machine can do the same job in $x + d$ hours. Together the two machines can manufacture N items in T hours. Show that

$$x = T - \frac{d}{2} + \sqrt{T^2 + \left(\frac{d}{2}\right)^2}.$$

C **65.** If air resistance is neglected, a projectile fired vertically upward with an initial velocity v_0 will be at a height of $v_0t - 4.9t^2$ meters t seconds later. If $v_0 = 500$ meters per second, how long will it take for the projectile to reach a height of 1000 meters?

C **66.** A cable television company plans to begin operations in a small town. The company foresees that about 600 people will subscribe to the service if the price per subscriber is \$20 per month, but that for each \$0.20 increase in the monthly subscription price, 4 of the original 600 people will decide not to subscribe. The company begins operations, and its total revenue for the first month is \$6000. How many people have subscribed to the service?

C **67.** A telephone company is placing telephone poles along a road. If the distance between successive poles were increased by 10 meters, 5 fewer poles per kilometer would be required. How many telephone poles is the company now placing along each kilometer of the road?

68. Carlos, who is training to run in the Boston Marathon, runs 18 miles every Saturday afternoon. His goal is to cut his running time by one-half hour; he figures that to accomplish this he will have to increase his average speed by 1.2 miles per hour. What is his current average speed for the 18-mile run?

69. In order to support a solar collector at the correct angle, the roof trusses for a house are designed as right triangles. Rafters form the right angle, and the base of the truss is the hypotenuse (Figure 2). If the rafter on the same side as the solar collector is 10 feet shorter

Figure 2

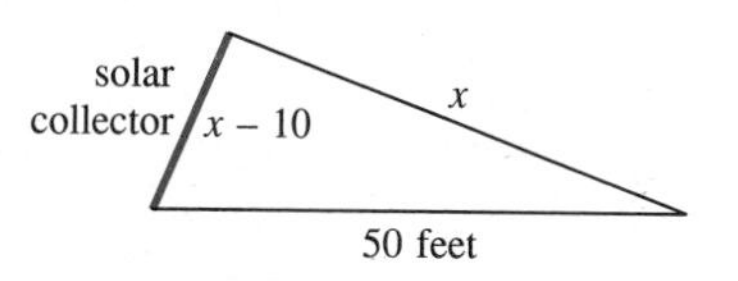

than the other rafter and if the base of each truss is 50 feet long, how long is each of the rafters? [*Hint:* Use the Pythagorean theorem, found in Appendix I.]

70. A **diagonal** of a polygon is a line segment between two nonadjacent vertices. (Figure 3 shows all of the diagonals of a polygon with five vertices.) It can be shown that a polygon with n vertices has a total of $\frac{1}{2}n(n-3)$ diagonals.

(a) How many vertices does a polygon with 14 diagonals have?

(b) Is there a polygon that has exactly 17 diagonals?

Figure 3

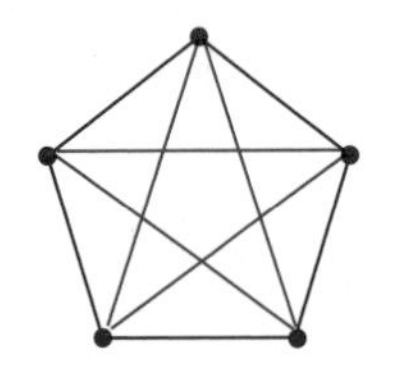

C **71.** The final value S dollars of a principal P dollars invested at a rate of interest r compounded annually for t years is given by the formula $S = P(1 + r)^t$. If $10,000 invested for 2 years at a rate of interest r compounded annually yields a final value of $12,544, find the interest rate r.

72. A standard "55 gallon" oil drum is a cylinder of height 34 inches with a closed top and bottom. Find the radius r of such a drum if its total surface area is 990π square inches.

73. Since ancient times, artists and artisans have recognized that figures whose proportions are in a certain ratio a/b, called the **golden ratio,** are especially pleasing. For instance, a rectangle whose width a and length b form the golden ratio is called a **golden rectangle.** The condition under which a/b is the golden ratio is that a is to b as b is to $a + b$.

(a) Find the exact value of the golden ratio.

C **(b)** The dimensions of a standard sheet of typing paper are 8.5 inches by 11 inches. Does such a sheet form a golden rectangle?

The Parthenon, built in Athens, Greece, in the fifth century B.C., is regarded as one of the world's most beautiful structures. The dimensions of its facade conform almost exactly to the golden ratio.

2.4 Miscellaneous Equations

In this section, we discuss special types of equations that can be solved using slight variations of the methods presented in previous sections. Here we consider only *real* roots of equations.

Equations of the Form $x^p = a$

If the exponent p is an integer, you can solve the equation $x^p = a$ as shown in the following example.

Example 1 Solve each equation.

(a) $x^3 = 5$ **(b)** $x^{-3} = 5$ **(c)** $x^4 = 5$ **(d)** $x^4 = -5$

Solution

(a) Since the exponent is odd, there is just one solution, $x = \sqrt[3]{5}$.

(b) We begin by rewriting the equation in the equivalent form

$$\frac{1}{x^3} = 5 \qquad \text{or} \qquad x^3 = \frac{1}{5}.$$

The solution is

$$x = \sqrt[3]{\frac{1}{5}} = \frac{1}{\sqrt[3]{5}} = \frac{\sqrt[3]{25}}{5}.$$

(c) Since the exponent is even, there are two solutions, $x = \sqrt[4]{5}$ and $x = -\sqrt[4]{5}$.

(d) A real number raised to an even power cannot be negative, so the equation $x^4 = -5$ has no real root. ■

The following example shows how to solve the equation $x^p = a$ when the exponent p is the reciprocal of an integer.

Example 2 Solve each equation.

(a) $x^{1/6} = 5$ **(b)** $x^{1/3} = -5$ **(c)** $x^{-1/3} = 5$ **(d)** $x^{1/6} = -5$

Solution

(a) The equation can be written as $\sqrt[6]{x} = 5$, so that $x = 5^6 = 15{,}625$.

(b) Here we have $\sqrt[3]{x} = -5$, so that $x = (-5)^3 = -125$.

(c) We begin by rewriting the equation in the equivalent form

$$\frac{1}{x^{1/3}} = 5 \qquad \text{or} \qquad x^{1/3} = \frac{1}{5};$$

that is, $\sqrt[3]{x} = \frac{1}{5}$. Thus $x = (\frac{1}{5})^3 = \frac{1}{125}$.

(d) The equation can be rewritten as $\sqrt[6]{x} = -5$. Since a principal sixth root cannot be negative, the equation has no real solution. ■

If n and m are positive integers and the fraction n/m is reduced to lowest terms, an equation of the form

$$x^{n/m} = a$$

can be rewritten as

$$\sqrt[m]{x^n} = a$$

and solved using the methods illustrated above.

Example 3 Solve each equation.

(a) $x^{3/2} = -8$ **(b)** $(t - 3)^{2/5} = 4$ **(c)** $x^{-5/2} = -\frac{1}{4}$

Solution

(a) The equation is equivalent to $\sqrt{x^3} = -8$. Because a principal square root of a real number cannot be negative, the equation has no real solution.

(b) The equation can be rewritten as $\sqrt[5]{(t-3)^2} = 4$ or $(t-3)^2 = 4^5$. Hence,

$$t - 3 = \pm\sqrt{4^5} = \pm(\sqrt{4})^5 = \pm 2^5 = \pm 32.$$

Therefore, $t = 3 \pm 32$, and the two roots are $t = -29$ and $t = 35$.

(c) We begin by rewriting the equation as

$$\frac{1}{x^{5/2}} = -\frac{1}{4} \qquad \text{or} \qquad x^{5/2} = -4.$$

The last equation is equivalent to $\sqrt{x^5} = -4$, so it has no real solution. ■

Radical Equations

An equation in which the unknown appears in a radicand is called a **radical equation.** For instance,

$$\sqrt[4]{x^2 - 5} = x + 1 \qquad \text{and} \qquad \sqrt{3t+7} + \sqrt{t+2} = 1$$

are radical equations.

To solve a radical equation, begin by isolating the most complicated radical expression on one side of the equation, and then eliminate the radical by raising both sides of the equation to a power equal to the index of the radical. You may have to repeat this technique in order to eliminate all radicals. When the equation is radical-free, simplify and solve it. Since extraneous roots may be introduced when both sides of an equation are raised to an even power, *all roots must be checked by substitution in the original equation whenever a radical with an even index is involved.*

In Examples 4 and 5, solve each equation.

Example 4 $\sqrt[3]{x-1} - 2 = 0$

Solution To isolate the radical, we add 2 to both sides of the equation:

$$\sqrt[3]{x-1} = 2.$$

Now we raise both sides to the power 3 and obtain

$$(\sqrt[3]{x-1})^3 = 2^3 \qquad \text{or} \qquad x - 1 = 8.$$

It follows that $x = 9$. Since we did not raise both sides of the equation to an even power, it isn't necessary to check our solution, but it's good practice to do so anyway. Substituting $x = 9$ in the left side of the original equation, we obtain

$$\sqrt[3]{9-1} - 2 = \sqrt[3]{8} - 2 = 2 - 2 = 0;$$

hence, the solution $x = 9$ is correct. ■

Example 5 $\sqrt{3t+7} + \sqrt{t+2} = 1$

Solution

We add $-\sqrt{t+2}$ to both sides of the equation to isolate $\sqrt{3t+7}$ on the left side. Thus,

$$\sqrt{3t+7} = 1 - \sqrt{t+2}.$$

Now we square both sides of the equation to obtain

$$(\sqrt{3t+7})^2 = (1 - \sqrt{t+2})^2$$
$$3t + 7 = 1 - 2\sqrt{t+2} + (t+2).$$

The equation still contains a radical, so we simplify and isolate this radical:

$$2t + 4 = -2\sqrt{t+2}$$
$$-(t+2) = \sqrt{t+2}.$$

Again, we square both sides, so

$$[-(t+2)]^2 = (\sqrt{t+2})^2$$
$$t^2 + 4t + 4 = t + 2$$
$$t^2 + 3t + 2 = 0$$
$$(t+2)(t+1) = 0$$

$$t + 2 = 0 \qquad\qquad t + 1 = 0$$
$$t = -2 \qquad\qquad t = -1.$$

Check: For $t = -2$,

$$\sqrt{3t+7} + \sqrt{t+2} = \sqrt{3(-2)+7} + \sqrt{-2+2} = \sqrt{1} + \sqrt{0} = 1;$$

hence, $t = -2$ is a solution. For $t = -1$,

$$\sqrt{3t+7} + \sqrt{t+2} = \sqrt{3(-1)+7} + \sqrt{-1+2} = \sqrt{4} + \sqrt{1} \neq 1.$$

Therefore, $t = -1$ is an extraneous root that was introduced by squaring both sides of the equation. The only solution is $t = -2$. ■

Equations of Quadratic Type

An equation with x as the unknown is said to be of **quadratic type** if it can be rewritten in the form

$$au^2 + bu + c = 0,$$

where u is an expression containing x. If such an equation is first solved for u, any resulting values can be used to solve for x; thus the solution of the original equation can be found.

Example 6 Solve the equation $x^{2/3} - x^{1/3} - 12 = 0$.

Solution

The given equation can be rewritten as

$$(x^{1/3})^2 - x^{1/3} - 12 = 0.$$

Let $u = x^{1/3}$. Then the equation becomes

$$u^2 - u - 12 = 0$$

$$(u - 4)(u + 3) = 0.$$

Therefore, $u = 4$ or $u = -3$. Since $u = x^{1/3}$, we have

$$x^{1/3} = 4 \qquad\qquad x^{1/3} = -3$$

$$x = 4^3 = 64 \qquad\qquad x = (-3)^3 = -27$$

Hence, the two solutions are $x = 64$ and $x = -27$. ■

Nonquadratic Equations That Can Be Solved by Factoring

The method of solving an equation by rewriting it as an equivalent equation with zero on the right side, then factoring the left side and equating each factor to zero, is not limited to quadratic equations. Indeed, this method works whenever the required factorization can be accomplished.

In Examples 7 and 8, solve each equation by factoring.

Example 7 $4x^3 - 8x^2 - x + 2 = 0$

Solution We begin by grouping the terms so the common factor $x - 2$ emerges:

$$4x^2(x - 2) - (x - 2) = 0$$

$$(x - 2)(4x^2 - 1) = 0$$

$$(x - 2)(2x - 1)(2x + 1) = 0.$$

Setting each factor equal to zero, we have

$$x - 2 = 0 \qquad 2x - 1 = 0 \qquad 2x + 1 = 0$$

$$x = 2 \qquad x = \tfrac{1}{2} \qquad x = -\tfrac{1}{2}.$$

Therefore, the three solutions are $x = 2$, $x = \frac{1}{2}$, and $x = -\frac{1}{2}$. ■

Example 8 $x^{5/6} - 3x^{1/3} - 2x^{1/2} + 6 = 0$

Solution Again, we group and factor:

$$x^{1/3}[x^{(5/6)-(1/3)} - 3] - 2[x^{1/2} - 3] = 0$$

$$x^{1/3}(x^{1/2} - 3) - 2(x^{1/2} - 3) = 0$$

$$(x^{1/2} - 3)(x^{1/3} - 2) = 0.$$

Setting each factor equal to zero, we have

$$x^{1/2} - 3 = 0 \qquad\qquad x^{1/3} - 2 = 0$$

$$x^{1/2} = 3 \qquad\qquad x^{1/3} = 2$$

$$x = 3^2 = 9 \qquad\qquad x = 2^3 = 8$$

Therefore, the two solutions are $x = 9$ and $x = 8$. ■

Applications

The techniques illustrated in this section are useful for solving problems that arise in applied mathematics.

C Example 9 An offshore oil well is located at point W, which is 13 kilometers from the nearest point Q on a straight shoreline. The oil is to be piped from W to a terminal at point T on the shoreline by piping it straight under water to a point P on the shoreline between Q and T, and then piping it straight along the shoreline to T (Figure 1). The distance from Q to T is 10 kilometers, it costs \$90,000 per kilometer to lay underwater pipe, and it costs \$60,000 per kilometer to lay pipe along the shoreline. If a total of \$1,500,000 has been allocated for this project, determine how the pipe is to be installed.

Figure 1

W, d, 13 km, 10 − x, P, x, T, Q, 10 km

Solution The problem is to locate the point P along the shoreline, and this will be accomplished if we can find the distance x kilometers between P and the terminal T. Letting d kilometers be the distance between W and P, and applying the Pythagorean theorem (see Appendix I) to the triangle WQP, we find that

$$d^2 = 13^2 + (10 - x)^2 = 169 + 100 - 20x + x^2 = 269 - 20x + x^2.$$

Therefore,

$$d = \sqrt{269 - 20x + x^2},$$

and the total cost of the pipe is given by the equation

$$90{,}000\sqrt{269 - 20x + x^2} + 60{,}000x = 1{,}500{,}000.$$

Dividing both sides of the last equation by 30,000, we obtain

$$3\sqrt{269 - 20x + x^2} + 2x = 50$$

or

$$3\sqrt{269 - 20x + x^2} = 50 - 2x.$$

Squaring both sides of this equation, we have

$$9(269 - 20x + x^2) = 2500 - 200x + 4x^2 \qquad \text{or} \qquad 5x^2 + 20x - 79 = 0.$$

By the quadratic formula,

$$x = \frac{-20 \pm \sqrt{400 - 4(5)(-79)}}{2(5)} = \frac{-20 \pm \sqrt{1980}}{10} = -2 \pm \sqrt{19.8}.$$

Rejecting the negative solution as being physically meaningless, we find that

$$x = \sqrt{19.8} - 2 \text{ kilometers} \approx 2.45 \text{ kilometers}.$$

(Assuming that the given information is correct to at most three significant digits, it seems reasonable to round off our answer accordingly.) Thus, the point P should be located approximately 2.45 kilometers from the terminal T. ■

Problem Set 2.4

In this problem set, we consider only real roots of equations.

In Problems 1 to 16, solve each equation.

1. $x^{-2} = 4$
2. $x^{-7} = 3$
3. $x^4 = -16$
4. $u^{-4} = \frac{1}{9}$
5. $x^7 = -4$
6. $t^{-3} = -\frac{1}{8}$
7. $x^{-5} = \frac{-1}{32}$
8. $(x - 1)^4 = 81$
9. $y^{1/3} = 2$
10. $x^{-1/2} = -4$
11. $z^{5/2} = 3$
12. $t^{-2/3} = 9$
13. $(2t - 1)^{-2/5} = 4$
14. $(3z - 7)^{4/3} = -1$
15. $(x^2 - 7x)^{4/3} = 16$
16. $(x^2 + 12x)^{2/3} = 9$

In Problems 17 to 30, solve each equation. Check for extraneous roots.

17. $\sqrt{16 - m} + 4 = m$
18. $\sqrt{u + \sqrt{u - 4}} = 2$
19. $\sqrt[3]{2x + 3} = -1$
20. $\sqrt[5]{3x - 1} = 2$
21. $\sqrt[4]{y^4 + 2y - 6} = y$
22. $\sqrt[3]{x^3 + 2x^2 - 9x - 26} = x + 1$
23. $\sqrt{3t + 12} - 2 = \sqrt{2t}$
24. $\sqrt{5y + 1} = 1 + \sqrt{3y}$
25. $\sqrt{2y - 5} - \sqrt{y - 2} = 2$
26. $\sqrt{3t + 8} = 3\sqrt{t} - 2\sqrt{2}$
27. $\sqrt{2x + 5} - \sqrt{5x - 1} = \sqrt{2x - 4}$
28. $x(\sqrt{x^2 + 1})^{-1} = (4 - x)(\sqrt{x^2 - 8x + 17})^{-1}$
29. $\sqrt{5t - 10} = \sqrt{8t + 2} - \sqrt{3t + 12}$
30. $\dfrac{\sqrt{x + 1} + \sqrt{x - 1}}{\sqrt{x + 1} - \sqrt{x - 1}} = \dfrac{4x - 1}{2}$

In Problems 31 to 40, solve each equation.

31. $x^4 - 7x^2 + 12 = 0$
32. $(y^2 + y - 2)^2 + (y^2 + y - 2) = 20$
33. $t^{1/3} - 4t^{1/6} + 3 = 0$
34. $y^{-2/3} + 2y^{-1/3} = 8$
35. $9x^{-4} - 145x^{-2} + 16 = 0$
36. $2x^2 - 5x + 7 - 8\sqrt{2x^2 - 5x + 7} + 7 = 0$
37. $y^2 + 2y - 3\sqrt{y^2 + 2y + 4} = 0$
38. $t^2 + 6t - 6\sqrt{t^2 + 6t - 2} + 3 = 0$
39. $\sqrt{t + 20} - 4\sqrt[4]{t + 20} + 3 = 0$
40. $3(1 - x)^{1/6} + 2(1 - x)^{1/3} = 2$

In Problems 41 to 46, solve each equation by factoring.

41. $x^3 + x^2 - x - 1 = 0$
42. $x^5 - 4x^3 - x^2 + 4 = 0$
43. $x^6 - x^4 + 4x^2 - 4 = 0$
44. $z^7 - 27z^4 - z^3 + 27 = 0$
45. $x^{5/6} + 2x^{1/2} - x^{1/3} - 2 = 0$
46. $x^{7/12} - 2x^{1/3} + 5x^{1/4} - 10 = 0$

C 47. A rectangular observation deck is to be built overlooking a scenic lake. Find the dimensions of the deck if its area is to be 60 square meters and its diagonal is 13 meters long.

C **48.** A conical tent with no floor is to have a height $h = 3.1$ meters (Figure 2). Find the radius r meters of the base of the tent if the amount of canvas to be used has an area of 68.9 square meters. [*Hint:* The lateral surface area A of a cone is given by $A = \pi r\sqrt{r^2 + h^2}$.]

Figure 2

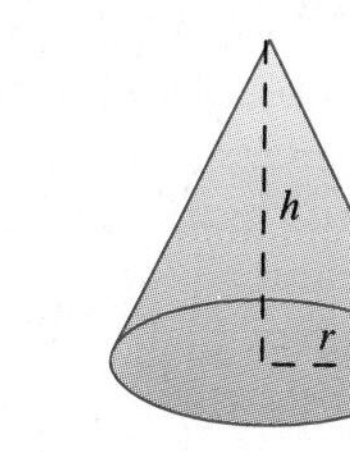

49. The master antenna for a cable television company is located at point A on the bank of a straight river 1 kilometer wide (Figure 3). A town T is situated on the opposite bank of the river, 3 kilometers downstream from A. The cable television company agrees to provide service to the town provided that it will pay one-half of the costs of making the cable connection. It costs \$30,000 per kilometer to lay the cable under water and \$18,000 per kilometer to lay the cable on land. The proposed cable route is shown in Figure 3. By issuing bonds, the town is able to raise \$39,000 for its share of the cost of the cable. Find the distance from P to T in Figure 3.

Figure 3

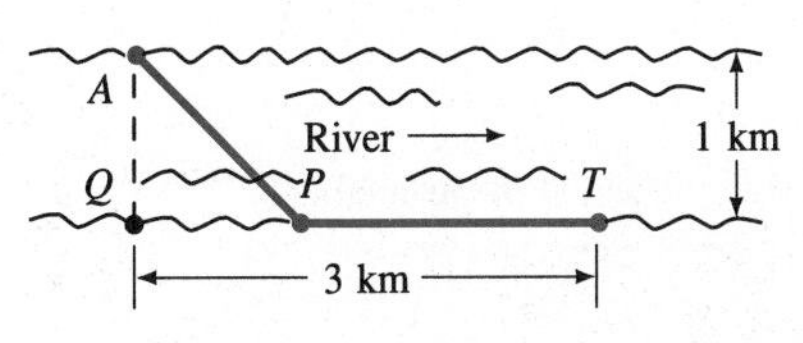

C **50.** If v is the velocity of sound in air and g is the acceleration of gravity, then the time t required for a person to hear the sound of a dropped object hitting a surface d units below is given by $t = d/v + \sqrt{2d/g}$. A spelunker (cave explorer) encounters a deep vertical shaft of unknown depth. A rock is dropped into the shaft, and the sound of the rock hitting the bottom is heard 3.7 seconds later. Taking $v = 340$ meters per second and $g = 9.8$ meters per second squared, find the depth d meters of the shaft.

51. In engineering mechanics, the deflection d of a beam of length l is given by $d = l\sqrt{a/(2l + a)}$, where a is a constant depending on the material from which the beam is made. Solve this equation for l in terms of a and d.

52. The frequency ω of a certain electronic circuit is given by

$$\omega = \frac{1}{2\pi\sqrt{\dfrac{LC_1C_2}{C_1 + C_2}}}.$$

Solve for C_2 in terms of ω, L, and C_1.

53. In meteorology, the heat loss H (wind chill) is given by

$$H = (A + B\sqrt{V} - CV)(S - T).$$

Solve for the wind speed V in terms of H, the neutral skin temperature S, the air temperature T, and the constants A, B, and C.

2.5 Inequalities

As important as it is to be able to deal with *equations* asserting that one quantity is equal to another, it is at least as important to be able to handle *inequalities* stating that one quantity is smaller than or larger than another. In this section, we use inequalities to define *intervals* of real numbers, study the basic properties of inequalities, and explain how to solve simple inequalities.

Inequalities and Intervals

Figure 1

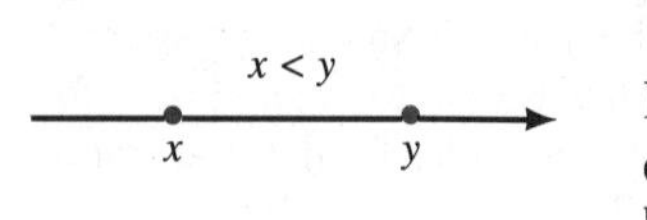

If the point on a number line with coordinate x lies to the left of the point with coordinate y (Figure 1), we say that y is **greater than** x (or equivalently, that x is **less than** y) and we write $y > x$ (or $x < y$). In other words, $y > x$ (or $x < y$) means that $y - x$ is positive.

Sometimes we know only that a certain inequality does *not* hold. If $x < y$ does not hold, then either $x > y$ or $x = y$. In this case, we say that x is **greater than or equal to** y, and we write $x \geq y$ (or $y \leq x$). If $y \leq x$, we say that y is **less than or equal to** x.

Statements of the form $y < x$, $x > y$, $y \leq x$, and $x \geq y$ are called **inequalities.** The inequalities $y < x$ and $x > y$ are said to be **strict,** and the inequalities $y \leq x$ and $x \geq y$ (which allow for the possibility of equality) are said to be **nonstrict.** The symbols $<$, $>$, $\leq$, and $\geq$ are called **inequality signs,** and the expressions on the left and right of these signs are called the **sides** or **members** of the inequality.

If we write $x < y < z$, we mean that $x < y$ *and* $y < z$. Likewise, $x \geq y > z$ means that $x \geq y$ *and* $y > z$. Notice that this notation for **combined** or **compound** inequalities is used only when the inequalities run in the *same direction.* If you are ever tempted to write something like $x < y > z$, resist the urge—such notation is improper and confusing.

Certain sets of real numbers, called **intervals,** have an important role to play in our study of inequalities.

Definition 1

Bounded Open or Closed Intervals

Let a and b be fixed real numbers with $a < b$.

(i) The **open interval** (a, b) with **endpoints** a and b is the set of all real numbers x such that $a < x < b$.

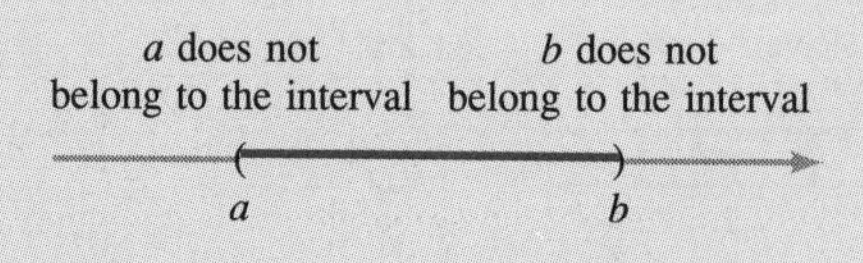

(ii) The **closed interval** $[a, b]$ with **endpoints** a and b is the set of all real numbers x such that $a \leq x \leq b$.

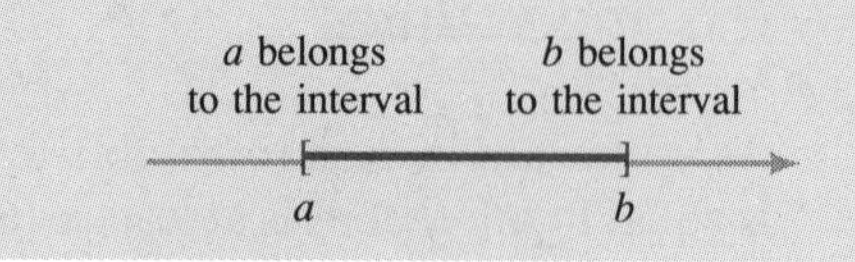

Notice that a closed interval contains its endpoints, but an open interval does not. A *half-open** interval, defined below, contains one of its endpoints, but not the other.

* Some authors use the term *half-closed* for these intervals.

Definition 2

Bounded Half-open Intervals

(i) The **half-open interval** $[a, b)$ with **endpoints** a and b is the set of all real numbers x such that $a \leq x < b$.

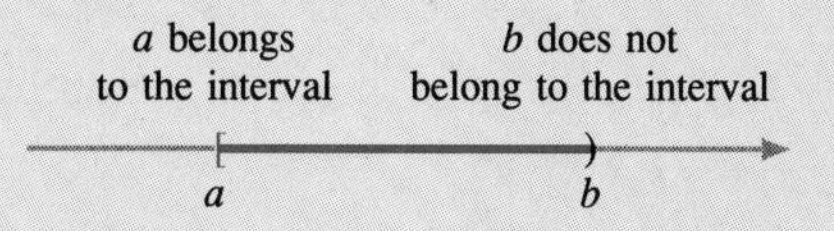

(ii) The **half-open interval** $(a, b]$ with **endpoints** a and b is the set of all real numbers x such that $a < x \leq b$.

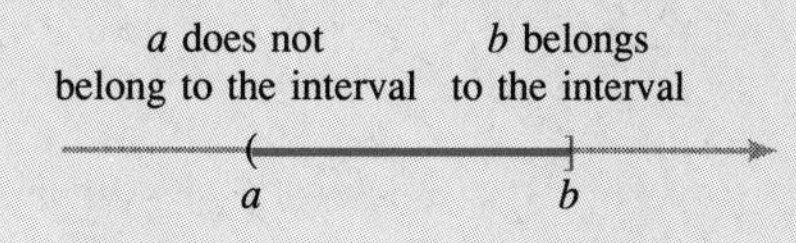

Unbounded intervals, which extend indefinitely to the right or left, are written with the aid of the special symbols $+\infty$ and $-\infty$, called **positive infinity** and **negative infinity.**

Definition 3

Unbounded Intervals

Let a be a fixed real number.

(i) $(a, +\infty)$ is the set of all real numbers x such that $a < x$.

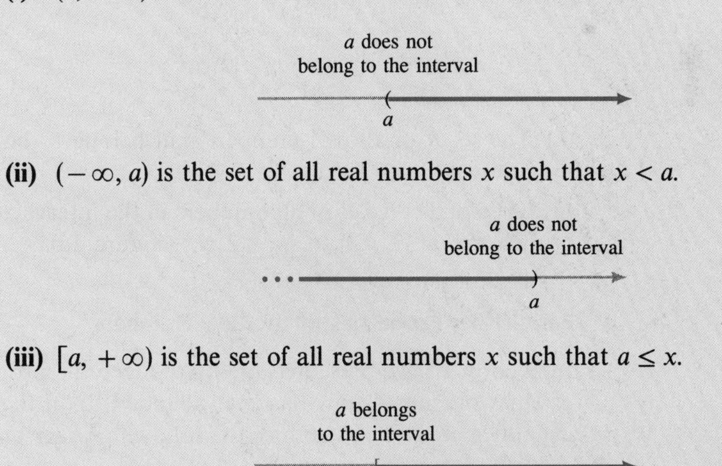

(ii) $(-\infty, a)$ is the set of all real numbers x such that $x < a$.

(iii) $[a, +\infty)$ is the set of all real numbers x such that $a \leq x$.

(iv) $(-\infty, a]$ is the set of all real numbers x such that $x \leq a$.

It must be emphasized that $+\infty$ and $-\infty$ are just convenient symbols—*they are not real numbers* and should not be treated as if they were. In the notation for unbounded intervals, we usually write ∞ rather than $+\infty$. For instance, $(5, \infty)$ denotes the set of all real numbers that are greater than 5.

Example 1 Illustrate each set on a number line.

(a) $(2, 5]$

(b) $(3, \infty)$

(c) The set A of all real numbers that belong to both intervals $(2, 5]$ and $(3, \infty)$.

(d) The set B of all real numbers that belong to at least one of the intervals $(2, 5]$ and $(3, \infty)$.

Solution **(a)** The interval $(2, 5]$ consists of all real numbers between 2 and 5, including 5, but excluding 2 (Figure 2a).

(b) The interval $(3, \infty)$ consists of all real numbers that are greater than 3 (Figure 2b).

Figure 2

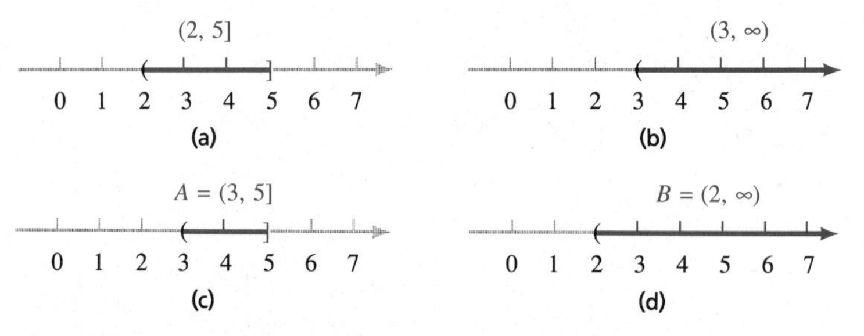

(c) The set A of all real numbers that belong to both intervals $(2, 5]$ and $(3, \infty)$ is the interval $(3, 5]$ (Figure 2c).

(d) The set B consists of all numbers in the interval $(2, 5]$ together with all numbers in the interval $(3, \infty)$, so $B = (2, \infty)$ (Figure 2d). ■

Basic Order Properties of the Real Numbers

In Section 1.1 we presented the basic algebraic properties of the real numbers. All of our work up to now has been founded upon these properties. To deal with inequalities, we must introduce the following **order properties:**

Let a, b, and c denote real numbers.

1. *The Trichotomy Property*

One and only one of the following is true:

$$a < b \quad \text{or} \quad b < a \quad \text{or} \quad a = b.$$

The trichotomy property has a simple geometric interpretation on the number line; namely, if a and b are distinct, then one must lie to the left of the other.

2. *The Transitivity Property*

If $a < b$ and $b < c$, then $a < c$.

The transitivity property means that if a lies to the left of b and b lies to the left of c, then a lies to the left of c (Figure 3).

Figure 3

3. *The Addition Property*

If $a < b$, then $a + c < b + c$.

The addition property permits us to add the same number c to both sides of an inequality $a < b$.

4. *The Multiplication Property*

If $a < b$ and $c > 0$, then $ac < bc$.

The multiplication property permits us to multiply both sides of an inequality $a < b$ by the same *positive* number c.

Example 2 State one basic order property of the real numbers to justify each statement.

(a) The discriminant of a quadratic equation is positive, negative, or zero.

(b) $-1 < 0$ and $0 < 1$, so it follows that $-1 < 1$.

(c) $5 < 6$, so it follows that $5 + (-11) < 6 + (-11)$.

(d) $4 < 7$ and $\frac{1}{2} > 0$, so it follows that $2 < \frac{7}{2}$.

Solution **(a)** The trichotomy property. **(b)** The transitivity property.

(c) The addition property. **(d)** The multiplication property. ■

Many of the important properties of inequalities can be *derived* from the basic order properties, although we shall not derive them here. Among the more important derived properties are the following.

5. *The Subtraction Property*

If $a < b$, then $a - c < b - c$.

6. *The Division Property*

If $a < b$ and $c > 0$, then $\frac{a}{c} < \frac{b}{c}$.

According to the multiplication and division properties, you can multiply or divide both sides of an inequality by a *positive* number. According to the following property, if you multiply or divide both sides of an inequality by a *negative* number, you must **reverse the inequality sign.**

7. *The Order-Reversing Properties*

(i) If $a < b$ and $c < 0$, then $ac > bc$.

(ii) If $a < b$ and $c < 0$, then $\frac{a}{c} > \frac{b}{c}$.

For instance, if we multiply both sides of the inequality $2 < 5$ by the negative number -3, we must reverse the inequality sign and write $-6 > -15$, or $-15 < -6$. (Indeed, on a number line, -15 lies to the left of -6.)

Statements analogous to Properties 2 to 7 can be made for nonstrict inequalities (those involving $\leq$ or $\geq$) and for compound inequalities. For instance, if you know that

$$-1 \leq 3 - \frac{x}{2} < 5,$$

you can multiply all members by -2, provided that you reverse the inequality signs:

$$2 \geq x - 6 > -10$$

or

$$-10 < x - 6 \leq 2.$$

Solving Inequalities

An inequality containing a variable is neither true nor false until a particular number is substituted for the variable. If a true statement results from such a substitution, we say that the substitution **satisfies** the inequality. If the substitution $x = a$ satisfies an inequality, we say that the real number a is a **solution** of the inequality. The set of all solutions of an inequality is called its **solution set.** Two inequalities are said to be **equivalent** if they have exactly the same solution set.

To **solve** an inequality—that is, to find its solution set—you proceed in much the same way as in solving equations, except of course that when you multiply or divide all members by a negative quantity, you must reverse all signs of inequality. The usual approach is to try to isolate the unknown on one side of the inequality.

Example 3 Solve each inequality and sketch its solution set on a number line.

(a) $5x + 3 < 18$ **(b)** $\frac{5}{6}x - \frac{3}{4} \geq \frac{1}{2}x - \frac{2}{3}$ **(c)** $-11 \leq 2x - 3 < 7$

Solution

(a)
$$
\begin{aligned}
5x + 3 &< 18 \\
5x &< 15 \qquad \text{(We subtracted 3 from both sides.)} \\
x &< 3 \qquad \text{(We divided both sides by 5.)}
\end{aligned}
$$

Since the last inequality is equivalent to the original one, we conclude that the solution of $5x + 3 < 18$ consists of all numbers less than 3. In other words, the solution set is the interval $(-\infty, 3)$ (Figure 4a).

Figure 4

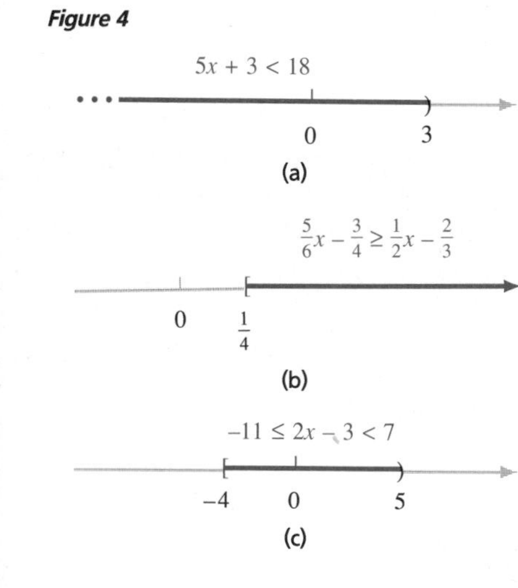

(b) We begin by multiplying both sides by 12, the LCD of the fractional coefficients:

$$
\begin{aligned}
10x - 9 &\geq 6x - 8 \\
4x - 9 &\geq -8 \qquad \text{(We subtracted } 6x \text{ from both sides.)} \\
4x &\geq 1 \qquad \text{(We added 9 to both sides.)} \\
x &\geq \tfrac{1}{4}. \qquad \text{(We divided both sides by 4.)}
\end{aligned}
$$

Therefore, the solution set is $[\frac{1}{4}, \infty)$ (Figure 4b).

(c) We begin by adding 3 to all members to help isolate x in the middle:

$$
\begin{aligned}
-8 &\leq 2x < 10 \\
-4 &\leq x < 5. \qquad \text{(We divided all members by 2.)}
\end{aligned}
$$

Therefore, the solution set is $[-4, 5)$ (Figure 4c). ■

To solve a compound inequality in which the unknown appears only in the middle member, you can always proceed as in Example 3c above. Otherwise, you can break the compound inequality up into simple inequalities and proceed as in the following example.

Example 4 Solve the inequality $3x - 10 \leq 5 < x + 3$ and sketch its solution set on a number line.

Solution The given inequality holds if and only if *both* the inequalities $3x - 10 \leq 5$ and $5 < x + 3$ hold.

Figure 5

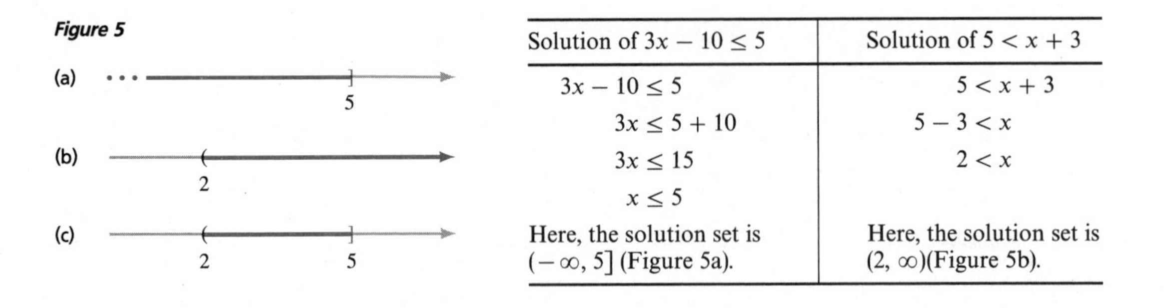

Solution of $3x - 10 \le 5$	Solution of $5 < x + 3$
$3x - 10 \le 5$	$5 < x + 3$
$3x \le 5 + 10$	$5 - 3 < x$
$3x \le 15$	$2 < x$
$x \le 5$	
Here, the solution set is $(-\infty, 5]$ (Figure 5a).	Here, the solution set is $(2, \infty)$(Figure 5b).

Therefore, the solution set of $3x - 10 \le 5 < x + 3$ consists of all real numbers x that belong to *both* the intervals $(-\infty, 5]$ and $(2, \infty)$; that is, the solution set is the interval $(2, 5]$ (Figure 5c). ■

The following example illustrates one of the many practical applications of inequalities.

Example 5 A technician determines that an electronic circuit fails to operate because the resistance between points A and B, 1200 ohms, exceeds the specifications, which call for a resistance of no less than 400 ohms and no greater than 900 ohms. The circuit can be made to satisfy the specifications by adding a shunt resistor of R ohms, $R > 0$ (Figure 6). After adding the shunt resistor, the resistance between points A and B will be $\dfrac{1200R}{1200 + R}$ ohms. What are the possible values of R?

Figure 6

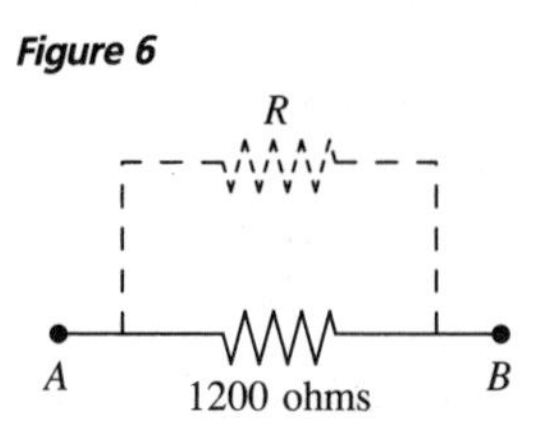

Solution To satisfy the specifications, we must have

$$400 \le \frac{1200R}{1200 + R} \le 900 \qquad \text{where } R > 0.$$

Because R is positive, it follows that $1200 + R$ is positive, so we can multiply all members of the inequality by $1200 + R$:

$$400(1200 + R) \le 1200R \le 900(1200 + R)$$

$$480{,}000 + 400R \le 1200R \le 1{,}080{,}000 + 900R.$$

Now we break the compound inequality into two simple inequalities.

$480{,}000 + 400R \le 1200R$	$1200R \le 1{,}080{,}000 + 900R$
$480{,}000 \le 1200R - 400R$	$1200R - 900R \le 1{,}080{,}000$
$480{,}000 \le 800R$	$300R \le 1{,}080{,}000$
$\dfrac{480{,}000}{800} \le R$	$R \le \dfrac{1{,}080{,}000}{300}$
$600 \le R$	$R \le 3600$

Therefore, we must have both $600 \le R$ and $R \le 3600$; that is,

$$600 \le R \le 3600.$$

The shunt resistance R must be no less than 600 ohms and no greater than 3600 ohms. ■

Problem Set 2.5

In Problems 1 to 6, illustrate each set on a number line.

1. **(a)** $(2, 5)$ **(b)** $[-1, 3)$ **(c)** $[-4, 0]$
   (d) $(-\infty, -3)$ **(e)** $[1, \infty)$ **(f)** $(-\frac{1}{2}, \infty)$
   (g) $[-\frac{3}{2}, \frac{5}{2}]$ **(h)** $(-4, 0]$

2. The set of real numbers x such that $-3 < x < 3$ and $2x$ is an integer.

3. The set A of all real numbers that belong to both of the intervals $(0, 3]$ and $(1, \infty)$.

4. The set B of all real numbers that belong to both of the intervals $(-1, \frac{1}{2}]$ and $[\frac{1}{2}, \frac{3}{4}]$.

5. The set C of all real numbers that belong to at least one of the intervals $[-2, -1]$ and $[1, 2]$.

6. The set D of all real numbers that belong to at least one of the intervals $(-1, 0]$ and $(0, 1]$.

7. State one basic order property of the real numbers to justify each statement.
   (a) $3 < 5$, so it follows that $3 + 2 < 5 + 2$.
   (b) $-3 < -2$ and $-2 < 2$, so it follows that $-3 < 2$.
   (c) $-3 < -2$ and $\frac{1}{3} > 0$, so it follows that $-1 < -\frac{2}{3}$.
   (d) We know that $\sqrt{2} \ne 1.414$, so it follows that either $\sqrt{2} < 1.414$ or else $\sqrt{2} > 1.414$.

8. State one derived order property of the real numbers to justify each statement.
   (a) $-1 < 0$, so it follows that $-2 < -1$.
   (b) $-5 < -3$ and $5 > 0$, so it follows that $-1 < -\frac{3}{5}$.
   (c) $3 < 5$ and $-2 < 0$, so it follows that $-6 > -10$.
   (d) $3 < 5$ and $-2 < 0$, so it follows that $-\frac{3}{2} > -\frac{5}{2}$.

9. Given that $x > -3$, what equivalent inequality is obtained if: **(a)** 4 is added to both sides? **(b)** 4 is subtracted from both sides? **(c)** Both sides are multiplied by 5? **(d)** Both sides are multiplied by -5?

10. Given that $-2 \le 3 - 4x < 1$, what equivalent inequality is obtained if: **(a)** 3 is subtracted from all members? **(b)** 2 is added to all members? **(c)** All members are divided by 2? **(d)** All members are divided by -4?

In Problems 11 to 30, solve each inequality and sketch its solution set on a number line.

11. $2x - 5 < 5$
12. $7 - 3x < 22$
13. $2x - 10 \le 15 - 3x$
14. $5x + 3 \le 8x - 6$
15. $7(x + 2) \ge 2(2x - 4)$
16. $21 - 3(x - 7) \le x + 20$
17. $\dfrac{x}{3} < 2 - \dfrac{x}{4}$
18. $\dfrac{2x}{3} > \dfrac{23}{24} - \dfrac{5x}{4}$
19. $\frac{1}{2}(x + 1) \ge \frac{1}{3}(x - 5)$
20. $\frac{1}{5}(x + 1) \ge \frac{1}{3}(x - 5)$
21. $\frac{1}{6}(x + 3) - \frac{1}{2}(x - 1) \le \frac{1}{3}$
22. $\dfrac{x + 5}{6} - \dfrac{10 - x}{-3} < \pi$
23. $5 \le 2x + 3 \le 13$
24. $14 \le 5 - 3x < 20$
25. $-3 < 5 - 4x \le 17$
26. $3 + x < 3(x - 1) \le x - 7$
27. $4x - 1 \le 3 \le 7 + 2x$
28. $3x + 1 \le 2 - x < 18 + 7x$
29. $1 - 2x < 5 - x \le 25 - 6x$
30. $2x - 1 \le 5 \le x + 2$

In Problems 31 to 34, use the method illustrated in the solution of Example 5 to solve the inequalities. Illustrate each solution set on a number line.

31. $1 \le \dfrac{10x}{10+x} \le 5$ where $x > 0$

32. $1 \le \dfrac{4x}{4+x} \le 3$ where $x > -4$

33. $\dfrac{4}{x+1} \le 3 \le \dfrac{6}{x+1}$ where $x > 0$

34. $\dfrac{8}{x+3} \le 2 \le \dfrac{13}{x+3}$ where $x > -3$

35. A student's scores on the first three tests in a physics course are 65, 78, and 84. What range of scores on a fourth test will give the student an average that is less than 80 but no less than 70 for the four tests?

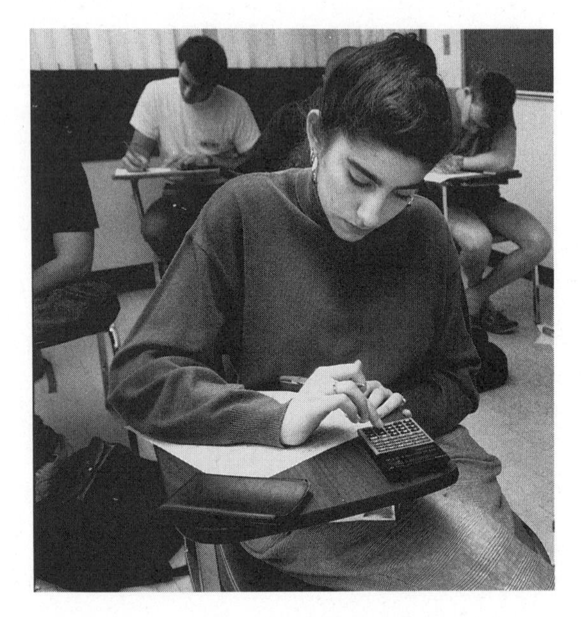

C **36.** Oven temperatures for baking various foods range between 320°F and 420°F, inclusive. Express this range in Celsius degrees by using the conversion formula $C = \frac{5}{9}(F - 32)$. Round off your answer to the nearest degree.

37. A shunt resistor of R ohms is to be added to a resistance of 600 ohms in order to reduce the resistance to a value strictly between 540 ohms and 550 ohms. After the shunt is added, the resistance is given by $600R/(600 + R)$. Find the possible values of R. (Assume that $R > 0$.)

38. An object thrown straight upward with an initial velocity of 100 feet per second will have a velocity v, given by $v = 100 - 32t$, exactly t seconds later. Over what interval of time will its velocity be greater than 50 feet per second but no greater than 60 feet per second?

39. A loupe is a small magnifying lens set in an eyepiece and used by jewelers, watchmakers, and hobbyists. If the focal length of the lens is f centimeters, then an object viewed through the lens at a distance of p centimeters from the lens will appear to be magnified by a factor $m = \dfrac{f}{f - p}$, provided that $p < f$. If $f = 5$ centimeters, what range of values of p will result in a magnification factor of between 2 and 5 inclusive?

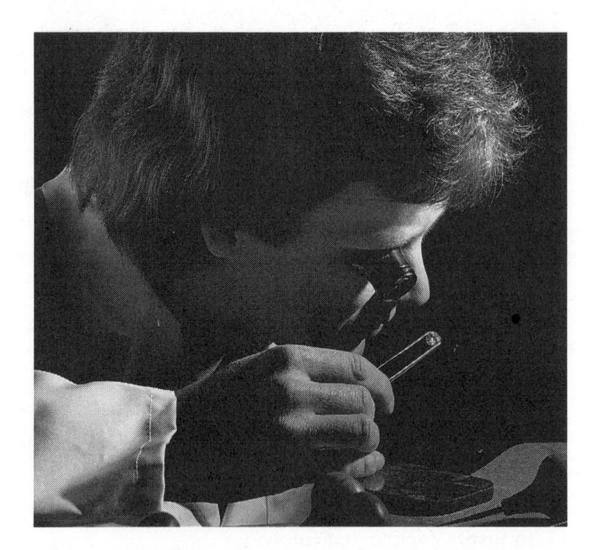

C **40.** A realtor charges the seller of a house a 7% commission on the price for which the house is sold. Suppose that the seller wants to clear at least \$60,000 after paying the commission, and that similar houses are selling for \$70,000 or less.

(a) What is the range of selling prices for the house?

(b) If the house is sold, what is the range of amounts for the realtor's commission?

41. A rectangular solar collector is to have a height of 1.5 meters, but its length is still to be determined. What is the range of values for this length if the collector provides 400 watts per square meter, and if it must provide a total of between 2000 and 3500 watts?

C 42. If families with an income, after deductions, of over \$29,750, but not over \$71,900, must pay an income tax of \$4463 plus 28% of the amount over \$29,750, what is the range of possible values for the income, after deductions, of families whose income tax liability is between \$5760 and \$8375?

43. Psychologists define the intelligence quotient (IQ) of a person to be 100 times the ratio of the person's mental age to his or her chronological age. A psychologist is studying a group of 13-year-olds who have an IQ range between 90 and 120, inclusive. What is the corresponding range of mental ages?

C 44. An operator-assisted station-to-station phone call from Boston, MA, to Tucson, AZ, costs \$2.25 plus \$0.38 for each additional minute after the first 3 minutes. A group of such calls were made, each costing between \$6.05 and \$8.71, inclusive. Give the range of minutes for the lengths of these calls.

45. In each case, determine the interval of values of x for which the given expression represents a real number.

(a) $\sqrt{2-3x}$ **(b)** $\dfrac{x}{\sqrt{3x+5}}$

46. If $c \neq 0$, prove that $c^2 > 0$. [*Hint:* Consider separately the two cases $c > 0$ and $c < 0$.]

47. Give an example to show that, in general, you *cannot* square both sides of an inequality and get an equivalent inequality.

48. If $c > 0$, show that $1/c > 0$. [*Hint:* If $1/c$ is not positive, it must be negative, since it cannot be zero.]

2.6 Equations and Inequalities Involving Absolute Values

Many programmable calculators have an *absolute-value* key marked $|x|$ or ABS. If you enter a *nonnegative* number and press this key, the same number appears in the display. However, if you enter a *negative* number and press the absolute-value key, the display shows the number without the negative sign. For instance,

Enter	Press ABS Key	Display Shows
3.2	$\longmapsto$	3.2
-3.2	$\longmapsto$	3.2

This feature is useful whenever you are interested only in the "magnitude" of a number, without regard to its algebraic sign.

On a calculator without an absolute-value key, you can obtain the absolute value $|x|$ of a number x by entering x, pressing the x^2 key, and then pressing the square-root key. In other words, if you square a number and take the principal square root of the result, you will remove any negative sign that may have been present, so that

$$\sqrt{x^2} = |x|.$$

For instance, $\sqrt{(-5)^2} = \sqrt{25} = 5 = |-5|$.

The absolute value is defined formally as follows.

Definition 1 **Absolute Value**

If x is a real number, then $|x|$, the **absolute value** of x, is defined by

$$|x| = \begin{cases} x, & \text{if } x \geq 0 \\ -x, & \text{if } x < 0. \end{cases}$$

For instance,

$$|5| = 5 \qquad \text{because } 5 \geq 0,$$
$$|0| = 0 \qquad \text{because } 0 \geq 0,$$
$$|-5| = -(-5) = 5 \qquad \text{because } -5 < 0.$$

Notice that $|x| \geq 0$ is always true.

On a number line, $|x|$ *is the number of units of distance between the point with coordinate* x *and the origin*, regardless of whether the point is to the right or left of the origin (Figure 1a). For instance, both the point with coordinate -3 and the point with coordinate 3 are 3 units from the origin (Figure 1b), so $|-3| = 3 = |3|$.

Figure 1

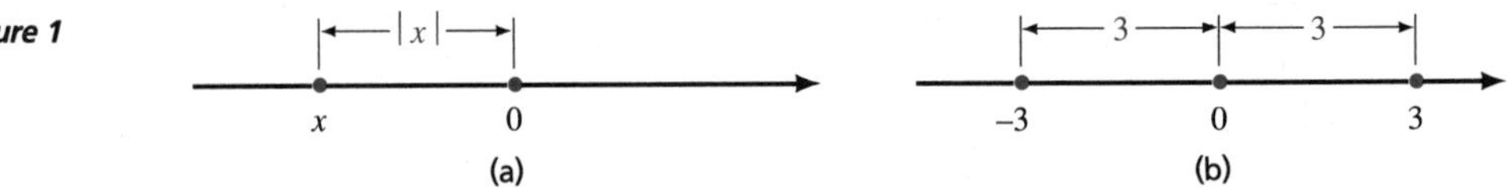

(a) (b)

More generally,

$|x - y|$ *is the number of units of distance between the point with coordinate* x *and the point with coordinate* y.

This statement holds no matter which point is to the left of the other (Figure 2a and b), and of course it also holds when $x = y$. For simplicity, we refer to the distance $|x - y|$ as the **distance between the numbers x and y.**

Figure 2

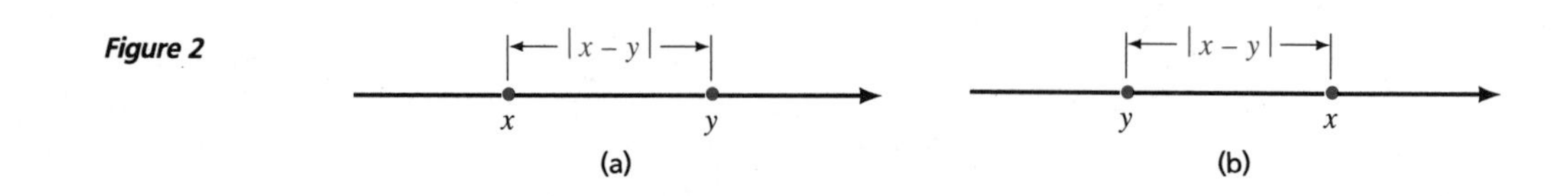

(a) (b)

Example 1 Find the distance between the numbers.

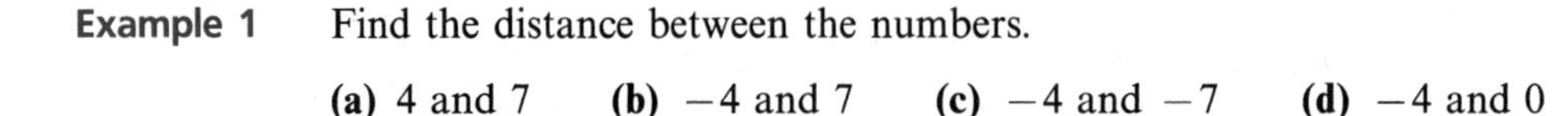

(a) 4 and 7 **(b)** -4 and 7 **(c)** -4 and -7 **(d)** -4 and 0

Solution

(a) $|4-7| = |-3| = 3$ **(b)** $|-4-7| = |-11| = 11$

(c) $|-4-(-7)| = |-4+7| = |3| = 3$ **(d)** $|-4-0| = |-4| = 4$ ■

The absolute value of a product of two numbers is equal to the product of their absolute values; that is,

$$|xy| = |x|\,|y|.$$

This rule can be derived as follows:

$$|xy| = \sqrt{(xy)^2} = \sqrt{x^2y^2} = \sqrt{x^2}\sqrt{y^2} = |x|\,|y|.$$

A similar rule applies to quotients; if $y \neq 0$,

$$\left|\frac{x}{y}\right| = \frac{|x|}{|y|}.$$

(See Problem 46.) However, there is no such rule for sums or differences:

$|x+y|$ *need not be the same as* $|x| + |y|$;

for instance,

$$|5+(-2)| = |3| = 3, \quad \text{but} \quad |5| + |-2| = 5 + 2 = 7.$$

Similarly,

$|x-y|$ *need not be the same as* $|x| - |y|$.

(See Problems 48 and 52.)

Example 2 Let $x = -3$ and $y = 5$. Find the value of each expression.

(a) $|6x+y|$ **(b)** $|6x| + |y|$ **(c)** $|xy|$ **(d)** $|x|\,|y|$

Solution

(a) $|6x+y| = |6(-3)+5| = |-18+5| = |-13| = 13$

(b) $|6x| + |y| = |6(-3)| + |5| = |-18| + |5| = 18 + 5 = 23$

(c) $|xy| = |(-3)5| = |-15| = 15$

(d) $|x|\,|y| = |-3|\,|5| = 3 \cdot 5 = 15$ ■

Equations Involving Absolute Values

Figure 3

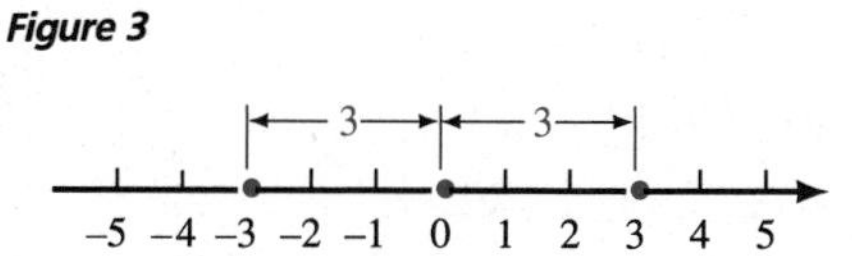

Geometrically, the equation $|x| = 3$ means that the point with coordinate x is 3 units from 0 on the number line. Obviously, the number line contains *two* points that are 3 units from the origin—one to the right of the origin and the other to the left (Figure 3). Thus, the equation $|x| = 3$ has two solutions, $x = 3$ and $x = -3$. This illustrates the following theorem, which simply expresses the same idea in a more general form.

Theorem 1 **Solution of the Equation $|u| = a$**

Let a be a real number.

(i) If $a > 0$, then $|u| = a$ if and only if $u = -a$ or $u = a$.

(ii) If $a < 0$, then the equation $|u| = a$ has no solution.

(iii) $|u| = 0$ if and only if $u = 0$.

In Examples 3 to 6, solve each equation.

Example 3 $|3t| - 1 = 17$

Solution The equation is equivalent to $|3t| = 18$. Thus, by part (i) of Theorem 1 with $u = 3t$, we have

$$3t = -18 \quad \text{or} \quad 3t = 18$$
$$t = -6 \quad \text{or} \quad t = 6.$$

An alternative solution is obtained by direct use of the definition of absolute value (Definition 1):

$$|3t| = \begin{cases} 3t & \text{if} \quad t \geq 0 \\ -3t & \text{if} \quad t < 0. \end{cases}$$

Hence, the equation $|3t| = 18$ yields

$$3t = 18 \quad \text{or} \quad -3t = 18;$$

so that, again, we obtain the two solutions $t = 6$ and $t = -6$. ■

Example 4 $|1 - 2x| = 7$

Solution The equation holds if and only if

$$1 - 2x = 7 \quad \text{or} \quad 1 - 2x = -7$$
$$-2x = 6 \quad \text{or} \quad -2x = -8$$
$$x = -3 \quad \text{or} \quad x = 4.$$
■

Example 5 $|x^2 + x - 6| = 0$

Solution By part (iii) of Theorem 1, the equation is equivalent to

$$x^2 + x - 6 = 0; \qquad \text{that is,} \qquad (x + 3)(x - 2) = 0.$$

The solutions are $x = -3$ and $x = 2$. ■

Example 6 $|7x + 1| = -3$

Solution By part (ii) of Theorem 1, the equation has no solution. ■

By applying Theorem 1 twice, you can solve an equation of the form $|u| = |v|$. However, the solution can be obtained directly by using the following theorem, which merely summarizes the result of such a procedure.

Theorem 2 **Solution of the Equation $|u| = |v|$**

$|u| = |v|$ if and only if $u = -v$ or $u = v$.

Example 7 Solve the equation $|p - 5| = |3p + 7|$.

Solution By Theorem 2 with $u = p - 5$ and $v = 3p + 7$, the given equation holds if and only if

$$
\begin{aligned}
p - 5 &= 3p + 7 &\quad\text{or}\quad& p - 5 = -(3p + 7)\\
-2p &= 12 &\quad\text{or}\quad& 4p = -2\\
p &= -6 &\quad\text{or}\quad& p = -\tfrac{1}{2}.
\end{aligned}
$$

■

Inequalities Involving Absolute Values

Referring once again to the number line (Figure 4), we see that the inequality $|x| < 3$ means that the point with coordinate x is less than 3 units from 0, but can be on either side of 0; that is, $-3 < x < 3$.

Figure 4

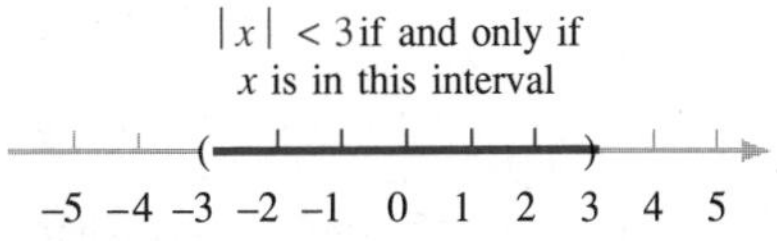

$|x| < 3$ if and only if x is in this interval

–5 –4 –3 –2 –1 0 1 2 3 4 5

More generally, we have the following theorem (Problem 54).

Theorem 3 **Solution of the Inequalities $|u| < a$ and $|u| \le a$**

If $a > 0$, then:

(i) $|u| < a$ if and only if $-a < u < a$.

(ii) $|u| \le a$ if and only if $-a \le u \le a$.

Example 8 Solve the inequality $|2x + 3| < 5$ and sketch the solution set on a number line.

Figure 5

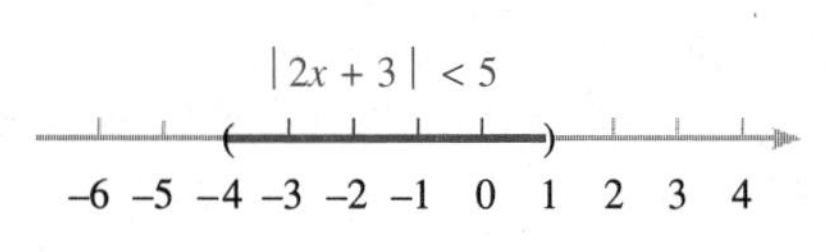

Solution

$$|2x + 3| < 5$$

$$-5 < 2x + 3 < 5 \qquad \text{[Part (i) of Theorem 3.]}$$

$$-8 < 2x < 2 \qquad \text{(We subtracted 3 from all members.)}$$

$$-4 < x < 1 \qquad \text{(We divided all members by 2.)}$$

Hence, the solution set is the interval $(-4, 1)$ (Figure 5). ■

Figure 6

$|x| > 3$ if and only if x belongs to one of these intervals

–5 –4 –3 –2 –1 0 1 2 3 4 5 6

On a number line, the inequality $|x| > 3$ means that the point with coordinate x is more than 3 units from 0 (Figure 6); that is, either $x > 3$ or $x < -3$.

More generally, we have the following theorem (Problem 56).

Theorem 4 **Solution of the Inequalities $|u| > a$ and $|u| \geq a$**

If $a > 0$, then:

(i) $|u| > a$ if and only if $u < -a$ or $u > a$.

(ii) $|u| \geq a$ if and only if $u \leq -a$ or $u \geq a$.

Example 9 Solve the inequality $|\frac{1}{2}x - 1| \geq 2$ and sketch its solution set on a number line.

Solution By part (ii) of Theorem 4 with $u = \frac{1}{2}x - 1$, we have

$$\tfrac{1}{2}x - 1 \leq -2 \quad \text{or} \quad \tfrac{1}{2}x - 1 \geq 2$$

$$\tfrac{1}{2}x \leq -1 \quad \text{or} \quad \tfrac{1}{2}x \geq 3$$

$$x \leq -2 \quad \text{or} \quad x \geq 6.$$

Thus, the solution set consists of all numbers that belong to either of the intervals $(-\infty, -2]$ or $[6, \infty)$ (Figure 7). ■

Figure 7

$|\frac{1}{2}x - 1| \geq 2$

–5 –4 –3 –2 –1 0 1 2 3 4 5 6 7 8 9 10

If A and B are sets, then the set consisting of all elements that belong to A or B (or to both) is called the **union** of A and B and is written as $A \cup B$. Thus, in Example 9, the solution set (Figure 7) is the union of the two intervals $(-\infty, -2]$ and $[6, \infty)$, and it can be written concisely as the set

$$(-\infty, -2] \cup [6, \infty).$$

Absolute, Relative, and Percent Error

As we mentioned in Section 1.2, we can write

$$t \approx a$$

to mean that the true value t is *approximated* by the number a. In practical work, approximations are of little use unless there is some indication of how much *error* might be involved. By definition, the **error** ε in the approximation $t \approx a$ is the difference between the true value and the approximate value; that is,

$$error = \varepsilon = t - a.$$

The idea of error enables us to rewrite the statement $t \approx a$ as the equation

$$t = a + \varepsilon$$

and thus to deal with it by the standard techniques of algebra.

Because the *magnitude* of an error may be of more concern than its algebraic sign, we define the **absolute error** in the approximation $t \approx a$ by

$$absolute\ error = |\varepsilon| = |t - a|.$$

Absolute error, by itself, may be a misleading indicator of accuracy. For instance, although an absolute error of 10 meters in measuring the perimeter of a parking lot might be troublesome, an absolute error of 10 meters in measuring the perimeter of an entire county is probably negligible. This leads us to the idea of **relative error** in the approximation $t \approx a$, defined by

$$relative\ error = \frac{|\varepsilon|}{|t|} = \frac{|t - a|}{|t|}$$

for $t \neq 0$. Relative error expressed as a percent is called **percent relative error,** or **percent error** for short. Thus,

$$percent\ error = \frac{|t - a|}{|t|} \times 100\%.$$

Example 10 In a guess-your-weight booth at a carnival, Cedric's weight is estimated to be 153 pounds. What is the percent error if Cedric actually weighs 150 pounds?

Solution Here $t = 150$ pounds, $a = 153$ pounds, $\varepsilon = t - a = -3$ pounds. Therefore, the absolute error is $|\varepsilon| = 3$ pounds, the relative error is $\frac{3}{150} = \frac{1}{50}$, and the percent error is

$$\tfrac{1}{50} \times 100\% = 2\%.$$

■

In working with numbers, it is important to be able to place a bound on the absolute error that results when a real number t is approximated by a rounded-off version a of t. This is accomplished by the following theorem, the proof of which we leave as an exercise (Problem 60).

Theorem 5 **Error Resulting from Rounding Off**

If t is a positive real number and a is the result of rounding off t to n decimal places, then

$$|t - a| \leq 5 \times 10^{-(n+1)}.$$

Many of the routines used by calculators and computers obtain successive approximations to a desired quantity by repeatedly executing a fixed procedure. Such a routine is said to be **iterative,** and the total number of repetitions of the fixed procedure used to obtain the final approximation is called the number of **iterations.** In deciding how many iterations will be required to obtain an approximation accurate to n decimal places, the following **rounding-off rule,** suggested by Theorem 5, is often employed:

If a computed number that is an approximation to a true value is to be rounded off to n decimal places, then the absolute error of the approximation should not exceed $5 \times 10^{-(n+1)}$.

Example 11 One of the simplest techniques for obtaining numerical solutions to equations is called the *bisection method**. After k iterations, this method yields an approximate solution with an absolute error of not more than $2^{-(k+1)}$. How many iterations will be required for an approximation that is correct to two decimal places?

Solution By the rounding-off rule with $n = 2$, the absolute error of the approximation should not exceed

$$5 \times 10^{-(2+1)} = 5 \times 10^{-3} = \tfrac{5}{1000} = \tfrac{1}{200}.$$

Thus, for the number of iterations k, we must have

$$2^{-(k+1)} \leq \tfrac{1}{200};$$

that is,

$$200 \leq 2^{k+1}.$$

By trial and error (raising 2 to successive positive integer powers), we find that the smallest integer power of 2 that is larger than 200 is $2^8 = 256$. Letting $k + 1 = 8$, we find that $k = 7$ iterations will be required. ■

* The bisection method is treated in Section 4.5.

Problem Set 2.6

1. Find the distance between each pair of numbers and illustrate on a number line.

   (a) -5 and 3 (b) $\frac{3}{4}$ and $\frac{5}{8}$

   C (c) -3.311 and 2.732 C (d) $-\sqrt{3}$ and $\sqrt{2}$

2. Rewrite each statement using absolute value.

   (a) The distance between x and 3 is 7.

   (b) The distance between -1 and x^3 is 0.01.

3. Find the value of each expression if $x = -4$ and $y = 6$.

   (a) $|5x + 2y|$ (b) $|5x| + |2y|$ (c) $|5x - 2y|$

   (d) $|5x| - |2y|$ (e) $||5x| - |2y||$ (f) $\left|\frac{5x}{2y}\right|$

   (g) $\left|\frac{5x}{2y}\right| - \frac{5|x|}{2|y|}$ (h) $\frac{|x|}{x} - \frac{x}{|x|}$

4. Let $m = \frac{x + y - |x - y|}{2}$ and $M = \frac{x + y + |x - y|}{2}$.

   Find m and M if

   (a) $x = 3$ and $y = -5$ (b) $x = -7$ and $y = -2$.

In Problems 5 to 22, solve each equation.

5. $|5y| + 2 = 17$
6. $|2t| - 3 = 5$
7. $|4 - y| = 2$
8. $|2p - 1| = 3$
9. $|x^2 - x - 20| = 0$
10. $|11 - 5x| = 16$
11. $|3p| = 3p$
12. $|7y - 3| = 3 - 7y$
13. $\left|\frac{3|x|}{5} + 1\right| = 0$
14. $\left|\frac{5c}{2} - 3\right| = -1$
15. $\left|\frac{x}{2} - \frac{3}{4}\right| = \frac{1}{12}$
16. $\left|\frac{13}{12} + \frac{2x}{3}\right| = \frac{11}{12}$
17. $|t + 1| = |4t|$
18. $|3x - 1| = x + 5$
19. $|1 - x| = |2x + 3|$
20. $|4 - 3y| = |10 - 5y|$
21. $|4x - 7| = |2 + x|$
22. $|x + 1| + |x - 1| = 4$

23. Rewrite each statement using absolute value.

    (a) The distance between x and 3 is at most 0.05.

    (b) The distance between x and a is less than δ. (The symbol δ is the small Greek letter *delta*.)

    (c) The distance between x and 2 is at least 5.

24. Describe each interval as the set of all real numbers x that satisfy an inequality containing an absolute value.

    (a) $(-1, 1)$ (b) $(-3, 7)$

    (c) $[-3, 3]$ (d) $[-2, 3]$

In Problems 25 to 42, solve each inequality and sketch its solution set on a number line.

25. $|5x| - 1 \le 9$
26. $|3t| > 15$
27. $|x - 3| < 2$
28. $|x| + 2 \ge 3$
29. $|4x - 3| > 9$
30. $|3y + 5| \le |-2|$
31. $|5x - 8| \le 2$
32. $|5 - 3p| \ge |-11|$
33. $|2x + 3| \ge \frac{1}{4}$
34. $|x + 5| \ge x + 1$
35. $|1 - \frac{7}{6}x| \le \frac{1}{3}$
36. $|3x + 6| > 0$
37. $|7 - 4x| \ge |-3|$
38. $|6y - 3| > -3$
39. $|\frac{11}{2} - 2x| > \frac{3}{2}$
40. $|3x + 5| < |2x + 1|$
41. $|2x - 5| \ge 0$
42. $|2x - 5| < -3$

43. In calculus, the Greek letter δ (*delta*) is often used to denote a small, positive number.

    (a) Solve the inequality $|x - 2| < \delta$.

    (b) Solve the inequality $0 < |x - 2| < \delta$.

    (c) How do your answers to parts (a) and (b) differ?

44. Show that if $a < b$, then the solution set of $|a + b - 2x| < b - a$ is the open interval (a, b).

45. In this problem, we illustrate an important technique often used in calculus for dealing with the absolute value of an expression containing a variable.

    (a) Using Definition 1, show that

    $$|2x - 6| = \begin{cases} 2x - 6 & \text{if } x \ge 3 \\ 6 - 2x & \text{if } x < 3. \end{cases}$$

    (b) Use the technique in part (a) to rewrite $|3x + 9|$.

46. If $y \neq 0$, show that $\left|\frac{x}{y}\right| = \frac{|x|}{|y|}$.

47. Show that $-|x| \le x \le |x|$ holds for all real numbers x.

48. Although the absolute value of a sum need not be the same as the sum of the absolute values, the inequality $|x + y| \le |x| + |y|$, called the **triangle inequality**, always holds. Prove the triangle inequality by considering the case in which $x + y \ge 0$ separately from the case in which $x + y < 0$. Use the facts that $-|x| \le x \le |x|$ and $-|y| \le y \le |y|$. (See Problem 47.)

49. If $a < b$, the expression $\frac{a + b}{2}$ represents the **midpoint** of the interval (a, b). Prove this by carrying out the following steps:

(**a**) Show that $a < \frac{a + b}{2} < b$.

(**b**) Show that the distance between a and $\frac{a + b}{2}$ is the same as the distance between $\frac{a + b}{2}$ and b.

50. Use the triangle inequality (Problem 48) to establish the following:

(**a**) If $|x - 2| < \frac{1}{2}$ and $|y + 2| < \frac{1}{3}$, then $|x + y| < \frac{5}{6}$.
(**b**) If $|x - 2| < \frac{1}{2}$ and $|z - 2| < \frac{1}{3}$, then $|x - z| < \frac{5}{6}$.

51. In this problem, we generalize the results in Problem 4.

(**a**) Show that the expression $\frac{1}{2}(x + y + |x - y|)$ always equals the larger of the two numbers x and y (or their common value if they are equal).

(**b**) Find an expression that is similar to the one given in part (a), but that gives the smaller of the two numbers x and y (or their common value if they are equal).

52. Using the triangle inequality (Problem 48), prove that $||x| - |y|| \le |x - y|$.

53. Statisticians often report their findings in terms of so-called *confidence intervals*—intervals that they believe (with a certain level of confidence) contain the unknown quantity with which they are concerned. Suppose that a statistician determines that (with a certain level of confidence) the average weight x pounds of newborn babies in a large city satisfies the inequality $|x - 6.8| < 0.85$. Find the confidence interval for x by solving this inequality.

54. Prove Theorem 3.

55. The diameter d, in millimeters, of a human capillary satisfies $|d - 10^{-2}| \le 5 \times 10^{-3}$. According to this inequality, what are the maximum and minimum diameters of a human capillary?

56. Prove Theorem 4.

57. Automakers in Detroit calculate that for each car they produce, they will have to replace a weight W of not less than 400 pounds and not more than 500 pounds of steel by light-weight materials, in order to meet mileage goals set by the government. Find numbers A and B such that W satisfies the inequality $|W - A| \le B$.

58. In calculus, it is necessary to show that a condition of the form $|x + y| < \varepsilon$ follows from the conditions $|x| < \varepsilon/2$ and $|y| < \varepsilon/2$. (ε is the Greek letter *epsilon.*) Use the triangle inequality (Problem 48) to make this argument.

59. A machine that fills one-liter milk cartons can make an error of at most 2%. If L is the number of liters of milk in a carton filled by this machine, find numbers A and B such that $|L - A| \le B$.

60. Prove Theorem 5. [*Hint*: Use Problem 50 on page 17.]

C **61.** In the SI (Système International) system of units the speed of light is exactly 299,792,458 meters per second. In 1766, the Danish astronomer, Olaf Roemer (1644–1710), made the first crude estimate of the speed of light by observing the periods of the satellites of Jupiter. He estimated that it takes light 1000 seconds to traverse the earth's orbit. Taking the modern value of 298,000,000 kilometers for the diameter of the earth's orbit, determine the percent error in Roemer's estimate. Round off your answer to the nearest tenth of a percent.

62. Let x be a positive number, let n be a positive integer, and let r be the result of rounding off x to n significant digits. Prove that the relative error $|x - r|/x$ made in estimating x by r cannot exceed 5×10^{-n}. [*Hint*: Use Problem 51 in Problem Set 1.2 together with Theorem 5.]

63. In Example 11, how many iterations will be required for an approximation that is accurate to three decimal places?

64. Using the result of Problem 62, formulate a rounding-off rule in terms of significant digits and relative error.

2.7 Polynomial and Rational Inequalities

Up to this point, we have dealt mainly with inequalities containing only first-degree polynomials in the unknown. In this section, we study inequalities containing higher-degree polynomials or rational expressions in the unknown.

Polynomial Inequalities

An inequality such as $4x^3 - 3x^2 + x - 1 > x^2 + 3x - x$, in which both sides are polynomials in the unknown, is called a **polynomial inequality.** Before solving such inequalities, we shall rewrite them in **standard form** with zero on the right side and a polynomial on the left side.

Suppose we have to solve the polynomial inequality

$$4x^2 - x - 8 < 3x^2 - 4x + 2.$$

We begin by subtracting $3x^2 - 4x + 2$ from both sides to bring it into the standard form

$$x^2 + 3x - 10 < 0.$$

Now, let's consider how the algebraic sign of the polynomial on the left changes as we vary the values of the unknown. As we move x along the number line, the quantity $x^2 + 3x - 10$ is sometimes positive, sometimes negative, and sometimes zero. To solve the inequality, we must find the values of x for which $x^2 + 3x - 10$ is negative.

It seems plausible (and is, in fact, true—see Problem 22) that *intervals where* $x^2 + 3x - 10$ *is positive are separated from intervals where it is negative by values of* x *for which it is zero.* To locate these intervals, we begin by solving the equation

$$x^2 + 3x - 10 = 0.$$

Factoring, we obtain $(x + 5)(x - 2) = 0$, so that $x = -5$ or $x = 2$. The two roots -5 and 2 divide the number line into three open intervals, $(-\infty, -5)$, $(-5, 2)$, and $(2, \infty)$ (Figure 1). Therefore, *the quantity* $x^2 + 3x - 10$ *will have a constant algebraic sign over each of these intervals.*

Figure 1

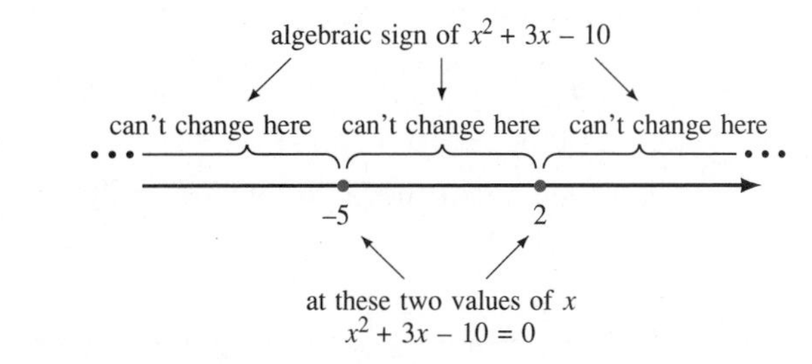

To find out whether $x^2 + 3x - 10$ is positive or negative over the first interval, $(-\infty, -5)$, we select any convenient test number in this interval, say, $x = -6$, and substitute it into $x^2 + 3x - 10$:

$$(-6)^2 + 3(-6) - 10 = 36 - 18 - 10 = 8 > 0.$$

Because $x^2 + 3x - 10$ is positive at one number in $(-\infty, -5)$ and cannot change its algebraic sign over this interval, we can conclude that $x^2 + 3x - 10 > 0$ for all values of x in $(-\infty, -5)$.

Similarly, to find the algebraic sign of $x^2 + 3x - 10$ over the interval $(-5, 2)$, we substitute a test number, say $x = 0$, to obtain

$$0^2 + 3(0) - 10 = -10 < 0,$$

and conclude that $x^2 + 3x - 10 < 0$ for all values of x in the interval $(-5, 2)$. Finally, using another test number, say, $x = 3$, from the interval $(2, \infty)$, we find that

$$3^2 + 3(3) - 10 = 9 + 9 - 10 = 8 > 0,$$

so that $x^2 + 3x - 10 > 0$ for all values of x in the interval $(2, \infty)$.

The information we now have about the algebraic sign of $x^2 + 3x - 10$ in each interval is summarized in Figure 2. Thus, $x^2 + 3x - 10 < 0$ if and only if $-5 < x < 2$; in other words, the solution set is the open interval $(-5, 2)$.

Figure 2

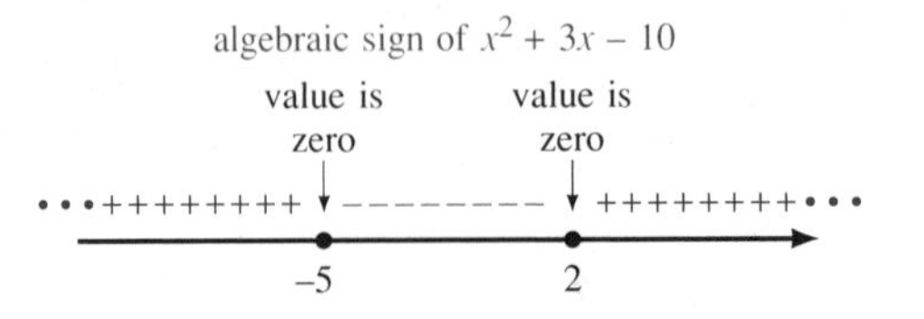

The method illustrated above can be used to solve any polynomial inequality in one unknown. This method is summarized in the following step-by-step procedure.

Procedure for Solving a Polynomial Inequality

Step 1. Bring the inequality into standard form, with zero on the right side and a polynomial on the left side.

Step 2. Set the polynomial equal to zero and find all real roots of the resulting equation.

Step 3. Arrange the roots obtained in step 2 in increasing order on a number line. These roots will divide the number line into open intervals.

Step 4. The algebraic sign of the polynomial cannot change over any of the intervals obtained in step 3. Determine this sign for each interval by selecting a test number in the interval and substituting it for the unknown in the polynomial. The algebraic sign of the resulting value is the sign of the polynomial over the entire interval.

Step 5. Draw a figure showing the information obtained in step 4 about the algebraic sign of the polynomial over the various open intervals. The solution set of the inequality can be read from this figure.

If it turns out in step 2 that the polynomial has no real roots, there is just one open interval in which to select the test number, namely, the entire number line $\mathbb{R}$.

In Examples 1 and 2, solve each polynomial inequality.

Example 1 $4x^2 + 8x \geq 5$

Solution We carry out the steps in the procedure above.

Step 1. $$4x^2 + 8x - 5 \geq 0$$

Step 2. $$4x^2 + 8x - 5 = 0$$

$$(2x + 5)(2x - 1) = 0$$

$$x = -\tfrac{5}{2} \quad \text{or} \quad x = \tfrac{1}{2}.$$

Step 3. The open intervals are $(-\infty, -\frac{5}{2})$, $(-\frac{5}{2}, \frac{1}{2})$, and $(\frac{1}{2}, \infty)$ (Figure 3).

Figure 3

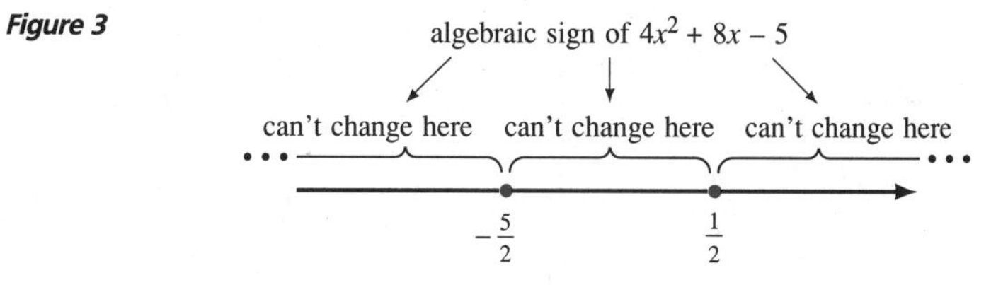

Step 4. Select, say, $x = -3$ in $(-\infty, -\frac{5}{2})$. Then,

$$4(-3)^2 + 8(-3) - 5 = 7 > 0,$$

so $$4x^2 + 8x - 5 > 0 \qquad \text{for all } x \text{ in } (-\infty, -\tfrac{5}{2}).$$

Select, say, $x = 0$ in $(-\frac{5}{2}, \frac{1}{2})$. Then,

$$4(0)^2 + 8(0) - 5 = -5 < 0,$$

so $$4x^2 + 8x - 5 < 0 \qquad \text{for all } x \text{ in } (-\tfrac{5}{2}, \tfrac{1}{2}).$$

Select, say, $x = 1$ in $(\frac{1}{2}, \infty)$. Then,

$$4(1)^2 + 8(1) - 5 = 7 > 0,$$

so $$4x^2 + 8x - 5 > 0 \qquad \text{for all } x \text{ in } (\tfrac{1}{2}, \infty).$$

Step 5. Figure 4 summarizes the information obtained in step 4. As the figure indicates, the solution set is $(-\infty, -\frac{5}{2}] \cup [\frac{1}{2}, \infty)$. (Note that the *nonstrict* inequality $4x^2 + 8x - 5 \geq 0$ requires that we include the values $-\frac{5}{2}$ and $\frac{1}{2}$ in the solution set.) ■

Figure 4

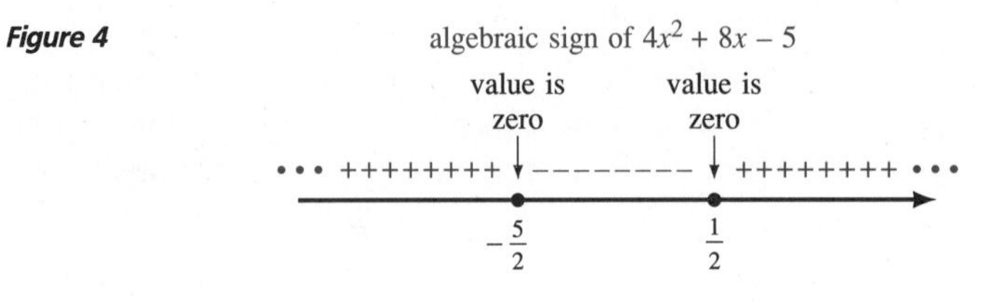

Example 2 $5x^2 + x + 16 < -4x^2 - 23x$

Solution The given inequality is equivalent to

$$9x^2 + 24x + 16 < 0; \qquad \text{that is,} \qquad (3x + 4)^2 < 0,$$

which is impossible. Therefore, the inequality has no solution. ■

C Example 3 High school students belonging to an amateur rocketry club are planning to launch a small experimental rocket straight upward. They have computed that the propellent will burn out when the rocket is 500 feet high and moving upward with a speed of 200 feet per second. The rocket will then coast to a highest point and begin to fall back down according to the equation $s = -16t^2 + 200t + 500$, where s feet is the height of the rocket above the ground and t seconds is the elapsed time since propellent burnout. When the rocket has fallen back to a height of 725 feet, a parachute will deploy and return it safely to the ground. For what time interval will the rocket be at least 725 feet above the ground?

Solution The values of t for which the rocket will be at least 725 feet high are solutions of the inequality

$$-16t^2 + 200t + 500 \geq 725;$$

that is,

$$-16t^2 + 200t - 225 \geq 0,$$

or

$$16t^2 - 200t + 225 \leq 0.$$

Solving the quadratic equation

$$16t^2 - 200t + 225 = 0$$

either by factoring or by using the quadratic formula, we find the two solutions $t = 1.25$ and $t = 11.25$. Test numbers chosen from the three intervals $(-\infty, 1.25)$, $(1.25, 11.25)$, and $(11.25, \infty)$ indicate the pattern of algebraic signs shown in Figure 5. Thus the rocket is at least 725 feet high during the time interval from 1.25 seconds to 11.25 seconds after propellant burnout. ■

Figure 5

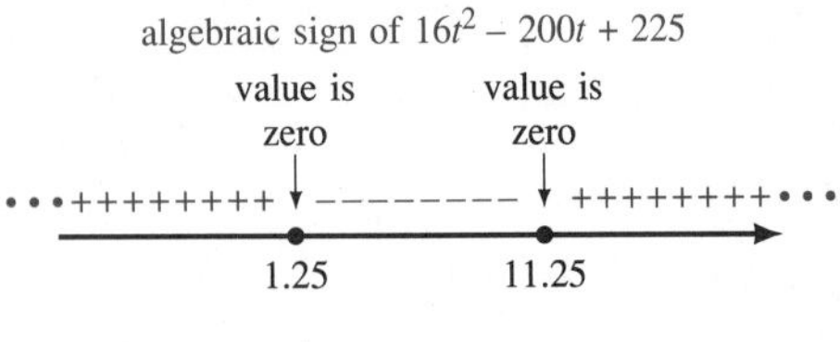

algebraic sign of $16t^2 - 200t + 225$

value is zero — value is zero

••• ++++++++ -------- ++++++++ •••

1.25 — 11.25

Rational Inequalities

An inequality such as

$$\frac{2x + 5}{x + 1} < \frac{x + 1}{x - 1},$$

in which all members are rational expressions in the unknown, is called a **rational inequality.** To rewrite such an inequality in **standard form,** subtract the expression on the right side from both sides and then simplify the left side. Be careful—*don't multiply both sides of an inequality by an expression containing the unknown unless you are certain that the multiplier can have only positive values.* (Why not?)

Example 4 Rewrite the rational inequality $\frac{2x + 5}{x + 1} < \frac{x + 1}{x - 1}$ in standard form.

Solution We begin by subtracting $\frac{x + 1}{x - 1}$ from both sides:

$$\frac{2x + 5}{x + 1} - \frac{x + 1}{x - 1} < 0.$$

Using the LCD $(x + 1)(x - 1)$, we combine the fractions on the left side to obtain

$$\frac{(2x + 5)(x - 1) - (x + 1)(x + 1)}{(x + 1)(x - 1)} < 0 \qquad \text{or} \qquad \frac{x^2 + x - 6}{(x + 1)(x - 1)} < 0.$$

Finally, factoring the numerator, we can write the inequality as

$$\frac{(x + 3)(x - 2)}{(x + 1)(x - 1)} < 0.$$

■

When you have rewritten a rational inequality in standard form, you can solve it by considering how the algebraic sign of the expression on the left side changes as the unknown is varied. Notice that a fraction can change its algebraic sign only if its numerator or denominator changes its algebraic sign. Hence:

> Intervals where a rational expression is positive are separated from intervals where it is negative by values of the variable for which the numerator or denominator* is zero.

It is thus possible to solve rational inequalities by a simple modification of the step-by-step procedure for solving polynomial inequalities. The method is illustrated in the following example.

* Of course, the fraction itself is *undefined* when its denominator is zero.

Example 5 Solve the rational inequality $\dfrac{(x+3)(x-2)}{(x+1)(x-1)} < 0.$

Solution The numerator of $\dfrac{(x+3)(x-2)}{(x+1)(x-1)}$ is zero when $x = -3$ and when $x = 2$; its denominator is zero when $x = -1$ and when $x = 1$. We arrange these four numbers in order along a number line to determine five open intervals over which the algebraic sign of $\dfrac{(x+3)(x-2)}{(x+1)(x-1)}$ does not change, namely, $(-\infty, -3)$, $(-3, -1)$, $(-1, 1)$, $(1, 2)$, and $(2, \infty)$ (Figure 6).

Figure 6

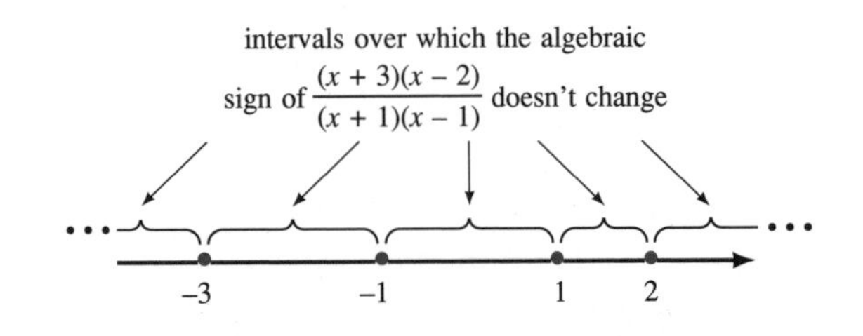

We select test numbers, one from each of these intervals, say, -4, -2, 0, $\frac{3}{2}$, and 3. Substituting these test numbers into $\dfrac{(x+3)(x-2)}{(x+1)(x-1)}$, we obtain

$$\frac{(-4+3)(-4-2)}{(-4+1)(-4-1)} = \frac{2}{5} > 0, \qquad \frac{(-2+3)(-2-2)}{(-2+1)(-2-1)} = -\frac{4}{3} < 0,$$

$$\frac{(0+3)(0-2)}{(0+1)(0-1)} = 6 > 0, \qquad \frac{(\frac{3}{2}+3)(\frac{3}{2}-2)}{(\frac{3}{2}+1)(\frac{3}{2}-1)} = -\frac{9}{5} < 0,$$

$$\frac{(3+3)(3-2)}{(3+1)(3-1)} = \frac{3}{4} > 0.$$

The algebraic signs obtained for the test numbers determine the algebraic signs of $\dfrac{(x+3)(x-2)}{(x+1)(x-1)}$ over each of the five open intervals, as shown in Figure 7. From this figure, we see that the given inequality $\dfrac{(x+3)(x-2)}{(x+1)(x-1)} < 0$ holds for values of x in the intervals $(-3, -1)$ and $(1, 2)$, and that it fails to hold for all other values of x. Therefore, the solution set is $(-3, -1) \cup (1, 2)$.

Figure 7

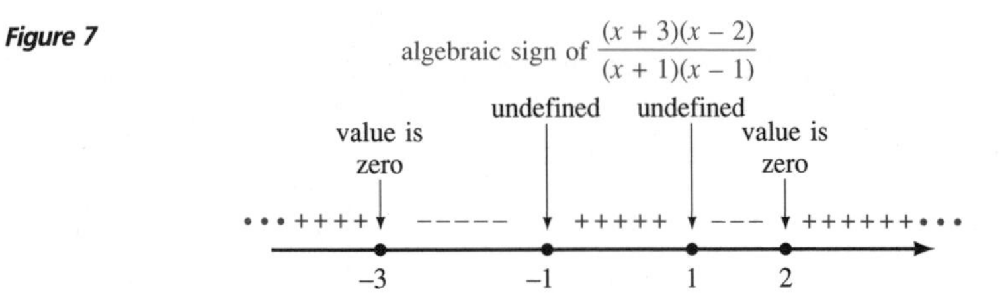

Problem Set 2.7

1. Rewrite each polynomial inequality in standard form.

   (a) $6x^2 - x + 3 < 5x^2 + 5x - 5$

   (b) $17x^2 - 12x + 33 > 14x^2 - 15x + 28$

   (c) $\frac{x^2}{2} + \frac{25}{2} \geq 5x$

   (d) $4x^3 + x \leq 3x^3 + 3x^2 - x$

2. In each case, write an inequality whose solution set consists of all values of k for which the condition is true.

   (a) $x^2 + kx + 1 = 0$ has real solutions.

   (b) $x^2 - 2kx + 3 - 2k = 0$ has no real solutions.

In Problems 3 to 20, solve each polynomial inequality.

3. $(x-2)(x-4) < 0$
4. $(x+4)(x+6) > 0$
5. $x^2 + 49 \leq 14x$
6. $x^2 + 32 \geq 12x + 6$
7. $6x^2 + 1 \leq 5x$
8. $2t^2 + 3t \leq 5$
9. $25y^2 + 4 \geq 20y$
10. $x(10 - 3x) < 8$
11. $1 - y \geq 2y^2$
12. $(x+2)^2 > (3x+1)^2$
13. $(x-1)(x+1)(x+2) \geq 0$
14. $x^2 - x + 1 > 0$
15. $x^3 > x$
16. $5x^3 \geq x^2 + 3x$
17. $4x^3 - 8x^2 + 2 < x$
18. $x^4 + 12 > 7x^2$
19. $(x+4)(x-2) \leq (2x-3)(x+1)$
20. $(2x+1)(3x-1)(x-2) \leq 0$
21. Rewrite each rational inequality in standard form.

    (a) $\frac{2x}{x+3} > \frac{1}{x+3}$ **(b)** $\frac{x^2-3}{x^2+5x+6} \leq \frac{1}{x+2}$

22. By examining the algebraic signs of the factors $x+5$ and $x-2$, explain why the intervals where $x^2 + 3x - 10$ is positive are separated from intervals where it is negative by values of x for which it is zero.

In Problems 23 to 34, solve the rational inequality.

23. $\frac{3x+6}{x+2} > 0$
24. $\frac{2x-1}{3x-2} < 0$
25. $\frac{3y-5}{y+3} \leq 0$
26. $\frac{(t-4)^2}{3t+1} \geq 0$
27. $\frac{(x-3)(x+3)}{(x+1)(x-2)} > 0$
28. $\frac{(x+1)(x-2)(x+2)}{(x-3)(x-4)(x+5)} \leq 0$
29. $\frac{x+1}{x-2} \leq 1$
30. $\frac{x+2}{x-3} \geq 1$
31. $\frac{6x+8}{x+2} > \frac{3x-1}{x-1}$
32. $\frac{x+1}{x+2} < \frac{x+3}{x+4}$
33. $\frac{(x-1)(x+2)(3-x)}{(x+3)(x^2-x-2)} \leq 0$
34. $\frac{(x-2)^2(x+1)(2x-1)}{(x+3)(x-1)^2(x+4)} \geq 0$

In Problems 35 and 36, determine the values of x for which the given expression represents a real number.

35. **(a)** $\sqrt{3x^2 + 22x + 35}$ **(b)** $\sqrt{\frac{x^2 - 2x - 15}{2x^2 + 3x - 5}}$

36. **(a)** $\sqrt{\frac{x^2 + 2x}{|x| - 4}}$ **(b)** $\sqrt{x^2 + 4|x| - 32}$

    [*Hint:* Consider separately the cases in which $x \geq 0$ and $x < 0$.]

C In Problems 37 and 38, solve each inequality with the aid of a calculator. Round off your answer to three decimal places.

37. **(a)** $23.8x^2 - 21.1x + 3.3 \geq 0$

    (b) $\frac{3.4x - 2.7}{2.1x^2 + 8.9x - 1.3} < 0$

38. **(a)** $2.12x^2 - 9.87x + 6.79 < 0$

    (b) $\frac{22.71x^2 + 11.11x - 8.23}{7.23x^2 - 41.02x + 6.66} \leq 0$

C 39. The sum of the first n positive integers is given by the formula

$$1 + 2 + 3 + \cdots + n = \frac{n(n+1)}{2}.$$

For what values of n will the sum be less than 625?

40. Wildlife biologists use the mathematical model $N = 50t^2 + 200t + 250$ to predict the population N of a certain endangered species after t years. In how many years will the population be at least 5050?

41. A projectile is fired straight upward from ground level with an initial velocity of 480 feet per second. Its distance above ground level t seconds later is given by the expression $480t - 16t^2$. For what time interval is the projectile more than 3200 feet above ground level?

C **42.** In economics, the *supply equation* for a commodity relates the selling price, p dollars per unit, to the total number of units q that the manufacturers would be willing to put on the market at that selling price. If the manufacturers' cost of producing one unit is c dollars, then, assuming that all units are sold, the total profit P dollars to the manufacturers is given by $P = pq - cq$. The supply equation for a small manufacturer of souvenirs in a resort area is $q = 8p + 90$, and each souvenir costs the manufacturer $c = \$4$ to produce. Assuming that all souvenirs are sold, what price per souvenir should be charged to bring in a total profit of at least \$315?

43. Let V and A denote the volume and the total surface area (top and bottom included) of a solid right circular cylinder of radius r centimeters and height $h = 20$ centimeters. For what values of r will the ratio $\frac{V}{A}$ be less than $\frac{1}{2}$?

C **44.** After being launched into space, the weight W of an object is given by the formula

$$W = W_0\left(\frac{3960}{3960 + h}\right)^2,$$

where h miles is the altitude of the object and W_0 is the weight of the object at the surface of the earth. For what altitudes is the weight of the object less than 4% of its weight on the surface of the earth?

45. In an experimental solar power plant, at least 60 mirrors that will focus sunlight on a boiler are to be arranged in rows, each row containing the same number of mirrors. For optimal operation, it is determined that the number of mirrors per row must be 2 more than twice the number of rows. At least how many rows must there be?

C **46.** In an apartment complex with 80 units, it is found that all units will be occupied if the rent is \$400 per month, but that for each \$20 increase in rent, one additional unit becomes vacant. Occupied units require \$40 per month for maintenance, while vacant units require none. Fixed costs for the building are \$20,000 per month. What rent should be charged in order for the profit from the complex to exceed \$6000 per month?

47. Doctors administering a new drug use the mathematical model

$$C = \frac{7t}{t^2 + 3t + 4}$$

to predict the concentration C milligrams per liter in the bloodstream of a patient t hours after the drug is taken orally. The drug is effective only when its concentration is at least 0.5 milligram per liter. During what period of time after the drug is taken will it be effective?

CHAPTER 2 Review Problem Set

In Problems 1 to 12, solve each equation.

1. $3(2y - 3) = 4(y + 1) - 3$

2. $5\left(\frac{t}{3} + \frac{2}{5}\right) = 2\left(\frac{4t}{5} + \frac{3}{2}\right)$

3. $\frac{1}{5}(6a - 7) = \frac{1}{2}(3a - 1)$

4. $\frac{5 + u}{6} - \frac{10 - u}{3} = 1$

5. $\frac{3 - y}{3y} + \frac{1}{4} = \frac{1}{2y}$

6. $\frac{5}{2x} = \frac{9 - 2x}{8x} + 3$

7. $\frac{n - 5}{n + 5} + \frac{n + 15}{n - 5} = \frac{25}{25 - n^2} + 2$

8. $\frac{2}{x - 2} + \frac{1}{x + 1} = \frac{1}{x^2 - x - 2}$

9. $\frac{2x}{x + 7} - 1 = \frac{x}{x + 3} + \frac{1}{x^2 + 10x + 21}$

10. $\dfrac{9}{y-3} - \dfrac{4}{y-6} = \dfrac{18}{y^2 - 9y + 18}$

11. $\dfrac{m}{x} + \dfrac{x-m}{x} - m = 1$ for x

12. $\dfrac{a-a^2}{x} = 1 - \dfrac{x-2a}{2x}$ for x

C In Problems 13 and 14, solve each equation with the aid of a calculator. Round off the answers to the correct number of significant digits.

13. $\dfrac{35x + 17.05}{-3x - 21.14} = 37.01$

14. $13.031 = 2\pi(11.076 + x)$

15. A pharmacist has 8 liters of a 15% solution of acid. How much distilled water must she add to it to reduce the concentration of acid to 10%?

16. A family has spent a total of \$3500 on a solar water heater and insulation for their home. Over the first year, they recover 30% of the cost of the solar water heater and 15% of the cost of the insulation in reduced energy bills. If the total reduction in energy bills amounts to \$825, how much was spent on the water heater and how much on the insulation?

17. A total of \$10,000 is invested, part at 7% simple annual interest and the other part at 8%. If a return of \$735 is expected at the end of one year, how much was invested at each rate?

18. In a certain district, 25% of the voters eligible for jury duty are members of minorities. Records show that during a 5-year period, 3% of all eligible people in that district were selected for jury duty, whereas 4% of the eligible minority members were selected. What percent of the eligible nonminority members in that district were selected for jury duty over the 5-year period?

19. Juanita and Pedro run a race from point A to point B and Juanita wins by 5 meters. She calculates that Pedro will have a fair chance to win a second race to point B if he starts at point A and she starts 6 meters behind him. What is the distance between point A and point B?

20. Plant A produces methane from biomass three times as fast as plant B. The two plants operating together yield a certain amount of methane in 12 hours. How long would it take plant A operating alone to yield the same amount of methane?

21. In each case, complete the square by adding a constant to the quantity.

(a) $x^2 + x$

(b) $x^2 - 24x$

(c) $x^2 - 9x$

(d) $x^2 + \sqrt{3}x$

22. Use the discriminant to determine the nature of the roots of each quadratic equation without solving it.

(a) $2x^2 - 3x + 1 = 0$

C **(b)** $169x^2 + 286x + 121 = 0$

(c) $4x^2 + 9x + 6 = 0$

(d) $\sqrt{3}x^2 + 2x - \sqrt{3} = 0$

In Problems 23 to 32, find all roots (real or complex) of each equation. (In Problems 27 and 28, round off your answers to three decimal places.)

23. $10y^2 = 3 + y$

24. $r(3r + 1) = 4$

25. $5x^2 - 7x - 2 = 0$

26. $x^2 + 2x = 5$

C **27.** $2.35t^2 + 6.42t - 0.91 = 0$

C **28.** $1.7y^2 + 0.33y = \dfrac{\pi}{3}$

29. $4x^2 + 3 = 3x$

30. $4s^2 + \frac{11}{2}s + 4 = 0$

31. $\dfrac{1-6x}{2x} = 2x$

32. $x^2 + \sqrt{2}x - 1 = 0$

33. A commercial jet could decrease the time needed to cover a distance of 2475 nautical miles by one hour if it were to increase its present speed by 100 knots (1 knot = 1 nautical mile per hour). What is its present speed?

C **34.** A biologist finds that the number N of water mites in a sample of lake water is related to the water temperature T in degrees Celsius by the mathematical model

$$N = 5.5T^2 - 19T.$$

At what temperature is $N = 2860$?

35. The diagonal of a rectangular television screen measures 20 inches, and the height of the screen is 4 inches less than its width. Find the dimensions of the screen.

36. All of the electricity in an alternative-energy home is supplied by batteries that are charged by a combination of solar panels and a wind-powered generator. Working together on a sunny, windy day, the solar panels and the generator can charge the batteries in 2.4 hours. Working alone on a windless, sunny day, it takes the solar panels 2 hours longer to charge the batteries than it takes the generator to do the job alone on a windy night. How long does it take the solar panels to charge the batteries on a windless, sunny day?

In Problems 37 to 60, find all real roots of each equation. Check for extraneous roots when necessary.

C **37.** $t + \sqrt{t+1} = 7$

38. $8 - \sqrt[3]{y-1} = 6$

39. $\sqrt[4]{x^2 + 5x + 6} = \sqrt{x+4}$

40. $\sqrt[3]{t+1} = \sqrt{t+1}$

41. $\sqrt{t+1} + \sqrt{2t+3} = 5$

42. $\sqrt{2r+5} - \sqrt{r+2} = \sqrt{3r-5}$

43. $t^4 - 5t^2 + 4 = 0$

44. $24v^4 - v^2 - 10 = 0$

45. $y - 2y^{1/2} - 15 = 0$

46. $t^{2/3} + 2t^{1/3} - 48 = 0$

47. $\left(x - \frac{8}{x}\right)^2 + \left(x - \frac{8}{x}\right) = 42$

48. $(t^2 + 2t)^2 - 2(t^2 + 2t) = 3$

49. $z^{-2} + z^{-1} = 6$

50. $x^{-2} = 2x^{-1} + 8$

51. $y + 2 + \sqrt{y+2} = 2$

52. $2x^2 + x - 4\sqrt{2x^2 + x + 4} = 1$

53. $x^5 - 2x^3 - x^2 + 2 = 0$

54. $(u-1)^3 - (u-1)^2 = u - 2$

55. $x^{-3/5} + x^{-2/5} - 2x^{-1/5} - 2 = 0$

56. $x^{7/12} + x^{1/4} - x^{1/3} = 1$

57. $\sqrt{5t + 2\sqrt{t^2 - t + 7}} - 4 = 0$

58. $\dfrac{\sqrt{v+16}}{\sqrt{4-v}} + \dfrac{\sqrt{4-v}}{\sqrt{v+16}} = \dfrac{5}{2}$

59. $\dfrac{\sqrt{1+x} + \sqrt{1-x}}{\sqrt{1+x} - \sqrt{1-x}} = 2$

60. $\sqrt{13 + 3\sqrt{y + 5 + 4\sqrt{y+1}}} = 5$

61. In engineering, the equation

$$L\sqrt[3]{F} = d\sqrt[3]{4bwY}$$

can be used to determine the bending of a beam with a rectangular cross section that is freely supported at the ends and loaded at the middle with a force F. Here L is the length of the beam, d is its depth, w is its width, b is the amount of bending in the middle, and Y is Young's modulus of elasticity for the material of the beam. Solve the equation for b in terms of L, F, d, w, and Y.

C **62.** The *equally tempered* musical scale, used by Bach in his "*Well-Tempered Clavier*" of 1722, consists of 13 notes, C, C♯, D, D♯, E, F, F♯, G, G♯, A, A♯, B, and high C. In this scale, the frequency of high C is twice the frequency of C, and the ratios of the frequencies of successive notes are equal. If the frequency of A is taken to be 440 Hz ("concert A"), find the frequencies of the remaining notes.

63. Government economists monitoring the rate of inflation calculate the percent of increase in the cost of living at the end of each month. If m is the percent increase in a given month, then the equivalent annual percent increase a is given by

$$a = 100\left(1 + \frac{m}{100}\right)^{12} - 100.$$

Solve this equation for m in terms of a.

C **64.** In Problem 63, find the monthly increase in the cost of living if the annual increase is 12%.

In Problems 65 and 66, find the distance between the numbers and illustrate on a number line.

65. **(a)** -3 and 4 **(b)** 0 and -5

(c) $-\frac{2}{3}$ and $\frac{5}{2}$ **(d)** -3.2 and 4.1

C **66.** **(a)** 1.42 and $\sqrt{2}$ **(b)** -2.735 and $-\pi$

In Problems 67 to 70, solve each inequality and sketch its solution set on a number line.

67. $-9t - 3 < -30$

68. $-8y - 4 \geq -32$

69. $\frac{4x}{3} - 2 > \frac{x}{4}$

70. $\frac{3x}{2} + 13 \leq \frac{x}{8}$

71. A student on academic probation must earn a C in his algebra course in order to avoid being expelled from college. Since he is lazy and has no interest in earning a grade any higher than necessary, he wants his final numerical grade for algebra to lie on the interval $[70, 80)$. This grade will be determined by taking three-fifths of his classroom average, which is 68, plus two-fifths of his score on the final exam. Since he has always had trouble with inequalities, he can't figure out the range in which his score on the final exam should fall to allow him to stay in school. Can you help him?

C **72.** A runner leaves a starting point and runs along a straight road at a steady speed of 8.8 miles per hour. After a while, she turns around and begins to run back to the starting point. How fast must she run on the way back so that her average speed for the entire run will be greater than 8 miles per hour but no greater than 8.5 miles per hour? (*Caution:* The average speed for her entire run *cannot* be found by adding 8.8 miles per hour to her return speed and dividing by 2.)

73. Ohm's law states that current I amperes, voltage E volts, and resistance R ohms in the simple circuit shown in Figure 1 are related by the equation $E = IR$.

(a) If the voltage ranges from 105 to 122 volts and the resistance is a constant 10 ohms, what is the corresponding range of values for the current?

(b) Find the range of values for the current if, in addition, the resistance varies from 8 to 12 ohms.

Figure 1

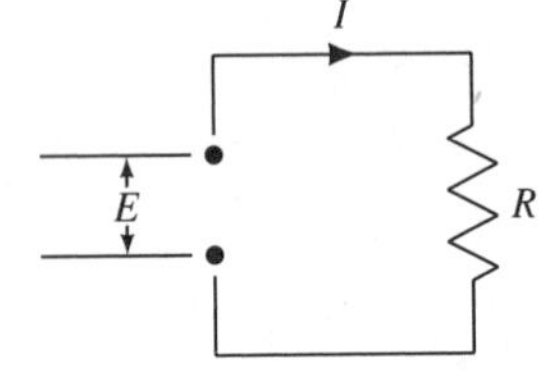

74. Which of the following equations or inequalities are true for all values of the variables?

(a) $|x^2 - 4| = |x - 2||x + 2|$

(b) $|x^2 - 4| = x^2 + 4$

(c) $|-x|^2 = x^2$ **(d)** $|x - y| = |y - x|$

(e) $|x^2 + 3x| \le x^2 + |3x|$ **(f)** $|x - y| \le |x| + |y|$

In Problems 75 to 92, solve each equation or inequality. Sketch the solution set of each inequality on a number line.

75. $|t - 1| = 3t$ **76.** $|2t + 3| = |t - 2|$

77. $|4x - 3| \ge 5$ **78.** $|2x + 1| < 3$

79. $|5 - 3z| < \frac{1}{2}$ **80.** $|x - 1| < |x + 1|$

81. $|2 - s| > |5 - 3s|$ **82.** $6x^2 - x \ge 2$

83. $6x^2 - 11x - 10 < 0$ **84.** $6x^2 \le 15x$

85. $6 - 5x - 6x^2 \ge 0$ **86.** $x(x + 3)(x + 5) < 0$

87. $\frac{3x - 1}{2x - 5} > 0$ **88.** $\frac{5x - 3}{7x + 1} \le 1$

89. $\frac{3x - 1}{5x - 7} \ge 2$ **90.** $\frac{1 - 5x}{9 - 3x} \le \frac{2x + 1}{x - 3}$

91. $\frac{x(x + 1)(x - 2)}{(x + 2)(x + 3)} < 0$ **92.** $\frac{x^2(x^2 - 1)}{x^2 - x - 2} \ge 0$

C **93.** A company specializes in buying used textbooks and selling them to students. The company has purchased 500 used copies of a calculus book at \$5 per copy. Suppose that all of these copies will be sold to students if the selling price is \$10 per copy, but that each \$1 increase in the selling price will result in 100 fewer copies being sold. If the profit to the company from the sale of these calculus books is to be between \$1000 and \$1100, inclusive, what will the selling price be? [*Hint:* Profit = revenue − cost.]

94. It is predicted that the number N of cars to be produced by the automobile industry for the coming year will satisfy the inequality $|N - 4{,}000{,}000| < 500{,}000$. Describe the anticipated production as an interval of real numbers.

95. It has been predicted that over the next decade, the number Q of quadrillions of Btu's of energy used per year in the United States will satisfy the inequality $|Q - 95| \le 10$. Find an interval containing Q according to this estimate.

96. Engineers estimate the percentage p of the heat stored in wood that will be delivered by a free-standing stove with good air control satisfies the inequality $|p - 65| \le 10$. Find an interval containing p according to this estimate.

C **97.** In calculus, a technique called *Simpson's rule* provides an estimate of the area of a region as a sum of $k + 1$ terms, where k must be even. The absolute error in this estimate does not exceed $M\frac{l^5}{180k^4}$, where M and l are certain constants. If $M = 24$ and $l = 2$, find the smallest positive even integer k for which the area estimate will be correct to three decimal places.

C **98.** The equation $8x^3 - 6x - 1 = 0$ is used in the proof that it is not possible to trisect a 60° angle with a straightedge and compass alone. One of the solutions r of this equation lies in the interval $[0.9, 1]$. If a is any number in this interval and if a is considered an approximation to r, it can be shown that the number $b = \frac{16a^3 + 1}{24a^2 - 6}$ is an even better approximation to r. Thus it is possible to construct successively better and better approximations to r by iteration of this procedure. After k iterations, starting with $a = 1$, it can be shown that the absolute error of the approximation does not exceed $\frac{1}{2}(\frac{1}{5})^{2^k}$. Find the smallest value of k for which the approximation will be correct to 10 decimal places.

CHAPTER 2 Test

1. Solve each equation.

   (a) $\frac{t-4}{7} = \frac{5}{t-2}$ (b) $\frac{5c}{2t} = 3c + 4$ for c

   (c) $|5x - 2| = 3$ (d) $r^2 = 4r - 6$

   (e) $3x^2 + 2x - 4 = 0$ (f) $(x + 2)^{3/4} = 27$

   (g) $x^{2/3} - 3x^{1/3} - 10 = 0$

   (h) $|1 - 2x| = |4x + 3|$ (i) $\sqrt{x+1} - 2\sqrt{x} = 1$

2. State a property of inequalities that justifies each statement.

   (a) If $x > -4$, then $-5x < 20$.

   (b) If $a - 2 \geq 3$, then $a \geq 5$.

   (c) If $5 < x$ and $x < t$, then $5 < t$.

   (d) If $\frac{x}{3} \leq -15$, then $x \leq -45$.

3. Solve each inequality and show the solution set on a number line.

   (a) $4(x - 3) \geq 3(x - 2)$ (b) $-11 \leq 2x - 3 \leq 7$

   (c) $|2x - 1| - 2 \leq 3$ (d) $|x - 3| > -2$

   (e) $x^2 + x - 2 > 0$ (f) $\frac{2x+1}{x-2} \geq 1$

4. Express $6.0\overline{2}$ as a quotient of integers.

5. Drew has a collection of 210 coins consisting of nickels, dimes, and quarters. He has twice as many dimes as nickels and 10 more quarters than dimes. How many coins of each kind does he have?

6. Kim has \$5000 to invest. If she invests \$3000 at 9.3% interest, at what rate must she invest the remainder so that the two investments together yield more than \$470 in yearly interest?

7. An electrician did a job for \$1080. It took him 6 hours longer than he expected, and therefore he earned \$6 per hour less than he had anticipated. How long did he expect that it would take to do the job?

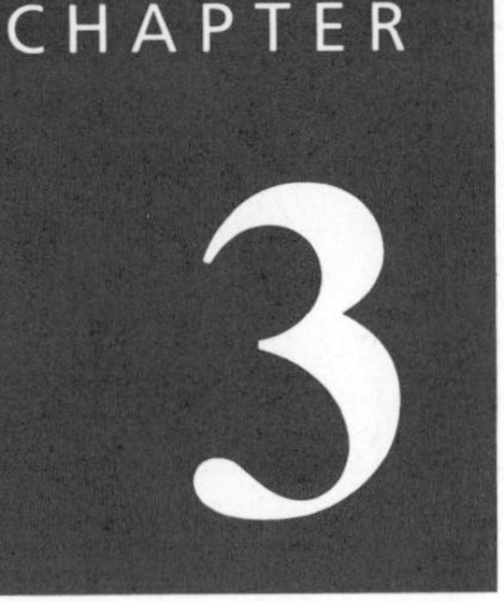

CHAPTER 3

Functions and Graphs

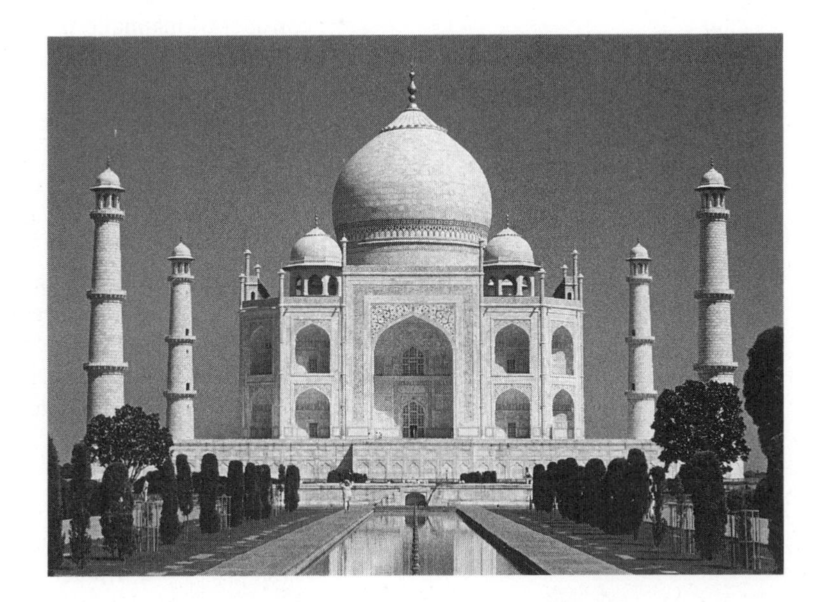

The Taj Mahal in Agra, India, provides a striking illustration of the idea of symmetry.

Ever since the Italian scientist Galileo Galilei (1564–1642) first used quantitative methods in the study of dynamics, advances in our understanding of the natural world have become more and more dependent on the use of mathematics. Important quantities have been identified, methods of measurement have been developed, and relationships among quantities have been formulated as scientific laws and principles. The idea of a *function* enables us to express relationships among observable quantities with efficiency and precision. In this chapter we discuss the Cartesian coordinate system, equations of straight lines and circles, graphs of functions, the algebra of functions, and inverse functions.

René Descartes

3.1 The Cartesian Coordinate System

In Section 1.1, we saw that a point P on a number line can be specified by a real number x called its *coordinate*. Similarly, by using a *Cartesian coordinate system*, named in honor of the French philosopher and mathematician René Descartes

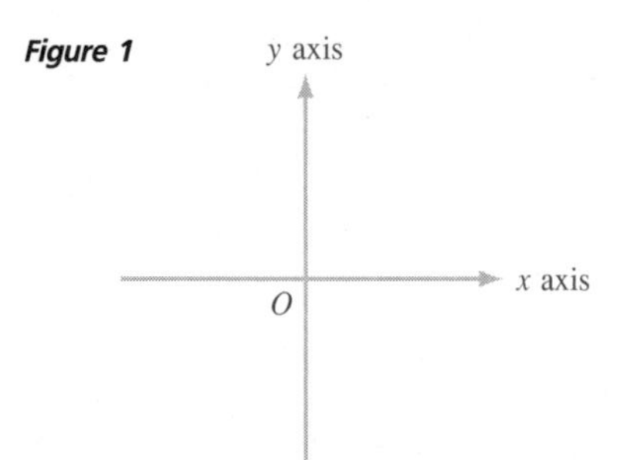

Figure 1

(1596–1650), we can specify a point P in the plane with *two* real numbers, also called *coordinates.*

A **Cartesian coordinate system** consists of two perpendicular number lines, called **coordinate axes,** which meet at a common origin O (Figure 1). Ordinarily, one of the number lines, called the **x axis,** is horizontal, and the other, called the **y axis,** is vertical. Numerical coordinates increase to the right along the x axis, and upward along the y axis. We usually use the same scale (that is, the same unit distance) on the two axes, although in some of our figures, space considerations make it convenient to use different scales.

If P is a point in the plane, the **coordinates** of P are the coordinates x and y of the points where perpendiculars from P meet the two axes (Figure 2). The x coordinate is called the **abscissa** of P, and the y coordinate is called the **ordinate** of P. The coordinates of P are traditionally written as an ordered pair (x, y) enclosed in parentheses, with the abscissa first and the ordinate second.*

To **plot** the point P with coordinates (x, y) means to draw Cartesian coordinate axes and to place a dot representing P at the point with abscissa x and ordinate y. You can think of the ordered pair (x, y) as the numerical "address" of P. The correspondence between P and (x, y) seems so natural that in practice we identify the point P with its "address" (x, y) by writing $P = (x, y)$. With this identification in mind, we call an ordered pair of real numbers (x, y) a **point,** and we refer to the set of all such ordered pairs as the **Cartesian plane** or the **xy plane.**

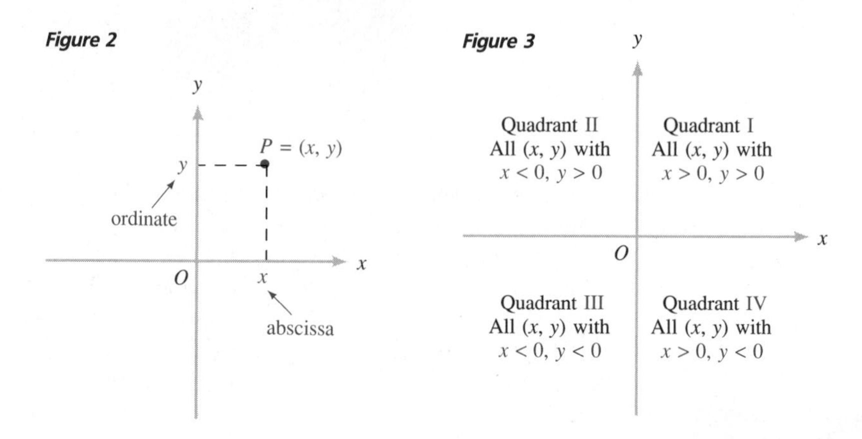

Figure 2

Figure 3

The x and y axes divide the plane into four regions called **quadrants I, II, III,** and **IV** (Figure 3). Quadrant I consists of all points (x, y) for which both x and y are positive, quadrant II consists of all points (x, y) for which x is negative and y is positive, and so forth, as shown in Figure 3. Notice that a point on a coordinate axis belongs to no quadrant.

* Unfortunately, this is the same symbolism as that used for an open interval; however, it is usually clear from the context what is intended.

Example 1 Plot each point and indicate which quadrant or coordinate axis contains the point.

(a) $(4, 3)$ **(b)** $(-3, 2)$ **(c)** $(-5, -1)$ **(d)** $(2, -4)$
(e) $(-3, 0)$ **(f)** $(0, 4)$ **(g)** $(0, -\frac{3}{2})$ **(h)** $(0, 0)$

Solution The points are plotted in Figure 4.

(a) $(4, 3)$ lies in quadrant I.
(b) $(-3, 2)$ lies in quadrant II.
(c) $(-5, -1)$ lies in quadrant III.
(d) $(2, -4)$ lies in quadrant IV.
(e) $(-3, 0)$ lies on the x axis.
(f) $(0, 4)$ lies on the y axis.
(g) $(0, -\frac{3}{2})$ lies on the y axis.
(h) $(0, 0)$, the origin, lies on both axes. ■

Figure 4

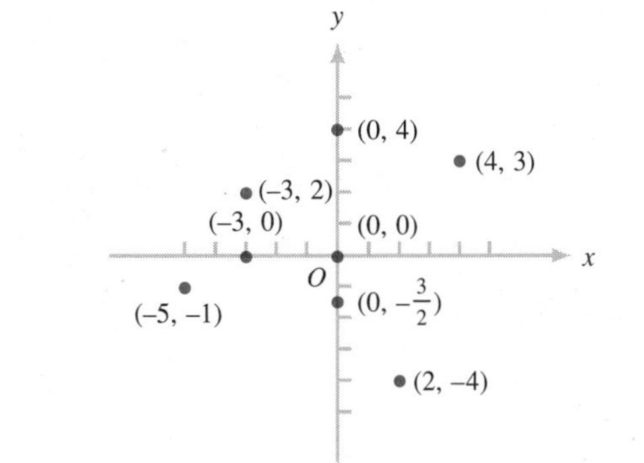

Example 2 Plot the point $P = (-4, 2)$ and determine the coordinates of the point Q if the line segment $\overline{PQ}$ is perpendicular to the x axis and is bisected by it.

Solution We begin by plotting the point $P = (-4, 2)$ (Figure 5). We see that P is 2 units directly above the point with coordinate -4 on the x axis; hence, Q is 2 units directly below the same point. Therefore, $Q = (-4, -2)$. ■

Figure 5

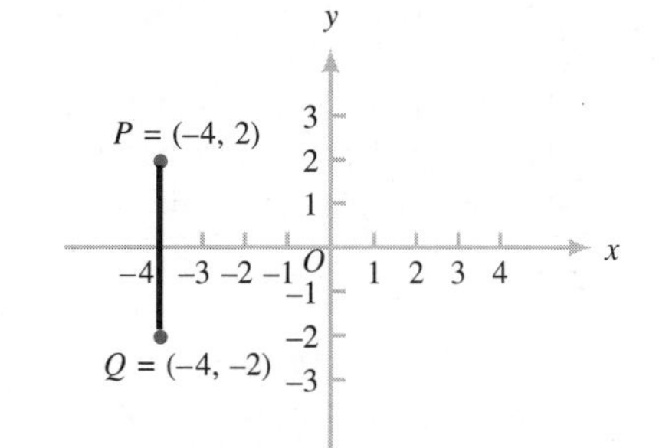

In Figure 5, the point $(-4, 0)$ is the midpoint of the line segment $\overline{PQ}$. More generally, the midpoint of a line segment can be obtained by using the following formula.

The Midpoint Formula

The point $\left(\frac{a+c}{2}, \frac{b+d}{2}\right)$ is the midpoint of the line segment joining the point (a, b) and the point (c, d).

We leave the proof of the midpoint formula as an exercise (Problem 28).

Example 3 Find the midpoint M of the line segment joining $R = (5, -4)$ and $S = (3, 6)$. Plot the points R, M, and S.

Solution By the midpoint formula, $M = \left(\frac{5+3}{2}, \frac{-4+6}{2}\right) = (4, 1)$. In Figure 6, we have plotted R, M, and S. ■

Figure 6

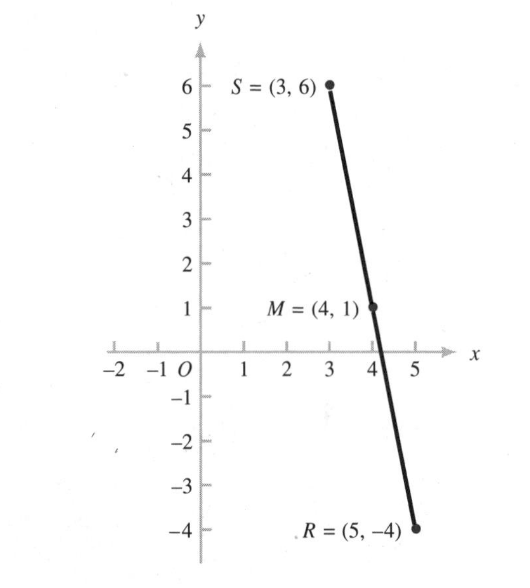

The Distance Formula

One of the attractive features of the Cartesian coordinate system is a simple formula that gives the distance between two points in terms of their coordinates. If P_1 and P_2 are two points in the Cartesian plane, we denote the distance between P_1 and P_2 by $|\overline{P_1P_2}|$.

The Distance Formula

If $P_1 = (x_1, y_1)$ and $P_2 = (x_2, y_2)$ are two points in the Cartesian plane, then the distance between P_1 and P_2 is given by

$$|\overline{P_1P_2}| = \sqrt{(x_2 - x_1)^2 + (y_2 - y_1)^2}.$$

Figure 7

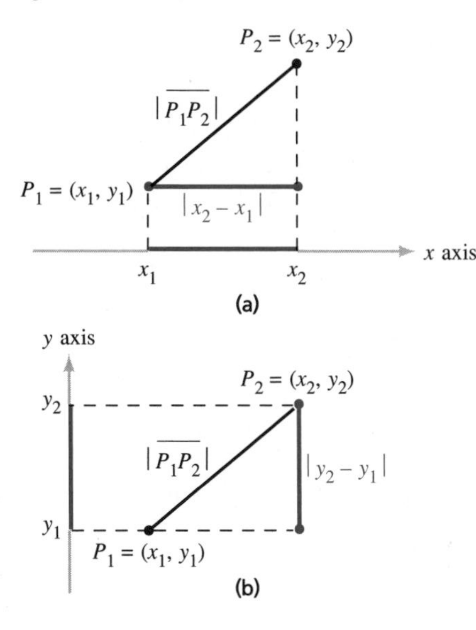

Figure 8

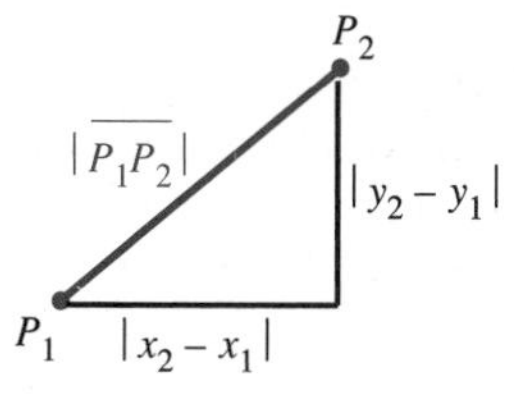

To derive the distance formula, we reason as follows: The *horizontal* distance between P_1 and P_2 is the same as the distance between the points with coordinates x_1 and x_2 on the x axis (Figure 7a). From the discussion of absolute value in Section 2.6, it follows that the horizontal distance between P_1 and P_2 is $|x_2 - x_1|$ units. Similarly, the *vertical* distance between P_1 and P_2 is $|y_2 - y_1|$ units (Figure 7b). If the line segment $\overline{P_1P_2}$ is neither horizontal nor vertical, it forms the hypotenuse of a right triangle with legs of lengths $|x_2 - x_1|$ and $|y_2 - y_1|$ (Figure 8). Therefore, by the Pythagorean theorem,*

$$\begin{aligned}|\overline{P_1P_2}|^2 &= |x_2 - x_1|^2 + |y_2 - y_1|^2 \\ &= (x_2 - x_1)^2 + (y_2 - y_1)^2,\end{aligned}$$

and it follows that

$$|\overline{P_1P_2}| = \sqrt{(x_2 - x_1)^2 + (y_2 - y_1)^2}.$$

We leave it to you to check that the formula works even if the line segment $\overline{P_1P_2}$ is horizontal or vertical (Problem 22).

Because $(x_2 - x_1)^2 = (x_1 - x_2)^2$ and $(y_2 - y_1)^2 = (y_1 - y_2)^2$, the distance formula can also be written as

$$|\overline{P_1P_2}| = \sqrt{(x_1 - x_2)^2 + (y_1 - y_2)^2}.$$

In other words, the order in which you subtract the abscissas or the ordinates does not affect the result.

Example 4 Let $A = (-2, -1)$, $B = (1, 3)$, $C = (-1, 2)$, and $D = (3, -2)$. Find:

(a) The distance $|\overline{AB}|$. **(b)** The distance $|\overline{CD}|$.

Solution By the distance formula, we have

(a) $|\overline{AB}| = \sqrt{[1 - (-2)]^2 + [3 - (-1)]^2} = \sqrt{3^2 + 4^2} = \sqrt{25} = 5$ units.

(b) $|\overline{CD}| = \sqrt{[3 - (-1)]^2 + (-2 - 2)^2} = \sqrt{32} = 4\sqrt{2}$ units. ■

C **Example 5** Let $P = (31.42, -17.04)$ and $Q = (13.75, 11.36)$. Using a calculator, find $|\overline{PQ}|$. Round off your answer to four significant digits.

Solution The distance between P and Q is given by

$$|\overline{PQ}| = \sqrt{(13.75 - 31.42)^2 + [11.36 - (-17.04)]^2} \approx 33.45 \text{ units.}$$ ■

* See Appendix I for a discussion of the Pythagorean theorem and its converse.

Example 6 Let $A = (-5, 3)$, $B = (6, 0)$, and $C = (5, 5)$. Show algebraically that ACB is a right triangle and find its area.

Figure 9

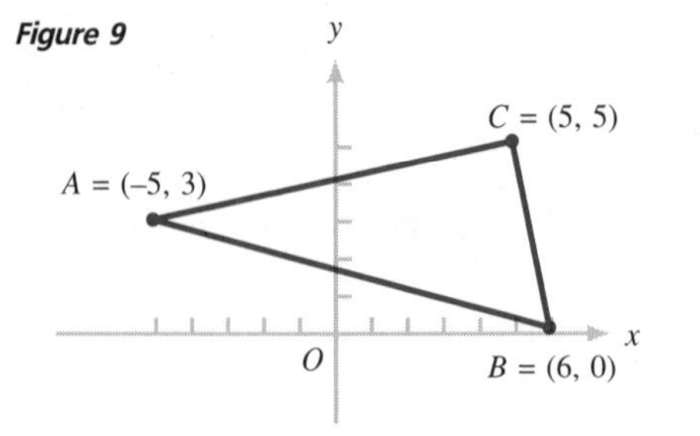

Solution We plot A, B, and C and draw the triangle in Figure 9. The figure leads us to suspect that the angle at vertex C is a right angle. To confirm this, we use the *converse* of the Pythagorean theorem; that is, we check to see if $|\overline{AC}|^2 + |\overline{BC}|^2 = |\overline{AB}|^2$. Now,

$$|\overline{AB}| = \sqrt{[6-(-5)]^2 + (0-3)^2} = \sqrt{11^2 + (-3)^2} = \sqrt{130}$$
$$|\overline{AC}| = \sqrt{[5-(-5)]^2 + (5-3)^2} = \sqrt{10^2 + 2^2} = \sqrt{104} = 2\sqrt{26}$$
$$|\overline{BC}| = \sqrt{(5-6)^2 + (5-0)^2} = \sqrt{(-1)^2 + 5^2} = \sqrt{26}.$$

Thus, $|\overline{AC}|^2 + |\overline{BC}|^2 = 104 + 26 = 130 = |\overline{AB}|^2$, and so ACB is, indeed, a right triangle.

Taking $|\overline{AC}| = 2\sqrt{26}$ as the base of the triangle and $|\overline{BC}| = \sqrt{26}$ as its altitude, we find that

$$\text{area} = \tfrac{1}{2}\text{ base} \times \text{altitude} = \tfrac{1}{2}(2\sqrt{26})\sqrt{26} = 26 \text{ square units.}$$ ■

Graphs in the Cartesian Plane

The graph of an equation or inequality in two unknowns x and y is defined to be the set of all points $P = (x, y)$ in the Cartesian plane whose coordinates x and y satisfy the equation or inequality. Many (but not all) equations in x and y have graphs that are smooth curves in the plane. For instance, consider the equation $x^2 + y^2 = 9$. We can rewrite this equation as $\sqrt{x^2 + y^2} = \sqrt{9}$ or as

$$\sqrt{(x-0)^2 + (y-0)^2} = 3.$$

Figure 10

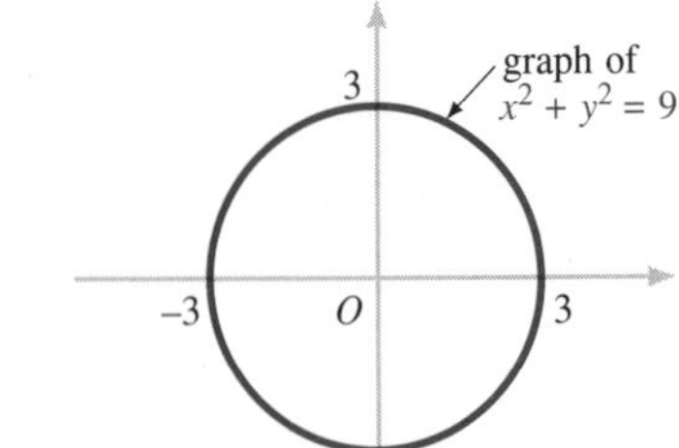

By the distance formula, the last equation holds if and only if the point $P = (x, y)$ is 3 units from the origin $O = (0, 0)$. Therefore, the graph of $x^2 + y^2 = 9$ is a circle of radius 3 units with its center at the origin O (Figure 10).

If we are given a curve in the Cartesian plane, we can ask whether there is an equation for which it is the graph. Such an equation is called an **equation for the curve** or an **equation of the curve.** For instance, $x^2 + y^2 = 9$ is an equation for the circle in Figure 10. Two equations or inequalities in x and y are said to be **equivalent** if they have the same graph. For example, the equation $x^2 + y^2 = 9$ is equivalent to the equation $\sqrt{x^2 + y^2} = 3$. We often use an equation for a curve to designate the curve; for instance, if we speak of "the circle $x^2 + y^2 = 9$," we mean "the circle for which $x^2 + y^2 = 9$ is an equation."

If $r > 0$, then the circle of radius r with center (h, k) consists of all points (x, y) such that the distance between (x, y) and (h, k) is r units (Figure 11). Using the

Figure 11

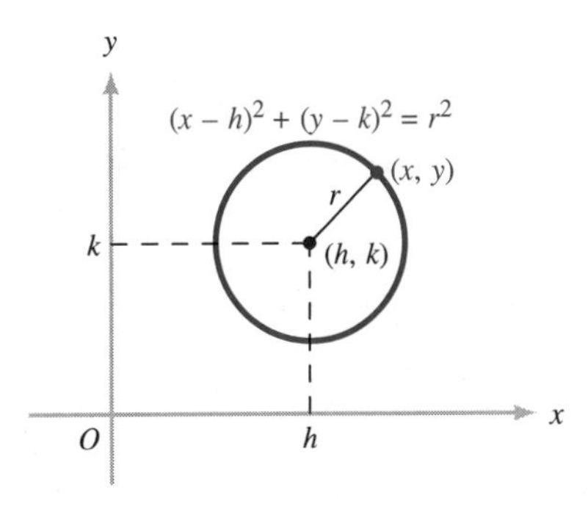

distance formula, we can write an equation for this circle as $\sqrt{(x-h)^2+(y-k)^2}=r$, or, equivalently,

$$(x-h)^2+(y-k)^2=r^2.$$

This last equation is called the **standard form** for the equation of a circle in the xy plane.

Example 7 Find an equation for the circle of radius 5 with center at the point $(3, -2)$ (Figure 12).

Solution Here $r = 5$ and $(h, k) = (3, -2)$, so, in standard form, the equation of the circle is

$$(x-3)^2+[y-(-2)]^2=5^2 \qquad \text{or} \qquad (x-3)^2+(y+2)^2=25.$$

Figure 12

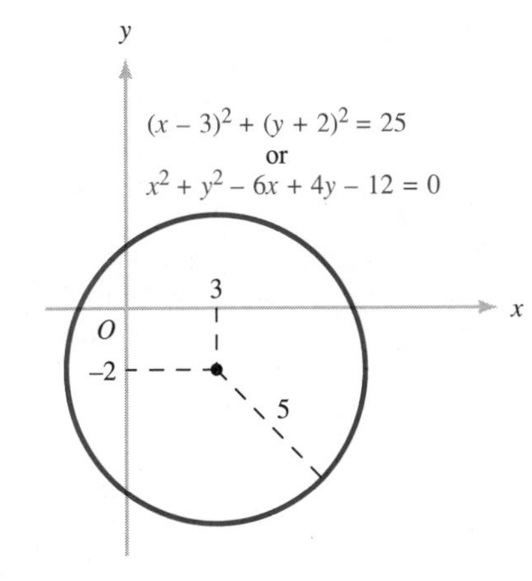

If desired, we can expand the squares, combine like terms, and rewrite the equation in the equivalent form

$$x^2+y^2-6x+4y-12=0.$$ ■

The last equation can be *restored* to standard form by completing the squares. (See page 99.) The work is arranged as follows:

$$\begin{aligned} x^2+y^2-6x+4y-12&=0\\ x^2-6x \quad +y^2+4y \quad &=12\\ x^2-6x+9+y^2+4y+4&=12+9+4\\ (x-3)^2+(y+2)^2&=25. \end{aligned}$$

Here, we added 9 to both sides of the equation to change x^2-6x to the perfect square x^2-6x+9, and we added 4 to both sides to change y^2+4y to the perfect square y^2+4y+4.

Figure 13

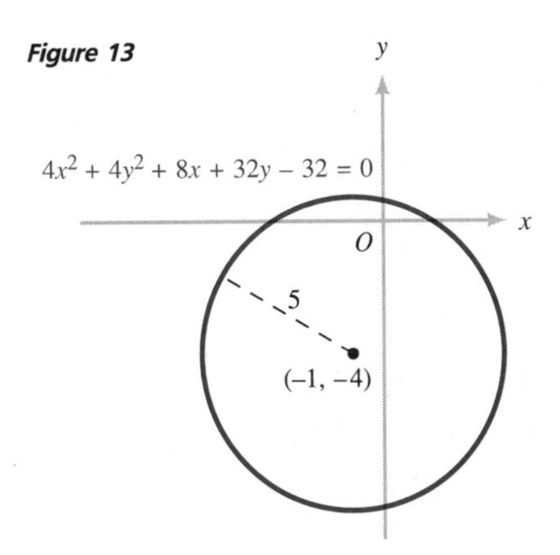

Example 8 Sketch the graph of $4x^2+4y^2+8x+32y-32=0$.

Solution We begin by dividing each side of the equation by 4. Then by completing the square, we have

$$\begin{aligned} x^2+2x \quad +y^2+8y \quad &=8\\ x^2+2x+1+y^2+8y+16&=8+1+16\\ (x+1)^2+(y+4)^2&=25. \end{aligned}$$

Therefore, the graph is a circle of radius $\sqrt{25}=5$ units with center $(-1, -4)$ (Figure 13). ■

Problem Set 3.1

In each problem set, problems with colored numbers constitute a good representation of the main ideas of the section. Note that some of the even-numbered problems may be considerably more challenging than the odd-numbered ones.

1. Plot each point and indicate which quadrant or coordinate axis contains it.

(a) $(1, 6)$ **(b)** $(-2, 3)$ **(c)** $(4, -1)$
(d) $(-1, -4)$ **(e)** $(0, 2)$ **(f)** $(\frac{5}{4}, 0)$
(g) $(0, -4)$ **(h)** $(\frac{3}{4}, -\frac{27}{5})$ **(i)** $(\pi, \sqrt{2})$

2. **(a)** Plot the points $A = (-5, -5)$, $B = (-1, -1)$, $C = (0, 0)$, $D = (\frac{1}{2}, \frac{1}{2})$, and $E = (\pi, \pi)$.

(b) Describe in words the set of all points of the form (x, x) for x a real number.

3. In each case, plot the point P and determine the coordinates of points Q, R, and S such that the line segment $\overline{PQ}$ is perpendicular to the x axis and is bisected by it, the line segment $\overline{PR}$ is perpendicular to the y axis and is bisected by it, and the line segment $\overline{PS}$ passes through the origin and is bisected by it.

(a) $P = (3, 2)$ **(b)** $P = (-1, 3)$

4. Describe in words the set of all points (x, y) satisfying each of the given conditions.

(a) $x = 1$ **(b)** $y = -2$
(c) $x \geq 0$ **(d)** $y < -2$

In Problems 5 to 14, **(a)** find the midpoint M of the line segment $\overline{RS}$, and **(b)** find the distance $d = |\overline{RS}|$ between R and S. In Problems 13 and 14, round off d to four significant digits.

5. $R = (1, 2)$, $S = (3, 4)$

6. $R = (2, 1)$, $S = (1, 4)$

7. $R = (-2, 3)$, $S = (8, 5)$

8. $R = (0, -9)$, $S = (-3, 7)$

9. $R = (0, 0)$, $S = (-8, -6)$

10. $R = (-\frac{1}{2}, -\frac{3}{2})$, $S = (3, -\frac{5}{2})$

11. $R = (a, b + 1)$, $S = (a + 1, b)$

12. $R = (s + t, s - t)$, $S = (t - s, t + s)$

C **13.** $R = (-2.714, 7.111)$, $S = (3.135, 4.982)$

C **14.** $R = (\pi, \frac{23}{4})$, $S = (-\sqrt{17}, \frac{211}{5})$

15. A **median** of a triangle is the line segment joining a vertex to the midpoint of the opposite side (Figure 14). In the triangle with vertices $A = (-4, 2)$, $B = (1, 4)$, and $C = (3, -1)$, find the exact length of the median from vertex A to the opposite side $\overline{BC}$.

Figure 14

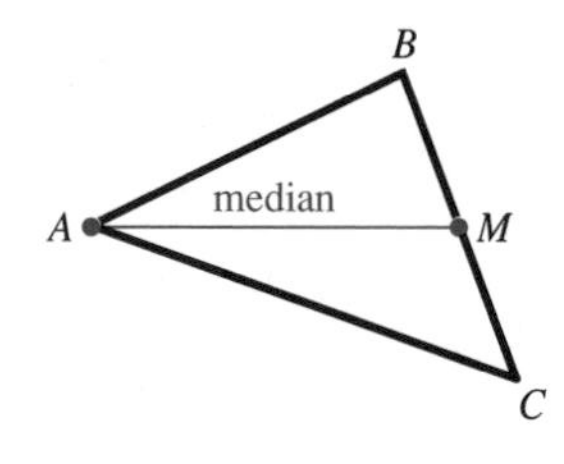

16. If $A = (3, 4)$, $B = (-5, 6)$, and $C = (1, 13)$, show algebraically that C is on the perpendicular bisector of $\overline{AB}$. [*Hint:* Find the midpoint M of $\overline{AB}$ and then use the converse of the Pythagorean theorem to show that triangle AMC is a right triangle.]

In Problems 17 and 18, use the distance formula and the converse of the Pythagorean theorem to show algebraically that triangle ABC is a right triangle and find the area of the triangle.

17. $A = (1, 1)$, $B = (5, 1)$, $C = (5, 7)$

18. $A = (-1, -2)$, $B = (3, -2)$, $C = (-1, -7)$

19. Show that the points $A = (-2, -3)$, $B = (3, -1)$, $C = (1, 4)$, and $D = (-4, 2)$ are the vertices of a square.

20. Show that the distance between the points (x_1, y_1) and (x_2, y_2) is the same as the distance between the point $(x_1 - x_2, y_1 - y_2)$ and the origin.

21. If $A = (-5, 1)$, $B = (-6, 5)$, and $C = (-2, 4)$, determine whether or not triangle ABC is isosceles.

22. Verify that the distance formula holds even if the line segment is horizontal or vertical.

23. Find all values of x for which the distance between the points $(x, 8)$ and $(-5, 3)$ is 13 units.

24. Find x if the distance between the points $(x, 0)$ and $(0, 3)$ is the same as the distance between the points $(x, 0)$ and $(7, 4)$.

25. Find all values of t for which the distance between the points $(-2, 3)$ and (t, t) is 5 units.

26. If P_1, P_2, and P_3 are points in the plane, then P_2 lies on the line segment $\overline{P_1P_3}$ if and only if $|\overline{P_1P_3}| = |\overline{P_1P_2}| + |\overline{P_2P_3}|$. Illustrate this geometric fact with diagrams.

27. In each case, determine whether or not P_2 lies on the line segment $\overline{P_1P_3}$ by checking to see if $|\overline{P_1P_3}| = |\overline{P_1P_2}| + |\overline{P_2P_3}|$ (see Problem 26).

(a) $P_1 = (1, 2)$, $P_2 = (0, \frac{5}{2})$, $P_3 = (-1, 3)$
(b) $P_1 = (2, 3)$, $P_2 = (3, -3)$, $P_3 = (-1, -1)$

28. Show that the point $P_2 = \left(\dfrac{a+c}{2}, \dfrac{b+d}{2}\right)$ is the midpoint of the line segment joining $P_1 = (a, b)$ and $P_3 = (c, d)$. [*Hint:* Use the condition in Problem 26 to show that P_2 actually belongs to the line segment $\overline{P_1P_3}$. Then show that $|\overline{P_1P_2}| = |\overline{P_2P_3}|$.]

29. Let $P_1 = (1, -2)$. Find P_2 such that the point $M = (2, 1)$ is the midpoint of the line segment $\overline{P_1P_2}$.

30. Figure 15 shows the midpoints of the sides of a triangle ABC. Find the coordinates of the vertices of the triangle.

Figure 15

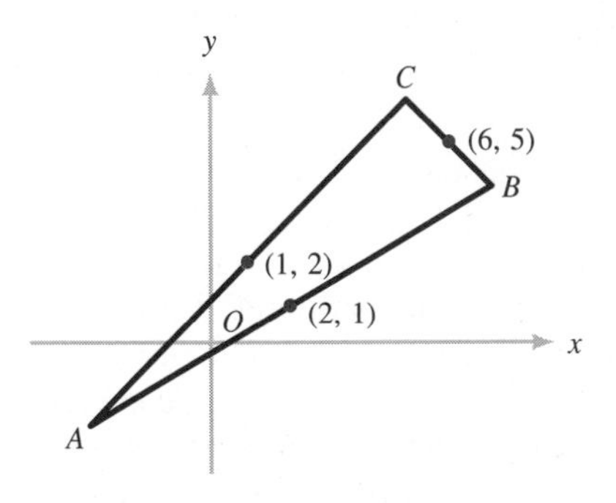

In Problems 31 to 40, find the equation in standard form of the circle satisfying the given conditions.

31. Radius 4, center $(0, 0)$.

32. Radius $\sqrt{3}$, center $(4, -\frac{3}{2})$.

33. Radius 2, center $(-1, 3)$.

34. Radius $\frac{2}{3}$, center $(\frac{3}{2}, -\sqrt{5})$.

35. Containing the point $(1, 2)$, center $(-4, 6)$.

36. Tangent to the y axis, center $(2, 0)$.

37. Tangent to the x axis, center $(1, 3)$.

38. Radius $\sqrt{17}$, center on the x axis, and containing the point $(0, 1)$. [There are two such circles.]

39. $(3, 2)$ and $(-2, 1)$ are endpoints of a diameter.

40. Tangent to both coordinate axes, center in quadrant II, radius 4.

In Problems 41 to 48, find the center and radius of each circle; then sketch the graph.

41. $x^2 + y^2 = 36$

42. $(x + 1)^2 + (y - 4)^2 = 4$

43. $(x - 3)^2 + (y - 5)^2 = 49$

44. $(x + 2)^2 + (y - 1)^2 = 64$

45. $x^2 - 4x + y^2 - 10y + 4 = 0$

46. $x^2 + y^2 - x - y - 1 = 0$

47. $5x^2 + 5y^2 + 10x + 20y + 20 = 0$

48. $4x^2 + 4y^2 + 8x - 4y + 1 = 0$

49. On a Cartesian coordinate grid, an aircraft carrier is detected by radar at point $A = (52, 71)$ and a submarine is detected by sonar at point $S = (47, 83)$. If distances are measured in nautical miles, how far is the carrier from a point on the surface of the water directly over the submarine?

50. Plans for a sausage-making machine require two sprockets with radii 2 inches, one of which is tangent to both coordinate axes and has its center in quadrant I. If the other sprocket has its center at the point (7, 6), find the length of the chain that fits tightly around the sprockets (Figure 16).

Figure 16

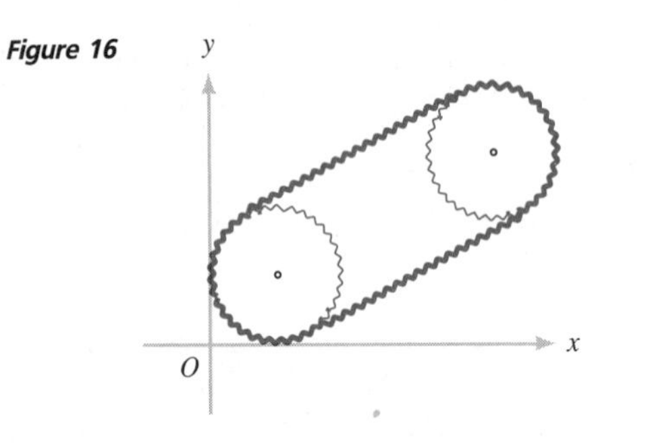

51. By a well-known proposition of Euclidean geometry, an angle inscribed in a semicircle is a right angle. Give a proof of this fact by using the converse of the Pythagorean theorem to show that triangle ABC in Figure 17 is a right triangle.

Figure 17

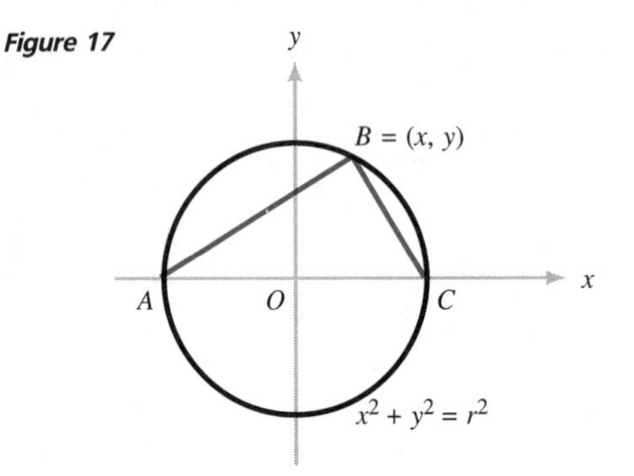

3.2 Lines and Their Slopes

Perhaps the simplest curves in the plane are straight lines (called simply *lines*) and circles. In Section 3.1, we derived equations for circles by using the distance formula. In this section, we derive equations for lines by using the idea of *slope.*

In ordinary language, the word "slope" refers to a steepness, an incline, or a deviation from the horizontal. For instance, we speak of a ski slope or the slope of a roof. In mathematics, the word "slope" has a precise meaning. Consider the line segment $\overline{AB}$ in Figure 1. The horizontal distance between A and B is called the run, and the vertical distance between A and B is called the rise. The ratio of rise to run is called the slope of the line segment $\overline{AB}$ and is traditionally denoted by the symbol m:

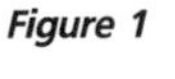
Figure 1

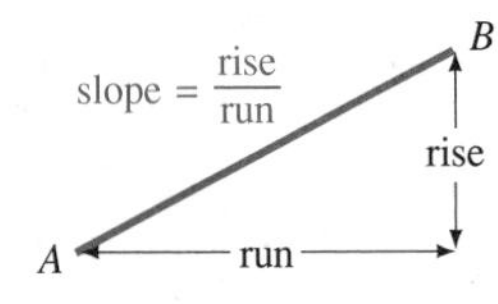

$$m = \text{the slope of } \overline{AB} = \frac{\text{rise}}{\text{run}}.$$

If the line segment $\overline{AB}$ is turned so that it becomes more nearly vertical, then the rise increases, the run decreases, and the slope $m = \text{rise/run}$ becomes larger. Therefore, the slope m really does give a numerical measure of the inclination or steepness of the line segment $\overline{AB}$—the greater the inclination, the greater the slope.

If the line segment $\overline{AB}$ is horizontal, its rise is zero, so its slope $m = \text{rise/run}$ is zero. Thus, *horizontal line segments have slope zero.* If $\overline{AB}$ slants downward to the right, as in Figure 2, its rise is considered to be negative; hence, its slope $m = \text{rise/run}$ is negative. (The run is always considered to be nonnegative.) Thus, *a line segment*

Figure 2

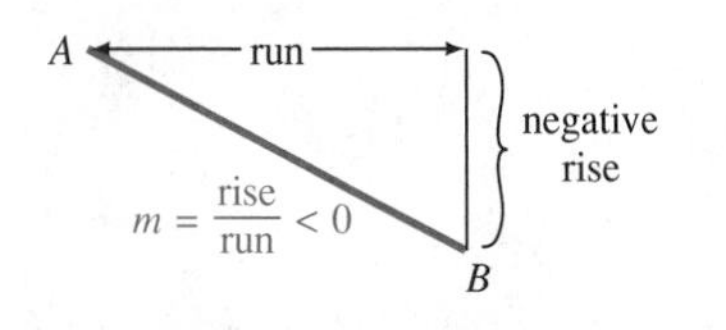

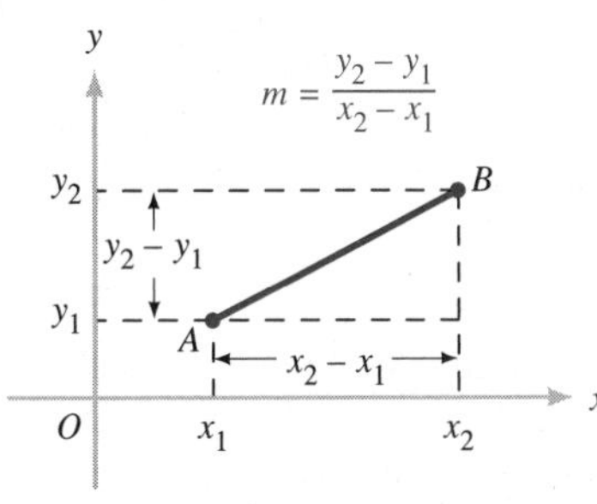

Figure 3

that slants downward from left to right has a negative slope. Notice that *the slope m = rise/run of a vertical line segment is undefined because the denominator is zero.*

Now let $A = (x_1, y_1)$ and $B = (x_2, y_2)$, and consider the line segment $\overline{AB}$ (Figure 3). If B is above and to the right of A, the line segment $\overline{AB}$ has rise $= y_2 - y_1$ and run $= x_2 - x_1$, so its slope is

$$m = \frac{y_2 - y_1}{x_2 - x_1}.$$

Even if B is not above and to the right of A, the slope m of $\overline{AB}$ is given by the same formula (Problem 4), and we have the following:

The Slope Formula

Let $A = (x_1, y_1)$ and $B = (x_2, y_2)$ be two points in the Cartesian plane. Then, if $x_1 \neq x_2$, the slope m of the line segment $\overline{AB}$ is

$$m = \frac{y_2 - y_1}{x_2 - x_1}.$$

Notice that $\frac{y_2 - y_1}{x_2 - x_1} = \frac{y_1 - y_2}{x_1 - x_2}$ (why?), so the slope of a line segment is the same regardless of which endpoint is called (x_1, y_1) and which is called (x_2, y_2).

Example 1 In each case, sketch the line segment $\overline{AB}$ and find its slope m by using the slope formula.

(a) $A = (-3, -2)$, $B = (4, 1)$ **(b)** $A = (-2, 3)$, $B = (5, 1)$

(c) $A = (-2, 4)$, $B = (5, 4)$ **(d)** $A = (3, -1)$, $B = (3, 6)$

Solution The line segments are sketched in Figure 4.

(a) $m = \frac{y_2 - y_1}{x_2 - x_1} = \frac{1 - (-2)}{4 - (-3)} = \frac{1 + 2}{4 + 3} = \frac{3}{7}$

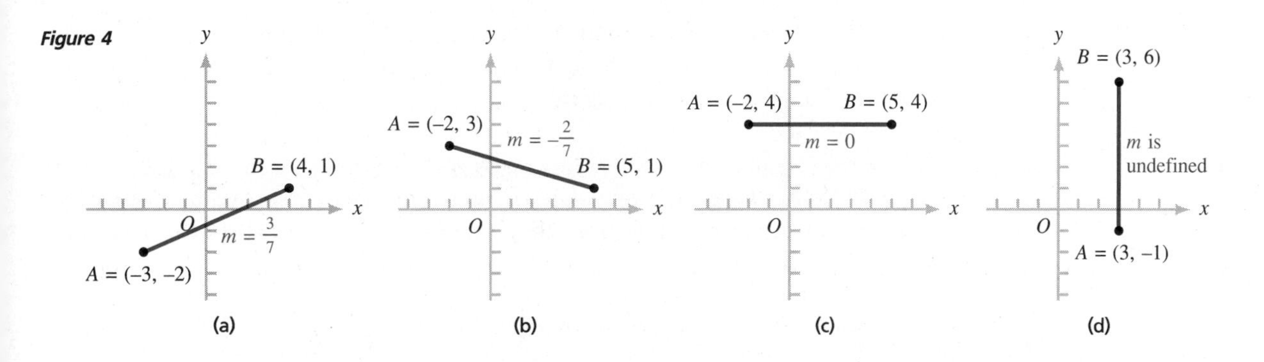

Figure 4

(b) $m = \dfrac{y_2 - y_1}{x_2 - x_1} = \dfrac{1 - 3}{5 - (-2)} = \dfrac{-2}{5 + 2} = -\dfrac{2}{7}$

(c) $m = \dfrac{y_2 - y_1}{x_2 - x_1} = \dfrac{4 - 4}{5 - (-2)} = \dfrac{0}{5 + 2} = 0$

(d) m is undefined, since $x_2 - x_1 = 0$. ■

From the similar triangles APB and CQD in Figure 5, you can see that *two parallel line segments $\overline{AB}$ and $\overline{CD}$ have the same slope.** Likewise, *if two line segments $\overline{AB}$ and $\overline{CD}$ lie on the same line L, then they have the same slope* (Figure 6). The common slope of all the segments lying on a line L is called the **slope** of L.

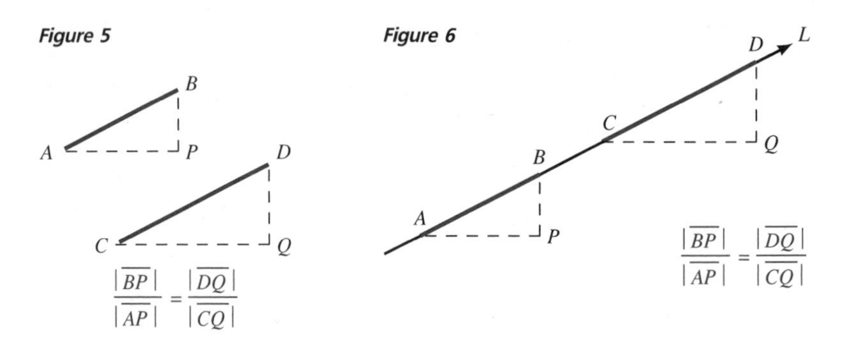

Example 2 Sketch the line L that contains the point $P = (1, 2)$ and has slope

(a) $m = \frac{2}{3}$ **(b)** $m = -\frac{2}{3}$.

Solution **(a)** The condition $m = \frac{2}{3}$ means that, for every 3 units we move to the right from a point on L, we must move up 2 units to get back to L. If we start at the point $P = (1, 2)$ on L, move 3 units to the right and 2 units up, we arrive at the point $Q = (1 + 3, 2 + 2) = (4, 4)$ on L. Because any two points on a line determine the line, we simply plot $P = (1, 2)$ and $Q = (4, 4)$ and use a straightedge to draw L (Figure 7a).

(b) The condition $m = -\frac{2}{3}$ means that, for every 3 units we move to the right from a point on L, we must move down 2 units to get back to L. If we start at the point $P = (1, 2)$ on L, move 3 units to the right and 2 units down, we arrive at the point $Q = (1 + 3, 2 - 2) = (4, 0)$ on L. Thus, we plot $P = (1, 2)$ and $Q = (4, 0)$ and use a straightedge to draw L (Figure 7b). ■

From the fact that two parallel line segments have the same slope, it follows that *two parallel lines have the same slope.* Conversely, you can prove by elementary geometry that *two distinct lines with the same slope are parallel* (Problem 12). Thus, we have the following:

* See Appendix I for the properties of triangles.

Figure 7

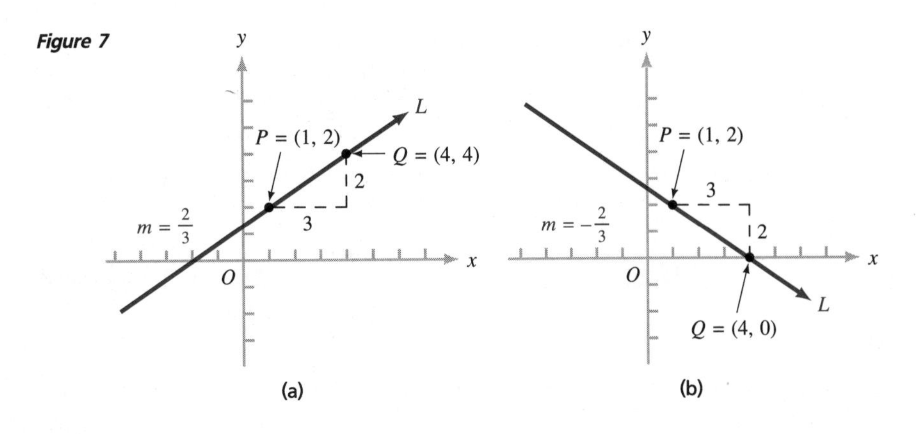

Parallelism Condition

> Two distinct nonvertical lines in the Cartesian plane are parallel if and only if they have the same slope.

For instance, the slope m_1 of the line containing the points $A = (-2, 5)$ and $B = (2, -1)$ is given by

$$m_1 = \frac{-1-5}{2-(-2)} = -\frac{3}{2},$$

and the slope m_2 of the line containing the points $C = (-4, 1)$ and $D = (0, -5)$ is given by

$$m_2 = \frac{-5-1}{0-(-4)} = -\frac{3}{2}.$$

Since these two lines have the same slope, they are parallel.

Now consider a nonvertical line L having slope m and containing the point $P_0 = (x_0, y_0)$* (Figure 8). If $P = (x, y)$ is any other point on L, then, by the slope formula,

Figure 8

$$m = \frac{y - y_0}{x - x_0};$$

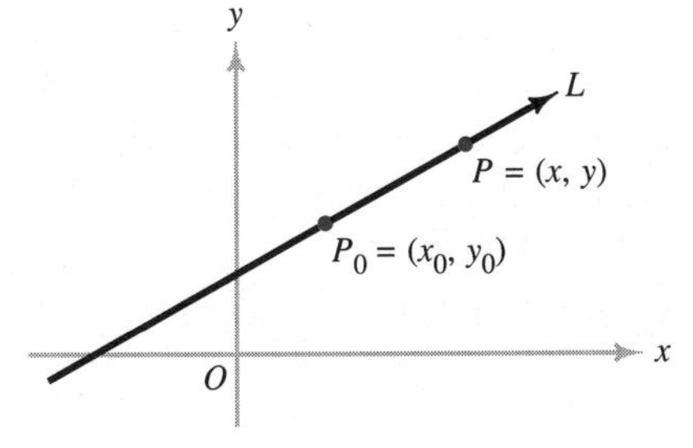

that is,

$$y - y_0 = m(x - x_0).$$

Notice that this last equation holds even if $P = P_0$, when it simply reduces to $0 = 0$. Using the parallelism condition and elementary geometry, you can see that, conversely, any point $P = (x, y)$ whose coordinates satisfy the equation $y - y_0 = m(x - x_0)$ lies on the line L (Problem 14). Therefore, we have the following:

* P_0 is read "P sub zero" or, more often, "P naught." Similarly, we refer to x_0 and y_0 as "x naught" and "y naught."

The Point-Slope Equation of a Straight Line

> In the Cartesian plane, the line L that contains the point $P_0 = (x_0, y_0)$ and has slope m is the graph of the equation
>
> $$y - y_0 = m(x - x_0).$$

You can write the equation $y - y_0 = m(x - x_0)$ as soon as you know the coordinates (x_0, y_0) of one point P_0 on the line L and the slope m of L. For this reason, the equation is called a **point-slope form.**

Example 3 Find an equation in point-slope form for the line that contains the point (3, 4) and is parallel to the line containing the points (1, −2) and (2, 3) (Figure 9).

Figure 9

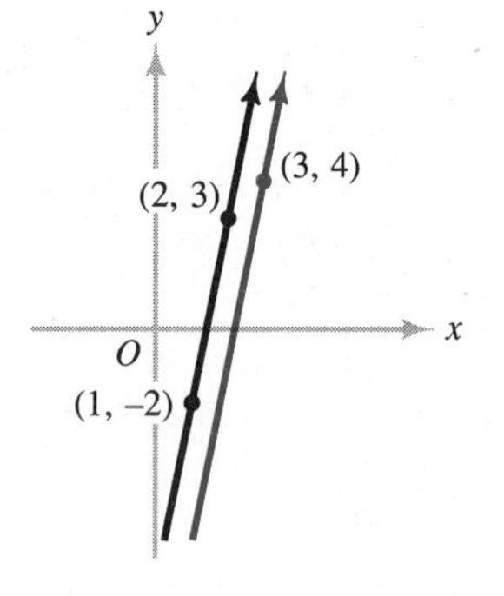

The line containing the points (1, −2) and (2, 3) has slope $(3 + 2)/(2 - 1) = 5$. Since the two lines are parallel, their slopes are equal; so $m = 5$.

Substituting $m = 5$, $x_0 = 3$, and $y_0 = 4$ in the equation

$$y - y_0 = m(x - x_0),$$

we obtain

$$y - 4 = 5(x - 3).$$

■

If L is a line that is not parallel to the y axis, then it must intersect this axis at some point $(0, b)$ (Figure 10). The ordinate b of the intersection point is called the **y intercept** of the line L. If m is the slope of L, then, since $(0, b)$ belongs to L, a point-slope equation of L is $y - b = m(x - 0)$. This equation simplifies to

$$y = mx + b$$

which is called the **slope-intercept form** of the equation for L. In this form, the coefficient of x is the slope and the constant term on the right is the y intercept.

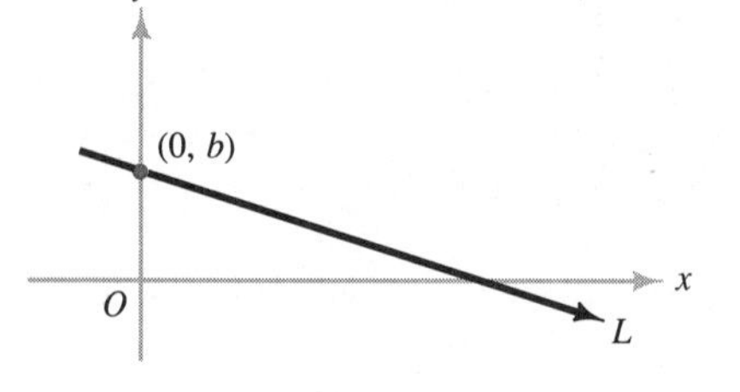

Example 4 Show that the graph of the equation $3x - 5y - 15 = 0$ is a straight line by rewriting the equation in slope-intercept form. Find the slope m and the y intercept b, and sketch the graph.

Solution We begin by solving the equation for y in terms of x:

$$3x - 5y - 15 = 0$$
$$-5y = -3x + 15$$
$$y = \tfrac{3}{5}x - 3.$$

Thus, we have the equation of a straight line in slope-intercept form with slope $m = \frac{3}{5}$ and y intercept $b = -3$. We can obtain the graph by drawing the line with slope $m = \frac{3}{5}$ through the point $(0, b) = (0, -3)$. An alternative method is to find the x intercept of the line, that is, the abscissa x of the point $(x, 0)$ where the line intersects the x axis. This is accomplished by putting $y = 0$ in the original equation $3x - 5y - 15 = 0$ and then solving for x:

$$3x - 15 = 0 \quad \text{or} \quad x = 5.$$

Thus, we obtain the graph by drawing the line through the point $(0, -3)$ on the y axis and the point $(5, 0)$ on the x axis (Figure 11). ■

Figure 11

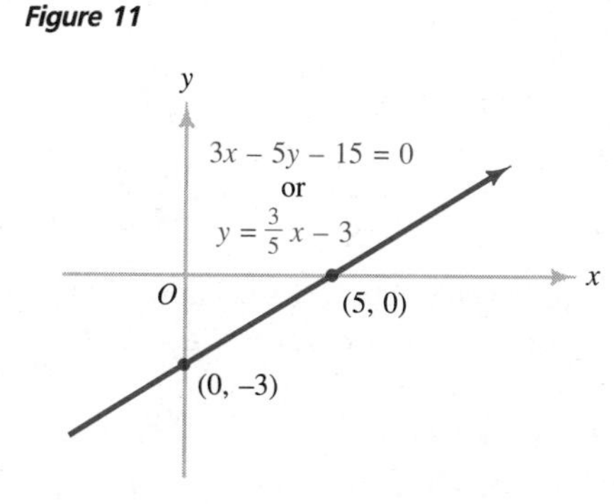

A horizontal line has slope $m = 0$; hence, in slope-intercept form, its equation is $y = 0(x) + b$, or simply $y = b$ (Figure 12). The equation $y = b$ places no restriction at all on the abscissa x of a point (x, y) on the horizontal line, but it requires that all of the ordinates y have the same value b.

Figure 12 Figure 13

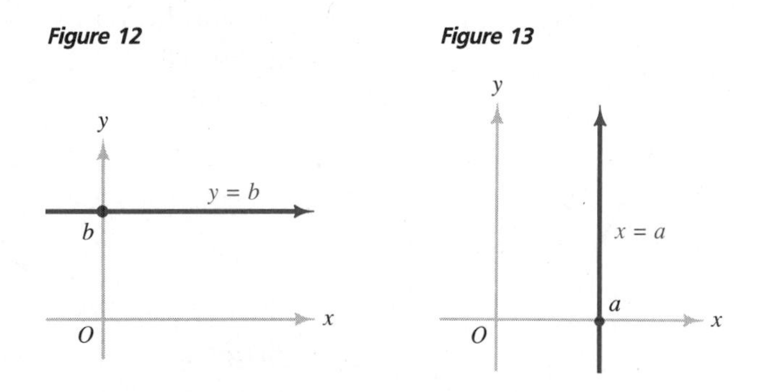

A vertical line has an undefined slope, so you can't write its equation in slope-intercept form. However, since all points on a vertical line have the same abscissa, say, a, an equation of such a line is $x = a$ (Figure 13).

If A, B, and C are constants and if A and B are not both zero, an equation of the form

$$Ax + By + C = 0$$

represents a straight line. If $B \neq 0$, the equation can be rewritten in the slope-intercept form as

$$y = \left(\frac{-A}{B}\right)x + \left(\frac{-C}{B}\right),$$

with slope $m = \frac{-A}{B}$ and y intercept $b = \frac{-C}{B}$. If $B = 0$, then $A \neq 0$ and the equation can be rewritten as

$$x = \frac{-C}{A},$$

the equation of a vertical line. The equation $Ax + By + C = 0$ is called the **general form** of the equation of a line.

The following condition for two lines to be perpendicular is quite useful.

Perpendicularity Condition

Two nonvertical lines in the Cartesian plane are perpendicular if and only if the slope of one of the lines is the negative of the reciprocal of the slope of the other.

Figure 14

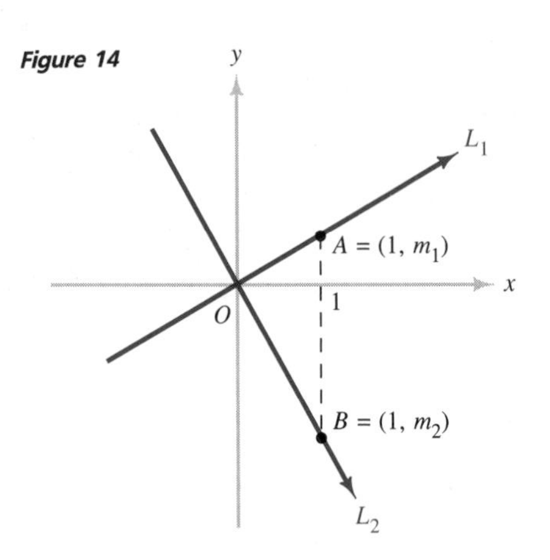

To establish this result, let the two lines be L_1 and L_2, with slopes m_1 and m_2, respectively. The condition that the slope of either one of the lines is the negative of the reciprocal of the slope of the other can be written as $m_1 m_2 = -1$. Neither the angle between the lines nor their slopes are affected if we place the origin O at the point where L_1 and L_2 intersect (Figure 14). Starting at O on L_1, we move 1 unit to the right and $|m_1|$ units vertically to arrive at the point $A = (1, m_1)$ on L_1. Likewise, the point $B = (1, m_2)$ is on line L_2. By the Pythagorean theorem and its converse, triangle AOB is a right triangle if and only if

$$|\overline{AB}|^2 = |\overline{OA}|^2 + |\overline{OB}|^2.$$

Using the distance formula, we find that

$$|\overline{AB}|^2 = (1-1)^2 + (m_1 - m_2)^2 = m_1^2 - 2m_1m_2 + m_2^2,$$

$$|\overline{OA}|^2 = (1-0)^2 + (m_1 - 0)^2 = 1 + m_1^2,$$

and $$|\overline{OB}|^2 = (1-0)^2 + (m_2 - 0)^2 = 1 + m_2^2.$$

Therefore, the condition $|\overline{AB}|^2 = |\overline{OA}|^2 + |\overline{OB}|^2$ is equivalent to

$$m_1^2 - 2m_1m_2 + m_2^2 = 1 + m_1^2 + 1 + m_2^2.$$

The last equation simplifies to $m_1 m_2 = -1$, and the result is established.

Example 5 Let L be the line that contains the point $(-1, 2)$ and is perpendicular to the line L_1 whose equation is $3x - 2y + 5 = 0$ (Figure 15). Find an equation of L in **(a)** point-slope form, **(b)** slope-intercept form, and **(c)** general form.

Solution We obtain the slope m_1 of L_1 by solving the equation $3x - 2y + 5 = 0$ for y in terms of x. The result is $y = \frac{3}{2}x + \frac{5}{2}$, so the slope of L_1 is $m_1 = \frac{3}{2}$. By the perpendicularity condition, the slope m of L is

Figure 15

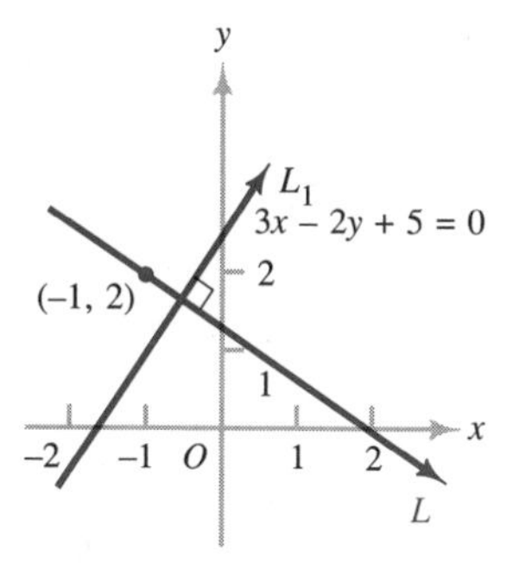

$$m = -\frac{1}{m_1} = -\frac{1}{\frac{3}{2}} = -\frac{2}{3}.$$

(a) Since L has slope $m = -\frac{2}{3}$ and contains the point $(-1, 2)$, its equation in point-slope form is

$$y - 2 = -\tfrac{2}{3}[x - (-1)]$$

or

$$y - 2 = -\tfrac{2}{3}(x + 1).$$

(b) Solving the equation in part (a) for y in terms of x, we obtain the equation of L in slope-intercept form:

$$y = -\tfrac{2}{3}x + \tfrac{4}{3}.$$

(c) Multiplying both sides of the equation in part (b) by 3 and rearranging terms, we obtain an equation of L in general form:

$$2x + 3y - 4 = 0.$$

■

Our work in Section 1.3 suggests the following natural extension of the idea of direct variation.

Definition 1 **Linear Variation**

If m and b are constants, and if x and y are variables related by an equation of the form

$$y = mx + b,$$

then we say that **y varies linearly as x** (or *with* x).

Example 6 One of the forms of depreciation for income tax purposes is the **straight-line method** in which the value V dollars of a property is assumed to vary linearly with the time t in years. An office complex originally purchased by an investment group for \$2.5 million is being depreciated by this method over a period of 20 years. At the end of this period, it is estimated that the complex will be worth \$800,000. Find out by how much the value of the complex will be depreciated each year, and find an equation relating the value V dollars of the complex to the time t years since it was purchased.

Solution The equation will have the form

$$V = mt + b,$$

where m and b are constants that have to be determined. Since $V =$ \$2,500,000 when $t = 0$ years, and $V =$ \$800,000 when $t = 20$ years, the points $(0, 2.5 \times 10^6)$ and $(20, 8 \times 10^5)$ belong to the graph of $V = mt + b$. Hence, the slope m is given by

$$m = \frac{800{,}000 - 2{,}500{,}000}{20 - 0} = -85{,}000 \text{ dollars per year.}$$

In other words, the value of the complex will depreciate \$85,000 each year over the 20-year period. Since $V = 2{,}500{,}000$ dollars when $t = 0$ years, we have $b = 2{,}500{,}000$; so, for $0 \le t \le 20$,

$$V = -85{,}000t + 2{,}500{,}000.$$

■

Problem Set 3.2

In Problems 1 and 2, sketch each line segment $\overline{AB}$ and find its slope using the slope formula.

1. **(a)** $A = (-1, 8)$, $B = (4, 3)$

(b) $A = (-2, 2)$, $B = (2, -2)$

(c) $A = (4, 0)$, $B = (0, -1)$

(d) $A = (-3, -\frac{1}{2})$, $B = (1, \frac{3}{2})$

C **2.** **(a)** $A = (2.6, -5.3)$, $B = (1.7, -1.1)$

(b) $A = (73.24, 31.53)$, $B = (1.71, 2.4)$

3. In each case, sketch the line L that contains the point P and has slope m.

(a) $P = (1, 1)$, $m = 2$

(b) $P = (-3, 2)$, $m = -\frac{2}{5}$

(c) $P = (-\frac{2}{3}, -3)$, $m = -2$

(d) $P = (-\frac{4}{3}, \frac{1}{2})$, $m = -\frac{3}{4}$

4. Show that the slope formula holds in all cases—even if point B is not above and to the right of point A.

5. Determine k so that the line containing the points $(k, 3)$ and $(-2, 1)$ is parallel to the line containing the points $(5, -2)$ and $(1, 4)$.

6. Use slopes to show that the quadrilateral with vertices $(-5, -2)$, $(1, -1)$, $(4, 4)$, and $(-2, 3)$ is a parallelogram. [*Hint:* Show that opposite sides have the same slope.]

7. Three points A, B, and C are said to be **collinear** if they lie on the same straight line. In each case, determine whether or not the three points are collinear by checking to see if the slope of line segment $\overline{AB}$ is the same as the slope of line segment $\overline{BC}$.

(a) $A = (0, 2)$, $B = (1, 1)$, $C = (5, -3)$

(b) $A = (7, -3)$, $B = (2, 3)$, $C = (-1, 5)$

(c) $A = (2, 1)$, $B = (3, 3)$, $C = (-5, -13)$

8. Let $A = (a_1, a_2)$, $B = (b_1, b_2)$, $C = (c_1, c_2)$, and $D = (d_1, d_2)$ be four points. Let P, Q, R, and S be the midpoints of the line segments $\overline{AB}$, $\overline{BC}$, $\overline{CD}$, and $\overline{DA}$, respectively. Prove that the line segments $\overline{PQ}$ and $\overline{RS}$ are parallel by showing that they both have the same slope, $m = (c_2 - a_2)/(c_1 - a_1)$.

9. In each case, sketch the line L that contains the point P and is parallel to the line segment $\overline{AB}$.

(a) $P = (4, -3)$, $A = (-2, 3)$, $B = (3, -7)$

(b) $P = (\frac{2}{3}, \frac{5}{3})$, $A = (\frac{1}{5}, \frac{3}{5})$, $B = (-\frac{2}{5}, \frac{4}{5})$

10. If $a > 0$ and $a + h > 0$, show that the slope of the line segment joining the points $(a, \sqrt{a})$ and $(a + h, \sqrt{a + h})$ is given by $m = \dfrac{1}{\sqrt{a + h} + \sqrt{a}}$.

11. If $a \neq 0$ and $a + h \neq 0$, show that the slope of the line segment joining the points $\left(a, \dfrac{1}{a}\right)$ and $\left(a + h, \dfrac{1}{a + h}\right)$ is given by $m = \dfrac{-1}{a(a + h)}$.

12. Show that two distinct lines with the same slope are necessarily parallel. [*Hint:* If they weren't parallel, they would meet at some point P.]

13. In each case, find an equation in point-slope form of the given line L.

    (a) L contains the point (3, 2) and has slope $m = \frac{3}{4}$.

    (b) L contains the points $(-3, 2)$ and $(4, 1)$.

    (c) L contains the point $(-3, 5)$ and is parallel to the line segment $\overline{AB}$, where $A = (3, 7)$ and $B = (-2, 2)$.

14. Let L be the line that contains the point (x_0, y_0) and has slope m. Suppose that (x, y) is a point such that $y - y_0 = m(x - x_0)$. Show that (x, y) lies on the line L.

In Problems 15 to 20, rewrite each equation in slope-intercept form, find the slope m and the y intercept b, and sketch the graph.

15. $3x - 2y = 6$

16. $5x - 2y - 10 = 0$

17. $y - 3x - 1 = 0$

18. $y + 1 = 0$

19. $x = -3y + 9$

20. $2x + y + 3 = 0$

In Problems 21 to 35, find an equation of the line L in **(a)** point-slope form, **(b)** slope-intercept form, and **(c)** general form.

21. L contains the point $(-5, 2)$ and has slope $m = 4$.

22. L contains the point $(3, -1)$ and has slope $m = 0$.

23. L has slope $m = -3$, and y intercept $b = 5$.

24. L has slope $m = \frac{4}{5}$ and intersects the x axis at $(-3, 0)$.

25. L intersects the x and y axes at (3, 0) and (0, 5).

26. L contains the points $(\frac{7}{2}, \frac{5}{3})$ and $(\frac{2}{5}, -6)$.

27. L contains the point $(3, -4)$ and is parallel to the x axis.

28. L contains the point $(3, -4)$ and is perpendicular to the x axis. [*Note:* In this case, there is neither a point-slope nor a slope-intercept form.]

29. L contains the point $(4, -4)$ and is parallel to the line that has the equation $2x - 5y + 3 = 0$.

30. L contains the point $(-3, \frac{2}{3})$ and is perpendicular to the y axis.

31. L contains the point $(-3, \frac{2}{3})$ and is perpendicular to the line that has the equation $5x + 3y - 1 = 0$.

32. L contains the point $(-6, -8)$ and has y intercept $b = 0$.

33. L contains the point $(\frac{2}{3}, \frac{5}{7})$ and is parallel to the line that has the equation $7x + 3y - 12 = 0$.

34. L is the line that bisects the right angle formed by the positive x and y axes.

35. L is the perpendicular bisector of the line segment $\overline{AB}$, where $A = (3, -2)$ and $B = (7, 6)$.

36. Suppose that the line L intersects the axes at $(a, 0)$ and $(0, b)$. Show that the equation of L can be written in **intercept form**

$$\frac{x}{a} + \frac{y}{b} = 1.$$

37. In each case, find an equation of the line L that contains the point P and is perpendicular to the line segment $\overline{AB}$ and sketch both L and $\overline{AB}$ on the same coordinate system.

    (a) $P = (1, 2)$, $A = (-7, -3)$, $B = (-5, 0)$

    (b) $P = (\frac{3}{2}, \frac{5}{2})$, $A = (4, -\frac{1}{3})$, $B = (\frac{1}{2}, 6)$

38. Consider the quadrilateral $ABCD$ with $A = (3, 1)$, $B = (2, 4)$, $C = (7, 6)$, and $D = (8, 3)$.

    (a) Use the concept of slope to determine whether or not the diagonals $\overline{AC}$ and $\overline{BD}$ are perpendicular.

    (b) Is the quadrilateral $ABCD$ a parallelogram? A rectangle? A square? A rhombus? Justify your answer.

39. Determine d so that the line containing the points $(d, 3)$ and $(-2, 1)$ is perpendicular to the line containing the points $(5, -2)$ and $(1, 4)$.

40. Find an equation of the tangent line to the circle with center at (0, 0) and radius 5 if the point of tangency is $(3, -4)$.

41. A car-rental company leases automobiles for a charge of \$22 per day plus \$0.20 per mile. Write an equation for the cost y dollars in terms of the distance x miles driven if the car is leased for N days. If $N = 3$, sketch a graph of the equation.

42. In 1989 the Solar Electric Company showed a profit of \$3.45 per share, and it expects this figure to increase annually by \$0.25 per share. If the year 1989 corresponds to $x = 0$, and successive years correspond to $x = 1, 2, 3$, and so on, find the equation $y = mx + b$ of the line that allows the company to predict its profit y dollars per share in future years. Sketch the graph

of this equation and find the predicted profit per share in 1997.

43. In 1990, tests showed that water in a lake was polluted with 7 milligrams of mercury compounds per 1000 liters of water. Cleaning up the lake became an immediate priority, and environmentalists determined that the pollution level would drop at the rate 0.75 milligram of mercury compounds per 1000 liters of water per year if all of their recommendations were followed. If 1990 corresponds to $x = 0$ and successive years correspond to $x = 1, 2, 3$, and so on, find the equation $y = mx + b$ of the line that allows the environmentalists to predict the pollution level y in future years if their recommendations are followed. Sketch the graph of the equation and determine when the lake will be free of mercury pollution according to this graph.

44. **(a)** The frequency n hertz of a musical tone is directly proportional to the speed v of sound in air and inversely proportional to the wavelength l of the sound wave. The speed of sound in air varies linearly as the temperature T in degrees Celsius. Write a formula for the frequency n in terms of l, T, and two constants.

C **(b)** Determine the numerical value of the two constants in part (a), given the following data: $n = 440$ hertz when $T = 0°C$ and $l = 0.75$ meter, and $n = 880$ hertz when $T = 20°C$ and $l = 0.39$ meter.

45. **(a)** Fahrenheit temperature F varies linearly as Celsius temperature C. Find a formula for F in terms of C if $F = 212°$ when $C = 100°$, and $F = 32°$ when $C = 0°$.

(b) To convert from degrees Celsius to absolute temperature K in kelvins, we add 273.15 to C. Find a formula for F in terms of K.

46. If a property is depreciated over a period of n years by the straight-line method, write a formula for its value V dollars t years after it was purchased at a cost of C dollars, assuming that its value at the end of the n-year period is S dollars.

47. A business firm is depreciating the value of a computer system originally purchased for \$20,000 over a period of 10 years by the straight-line method. If the system will have a salvage value of \$2000 at the end of the 10-year period, find an equation relating the time t years since the system was purchased to the value V dollars of the system at time t. Also, determine by how many dollars the system depreciates in one year.

48. In exercise physiology, a mathematical model used to relate a jogger's heart rate N beats per minute and the jogger's speed V feet per second states that N varies linearly as V. If a jogger's heart rate is 75 beats per minute at a speed of 10 feet per second, and 80 beats per minute at a speed of 12 feet per second, **(a)** find an equation expressing N in terms of V, and **(b)** find V if $N = 90$ beats per minute.

49. A communication satellite handled 1.3 million messages during its first year of operation, and 9.7 million messages during its eighth year of operation. Assuming that the number N million of messages handled during

the tth year of operation varies linearly as t, predict the number of messages that will be handled during the ninth year of operation.

50. A mathematical model used to relate the weight W metric tons of an adult whale shark to its length l meters states that W varies linearly as l.* A 12-meter whale shark weighs 11.8 metric tons, and an 18-meter whale shark weighs 17.7 metric tons.

(a) Write an equation relating W to l according to this model.

(b) Use the equation in part (a) to predict the weight of a whale shark of length 19.5 meters.

* This linear model is probably rather crude. For a discussion of such matters, see T. McMahon, "Size and Shape in Biology," Science, Vol. 179, 1973.

3.3 Functions

The idea of a *function*, which we introduce in this section, is nicely illustrated by a calculator or computer. Such a device can accept a number x as *input*, process the input according to definite instructions, and return a unique *output* y. The fact that the input x determines the output y is symbolized by the **mapping notation**

$$x \longmapsto y$$

indicating that each value of x is "transformed into" or "mapped into" a corresponding value of y.

For instance, if you enter a number x into a calculator and touch the $\sqrt{x}$ key, you obtain a vivid demonstration of the mapping

$$x \longmapsto \sqrt{x}$$

as the display changes from the input x to the output $\sqrt{x}$. For example:

Enter	Press $\sqrt{x}$ Key	Display Shows
4	$\longmapsto$	2
25	$\longmapsto$	5
2	$\longmapsto$	$1.414213562 \approx \sqrt{2}$

Another name for a mapping is a *function*.

Definition 1 **Function***

A **function** is a rule, correspondence, or mapping

$$x \longmapsto y$$

that assigns to each real number x (**the input**) in a certain set D one and only one real number y (**the output**).

In Definition 1, the set D is called the **domain** of the function, and y is called the **dependent variable,** since its value depends on the value of x. Because x can be assigned any value in the domain D, we refer to x as the **independent variable.** The set of values assumed by y as x runs through all values in D is called the **range** of the function.

Most calculators have special keys for some of the more important functions such as $x \longmapsto \sqrt{x}$ and $x \longmapsto x^2$.

Programmable calculators and microcomputers also have "user-definable" keys that can be programmed for whatever function $x \longmapsto y$ may be required. The program for the required function is the actual rule whereby the output y is to be calculated from the input x. Each user-definable key is marked with a letter of the alphabet or other symbol, so that, after the key has been programmed for a particular function, the letter or symbol can be used as the "name" of the function.

The use of letters of the alphabet to designate functions is not restricted exclusively to calculating machines. Although any letters of the alphabet can be used to designate functions, the letters f, g, and h as well as F, G, and H are most common. (Letters of the Greek alphabet are also used.) For instance, if we wish to designate the square-root function $x \longmapsto \sqrt{x}$ by the letter f, we write

$$x \overset{f}{\longmapsto} \sqrt{x} \quad \text{or} \quad f\colon x \longmapsto \sqrt{x}.$$

If $f\colon x \longmapsto y$ is a function, it is customary to write the value of y that corresponds to x as $f(x)$, read "f of x." In other words, $f(x)$ is the output produced when the function f is applied to the input x.

For instance, if $f\colon x \longmapsto \sqrt{x}$ is the square-root function, then

$$f(4) = \sqrt{4} = 2,$$

$$f(25) = \sqrt{25} = 5,$$

$$f(2) = \sqrt{2} \approx 1.414,$$

and, in general, for any nonnegative value of x,

$$f(x) = \sqrt{x}.$$

Note carefully that $f(x)$ does *not* mean f times x.

* It is also possible to define a complex-valued function of a complex variable or, more generally, a function from any set D into any set R.

If $f: x \longmapsto y$, then, for every value of x in the domain of f, we have

$$y = f(x),$$

an equation relating the dependent variable y to the independent variable x. Conversely, if an equation of the form

$$y = \text{an expression involving } x$$

determines a function $f: x \longmapsto y$, we say that the function f is **defined by** or **given by** the equation. For instance, the equation

$$y = 3x^2 - 1$$

defines a function $f: x \longmapsto y$, so that

$$y = f(x) = 3x^2 - 1$$

or simply

$$f(x) = 3x^2 - 1.$$

In defining a function $f: x \longmapsto y$ by an equation, you must pay attention to the requirement in Definition 1 that each value of x in the domain determines *one and only one* corresponding value of y. For instance,

$$y = \pm\sqrt{x}$$

does *not* define a function.

When a function f is defined by an equation, you can determine, by substitution, the output $f(a)$ corresponding to a particular input value $x = a$. For instance, if f is defined by

$$f(x) = 3x^2 - 1,$$

then

$$f(2) = 3(2)^2 - 1 = 11,$$

$$f(0) = 3(0)^2 - 1 = -1,$$

$$f(-1) = 3(-1)^2 - 1 = 2,$$

and

$$f(t + 1) = 3(t + 1)^2 - 1 = 3(t^2 + 2t + 1) - 1$$
$$= 3t^2 + 6t + 2.$$

Example 1 Let g be the function defined by $g(x) = 5x^2 + 3x$. Find the indicated values.

(a) $g(1)$ **(b)** $g(-2)$ **(c)** $g(t^3)$
(d) $[g(-1)]^2$ **(e)** $g(t + h)$ **(f)** $g(-x)$

Solution

(a) $g(1) = 5(1)^2 + 3(1) = 5 + 3 = 8$
(b) $g(-2) = 5(-2)^2 + 3(-2) = 20 - 6 = 14$
(c) $g(t^3) = 5(t^3)^2 + 3(t^3) = 5t^6 + 3t^3$
(d) $[g(-1)]^2 = [5(-1)^2 + 3(-1)]^2 = (5 - 3)^2 = 2^2 = 4$

(e) $g(t+h) = 5(t+h)^2 + 3(t+h) = 5t^2 + 10th + 5h^2 + 3t + 3h$

(f) $g(-x) = 5(-x)^2 + 3(-x) = 5x^2 - 3x$ ■

Many problems in applied mathematics, and especially in calculus, involve the *change* $f(b) - f(a)$ in the function value $f(x)$ caused by changing the independent variable from $x = a$ to $x = b$. For instance, if $y = f(x)$ expresses the profit y of a company as a function of the production level x, then $f(b) - f(a)$ is the change in profit caused by changing the production level from $x = a$ to $x = b$. Expressions such as $f(b) - f(a)$ arise so frequently that the special notation $f(x)\big|_{x=a}^{x=b}$, or simply $f(x)\big|_a^b$, is often used for them. Thus, by definition, if a and b are in the domain of the function f, we write

$$f(x)\Big|_a^b = f(b) - f(a).$$

The symbolism $f(x)\big|_a^b$ is called the **evaluation bar** notation.

Example 2 Evaluate $\frac{1}{10}(1 + 2x^2)^{3/2}\big|_0^2$.

Solution

$$\begin{aligned}\tfrac{1}{10}(1 + 2x^2)^{3/2}\Big|_0^2 &= \tfrac{1}{10}(1 + 8)^{3/2} - \tfrac{1}{10}(1 + 0)^{3/2} \\ &= \tfrac{1}{10}(9^{3/2}) - \tfrac{1}{10}(1^{3/2}) \\ &= \tfrac{1}{10}(27 - 1) = \tfrac{13}{5}\end{aligned}$$ ■

Whenever a function $f: x \longmapsto y$ is defined by an equation, you may assume (unless we say otherwise) that *its domain is the set of all values of x for which the equation makes sense and determines a corresponding real number y.* The range of the function is then automatically determined, since it consists of the set of all values of y that correspond, by the equation that defines the function, to values of x in the domain. In Section 3.4, we give a useful graphical method for determining the range of a function.

Example 3 Find the domain of the function defined by each equation.

(a) $h(x) = \dfrac{1}{x - 1}$

(b) $G(x) = \sqrt{4 - x}$

(c) $F(x) = 3x - 5$

Solution

(a) The domain of h is the set of all real numbers except 1; that is, it is the set $(-\infty, 1) \cup (1, \infty)$.

(b) The expression $\sqrt{4 - x}$ represents a real number if and only if $4 - x \geq 0$; that is, if and only if $x \leq 4$. Therefore, the domain of G is the interval $(-\infty, 4]$.

(c) Since the expression $3x - 5$ is defined for all real values of x, the domain of F is the set $\mathbb{R}$ of all real numbers. ■

Functions that arise in applied mathematics may have restrictions imposed on their domains by physical or geometrical circumstances. For instance, the function $x \longmapsto \pi x^2$ that expresses the correspondence between the radius x and the area πx^2 of a circle would have its domain restricted to the interval $(0, \infty)$, since a circle must have a positive radius.

For a function f, the expression

$$\frac{f(x+h)-f(x)}{h}, \qquad h \neq 0,$$

called the **difference quotient,** plays an important role in calculus.

Example 4 Find the difference quotient for the function f defined by $f(x) = \sqrt{x}$ and simplify the result.

Solution Assuming that $h \neq 0$, $x \geq 0$, and $x + h \geq 0$, we have

$$\frac{f(x+h)-f(x)}{h} = \frac{\sqrt{x+h}-\sqrt{x}}{h}.$$

In calculus, this is "simplified" by rationalizing the numerator, so that

$$\begin{aligned}\frac{f(x+h)-f(x)}{h} &= \frac{(\sqrt{x+h}-\sqrt{x})(\sqrt{x+h}+\sqrt{x})}{h(\sqrt{x+h}+\sqrt{x})}\\ &= \frac{(x+h)-x}{h(\sqrt{x+h}+\sqrt{x})} = \frac{h}{h(\sqrt{x+h}+\sqrt{x})}\\ &= \frac{1}{\sqrt{x+h}+\sqrt{x}}.\end{aligned}$$

■

In dealing with a function f, it is important to distinguish among the *function itself*

$$f: x \longmapsto y,$$

which is a rule or correspondence; the *output*

$$y \qquad \text{or} \qquad f(x),$$

which is a number depending on the input x; and the *equation*

$$y = f(x),$$

which relates the dependent variable y to the independent variable x. Nevertheless, people tend to take shortcuts and speak, incorrectly, of "the function $f(x)$" or "the function $y = f(x)$." Similarly, in applied mathematics, people often say that "y is a function of x," for instance, "current is a function of voltage." Although we avoid these practices when absolute precision is required, we indulge in them whenever it seems convenient and harmless.

The particular letters used to denote the dependent and independent variables are of no importance in themselves—the important thing is the rule by which a definite value of the dependent variable is assigned to each value of the independent variable. In applied work, variables other than x and y are often used because physical and geometrical quantities are designated by conventional symbols.

Example 5 A wafer cone is to be filled to the bottom with frozen yogurt and topped off with a frozen yogurt hemisphere (Figure 1). If the height of the cone is 12 centimeters, express the total volume V cubic centimeters of frozen yogurt as a function of the radius r centimeters of the hemisphere.

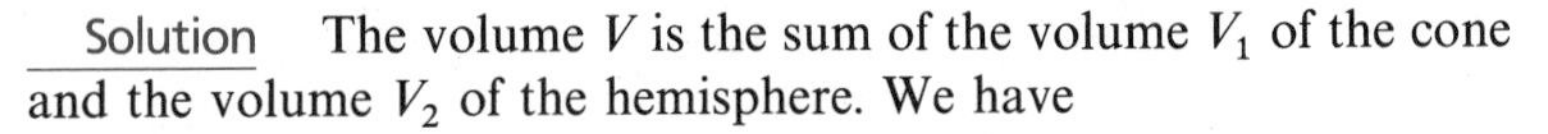

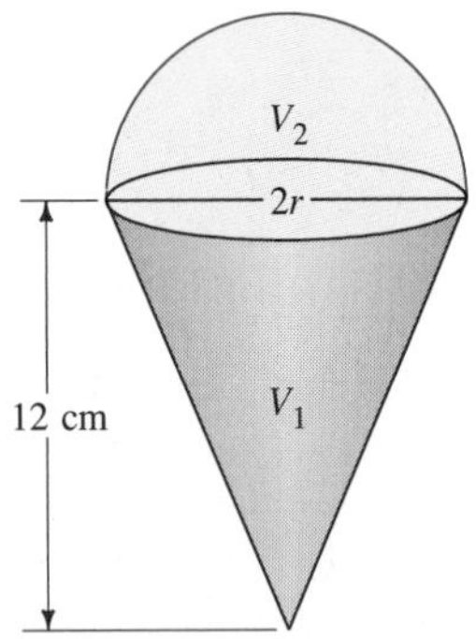

Solution The volume V is the sum of the volume V_1 of the cone and the volume V_2 of the hemisphere. We have

Figure 1

$$V_1 = \tfrac{1}{3}\pi r^2(12) = 4\pi r^2$$

and

$$V_2 = \tfrac{1}{2}\left(\tfrac{4}{3}\pi r^3\right) = \tfrac{2}{3}\pi r^3.$$

Thus, if we denote by, say, v the function that gives V in terms of r, we have

$$V = v(r) = 4\pi r^2 + \tfrac{2}{3}\pi r^3,$$

where the domain of v is the interval $(0, \infty)$. ■

The Graph of a Function

The **graph** of a function f is defined to be the graph of the corresponding equation $y = f(x)$. In other words, the graph of f is the set of all points (x, y) in the Cartesian plane such that x is in the domain of f and $y = f(x)$.

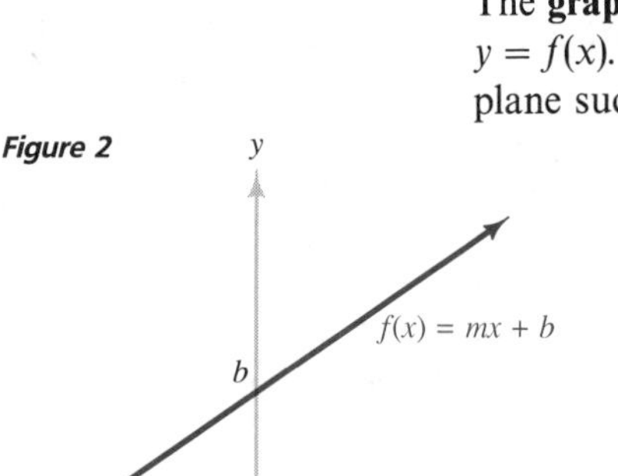

Figure 2

For instance, if m and b are constants, then the graph of the function

$$f(x) = mx + b$$

is the same as the graph of the equation

$$y = mx + b,$$

a line with slope m and y intercept b (Figure 2). For this reason, a function of the form $f(x) = mx + b$ is called a **linear function.**

Example 6 A mathematical model for the growth of a human fetus proposes that the length L centimeters of the fetus is a linear function of the time t weeks after conception,

$$L = l(t) \qquad \text{for } 9 \le t \le 38.$$

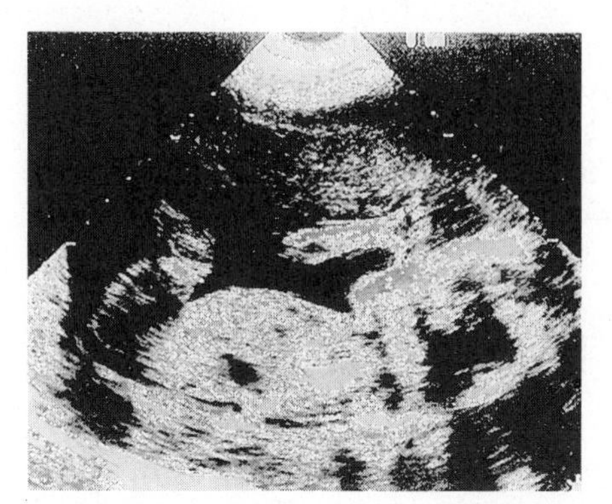
Ultrasound of human fetus at 17 weeks

Measurements by ultrasound show that $L = 7$ centimeters when $t = 9$ weeks and $L = 13$ centimeters when $t = 17$ weeks.

(a) Draw a graph of the function l.

(b) Predict the length of the baby when it is born at $t = 38$ weeks.

Solution The function l will be given by an equation of the form

$$L = l(t) = mt + b.$$

To find m, we use the fact that (9, 7) and (17, 13) are two points on the graph, so that

$$m = \frac{13 - 7}{17 - 9} = \frac{3}{4}.$$

Substituting $m = \frac{3}{4}$, $t = 9$, and $L = 7$ into the equation $L = mt + b$ and solving for b, we find that

$$7 = (\tfrac{3}{4})9 + b,$$

$$b = 7 - \tfrac{27}{4} = \tfrac{1}{4}.$$

Thus, $$L = l(t) = \tfrac{3}{4}t + \tfrac{1}{4} \qquad \text{for } 9 \le t \le 38.$$

(a) A graph of the linear function l is shown in Figure 3.

(b) When $t = 38$, we have

$$L = l(38) = (\tfrac{3}{4})38 + \tfrac{1}{4} = 28.75 \text{ centimeters.}$$ ■

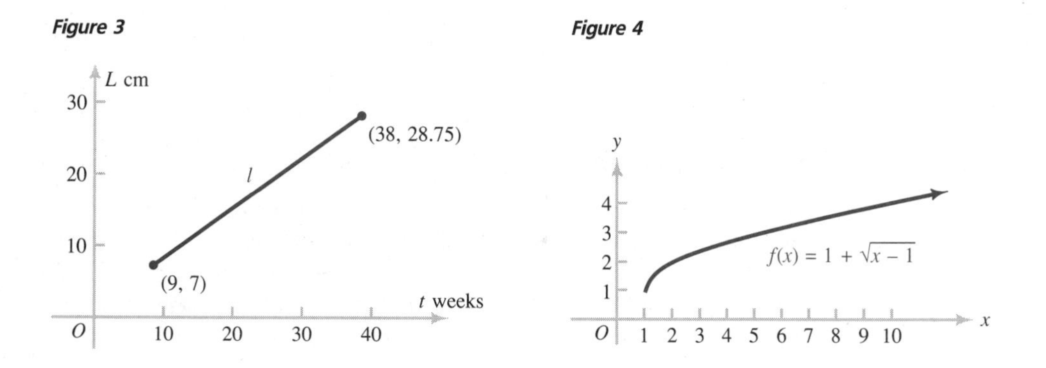

Graphs of functions that are not linear are often (but not always) smooth curves in the Cartesian plane. For instance, the graph of $f(x) = 1 + \sqrt{x - 1}$ is shown in Figure 4. Sketching such a graph "by hand" can be considerably more challenging than sketching a straight line, although the use of a calculator to determine several points on the graph will often give a good indication of its general shape.

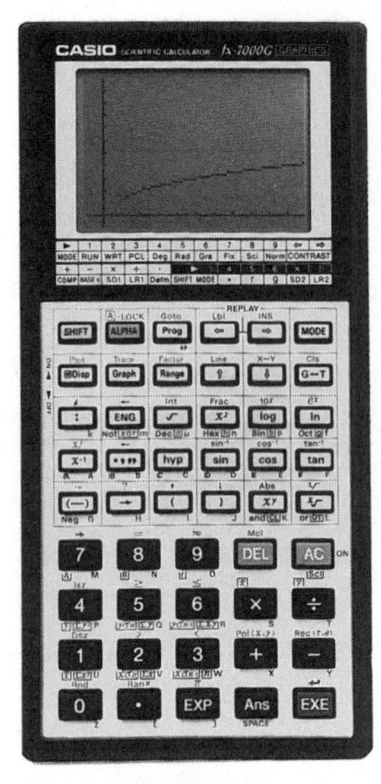

Casio fx-7000G showing the graph of Figure 4.

Today many students own or have access to *graphing calculators* that offer an almost effortless way to obtain graphs of general functions. Graphing calculators are available from manufacturers such as Casio, Hewlett-Packard, Sharp, and Texas Instruments, among others. Also, numerous *mathematical software* packages are available that enable microcomputers to generate graphs.* In this textbook, we use the symbol gc to mark examples and problems for which a graphing calculator (or a computer) may be used. In Sections 3.4 and 3.5, we consider graph sketching in more detail.

In scientific work, a graph showing the relationship between two variable quantities is often obtained by means of actual measurement. For instance, Figure 5 shows the blood pressure P (in millimeters of mercury) in an artery of a healthy person plotted against time t in seconds. Such a curve may be regarded as the graph of a function $P = f(t)$, even though it may not be clear how to write a "mathematical formula" giving P in terms of t.

Figure 5

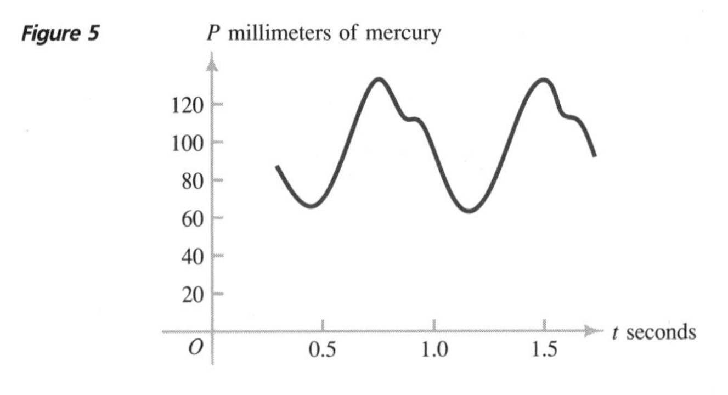

The Vertical-Line Test

It is important to realize that *not every curve in the Cartesian plane is the graph of a function.* Indeed, the definition of a function (Definition 1) requires that there be one and *only one* value of y corresponding to each value of x in the domain. Thus, on the graph of a function, *we cannot have two points* (x, y_1) *and* (x, y_2) *with the same abscissa* x *and different ordinates* y_1 *and* y_2. Hence, we have the following test.

The Vertical-Line Test

A set of points in the Cartesian plane is the graph of a function if and only if no vertical straight line intersects the set more than once.

* Among these are Derive, Maple, Mathematica, and Microcalc.

Example 7 Which of the curves in Figure 6 is the graph of a function?

Solution By the vertical-line test, the curve in Figure 6a is the graph of a function, but the curve in Figure 6b is not. ■

Figure 6

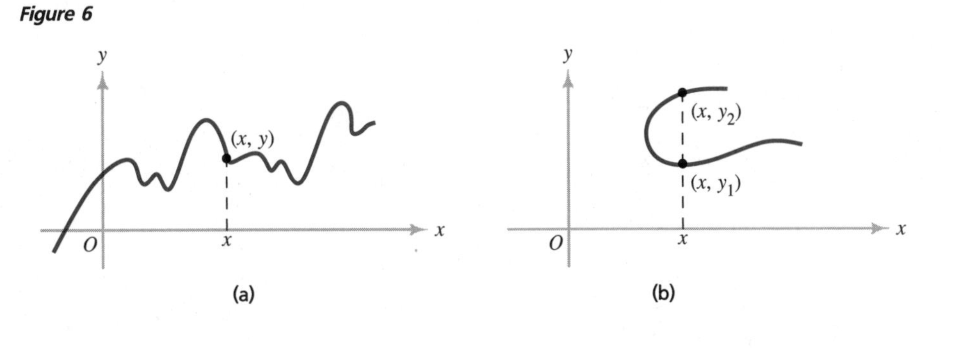

Problem Set 3.3

1. Let $f(x) = x^2 - 3x - 4$. Find:

   (a) $f(-2)$ **(b)** $f(0)$
   (c) $f(-b)$ **(d)** $f(a + b)$

2. Let $h(x) = \sqrt{3x + 5}$. Find:

   (a) $h(-\frac{1}{3})$ **(b)** $\sqrt{h(1)}$
   (c) $h(2k - 1)$ **(d)** $h(-p)$

3. Let $g(x) = \sqrt[3]{x^3 - 4}$. Find:

   (a) $g(\sqrt[3]{31})$ **(b)** $g(-5)$
   (c) $g(x + 1)$ **(d)** $g(-x)$

4. Let $F(t) = \dfrac{t - 2}{3t + 7}$. Find:

   (a) $F(\frac{7}{3})$ **(b)** $F(-\frac{1}{3})$
   (c) $F(-x)$ **(d)** $\dfrac{1}{F(x)}$

In Problems 5 and 6, evaluate each expression.

5. **(a)** $\sqrt{2x^2 + 1}\Big|_0^2$ **(b)** $\dfrac{x}{x + 1}\Big|_3^5$ **(c)** $|2 - 5x|\Big|_{-2}^2$

6. **(a)** $\sqrt{x + 1}\Big|_a^{a+h}$ **(b)** $\dfrac{1}{x}\Big|_a^{a+h}$ **(c)** $\dfrac{f(x)}{h}\Big|_x^{x+h}$

In Problems 7 and 8, find the domain of each function.

7. **(a)** $f(x) = \dfrac{1}{x}$ **(b)** $f(x) = 1 - 4x^2$
   (c) $h(x) = \sqrt{x}$ **(d)** $p(x) = \sqrt{9 - x^2}$
   (e) $F(x) = \dfrac{x^3 - 8}{x^2 - 4}$

8. **(a)** $g(x) = |x|$ **(b)** $g(x) = (x + 2)^{-1}$
   (c) $K(x) = (4 - 5x)^{-1/2}$ **(d)** $h(x) = \dfrac{1}{x + |x|}$
   (e) $h(x) = \sqrt{\dfrac{x - 2}{x - 4}}$

In Problems 9 to 14, find the difference quotient $\dfrac{f(x + h) - f(x)}{h}$ and simplify the result.

9. $f(x) = 4x - 1$ **10.** $f(x) = 5$

11. $f(x) = x^2 + 3$

12. $f(x) = mx + b, \quad m \neq 0$

13. $f(x) = 1/\sqrt{x}$

14. $f(x) = 1/x$

In Problems 15 to 20, sketch the graph of the function.

15. $f(x) = \frac{2}{3}x - 5$

16. $f(x) = 4$

17. $f(x) = x$

18. $f(x) = \sqrt{9 - x^2}$

19. $f(x) = -2x + 1$

20. $f(x) = -\sqrt{9 - x^2}$

21. Use the vertical-line test to determine which of the curves in Figure 7 are graphs of functions.

Figure 7

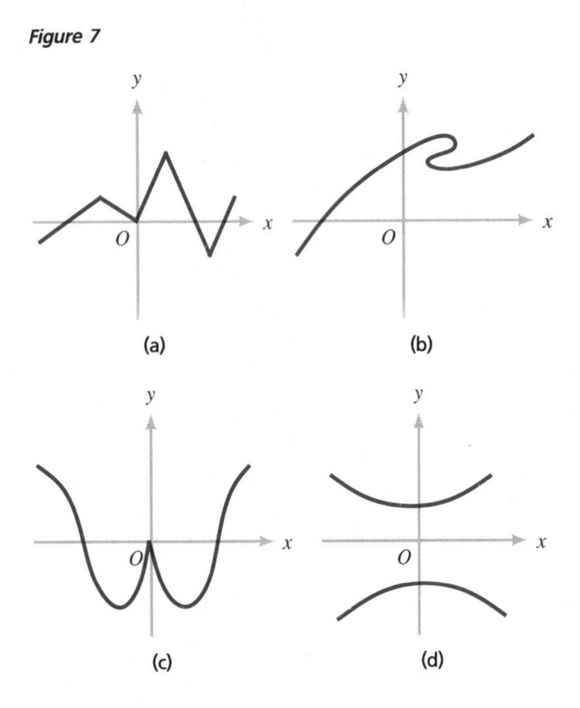

22. Is the graph of the equation $x^2 + y^2 = 9$ the graph of a function? Why or why not?

23. The function $f: C \longmapsto F$ given by the equation $F = \frac{9}{5}C + 32$ converts the temperature C in degrees Celsius to the corresponding temperature F in degrees Fahrenheit. Find $f(0)$, $f(15)$, $f(-10)$, and $f(55)$ and write the results using both function and mapping notation.

24. A small company finds that its profit depends on the amount of money it spends on advertising. If x dollars are spent on advertising, the corresponding profit $p(x)$ dollars is given by

$$p(x) = \frac{x^2}{100} - \frac{x}{50} + 100.$$

(a) Find $p(x)\big|_{50}^{100}$.

(b) Find $p(x)\big|_{100}^{300}$.

(c) Give an interpretation of the quantities found in parts (a) and (b).

25. Assuming that a, b, and c are in the domain of the function f, show that $f(x)\big|_a^c = f(x)\big|_a^b + f(x)\big|_b^c$.

26. Let $A(x)$ denote the area shown in Figure 8. Give a geometric interpretation of $A(x)\big|_2^3$.

Figure 8

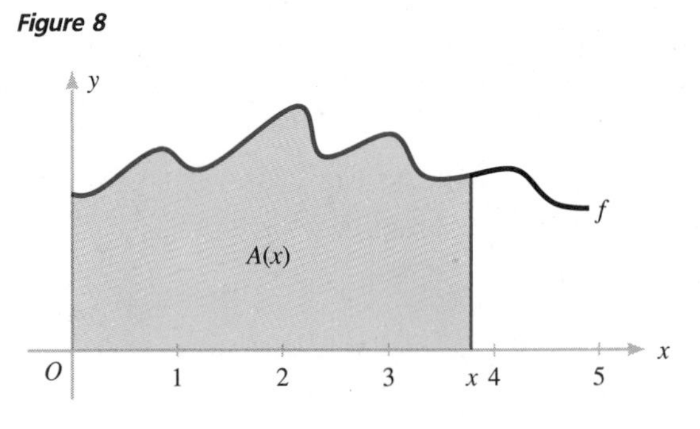

27. An airline chart shows that the temperature T in degrees Fahrenheit at an altitude of $h = 15{,}000$ feet is $T = 5°$. At an altitude of $h = 20{,}000$ feet, $T = -15°$. Supposing that T is a linear function of h, obtain an equation that defines this function, sketch its graph, and find the temperature at an altitude $h = 30{,}000$ feet.

C **28.** In physics, the (absolute) pressure P in newtons per square meter at a point h meters below the surface of a body of water is shown to be a linear function of h. When $h = 0$, $P = 1.013 \times 10^5$ newtons per square meter. When $h = 1$ meter, $P = 2.003 \times 10^5$ newtons per square meter. Obtain the equation that defines P as a function of h, and use it to find the pressure at a depth of $h = 100$ meters.

29. An outdoor track is to be constructed in the shape shown in Figure 9.

    (a) Express the total area A enclosed by the track as a function of x.

    (b) Express the perimeter P of the track as a function of x.

    (c) Express A as a function of P.

Figure 9

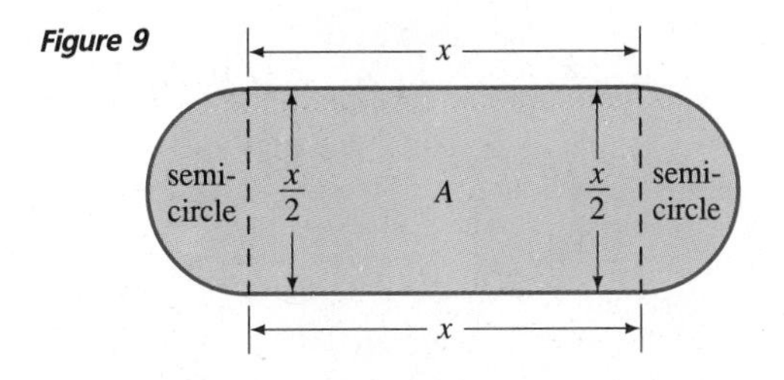

30. An ecologist investigating the effect of air pollution on plant life finds that the percentage, $p(x)$ percent, of diseased trees and shrubs at a distance of x kilometers from an industrial city is given by $p(x) = 32 - (3x/50)$ for $50 \leq x \leq 500$. Sketch a graph of the function p and find $p(50)$, $p(100)$, $p(200)$, $p(400)$, and $p(500)$.

C 31. Anthropologists often find only parts of a human skeleton, from which they must reconstruct the appearance of the original human. Suppose that the height h centimeters of a female human is a linear function of the length l centimeters of her humerus (Figure 10). Data from two skeletons, found intact, show that $h = 151$ centimeters when $l = 28$ centimeters and that $h = 162$ centimeters when $l = 31.5$ centimeters.

Figure 10

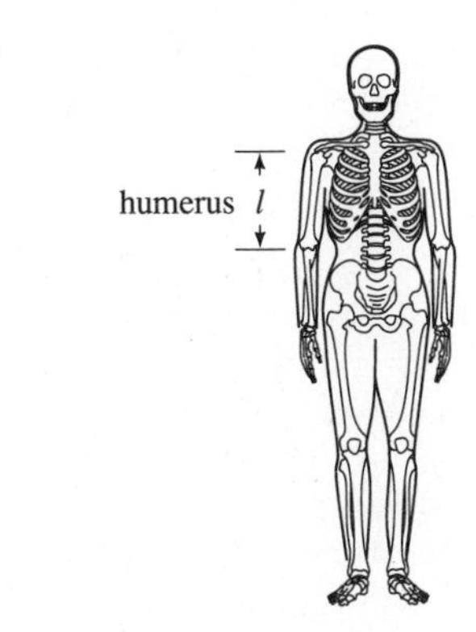

(a) Express h as a function of l.

(b) A female humerus of length $l = 33$ centimeters is discovered. Estimate the height h.

C 32. The manufacturers of modern scientifically designed walking shoes must conduct a detailed analysis of the mechanics of walking. One manufacturer uses the mathematical model $T(l) = 2\pi\sqrt{0.068l}$ for the period $T(l)$ seconds of the swing of a leg of length l meters (Figure 11). Find:

    (a) $T(0.75)$

    (b) $T(0.98)$

    (c) $T(l)\big|_{0.75}^{0.98}$.

    Round off your answers to two significant digits.

Figure 11

33. Small squares are cut from the four corners of a rectangular piece of cardboard that originally measures 10 inches by 14 inches, and an open box is then formed by folding up the flaps (Figure 12). If x inches is the length of the sides of the removed squares, express the volume $V(x)$ cubic inches of the box in terms of x.

Figure 12

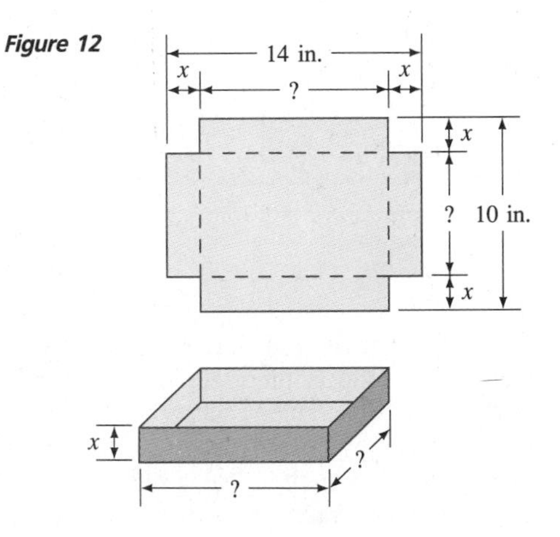

34. A first ship is steaming due north at 24 knots (nautical miles per hour), and a second ship is steaming due east at 30 knots. At 1:00 A.M., the second ship is 150 nautical miles due north of the first ship (Figure 13). Express the distance $d(t)$ nautical miles between the two ships as a function of the elapsed time t hours since 1:00 A.M.

Figure 13

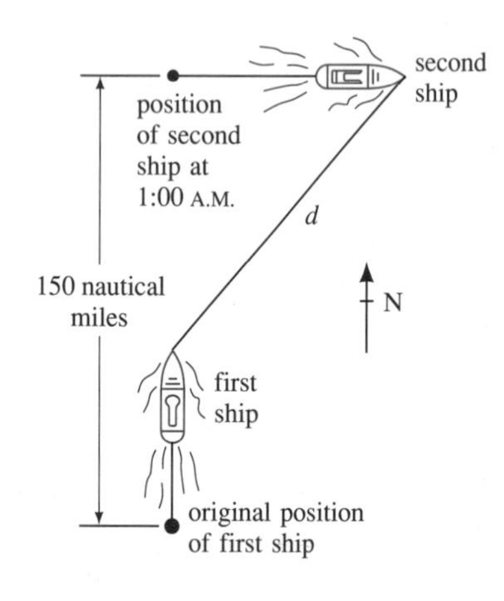

35. A rod-shaped bacterium has the form of a circular cylinder closed by two hemispherical caps (Figure 14). If the length l of such a bacterium is $l = 5.7 \times 10^{-6}$ meter, express the volume V cubic meters of the bacterium as a function of its radius r meters.

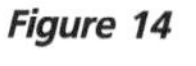

Figure 14

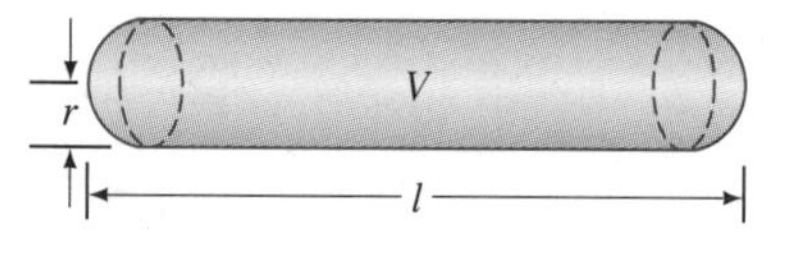

36. The reentry vehicles used during projects Mercury, Gemini, and Apollo in the 1960s were shaped like a frustum of a right circular cone, a figure formed by truncating a cone by a plane parallel to its base (Figure 15). If the base of a frustum has a fixed radius a and the top has a fixed radius b, find the volume $V(h)$ of the frustum as a function of its height h.

Figure 15

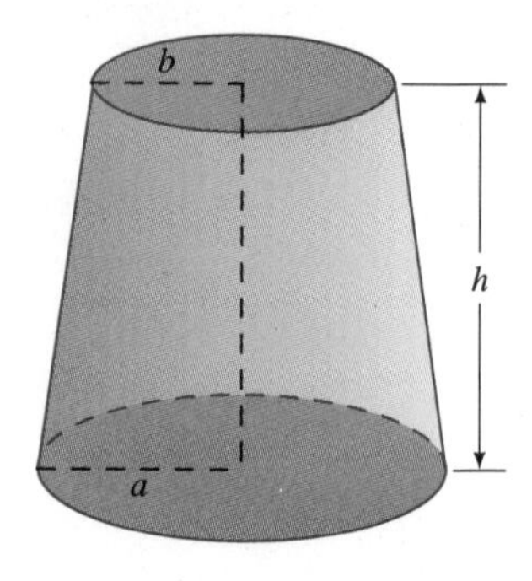

3.4 Graph Sketching and Properties of Graphs

In this section and the next, we consider methods for sketching graphs, properties of graphs, and graphs of some important functions. This material not only provides insight into the relationship between properties of functions and the shapes of their graphs, but it is also essential for proficient use of a graphing calculator or computer.

The basic graph-sketching procedure is as follows.

The Point-Plotting Method

To sketch the graph of $y = f(x)$, select several values of x in the domain of f, calculate the corresponding values of $f(x)$, plot the resulting points, and connect the points with a smooth curve. The more points you plot, the more accurate your sketch will be.

All graphing calculators operate on the basis of the point-plotting method. They plot a large number of points (actually pixels) that seem to blend together to form a curve.

Example 1 Use the point-plotting method to sketch the graph of $f(x) = x^2$, where the domain of f is restricted by the condition that $x > 0$.

Solution The domain of f is the interval $(0, \infty)$, so we begin by selecting several values of x in this interval and calculating the corresponding values of $f(x) = x^2$, as in the table in Figure 1. We then plot the points $(x, f(x))$ from the table and connect them with a smooth curve (Figure 1). Because the domain of f consists only of positive numbers, the point $(0, 0)$ is excluded from the graph. This excluded point is indicated by a small open circle. ■

If you have access to a graphing calculator, you may wish to use it to confirm the correctness of the graph in Figure 1.

Figure 1

x	$f(x) = x^2$
1	1
2	4
3	9
4	16

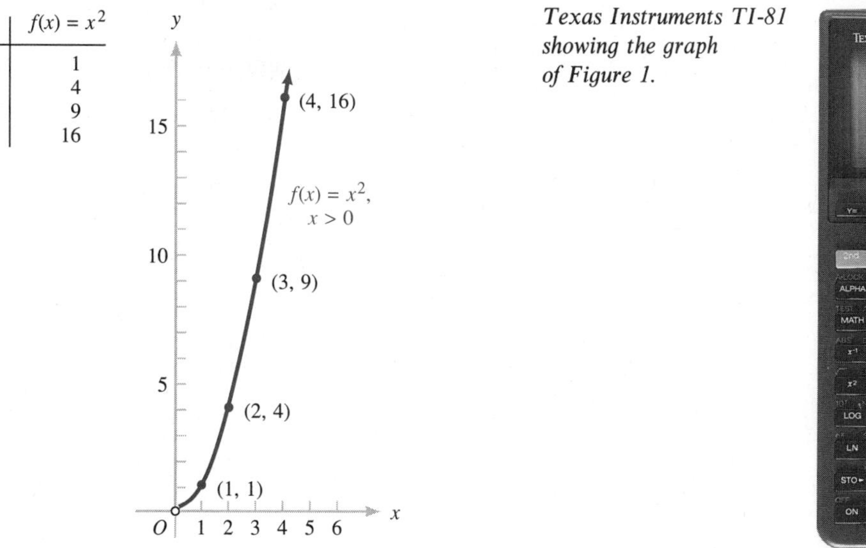

Texas Instruments TI-81 showing the graph of Figure 1.

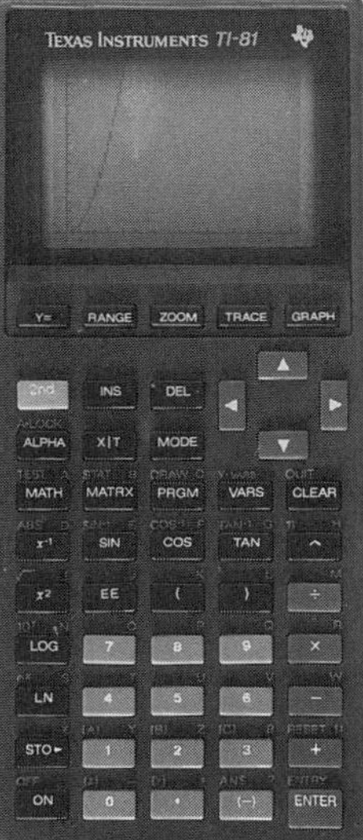

The point-plotting method requires us to *guess* the shape of the graph between or beyond known points, and *therefore must be used with caution* (see Problems 41 and 42). If the function is fairly simple, the point-plotting method usually works pretty well; however, more complicated functions may require the more advanced methods studied in calculus.

Geometric Properties of Graphs

The graph in Figure 2a is always *rising* as we move to the right, a geometric indication that the function f is **increasing;** that is, as x increases, so does the value of $f(x)$. On the other hand, the graph in Figure 2b is always *falling* as we move to the right, indicating that the function g is **decreasing;** that is, as x increases, the value of $g(x)$ decreases. In Figure 2c, the graph doesn't rise or fall, indicating that h is a **constant function** whose values $h(x)$ do not change as we increase x. The following definition makes these ideas more precise.

Figure 2

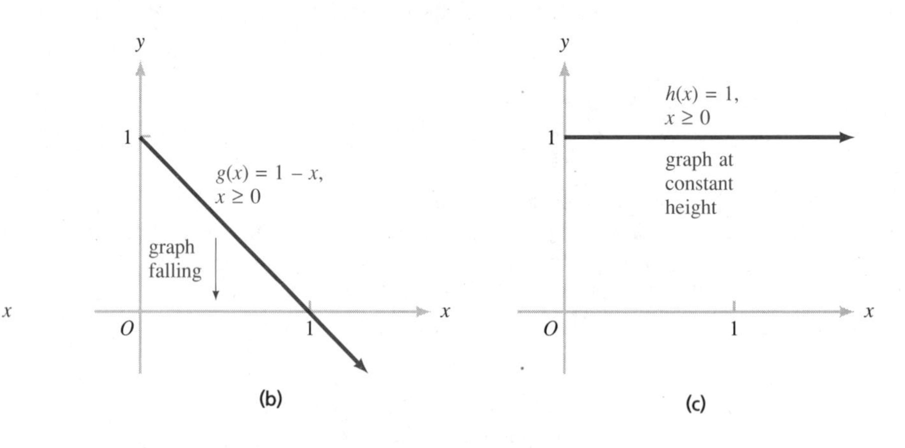

Figure 3

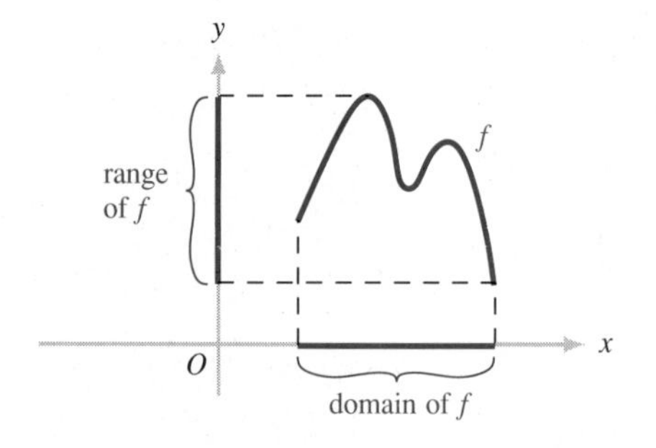

Definition 1 Increasing, Decreasing, and Constant Functions

Let the interval I be contained in the domain of the function f.

(i) f is **increasing** on I if for every two numbers a and b in I with $a < b$ we have $f(a) < f(b)$.

(ii) f is **decreasing** on I if for every two numbers a and b in I with $a < b$ we have $f(a) > f(b)$.

(iii) f is **constant** on I if for every two numbers a and b in I we have $f(a) = f(b)$.

As Figure 3 illustrates, the domain and range of a function are easily found from its graph.

The domain of a function is the set of all abscissas of points on its graph, and the range is the set of all ordinates of points on its graph.

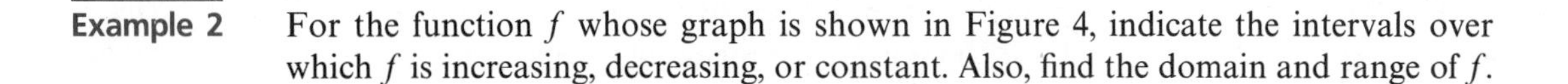

Example 2 For the function f whose graph is shown in Figure 4, indicate the intervals over which f is increasing, decreasing, or constant. Also, find the domain and range of f.

Figure 4

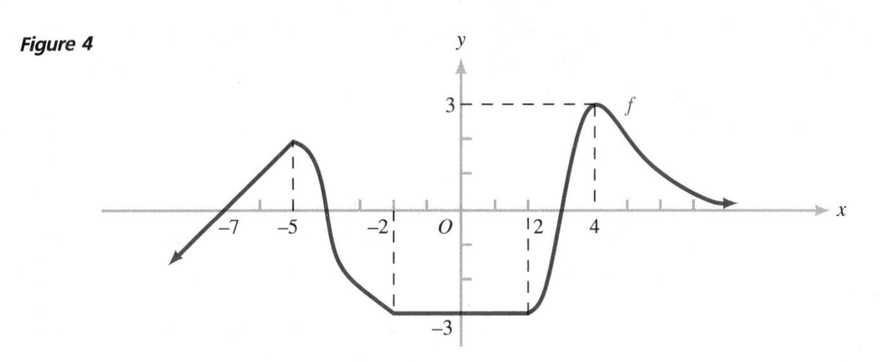

Solution As we move from left to right, the function f is increasing over intervals where the graph is rising, and decreasing over intervals where it is falling. Thus, assuming that the graph continues indefinitely to the left and right in the directions indicated by the arrowheads, we conclude that f is increasing on the intervals $(-\infty, -5]$ and $[2, 4]$ and that f is decreasing on the intervals $[-5, -2]$ and $[4, \infty)$. On the interval $[-2, 2]$, f is constant.

The graph "covers" the entire x axis, so the domain of f is the set $\mathbb{R}$ of all real numbers. We assume that the graph keeps dropping as we move to the left of -5 on the x axis, but that it never climbs any higher than $y = 3$; hence, the range of f is the interval $(-\infty, 3]$. ■

Consider the graphs in Figure 5. The graph of f (Figure 5a) is **symmetric about the y axis;** that is, the portion of the graph to the right of the y axis is the mirror image of the portion to the left of it. Specifically, if the point (x, y) belongs to the

Figure 5

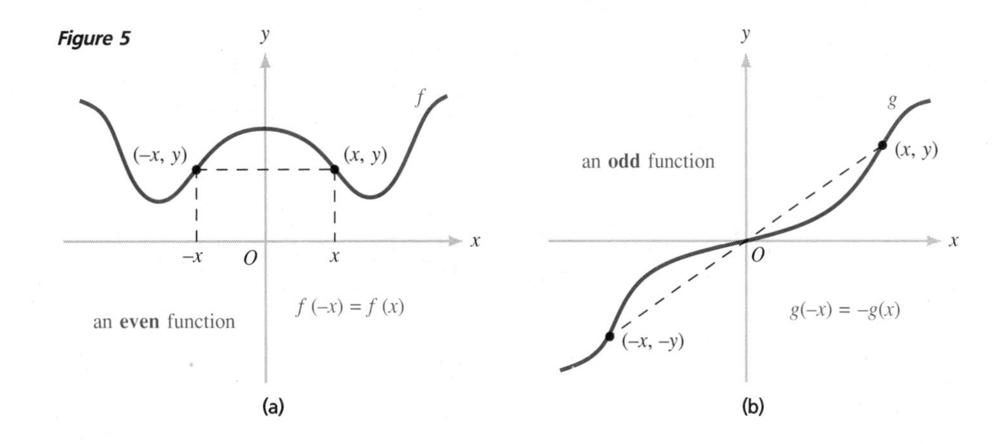

graph of f, then so does the point $(-x, y)$, so that $f(-x) = f(x)$. Similarly, the graph of g (Figure 5b) is **symmetric about the origin** because, if the point (x, y) belongs to the graph, then so does the point $(-x, -y)$; that is, $g(-x) = -g(x)$.

A function whose graph is symmetric about the y axis is called an **even function;** a function whose graph is symmetric about the origin is called an **odd function.** This is stated more formally in the following definition.

Definition 2 **Even and Odd Functions**

(i) A function f is said to be **even** if, for every number x in the domain of f, $-x$ is also in the domain of f and

$$f(-x) = f(x).$$

(ii) A function f is said to be **odd** if, for every number x in the domain of f, $-x$ is also in the domain of f and

$$f(-x) = -f(x).$$

Of course, there are many functions that are neither even nor odd.

Example 3 In each case, indicate the type of symmetry, if any, shown by the computer-generated graph in Figure 6 and confirm your answer by using Definition 2.

(a) $f(x) = x^4 - 3x^2$

(b) $g(x) = x^3 - 2x$

(c) $h(x) = x^4 - 3x$

Figure 6

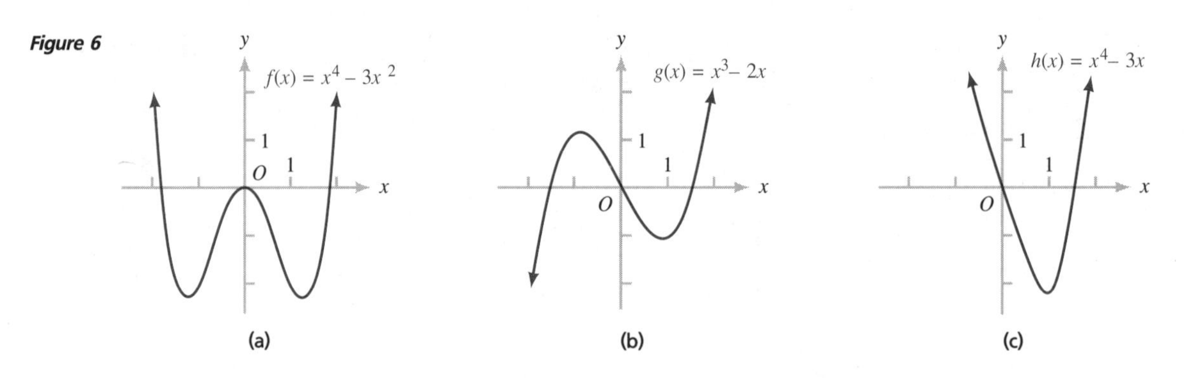

Solution **(a)** The graph appears to be symmetric about the y axis, suggesting that f is an even function. To confirm this, we must check to see if $f(-x) = f(x)$. We have

$$f(-x) = (-x)^4 - 3(-x)^2 = x^4 - 3x^2 = f(x),$$

so f is indeed an even function.

(b) The graph seems to be symmetric about the origin, suggesting that g is an odd function. Here we have

$$g(-x) = (-x)^3 - 2(-x) = -x^3 + 2x = -g(x),$$

showing that g really is an odd function.

(c) The graph is not symmetric about the y axis or about the origin. In this case, we have

$$h(-x) = (-x)^4 - 3(-x) = x^4 + 3x,$$

whereas

$$h(x) = x^4 - 3x \qquad \text{and} \qquad -h(x) = -x^4 + 3x.$$

Thus, both of the conditions $h(-x) = h(x)$ and $h(-x) = -h(x)$ fail, so h is neither even nor odd. ■

Graphs of Some Particular Functions

The following examples illustrate the ideas discussed above and exhibit the graphs of some important functions.

In Examples 4 to 8, find the domain of f, sketch each graph, discuss the symmetry of the graph, indicate the intervals where f is increasing or decreasing, and find the range of f. (Do these in whatever order seems most convenient.)

Example 4 $f(x) = x$ (the **identity** function)

Solution This is a special case of a linear function $f(x) = mx + b$ with slope $m = 1$ and y intercept $b = 0$ (Figure 7). Notice that the graph consists of all points for which the abscissa equals the ordinate. The set $\mathbb{R}$ of all real numbers is both the domain and the range of f, and f is increasing over $\mathbb{R}$. Since $f(-x) = -x = -f(x)$, it follows that f is an odd function and that its graph is symmetric about the origin. ■

Figure 7

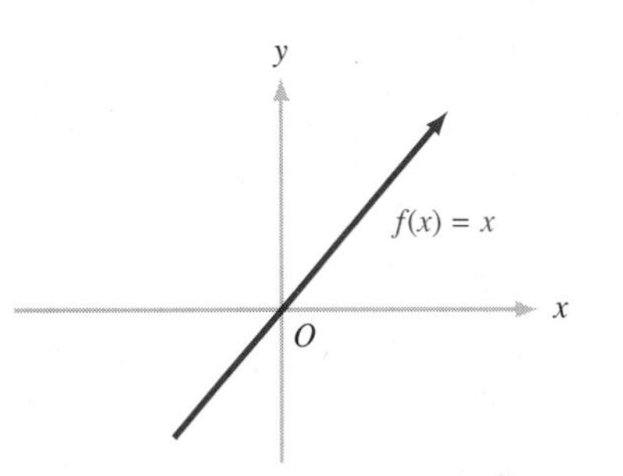

Example 5 $f(x) = x^2$ (the **squaring** function)

Solution The domain of f is the set $\mathbb{R}$. Because $f(-x) = (-x)^2 = x^2 = f(x)$, the function f is even and its graph is symmetric about the y axis. We have already sketched the portion of this graph for $x > 0$ in Figure 1. The full graph includes the mirror image of Figure 1 on the other side of the y axis, and the point (0, 0) (Figure 8). From the graph we see that the range of f is the interval $[0, \infty)$, and that f is decreasing on the interval $(-\infty, 0]$ and increasing on the interval $[0, \infty)$. ■

Figure 8

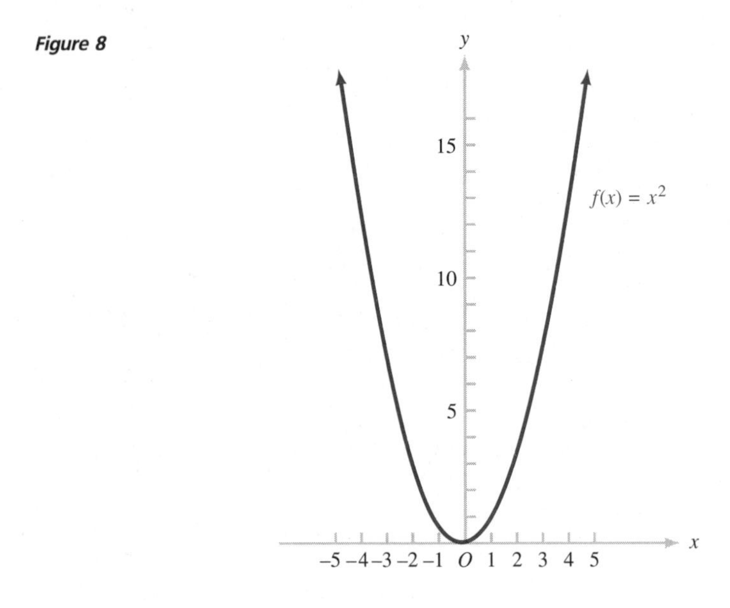

Example 6 $f(x) = \sqrt{x}$ (the **square-root** function*)

Solution Here the domain of f is the interval $[0, \infty)$. We tabulate some points $(x, f(x))$, plot them, and connect them by a smooth curve to obtain a sketch of the graph (Figure 9).

* In BASIC programming language, the square-root function is written SQR(X).

Figure 9

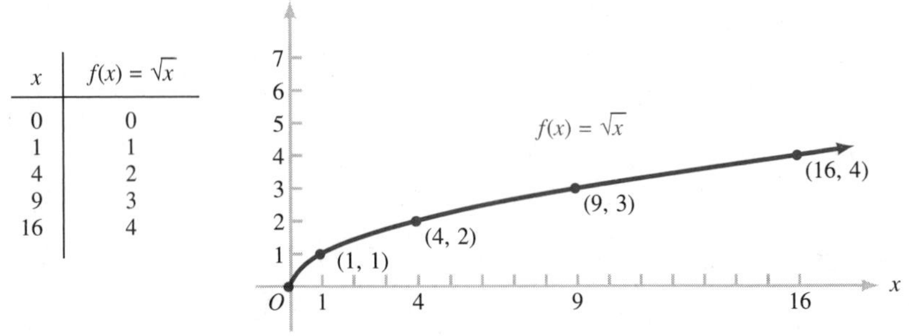

x	$f(x) = \sqrt{x}$
0	0
1	1
4	2
9	3
16	4

The function f is increasing on the interval $[0, \infty)$, and its graph rises higher and higher without bound, so the range of f is the interval $[0, \infty)$. The graph is symmetric about neither the y axis nor the origin, so f is neither even nor odd. ■

Example 7 $f(x) = b$ (a **constant** function)

Solution This is a special case of the linear function $f(x) = mx + b$, with slope $m = 0$ and y intercept b. Its graph is a line parallel to the x axis and containing the point $(0, b)$ on the y axis. All points on this line have the same ordinate b (Figure 10). The domain of f is the set $\mathbb{R}$ of all real numbers, and the range is the set $\{b\}$. The constant function f is neither increasing nor decreasing. Since $f(-x) = b = f(x)$, the function f is even and its graph is symmetric about the y axis. ■

Figure 10

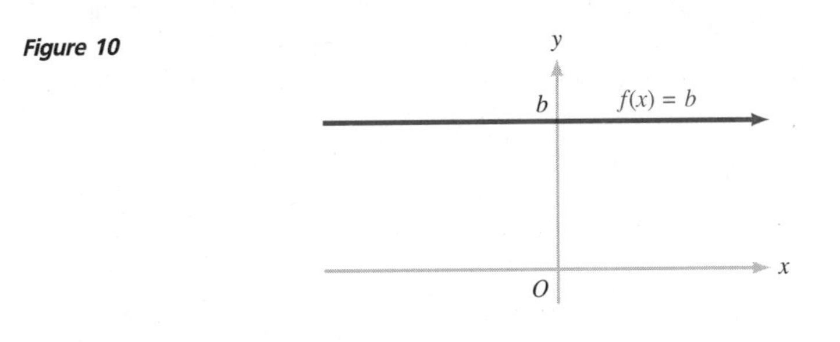

Example 8 $f(x) = |x|$ (the **absolute-value** function*)

Solution The domain of f consists of all real numbers $\mathbb{R}$. For $x \geq 0$, $f(x) = x$, so the portion of the graph to the right of the y axis is the same as the graph of the identity function (Example 4). Since $f(-x) = |-x| = |x| = f(x)$, the function f is even and its graph is symmetric about the y axis. Reflecting the portion of the graph of the identity function for $x \geq 0$ across the y axis, we obtain the V-shaped graph shown in Figure 11. Evidently, the range of f is the interval $[0, \infty)$; f is decreasing on the interval $(-\infty, 0]$ and increasing on the interval $[0, \infty)$. ■

Figure 11

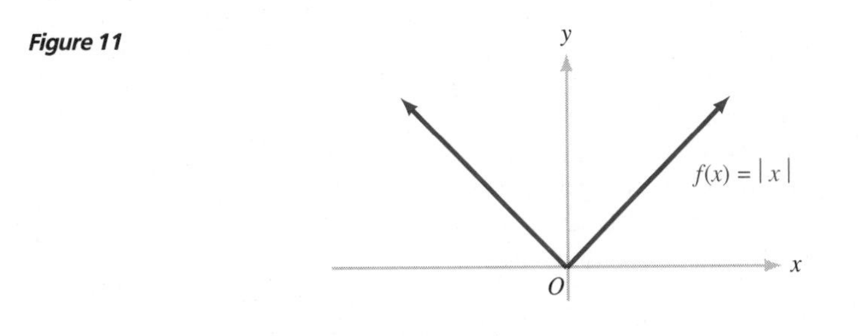

Sometimes a function is defined by using different equations on different intervals. Such a **piecewise-defined** function is illustrated in the following example.

* In BASIC programming language, the absolute-value function is written ABS(X).

Example 9 Sketch the graph of the function f defined by

$$f(x) = \begin{cases} -2 & \text{if } x < 0 \\ x^2 & \text{if } 0 \le x < 2 \\ x & \text{if } x \ge 2. \end{cases}$$

Solution For $x < 0$, we have $f(x) = -2$, so this portion of the graph of f will look like the graph of a constant function (Figure 10 with $b = -2$). In sketching this portion of the graph, we must be careful to leave out the point $(0, -2)$ on the y axis because of the strict inequality $x < 0$ (Figure 12).

For $0 \le x < 2$, we have $f(x) = x^2$; this part of our graph coincides with the graph of the squaring function (Figure 8), starting at the origin and ending at the point $(2, 4)$, which does not belong to the graph of f (Figure 12).

Finally, for $x \ge 2$, we have $f(x) = x$; this part of our graph will coincide with the graph of the identity function (Figure 7), starting at the point $(2, 2)$, which does belong to the graph, and continuing indefinitely to the right (Figure 12). ■

Figure 12

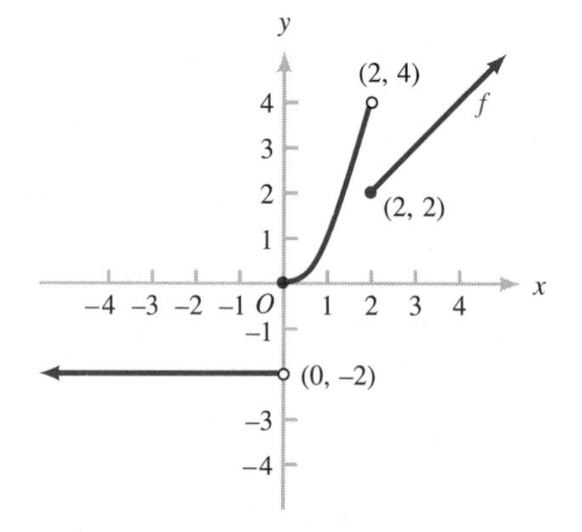

Problem Set 3.4

In Problems 1 to 8, use the point-plotting method to sketch the graph of each function. [c] You may wish to improve the accuracy of your sketch by using a calculator to determine the coordinates of some points on the graph. [gc] Alternatively, you may use a graphing calculator.

1. $f(x) = x^3$ for $x \ge 0$

2. $g(x) = \sqrt[3]{x}$ for $-8 \le x \le 8$

3. $h(x) = \dfrac{1}{x}$ for $x \ge 1$

4. $F(x) = \dfrac{x-1}{x+1}$ for $x \ge 0$

5. $G(x) = x^2 + x + 1$

6. $H(x) = \dfrac{1}{x^2 + 1}$

7. $p(x) = \sqrt{25 - x^2}$

8. $q(x) = x + \sqrt{x}$

9. For each function in Figure 13, find the domain and range; indicate the intervals over which the function is increasing, decreasing, or constant; and determine whether the function is even, odd, or neither.

Figure 13

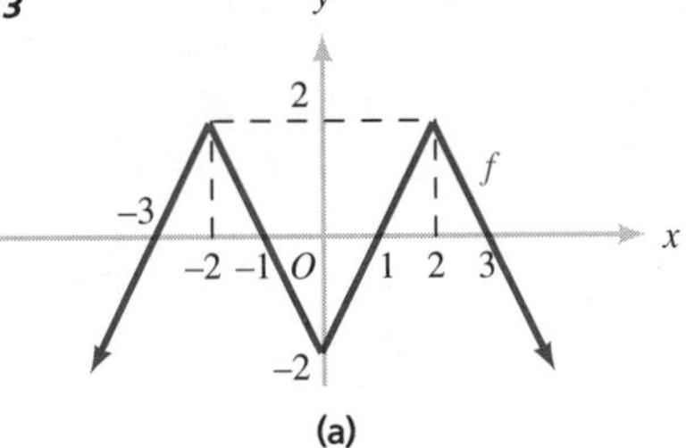

(a)

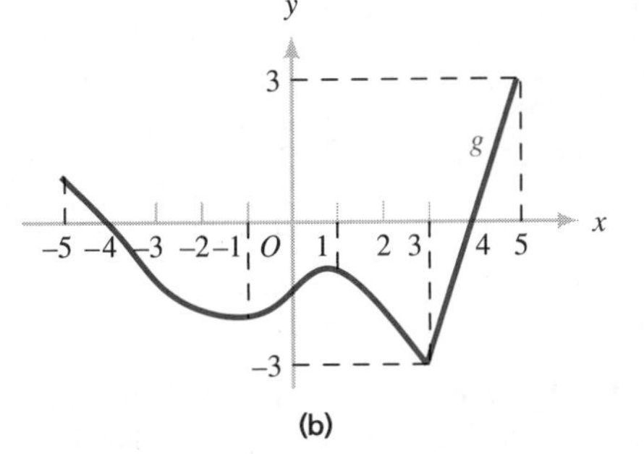

(b)

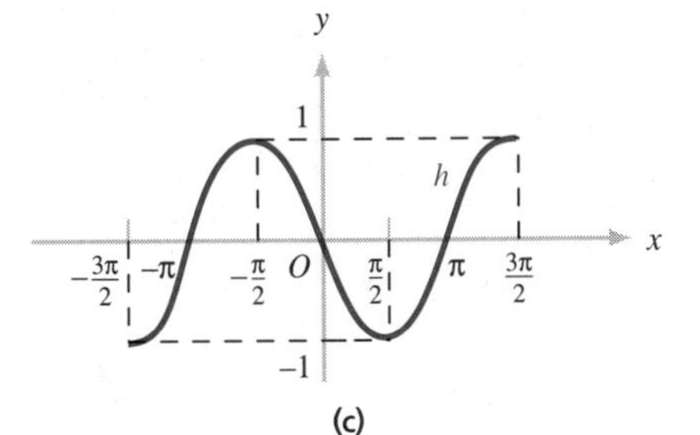

(c)

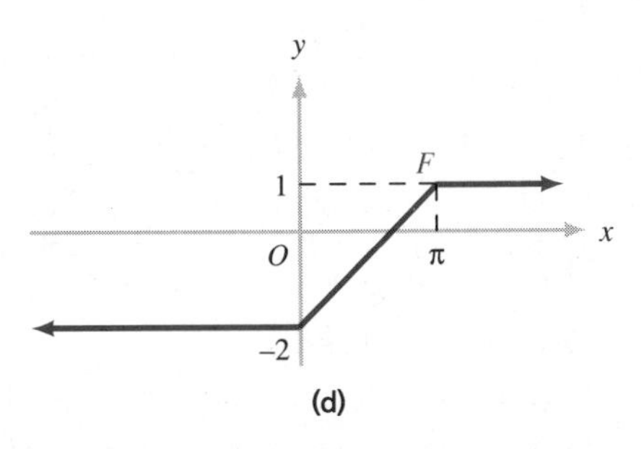

(d)

10. Figure 14 shows a computer-generated graph of the function $f(x) = x^{1/3}(x - 1)^{2/3}$. Indicate the intervals over which f is increasing or decreasing and determine whether f is even, odd, or neither.

Figure 14

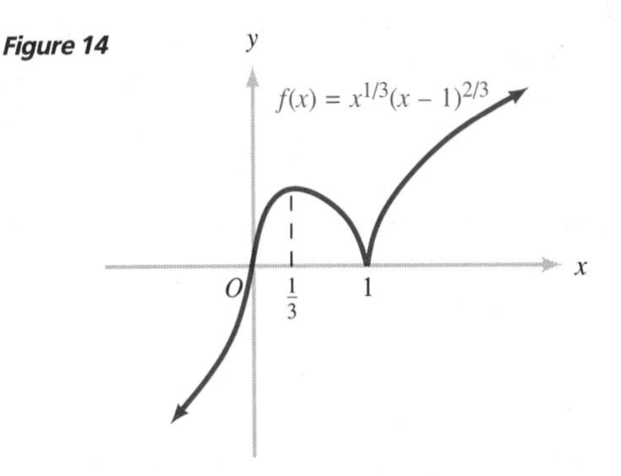

In Problems 11 to 19, determine without drawing a graph whether the function is even, odd, or neither, and discuss any symmetry of the graph.

11. $f(x) = x^4 + 3x^2$

12. $g(x) = -5x^2 + 4$

13. $h(x) = 4x^3 - 5x$

14. $F(x) = x^2 + |x|$

15. $G(x) = \sqrt{8x^4 + 1}$

16. $H(x) = 5x^2 - x^3$

17. $q(x) = \sqrt[3]{x^3 - 9}$

18. $p(x) = \dfrac{\sqrt{x^2 + 1}}{|x|}$

19. $r(x) = \dfrac{1}{x}$

20. Let $f(x) = 3x^4 - 2x^3 + x^2 - 1$. Define the functions g and h by $g(x) = \frac{1}{2}[f(x) + f(-x)]$ and $h(x) = \frac{1}{2}[f(x) - f(-x)]$.

(a) Show that f is neither even nor odd.

(b) Show that g is even.

(c) Show that h is odd.

(d) Show that $f(x) = g(x) + h(x)$ holds for all values of x.

In Problems 21 to 34, find the domain of the function, sketch its graph, discuss the symmetry of the graph, indicate the intervals where the function is increasing or decreasing, and find the range of the function. Do these things in any convenient order. gc Use a graphing calculator if you wish.

21. $f(x) = 3x + 1$

22. $g(x) = -4x + 3$

23. $h(x) = 5$

24. $F(x) = -2\sqrt{x}$

25. $H(x) = 2 - x^2$

26. $G(x) = -7$

27. $f(x) = x|x|$

28. $g(x) = \sqrt{x - 4} + x$

29. $F(x) = 1 + \sqrt{x - 1}$

30. $h(x) = x^2 - 4x$

31. $H(x) = \dfrac{x}{|x|}$

32. $g(x) = x - |\tfrac{1}{2}x|$

33. $f(x) = x^3 + 2x$

34. $f(x) = |x| - x$

In Problems 35 to 40, sketch the graph of each piecewise-defined function.

35. $f(x) = \begin{cases} 2 & \text{if } x < 0 \\ -1 & \text{if } x \geq 0 \end{cases}$

36. $g(x) = \begin{cases} x & \text{if } x \geq 0 \\ 2 & \text{if } x < 0 \end{cases}$

37. $F(x) = \begin{cases} 5 + x & \text{if } x \leq 3 \\ 9 - \dfrac{x}{3} & \text{if } x > 3 \end{cases}$

38. $h(x) = \begin{cases} x^2 & \text{if } x < 0 \\ -x^2 & \text{if } x \geq 0 \end{cases}$

39. $g(x) = \begin{cases} -x & \text{if } x < 0 \\ x^2 & \text{if } 0 \leq x < 2 \\ 8 - 2x & \text{if } x \geq 2 \end{cases}$

40. $G(x) = \begin{cases} 0 & \text{if } \quad x < -2 \\ 3 + x^2 & \text{if } -2 \leq x < 2 \\ 0 & \text{if } \quad x \geq 2 \end{cases}$

c 41. A student sketches the graph of the function $f(x) = 4x^4 - 14x^3 + 20x^2 - 5x$ for $x \geq 0$ by plotting five points and connecting them with a smooth curve, as shown in Figure 15. However, the graph is not correct as shown. Find the error. [*Hint:* Plot more points for x between 0 and 0.5, or gc use a graphing calculator.]

Figure 15

x	$f(x)$
0	0
0.5	1
1	5
1.5	10.5
2	22

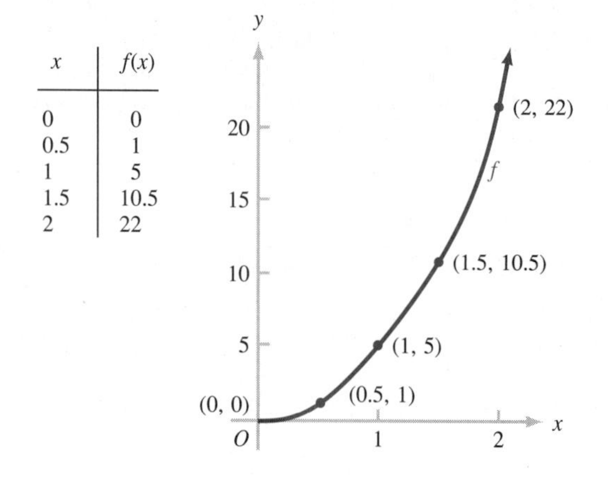

42. A student sketches the graph of $f(x) = x^3 + 3x^2$ by plotting seven points and connecting them with a smooth curve, as shown in Figure 16. Criticize the student's work. gc Use a graphing calculator if you wish.

Figure 16

x	$f(x)$
–2	4
–1.5	3.38
–1	2
–0.5	0.63
0	0
0.5	0.88
1	4

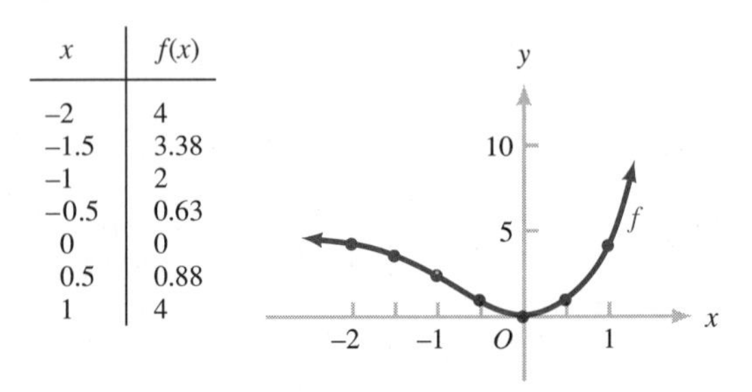

43. The management of a city parking lot charges its customers \$0.50 per hour or part of an hour for up to 12 hours, with a minimum charge of \$2.00 and a maximum charge of \$5.00. If $C(x)$ denotes the charge in dollars for parking a car for x hours, sketch a graph of the function C for $0 \leq x \leq 12$ hours.

44. Explain how to use the graph of the squaring function (Figure 8) to see that if $a < b < 0$, then $a^2 > b^2$.

45. Surveys conducted by the department of social work in one state indicate that payment of child support tends to decline with time elapsed after the divorce decree. One survey shows that the percentage p of cases in which payments are made is given by a linear function, $p = f(t)$, of the time t years since the decree. Specifically, after 3 years, payments are made in only 35% of the cases, and after 10 years, payments are maintained in a mere 7% of the cases.

(a) Write an equation that determines the function f over the domain [0, 10.5] and sketch the graph of f.

(b) Find the range of f.

(c) Interpret the slope and the p intercept of f.

(d) In what percentage of the cases is child support paid after 7 years?

46. The symbol $[\![x]\!]$ is often used to denote the *greatest integer* less than or equal to x; that is, $[\![x]\!]$ is the one and only integer such that $[\![x]\!] \leq x < [\![x]\!] + 1$. For instance, $[\![1.7]\!] = 1$, $[\![3.14]\!] = 3$, $[\![2]\!] = 2$, $[\![-1.3]\!] = -2$,

$[\![-3]\!] = -3$, and so forth. The function defined by $f(x) = [\![x]\!]$ is called the **greatest integer function.*** Sketch a graph of the greatest integer function for $-4 \leq x \leq 4$.

47. Figure 17 shows computer-generated graphs of $y = 3x - x^2$ and $y = x$. Find an equation that defines the function $A(x)$ giving the area of the indicated square in terms of x.

Figure 17

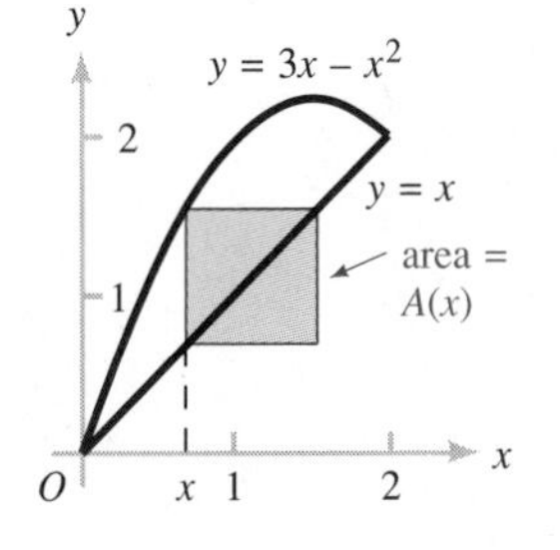

48. A point P is moving along the s axis in such a way that its distance s from the origin at time t is given by $s = f(t)$ (Figure 18). If $a < b$, then the *average speed* of the moving point over the time interval from $t = a$ to $t = b$ is $\dfrac{f(b) - f(a)}{b - a}$. Show that the average speed is the slope of the line segment from the point $(a, f(a))$ to the point $(b, f(b))$ on the graph of f and illustrate with a diagram.

Figure 18

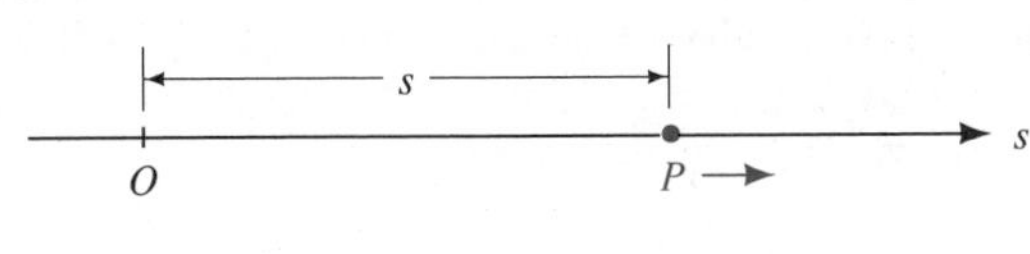

* In BASIC the greatest integer function is written INT(X).

49. The function $f(t) = 16t^2$ gives the number of feet that an object dropped from rest will fall in t seconds (Figure 19). If the object is dropped from a building 100 feet high, sketch a graph of the function f. gc Use a graphing calculator if you wish.

Figure 19

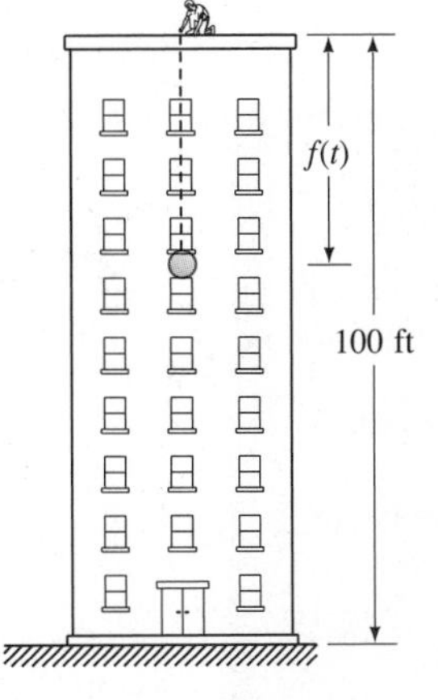

C **50.** For the falling object in Problem 49, use the formula in Problem 48 to find the average speed over each time interval.

(a) $a = 2$ to $b = 2.1$

(b) $a = 2$ to $b = 2.01$

(c) $a = 2$ to $b = 2.001$

(d) $a = 2$ to $b = 2.0001$.

51. Find the domain and the range of the function f in Problem 49.

52. In Problem 50, what value does the average speed seem to be approaching as the time interval gets shorter and shorter? This value is called the *instantaneous speed* of the falling object at the instant when $t = 2$.

3.5 Shifting, Stretching, and Reflecting Graphs

By altering a function f in certain ways we can obtain a new function F whose graph is closely related to the graph of the original function. In this section we study some of the basic ways in which functions can be altered and the resulting effects on their graphs.

Example 1 Sketch the graphs of $f(x) = x^2$ and $F(x) = x^2 + 1$ on the same coordinate system.

Solution The graph of $f(x) = x^2$ was shown in Figure 8 of Section 3.4. The graph of $F(x) = x^2 + 1$ is obtained by adding 1 to the ordinate of each point on the graph of f. Thus, the graph of $F(x) = x^2 + 1$ is just the graph of $f(x) = x^2$ shifted upward by 1 unit (Figure 1). A graphing calculator may be used to confirm the relationship between the graphs in Figure 1. ■

Figure 1

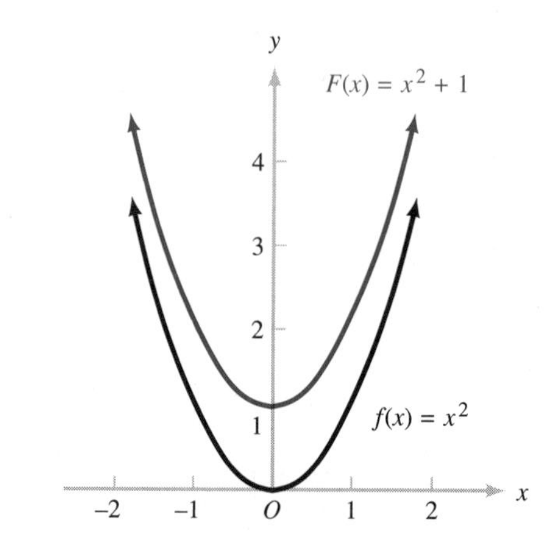

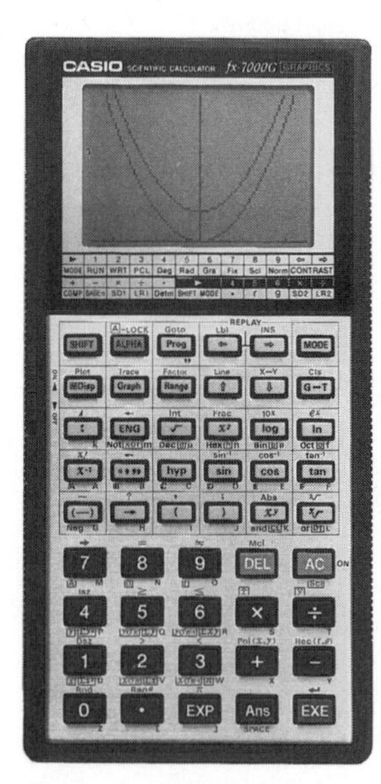

Casio fx-7000G showing the graph in Figure 1.

More generally, we have the following rule.

Vertical Shifting Rule

If f is a function and k is a constant, then the graph of the function F defined by

$$F(x) = f(x) + k$$

is obtained by **shifting** the graph of f **vertically** by $|k|$ units, **upward** if $k > 0$, and **downward** if $k < 0$ (Figure 2).

Figure 2

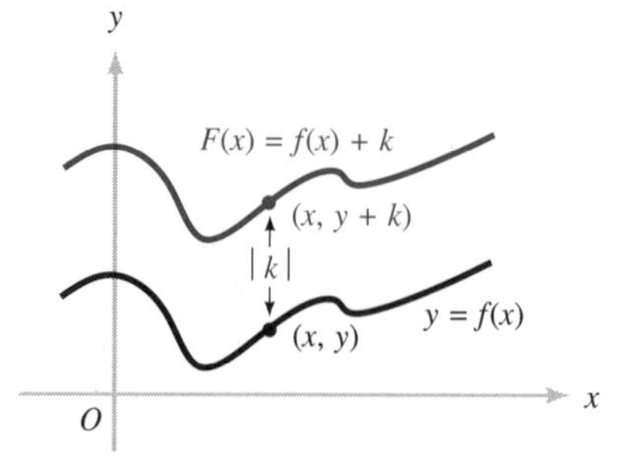

Example 2 Sketch the graphs of $F(x) = |x| + 2$ and $G(x) = |x| - 3$ on the same coordinate system.

Solution

The graph of

$$f(x) = |x|$$

was shown in Figure 11 of Section 3.4. We shift this graph upward by 2 units to obtain the graph of $F(x) = |x| + 2$; we shift it downward by 3 units to obtain the graph of $G(x) = |x| - 3$ (Figure 3). ■

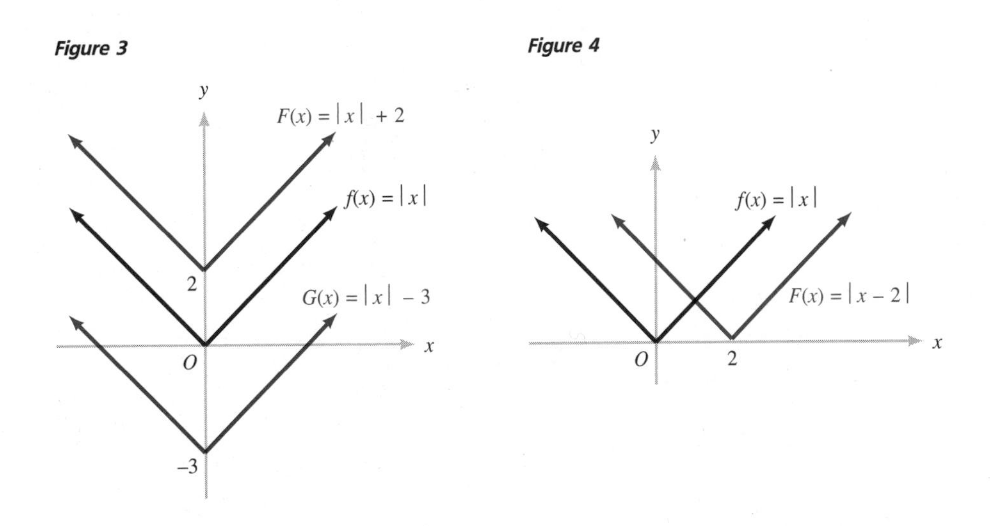

Example 3 Sketch the graphs of $f(x) = |x|$ and $F(x) = |x - 2|$ on the same coordinate system.

Solution

A point (x, y) belongs to the graph of F if and only if $y = |x - 2| = f(x - 2)$, that is, if and only if $(x - 2, y)$ belongs to the graph of f. For instance, the point $(2, 0)$ belongs to the graph of F because the point $(2 - 2, 0) = (0, 0)$ belongs to the graph of f. Because each point (x, y) on the graph of F is 2 units to the right of the point $(x - 2, y)$ on the graph of f, it follows that the graph of F is obtained by shifting the graph of f horizontally by 2 units to the right (Figure 4). ■

More generally, we have the following rule.

Horizontal Shifting Rule

If f is a function and h is a constant, then the graph of the function F defined by

$$F(x) = f(x - h)$$

is obtained by **shifting** the graph of f **horizontally** by $|h|$ units, to the **right** if $h > 0$ and to the **left** if $h < 0$.

To understand the horizontal shifting rule, notice that a point (x, y) belongs to the graph of $F(x) = f(x - h)$ if and only if $y = f(x - h)$, that is, if and only if the point $(x - h, y)$ belongs to the graph of f (Figure 5).

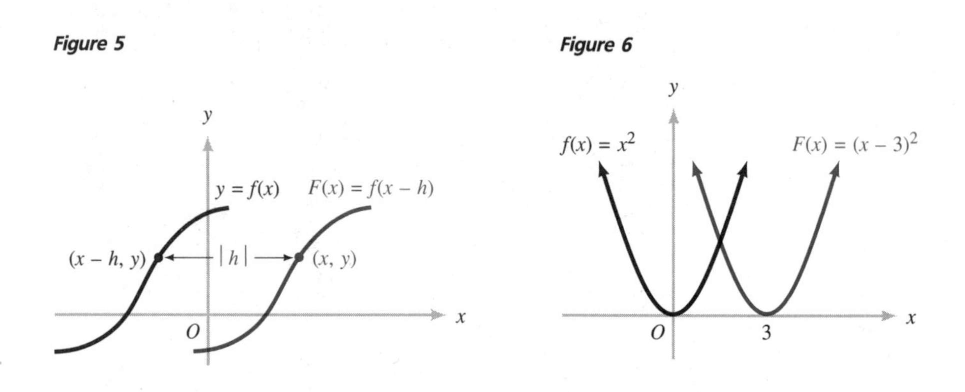

Example 4 Sketch the graph of $F(x) = (x - 3)^2$.

Solution We begin by sketching the graph of $f(x) = x^2$. By shifting this graph 3 units to the right, we obtain the graph of $F(x) = (x - 3)^2$ (Figure 6). ■

Of course, we can combine vertical and horizontal shifts.

Combined Graph-Shifting Rule

> If h and k are constants and f is a function, then the graph of
>
> $$F(x) = f(x - h) + k$$
>
> is obtained by shifting the graph of f horizontally by $|h|$ units and vertically by $|k|$ units. The horizontal shift is to the right if $h > 0$, and to the left if $h < 0$. The vertical shift is upward if $k > 0$, and downward if $k < 0$.

Example 5 Sketch the graph of $F(x) = |x + 2| - 3$.

Solution We begin by sketching the graph of $f(x) = |x|$. In the combined graph-shifting rule we take $F(x) = f(x - h) + k$ with $h = -2$ and $k = -3$. Therefore, the graph of $F(x) = |x + 2| - 3$ is obtained by shifting the graph of $f(x) = |x|$ to the left by 2 units and downward by 3 units (Figure 7). ■

The effect of multiplying a function by a constant is to multiply the ordinate of each point on its graph by that constant.

Figure 7

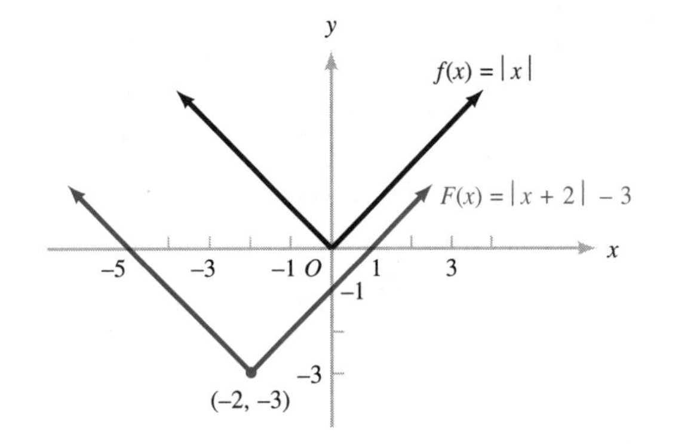

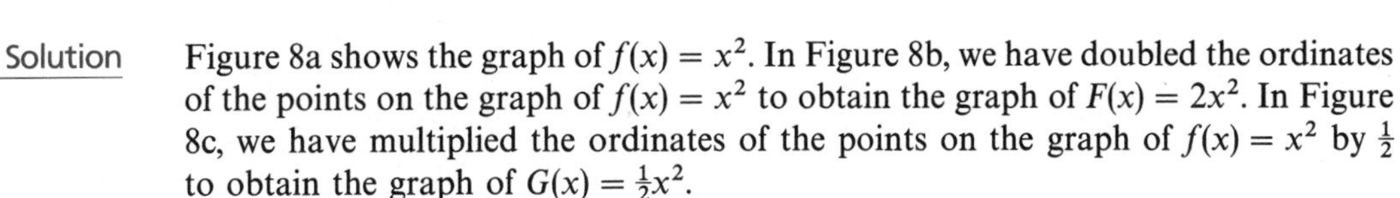

Example 6 Sketch the graphs of $F(x) = 2x^2$ and $G(x) = \frac{1}{2}x^2$.

Solution Figure 8a shows the graph of $f(x) = x^2$. In Figure 8b, we have doubled the ordinates of the points on the graph of $f(x) = x^2$ to obtain the graph of $F(x) = 2x^2$. In Figure 8c, we have multiplied the ordinates of the points on the graph of $f(x) = x^2$ by $\frac{1}{2}$ to obtain the graph of $G(x) = \frac{1}{2}x^2$. ■

Figure 8

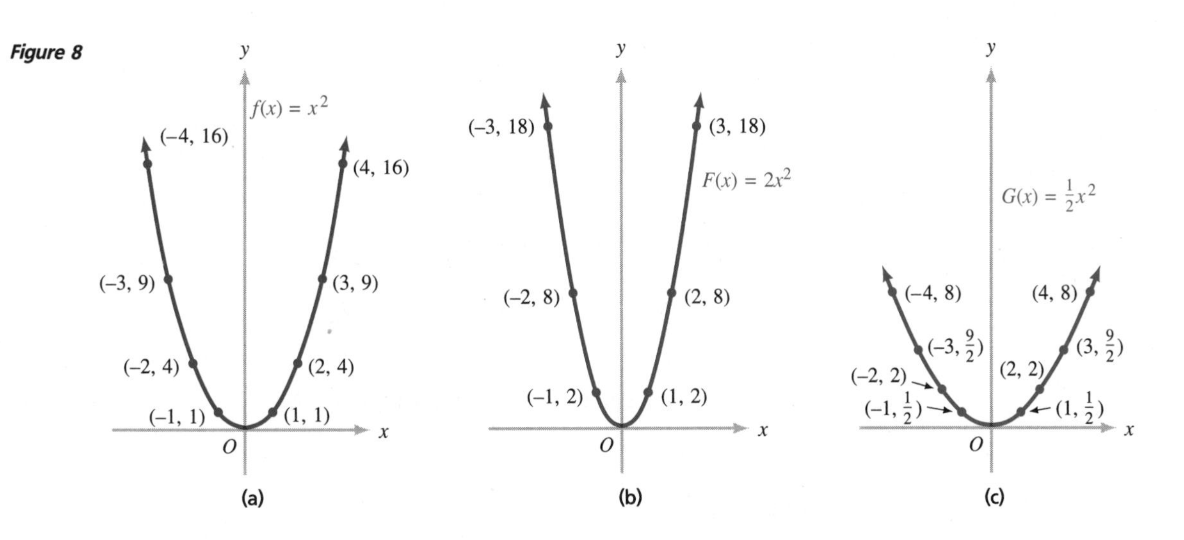

More generally, we have the following rule.

Vertical Stretching and Flattening Rule

If f is a function and c is a constant, then the graph of the function F defined by

$$F(x) = cf(x)$$

is obtained from the graph of f by multiplying each ordinate by c.

If $c > 1$, the effect is to "stretch" the graph vertically by a factor of c. If $0 < c < 1$, the effect is to "flatten" the graph. The case in which $c < 0$ is addressed in Example 8.

Example 7 Sketch the graphs of $F(x) = 3|x|$ and $G(x) = \frac{1}{3}|x|$ on the same coordinate system.

Solution In Figure 9 we have used the stretching and flattening rule to obtain the graphs of F and G from the graph of $f(x) = |x|$. ■

Figure 9

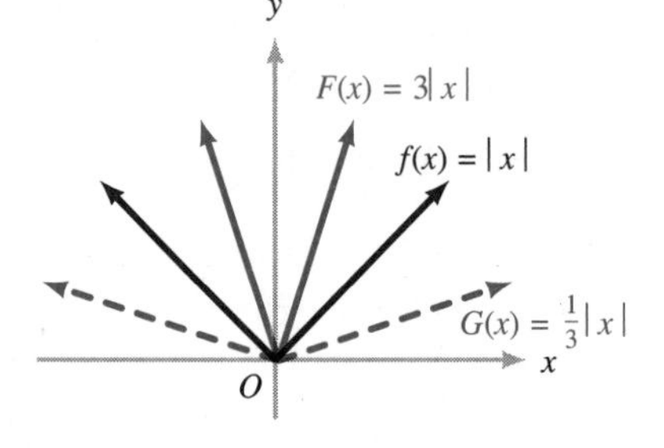

Figure 10

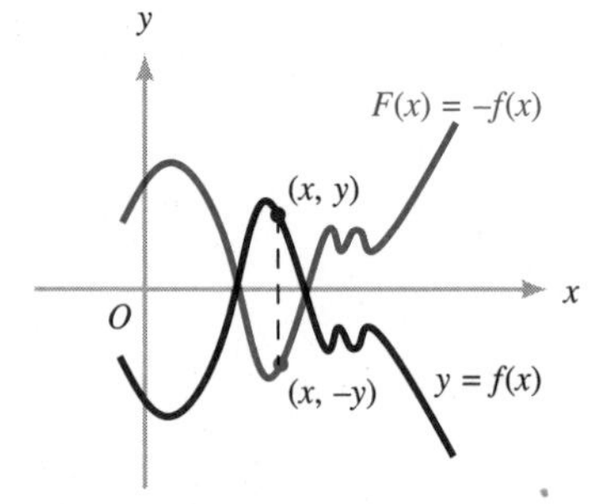

Figure 10 shows the effect of multiplying a function by -1 and illustrates the following rule.

x Axis Reflection Rule

If f is a function, then the graph of the function

$$F(x) = -f(x)$$

is obtained by **reflecting** the graph of f across the ***x* axis.**

Example 8 Sketch the graphs of $F(x) = -|x|$ and $G(x) = 2 - |x|$.

Solution We obtain the graph of $F(x) = -|x|$ by reflecting the graph of $f(x) = |x|$ across the x axis (Figure 11a). By shifting the graph of $F(x) = -|x|$ upward by 2 units, we obtain the graph of $G(x) = 2 - |x|$ (Figure 11b). ■

Figure 11

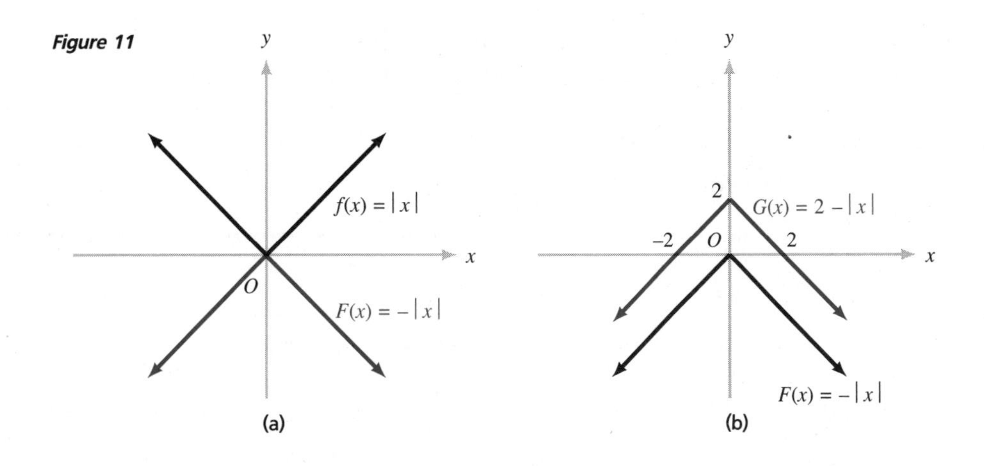

In Problem 43, we ask you to justify the following.

y Axis Reflection Rule

> If f is a function, then the graph of the function
> $$G(x) = f(-x)$$
> is obtained by **reflecting** the graph of f across the **y axis.**

Example 9 Use the graph of the function f in Figure 12a to obtain the graph of $G(x) = f(-x)$.

Solution We obtain the graph of $G(x) = f(-x)$ by reflecting the graph of f across the y axis (Figure 12b). ■

Figure 12

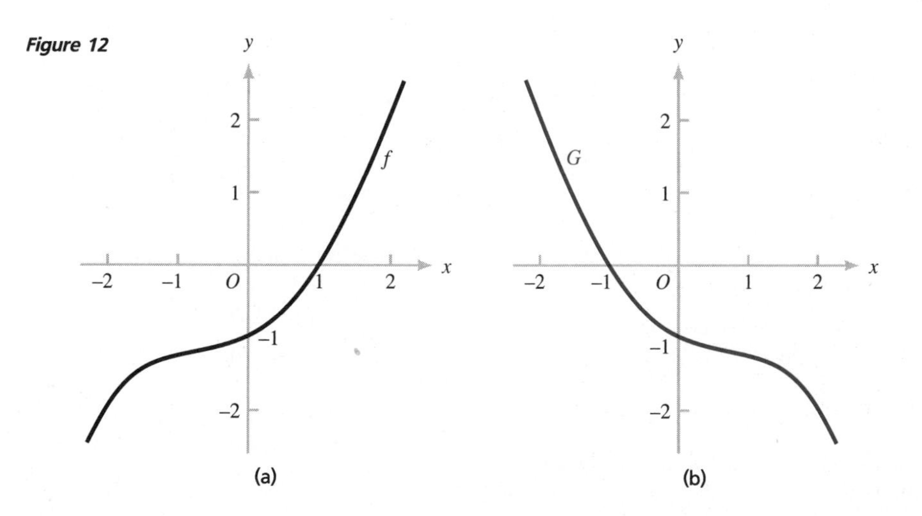

The techniques of shifting, stretching, and reflecting are often useful for determining the "general shape" of a graph before you begin to sketch it. Using this information and plotting a few points, you can rapidly sketch a reasonably accurate graph.

Problem Set 3.5

In Problems 1 to 8, the graph of the function F is obtained from the graph of the given function f as indicated. Find an equation that defines F and sketch graphs of f and F on the same coordinate system. gc If you wish, you may use a graphing calculator to check your work.

1. Shift the graph of $f(x) = x$ vertically upward 2 units.
2. Shift the graph of $f(x) = x^2$ vertically downward 3 units.
3. Shift the graph of $f(x) = |x|$ vertically downward 2 units.
4. Shift the graph of $f(x) = \sqrt{x}$ horizontally 3 units to the left.
5. Shift the graph of $f(x) = \sqrt{x}$ horizontally 2 units to the right.
6. Reflect the graph of $f(x) = x^2$ across the x axis and then vertically stretch it by a multiple of 3.
7. Shift the graph of $f(x) = x^2$ vertically upward 2 units and horizontally to the left 3 units and then flatten it by a multiple of $\frac{2}{3}$.
8. Reflect the graph of $f(x) = x^2$ across the x axis, then shift it $\frac{3}{2}$ units to the right, and then reflect the resulting graph across the y axis.

In Problems 9 to 12, sketch on the same coordinate system the graph of the given functions for the indicated values of the constant k.

9. $f(x) = |x| + k$; $k = -3, 0, 3$
10. $f(x) = (x - k)^2$; $k = -3, 0, 3$
11. $f(x) = k\sqrt{x} + 2$; $k = -3, 0, 3$
12. $f(x) = (x - k)^2 + k$; $k = -3, 0, 3$

In Problems 13 to 20, compare the graph of the first function with the graph of the second by sketching them both on the same coordinate system and explaining in words how they are related.

13. $f(x) = x$ and $F(x) = x + 4$
14. $g(x) = |x|$ and $G(x) = |x| - 5$
15. $p(x) = x^2$ and $P(x) = 1 - x^2$
16. $q(x) = |x|$ and $Q(x) = |x - 3| + 3$
17. $r(x) = |x|$ and $R(x) = 2|x| + 1$
18. $s(x) = x^2$ and $S(x) = -\frac{1}{2}(x - 1)^2$
19. $t(x) = \sqrt{x}$ and $T(x) = 2\sqrt{x - 2} + 4$
20. $f(x) = x^3$ and $F(x) = 1 - \frac{2}{3}(x - 1)^3$

In Problems 21 to 38, use the techniques of shifting, stretching, and reflecting to sketch the graph of each function.

21. $F(x) = x^2 + 2$
22. $G(x) = x^2 - 1$
23. $H(x) = |x - 3|$
24. $f(x) = \sqrt{x - 1}$
25. $g(x) = (x - 2)^2$
26. $p(x) = \sqrt{x + 4} - 2$
27. $q(x) = |x - 1| + 1$
28. $F(x) = 3\sqrt{x - 1}$
29. $G(x) = 2|x|$
30. $H(x) = 3|x - 1| + 1$
31. $Q(x) = \frac{3}{4}x^2 + 1$
32. $R(x) = \frac{2}{3}(x - 1)^2 + 2$
33. $F(x) = -x^2$
34. $G(x) = 3 - \sqrt{x}$
35. $H(x) = 1 - 3|x|$

36. $f(x) = 1 - 2|x + 4|$

37. $g(x) = 1 - 2x^2$

38. $q(x) = 2 - \frac{3}{2}\sqrt{x + 4}$

In Problems 39 to 42, use the graph of the function f in the figure to obtain the graph of **(a)** $F(x) = f(x) + 2$, **(b)** $G(x) = f(x - 2)$, **(c)** $H(x) = 2f(x)$, **(d)** $Q(x) = -f(x)$, and **(e)** $R(x) = f(-x)$.

39.

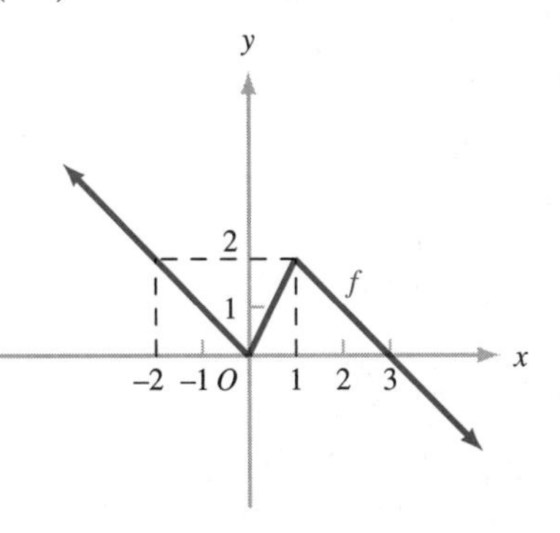

40.

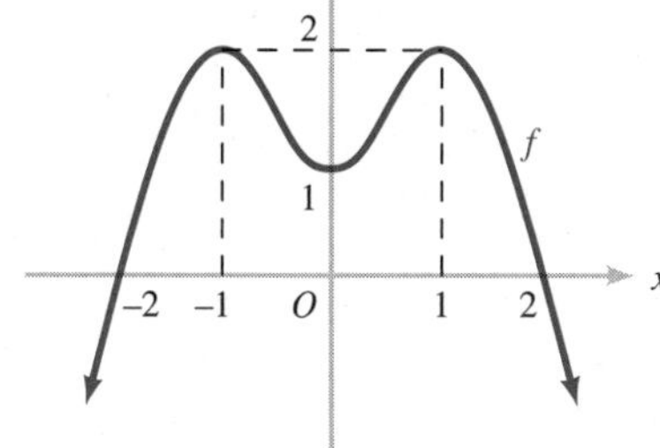

41.

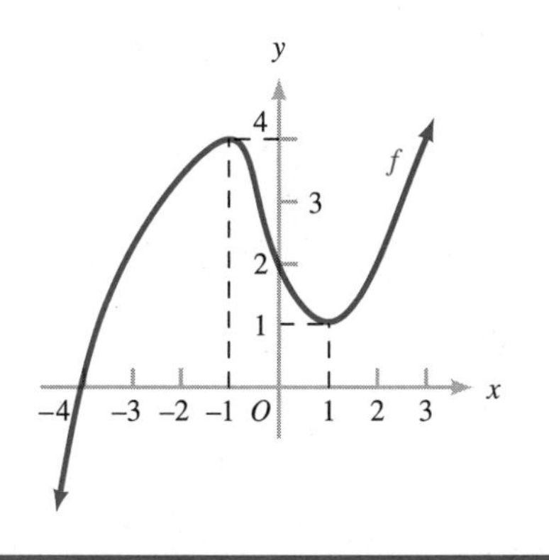

42.

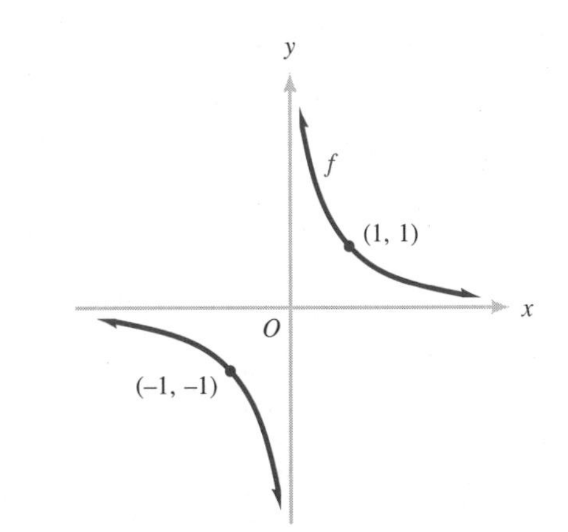

43. Explain why the y axis reflection rule works.

44. **(a)** For $x \neq 8$ and $x > 4$, show that $\dfrac{8 - x}{2 - \sqrt{x - 4}} = 2 + \sqrt{x - 4}$.

(b) Use the result of part (a) to sketch a graph of $f(x) = \dfrac{8 - x}{2 - \sqrt{x - 4}}$.

45. **(a)** A graph will be shifted to the right by h units and it will be shifted vertically by k units. Does it matter in what order these shifts are performed?

(b) A graph will be shifted vertically by k units and reflected across the x axis. Does it matter in what order these operations are performed?

46. If c is a constant and $c > 1$, the graph of $F(x) = cf(x/c)$ is said to be obtained from the graph of f by *magnification*, and c is called the *magnification factor*.

(a) Sketch the graph obtained if the graph of $f(x) = x^2$ is magnified by a factor of 2.

(b) What happens if you magnify the graph of a linear function $f(x) = mx + b$ by a factor of c?

47. If c is a positive constant, how is the graph of $F(x) = f(cx)$ related to the graph of f?

48. Hugo has developed a mathematical model $p = f(t)$ for the annual profit p dollars of the Widget Company t years after it was founded in 1979 and is confident that this model is reasonably accurate for $0 \leq t \leq 25$ years. Hugo's boss, Emily, would prefer to have a mathematical model $p = F(t)$ for the annual profit t years after the company was reorganized in 1987. Hugo is able to write an equation expressing F in terms of f. What is this equation and what is the domain of F?

3.6 Algebra of Functions and Composition of Functions

In the business world, the *profit* P from the sale of goods is related to the *revenue* R from the sale and to the *cost* C of producing the goods by the equation

$$P = R - C.$$

In other words, $R = P + C$. Often, the quantities P, R, and C are variables that depend on another variable, for instance, on the number x of units of goods produced; that is, P, R, and C are functions of x. Thus, business applications of mathematics may involve the *addition and subtraction of functions;* other applications require the *multiplication and division of functions.* For example, the function that describes an amplitude-modulated (AM) radio signal is the product of the function that describes the audio signal and the function that describes the carrier wave. Therefore, we make the following definition.

Definition 1 **Sum, Difference, Product, and Quotient of Functions**

Let f and g be functions whose domains overlap. We define functions $f + g$, $f - g$, $f \cdot g$, and f/g as follows: For all values of x in the domain of both f and g,

$$(f + g)(x) = f(x) + g(x)$$
$$(f - g)(x) = f(x) - g(x)$$
$$(f \cdot g)(x) = f(x) \cdot g(x)$$
$$\left(\frac{f}{g}\right)(x) = \frac{f(x)}{g(x)}, \qquad g(x) \neq 0.$$

Example 1 Let $f(x) = 2x^3 + 1$ and $g(x) = x^2 - 5$. Find:

(a) $(f + g)(x)$ **(b)** $(f - g)(x)$ **(c)** $(f \cdot g)(x)$

(d) $\left(\frac{f}{g}\right)(x)$ **(e)** The domain of $\frac{f}{g}$

Solution

(a) $(f + g)(x) = f(x) + g(x) = (2x^3 + 1) + (x^2 - 5) = 2x^3 + x^2 - 4$

(b) $(f - g)(x) = f(x) - g(x) = (2x^3 + 1) - (x^2 - 5) = 2x^3 - x^2 + 6$

(c) $(f \cdot g)(x) = f(x) \cdot g(x) = (2x^3 + 1)(x^2 - 5) = 2x^5 - 10x^3 + x^2 - 5$

(d) $\left(\frac{f}{g}\right)(x) = \frac{f(x)}{g(x)} = \frac{2x^3 + 1}{x^2 - 5}$

(e) The domain of $\frac{f}{g}$ is the set of all real numbers except $\sqrt{5}$ and $-\sqrt{5}$. ■

Example 2 The Molar Brush Company finds that the total production cost for manufacturing x toothbrushes is given by the function

$$C(x) = 5000 + 30\sqrt{x} \text{ dollars.}$$

These toothbrushes sell for \$1.50 each.

(a) Write a formula for the revenue $R(x)$ dollars to the company if x toothbrushes are sold.

(b) Write a formula for the profit $P(x)$ dollars if x toothbrushes are sold.

(c) Find $P(40{,}000)$.

Solution **(a)** If x toothbrushes are sold at \$1.50 each, $R(x) = 1.50x = 1.5x$ dollars.

(b) $P(x) = R(x) - C(x) = 1.5x - (5000 + 30\sqrt{x})$ dollars.

(c) $P(40{,}000) = 1.5(40{,}000) - (5000 + 30\sqrt{40{,}000}) = 49{,}000$ dollars. ■

Geometrically, the graph of the sum, difference, product, or quotient of f and g has at each point an ordinate that is the sum, difference, product, or quotient, respectively, of the ordinates of f and g at the corresponding points. Thus, we can sketch graphs of $f + g$, $f - g$, $f \cdot g$, or f/g by the method of **adding, subtracting, multiplying,** or **dividing ordinates,** respectively. The following example illustrates the method of adding ordinates.

Example 3 Use the method of adding ordinates to sketch the graph of $h = f + g$ if $f(x) = x^2$ and $g(x) = x - 1$.

Solution We begin by sketching the graph of f (Figure 1a) and the graph of g (Figure 1b). Then we add the ordinates $f(x)$ and $g(x)$ corresponding to selected values of x

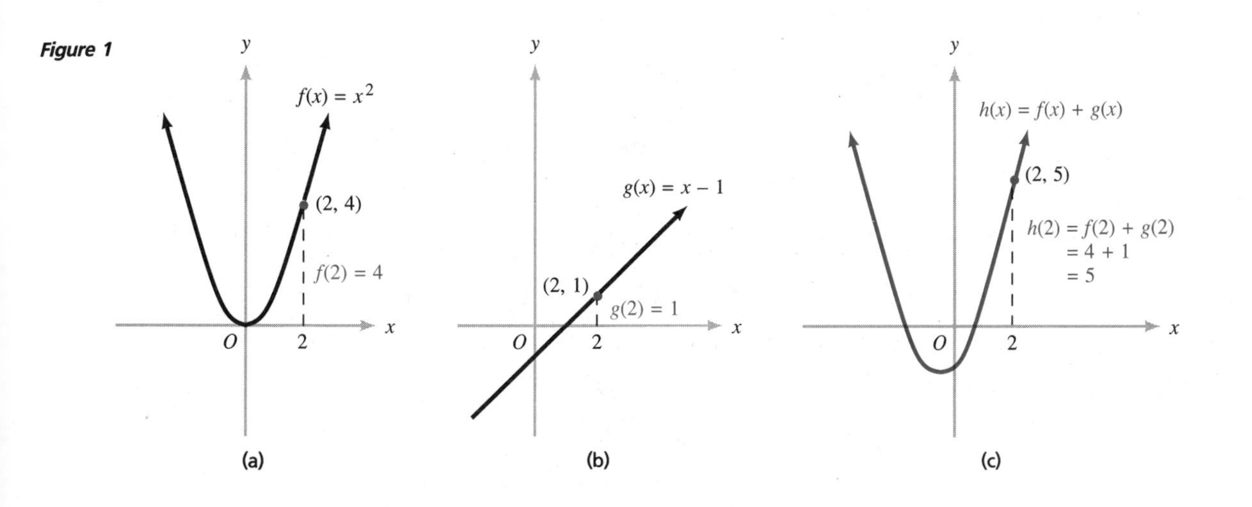

in order to obtain the corresponding ordinates $h(x) = f(x) + g(x)$. For instance, $f(2) = 2^2 = 4$ and $g(2) = 2 - 1 = 1$, so $h(2) = f(2) + g(2) = 4 + 1 = 5$. By calculating and plotting several such points and connecting them with a smooth curve, we obtain the graph of h (Figure 1c). ■

Composition of Functions

Suppose that we have three variable quantities y, t, and x. If y depends on t and if t, in turn, depends on x, then it is clear that y depends on x. In other words, if y is a function of t and t is a function of x, then y is a function of x. For instance, suppose that

$$y = t^2 \qquad \text{and} \qquad t = 3x - 1.$$

Then, substituting the value of t from the second equation into the first equation, we find that

$$y = (3x - 1)^2.$$

More generally, if

$$y = f(t) \qquad \text{and} \qquad t = g(x),$$

then, substituting t from the second equation into the first equation, we find that

$$y = f(g(x)).$$

To avoid a pileup of parentheses, we often replace the outside parentheses in the last equation by square brackets and write

$$y = f[g(x)].$$

If f and g are functions, the equation $y = f[g(x)]$ defines a new function h where

$$h(x) = f[g(x)].$$

The function h, obtained by "chaining" f and g together this way, is called the **composition** of f and g and is written as

$$h = f \circ g,$$

read, "h equals f composed with g." Thus, by definition,

$$(f \circ g)(x) = f[g(x)].$$

The domain of $f \circ g$ is the set of all numbers x in the domain of g such that $g(x)$ is in the domain of f.

Example 4 Let $f(x) = 3x - 1$, $g(x) = x^3$, and $h(x) = \sqrt{x}$. Find:

(a) $(f \circ g)(2)$ **(b)** $(g \circ f)(2)$ **(c)** $(f \circ g)(x)$
(d) $(g \circ f)(x)$ **(e)** $(f \circ f)(x)$ [C] **(f)** $(f \circ g)(3.007)$
(g) $[f \circ (g \circ h)](x)$ **(h)** The domain of $f \circ (g \circ h)$

Solution

(a) $(f \circ g)(2) = f[g(2)] = f(2^3) = f(8) = 3(8) - 1 = 23$

(b) $(g \circ f)(2) = g[f(2)] = g[3(2) - 1] = g(5) = 5^3 = 125$

(c) $(f \circ g)(x) = f[g(x)] = f(x^3) = 3x^3 - 1$

(d) $(g \circ f)(x) = g[f(x)] = g(3x - 1) = (3x - 1)^3$

(e) $(f \circ f)(x) = f[f(x)] = f(3x - 1) = 3(3x - 1) - 1 = 9x - 4$

(f) $(f \circ g)(3.007) = f[g(3.007)] = 3(3.007)^3 - 1 \approx 80.568324$

(g) $[f \circ (g \circ h)](x) = f[(g \circ h)(x)] = f[g(h(x))] = f[g(\sqrt{x})] = f[(\sqrt{x})^3]$
$= f(x^{3/2}) = 3x^{3/2} - 1$

(h) The domain of $f \circ (g \circ h)$ is the set of real numbers x with $x \geq 0$. ■

Although the symbolism $f \circ g$ looks vaguely like some kind of "product," you must not confuse it with the actual product $f \cdot g$ of f and g. For instance, whereas $f \cdot g = g \cdot f$, note (in Example 4) that $f \circ g \neq g \circ f$. However, it does turn out that function composition is **associative;** that is, for any three functions f, g, and h,

$$f \circ (g \circ h) = (f \circ g) \circ h$$

(Problem 38). Because of the associative property of composition, parentheses aren't really necessary when three or more functions are to be composed. Thus, we may simply write $f \circ g \circ h$, $f \circ g \circ h \circ F$, and so forth.

Using a programmable calculator, you can see a vivid demonstration of function composition. Suppose that the f and g keys are programmed with functions of your choice. If you enter a number x and touch the g key, you see the mapping

$$x \longmapsto g(x)$$

take place, and the number $g(x)$ appears in the display. Now, if you touch the f key, the mapping

$$g(x) \longmapsto f[g(x)]$$

will take place. Therefore, after you enter the number x, you can perform the composite mapping

$$x \longmapsto (f \circ g)(x)$$

by *touching first the g key, then the f key*. This fact is represented by the diagram in Figure 2.

Figure 2

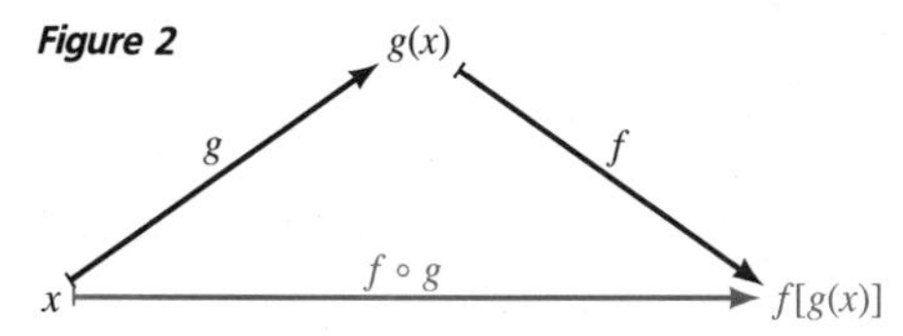

In using a programmable calculator or a computer, it is important to be able to tell when a complicated function can be obtained as a composition of simpler functions. The same skill is essential in calculus. If

$$h(x) = (f \circ g)(x) = f[g(x)],$$

let's agree to call g the "inside function" and f the "outside function" because of the positions they occupy in the expression $f[g(x)]$. In order to see that h can be obtained as a composition $h = f \circ g$, you must be able to recognize the inside function g and the outside function f in the equation that defines h.

Example 5 Express the function $h(x) = (3x + 2)^2$ as a composition $h = f \circ g$ of two functions f and g.

Solution Here we can take the inside function to be $g(x) = 3x + 2$ and the outside function to be $f(t) = t^2$, so that

$$(f \circ g)(x) = f[g(x)] = f(3x + 2) = (3x + 2)^2 = h(x).$$

■

In the solution of Example 5, we have chosen to write the squaring function as $f(t) = t^2$ rather than as $f(x) = x^2$, since x was used for the independent variable of the function $g(x) = 3x + 2$. Of course, we could just as well have written $f(x) = x^2$, since it is the *rule* f that is important, and no special significance is attached to the symbols used to denote dependent and independent variables. We would still obtain $(f \circ g)(x) = h(x)$.

Example 6 A conveyer dumps sand into a pile which maintains the form of a right circular cone whose height h meters is always the same as the radius of the base (Figure 3). The function g defined by

$$g(V) = \left(\frac{3V}{\pi}\right)^{1/3}$$

expresses the height h meters of the pile in terms of the volume V cubic meters of sand in the pile, so that

$$h = g(V)$$

holds for $V \geq 0$. If the conveyer delivers sand to the pile at a constant rate of r cubic meters per minute, and if the original volume of the conical pile at time $t = 0$ minutes was k cubic meters, then the function f defined by

$$f(t) = rt + k$$

expresses the volume V cubic meters of sand in the pile at time t minutes, so that

$$V = f(t)$$

holds for $t \geq 0$. **(a)** Interpret the composite function $g \circ f$. **(b)** Find a formula for the height h of the pile in terms of the time t.

Figure 3

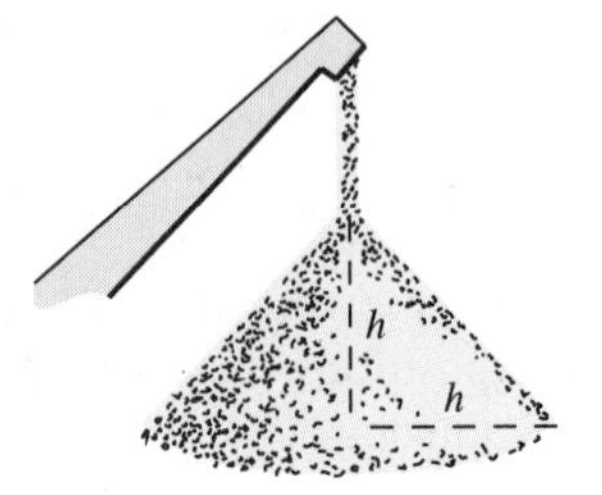

Solution **(a)** We have $h = g(V)$ and $V = f(t)$, so that

$$h = g[f(t)] = (g \circ f)(t).$$

Thus, the composite function $g \circ f$ may be used to express the height of the pile as a function of the time t.

(b) By part (a), we have

$$h = g[f(t)] = \left[\frac{3 \cdot f(t)}{\pi}\right]^{1/3} = \left[\frac{3(rt + k)}{\pi}\right]^{1/3}.$$

■

Iteration and Function Composition

The idea of an *iterative* routine for computing successively better and better approximations to a desired quantity q was first mentioned in Section 2.6. The fixed procedure to be repeated or iterated can be regarded as a function f that maps an approximation of q into an even better approximation. Thus, if $a \approx q$, then

$$a \longmapsto f(a) \approx q$$

produces an improved approximation. Consequently,

$$f(a) \longmapsto f[f(a)] = (f \circ f)(a) \approx q$$

yields an even better approximation, which in turn can be improved upon,

$$(f \circ f)(a) \longmapsto f[(f \circ f)(a)] = (f \circ f \circ f)(a) \approx q,$$

and so forth.

The functions

$$f, \quad f \circ f, \quad f \circ f \circ f, \quad f \circ f \circ f \circ f, \quad \ldots$$

obtained by repeatedly composing f with itself are called the **iterates** of f. In an iterative routine, the numbers

$$a, \quad f(a), \quad (f \circ f)(a), \quad (f \circ f \circ f)(a), \quad (f \circ f \circ f \circ f)(a), \quad \ldots$$

obtained by applying the iterates of f to an initial approximation $a \approx q$, produce better and better approximations to q. Thus, on a calculator or computer, we can begin with $a \approx q$, compute $f(a)$ to obtain a better approximation, compute $f[f(a)]$ for an even better approximation, and continue in this way until we finally arrive at a number p that no longer changes when we apply f, so that

$$f(p) = p.$$

Then the approximation $p \approx q$ will be correct to within the limits of accuracy of our calculating machine. A number p such that $f(p) = p$ is called a **fixed point** of the function f.

C Example 7 The ancient Babylonians discovered an iterative method for approximating the square root of a positive number c. The **Babylonian method** is based on the observation that if $a \approx \sqrt{c}$, then $c/a \approx \sqrt{c}$. Furthermore, if, say, $a > \sqrt{c}$, then $c/a < \sqrt{c}$, so the average $\frac{1}{2}\left(a + \frac{c}{a}\right)$ of a and c/a should be a better approximation to $\sqrt{c}$ than either a or c/a. Thus, if we define

$$f(x) = \frac{1}{2}\left(x + \frac{c}{x}\right)$$

for $x > 0$ and apply the successive iterates of f to an initial approximation a for $\sqrt{c}$, we should obtain better and better approximations to $\sqrt{c}$. Use this method, starting with the initial approximation $a = 1$, to obtain an approximation to $\sqrt{2}$.

Solution

Taking $c = 2$, we have

$$f(x) = \frac{1}{2}\left(x + \frac{2}{x}\right)$$

for $x > 0$. Using a 10-digit calculator, we find that

$$f(1) = 1.5$$
$$f(1.5) = 1.416666667$$
$$f(1.416666667) = 1.414215686$$
$$f(1.414215686) = 1.414213562$$
$$f(1.414213562) = 1.414213562$$

Thus,

$$\sqrt{2} \approx 1.414213562.$$

■

We note that the calculation in Example 7 gives $\sqrt{2}$ correct to nine decimal places (the limit of accuracy of our calculator) after only four iterations.

Problem Set 3.6

In Problems 1 to 8, find **(a)** $(f + g)(x)$, **(b)** $(f - g)(x)$, **(c)** $(f \cdot g)(x)$, **(d)** $\left(\frac{f}{g}\right)(x)$, and **(e)** the domain of $\frac{f}{g}$.

1. $f(x) = 5x + 2$, $g(x) = 2x - 5$

2. $f(x) = |x|$, $g(x) = x - |x|$

3. $f(x) = x^2$, $g(x) = 4$

4. $f(x) = ax + b$, $g(x) = cx + d$

5. $f(x) = 2x - 5$, $g(x) = x^2 + 1$

6. $f(x) = 1 + \frac{1}{x}$, $g(x) = 1 - \frac{1}{x}$

7. $f(x) = \frac{5x}{2x - 1}$, $g(x) = \frac{3x + 1}{2x - 1}$

8. $f(x) = \sqrt{x - 3}$, $g(x) = \frac{1}{\sqrt{x - 3}}$

In Problems 9 and 10, use the method of adding or subtracting ordinates to sketch the graph of h.

9. $f(x) = x^2 - 2$, $g(x) = 1 - \frac{x}{2}$, $h = f + g$

10. $f(x) = 1 - x^2$, $g(x) = x - 1$, $h = f - g$

In Problems 11 and 12, find the indicated values if $f(x) = x - 3$ and $g(x) = x^2 + 4$.

11. **(a)** $(f \circ g)(4)$ **(b)** $(f \circ g)(\sqrt{2})$
(c) $(g \circ f)(4.73)$ **(d)** $(g \circ f)(-2.08)$

12. **(a)** $(f \circ f)(3)$ **(b)** $(g \circ g)(-3)$
(c) $[f \circ (g \circ f)](2)$ **(d)** $[(f \circ g) \circ f](2)$

In Problems 13 to 20, find **(a)** $(f \circ g)(x)$, **(b)** $(g \circ f)(x)$, and **(c)** $(f \circ f)(x)$. In each case, specify the domain of the composite function.

13. $f(x) = 3x$, $g(x) = x + 1$

14. $f(x) = ax + b$, $g(x) = cx + d$

15. $f(x) = x^2$, $g(x) = \sqrt{x}$

16. $f(x) = x^3 + 1$, $g(x) = \sqrt[3]{x - 1}$

17. $f(x) = x + x^{-1}$, $g(x) = \sqrt{x - 1}$

18. $f(x) = 1 - 5x$, $g(x) = |2x + 3|$

19. $f(x) = \dfrac{1}{2x - 3}$, $g(x) = 2x - 3$

20. $f(x) = \dfrac{Ax + B}{Cx + D}$, $g(x) = \dfrac{ax + b}{cx + d}$

In Problems 21 and 22, let $f(x) = 4x$, $g(x) = x^2 - 3$, and $h(x) = \sqrt{x}$. Express each function as a composition of functions chosen from f, g, and h.

21. **(a)** $F(x) = \sqrt{x^2 - 3}$ **(b)** $G(x) = (\sqrt{x})^2 - 3$
(c) $H(x) = 2\sqrt{x}$ **(d)** $K(x) = 4x^2 - 12$

22. **(a)** $Q(x) = 4\sqrt{x}$ **(b)** $q(x) = 16x^2 - 3$
(c) $r(x) = \sqrt[4]{x}$ **(d)** $s(x) = x^4 - 6x^2 + 6$

In Problems 23 to 30, express each function h as a composition $h = f \circ g$ of two simpler functions f and g.

23. $h(x) = (2x^2 - 5x + 1)^{-7}$

24. $h(x) = |2x^2 - 3|$

25. $h(x) = \left(\dfrac{1 + x^2}{1 - x^2}\right)^5$

26. $h(x) = \dfrac{3(x^2 - 1)^2 + 4}{1 - (x^2 - 1)^2}$

27. $h(x) = \sqrt{\dfrac{x + 1}{x - 1}}$

28. $h(x) = \dfrac{1}{(4x + 5)^5}$

29. $h(x) = \dfrac{|x + 1|}{x + 1}$

30. $h(x) = \sqrt{1 - \sqrt{x - 1}}$

31. Let $f(x) = 2x + 3$, $g(x) = x^2$, and $F(x) = \frac{1}{2}(x - 3)$. Find:

(a) $(f \circ F)(x)$ **(b)** $(F \circ f)(x)$
(c) $[f \circ (g \circ F)](x)$ **(d)** $[(f \circ g) \circ F](x)$
(e) $[(f \circ g) + (f \circ F)](x)$

32. Suppose that user-definable keys f and g on a programmable calculator are programmed so that $f: x \longmapsto 3x^2 - 2$ and $g: x \longmapsto 5x - 3$. Draw a mapping diagram (see Figure 2, page 201) to show the effect of entering a number x and **(a)** touching first the g key, then the f key; **(b)** touching first the f key, then the g key; **(c)** touching the g key twice in succession.

33. A company manufacturing integrated circuits finds that its production cost in dollars for manufacturing x of these circuits is given by the function $C(x) = 50{,}000 + 10{,}000\sqrt[3]{x + 1}$. It sells the integrated circuits to distributors for \$10 each, and the demand is so high that all the manufactured circuits are sold.

(a) Write a formula for the revenue $R(x)$ in dollars to the company if x integrated circuits are sold.

(b) Write a formula for the profit function $P(x)$ in dollars.

(c) Find the profit if $x = 46{,}655$.

34. The labor force $F(x)$ persons required by a certain industry to manufacture x units of a product is given by $F(x) = \frac{1}{2}\sqrt{x}$. At present, there is a demand for 40,000 units of the product, and this demand is increasing at a constant rate of 10,000 units per year.

(a) Write an equation for the function $u(t)$ that gives the number of units that will be demanded t years from now.

(b) Write a formula for $(F \circ u)(t)$ and interpret the resulting expression.

35. A baseball diamond is a square 90 feet on each side (Figure 4). Suppose a ball is hit directly toward third

Figure 4

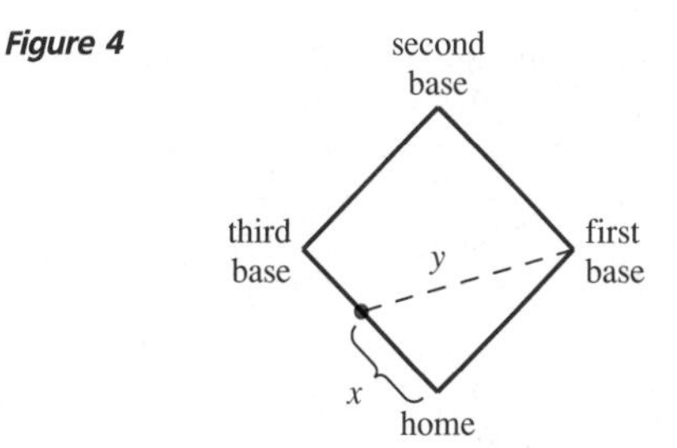

base at the rate of 50 feet per second. Let y denote the distance in feet of the ball from first base, let x denote its distance from home plate, and let t denote the elapsed time in seconds since the ball was hit. Here y is a function of x, say $y = f(x)$, and x is a function of t, say $x = g(t)$.

(a) Find formulas for $f(x)$ and $g(t)$.

(b) Find a formula for $(f \circ g)(t)$.

(c) Explain why $y = (f \circ g)(t)$.

36. An offshore oil well begins to leak, and the oil slick begins to spread on the surface of the water in a circular pattern with a radius $r(t)$ kilometers. The radius depends on the time t in hours since the beginning of the leak, according to the equation $r(t) = 0.5 + 0.25t$.

(a) If $A(x) = \pi x^2$, obtain a formula for $(A \circ r)(t)$ and interpret the resulting expression.

C **(b)** Find the area of the oil slick 4 hours after the beginning of the leak. Round off your answer to two decimal places.

37. The area A of an equilateral triangle depends on its side length x, and the side length x depends on the perimeter p of the triangle. In fact, $A = f(x)$ and $x = g(p)$, where $f(x) = (\sqrt{3}/4)x^2$ and $g(p) = (1/3)p$. Obtain a formula for $(f \circ g)(p)$ and interpret the resulting expression.

38. If f, g, and h are three functions, show that $(f \circ g) \circ h$ is the same function as $f \circ (g \circ h)$.

39. Let $f(x) = 5x + 3$ and $g(x) = 3x + k$, where k is a constant. Find a value of k for which $f \circ g$ and $g \circ f$ are the same function.

40. **(a)** Is $f \cdot (g + h)$ always the same as $(f \cdot g) + (f \cdot h)$?

(b) Is the function $f \circ (g + h)$ always the same as the function $(f \circ g) + (f \circ h)$?

41. Prove that the composition of two linear functions is again a linear function.

42. **(a)** If $f(x) = mx + b$ with $m \neq 0$, find a function F such that $F \circ f$ is the same as the identity function.

(b) Let $f(x) = cx + 1$, where c is a constant. Find a value of c so that $f \circ f$ is the same as the identity function.

43. Show that $(f \circ g)(x)$ can be obtained graphically as follows: Start at the point $(x, 0)$ on the x axis. Move vertically to the graph of g, then horizontally to the graph of $y = x$, then vertically to the graph of f, and finally horizontally to the point $(0, y)$ on the y axis (Figure 5). Conclude that $y = (f \circ g)(x)$.

Figure 5

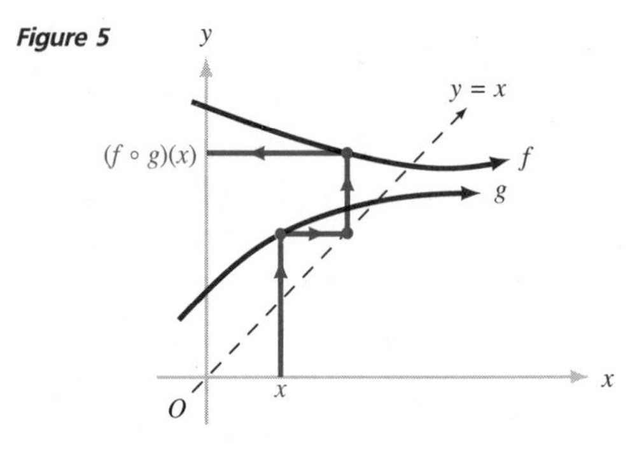

44. Show how to modify the procedure in Problem 43 to find the values $f(x)$, $(f \circ f)(x)$, $(f \circ f \circ f)(x)$, $(f \circ f \circ f \circ f)(x)$, and so on, of the successive iterates of f at x.

C **45.** Use the Babylonian method and a calculator to find an approximation to $\sqrt{3}$.

C **46.** In the life sciences, a **simple recursive model** for population growth is a function f that predicts the population $f(p)$ of an organism 1 unit of time later if the current population is p. (For instance, the unit of time might be the gestation period of the organism.) Thus, according to this model, if p is the current population, then $f(p)$ is the population 1 unit of time later, $f[f(p)] = (f \circ f)(p)$ is the population 2 units of time later, $f[(f \circ f)(p)] = (f \circ f \circ f)(p)$ is the population 3 units of time later, and so forth. One of the most popular simple recursive growth models is the **discrete logistic model** $f(p) = [\![Bp - Ap^2]\!]$, where A and B are nonnegative constants depending on the birth and death rates of the organism and $[\![x]\!]$ denotes the greatest integer less than or equal to x. Suppose that $B = 2.6$ and $A = 0.08$ for a certain organism in a particular habitat. Starting with $p = 4$ organisms, use the discrete logistic model to predict the population 1, 2, 3, 4, 5, and 6 units of time later.

47. Let c be a positive constant and let $f(x) = \frac{1}{2}\left(x + \frac{c}{x}\right)$ be the function used in Example 7. If p is a fixed point of f, prove that $p = \sqrt{c}$.

48. If f is a simple recursive model for population growth as in Problem 46, then a positive fixed point p for f is called a **stable population** for the organism in question. Find the value of the stable population for the discrete logistic model $f(p) = [\![Bp - Ap^2]\!]$ if $B = 2.6$ and $A = 0.08$.

3.7 Inverse Functions

Many calculators have both a squaring key (marked x^2) and a square-root key (marked $\sqrt{x}$). For *nonnegative numbers* x, the functions represented by these keys "undo each other" in the sense that

$$x \xmapsto{\text{squaring function}} x^2 \xmapsto{\text{square-root function}} \sqrt{x^2} = x$$

and

$$x \xmapsto{\text{square-root function}} \sqrt{x} \xmapsto{\text{squaring function}} (\sqrt{x})^2 = x.$$

Two functions related in such a way that each "undoes" what the other "does" are said to be **inverses** of each other. Most scientific calculators not only have keys corresponding to a number of important functions, but they can also calculate the values of the inverses of these functions.

In order that two functions f and g be inverses of one another, we must have the following: For every value of x in the domain of f, $f(x)$ is in the domain of g and

$$x \xmapsto{f} f(x) \xmapsto{g} g[f(x)] = x;$$

likewise, for every value of x in the domain of g, $g(x)$ is in the domain of f and

$$x \xmapsto{g} g(x) \xmapsto{f} f[g(x)] = x.$$

This leads us to the following definition.

Definition 1 **Inverse Functions**

The functions f and g are said to be **inverses** of each other if and only if

$$g[f(x)] = x$$

for every value of x in the domain of f, and

$$f[g(x)] = x$$

for every value of x in the domain of g.

A function f for which such a function g exists is said to be **invertible.**

In Examples 1 and 2, show that the functions f and g are inverses of each other.

Example 1 $f(x) = 5x - 1$ and $g(x) = \dfrac{x+1}{5}.$

Solution The domain of f is the set $\mathbb{R}$ of all real numbers. For every real number x, we have

$$g[f(x)] = g(5x-1) = \frac{(5x-1)+1}{5} = \frac{5x}{5} = x.$$

The domain of g is also the set $\mathbb{R}$ of all real numbers, and for every real number x, we have

$$f[g(x)] = f\left(\frac{x+1}{5}\right) = 5\left(\frac{x+1}{5}\right) - 1 = x + 1 - 1 = x.$$ ■

Example 2 $f(x) = \dfrac{2x-1}{x}$ and $g(x) = \dfrac{1}{2-x}.$

Solution The domain of f is the set of all nonzero real numbers. If $x \neq 0$, we have

$$g[f(x)] = g\left(\frac{2x-1}{x}\right) = \frac{1}{2-\left(\dfrac{2x-1}{x}\right)} = \frac{x}{\left[2-\left(\dfrac{2x-1}{x}\right)\right]x}$$

$$= \frac{x}{2x-(2x-1)} = \frac{x}{1} = x.$$

The domain of g is the set of all real numbers except 2. If $x \neq 2$, we have

$$f[g(x)] = f\left(\frac{1}{2-x}\right) = \frac{2\left(\dfrac{1}{2-x}\right) - 1}{\left(\dfrac{1}{2-x}\right)} = \left[2\left(\frac{1}{2-x}\right) - 1\right](2-x)$$

$$= 2 - (2-x) = x.$$ ■

Figure 1

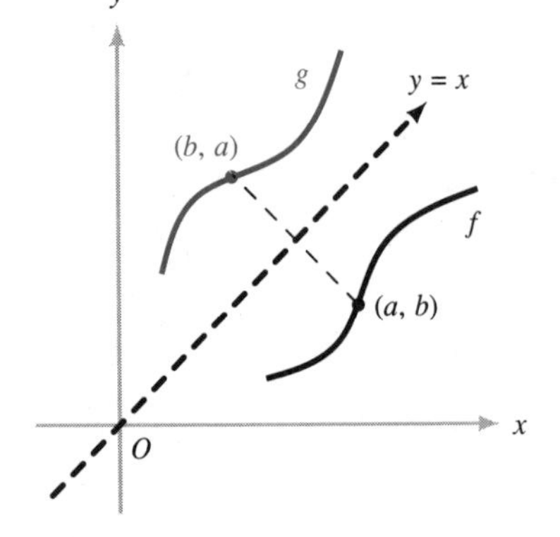

Geometrically, f and g are inverses of each other if and only if the graph of g is the mirror image of the graph of f across the straight line $y = x$ (Figure 1).

This will be clear if you notice first that the mirror image of a point (a, b) across the line $y = x$ is the point (b, a). But, if f and g are inverses of each other and if (a, b) belongs to the graph of f, then $b = f(a)$, so $g(b) = g[f(a)] = a$; that is, (b, a) belongs to the graph of g. Similarly, you can show that if (b, a) belongs to the graph of g, then (a, b) belongs to the graph of f (Problem 10).

Figure 2

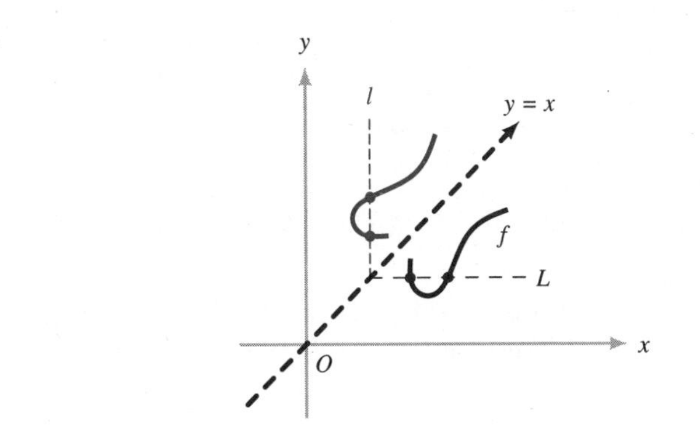

Not every function is invertible. Indeed, consider the function f whose graph appears in Figure 2. The mirror image of the graph of f across the line $y = x$ isn't the graph of a function, because there is a vertical line l that intersects it more than once. (Recall the vertical-line test, page 174.) Notice that the horizontal line L obtained by reflecting l across the line $y = x$ intersects the graph of f more than once. These considerations provide the basis of the following test (see Problem 24a).

Horizontal-Line Test

A function f is invertible if and only if no horizontal straight line intersects its graph more than once.

Example 3 Use the horizontal-line test to determine whether or not the functions graphed in Figure 3 are invertible.

Figure 3

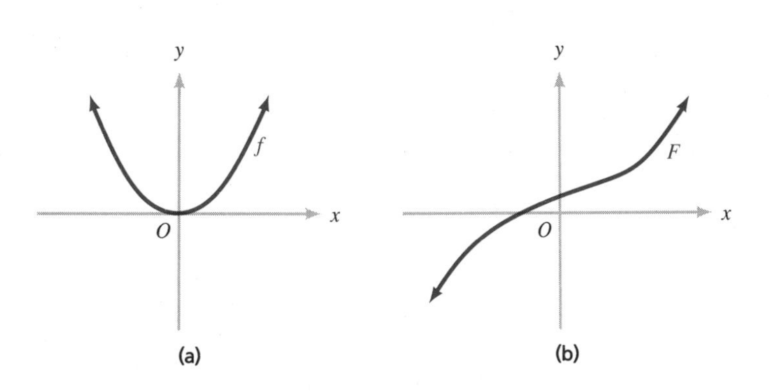

Solution **(a)** Any horizontal line drawn above the origin will intersect the graph of f *twice.* Therefore, f is not invertible.

(b) No horizontal line intersects the graph of F more than once; hence, F is invertible. ■

The horizontal-line test can be restated in terms of the following condition.

Definition 2 **One-to-One Function**

A function f is said to be **one-to-one** if, whenever a and b are in the domain of f and $f(a) = f(b)$, it follows that $a = b$.

In Problem 24b, we ask you to use the horizontal-line test to show that *a function f is invertible if and only if it is one-to-one.*

Example 4 Show that $f(x) = 3x - 7$ is a one-to-one function.

Solution Suppose that a and b are real numbers and $f(a) = f(b)$. Then

$$3a - 7 = 3b - 7$$

$$3a = 3b \qquad \text{(We added 7 to both sides.)}$$

$$a = b \qquad \text{(We divided both sides by 3.)}$$

Therefore, $f(x) = 3x - 7$ is one-to-one.

If a function f is invertible, there is *exactly one* function g such that f and g are inverses of each other; indeed, g is the one and only function whose graph is the mirror image of the graph of f across the line $y = x$. We call g the **inverse** of the function f, and we write $g = f^{-1}$. The notation f^{-1} is read "f inverse."* If you imagine that the graph of f is drawn with wet ink, then the graph of f^{-1} would be the imprint obtained by folding the paper along the line $y = x$ (Figure 4). If f is invertible, then *the domain of f^{-1} is the range of f and the range of f^{-1} is the domain of f.* (Do you see why?)

Figure 4

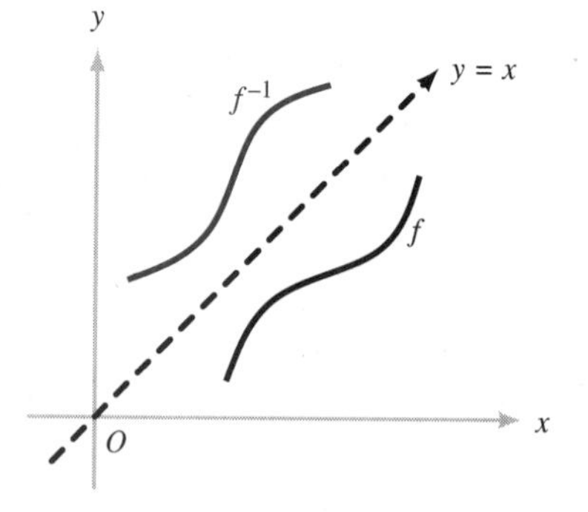

The fact that the graph of f^{-1} is the set of all points (x, y) such that (y, x) belongs to the graph of f provides the basis for the following procedure.

* Note that -1, used in this way to denote an inverse function, must not be confused with an exponent.

Algebraic Method for Finding f^{-1}

Step 1. Write the equation $y = f(x)$ that defines f.

Step 2. Interchange x and y in the equation obtained in step 1.

Step 3. Solve the equation in step 2 for y in terms of x to get $y = f^{-1}(x)$.

In step 2, you must change y to x and change all x's to y's so that the equation becomes

$$x = f(y).$$

In step 3, if the equation $x = f(y)$ has more than one solution, then the function f is not one-to-one and therefore is not invertible. However, if you can solve the equation for a *unique* y, then the resulting equation $y = f^{-1}(x)$ will define the inverse function f^{-1}. The interchange of variables in step 2 enables you to graph both f and f^{-1} on the same xy coordinate system.

Example 5 **(a)** Use the algebraic method to find the inverse of the function $f(x) = x^2$, for $x \geq 0$, and sketch the graphs of f and f^{-1} on the same coordinate system.
(b) Check that $f^{-1}[f(x)] = x$ for all values of x in the domain of f.
(c) Check that $f[f^{-1}(x)] = x$ for all values of x in the domain of f^{-1}.

Figure 5

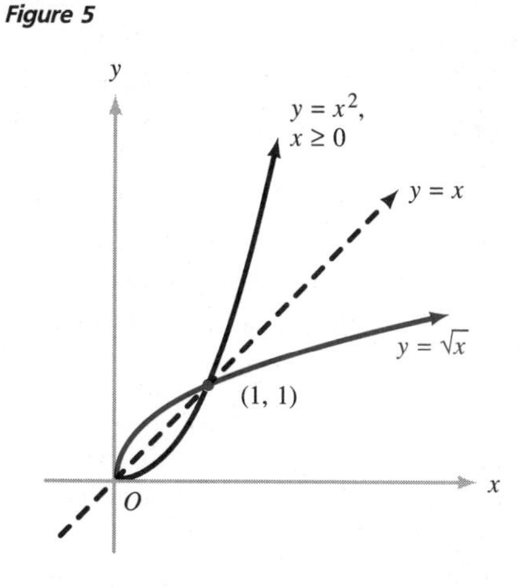

Solution

(a) We carry out the algebraic procedure:

Step 1. $y = x^2$ for $x \geq 0$

Step 2. $x = y^2$ for $y \geq 0$

Step 3. Solving the equation $x = y^2$ for y and using the condition that $y \geq 0$, we obtain $y = \sqrt{x}$; therefore, $f^{-1}(x) = \sqrt{x}$ (Figure 5).

(b) For $x \geq 0$, we have

$$f^{-1}[f(x)] = f^{-1}(x^2) = \sqrt{x^2} = x.$$

(c) For $x \geq 0$, we have

$$f[f^{-1}(x)] = f(\sqrt{x}) = (\sqrt{x})^2 = x.$$

■

Example 6 Use the algebraic method to find f^{-1} if $f(x) = 2x + 1$.

Solution We carry out the algebraic procedure:

Step 1. $y = 2x + 1$

Step 2. $x = 2y + 1$

Step 3. $2y + 1 = x$

$$2y = x - 1$$

$$y = \tfrac{1}{2}(x - 1); \text{ therefore, } f^{-1}(x) = \tfrac{1}{2}(x - 1).$$

It is easy to check that $f^{-1}[f(x)] = x$ and $f[f^{-1}(x)] = x$ for all real values of x (see Problem 1). ■

In working with inverses of functions, *you must be careful not to confuse* $f^{-1}(x)$ *and* $[f(x)]^{-1}$. Notice that $f^{-1}(x)$ is the value of the function f^{-1} at x, while $[f(x)]^{-1} = \dfrac{1}{f(x)}$ is the reciprocal of the value of the function f at x. For instance, in Example 6, above, $f^{-1}(x) = \tfrac{1}{2}(x - 1)$, whereas $[f(x)]^{-1} = \dfrac{1}{2x + 1}$.

Our discussion of the inverse of a function can be summarized as follows:

> Let f be a function. Then, if any one of the following statements is true, they are all true:
>
> **(i)** There is a function g such that $g[f(x)] = x$ for every value of x in the domain of f, and $f[g(x)] = x$ for every value of x in the domain of g.
>
> **(ii)** f is invertible and has a unique inverse function, f^{-1}.
>
> **(iii)** No horizontal straight line intersects the graph of f more than once.
>
> **(iv)** f is one-to-one.
>
> **(v)** The equation $x = f(y)$ can be solved uniquely for y in terms of x.

When conventional symbols are used to represent quantities in applied mathematics, it is necessary to modify the algebraic method for finding inverses by omitting the second step in which the variables are interchanged. The meaning of the symbols would be lost if they were interchanged. The technique is illustrated in the following example.

Example 7 In economics, it is often assumed that there is a functional relationship between the price p charged for a commodity and the quantity q of the commodity demanded in the marketplace. Typically, as the price p is increased, the demand q decreases. This situation is represented by a mathematical model $q = f(p)$ in which f is a decreasing function. Such a function is necessarily one-to-one (Problem 30), and the inverse function f^{-1} may then be used to express the price p in terms of the demand q. If

$$q = f(p) = -\frac{p}{2} + 100$$

find p in terms of q and find f^{-1}.

Solution Because of the economic interpretation, $0 \le p$ and $0 \le q$. Therefore,

$$0 \le p \qquad \text{and} \qquad 0 \le -\frac{p}{2} + 100;$$

that is, $0 \le p \le 200$, so the domain of the function f is the interval $[0, 200]$. Solving the given equation for p in terms of q, we find that

$$p = 200 - 2q,$$

and therefore

$$f^{-1}(q) = 200 - 2q.$$

Since $0 \le p = f^{-1}(q)$, we must have $q \le 100$, so the domain of f^{-1} is the interval $[0, 100]$. ■

Problem Set 3.7

In Problems 1 to 6, show that the functions f and g are inverses of each other.

1. $f(x) = 2x + 1$ and $g(x) = \dfrac{x - 1}{2}$
2. $f(x) = x^3$ and $g(x) = \sqrt[3]{x}$
3. $f(x) = \dfrac{1}{x}$ and $g(x) = \dfrac{1}{x}$
4. $f(x) = \dfrac{2x - 3}{3x - 2}$ and $g(x) = \dfrac{2x - 3}{3x - 2}$
5. $f(x) = \sqrt[3]{x + 8}$ and $g(x) = x^3 - 8$
6. $f(x) = x^2 + 1$ for $x \ge 1$ and $g(x) = \sqrt{x - 1}$
7. Use the horizontal-line test to determine whether or not each of the functions graphed in Figure 6 is invertible.

Figure 6

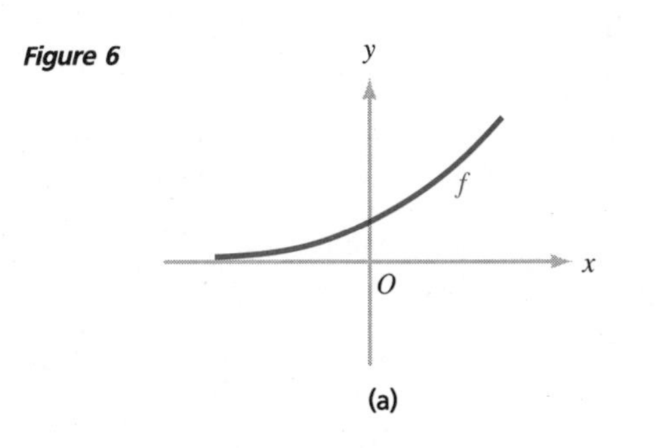

(a)

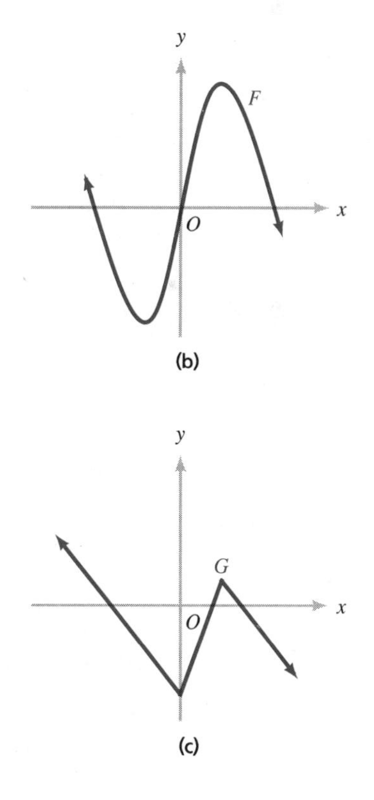

(b)

(c)

8. (a) Under what conditions is a linear function $f(x) = mx + b$ one-to-one?

   (b) If a linear function f is one-to-one, show that f^{-1} is also a linear function.

9. The three functions graphed in Figure 7 are invertible. In each case, obtain the graph of the inverse function by reflecting the given graph across the line $y = x$.

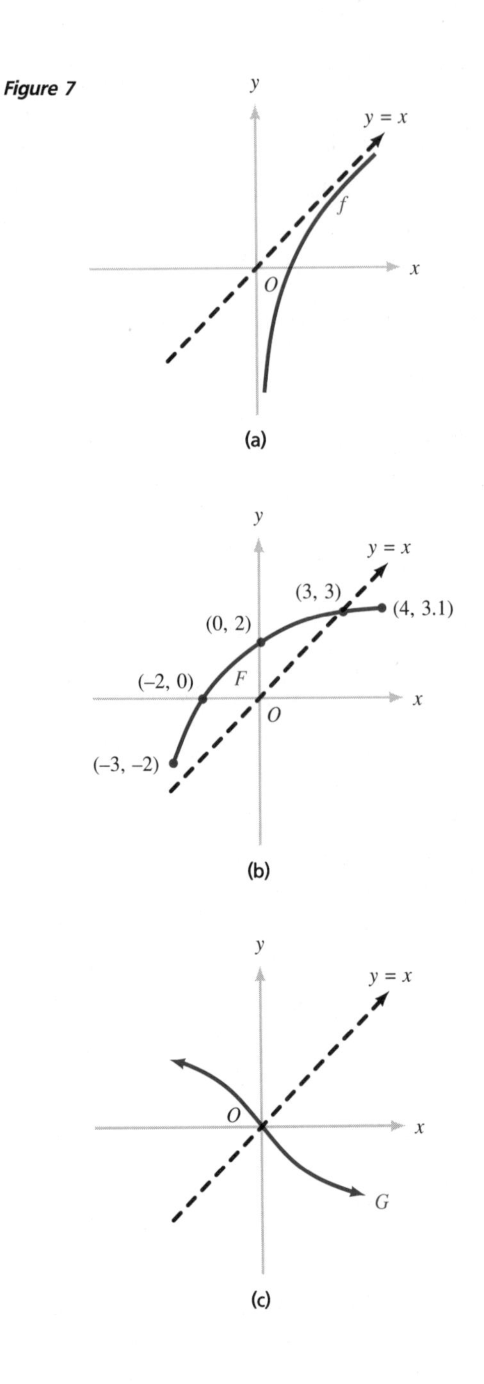

Figure 7

10. **(a)** If f and g are inverses of each other and $P = (b, a)$ belongs to the graph of g, show that $Q = (a, b)$ belongs to the graph of f.

    (b) Show that the graph of $y = x$ is the perpendicular bisector of the line segment $\overline{PQ}$.

In Problems 11 to 22, **(a)** show that f is one-to-one, **(b)** use the algebraic method to find $f^{-1}(x)$, **(c)** check that $f^{-1}[f(x)] = x$ for all values of x in the domain of f, and **(d)** check that $f[f^{-1}(x)] = x$ for all x in the domain of f^{-1}. Also, **(e)** in Problems 11 to 16, sketch the graphs of f and f^{-1} on the same coordinate system. gc If you wish, you may use a graphing calculator to check your work.

11. $f(x) = 7x - 13$

12. $f(x) = -\frac{1}{2}x + 3$

13. $f(x) = -x^2$ for $x \leq 0$

14. $f(x) = 4 - x^2$ for $x \geq 0$

15. $f(x) = 1 + \sqrt{x}$

16. $f(x) = (x - 2)^2$ for $x \geq 2$

17. $f(x) = (x - 1)^2 + 1$ for $x \geq 1$

18. $f(x) = (x + 1)^3 - 2$

19. $f(x) = 1 - 2x^3$

20. $f(x) = -\dfrac{1}{x} - 1$

21. $f(x) = \dfrac{3x - 7}{x + 1}$

22. $f(x) = (x + 3)^5 - 2$

23. Sketch the graph of $f(x) = |x - 1|$ and determine whether or not f is one-to-one.

24. **(a)** Complete the argument to justify the horizontal-line test by showing that, if no horizontal line intersects the graph of f more than once, then f is invertible.

    (b) By using the horizontal-line test, show that f is invertible if and only if it is one-to-one.

25. Figure 8 shows the graph of a function f. Sketch the graph of the function defined by each equation.

    (a) $y = f^{-1}(x)$ **(b)** $y = f^{-1}(x) - 2$

    (c) $y = f^{-1}(x - 1)$ **(d)** $y = -f^{-1}(x)$

Figure 8

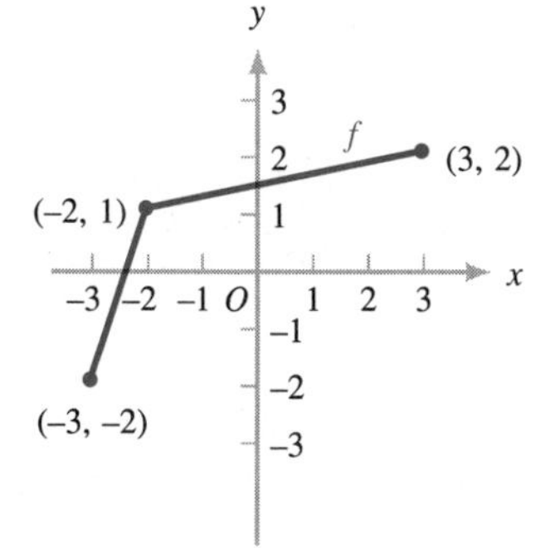

26. Use the horizontal-line test to determine whether or not each function is one-to-one.

(a) $f(x) = \begin{cases} x^2 & \text{if } -2 \le x \le 0 \\ x^2 + 2 & \text{if } x > 0 \end{cases}$

(b) $g(x) = \begin{cases} -x^2 & \text{if } -2 \le x \le 0 \\ x^2 + 2 & \text{if } x > 0 \end{cases}$

27. If a, b, c, and d are constants such that $ad \ne bc$, use the algebraic method to find the inverse of

$$f(x) = \frac{ax + b}{cx + d}.$$

28. If f is invertible, show that the domain of f coincides with the range of f^{-1}.

29. The linear function $f(x) = x$ is its own inverse. Find all linear functions that are their own inverses.

30. Suppose that f is a function and that the domain of f is an interval I. If f is either increasing on I or decreasing on I, show that f is one-to-one and therefore invertible.

31. Find all values of the constants a, b, c, and d in Problem 27 for which the function f is its own inverse.

32. If $f(x) = \sqrt{4 - x^2}$ for $0 \le x \le 2$, show that f is its own inverse.

33. Sketch the graph of any invertible function f; then turn the paper over, rotate it 90° clockwise, and hold it up to the light. Through the paper you will see the graph of f^{-1}. Why does this procedure work?

34. Show that the functions f and g are inverses of each other if and only if $(g \circ f)(x) = x$ for every x in the domain of f and $(f \circ g)(x) = x$ for every x in the domain of g.

35. Let A, B, and C be constants with $A > 0$ and suppose that

$f(x) = Ax^2 + Bx + C$ for $x \ge \dfrac{-B}{2A}$. Find $f^{-1}(x)$.

36. Is there any invertible function f such that $f^{-1}(x) = [f(x)]^{-1}$ for all values of x in the domain of f?

37. The equation $4p + 10q = 12{,}500$ expresses the relationship between the price p dollars of a product and the number of units q of the product demanded.

(a) Find functions f and g such that $q = f(p)$ and $p = g(q)$.

(b) Show that f and g are inverse functions of each other.

38. The volume V of a sphere of radius r is given by the function $V = f(r) = \frac{4}{3}\pi r^3$ (Figure 9).

(a) Find the domain and the range of f.

(b) Find f^{-1} and use f^{-1} to express a relationship between V and r.

C (c) Sketch graphs of $y = f(x)$ and $y = f^{-1}(x)$ on the same xy coordinate system.

Figure 9

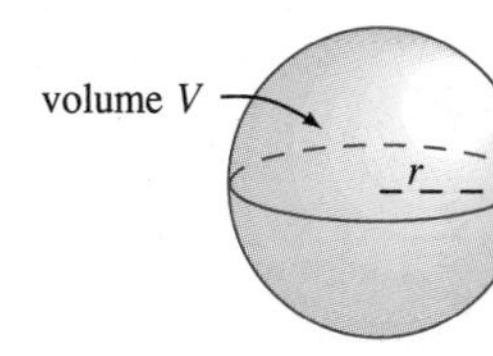

39. Suppose that the concentration C of an anesthetic in body tissues satisfies an equation of the form $t = f(C)$, where t is the elapsed time since the anesthetic was administered. If f is invertible, write an equation for C in terms of t.

40. If bacteria are allowed to reproduce in a culture, the number N of bacteria at time t satisfies an equation of the form $N = f(kt + c)$, where k and c are constants, $k \ne 0$, and f is an invertible function. Write an equation involving f^{-1} that gives t in terms of N.

CHAPTER 3 Review Problem Set

In Problems 1 to 4, find **(a)** the distance $d = |\overline{PQ}|$ between the points P and Q, **(b)** the slope m of $\overline{PQ}$, **(c)** the coordinates of the midpoint M of $\overline{PQ}$, **(d)** an equation of the line containing P and Q, and **(e)** an equation (in standard form) of the circle with center at M and with diameter d.

1. $P = (1, 1), Q = (4, 5)$

2. $P = (-1, 2), Q = (5, -7)$

3. $P = (-2, -5), Q = (\frac{1}{2}, \frac{2}{3})$

C **4.** $P = (4.71, -3.22), Q = (0, 2.71)$

C **5.** What is the perimeter of the triangle with vertices $(-1, 5)$, $(8, -7)$, and $(4, 1)$? Round off your answer to two decimal places.

6. Are the points $(-5, 4)$, $(7, -11)$, $(12, -11)$, and $(0, 4)$ the vertices of a parallelogram?

7. Sketch the line L that contains the point P and has slope m, and find an equation in point-slope form for L.

(a) $P = (5, 2)$ and $m = -\frac{3}{5}$

(b) $P = (-\frac{2}{3}, \frac{1}{2})$ and $m = \frac{3}{2}$

8. In calculus, it is shown that the slope m of the line tangent to the graph of $y = x^2 + 4$ at the point (a, b) is given by $m = 2a$. **(a)** Write an equation of this tangent line. **(b)** Find an equation of the tangent line at $(2, 8)$. **(c)** Sketch the graph of $y = x^2 + 4$ and show the tangent line at $(2, 8)$.

In Problems 9 and 10, find an equation of the line L in **(a)** point-slope form, **(b)** slope-intercept form, and **(c)** general form.

9. L contains the point $(1, -2)$ and is parallel to the line whose equation is $7x - 3y + 2 = 0$.

10. L contains the point $(3, -4)$ and is perpendicular to the line $2x - 5y + 4 = 0$.

11. If (a, b) is a point on the circle $x^2 + y^2 = 9$, find an equation of the line that is tangent to the circle at (a, b).

[*Hint*: The line tangent to a circle at a point is perpendicular to the radius drawn from the center to the point.]

12. Show that the line containing the two points $P = (a, b)$ and $Q = (c, d)$ has an equation of the form $(b - d)x - (a - c)y + ad - bc = 0$.

13. If $f(x) = 3x^2 - 4$, $g(x) = 6 - 5x$, and $h(x) = 1/x$, find the indicated values.

(a) $f(x) - f(2)$ **(b)** $f(x + k) - f(x)$

(c) $f[g(x)]$ **(d)** $g\left(\frac{1}{4 + k}\right)$

(e) $g(x) + g(-x)$ **(f)** $\sqrt{f(-|x|)}$

(g) $\frac{h(x + k) - h(x)}{k}$ **(h)** $\frac{1}{h(4 + k)}$

14. Find the difference quotient $\frac{f(x + h) - f(x)}{h}$ and simplify the resulting expression.

(a) $f(x) = 3x^2 - 2x + 1$

(b) $f(x) = x^{-1/3}$

15. Find the domain of each function.

(a) $f(x) = \frac{1}{x - 1}$

(b) $g(x) = \frac{1}{\sqrt{4 - x^2}}$

(c) $h(x) = \sqrt{1 + x}$

(d) $F(x) = \frac{3}{|x| - x}$

16. Let $g(x) = |x - 5| - |x + 5|$.

(a) Sketch the graph of g.

(b) Determine the intervals over which g is increasing or decreasing.

[*Hint:* Express $g(x)$ without absolute value symbols for the three cases $x \leq -5$, $-5 < x < 5$, and $5 \leq x$.]

17. For each function in Figure 1, find the domain and range; indicate the intervals over which the function is increasing, decreasing, or constant; and determine whether the function is even, odd, or neither.

Figure 1

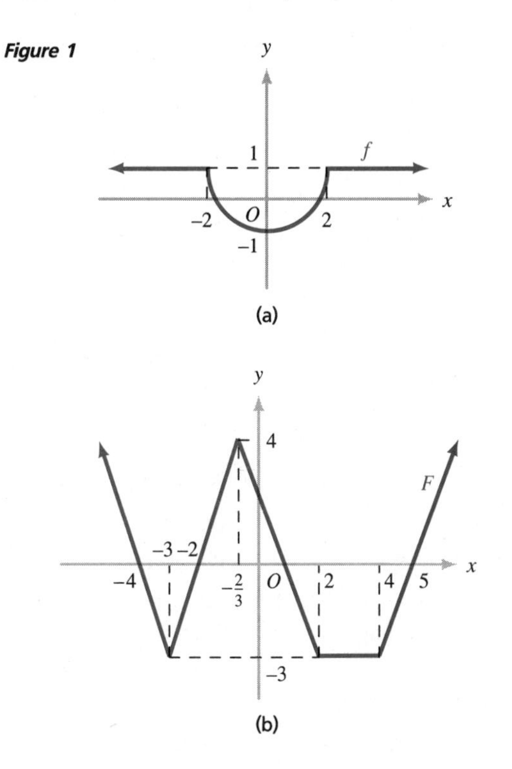

18. If $g(x) = \dfrac{2+x}{2-x}$, find and simplify $\dfrac{g(t) - g(-t)}{1 + g(t)g(-t)}$.

[c] **19.** The function $f: x \longmapsto y$ given by the equation $f(x) = 71.88 + 0.37x$ relates the atmospheric pressure x in centimeters of mercury to the boiling temperature $f(x)$ of water in degrees Celsius at that pressure. Find $f(74)$, $f(75)$, and $f(76)$. Round off your answers to two decimal places.

20. Determine without drawing a graph whether each function is even, odd, or neither, and discuss any symmetry of the graph.

(a) $f(x) = 5x^5 + 3x^3 + x$

(b) $g(x) = (x^4 + x^2 + 1)^{-1}$

(c) $h(x) = (x + 1)x^{-1}$

(d) $F(x) = -x^3|x|$

(e) $G(x) = x^{80} - 5x^6 + 9$

(f) $H(x) = \dfrac{\sqrt{x}}{1+x}$

In Problems 21 to 24, find the domain of the function, sketch its graph, discuss any symmetry of the graph, indicate the intervals where the function is increasing or decreasing, and find the range of the function. Do these things in any convenient order. [c] Use a calculator if you wish. [gc] Alternatively, use a graphing calculator.

21. $F(x) = 2\sqrt{x-2}$

22. $G(x) = x^{1/3}$

23. $h(x) = \begin{cases} x^2 & \text{if } x > 0 \\ -x^2 & \text{if } x \le 0 \end{cases}$

24. $H(x) = \begin{cases} x & \text{if } x < 0 \\ 2x & \text{if } 0 \le x \le 1 \\ 3x^3 - 1 & \text{if } x > 1 \end{cases}$

25. The graph of a function f is shown in Figure 2. Sketch the graph of the function F defined by:

(a) $F(x) = f(x) + 1$

(b) $F(x) = f(x) - 2$

(c) $F(x) = f(x - 1)$

(d) $F(x) = f(x + 2)$

(e) $F(x) = -f(x)$.

Figure 2

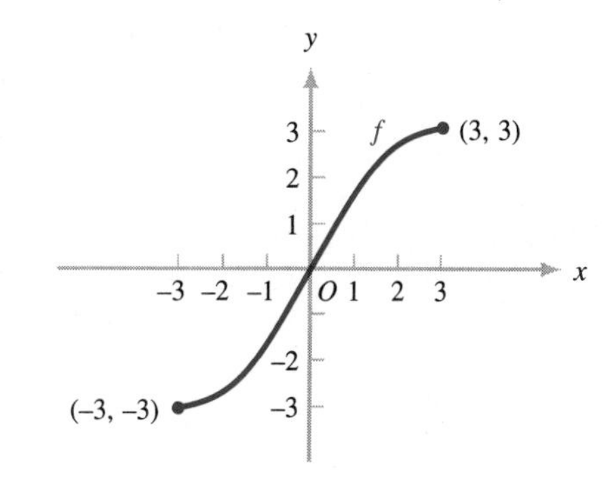

26. The graph in Figure 3a is the graph of a function, but the graph in Figure 3b is not. Explain.

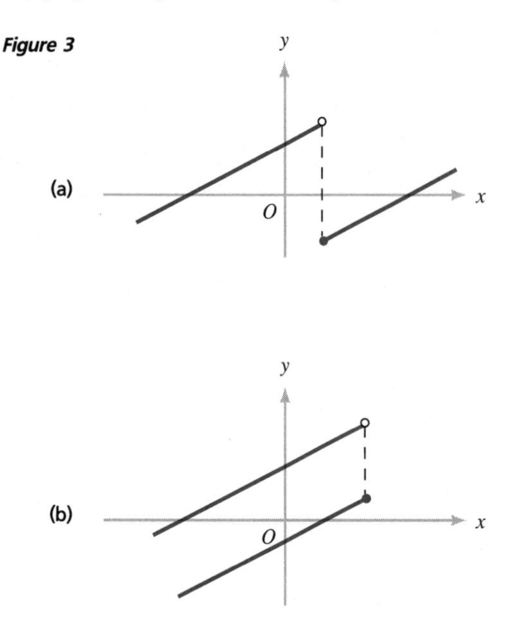

Figure 3

27. Compare the graph of the first function with the graph of the second by sketching them both on the same coordinate system and explaining in words how they are related.

(a) $F(x) = 3x + 2$, $f(x) = 3x - 1$
(b) $G(x) = x^2 - 5$, $g(x) = x^2$
(c) $H(x) = 1 - |x|$, $h(x) = |x|$
(d) $K(x) = \sqrt{x - 1} + 2$, $k(x) = \sqrt{x}$
(e) $Q(x) = \frac{1}{2}(x + 2)^2$, $q(x) = x^2$
(f) $R(x) = 1 - \frac{1}{2}|x + 1|$, $r(x) = |x|$

28. What is wrong with the following proposed definition of a function:

$$f(x) = \begin{cases} x^2 + 1 & \text{if } x \le 0 \\ x^2 - 2 & \text{if } x \ge 0 \end{cases} ?$$

In Problems 29 to 32, find **(a)** $(f + g)(x)$, **(b)** $(f - g)(x)$, **(c)** $(f \cdot g)(x)$, **(d)** $\left(\frac{f}{g}\right)(x)$, and **(e)** $(f \circ g)(x)$.

29. $f(x) = \dfrac{1}{x - 1}$, $g(x) = \dfrac{1}{x + 1}$

30. $f(x) = \dfrac{x + 3}{x - 2}$, $g(x) = \dfrac{x}{x - 2}$

31. $f(x) = x^4$, $g(x) = \sqrt{x + 1}$

32. $f(x) = \dfrac{|x|}{x}$, $g(x) = \dfrac{-x}{|x|}$

33. If $f(x) = x^2 + 1$, $g(x) = x^2 - 1$, and $h(x) = \sqrt{x}$, find:

(a) $[(f + g) \circ h](x)$ **(b)** $[(f - g) \circ h](x)$
(c) $[(f \circ h) + (g \circ h)](x)$ **(d)** $[(f \circ h) - (g \circ h)](x)$

34. Find $(f \circ g)(x)$ if

$$f(x) = \begin{cases} x & \text{if } x \ge 0 \\ 2x & \text{if } x < 0 \end{cases}$$

and

$$g(x) = \begin{cases} -2x & \text{if } x \ge 0 \\ 4x & \text{if } x < 0. \end{cases}$$

35. Let $f(x) = x^2$, $g(x) = \sqrt{x}$, and $h(x) = x + 1$. Express each function as a composition of functions chosen from f, g, and h.

(a) $F(x) = \sqrt{x + 1}$ **(b)** $G(x) = |x|$
(c) $H(x) = \sqrt{x} + 1$ **(d)** $K(x) = x^4$
(e) $P(x) = x^2 + 2x + 1$ **(f)** $p(x) = x + 2$

C **36.** Let $f(x) = 1 + x^5$ and let $g(x) = 1 - \sqrt{x}$. Use a calculator to find each value. Round off to four decimal places.

(a) $(f \circ g)(2.7746)$ **(b)** $[(f \circ g) \circ f](\pi)$
(c) $(g \circ f)(2.7746)$ **(d)** $(g \circ g)(0.0007)$

37. Express each function h as a composition $h = f \circ g$ of two simpler functions f and g.

(a) $h(x) = (4x^3 - 2x + 5)^{-3}$

(b) $h(x) = \sqrt[3]{\dfrac{4 + x^3}{4 - x^3}}$

(c) $h(x) = \dfrac{2(x^2 + 1)^2 + x^2 + 1}{\sqrt{x^2 + 1}}$

38. Which of the following functions is one-to-one?

(a) $f(x) = x$ **(b)** $f(x) = 1/x$
(c) $f(x) = 3x + 5$ **(d)** $f(x) = 3x^2 + 5$

In Problems 39 to 42, find $f^{-1}(x)$ for each function f, verify that $f^{-1}[f(x)] = x$ and $f[f^{-1}(x)] = x$, and sketch the graphs of f and f^{-1} on the same coordinate system. gc If you wish, you may use a graphing calculator to check your work.

39. $f(x) = \frac{1}{5}x + 5$

40. $f(x) = (x - 1)^{-1}$

41. $f(x) = 2\sqrt{x} - 1$

42. $f(x) = \frac{1}{4}x^3$

43. Which of the functions whose graphs are shown in Figure 4 are invertible?

Figure 4

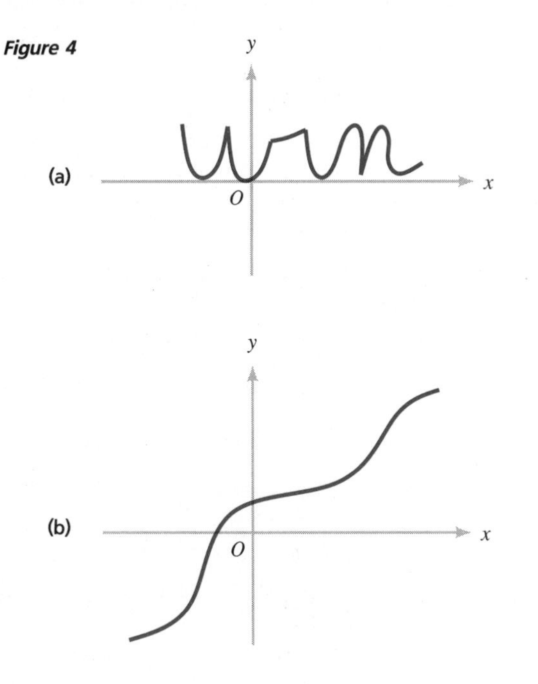

44. If f is invertible, show that f^{-1} is also invertible and that $(f^{-1})^{-1}$ is the same as f.

45. Find the inverse of each function.

(a) $f(x) = \dfrac{x - 1}{x}$ **(b)** $g(x) = x^2 - 3x + 2$ for $x \leq \frac{3}{2}$

46. Suppose that f is invertible, that the domain of f is $\mathbb{R}$, and that the range of f is $\mathbb{R}$. What function is $f^{-1} \circ f$?

C **47.** The power delivered by a wind-powered generator is given by

$$P(x) = kx^3 \text{ horsepower,}$$

where x is the speed of the wind in miles per hour and k is a constant depending on the size of the generator and its efficiency. For a certain wind-powered generator, the constant $k = 3.38 \times 10^{-4}$. Sketch the graph of the function P for this generator and determine how many horsepower are generated when the wind speed is 35 miles per hour.

48. Solar cells and a wind-powered generator charge batteries that supply electric power to an alternative-energy home. The batteries have a constant internal resistance of r ohms and provide a fixed voltage of E volts. The current I amperes drawn from the batteries depends on the net resistance R ohms of the appliances being used in the home according to the equation $I = \dfrac{E}{r + R}$. The electric power P watts being consumed by these appliances is given by the equation $P = I^2R$. Find a function f such that $P = f(R)$.

49. In economics, it is often assumed that the demand for a commodity is a function of selling price; that is, $q = f(p)$, where p dollars is the selling price per unit of the commodity and q is the number of units that will sell at that price.

(a) Write an equation for the total amount of money in dollars $F(p)$ spent by consumers for the commodity if the selling price per unit is p dollars.

(b) What would it mean to say that there is a value p_0 dollars for which $f(p_0) = 0$?

50. In economics, it is often assumed that the number of units s of a commodity that producers will supply to the marketplace is a function of the selling price p dollars per unit; that is, $s = g(p)$.

(a) Would you expect g to be an increasing or a decreasing function?

(b) Write an equation for the total amount of money in dollars $G(p)$ spent by consumers for the commodity if the selling price per unit is p dollars and all supplied units are purchased.

(c) What would it mean to say that there is a value p_1 dollars for which $g(p_1) = 0$?

51. A manufacturer finds that 100,000 programmable calculators are sold per month at a price of \$50 each, but that only 60,000 are sold per month if the price is \$75 each. Suppose that the demand function f for these calculators is linear (see Problem 49).

(a) Find a formula for $f(p)$, where p dollars is the selling price per calculator and $f(p)$ is the number that will sell at that price.

(b) Find the selling price per calculator if the monthly demand for calculators is 80,000.

(c) What price would be so high that no calculators would be sold?

(d) Find f^{-1} and give an economic interpretation of the equation $p = f^{-1}(q)$.

52. Two college students earn extra money on weekends by delivering firewood in their pickup truck. They have found that they can sell x cords per weekend at a price of p dollars per cord, where $x = 75 - \frac{3}{5}p$. The students buy the firewood from a supplier who charges them C dollars for x cords according to the equation $C = 500 + 15x + \frac{1}{5}x^2$.

(a) Find a function f such that $P = f(p)$, where P dollars is the profit per weekend for the students if they charge p dollars per cord.

(b) Find the profit P dollars if $p = \$95$.

53. An advertising agency claims that a furniture store's revenue will increase by \$20 per month for each additional dollar spent on advertising. The current average monthly sales revenue is \$140,000, with an expenditure of \$100 per month for advertising. Find the equation that relates the store's expected average monthly sales revenue y to the total expenditure x for advertising. Find the value of y if \$400 per month is spent for advertising.

54. Two ships sail from the same port. The first ship leaves at 1:00 A.M. and travels east at a rate of 14 knots (nautical miles per hour). The second ship leaves at 2:00 A.M. and travels north at a rate of 11 knots (Figure 5).

(a) If t is the time in hours after 2:00 A.M. and d is the distance between the ships at time t, find a function f such that $d = f(t)$.

(b) Explain why the function f is one-to-one.

Figure 5

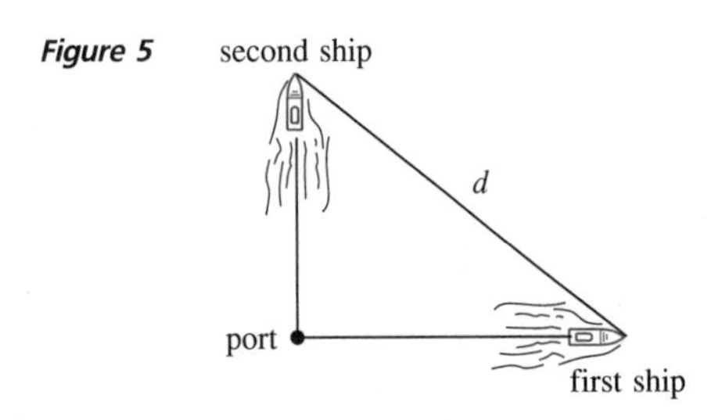

55. A meteorologist is inflating a spherical weather balloon with helium gas in such a way that at the end of t seconds, the radius of the balloon is given by $r = 6\sqrt{t} + 10$ centimeters, $0 \leq t \leq 4$.

(a) If A is the surface area in square centimeters of the balloon at time t, find a function f such that $A = f(t)$.

(b) Explain why the function f is one-to-one.

56. An oil tank consists of a right circular cylinder of radius r meters and height 10 meters on top of a right circular cone of radius r meters and height r meters (Figure 6).

(a) If V cubic meters is the volume of the tank, find a function f such that $V = f(r)$.

(b) Explain why the function f is one-to-one.

Figure 6

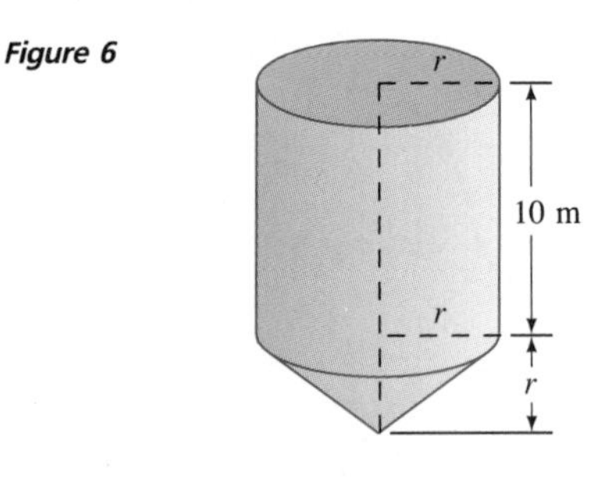

57. A Coast Guard helicopter reports that oil spilling from a ruptured tanker in the Gulf of Alaska is spreading in a circular pattern whose radius is increasing at the rate of 250 meters per hour.

(a) If A is the surface area in square meters of the spill t hours after the rupture occurred, find a function f such that $A = f(t)$.

(b) Explain why the function f is one-to-one.

(c) Find f^{-1}.

CHAPTER 3 Test

1. Let $P = (-1, 5)$ and $Q = (3, -7)$. Find:
   (a) The slope m of the line segment $\overline{PQ}$.
   (b) The midpoint M of the line segment $\overline{PQ}$.
   (c) An equation in slope-intercept form for the line containing P and Q.
   (d) The distance d between P and Q.
   (e) An equation in standard form for the circle with center M that contains the two points P and Q.

2. Find an equation for the line tangent to the circle $(x - 1)^2 + (y - 2)^2 = 25$ at the point $(4, 6)$ by using the fact that the tangent line is perpendicular to the radius drawn from the center of the circle to the point.

3. Let $f(x) = x^2 + x$.
   (a) Evaluate $\dfrac{f(-3)}{f(3)}$.
   (b) Find and simplify $\dfrac{f(a + h) - f(a)}{h}$ for $h \neq 0$.

4. Find the domain of each function.
   (a) $f(x) = \dfrac{1 + x}{\sqrt{x}}$
   (b) $g(x) = \sqrt{x^2 - 1}$
   (c) $h(x) = \dfrac{4}{|x| - x}$

5. Let $f(x) = \begin{cases} 5 - x & \text{if } -1 \leq x \leq 1 \\ x + 3 & \text{if } 1 < x \leq 2. \end{cases}$
   (a) Sketch the graph of f.
   (b) Find the domain and the range of f.
   (c) Determine the intervals on which f is increasing and decreasing.
   (d) Is f a one-to-one function? Why or why not?

6. Determine whether each function is even, odd, or neither and discuss the symmetry, if any, of its graph.
   (a) $f(x) = x^2\sqrt{x^4 + 7}$
   (b) $g(x) = \dfrac{1}{x(x^2 + 1)}$
   (c) $h(x) = 1 - x|x|$.

7. Let $f(x) = 3x - 7$ and $g(x) = x^2$. Find:
   (a) $(f + g)(x)$ **(b)** $(f \cdot g)(x)$
   (c) $\left(\dfrac{f}{g}\right)(x)$ **(d)** $(f \circ g)(x)$
   (e) $f^{-1}(x)$

8. Use the graph of $y = |x|$ and the techniques of shifting and stretching to sketch the graph of each function.
   (a) $f(x) = |x| - 1$ **(b)** $g(x) = 2 - |x|$
   (c) $h(x) = |x - 1|$ **(d)** $p(x) = 2|x + 1| - 1$

9. Fahrenheit temperature F and Kelvin (or absolute) temperature K are related by the equation $9K - 5F = 2297$.
   (a) Find a function f such that $F = f(K)$.
   (b) Show that f is one-to-one.
   (c) Find f^{-1} and use it to express K in terms of F.

10. Let $f(x) = \dfrac{2x - 1}{3x + 2}$.
   (a) Find the domain of f.
   (b) Prove that f is one-to-one by showing algebraically that if a and b are in the domain of f and $f(a) = f(b)$, then $a = b$.
   (c) Find f^{-1}.

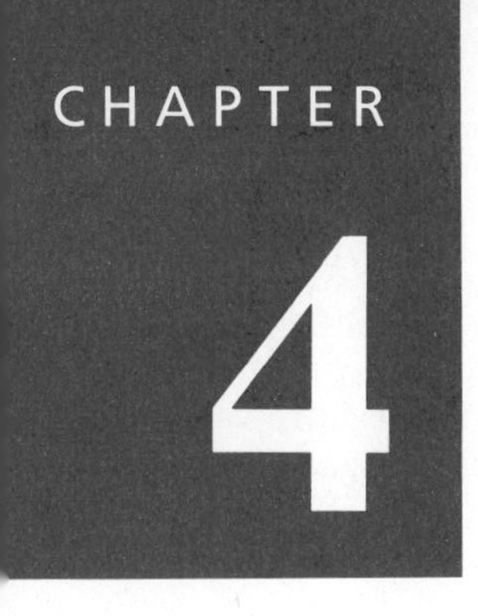

Polynomial and Rational Functions

This illuminated fountain in St. Louis, Missouri, shows parabolic jets of water.

In this chapter, we continue the study of functions and their graphs. Here we explore polynomial and rational functions, their "zeros," and some of the properties of their graphs. The chapter also includes the remainder theorem, synthetic division, the rational-zeros theorem, Descartes' rule of signs, and a section on complex polynomials.

4.1 Quadratic Functions

A function f of the form

$$f(x) = ax^2 + bx + c,$$

where a, b, and c are constants and $a \neq 0$, is called a **quadratic function.** Such functions often arise in applied mathematics. For instance, the height of a projectile is a quadratic function of time; the velocity of blood flow is a quadratic function of the distance from the center of the blood vessel; and the force exerted by the wind on the blades of a wind-powered generator is a quadratic function of the wind speed.

The simplest quadratic function is the squaring function $f(x) = x^2$, whose graph was sketched in Section 3.4 (page 184). As we saw in Section 3.5, the graph of $f(x) = ax^2$ is obtained from the graph of $y = x^2$ by vertical stretching, if $a > 1$, or flattening, if $0 < a < 1$. Furthermore, the graph of $f(x) = ax^2$ for negative values of a is obtained by reflecting the graph $y = |a|x^2$ across the x axis. Figure 1 shows the graph of $f(x) = ax^2$ for various values of a.

Figure 1

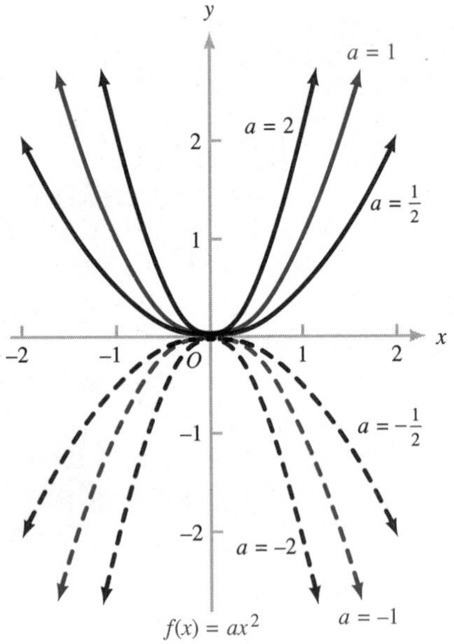

Figure 2

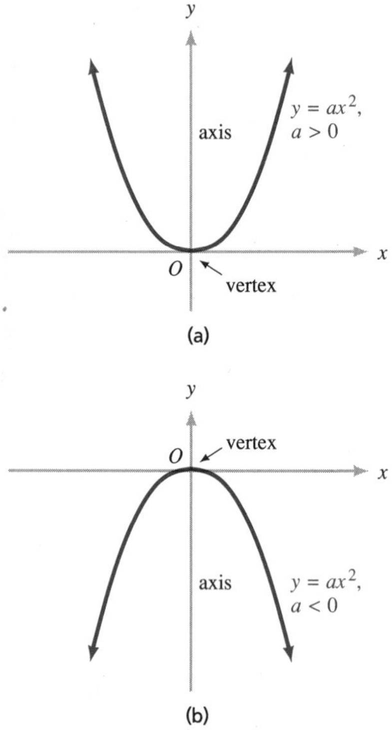

The graphs of equations of the form $y = ax^2$ are examples of curves called **parabolas.** These parabolas* are symmetric about the y axis; they **open upward** and have a lowest point at (0, 0) if $a > 0$ (Figure 2a), and they **open downward** and have a highest point at (0, 0) if $a < 0$ (Figure 2b). The highest or lowest point of the graph of $y = ax^2$ is called the **vertex** of the parabola, and its line of symmetry (in this case the y axis) is called the **axis of symmetry** or simply the **axis** of the parabola.

By the combined graph-shifting rule (Section 3.5, page 192), the graph of

$$f(x) = a(x - h)^2 + k$$

is obtained by shifting the parabola $y = ax^2$ horizontally by $|h|$ units and vertically by $|k|$ units. Hence its graph is a parabola with vertex at (h, k) (Figure 3). The parabola opens upward if $a > 0$ and downward if $a < 0$, and its axis is the vertical line $x = h$.

* We continue our study of parabolas in a later chapter on analytical geometry.

Figure 3

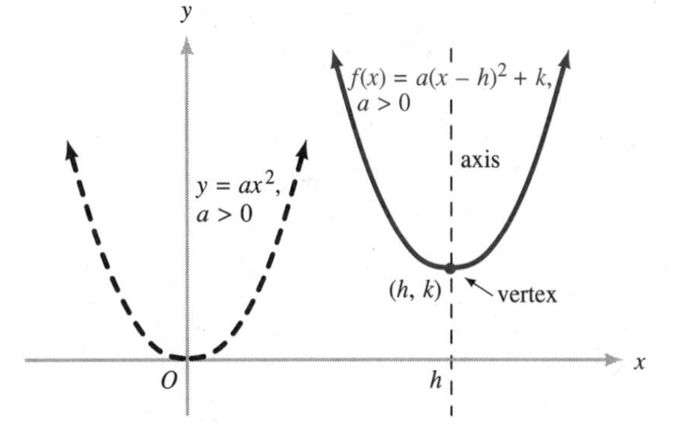

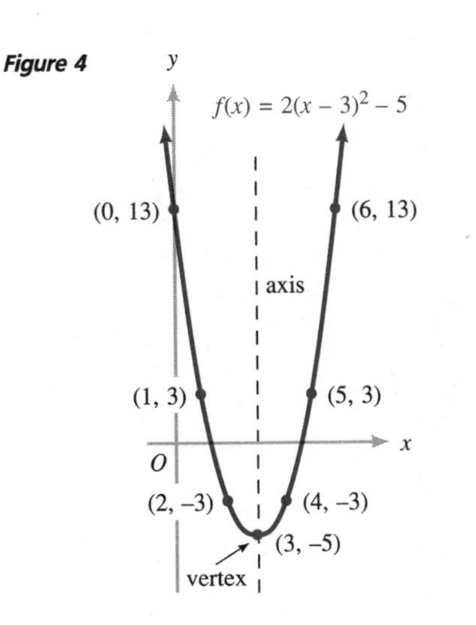

Figure 4

Example 1 Sketch the graph of $f(x) = 2(x - 3)^2 - 5$.

Solution The function has the form $f(x) = a(x - h)^2 + k$, with $a = 2$, $h = 3$, and $k = -5$; hence, its graph is a parabola, opening upward, with vertical axis $x = 3$ and vertex $(h, k) = (3, -5)$. This information indicates the general appearance of the graph. By plotting a few points, we can obtain a reasonably accurate sketch of the parabola (Figure 4). In Figure 4 we have used different scales on the x and y axes to obtain a graph with reasonable proportions. ■

By completing the square as shown in the following example, you can always rewrite a quadratic function $f(x) = ax^2 + bx + c$ in the form $f(x) = a(x - h)^2 + k$. Then, using the technique illustrated in Example 1, you can sketch the graph of f.

Example 2 Rewrite $f(x) = 2x^2 + 12x + 17$ in the form $f(x) = a(x - h)^2 + k$ and sketch the graph of f.

Solution To prepare for completing the square (Section 2.3, page 99), we factor the coefficient 2 of x^2 out of the first two terms:

$$f(x) = 2x^2 + 12x + 17 = 2(x^2 + 6x \qquad) + 17.$$

By adding $(\frac{6}{2})^2 = 9$ to $x^2 + 6x$, we obtain the perfect square

$$x^2 + 6x + 9 = (x + 3)^2.$$

Therefore, we have

$$f(x) = 2(x^2 + 6x + 9) + 17 - 2(9),$$

where we have subtracted 2(9) to compensate for the addition of 9 to $x^2 + 6x$. [Notice that the 9 added will be multiplied by 2, which is why we subtracted 2(9).]

Figure 5

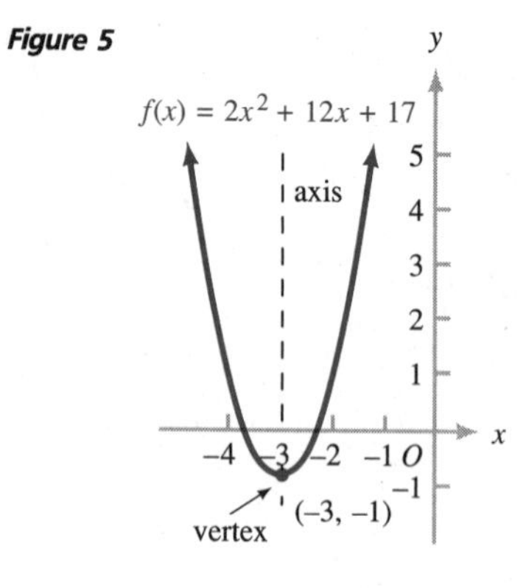

It follows that

$$f(x) = 2(x + 3)^2 - 1,$$

which has the form

$$f(x) = a(x - h)^2 + k$$

with $a = 2$, $h = -3$, and $k = -1$. Thus, the graph is a parabola, opening upward, with vertex $(h, k) = (-3, -1)$ and vertical axis $x = -3$ (Figure 5). ■

More generally, we have the following result.

Theorem 1 **The Quadratic Function Theorem**

The quadratic function

$$f(x) = ax^2 + bx + c, \qquad a \neq 0$$

can be rewritten in the form

$$f(x) = a(x - h)^2 + k,$$

where $$h = -\frac{b}{2a} \quad \text{and} \quad k = f(h).$$

Proof We begin by writing

$$f(x) = ax^2 + bx + c = a\left(x^2 + \frac{b}{a}x \qquad \right) + c.$$

By adding $\left[\frac{1}{2}\left(\frac{b}{a}\right)\right]^2 = \frac{b^2}{4a^2}$ to $x^2 + \frac{b}{a}x$, we obtain the perfect square

$$x^2 + \frac{b}{a}x + \frac{b^2}{4a^2} = \left(x + \frac{b}{2a}\right)^2.$$

Therefore,

$$\begin{aligned} f(x) &= a\left(x^2 + \frac{b}{a}x + \frac{b^2}{4a^2}\right) + c - a\left(\frac{b^2}{4a^2}\right) \\ &= a\left(x + \frac{b}{2a}\right)^2 + c - \frac{b^2}{4a} = a(x - h)^2 + k, \end{aligned}$$

where $h = -\frac{b}{2a}$ and $k = c - \frac{b^2}{4a}$. But, from the equation

$$f(x) = a(x - h)^2 + k,$$

we have

$$f(h) = a(h - h)^2 + k = a(0)^2 + k = k;$$

that is, $k = f(h)$. ■

As a consequence of Theorem 1, we have the following formula for the coordinates of the vertex of a parabola.

The Vertex Formula

If $a \neq 0$, then the graph of

$$f(x) = ax^2 + bx + c$$

is a parabola with vertex (h, k) where

$$h = -\frac{b}{2a} \quad \text{and} \quad k = f(h) = f\left(-\frac{b}{2a}\right).$$

If $a > 0$, the parabola opens upward and the vertex (h, k) is its lowest point; if $a < 0$, the parabola opens downward and the vertex (h, k) is its highest point.

It follows from the vertex formula that *the vertical line $x = h$ is the axis of symmetry of the graph of $f(x) = ax^2 + bx + c$.*

Example 3 Find the vertex and axis of symmetry of the graph of $f(x) = -3x^2 - 12x - 1$ and sketch the graph.

Solution By the vertex formula, with $a = -3$ and $b = -12$, we have

$$h = -\frac{b}{2a} = -\frac{(-12)}{2(-3)} = -2$$

and

$$k = f(h) = f(-2)$$
$$= -3(-2)^2 - 12(-2) - 1 = 11.$$

Therefore, the graph is a parabola, opening downward, with vertex $(-2, 11)$ and vertical axis $x = -2$ (Figure 6). ■

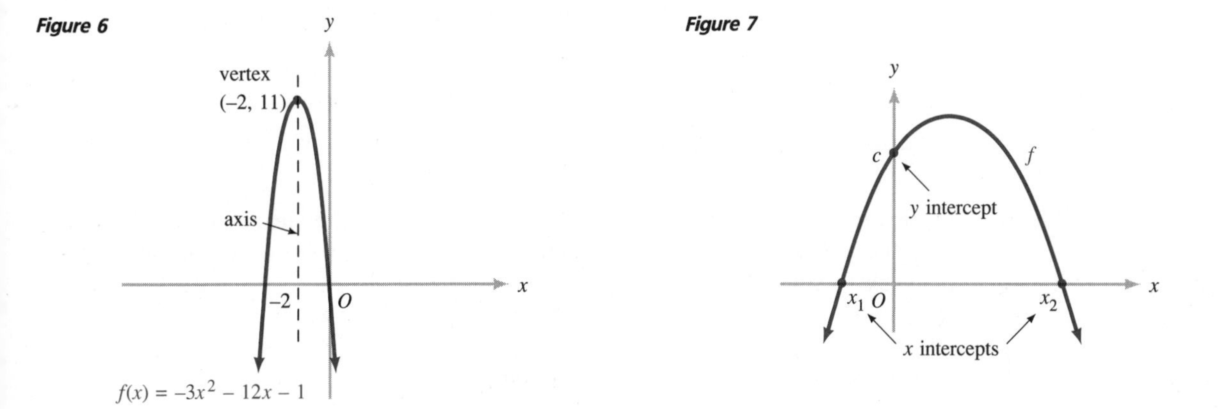

Figure 6

Figure 7

Notice that the graph of $f(x) = ax^2 + bx + c$ intersects the y axis at the point $(0, c)$ (Figure 7).

> We call c the **y intercept** of the graph. If the graph intersects the x axis at the points $(x_1, 0)$ and $(x_2, 0)$, we call x_1 and x_2 its ***x* intercepts.**

Note that x_1 and x_2, if they exist, are the real roots of the quadratic equation $ax^2 + bx + c = 0$. (Do you see why?)

gc Example 4 Find the vertex and the y and x intercepts of the graph of $f(x) = -2x^2 - 5x + 3$, determine whether the graph opens upward or downward, sketch the graph, and find the domain and range of f.

Solution Here we have $a = -2$, $b = -5$, and $c = 3$; hence, by the vertex formula,

$$h = -\frac{b}{2a} = -\frac{(-5)}{2(-2)} = -\frac{5}{4}$$

and

$$k = f(h) = f(-\tfrac{5}{4})$$
$$= -2(-\tfrac{5}{4})^2 - 5(-\tfrac{5}{4}) + 3 = \tfrac{49}{8}.$$

Therefore, the vertex is

$$(h, k) = (-\tfrac{5}{4}, \tfrac{49}{8}).$$

Figure 8

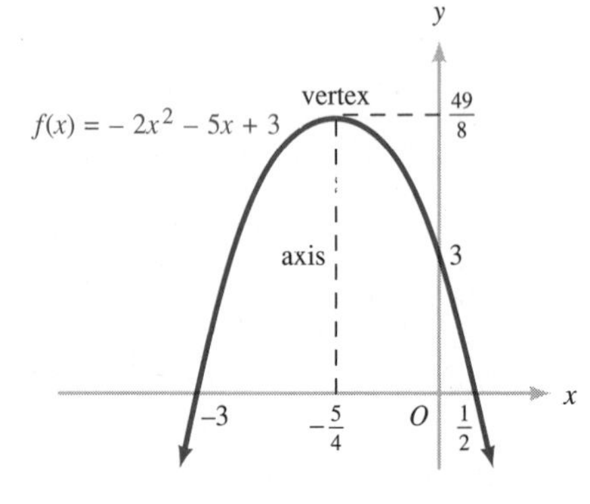

The y intercept is $c = 3$, and the x intercepts are the solutions of the quadratic equation

$$-2x^2 - 5x + 3 = 0 \qquad \text{or} \qquad (x + 3)(-2x + 1) = 0.$$

It follows that the x intercepts are -3 and $\frac{1}{2}$. Because $a = -2 < 0$, the graph opens downward. Using this information, we can sketch the parabola (Figure 8). (If a more accurate sketch is desired, a calculator can be used to determine additional points on the graph or a graphing calculator can be used.) The domain of f is $\mathbb{R}$ and, as Figure 8 shows, its range is the interval $(-\infty, \frac{49}{8}]$. ■

Example 5 Use the graph of $f(x) = -2x^2 - 5x + 3$ to solve the quadratic inequalities:

(a) $-2x^2 - 5x + 3 < 0$ **(b)** $-2x^2 - 5x + 3 \geq 0$

Solution **(a)** From the graph of $f(x) = -2x^2 - 5x + 3$, already sketched in Figure 8, we see that $f(x) < 0$ when $x < -3$ and when $x > \frac{1}{2}$. Hence, the solution set of the inequality $-2x^2 - 5x + 3 < 0$ is the union

$$(-\infty, -3) \cup (\tfrac{1}{2}, \infty)$$

of the two open intervals $(-\infty, -3)$ and $(\frac{1}{2}, \infty)$.

(b) The graph of $f(x)$ in Figure 8 shows that the solution set of the inequality $-2x^2 - 5x + 3 \geq 0$ is the closed interval $[-3, \frac{1}{2}]$. ■

Maximum and Minimum Values of Quadratic Functions

Using the vertex formula, we can easily locate the maximum and minimum values of a quadratic function $f(x) = ax^2 + bx + c$. Indeed, if $h = -b/(2a)$, then the vertex $(h, f(h))$ is the *lowest point* on the graph if $a > 0$ (Figure 9a) and the *highest point* on the graph if $a < 0$ (Figure 9b).

Figure 9

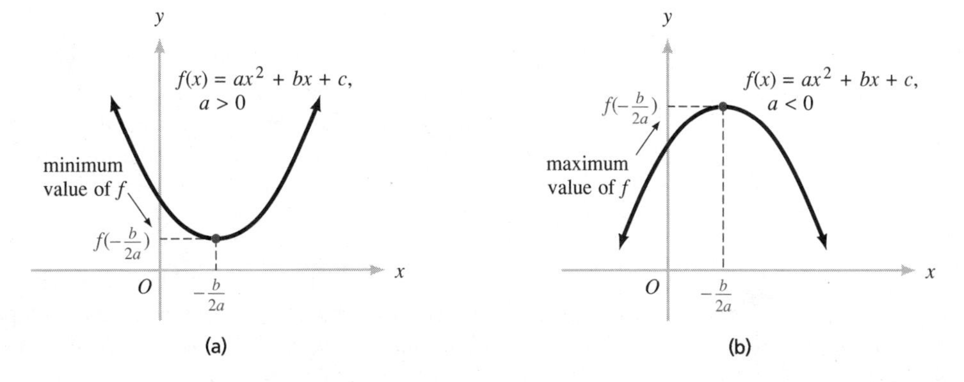

Therefore, we have the following useful fact.

Maximum and Minimum Values of Quadratic Functions

Let $f(x) = ax^2 + bx + c$. If $a > 0$, the number

$$f\left(-\frac{b}{2a}\right)$$

is the **minimum** (smallest) value of the function f, and if $a < 0$, it is the **maximum** (largest) value of f.

Example 6 When a single-stage rocket is launched straight upward, it accelerates until propellent burnout, after which it coasts upward to its highest point. The height $h(t)$ meters of the rocket above the ground t seconds after burnout is given by the function

$$h(t) = -\tfrac{1}{2}gt^2 + v_0t + h_0,$$

where $g = 9.8$ meters per second squared is the acceleration of gravity, v_0 meters per second is the velocity of the rocket at burnout, and h_0 meters is its height at burnout. If $v_0 = 343$ meters per second and $h_0 = 2200$ meters, find the maximum height attained by the rocket.

Solution The quadratic function h has the form

$$h(t) = at^2 + bt + c,$$

where $a = -\tfrac{1}{2}g = -\tfrac{1}{2}(9.8) = -4.9$, $b = v_0 = 343$ and $c = h_0 = 2200$.

Using the vertex formula, we find that the rocket reaches its maximum height when

$$t = -\frac{b}{2a} = -\frac{343}{2(-4.9)} = 35 \text{ seconds}$$

after burnout. Therefore, the maximum height of the rocket is

$$h(35) = (-4.9)(35)^2 + 343(35) + 2200 = 8202.5 \text{ meters.}$$

■

Example 7 An orchard contains 30 apple trees, each of which yields approximately 400 apples over the growing season. The owner plans to add more trees to the orchard, but the State Agricultural Service advises that because of crowding, each new tree will reduce the average yield per tree by about 10 apples over the growing season. How many trees should be added to maximize the total yield of apples, and what is the maximum yield?

Solution Let x denote the number of trees added. After x trees are added, the orchard will contain $30 + x$ trees, but the average yield of each tree per season will be reduced from 400 apples to $400 - 10x$ apples. Therefore, the total yield of apples per season is given by

$$\begin{aligned} f(x) &= (30 + x)(400 - 10x) = 12{,}000 + 100x - 10x^2 \\ &= ax^2 + bx + c, \end{aligned}$$

where $a = -10$, $b = 100$, and $c = 12{,}000$. Thus, $f(x)$ attains its maximum value when

$$x = -\frac{b}{2a} = -\frac{100}{2(-10)} = 5 \text{ trees.}$$

If 5 trees are added to the orchard, the total yield per growing season will be

$$\begin{aligned} f(5) &= 12{,}000 + 100(5) - 10(5)^2 \\ &= 12{,}250 \text{ apples.} \end{aligned}$$

■

Problem Set 4.1

In each problem set, problems with colored numbers constitute a good representation of the main ideas of the section. Note that some of the even-numbered problems may be considerably more challenging than the odd-numbered ones.

1. Sketch the graph of $f(x) = ax^2$ for **(a)** $a = 3$, **(b)** $a = -\frac{1}{3}$, and **(c)** $a = -3$.

2. Sketch the graph of $f(x) = 2x^2 + k$ for **(a)** $k = 0$, **(b)** $k = 1$, and **(c)** $k = -1$.

In Problems 3 and 4, sketch the graphs of both quadratic functions on the same coordinate system.

3. **(a)** $f(x) = 2(x - 1)^2 + 3$
   (b) $g(x) = -2(x + 2)^2 - 3$

4. **(a)** $h(x) = -\frac{1}{2}(x + 1)^2 - 4$
   (b) $p(x) = \frac{1}{2}(x + 1)^2 - \frac{1}{2}$

In Problems 5 to 12, use the process of completing the square as in Example 2 to rewrite each quadratic function

f in the form $f(x) = a(x - h)^2 + k$. *Do not use the vertex formula*—show the work of completing the square in each case.

5. $f(x) = x^2 - 4x - 1$
6. $f(x) = -6x^2 + 12x + 5$
7. $f(x) = -x^2 - 2x + 8$
8. $f(x) = 8x - x^2$
9. $f(x) = 2x^2 - 4x + 1$
10. $f(x) = \frac{3}{2}x^2 - 6x - 7$
11. $f(x) = -4x^2 + 8x - 5$
12. $f(x) = (2x - 1)(3x + 2)$

In Problems 13 to 20, find the vertex and the intercepts of the graph of each function, determine whether the graph opens upward or downward, sketch the graph, and find the domain and range of the function. c You may use a calculator to determine additional points on the graph if you want to obtain a more accurate sketch. gc Alternatively, you may use a graphing calculator.

13. $f(x) = -3x^2 + 12x + 15$
14. $H(x) = 30 - x(1 + 14x)$
15. $Q(x) = -2x^2 + x - 15$
16. $g(x) = 4x - 12x^2 + 21$
17. $F(x) = \frac{1}{2}x^2 + x + 2$
18. $G(t) = t - \frac{2}{3}t^2 - \frac{1}{3}$
19. $f(x) = 2x^2 - 20x + 57, 0 \le x \le 10$
20. $g(x) = -3x^2 + 24x - 50, 2 < x < 6$

In Problems 21 to 24, find the equation in the form $f(x) = a(x - h)^2 + k$ of a quadratic function whose graph satisfies the given conditions and sketch the graph. gc If you wish, use a graphing calculator to check your work.

21. The graph contains the origin and has its vertex at the point (1, 1).
22. The graph has y intercept 8, and its x intercepts are -4 and 2.
23. The graph has a vertex at the point (1, 2), and one of the x intercepts is 2.
24. The graph has a vertex at the point $(-2, -1)$, and its y intercept is 7.
25. If a baseball is thrown straight upward, then its height $h(t)$ feet t seconds after being released from the thrower's hand is given by the function

$$h(t) = -16t^2 + v_0 t + h_0,$$

where v_0 feet per second is the speed of the baseball at the instant of release and h_0 feet is the height above the ground at which it is released. If $v_0 = 25$ feet per second and $h_0 = 7$ feet, find the maximum height reached by the baseball.

26. In Problem 25, determine the time t seconds at which the baseball returns to the height (7 feet) from which it was originally released.
27. Suppose that the distance d kilometers that a certain car can travel on one tank of gasoline at a speed of v kilometers per hour is given by $d = 12v - (v/4)^2$. What speed maximizes the distance d and hence minimizes fuel consumption?
28. When an object is projected straight upward from an initial height h_0 with an initial velocity v_0, its height $h(t)$ after t units of time have passed is given by $h(t) = -\frac{1}{2}gt^2 + v_0 t + h_0$. Here, g is the acceleration of gravity, and air resistance is considered negligible.
    - **(a)** Find the time at which the object reaches its maximum height.
    - **(b)** Find the maximum height reached by the object.
    - **(c)** Find the time at which the object hits the ground.
29. A manufacturer of synfuel (synthetic fuel) from coal estimates that the cost $f(x)$ in dollars per barrel for a production run of x thousand barrels is given by $f(x) = 9x^2 - 180x + 940$. How many thousands of barrels should be produced during each run to minimize the cost per barrel, and what is the minimum cost per barrel of the synfuel?
30. In physics it is shown that if air resistance is neglected, the stream of water projected from a fire hose satisfies the equation

$$y = mx - \frac{g}{2}(1 + m^2)\left(\frac{x}{v}\right)^2,$$

where m is the slope of the nozzle, v is the velocity of the stream at the nozzle, y is the height of the stream x units from the nozzle, and g is the acceleration of gravity (Figure 10). Assume that v and g are positive constants.

(a) For a fixed value of m, find the value of x for which the height y of the stream is maximum.

(b) For a fixed value of m, find the distance d from the nozzle at which the stream hits the ground.

(c) Find the value of m for which the water reaches the greatest height on a vertical wall x units from the nozzle.

Figure 10

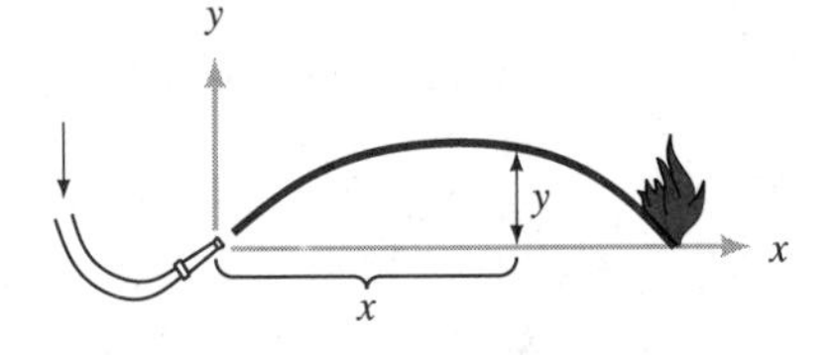

31. In medicine, it is often assumed that a patient's *reaction* $R(x)$ to a drug dose of size x is given by an equation of the form $R(x) = Ax^2(B - x)$, where A and B are positive constants. It can then be shown that the body's *sensitivity* $S(x)$ to a dose of size x is given by $S(x) = 2ABx - 3Ax^2$. Find the reaction to the dose for which the sensitivity is maximum.

32. Find the minimum value of $g(x) = (ax^2 + bx + c)^2$ in terms of the constants a, b, and c.

33. If a manufacturer produces x thousand tons of a new lightweight alloy for engine blocks, each block will cost $3x^2 - 600x + 30{,}090$ dollars. How many thousand tons of the alloy should be produced to minimize the cost of the engine blocks, and what is the resulting minimum cost of one block?

34. The management of a health spa foresees that 120 members will join the spa if each membership is \$10 per month, but that for each \$1 increase in the membership price per month, 8 of the 120 potential members will decide not to join. The cost to the spa per member is estimated to be \$7 per month. What membership price per month will bring in the maximum profit for the spa?

35. A rancher has 40 meters of fencing to enclose a rectangular pen next to a barn (Figure 11), with the barn wall forming one side of the pen. What dimensions of the pen will produce a maximum enclosed area?

Figure 11

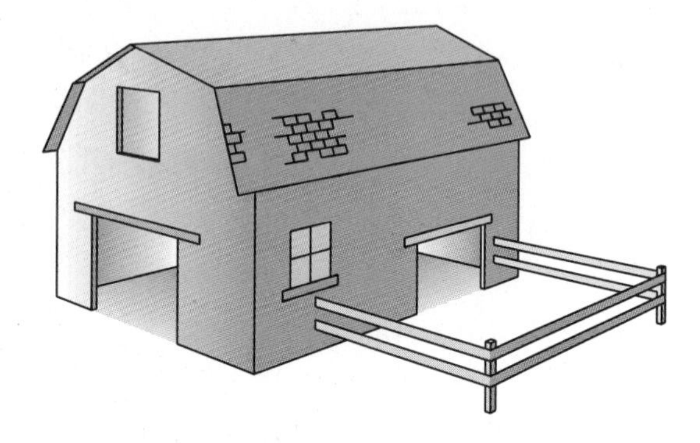

36. Ship A is 65 nautical miles due east of ship B and is sailing south at 15 knots (nautical miles per hour), while ship B is sailing east at 10 knots. Figure 12 shows the original positions P and Q of ships A and B, their positions t hours later, and the distance d between the ships at time t. Find the minimum distance between the ships and the time when it occurs. [*Hint:* Use the Pythagorean theorem and the fact that d is minimum when d^2 is minimum.]

Figure 12

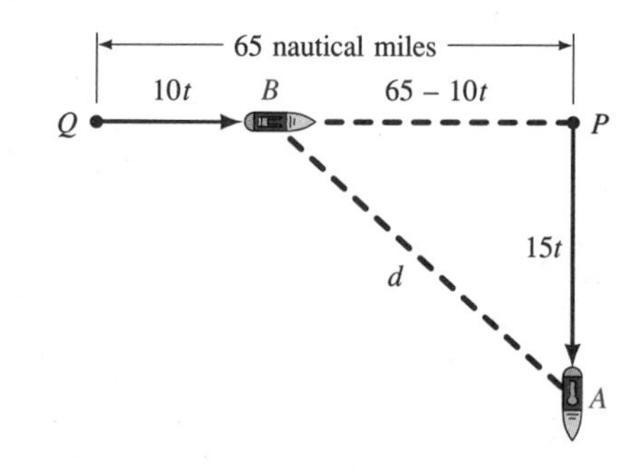

37. In order to reduce highway congestion, rapid commuter train service will be established between two nearby cities. It is projected that 2400 passengers will ride the train each day if the fare is \$4.00, and that for every \$0.20 decrease in the fare per passenger, 200 more passengers will decide to take the train rather than to drive.

(a) What fare should be charged to maximize the revenue to the railroad company?

(b) What fare will maximize the revenue if the train can carry at most 3000 passengers?

38. A real estate company manages an apartment complex containing 160 units. When the rent for each unit is \$500 per month, all apartments are occupied. However, for each \$20 increase in monthly rent per unit, one of the units becomes vacant. Each vacant unit costs the management \$30 per month for taxes and upkeep, and each occupied unit costs the management \$130 per month for taxes, service, upkeep, and water. What rent should be charged to obtain a maximum profit for the management company?

39. A rain gutter with vertical sides and no top is to be constructed by bending up equal sides of a rectangular piece of metal that is 30 centimeters wide (Figure 13). How many centimeters should be turned up to give the gutter its maximum capacity?

Figure 13

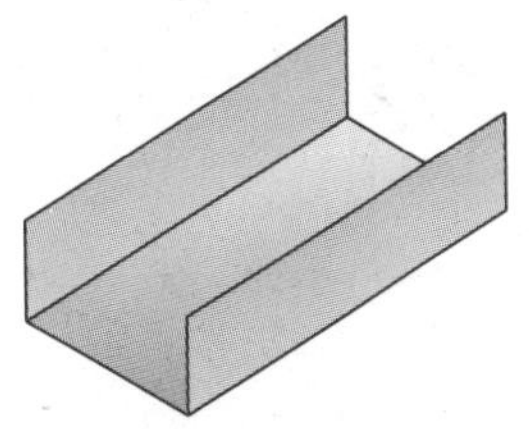

40. One section of a suspension bridge has its weight uniformly distributed between twin towers that are 300 meters apart and extend 80 meters above the horizontal roadway. A suspension cable runs between the tops of the towers and has the shape of a parabola tangent to the roadway at its vertex midway between the two towers. Imagine an xy coordinate system with the x axis extending along the roadway, the origin at the vertex of the parabola, and the y axis straight upward.

(a) Find an equation for the parabola.

(b) Find the height of the cable at a point 100 meters from the midpoint between the towers.

41. Find the point on the graph of $y = \sqrt{2x}$ that is nearest to the point (4, 0).

4.2 Polynomial Functions

We have now discussed constant functions

$$f(x) = b,$$

linear functions

$$f(x) = ax + b,$$

and quadratic functions

$$f(x) = ax^2 + bx + c.$$

These are all special cases of **polynomial functions,** that is, functions whose values are given by polynomials in the independent variable.

A polynomial function f of the form

$$f(x) = ax^n,$$

where $a \neq 0$ and n is a positive integer, is called a **power function of degree *n*.** For $n = 1$, the graph of f is a straight line that has slope a and contains the origin; for $n = 2$, the graph of f is a parabola that has its vertex at the origin and opens upward if $a > 0$ and downward if $a < 0$.

Figure 1

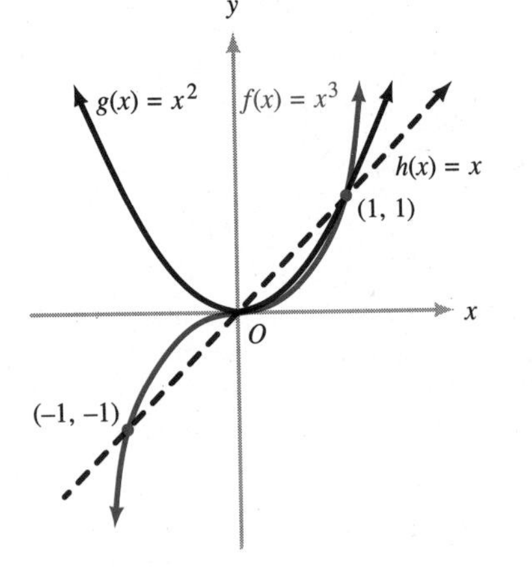

Figure 1 shows the graph of $f(x) = x^3$ and, for comparison, the graphs of $g(x) = x^2$ and $h(x) = x$. Notice that if x is less than 1 and positive, x^2 is smaller than x, and x^3 is even smaller than x^2. (For instance, if $x = \frac{1}{4}$, then $x^2 = \frac{1}{16} < \frac{1}{4}$ and $x^3 = \frac{1}{64} < \frac{1}{16}$.) It follows that on the open interval $(0, 1)$, the graph of $g(x) = x^2$ lies below the graph of $h(x) = x$, while the graph of $f(x) = x^3$ lies below the graph of $g(x) = x^2$. However, on the interval $(1, \infty)$, $x^3 > x^2 > x$, so the graph of $f(x) = x^3$ is above the graph of $g(x) = x^2$, which in turn is above the graph of $h(x) = x$. Notice that all three graphs contain the origin and the point $(1, 1)$. Since $f(x) = x^3$ is an odd function (Definition 2, page 182), its graph is symmetric about the origin.

If n is an *even* positive integer, then the power function $f(x) = x^n$ is even (why?) and its graph is symmetric about the y axis (Figure 2). The graph contains the origin and the points $(-1, 1)$ and $(1, 1)$. It never falls below the x axis. If $n = 2$, the graph is a parabola; but, for $n > 2$, the graph of $f(x) = x^n$ falls below the parabola $y = x^2$ over the open intervals $(-1, 0)$ and $(0, 1)$ and rises above the parabola over the intervals $(-\infty, -1)$ and $(1, \infty)$.

Figure 2

Figure 3

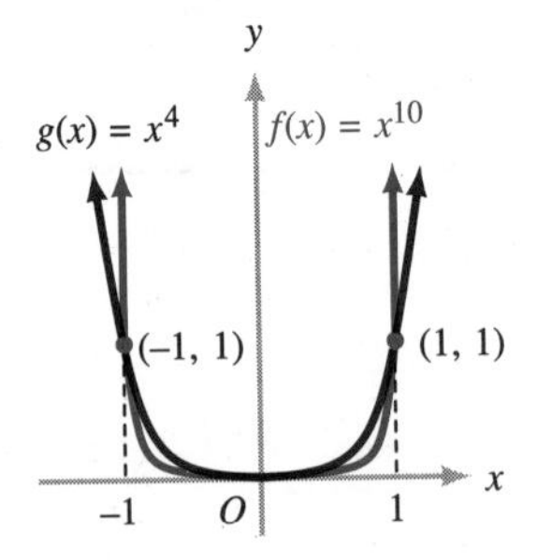

Figure 4

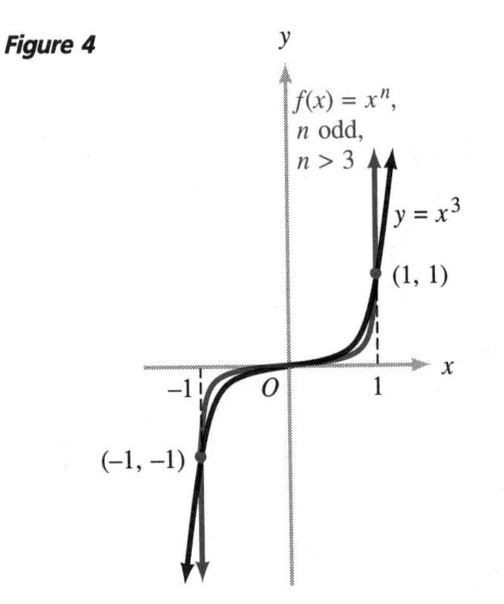

As the even integer n becomes larger and larger, the graph of $f(x) = x^n$ becomes flatter and flatter on both sides of the origin and rises more and more sharply through the two points $(-1, 1)$ and $(1, 1)$. Figure 3 shows the graph of $f(x) = x^{10}$ and contrasts it with the graph of $g(x) = x^4$. For large values of n, the graph of $f(x) = x^n$ may come so close to the x axis on both sides of the origin that it appears to coincide with a segment of the x axis. In reality, of course, the graph of $f(x) = x^n$ touches the x axis only at the origin.

If n is an *odd* positive integer, then the power function $f(x) = x^n$ is odd (why?) and its graph is symmetric about the origin (Figure 4). The graph contains the origin and the points $(-1, -1)$ and $(1, 1)$. For odd $n > 3$, the graph is similar to the graph of $y = x^3$ but is flatter near the origin, rises more sharply to the right of $x = 1$, and drops more rapidly to the left of $x = -1$ (Figure 4).

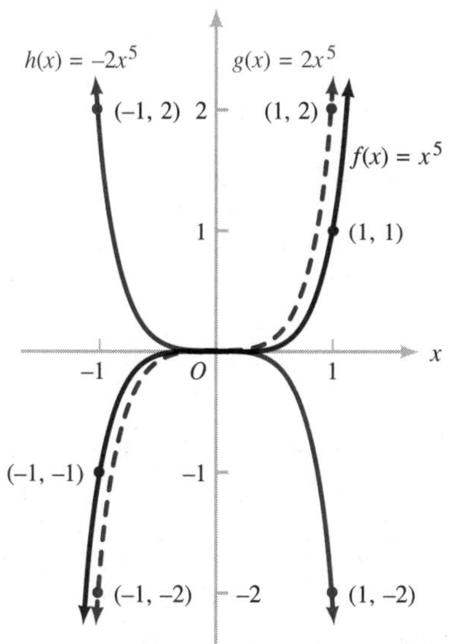

Figure 5

Example 1 Sketch each of the graphs of $f(x) = x^5$, $g(x) = 2x^5$, and $h(x) = -2x^5$ on the same coordinate system.

Solution We begin by sketching the graph of $f(x) = x^5$. Then, by doubling the ordinates, we obtain the graph of $g(x) = 2x^5$. Finally, we reflect the graph of $g(x) = 2x^5$ across the x axis to obtain the graph of $h(x) = -2x^5$ (Figure 5). ■

In Section 4.1 we studied the intercepts of the graph of a quadratic function. More generally, we have the following definition.

Definition 1 Intercepts of a Graph

The y **intercept** of the graph of a function f is the ordinate $f(0)$ of the point $(0, f(0))$ where the graph intersects the y axis. Similarly, the abscissa of a point where the graph of f intersects the x axis is called an x **intercept** of the graph (Figure 6).

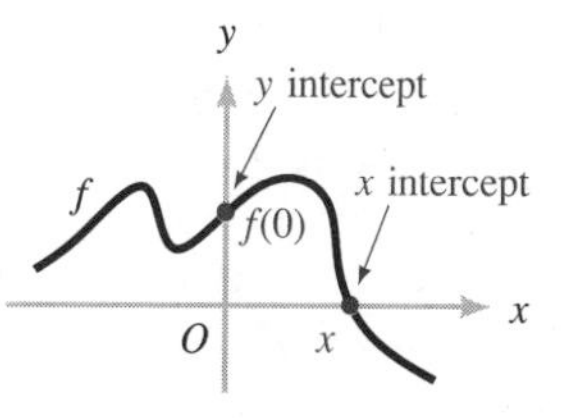

Figure 6

Thus, *the x intercepts of the graph of f are the real roots (if any) of the equation* $f(x) = 0$.

In Examples 2 and 3, find the y and x intercepts and sketch the graph of each function.

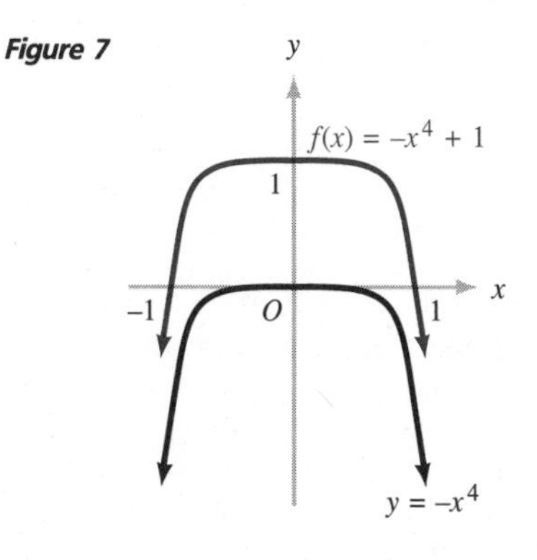

Figure 7

Example 2 $f(x) = -x^4 + 1$

Solution The y intercept is $f(0) = 1$, and the x intercepts are the real roots of the equation $f(x) = 0$; that is, $-x^4 + 1 = 0$ or $x^4 = 1$. Hence, the x intercepts are $x = \sqrt[4]{1} = 1$ and $x = -\sqrt[4]{1} = -1$. By reflecting the graph of $y = x^4$ (Figure 3) across the x axis and then shifting it 1 unit upward, we obtain the graph of $f(x) = -x^4 + 1$ (Figure 7). ■

Example 3 $G(x) = -\frac{1}{4}(x - 1)^5 + 8$

Solution The y intercept is $G(0) = -\frac{1}{4}(0 - 1)^5 + 8 = \frac{1}{4} + 8 = \frac{33}{4}$, and the x intercept is the real root of the equation $G(x) = 0$; that is,

$$-\tfrac{1}{4}(x - 1)^5 + 8 = 0 \qquad \text{or} \qquad (x - 1)^5 = 32.$$

Figure 8

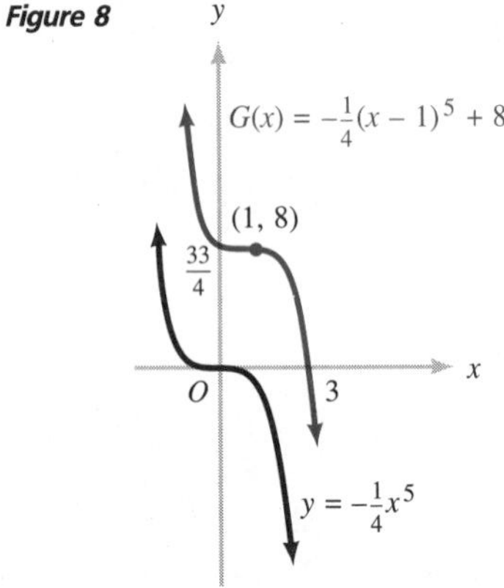

Thus, $x - 1 = \sqrt[5]{32} = 2$, so the x intercept is given by $x = 3$. By the combined graph-shifting rule (page 192), the graph of $G(x) = -\frac{1}{4}(x - 1)^5 + 8$ is obtained by shifting the graph of $y = -\frac{1}{4}x^5$ upward by 8 units and to the right by 1 unit (Figure 8). ■

In calculus, it is shown that the graph of every polynomial function is a smooth curve with no jumps or breaks. Although the graph of a polynomial function f can wiggle up and down as in Figure 9a, it cannot have the sharp corners or jumps illustrated in Figure 9b. The x intercepts of the graph of f divide the x axis into open intervals. You can determine the algebraic signs of $f(x)$ over these open intervals by using convenient test numbers, just as you did in Section 2.7. The following example illustrates how you can use this information to help sketch the graph of f.

Figure 9

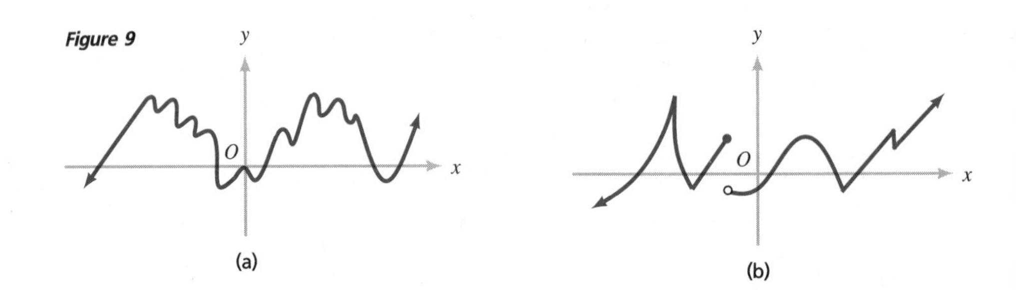

gc **Example 4** Let $f(x) = (2x - 1)(x^2 - x - 2)$.

(a) Find the x intercepts of the graph of f.

(b) Sketch the graph of f.

(c) Use the graph to solve the inequality $f(x) < 0$.

Solution **(a)** The x intercepts are the real roots of $f(x) = 0$; that is, $(2x - 1)(x^2 - x - 2) = 0$ or $(2x - 1)(x + 1)(x - 2) = 0$. Setting each factor equal to zero, we obtain

$2x - 1 = 0$	$x + 1 = 0$	$x - 2 = 0$
$x = \frac{1}{2}$	$x = -1$	$x = 2.$

Thus, in increasing order, the x intercepts are -1, $\frac{1}{2}$, and 2.

Figure 10

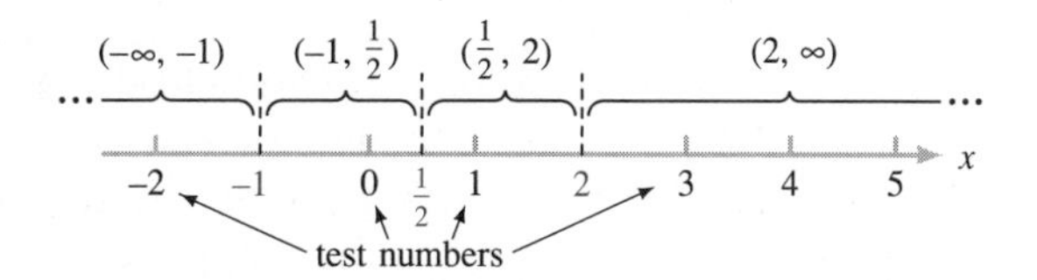

(b) The x intercepts divide the x axis into four open intervals, as shown in Figure 10. The graph of f touches the x axis only at the endpoints of these intervals; over each interval, the graph is either entirely above the x axis $[f(x) > 0]$ or entirely below it $[f(x) < 0]$. To see which is the case, we select convenient test numbers on each open interval (Figure 10) and evaluate $f(x) = (2x - 1)(x^2 - x - 2)$ at each test number. We obtain

$$f(-2) = -20, \qquad f(0) = 2, \qquad f(1) = -2, \qquad \text{and} \qquad f(3) = 20.$$

Since $f(-2) = -20$, the point $(-2, -20)$ belongs to the graph of f. This point is below the x axis, so the graph of f stays below the x axis over the first interval $(-\infty, -1)$. We apply similar reasoning to the remaining intervals and obtain the results in Figure 11a. Plotting the points $(-1, 0)$, $(\frac{1}{2}, 0)$, and $(2, 0)$ corresponding to the x intercepts, and the points $(-2, -20)$, $(0, 2)$, $(1, -2)$, and $(3, 20)$ corresponding to the test numbers, and using the information in Figure 11a, we can sketch the graph of f (Figure 11b). A graphing calculator can be used to confirm the correctness of the graph.

(c) From the graph of f (Figure 11b), it is clear that $f(x) < 0$ only for values of x in the two intervals $(-\infty, -1)$ and $(\frac{1}{2}, 2)$, so the solution set of the inequality $f(x) < 0$ is the union $(-\infty, -1) \cup (\frac{1}{2}, 2)$. ■

Figure 11

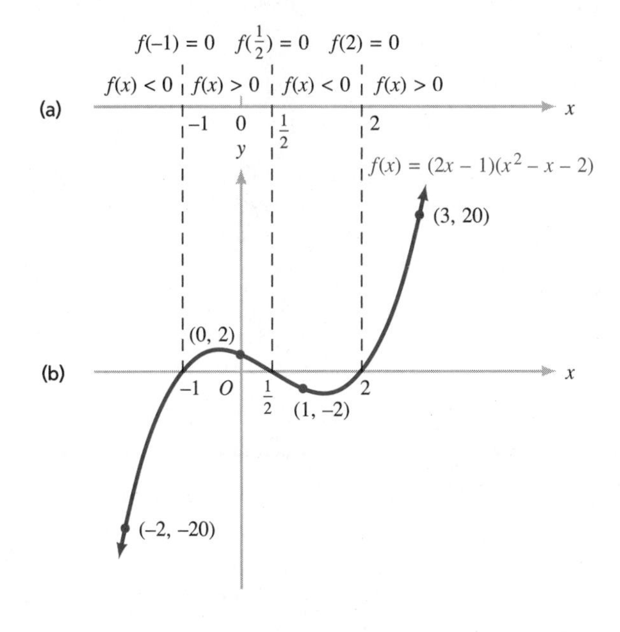

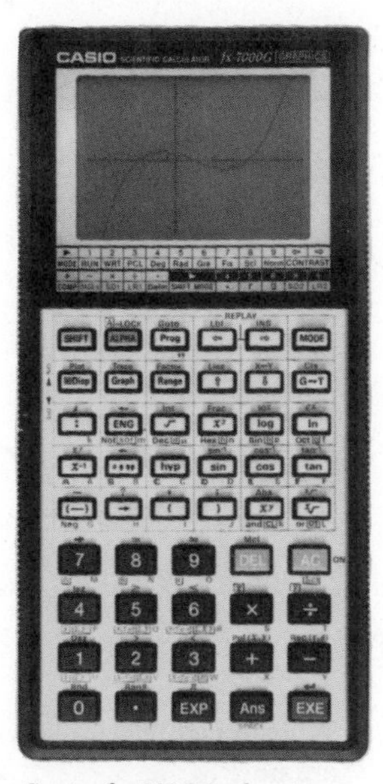

Casio fx-7000G showing the graph in Figure 11b.

When you use the method illustrated in Example 4, you should always ask whether the graph contains hidden peaks and valleys that are not indicated by the plotted points. Although methods studied in calculus are required to answer this question, the polynomial functions that we consider will be fairly simple, so that all peaks and valleys on their graphs can be detected by plotting a reasonable number of points. As always, proper use of a graphing calculator, especially one with a "zoom" feature, will usually reveal most of the interesting features of the graph.

Example 5 Figure 12a shows a computer-generated graph of

$$f(x) = (2x + 1)(x - 1)(x - 3)(x + 2).$$

Find the x and y intercepts, label them on the graph, and then use the graph to solve the inequality $f(x) > 0$.

Solution By setting each of the factors $2x + 1$, $x - 1$, $x - 3$, and $x + 2$ equal to zero and solving for x, we find that the x intercepts (rewritten in increasing order) are -2, $-\frac{1}{2}$, 1, and 3. The y intercept is given by

$$f(0) = (1)(-1)(-3)(2) = 6.$$

In Figure 12b, we have labeled these intercepts on the graph. The graph rises above the x axis over the intervals $(-\infty, -2)$, $(-\frac{1}{2}, 1)$, and $(3, \infty)$, so the solution set of $f(x) > 0$ is the union

$$(-\infty, -2) \cup (-\tfrac{1}{2}, 1) \cup (3, \infty).$$

■

Figure 12

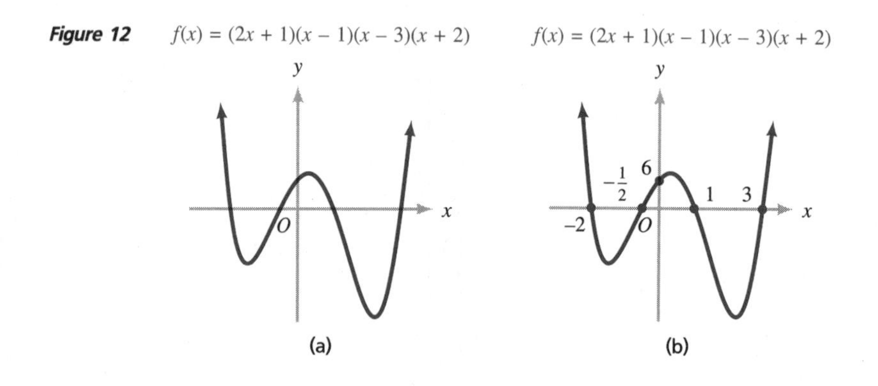

Problem Set 4.2

1. Which functions are polynomial functions and which are not?

   (a) $f(x) = 2x^2 - 16x + 29$

   (b) $g(x) = \pi x - \sqrt{3}x^5 + \frac{1}{2}x^3$

   (c) $h(x) = \sqrt{x}$

   (d) $F(x) = (2x^2 - 1)(3x^3 + 2)$

   (e) $G(x) = 2x^{-3} + 7$ (f) $H(x) = \dfrac{1}{x}$

2. Complete the sentence: All polynomial functions have the same domain, namely, the set . . . .

In each of Problems 3 to 8, sketch graphs of the functions f, g, and h on the same coordinate system. [c] You may use a calculator to determine additional points on the graph if you want to obtain a more accurate sketch. [gc] Alternatively, you may use a graphing calculator.

3. $f(x) = x^4$, $g(x) = 2x^4$, $h(x) = -2x^4$

4. $f(x) = x^6$, $g(x) = \dfrac{x^6}{3}$, $h(x) = -\dfrac{x^6}{3}$

5. $f(x) = x^5$, $g(x) = \frac{1}{2}x^5$, $h(x) = -\frac{1}{2}x^5$

6. $f(x) = x^7$, $g(x) = \dfrac{x^7}{9}$, $h(x) = -\dfrac{x^7}{9}$

7. $f(x) = x^4 - 1$, $g(x) = x^4 + 1$, $h(x) = (x - 1)^4$

8. $f(x) = x^5 - 2$, $g(x) = x^5 + 3$, $h(x) = (x + 2)^5$

[c] In Problems 9 to 16, find the y and x intercepts and sketch the graph of each function. [gc] Use a graphing calculator if you wish.

9. $h(x) = 3x^5 - 1$

10. $F(x) = \frac{1}{16}(x + 2)^9$

11. $f(x) = -x^8 - 1$

12. $g(x) = -\frac{2}{3}x^7 + 1$

13. $h(x) = (x + \frac{1}{2})^3 - 1$

14. $F(x) = 7(x + \frac{1}{2})^3 + 2$

15. $G(x) = 2(x - 6)^4 + 1$

16. $H(x) = -20(x - 8)^8 + 8$

In Problems 17 to 22, find all x intercepts of the graph of each function.

17. $h(x) = (x - 1)(4x^2 - 12x + 9)$

18. $k(x) = (3x + 1)(3x^2 - 4x)$

19. $p(x) = (4x^2 - 1)(6x^2 - 5x + 1)$

20. $H(x) = (9x^2 - 25)(x^2 - 5x - 14)$

21. $F(x) = x^6 - x^4$

22. $f(x) = x^5 + 6x^3 + 9x$

[c] In Problems 23 to 26, **(a)** find the x intercepts of the graph of the function, **(b)** sketch the graph, and **(c)** use the graph to solve the given inequality. [gc] Use a graphing calculator if you wish.

23. $f(x) = x(x - 1)(x + 1)$; $f(x) > 0$

24. $g(x) = (x^2 - x - 2)(x + 3)$; $g(x) \geq 0$

25. $p(x) = -x(x + 2)(x - 1)$; $p(x) \geq 0$

26. $F(x) = -(2x + 1)(x - 1)(3x + 4)$; $F(x) < 0$

In Problems 27 to 32, the indicated figure shows a computer-generated graph of the given function. In each case, find and label the x and y intercepts and then use the graph to solve the given inequality.

27. $f(x) = x^3 + x^2 - 12x < 0$ (Figure 13)

Figure 13

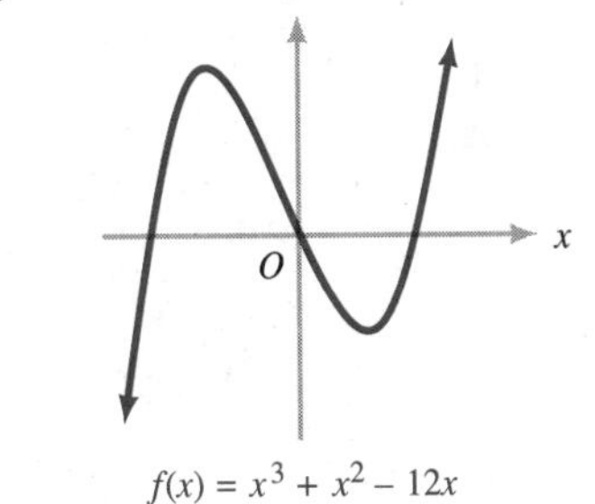

28. $g(x) = (2x - 5)(x - 1)^2 \geq 0$ (Figure 14)

Figure 14

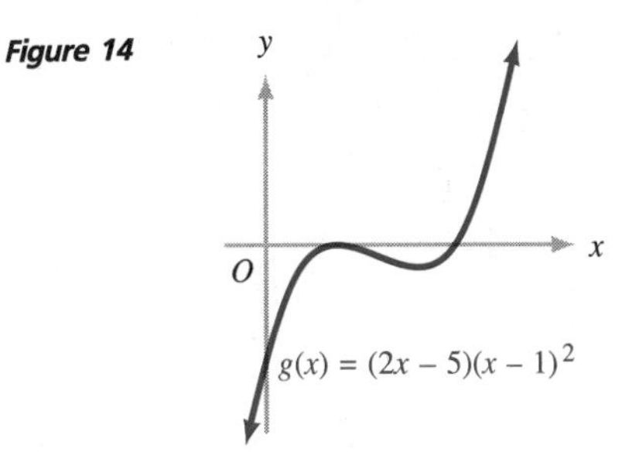

29. $h(x) = -2x^4 + 2x^2 > 0$ (Figure 15)

Figure 15

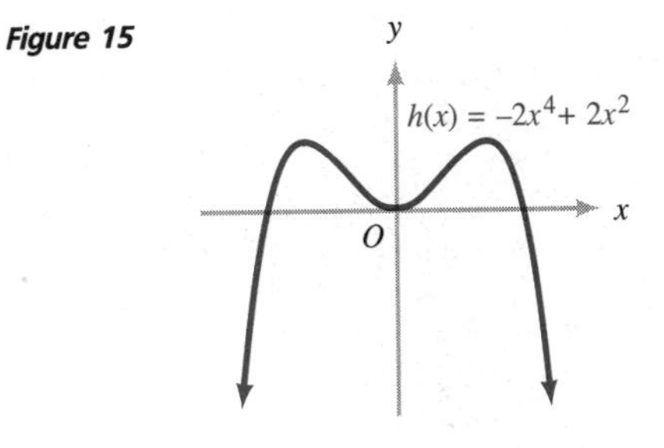

30. $F(x) = -x^5 + 3x^3 + 4x \leq 0$ (Figure 16)

Figure 16

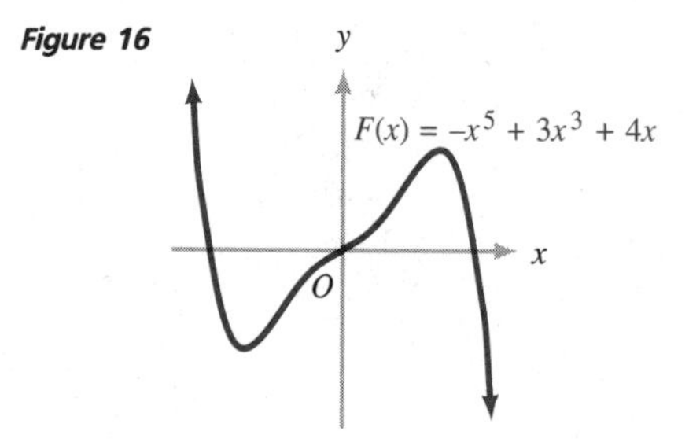

31. $G(x) = x^3(x + 1)(x - 2) \geq 0$ (Figure 17)

Figure 17

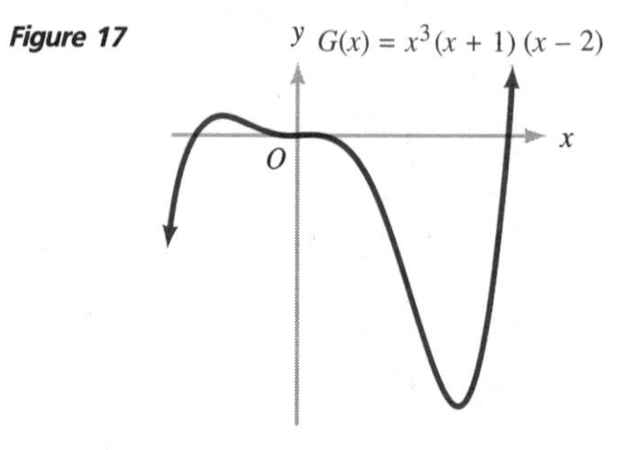

32. $H(x) = x^4(3x - 5) \leq 0$ (Figure 18)

Figure 18

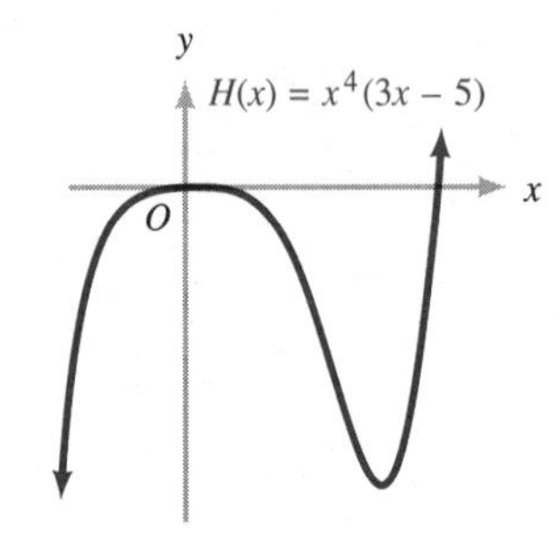

33. Equal squares are cut off at each corner of a rectangular piece of cardboard 8 inches wide by 15 inches long, and an open-topped box is formed by turning up the sides (Figure 19). If x is the length of the sides of the cutoff squares, then the volume $V(x)$ of the resulting box is given by $V(x) = x(8 - 2x)(15 - 2x)$ cubic inches. [c] Sketch the graph of the polynomial function V for $x > 0$, [gc] using a graphing calculator if you wish. Use the graph to determine the approximate value of x for which $V(x)$ is maximum.

Figure 19

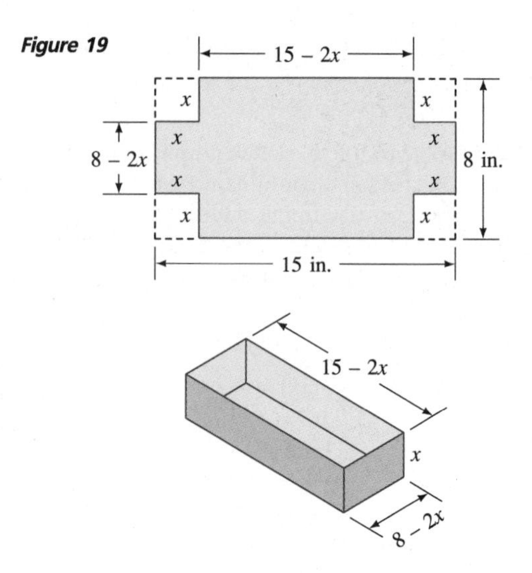

34. A child's sandbox is to be made by cutting equal squares from the corners of a square sheet of galvanized iron and turning up the sides. Suppose each side of the sheet of galvanized iron is 2 meters long.

(a) Find a formula for the volume $V(x)$ of the sandbox if x is the length in meters of the sides of the cutoff squares.

[c] **(b)** Sketch the graph of $V(x)$ for $x > 0$, or [gc] use a graphing calculator.

(c) Use the graph to determine the approximate value of x for which $V(x)$ is a maximum.

35. A box with a square end x inches on a side is to be mailed. The U.S. Postal Service will accept the box for domestic shipment only if the length L of the box plus its girth $4x$ is less than 108 inches. In attempting to maximize the volume of the box, it seems reasonable to take $L = 108 - 4x$ (Figure 20).

Figure 20

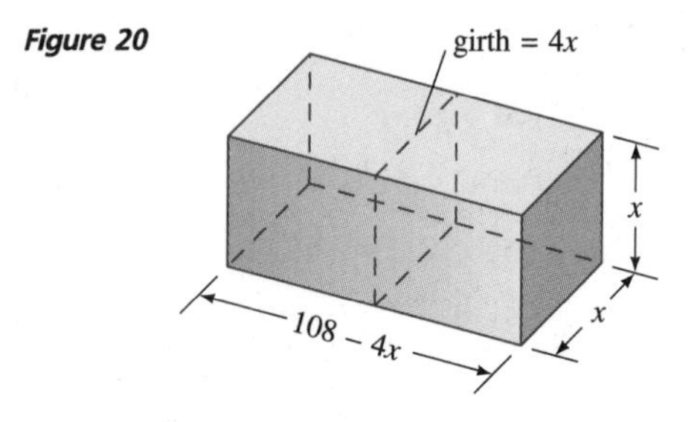

(a) Find a formula for the volume $V(x)$ of the box.

C (b) Sketch the graph of $V(x)$ for $0 \leq x \leq 27$, or gc use a graphing calculator.

(c) Use the graph to determine the approximate value of x for which $V(x)$ is maximum.

36. The **Legendre polynomials** are of importance in applied mathematics for solving problems involving electrical potential, diffusion, heat transfer, and so on. For each integer $n \geq 0$ there is a corresponding Legendre polynomial $P_n(x)$ of degree n. For $n = 0, 1, 2$, the Legendre polynomials are

$$P_0(x) = 1, \quad P_1(x) = x, \quad \text{and} \quad P_2(x) = \tfrac{1}{2}(3x^2 - 1).$$

For each positive integer value of n, the Legendre polynomials satisfy the following relation (called the *recurrence relation*):

$$(n + 1)P_{n+1}(x) + nP_{n-1}(x) = (2n + 1)xP_n(x).$$

(a) Find the Legendre polynomial $P_3(x)$.

C (b) Sketch the graph of $P_3(x)$, or gc use a graphing calculator.

37. An infectious disease is spreading through a certain city. Health officials have determined that antibodies for the infection develop so rapidly that the disease is self-limiting and will run its course in a period of T days. They have developed the mathematical model $N(t) = -t^3 + 30t^2 + 400t$, which they believe will give an accurate prediction of the number of people afflicted by the disease after t days for $0 \leq t \leq T$.

(a) Using the facts that $T > 0$ and $N(T) = 0$, find T.

C (b) Sketch a graph of $N(t)$ for $0 \leq t \leq T$, or gc use a graphing calculator.

(c) Use the graph to estimate the value of t between 0 and T for which $N(t)$ is maximum.

4.3 Division of One Polynomial by Another

In elementary arithmetic, you learned to divide one integer by another to obtain a quotient and a remainder; for instance,

```
                   71 ←—— quotient
divisor ——→ 32 | 2277 ←—— dividend
                224
                 37
                 32
                  5 ←—— remainder.
```

The result of this calculation can be expressed as

$$2277 = (32)(71) + 5;$$

that is,

dividend = (divisor)(quotient) + remainder.

In this section, we study a similar procedure, called **long division,** for dividing one polynomial by another, and we introduce a useful shortcut called **synthetic division.** These procedures will be used extensively in Section 4.4, where we continue our study of polynomial functions and their graphs.

When you divide one positive integer by another, you continue the procedure until you obtain a remainder that is less than the divisor. Likewise, when you divide one polynomial by another, you should continue the long-division procedure until the remainder is either the zero polynomial or a polynomial of lower degree than the divisor.

In Examples 1 and 2, perform the indicated long division to find a quotient polynomial and a remainder polynomial. Be sure that the remainder is either the zero polynomial or a polynomial of lower degree than the divisor. Check your work by verifying that

$$\textit{dividend} = (\textit{divisor})(\textit{quotient}) + \textit{remainder}.$$

Example 1 $2x^2 + x - 1 \overline{)\, 6x^4 + x^3 + 4x + 4}$

Solution Notice that both the divisor $2x^2 + x - 1$ and the dividend $6x^4 + x^3 + 4x + 4$ are arranged in descending powers of x. (If they weren't, we would begin by rewriting them so they were.) We divide $6x^4$, the leading term of the dividend, by $2x^2$, the leading term of the divisor, to obtain

$$\frac{6x^4}{2x^2} = 3x^2,$$

the first term of the quotient. Thus we write

$$\begin{array}{r} 3x^2 \leftarrow \text{first term of the quotient} \\ 2x^2 + x - 1 \overline{)\, 6x^4 + x^3 + 4x + 4.} \end{array}$$

Now, we multiply $3x^2$ by the divisor $2x^2 + x - 1$, write this product under the dividend, and subtract to obtain a first trial remainder:

$$\begin{array}{rl} & 3x^2 \\ 2x^2 + x - 1 & \overline{)\, 6x^4 + x^3 \qquad\quad + 4x + 4} \\ \text{subtract} \rightarrow & \underline{6x^4 + 3x^3 - 3x^2} \\ & \quad -2x^3 + 3x^2 + 4x + 4. \leftarrow \text{first trial remainder} \end{array}$$

(To make the subtraction easier, we have spaced the dividend so that like terms are aligned vertically.) Because the degree of our first trial remainder isn't less than the degree of the divisor, we must repeat the procedure. Thus, we divide $-2x^3$, the leading term of the first trial remainder, by $2x^2$, the leading term of the divisor, to obtain $-x$, the second term of the quotient. We multiply $-x$ by the divisor $2x^2 + x - 1$, write this product under the first trial remainder, and subtract to obtain a second trial remainder:

$$\begin{array}{rl} & 3x^2 - x \leftarrow \text{second term of the quotient} \\ 2x^2 + x - 1 & \overline{)\, 6x^4 + x^3 \qquad\quad + 4x + 4} \\ & \underline{6x^4 + 3x^3 - 3x^2} \\ & \quad -2x^3 + 3x^2 + 4x + 4 \\ \text{subtract} \rightarrow & \quad \underline{-2x^3 - x^2 + x} \\ & \qquad\quad 4x^2 + 3x + 4. \leftarrow \text{second trial remainder} \end{array}$$

The degree of the second trial remainder still isn't less than the degree of the divisor, so we divide its leading term $4x^2$ by $2x^2$ to obtain 2, the third term of the quotient. We multiply 2 by the divisor $2x^2 + x - 1$, write this product under the second trial remainder, and subtract to obtain a third trial remainder:

$$
\begin{array}{r|l}
 & 3x^2 - x + 2 \\
\cline{2-2}
2x^2 + x - 1 & 6x^4 + x^3 + 4x + 4 \\
 & 6x^4 + 3x^3 - 3x^2 \\
\cline{2-2}
 & -2x^3 + 3x^2 + 4x + 4 \\
 & -2x^3 - x^2 + x \\
\cline{2-2}
 & 4x^2 + 3x + 4 \\
 & 4x^2 + 2x - 2 \\
\cline{2-2}
 & x + 6.
\end{array}
$$

third term of the quotient ($+2$)

subtract ($4x^2 + 2x - 2$)

third trial remainder (the remainder) ($x + 6$)

The third trial remainder is actually the remainder, since its degree is less than the degree of the divisor. We conclude that

$3x^2 - x + 2$ is the quotient polynomial

and $x + 6$ is the remainder polynomial.

To check, we calculate

$$
\begin{aligned}
\text{(divisor)(quotient)} + \text{remainder} &= (2x^2 + x - 1)(3x^2 - x + 2) + (x + 6) \\
&= 6x^4 + x^3 + 3x - 2 + x + 6 \\
&= 6x^4 + x^3 + 4x + 4 \\
&= \text{the dividend,}
\end{aligned}
$$

so our work is correct. ■

Example 2 $x^3 - 2x^2 + 3x - 4 \overline{\smash{)}\, 2x^5 - 7x^4 + 13x^3 - 19x^2 + 15x - 4}$

Solution

$$
\begin{array}{r|l}
 & 2x^2 - 3x + 1 \\
\cline{2-2}
x^3 - 2x^2 + 3x - 4 & 2x^5 - 7x^4 + 13x^3 - 19x^2 + 15x - 4 \\
 & 2x^5 - 4x^4 + 6x^3 - 8x^2 \\
\cline{2-2}
 & -3x^4 + 7x^3 - 11x^2 + 15x - 4 \\
 & -3x^4 + 6x^3 - 9x^2 + 12x \\
\cline{2-2}
 & x^3 - 2x^2 + 3x - 4 \\
 & x^3 - 2x^2 + 3x - 4 \\
\cline{2-2}
 & 0.
\end{array}
$$

quotient polynomial ($2x^2 - 3x + 1$)

subtract ($2x^5 - 4x^4 + 6x^3 - 8x^2$); first trial remainder ($-3x^4 + 7x^3 - 11x^2 + 15x - 4$)

subtract ($-3x^4 + 6x^3 - 9x^2 + 12x$); second trial remainder ($x^3 - 2x^2 + 3x - 4$)

subtract ($x^3 - 2x^2 + 3x - 4$); third trial remainder (the remainder) (0)

The quotient polynomial is $2x^2 - 3x + 1$, and the remainder is the zero polynomial. To check, we calculate

$$
\begin{aligned}
\text{(divisor)(quotient)} + \text{remainder} &= (x^3 - 2x^2 + 3x - 4)(2x^2 - 3x + 1) + 0 \\
&= 2x^5 - 7x^4 + 13x^3 - 19x^2 + 15x - 4 \\
&= \text{the dividend,}
\end{aligned}
$$

and our work is correct. ■

A systematic procedure that is guaranteed to work in a finite number of steps is called an **algorithm.** Long division is an algorithm because, as you carry it out, you accumulate the quotient polynomial term by term and you reduce the degree of the trial remainder at each stage. The process comes to an end when the trial remainder is either the zero polynomial or a polynomial of lower degree than the divisor. This is summarized by the following noteworthy theorem.

Theorem 1 **The Division Algorithm**

Let f and g be polynomial functions and suppose that g is not the constant zero polynomial. Then there exist unique polynomial functions q and r such that

$$f(x) = g(x)q(x) + r(x)$$

holds for all values of x, and r is either the constant zero polynomial or a polynomial of degree lower than g.

In the division algorithm, $f(x)$ is the dividend polynomial, $g(x)$ the divisor polynomial, $q(x)$ the quotient polynomial, and $r(x)$ the remainder polynomial. The identity

$$f(x) = g(x)q(x) + r(x)$$

expresses the fact that

$$\text{dividend} = (\text{divisor})(\text{quotient}) + \text{remainder}.$$

We often refer to the function f/g or to a value $f(x)/g(x)$ of such a function as a *quotient*, but the word used in this way must not be confused with the *quotient polynomial* $q(x)$ in the division algorithm. The relationship between the two types of quotients is expressed in the following theorem.

Theorem 2 **The Quotient Theorem**

Let $q(x)$ be the quotient polynomial and $r(x)$ be the remainder polynomial obtained by long division of the polynomial $f(x)$ by the nonzero polynomial $g(x)$. Then, for all values of x such that $g(x) \neq 0$,

$$\frac{f(x)}{g(x)} = q(x) + \frac{r(x)}{g(x)}.$$

Proof Divide both sides of the identity $f(x) = g(x)q(x) + r(x)$ by $g(x)$. ■

The quotient theorem is often used to write a rational expression $\frac{f(x)}{g(x)}$ as the sum of a polynomial $q(x)$ and a rational expression $\frac{r(x)}{g(x)}$ that is **proper** in the sense that its numerator is of lower degree than its denominator.

Example 3 Rewrite the rational expression $\dfrac{2x^3 - x^2 - 7}{x - 2}$ as the sum of a polynomial and a proper rational expression.

Solution Let $f(x) = 2x^3 - x^2 - 7$ and $g(x) = x - 2$. By long division, we have

$$
\begin{array}{r|l}
 & 2x^2 + 3x + 6 \qquad = q(x) \\
\hline
x - 2 & 2x^3 - x^2 \qquad - 7 \\
 & 2x^3 - 4x^2 \\
\hline
 & \quad 3x^2 \qquad - 7 \\
 & \quad 3x^2 - 6x \\
\hline
 & \qquad 6x - 7 \\
 & \qquad 6x - 12 \\
\hline
 & \qquad\quad 5 = r(x).
\end{array}
$$

By Theorem 2, $\dfrac{f(x)}{g(x)} = q(x) + \dfrac{r(x)}{g(x)}$; that is,

$$\frac{2x^3 - x^2 - 7}{x - 2} = 2x^2 + 3x + 6 + \frac{5}{x - 2}. \quad ■$$

In Example 3 above, the divisor is a *first-degree* polynomial of the form $g(x) = x - c$. In all such cases, the remainder will either be the zero polynomial or it will have degree 0—in other words, the remainder will be a *constant* R.

Theorem 3 **The Remainder Theorem**

When a polynomial function f is divided by $x - c$, the remainder is $f(c)$.

Proof By the division algorithm (Theorem 1) with $g(x) = x - c$, we have

$$f(x) = (x - c)q(x) + R,$$

where R is a constant. Substituting $x = c$ into the last equation, we find that

$$f(c) = (c - c)q(c) + R = 0q(c) + R = R. \quad ■$$

Example 4 If $f(x) = 2x^3 - x^2 - 7$, use the remainder theorem to find $f(2)$.

Solution In Example 3 we used long division to divide $f(x) = 2x^3 - x^2 - 7$ by $g(x) = x - 2$ to obtain a quotient $q(x) = 2x^2 + 3x + 6$ and a remainder $R = 5$. Therefore, by the remainder theorem, we have

$$f(2) = 5.$$

That this is correct may be confirmed by a direct calculation:

$$f(2) = 2(2)^3 - 2^2 - 7 = 16 - 4 - 7 = 5.$$

One of the interesting consequences of the remainder theorem is the following.

Theorem 4 **The Factor Theorem**

Let f be a polynomial function and let c be a constant. Then $f(c) = 0$ if and only if $x - c$ is a factor of $f(x)$.

Proof

We have to prove that **(i)** if $f(c) = 0$, then $x - c$ is a factor of $f(x)$, and **(ii)** if $x - c$ is a factor of $f(x)$, then $f(c) = 0$.

(i) Suppose that $f(c) = 0$. Then we know by the remainder theorem that 0 is the remainder when $f(x)$ is divided by $x - c$; hence, $x - c$ is a factor of $f(x)$.

(ii) Suppose that $x - c$ is a factor of $f(x)$. Then, we can write

$$f(x) = (x - c)q(x),$$

and it follows that

$$f(c) = (c - c)q(c) = 0q(c) = 0.$$

■

Example 5 Let $f(x) = x^3 + 2x^2 - 5x - 6$. Use the factor theorem to determine whether:

(a) $x + 1$ is a factor of $f(x)$

(b) $x - 3$ is a factor of $f(x)$

Solution

(a) We have $x + 1 = x - (-1)$, which has the form $x - c$ with $c = -1$. Since

$$f(-1) = (-1)^3 + 2(-1)^2 - 5(-1) - 6 = -1 + 2 + 5 - 6 = 0,$$

it follows from the factor theorem that $x + 1$ is a factor of $f(x)$.

(b) Here $x - 3$ has the form $x - c$ with $c = 3$. Since

$$f(3) = 3^3 + 2(3)^2 - 5(3) - 6 = 27 + 18 - 15 - 6 = 24 \neq 0,$$

it follows from the factor theorem that $x - 3$ is not a factor of $f(x)$.

■

In using the remainder theorem to find $f(c)$, it is necessary to divide $f(x)$ by $g(x) = x - c$. This particular type of division can be carried out by a shortcut called **synthetic division.**

Procedure for Synthetic Division

To find the quotient polynomial $q(x)$ and the remainder $r(x) = R$ when a dividend polynomial $f(x)$ of degree $n \geq 1$ is divided by a *first-degree* polynomial $g(x) = x - c$, do the following:

Step 1. *Arrange the polynomial $f(x)$ in descending powers of x:*

$$f(x) = a_n x^n + a_{n-1}x^{n-1} + a_{n-2}x^{n-2} + \cdots + a_1 x + a_0.$$

Represent all missing powers by using zero coefficients.

Step 2. *Write down the value of c, then draw a vertical line and after it list the coefficients of $f(x)$:*

$$\begin{array}{c|cccccc} c & a_n & a_{n-1} & a_{n-2} & \cdots & a_1 & a_0 \end{array}$$

Step 3. *Leave some space below the row of coefficients, draw a horizontal line, and copy the leading coefficient a_n below the line:*

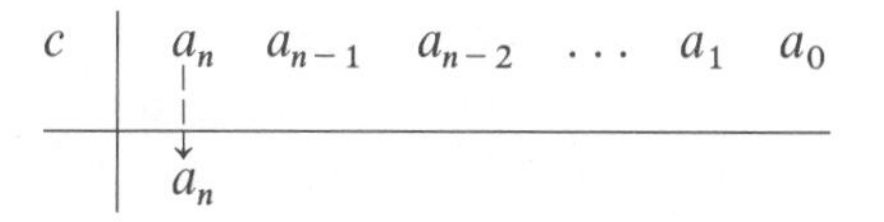

Step 4. *Multiply a_n by c and write the product above the horizontal line under the second coefficient a_{n-1}; then add a_{n-1} to this product and write the result s_1 below the line:*

$$\begin{array}{c|cccccc} c & a_n & a_{n-1} & a_{n-2} & \cdots & a_1 & a_0 \\ & & ca_n & & & & \\ \hline & a_n & s_1 & & & & \end{array}$$

Now multiply s_1 by c and write the product above the line under the third coefficient a_{n-2}; then add a_{n-2} to this product and write the result s_2 below the line:

$$\begin{array}{c|cccccc} c & a_n & a_{n-1} & a_{n-2} & \cdots & a_1 & a_0 \\ & & ca_n & cs_1 & & & \\ \hline & a_n & s_1 & s_2 & & & \end{array}$$

Continue in this way, multiplying each newly obtained number below the line by c, writing the product above the line under the next coefficient, and adding to produce the next number below the line. Do this until you have numbers below the line for every coefficient. Isolate the very last sum by drawing a short vertical line:

$$\begin{array}{c|ccccc|c} c & a_n & a_{n-1} & a_{n-2} & \cdots & a_1 & a_0 \\ & & ca_n & cs_1 & \cdots & cs_{n-2} & cs_{n-1} \\ \hline & a_n & s_1 & s_2 & \cdots & s_{n-1} & s_n \end{array}$$

Step 5. *Conclude that the numbers $a_n, s_1, s_2, \ldots, s_{n-1}$ are the coefficients of the quotient polynomial*

$$q(x) = a_n x^{n-1} + s_1 x^{n-2} + s_2 x^{n-3} + \cdots + s_{n-2}x + s_{n-1},$$

and that $s_n = R$, the remainder. [Note that $q(x)$ has degree 1 less than $f(x)$.]

Example 6 Use synthethic division to obtain the quotient polynomial $q(x)$ and the remainder R upon division of $f(x) = 2x^3 - x^2 - 7$ by $g(x) = x - 2$.

Solution The dividend $f(x) = 2x^3 - x^2 + 0x - 7$ has coefficients 2, -1, 0, and -7, and the divisor has the form $g(x) = x - c$ with $c = 2$. By synthetic division:

$$\begin{array}{r|rrrr} 2 & 2 & -1 & 0 & -7 \\ & & 4 & 6 & 12 \\ \hline & 2 & 3 & 6 & 5 \end{array}$$

Hence, the quotient polynomial is $q(x) = 2x^2 + 3x + 6$ and the remainder is $R = 5$. Note that this is the same problem that we did by long division in Example 3. Do you see how the shortcut works? ■

A detailed proof that synthetic division always works is a bit tedious, so we shall not give it here. However, if you work a few more examples by both long division and synthetic division, you will see clearly how every bit of the arithmetic required in the long division is accounted for in the synthetic division (see Problems 21 and 22).

Example 7 Use synthethic division and the remainder theorem to find $f(-2)$ if $f(x) = 3x^4 - 2x^3 - 5x$.

Solution We must divide $f(x) = 3x^4 - 2x^3 + 0x^2 - 5x + 0$ by $g(x) = x - c$ with $c = -2$. By synthetic division:

$$\begin{array}{r|rrrrr} -2 & 3 & -2 & 0 & -5 & 0 \\ & & -6 & 16 & -32 & 74 \\ \hline & 3 & -8 & 16 & -37 & 74 \end{array}$$

Thus, $f(-2) = 74$. ■

The technique for evaluating a polynomial illustrated in Example 7, which is sometimes called **Horner's method,** is readily carried out on a calculator or computer. The instruction booklet furnished with your calculator may contain efficient procedures for carrying out Horner's method using the particular memory features of that calculator.

C Example 8 If $f(x) = x^4 - 4.5x^3 - 7.14x^2 + 19.06x + 20.7$, show that $f(-1.8) = 0$.

Solution Using Horner's method, we have:

$$\begin{array}{r|rrrrr} -1.8 & 1 & -4.5 & -7.14 & 19.06 & 20.7 \\ & & -1.8 & 11.34 & -7.56 & -20.7 \\ \hline & 1 & -6.3 & 4.20 & 11.50 & 0 \end{array}$$

Therefore, $f(-1.8) = 0$. ■

Problem Set 4.3

In Problems 1 to 18, perform the indicated long division to find a quotient polynomial and a remainder polynomial. Be sure that the remainder is either the zero polynomial or a polynomial of lower degree than the divisor. Check your work by verifying the fact that dividend = (divisor)(quotient) + remainder.

1. $x - 5 \overline{\smash{)}\, x^2 + 3x - 10}$

2. $3x^2 - 1 \overline{\smash{)}\, 6x^4 + 10x^2 + 7}$

3. $x - 1 \overline{\smash{)}\, x^3 - 1}$

4. $4x - 3 \overline{\smash{)}\, 4x^6 + 5x^3 - 6}$

5. $2x^2 - 4x + 1 \overline{\smash{)}\, 6x^4 - 31x^2 + 26x - 6}$

6. $x + 1 \overline{\smash{)}\, x^5 - 1}$

7. $2x - 3 \overline{\smash{)}\, 4x^4 - 12x^3 + 15x^2 - 17x}$

8. $2x^4 - 3x - 1 \overline{\smash{)}\, -6x^4 - 7x^3 + 14x^2 - 5x + 1}$

9. $x^2 + 3 \overline{\smash{)}\, x^3 + 3x^2 + 2x - 4}$

10. $1 - 4x - 2x^3 \overline{\smash{)}\, x^2 - 3x^4 - x^3 + 5}$

11. $x^3 + x^2 \overline{\smash{)}\, x^5 - 1}$

12. $x - c \overline{\smash{)}\, x^2 - 2cx + 2}$

13. $2x - 1 \overline{\smash{)}\, x^2 + \frac{1}{2}x + 1}$

14. $t^2 - t + 1 \overline{\smash{)}\, 2t^4 + t^2 - 1}$

15. $3x^2 + 2x + 1 \overline{\smash{)}\, x^3 + x^2 + 1}$

16. $x - c \overline{\smash{)}\, x^3 - c^3}$

17. $5t^2 - t + 4 \overline{\smash{)}\, 10t^3 + 13t^2 + 5t + 2}$

18. $x^2 + x + 1 \overline{\smash{)}\, ax^2 + bx + c}$

In Problems 19 and 20, use the quotient theorem (Theorem 2) to rewrite each rational expression as the sum of a polynomial and a proper rational expression.

19. (a) $\dfrac{5x^3 + 3x^2 - x + 2}{x - 4}$ **(b)** $\dfrac{5x^3 - 6x^2 - 68x - 16}{x^3 - 2x^2 - 8x}$

20. (a) $\dfrac{x^2 - x + 1}{x^2 + x - 1}$ **(b)** $\dfrac{ax + b}{cx + d}$

In Problems 21 and 22, find the quotient and remainder **(a)** by long division and **(b)** by synthetic division.

21. $x - 3 \overline{\smash{)}\, 5x^3 - 11x^2 - 14x - 10}$

22. $x + 3 \overline{\smash{)}\, 2x^3 + 3x^2 - 5x + 12}$

In Problems 23 to 28, use synthetic division to obtain the quotient polynomial $q(x)$ and the remainder R upon division of the polynomial $f(x)$ by the first-degree polynomial $g(x)$.

23. $f(x) = 3x^3 - 2x^2 - x + 4$; $g(x) = x - 2$

24. $f(x) = x^6 - x^5 - x^2 - x - 1$; $g(x) = x - 2$

25. $f(x) = x^5 - 5x^3 + x - 16$; $g(x) = x + 2$

26. $f(x) = 5x^3 - 7x^2 + 3x - 2$; $g(x) = x - 1$

27. $f(x) = -16x^3 - 12x^2 + 2x + 7$; $g(x) = x - \frac{1}{2}$

28. $f(x) = x^2 + 2x + 1$; $g(x) = x - c$

C In Problems 29 to 36, use Horner's method to find the value of $f(c)$ for the given polynomial function f and the indicated value of c.

29. $f(x) = x^3 - 9x^2 + 23x - 15$; $c = 1$

30. $f(x) = 2x^3 - 2x^2 - x - 2$; $c = -1$

31. $f(x) = 8x^3 - 25x^2 + 4x - 3$; $c = -3$

32. $f(x) = 3x^4 - x^3 + 2x - 10$; $c = 2$

33. $f(x) = x^5 - 2x^4 + x^3 - 3x^2 + 8$; $c = 2$

34. $f(x) = x^5 + 2x^3 - 4x + 5$; $c = -2$

35. $f(x) = 2x^4 + 3x^2 - 2x + 1$; $c = 1.2$

36. $f(x) = 5x^5 - 20x^3 + 2x - 1$; $c = -2.13$

C In Problems 37 to 42, use the factor theorem to determine (without actually dividing) whether or not the indicated binomial is a factor of the given polynomial. (You may want to use a calculator and Horner's method to find values of the polynomial functions.)

37. $f(x) = 4x^4 + 13x^3 - 13x^2 - 40x + 12$; $x + 2$

38. $g(x) = x^4 - 9x^3 + 18x^2 - 3$; $x + 1$

39. $H(x) = x^5 - 17x^3 + 75x + 9$; $x - 3$

40. $F(x) = 2x^4 - x^3 + x^2 + x - 3$; $x + 1$

41. $h(x) = 30x^3 - 20x^2 - 100x + 1000$; $x - 10$

42. $G(t) = t^4 + 2t^3 - 6t^2 - 14t - 7$; $t - 7$

C In Problems 43 and 44, use a calculator and Horner's method to show that $f(c) = 0$ for the given polynomial function f and the indicated value of c.

43. $f(x) = 5x^3 - 11x^2 - 46x + 96$; $c = 3.2$

44. $f(x) = x^3 - 0.43x^2 - 2.57x - 1.14$; $c = -0.57$

In Problems 45 to 50, use the factor theorem to answer each question.

45. For what value of k will $x + 5$ be a factor of $x^3 + kx + 125$?

46. For what positive integer values of n is $x + a$ a factor of $x^n + a^n$?

47. For what value of k will $x - \frac{1}{3}$ be a factor of $3x^3 - x^2 + kx - 5$?

48. For what positive integer values of n is $x + a$ a factor of $x^n - a^n$?

49. For what positive integer values of n is $x - a$ a factor of $x^n - a^n$?

50. Are there any real numbers k for which $x - k$ is a factor of $x^4 + 3x^2 + 1$? Why or why not?

C 51. If $0 \le h \le 2$, the equation

$$P = 15 - 6h + 1.2h^2 - 0.16h^3$$

gives a good approximation of the atmospheric pressure P in pounds per square inch at a height of h thousand feet above sea level. Use a calculator and Horner's method to find the approximate value of P for a hot-air balloon at a height of $h = 1.3$ thousand feet.

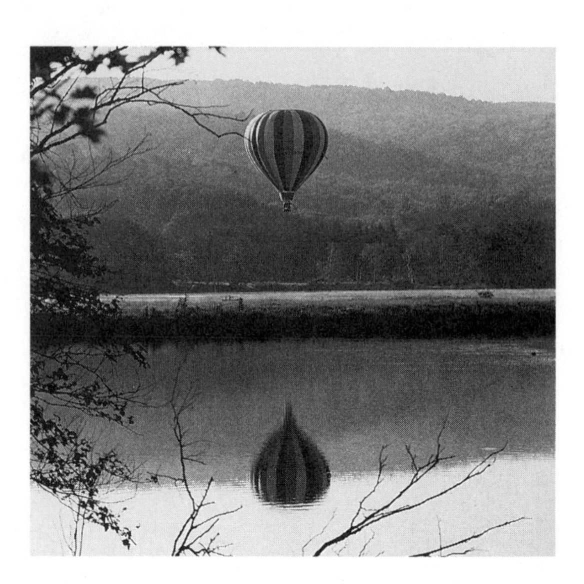

52. Suppose that the polynomial $f(x)$ is divided by the nonzero polynomial $g(x)$ to produce a quotient polynomial $q(x)$ and a remainder polynomial $r(x)$. Now, suppose that $q(x)$ is divided by the nonzero polynomial $G(x)$ to produce a quotient polynomial $Q(x)$ and a remainder polynomial $R(x)$. Show that, if $f(x)$ is divided by the product $g(x)G(x)$, the quotient polynomial is $Q(x)$ and the remainder polynomial is $r(x) + g(x)R(x)$.

C 53. Let $f(x) = 1 + 0.5x - 0.125x^2 + 0.0625x^3$. For small values of x, the approximation $\sqrt{1 + x} \approx f(x)$ is fairly accurate. How much error is involved in using this approximation for **(a)** $x = -0.2$, **(b)** $x = 0.1$, **(c)** $x = 0.5$?

54. If f is a function and a is a number in the domain of f, then the expression $\dfrac{f(x) - f(a)}{x - a}$ plays an important role in calculus. If f is a polynomial function, show that there exists a polynomial function q such that, for $x \neq a$,

$$\frac{f(x) - f(a)}{x - a} = q(x).$$

4.4 Real Zeros of Polynomial Functions

In this section, we continue our study of polynomial functions and their graphs, concentrating our attention on techniques for finding *real zeros* of such functions.

If f is a function, then a number c in the domain of f is called a **zero** of f if

$$f(c) = 0.$$

In particular, if f is a polynomial function and c is a real number, then the following statements are equivalent:

(i) c is a *zero* of f.

(ii) $x = c$ is a *solution* or **root** of the equation $f(x) = 0$.

(iii) $x - c$ is a *factor* of $f(x)$.

(iv) c is an x *intercept* of the graph of f.

For instance, a computer-generated graph of the polynomial function

$$f(x) = x^3 - 4x$$

(Figure 1) shows that the real zeros of f are -2, 0, and 2.

Figure 1

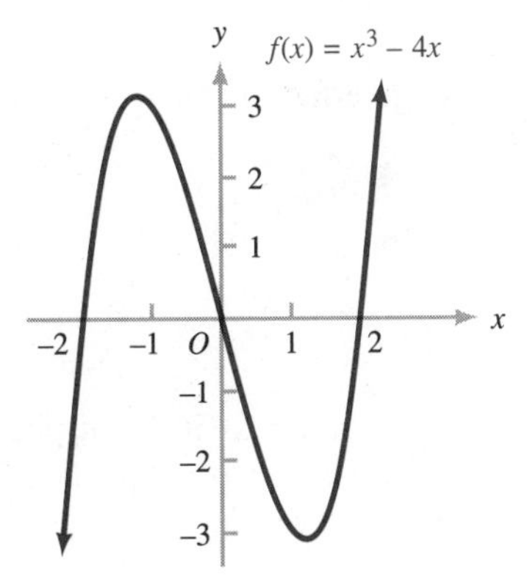

As indicated above, each zero of a polynomial function f corresponds to a first-degree factor $(x - c)$ of $f(x)$. When $f(x)$ is factored completely, the same factor $(x - c)$ may occur more than once, in which case c is called a **repeated** or **multiple** zero of f.

If $(x - c)^m$ is a factor of $f(x)$, but $(x - c)^{m+1}$ is not, we say that c is a zero of multiplicity m.

For instance, if $(x - 5)^2$ is a factor of $f(x)$, but $(x - 5)^3$ is not, we say that 5 is a zero of multiplicity 2. At a zero that is not repeated, the graph of a polynomial function f simply cuts across the x axis (Figure 2a); however, the graph of f is very "flat" near a zero of multiplicity $m > 1$ (Figures 2b and 2c).*

Figure 2

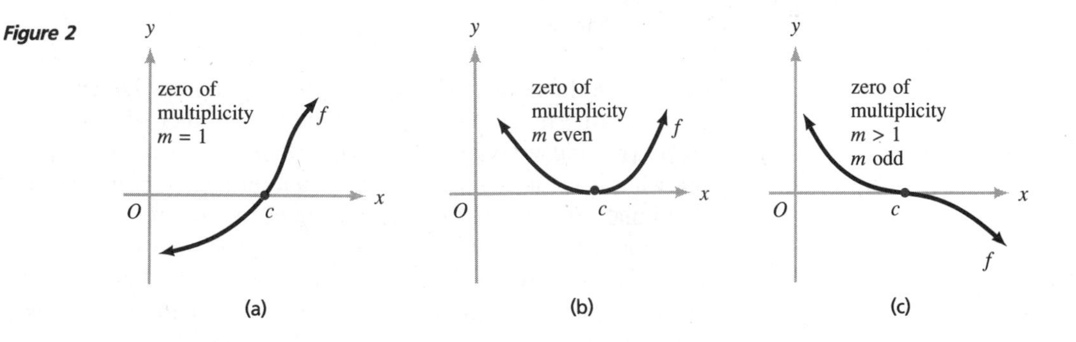

Example 1 Given that 2 is a zero of

$$f(x) = x^5 - 6x^4 + 11x^3 - 2x^2 - 12x + 8,$$

determine its multiplicity.

* At a zero of multiplicity $m > 1$, the x axis is actually *tangent* to the graph of f.

Solution

Using synthetic division, we divide $f(x)$ by $x - 2$ to obtain

$$f(x) = (x - 2)(x^4 - 4x^3 + 3x^2 + 4x - 4).$$

Again, we divide $x^4 - 4x^3 + 3x^2 + 4x - 4$ by $x - 2$ and find that

$$x^4 - 4x^3 + 3x^2 + 4x - 4 = (x - 2)(x^3 - 2x^2 - x + 2).$$

Combining the last two equations, we obtain

$$f(x) = (x - 2)(x - 2)(x^3 - 2x^2 - x + 2) = (x - 2)^2(x^3 - 2x^2 - x + 2).$$

Now we divide $x^3 - 2x^2 - x + 2$ by $x - 2$ and find that

$$x^3 - 2x^2 - x + 2 = (x - 2)(x^2 - 1).$$

Therefore, $$f(x) = (x - 2)^2(x - 2)(x^2 - 1) = (x - 2)^3(x^2 - 1).$$

Because $x - 2$ is not a factor of $x^2 - 1$, our successive divisions terminate, and it follows that 2 is a zero of multiplicity 3 for the polynomial $f(x)$. ■

Because a polynomial $f(x)$ can't have more first-degree factors than its degree, we have the following result.

Theorem 1

The Maximum Number of Zeros of a Polynomial Function

A polynomial function cannot have more zeros than its degree.

In using Theorem 1, you must count each zero as many times as its multiplicity. For instance, we found in Example 1 that 2 was a zero of multiplicity 3 for the fifth-degree polynomial function f; hence, we can conclude that f can have at most two more zeros. Indeed, f has exactly two more zeros: 1 and -1.

Descartes' Rule of Signs

The seventeenth-century French mathematician René Descartes discovered a simple and useful rule that helps to determine the number of positive and negative zeros of a polynomial with real coefficients. Before stating Descartes' rule, we must explain what is meant by a *variation of sign* for such a polynomial. If the terms of the polynomial are arranged in order of descending powers, we say that a **variation of sign** occurs whenever two successive terms have opposite signs. Missing terms (with zero coefficients) are ignored when counting the total number of variations of sign. In using Descartes' rule, it is necessary to count the variations of sign both for a polynomial $f(x)$ and for the related polynomial $f(-x)$.

Example 2

If $f(x) = 7x^5 - 3x^4 + 5x^2 + x - 2$, determine the total number of variations of sign for:

(a) $f(x)$ **(b)** $f(-x)$

Solution

(a) $f(x) = 7x^5 - 3x^4 + 5x^2 + x - 2$ has three variations of sign.

(b) $f(-x) = 7(-x)^5 - 3(-x)^4 + 5(-x)^2 + (-x) - 2$

$$= -7x^5 - 3x^4 + 5x^2 - x - 2$$

has two variations of sign. ■

We can now state **Descartes' rule of signs** for a polynomial $f(x)$ with real coefficients.

Theorem 2 **Descartes' Rule of Signs**

Suppose that $f(x)$ is arranged in descending powers of x. Then:

(i) The number of *positive zeros* of $f(x)$ is either equal to the number of variations of sign for $f(x)$, or it is less than that number by an even integer.

(ii) The number of *negative zeros* of $f(x)$ is either equal to the number of variations of sign for $f(-x)$, or it is less than that number by an even integer.

In using Descartes' rule, zeros of multiplicity m are to be counted m times. The proof of part (i) of the rule is somewhat technical and will be omitted here.* Part (ii) of the rule can be derived from part (i) by applying the results of Problems 53 and 55.

Example 3 Use Descartes' rule of signs to determine the possible number of positive and negative zeros of

$$f(x) = 2x^5 + 3x^4 - 7x^3 + 2x^2 + 8x - 5.$$

Solution There are three variations of sign for $f(x)$, so $f(x)$ has either 3 or 1 positive zeros. There are two variations of sign for

$$f(-x) = -2x^5 + 3x^4 + 7x^3 + 2x^2 - 8x - 5,$$

so $f(x)$ has either 2 or 0 negative zeros. ■

Figure 3

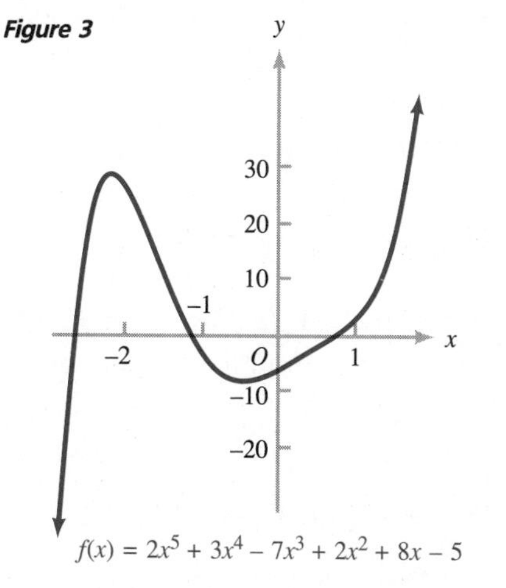

$f(x) = 2x^5 + 3x^4 - 7x^3 + 2x^2 + 8x - 5$

As a matter of fact, the function $f(x)$ in Example 3 has 2 negative zeros and 1 positive zero, as shown by the computer-generated graph in Figure 3. It also has 2 *complex* zeros. (We study complex zeros in Section 4.6.)

* A proof of Descartes' rule of signs can be found in *A Modern Course on the Theory of Equations* by David E. Dobbs and Robert Hanks (Polygonal Publishing House, Passaic, N.J., 1980).

The Upper-and-Lower-Bound Rule

A second rule, which can be used efficiently in conjunction with Descartes' rule of signs, determines a closed interval that contains all of the real zeros of a polynomial function $f(x)$. This rule involves the idea of *upper and lower bounds.* Any number that is greater than or equal to all the real zeros of $f(x)$ is called an **upper bound** for these zeros. Likewise, any number that is less than or equal to all of the real zeros of $f(x)$ is called a **lower bound** for these zeros. Notice that such upper and lower bounds are not unique—any number greater than an upper bound is again an upper bound, and any number less than a lower bound is again a lower bound. However, if L is a lower bound and U is an upper bound for the real zeros of $f(x)$, then we can be certain that all of these zeros belong to the closed interval $[L, U]$.

We can now state the following theorem.

Theorem 3 **The Upper-and-Lower-Bound Rule for Real Zeros**

Suppose that $f(x)$ is a polynomial with real coefficients and that the *leading coefficient** of $f(x)$ is positive. Let $f(x)$ be divided by $x - c$ using synthetic division.

(i) If $c > 0$ and all numbers in the last (quotient) row of the synthetic division are positive or zero, then c is an upper bound for the zeros of $f(x)$.

(ii) If $c < 0$ and the numbers in the last (quotient) row of the synthetic division alternate in sign, then c is a lower bound for the zeros of $f(x)$.

If 0 appears in one or more places in the quotient row, it can be regarded as being either positive or negative for purposes of applying part (ii) *of this rule.* Problem 54 outlines a proof of part (i) of the rule. Part (ii) can be derived from part (i) by applying the results of Problems 53 and 55.

Example 4 Use the upper-and-lower-bound rule to find integers that are upper and lower bounds for the real zeros of

$$f(x) = 6x^4 - 19x^3 + 13x^2 + 4x - 4.$$

Solution To locate an upper bound for the real zeros of $f(x)$ we use part (i) of the upper-and-lower-bound rule to test successive values of $c = 1, 2, 3, \ldots,$ until the last row of the synthetic division becomes nonnegative:

$$\begin{array}{r|rrrr|r} 1 & 6 & -19 & 13 & 4 & -4 \\ & & 6 & -13 & 0 & 4 \\ \hline & 6 & -13 & 0 & 4 & 0 \end{array} \qquad \begin{array}{r|rrrr|r} 2 & 6 & -19 & 13 & 4 & -4 \\ & & 12 & -14 & -2 & 4 \\ \hline & 6 & -7 & -1 & 2 & 0 \end{array}$$

$$\begin{array}{r|rrrr|r} 3 & 6 & -19 & 13 & 4 & -4 \\ & & 18 & -3 & 30 & 102 \\ \hline & 6 & -1 & 10 & 34 & 98 \end{array} \qquad \begin{array}{r|rrrr|r} 4 & 6 & -19 & 13 & 4 & -4 \\ & & 24 & 20 & 132 & 544 \\ \hline & 6 & 5 & 33 & 136 & 540 \end{array}$$

* The leading coefficient of a polynomial is the coefficient of its term of highest degree.

The first value of c for which the last row becomes nonnegative is $c = 4$; hence, 4 is an upper bound for the real zeros of $f(x)$. Now we use part (ii) of the rule to test successive values of $c = -1, -2, -3, \ldots$, until the last row of the synthetic division alternates in sign:

$$\begin{array}{r|rrrrr} -1 & 6 & -19 & 13 & 4 & -4 \\ & & -6 & 25 & -38 & 34 \\ \hline & 6 & -25 & 38 & -34 & \multicolumn{1}{|r}{30} \end{array}$$

Here, on the very first trial with $c = -1$, we obtain alternating signs, and it follows that -1 is a lower bound for the real zeros of $f(x)$. ■

By using the upper-and-lower-bound rule as in Example 4, we can conclude that all the real zeros of the function

$$f(x) = 6x^4 - 19x^3 + 13x^2 + 4x - 4$$

lie on the interval $[-1, 4]$. This is quite correct, but, as the computer-generated graph in Figure 4 shows, all the real zeros of f actually lie on the smaller interval $[-1, 2]$. Thus, although the upper-and-lower-bound rule will give an interval that is guaranteed to contain all of the real zeros of a polynomial function, it does not necessarily produce the *smallest* such interval.

Figure 4

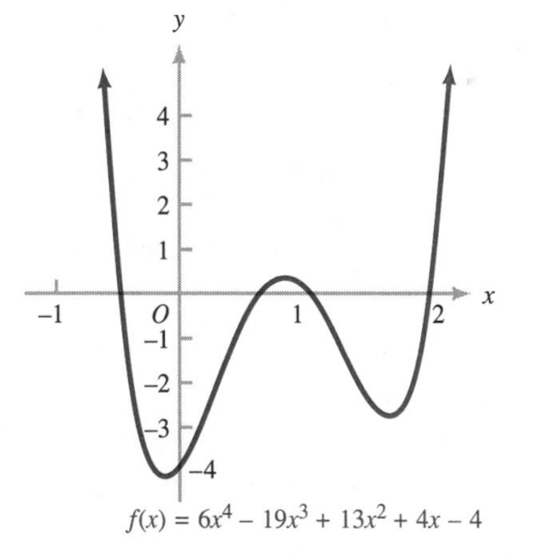

Rational Zeros of Polynomial Functions

Suppose that all the coefficients of a polynomial function are rational numbers. If you multiply the polynomial by the least common denominator (LCD) of these coefficients, you will obtain a polynomial function with *integer* coefficients and with the same zeros as the original function. Then, the following theorem can be used to find all the rational zeros of the polynomial function.

Theorem 4 **The Rational-Zeros Theorem**

Let $f(x) = a_n x^n + a_{n-1}x^{n-1} + \cdots + a_1 x + a_0$ be a polynomial function of degree $n > 0$ with integers as coefficients and suppose that $a_0 \neq 0$. Then, if p/q is any rational zero of f and if the fraction p/q is reduced to lowest terms, p must be a factor of a_0 and q must be a factor of a_n.

A rigorous proof* of the rational-zeros theorem can be found on page 280 of *A First Undergraduate Course in Abstract Algebra* by A.P. Hillman and G.L. Anderson (Wadsworth Publishing Co., Belmont, Calif., 1973).

In the following procedure, based on Theorem 4, we assume that $a_0 \neq 0$. [If $a_0 = 0$, then 0 is a rational root of $f(x)$.]

Procedure for Finding all Rational Zeros of a Polynomial Function

Step 1. Check to see that all coefficients of $f(x)$ are integers. (If not, multiply by the LCD of the coefficients to clear fractions.) Find all factors p of a_0 (the constant term) and all factors q of a_n (the leading coefficient).

Step 2. Form all possible ratios p/q. These rational numbers are all possible rational zeros of f.

Step 3. Using synthetic division, check each† of the rational numbers p/q to see whether it is a zero of f. If none of them works, conclude that f has no rational zeros. If a rational zero p/q is found, proceed to step 4.

Step 4. If $c = p/q$ is the rational zero produced in step 3, then, by the factor theorem, $f(x) = (x - c)Q(x)$, where the coefficients of $Q(x)$ were determined when you performed the synthetic division. The remaining zeros of f are the zeros of Q.

If the polynomial $Q(x)$ in step 4 is quadratic, you can use the quadratic formula (or factoring) to find its zeros. If $Q(x)$ has degree 3 or more, just repeat the whole procedure, starting with Q this time. To find all rational roots of the original polynomial may require several cycles through the procedure. If you encounter the same rational zero more than once, this just means that the zero has multiplicity greater than one.

* The proof, although elementary, makes use of facts concerning relatively prime integers that are usually proved only in more advanced courses.

† For greater efficiency, you can use Descartes' rule of signs and the upper-and-lower-bound rule to reject all ratios p/q that cannot be zeros of f.

Example 5 Find all rational zeros of $f(x) = 9x^3 + 6x^2 - 5x - 2$.

Solution We follow the procedure above.

Step 1. All coefficients are integers, so there is no need to clear fractions. Here $a_0 = -2$ and $a_3 = 9$. The factors of a_0 are

$$p: \pm 1, \pm 2,$$

and the factors of a_3 are

$$q: \pm 1, \pm 3, \pm 9.$$

Step 2. The possible zeros p/q are

$$\frac{p}{q}: \pm 1, \pm 2, \pm\tfrac{1}{3}, \pm\tfrac{2}{3}, \pm\tfrac{1}{9}, \pm\tfrac{2}{9}.$$

Step 3. Of the twelve possible rational zeros in step 2, three at most can be zeros of $f(x)$ (Theorem 1), but it may be that none of them is. (The possibilities can be narrowed down by using Descartes' rule of signs and the upper-and-lower-bound rule, but we shall not do so here.) By checking the possibilities one at a time, using synthetic division, we find, for instance, that $\frac{2}{3}$ works.

$$\begin{array}{c|cccc} \frac{2}{3} & 9 & 6 & -5 & -2 \\ & & 6 & 8 & 2 \\ \hline & 9 & 12 & 3 & \vert\ 0 = R. \end{array}$$

Step 4. The synthetic division just performed yields the coefficients of the quotient polynomial $Q(x) = 9x^2 + 12x + 3$. We could now apply the same procedure to the polynomial function $Q(x)$, but it's easier to factor:

$$9x^2 + 12x + 3 = 3(x + 1)(3x + 1).$$

We conclude that the remaining zeros are -1 and $-\frac{1}{3}$. Therefore, the rational zeros of f are $\frac{2}{3}$, -1, and $-\frac{1}{3}$. ■

Figure 5

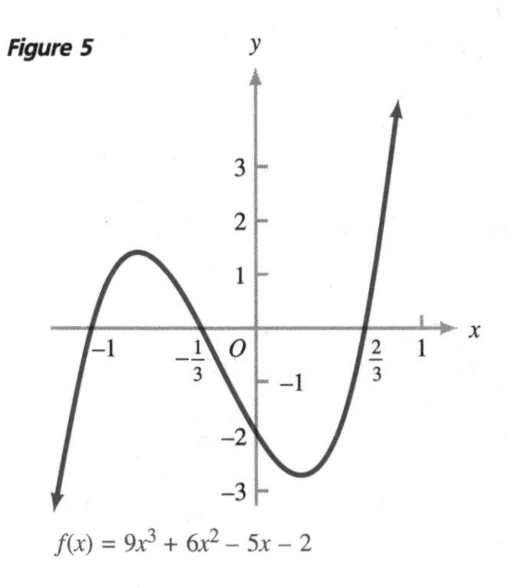

$f(x) = 9x^3 + 6x^2 - 5x - 2$

By using the rational-zeros theorem and the factor theorem, you can often factor a polynomial function completely into prime factors. For instance, as a consequence of Example 5,

$$\begin{aligned} f(x) &= 9x^3 + 6x^2 - 5x - 2 \\ &= (x - \tfrac{2}{3})(9x^2 + 12x + 3) \\ &= (x - \tfrac{2}{3})(3)(x + 1)(3x + 1) \\ &= (3x - 2)(x + 1)(3x + 1). \end{aligned}$$

If you succeed in factoring a polynomial function completely, you can use the techniques discussed in Section 4.2 to sketch its graph. For instance, the graph of $f(x) = 9x^3 + 6x^2 - 5x - 2$ (Example 5) is shown in Figure 5.

Example 6 Let $f(x) = 4x^4 - 4x^3 - 25x^2 + x + 6$.

(a) Factor $f(x)$ completely into prime factors.

(b) Sketch the graph of f. gc Use a graphing calculator if you wish.

Solution **(a)** We begin by applying the procedure for finding rational zeros to the function f. Here $a_0 = 6$ and $a_4 = 4$. Thus,

$$p: \ \pm 1, \pm 2, \pm 3, \pm 6 \qquad \text{(factors of 6)}$$

$$q: \ \pm 1, \pm 2, \pm 4 \qquad \text{(factors of 4)}$$

$$\frac{p}{q}: \ \pm 1, \pm 2, \pm 3, \pm 6, \pm \tfrac{1}{2}, \pm \tfrac{3}{2}, \pm \tfrac{1}{4}, \pm \tfrac{3}{4}.$$

We test each* of the possible rational zeros p/q, one by one, using synthetic division. The first one that works is -2:

$$\begin{array}{r|rrrr|r} -2 & 4 & -4 & -25 & 1 & 6 \\ & & -8 & 24 & 2 & -6 \\ \hline & 4 & -12 & -1 & 3 & 0 = R. \end{array}$$

Therefore, -2 is a rational zero of f and the quotient polynomial is given by $Q(x) = 4x^3 - 12x^2 - x + 3$. Hence,

$$f(x) = (x + 2)(4x^3 - 12x^2 - x + 3).$$

Now, we repeat the procedure for $Q(x) = 4x^3 - 12x^2 - x + 3$. Here $a_0 = 3$ and $a_3 = 4$. Thus,

$$p: \ \pm 1, \pm 3 \qquad \text{(factors of 3)}$$

$$q: \ +1, \pm 2, \pm 4 \qquad \text{(factors of 4)}$$

$$\frac{p}{q}: \ \pm 1, \pm 3, \pm \tfrac{1}{2}, \pm \tfrac{1}{4}, \pm \tfrac{3}{2}, \pm \tfrac{3}{4}.$$

We test the possible rational zeros p/q of $Q(x)$, one by one, using synthetic division. The first one that works is 3:

$$\begin{array}{r|rrr|r} 3 & 4 & -12 & -1 & 3 \\ & & 12 & 0 & -3 \\ \hline & 4 & 0 & -1 & 0 = R. \end{array}$$

Therefore, 3 is a rational zero of $Q(x)$, and

$$\begin{aligned} Q(x) = 4x^3 - 12x^2 - x + 3 &= (x - 3)(4x^2 + 0x - 1) \\ &= (x - 3)(4x^2 - 1) = (x - 3)(2x - 1)(2x + 1). \end{aligned}$$

* By Descartes' rule of signs, f has either two or no positive zeros and either two or no negative zeros—although Descartes' rule alone gives no information about whether these zeros are rational numbers. By the upper-and-lower-bound rule, all zeros of f lie on the interval $[-3, 4]$. Thus, we can rule out ± 6 as possible zeros without bothering to test them by synthetic division.

It follows that

$$f(x) = (x + 2)Q(x) = (x + 2)(x - 3)(2x - 1)(2x + 1).$$

(b) Using the techniques discussed in Section 4.2, we can now sketch the graph of f (Figure 6). As usual, accuracy is enhanced if a calculator is used to help determine points on the graph, or if a graphing calculator is used. ■

Figure 6

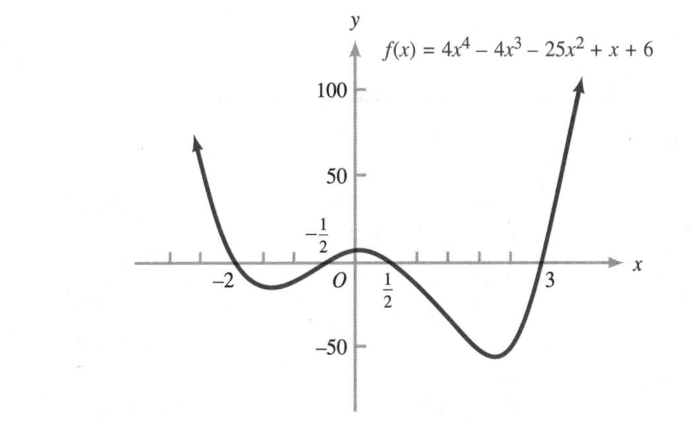

Example 7 A storage tank on a milk truck has the shape of a right circular cylinder with hemispheres on each end (Figure 7). The total length of the tank is 14 feet, and its volume is 108π cubic feet. Find the radius x of the tank.

Solution The volume of the cylindrical part of the tank is

$$\pi x^2(14 - 2x)$$

cubic feet, and the volume of the two hemispherical ends is

$$\frac{4}{3}\pi x^3$$

cubic feet. Therefore,

$$\pi x^2(14 - 2x) + \tfrac{4}{3}\pi x^3 = 108\pi.$$

Figure 7

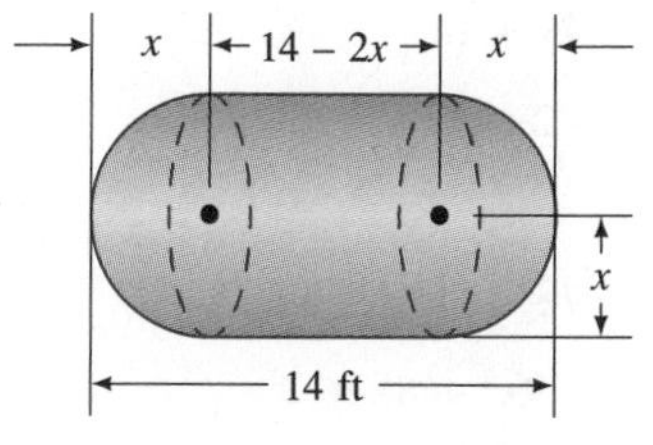

Simplifying this equation, we obtain

$$x^3 - 21x^2 + 162 = 0.$$

Using the procedure for finding rational zeros, we find that $x = 3$ is a solution of the last equation. Indeed,

$$\begin{array}{r|rrrr} 3 & 1 & -21 & 0 & 162 \\ & & 3 & -54 & -162 \\ \hline & 1 & -18 & -54 & 0 \end{array}$$

Therefore, $x^3 - 21x^2 + 162 = (x - 3)(x^2 - 18x - 54) = 0.$

By the quadratic formula, the solutions of the equation

$$x^2 - 18x - 54 = 0$$

are $x = 9 \pm 3\sqrt{15}$. Since the tank cannot have a negative radius, we can reject the solution $x = 9 - 3\sqrt{15}$. Also, the length of the cylindrical part of the tank is $14 - 2x$ feet, and if $x = 9 + 3\sqrt{15}$, then $14 - 2x$ is negative, so we can reject the solution $x = 9 + 3\sqrt{15}$. Therefore, the radius of the tank is 3 feet. ■

Problem Set 4.4

In Problems 1 to 6, the indicated number is a zero of the polynomial function $f(x)$. Determine the multiplicity of this zero.

1. $1; f(x) = x^3 + x^2 - 5x + 3$

2. $1; f(x) = 3x^3 - 8x^2 + 7x - 2$

3. $2; f(x) = x^4 - 11x^3 + 42x^2 - 68x + 40$

4. $-2; f(x) = x^5 + 2x^4 - 9x^3 - 22x^2 + 4x + 24$

5. $-1; f(x) = 4x^4 + 9x^3 + 3x^2 - 5x - 3$

6. $-3; f(x) = x^5 + 10x^4 + 37x^3 + 63x^2 + 54x + 27$

In Problems 7 to 12, determine the total number of variations in sign for **(a)** $f(x)$ and **(b)** $f(-x)$.

7. $f(x) = 3x^3 - 2x^2 - 5x + 7$

8. $f(x) = 2x^4 + 5x^3 - 8x^2 - 5x + 4$

9. $f(x) = 12x^5 - 17x^4 + 7x^2 + 3x - 2$

10. $f(x) = -2x^5 - x^4 + 5x^3 + x^2$

11. $f(x) = 7x^7 - 8x^5 - 5x^3 + 11x + 6$

12. $f(x) = x^6 - 9x^4 - 3x^2 + 13$

In Problems 13 to 20, use Descartes' rule of signs to determine the possible number of positive and negative zeros of the polynomial function $f(x)$.

13. $f(x) = 2x^3 - 3x^2 - 4x - 5$

14. $f(x) = 2x^3 - 7x^2 + 12x - 4$

15. $f(x) = 2x^3 - 7x^2 + 17x - 4$

16. $f(x) = 3x^4 - 4x^3 - 5x - 1$

17. $f(x) = x^4 - 2x^3 - 2x^2 + 8x - 1$

18. $f(x) = 2x^3 - 7x^2 - x - 21$

19. $f(x) = x^6 + x^2 + x + 1$

20. $f(x) = 2x^6 - x^4 + 3x + 13$

In Problems 21 to 28, use the upper-and-lower-bound rule to find integers that are upper and lower bounds for the real zeros of the polynomial function $f(x)$.

21. $f(x) = 2x^3 - 5x^2 - 4x + 3$

22. $f(x) = 2x^3 + 9x^2 - 5x - 41$

23. $f(x) = x^4 - x^3 - 10x^2 - 2x + 12$

24. $f(x) = 2x^5 - 2x^2 + x - 2$

25. $f(x) = x^4 - x^3 + x - 3$

26. $f(x) = 2x^4 - 5x^3 - 8x^2 + 25x - 10$

27. $f(x) = 4x^4 - 8x^3 - 43x^2 + 28x + 60$

28. $f(x) = 8x^5 - 44x^4 + 86x^3 - 73x^2 + 28x - 4$

In Problems 29 to 38, find all rational zeros of each polynomial function.

29. $f(x) = x^3 - 9x^2 + 23x - 15$

30. $g(x) = x^3 - 3x^2 - 4x + 12$

31. $F(x) = 2x^3 - 3x^2 + 2x + 2$

32. $G(x) = x^3 - 2x^2 - 13x - 10$

33. $f(x) = x^4 - 4x^3 + 7x^2 - 12x + 12$

34. $H(x) = 4x^4 + 4x^3 + 9x^2 + 8x + 2$

35. $f(x) = x^4 - 5x^3 + \frac{1}{4}x^2 + \frac{9}{2}x + \frac{3}{2}$

36. $g(x) = x^5 - x^3 + 27x^2 - 27$

37. $Q(x) = x^6 - 17x^4 + 75x^2 - 9x$

38. $f(x) = \frac{1}{2}x^5 + 2x^4 + \frac{1}{2}x^2 - \frac{3}{2}x - 14$

In Problems 39 to 42, find all real numbers that are solutions of the given equation. [*Hint:* Begin by finding the rational solutions.]

39. $8x^3 - 25x^2 + 4x - 3 = 0$

40. $2x^3 - 15x^2 + 27x - 10 = 0$

41. $20x^5 - 9x^4 - 74x^3 + 30x^2 + 42x - 9 = 0$

42. $x^5 + x^4 - \frac{5}{4}x^3 + \frac{25}{4}x^2 - 21x + 9 = 0$

C In Problems 43 to 46, **(a)** factor each polynomial function completely into prime factors and **(b)** sketch its graph. gc Use a graphing calculator if you wish.

43. $f(x) = x^3 - 6x^2 - x + 6$

44. $g(x) = x^3 - 8x^2 + 2x - 16$

45. $Q(x) = 2x^4 - 3x^3 - 7x^2 + 12x - 4$

46. $G(x) = 16x^4 - 40x^2 + 9$

47. A box is to be constructed so that the length is twice the width and the depth exceeds the width by 2 inches (Figure 8). Find the box dimensions if its volume is 350 cubic inches. [*Hint:* Let x denote the box width, express the length and depth in terms of x, and solve the equation length × depth × width = 350.]

Figure 8

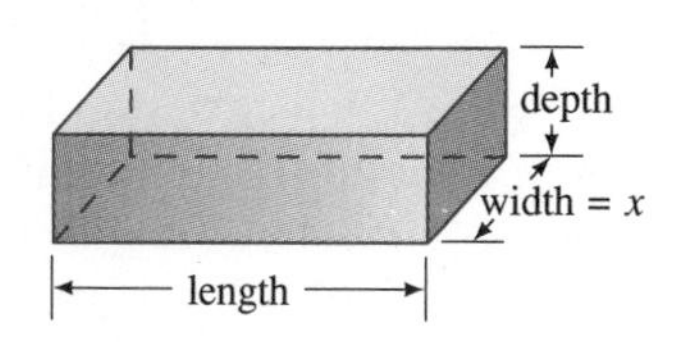

48. A plot of land has the shape of a right triangle with a hypotenuse 1 kilometer longer than one of the sides. Find the lengths of the sides of the plot of land if its area is 6 square kilometers.

49. A cistern in the shape of a right circular cylinder with a hemisphere at the bottom (Figure 9) is to have a depth of 12 meters. **(a)** Determine the volume V of the cistern as a function of its radius r. **(b)** Find the radius r of the cistern if its volume is 360π cubic meters.

Figure 9

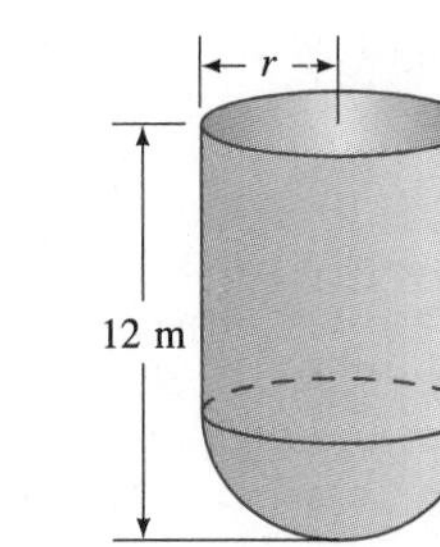

50. The capstone for a monument has the shape of a cube topped by a pyramid (Figure 10). Find the length x of a side of the capstone if its height is 4 feet and its volume is 5.25 cubic feet. [*Hint:* The volume of a pyramid is one-third its height times the area of its base.]

Figure 10

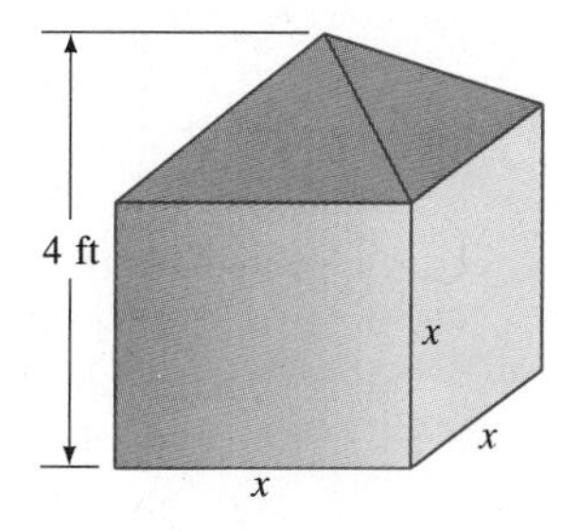

51. In computer science, a class of problems is said to be solvable in *polynomial time* if there is a polynomial function $f(x)$ such that the time t seconds required by a computer to solve a problem in this class is given by $t = f(n)$, where n is a measure of the size or complexity of the problem. Suppose that, for such a class, $f(x) = x^4 + 3x - 3$. Find the size n of a problem in this class that requires 6585 seconds for its solution. [*Hint:* $6588 = 4 \cdot 27 \cdot 61$.]

52. **(a)** Suppose that c is a real zero of a polynomial function $f(x)$, the coefficients of $f(x)$ are integers, and the leading coefficient of $f(x)$ is 1. Using the rational-zeros theorem, show that either c is an integer, or else c is an irrational number.

(b) Using the result of part (a), show that $\sqrt{2}$ is an irrational number.

53. How are the zeros of $f(x)$ related to the zeros of $f(-x)$?

54. Suppose that $c > 0$, that the polynomial $f(x)$ is divided by $x - c$ to produce a quotient polynomial $q(x)$ and a remainder R, and that R and all the coefficients of $q(x)$ are nonnegative. Show that it is impossible for $f(x)$ to have a zero z with $z > c$. [*Hint:* Show that $f(z) > 0$ by using the fact that $f(x) = (x - c)q(x) + R$.]

55. How are the coefficients of $f(x)$ related to the coefficients of $f(-x)$?

56. Suppose $f(x) = a_n x^n + a_{n-1}x^{n-1} + \cdots + a_1 x + a_0$ is a polynomial function such that $f(x) = 0$ for all values of x. Prove that all the coefficients $a_n, a_{n-1}, \ldots, a_1, a_0$ must be zero. [*Hint:* Use Theorem 1.]

57. If p and q are zeros of the polynomial function $f(x) = x^2 + bx + c$, show that $b = -(p + q)$ and $c = pq$.

58. Suppose that f and g are polynomial functions such that $f(x) = g(x)$ for all values of x. Prove that the coefficients of f are the same as the coefficients of g. [*Hint:* Use Problem 56.]

4.5 Approximation of Zeros of Polynomial Functions

The technique used in Section 4.4 for finding zeros of polynomial functions has two limitations: It works only for polynomials with rational coefficients, and it locates only the rational zeros of such polynomials. Most of the practical methods for finding zeros in situations not covered by the techniques already illustrated involve the idea of **successive approximation;** that is, starting from a rough first approximation, they produce successively better and better approximations to a zero. Some of these methods are based on the idea of a *change of sign* of a function.

A function f is said to **change sign** on an interval I if there are numbers a and b in I such that $f(a)$ and $f(b)$ have opposite algebraic signs. In Section 2.7, we mentioned that *intervals where a polynomial with real coefficients is positive are separated from intervals where it is negative by values of the variable for which it is zero.* This fact can be restated as follows.

Figure 1

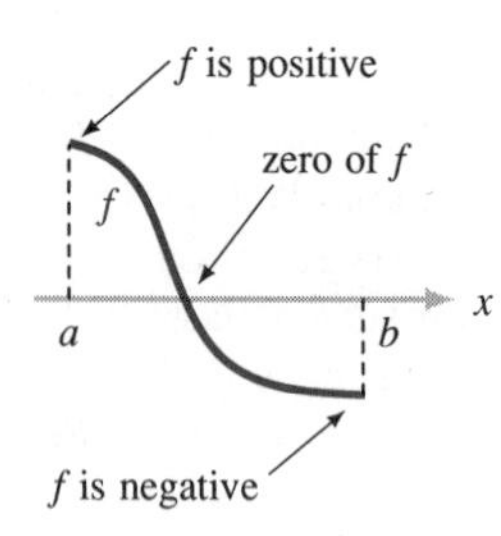

The Change-of-Sign Property

If a polynomial function with real coefficients changes sign on an interval, then it has a zero in that interval.

The change-of-sign property is illustrated in Figure 1, which shows the graph of a polynomial function f that changes sign on an interval $[a, b]$. Because the graph is a continuous curve, with no jumps or breaks, it must cut across the x axis at least once between a and b.

Example 1 Show that the polynomial function $f(x) = x^3 + x - 1$ has a zero in the interval $[0, 1]$.

Solution Because $f(0) = -1$ and $f(1) = 1$, the polynomial function f **changes** its sign on the interval $[0, 1]$. Therefore, by the change-of-sign property, f has at least one zero in the interval $[0, 1]$. ■

Figure 2

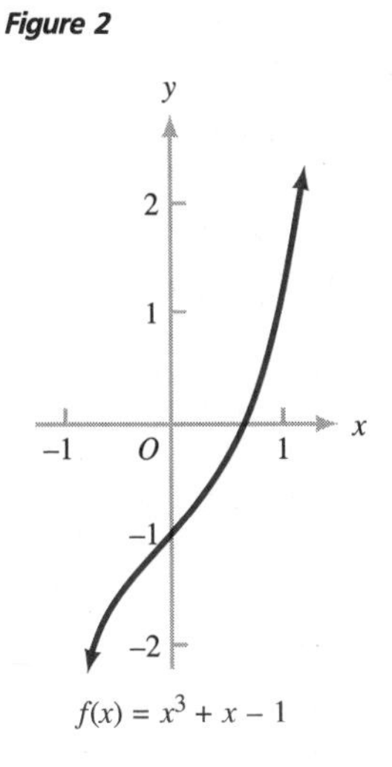

$f(x) = x^3 + x - 1$

The computer-generated graph of $f(x) = x^3 + x - 1$ in Figure 2 shows that f has exactly one zero in the interval $[0, 1]$.

The change-of-sign property is used in the following **bisection method** for finding zeros of a polynomial function f with real coefficients.

The Bisection Method

Step 1. By sketching a rough graph, or by trial and error using synthetic division, or by using a graphing calculator or a computer-generated graph, find a closed interval $[a, b]$ such that $f(a)$ and $f(b)$ have opposite algebraic signs. By the change-of-sign property, f must have a zero in this interval.

Step 2. Using the midpoint $c = (a + b)/2$, divide the interval $[a, b]$ into two subintervals $[a, c]$ and $[c, b]$.

Step 3. Calculate $f(c)$. If $f(a)$ and $f(c)$ have opposite algebraic signs, then f must have a zero in the interval $[a, c]$. If $f(c) = 0$, then c itself is a zero of f. The only other possibility is that $f(c)$ and $f(b)$ have opposite algebraic signs, in which case f must have a zero in the interval $[c, b]$.

Each time you apply the bisection method, either you find the exact value of a zero of f, or else you isolate a zero of f in an interval half the size of the preceding one. Using the bisection method over and over again, you can locate a zero of f as accurately as you wish by confining it in successively shorter and shorter intervals.

C Example 2 In Example 1, we used the change-of-sign property to show that the polynomial function

$$f(x) = x^3 + x - 1$$

has a zero between 0 and 1. Starting with the interval $[0, 1]$, use the bisection method twice in succession to locate a zero of f with more accuracy.

Solution The midpoint of the interval $[0, 1]$ is $(0 + 1)/2 = 0.5$. A calculation reveals that

$$f(0) = -1 \quad \text{and} \quad f(0.5) = -0.375,$$

so $f(0)$ and $f(0.5)$ have the same algebraic sign. However, $f(0.5)$ and $f(1)$ have opposite algebraic signs:

$$f(0.5) = -0.375 \quad \text{and} \quad f(1) = 1,$$

so f has a zero between 0.5 and 1. Thus, for our second application of the bisection method, we start with the interval $[0.5, 1]$. The midpoint of this interval is $(0.5 + 1)/2 = 0.75$. Now we have

$$f(0.5) = -0.375 \quad \text{and} \quad f(0.75) = 0.171875,$$

so $f(0.5)$ and $f(0.75)$ have opposite algebraic signs. It follows that f has a zero between 0.5 and 0.75. ■

After the bisection method is applied several times in succession, the midpoint of the final interval provides a numerical approximation to a zero of the function. For instance, in the solution of Example 2, we could take the midpoint

$$\frac{0.5 + 0.75}{2} = 0.625$$

of the interval $[0.5, 0.75]$ as an approximation to a zero of $f(x) = x^3 + x - 1$. The following rule governs the accuracy of such an approximation.

> The midpoint of the interval obtained after applying the bisection method k times in succession, starting from an interval of length one unit, can be considered accurate to n decimal places, where n is the greatest integer less than or equal to $3k/10$.

C Example 3 Use the bisection method to find a zero of $f(x) = x^3 + x - 1$ that is accurate to 2 decimal places. Start with the interval $[0, 1]$.

Solution The smallest value of k for which $3k/10 \geq 2$ is $k = 7$, so we will need 7 successive applications of the bisection method. The first two applications were already carried out in Example 2, so we have five more to go. Our work is shown in the following table.

Number of Applications	Interval	Midpoint c	$f(c)$
0	[0, 1]	0.5	−0.375
1	[0.5, 1]	0.75	0.171875
2	[0.5, 0.75]	0.625	−0.130859375
3	[0.625, 0.75]	0.6875	0.012451172
4	[0.625, 0.6875]	0.65625	−0.061126709
5	[0.65625, 0.6875]	0.671875	−0.024829865
6	[0.671875, 0.6875]	0.6796875	−0.006313801
7	[0.6796875, 0.6875]	0.68359375	0.003037393

Rounding off the last midpoint to two decimal places, we obtain 0.68 as an approximation to a zero of $f(x) = x^3 + x - 1$. ■

As Example 3 illustrates, the computation of real zeros of a polynomial function by the bisection method, even with the aid of a calculator, can involve tedious repetition. However, the method is easily adapted to a programmable calculator or computer* and thus provides a reasonably efficient technique for finding real

* Software that uses the bisection method to approximate a zero of a polynomial function is available for IBM PC and its compatible computers. See Appendix III for a BASIC program for the bisection method.

zeros. For instance, after 24 successive applications of the bisection method, a programmable calculator quickly produces 0.6823278 as an approximation, accurate to seven decimal places, for a zero of the polynomial function in Example 3.

The Fixed-Point Method

Some of the most efficient techniques for finding or approximating a zero of a function f with a calculator or a computer use the following **fixed-point method.**

Step 1. Find a function g such that the equation

$$f(x) = 0$$

is equivalent to the equation

$$g(x) = x.$$

Step 2. By sketching a rough graph, or by using a computer-generated graph of f, find a first approximation a to a zero of f.

Step 3. Compute the values of the successive iterates[†]

$$g(a), \qquad g(g(a)), \qquad g(g(g(a))),$$

and so on, continuing until you arrive at a number p that no longer changes when you apply g.

Step 4. Conclude that, to within the limits of accuracy of your computing device, p is an approximation to a zero of f.

In step 1, there are many possible choices for the function g. For instance, if h is any function, then

$$g(x) = x - \frac{f(x)}{h(x)}$$

will do, provided there is no difficulty with a zero denominator and that the values obtained in step 3 converge toward a fixed point p of the function g. For polynomial functions of degrees 3 and 4, the following suggested choices* for h usually work quite well.

For $f(x) = ax^3 + bx^2 + cx + d$, try $h(x) = 3ax^2 + 2bx + c$.
For $f(x) = ax^4 + bx^3 + cx^2 + dx + k$, try $h(x) = 4ax^3 + 3bx^2 + 2cx + d$.

[†] See Section 3.6.

* These choices are based on **Newton's method,** which is discussed on pages 366 and 367 of *After Calculus: Analysis* by D. Foulis and M. Munem (Dellen/Macmillan Publishing Company, Riverside, N.J., 1989).

C Example 4 Use the fixed-point method to find a zero of $f(x) = x^3 + x - 1$ that is accurate to as many decimal places as are available on your calculator.

Solution The function f is a third-degree polynomial function

$$f(x) = ax^3 + bx^2 + cx + d$$

with $a = 1$, $b = 0$, $c = 1$, and $d = -1$. Following the suggestion above, we let

$$h(x) = 3ax^2 + 2bx + c = 3x^2 + 1.$$

Thus,

$$g(x) = x - \frac{f(x)}{h(x)} = x - \frac{x^3 + x - 1}{3x^2 + 1}$$

$$= \frac{2x^3 + 1}{3x^2 + 1}.$$

Again, we use the computer-generated graph in Figure 2 to see that, in step 2,

$$a = 0.7$$

is a good approximation to a zero of f. Values of successive iterates $g(a)$, $g(g(a))$, $g(g(g(a)))$, and so on, are

$$g(0.7) = 0.682591093$$

$$g(0.682591093) = 0.682327863$$

$$g(0.682327863) = 0.682327804$$

$$g(0.682327804) = 0.682327804.$$

Therefore, 0.682327804 is an approximate zero of $f(x) = x^3 + x - 1$. ■

As you can see by comparing Examples 3 and 4, the fixed-point method can be considerably more efficient than the bisection method for finding approximate zeros of functions.

Problem Set 4.5

In Problems 1 to 6, show that the polynomial function has a zero in the interval by using the change-of-sign property. (You need not try to find the value of the zero.)

1. $f(x) = x^5 - 2x^3 - 1$ in $[1, 2]$

C **2.** $g(x) = x^4 + 6x^3 - 18x^2$ in $[2.1, 2.2]$

C **3.** $f(x) = x^5 - 2x^3 - 1$ in $[1.5, 1.6]$

C **4.** $g(x) = x^4 + 6x^3 - 18x^2$ in $[-8.2, -8.1]$

5. $h(x) = x^5 - 2x + 3$ in $[-2, -1]$

C **6.** $p(x) = 6x^4 + 19x^3 - 7x^2 - 26x + 12$ in $[-1.54, -1.53]$

C **7.** In Problem 1, use the bisection method twice in succession, starting with the interval [1, 2], to locate a zero of $f(x) = x^5 - 2x^3 - 1$ with greater accuracy.

C **8.** In Problem 2, use the bisection method twice in succession, starting with the interval [2.1, 2.2], to locate a zero of $g(x) = x^4 + 6x^3 - 18x^2$ with greater accuracy.

C **9.** In Problem 1, start with the interval [1, 2] and use the bisection method to find a zero of $f(x) = x^5 - 2x^3 - 1$ that is accurate to two decimal places.

C **10.** In Problem 5, start with the interval $[-2, -1]$ and use the bisection method to find a zero of $h(x) = x^5 - 2x + 3$ that is accurate to two decimal places.

C The computer-generated graph in Figure 3 shows that the polynomial function $f(x) = x^3 - 3x + 1$ has three real zeros. Use this information in Problems 11 to 16.

Figure 3

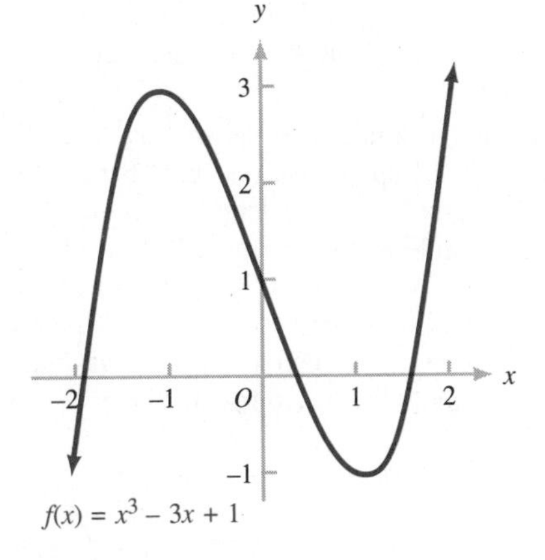

11. Locate each of the zeros of f on intervals of length 0.25 unit by using the bisection method.

12. Use the bisection method to find the largest of the three zeros of f with an accuracy of two decimal places.

13. Use the bisection method to find the zero of f that is negative with an accuracy of two decimal places.

14. Use the bisection method to find the smallest positive zero of f with an accuracy of two decimal places.

15. Use the fixed-point method to find the largest of the three zeros of f as accurately as you can with your calculator.

16. Use the fixed-point method to find the negative zero of f as accurately as you can with your calculator.

C In Problems 17 to 20, use the information in the computer-generated graph in the indicated figure and the bisection method to find the largest zero of each function with an accuracy of two decimal places.

17. $f(x) = 2x^4 - 2x - 3$ (Figure 4)

Figure 4

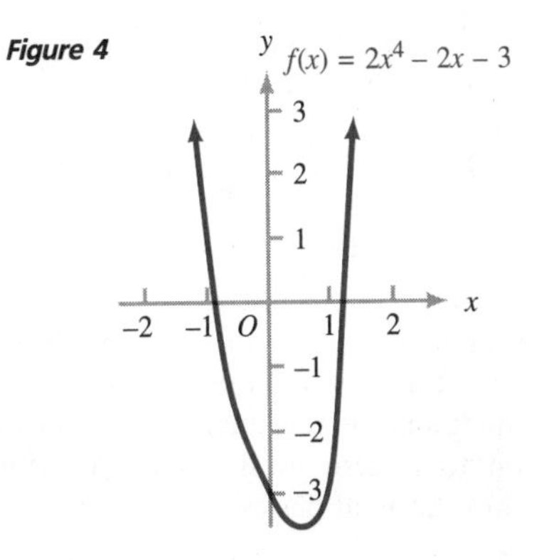

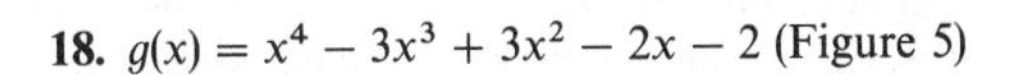

18. $g(x) = x^4 - 3x^3 + 3x^2 - 2x - 2$ (Figure 5)

Figure 5

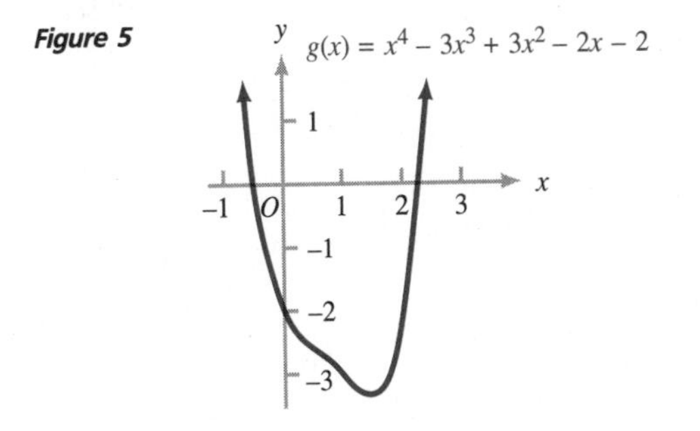

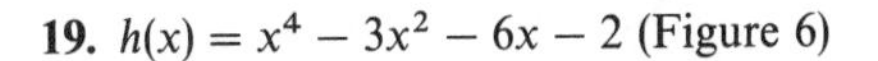

19. $h(x) = x^4 - 3x^2 - 6x - 2$ (Figure 6)

Figure 6

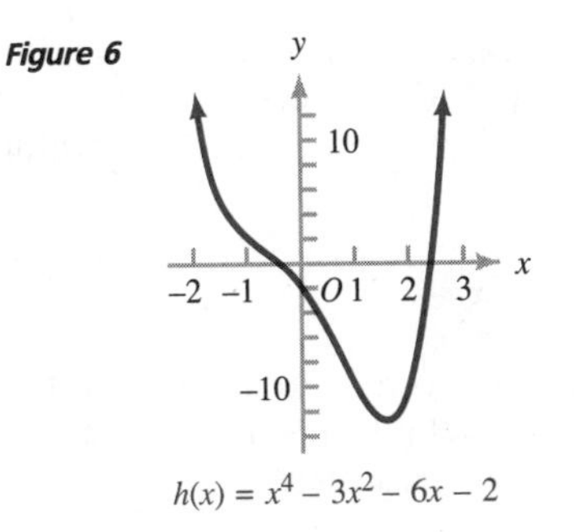

20. $p(x) = 2x^5 - 3x^4 + x^2 - 3x + 1$ (Figure 7)

Figure 7

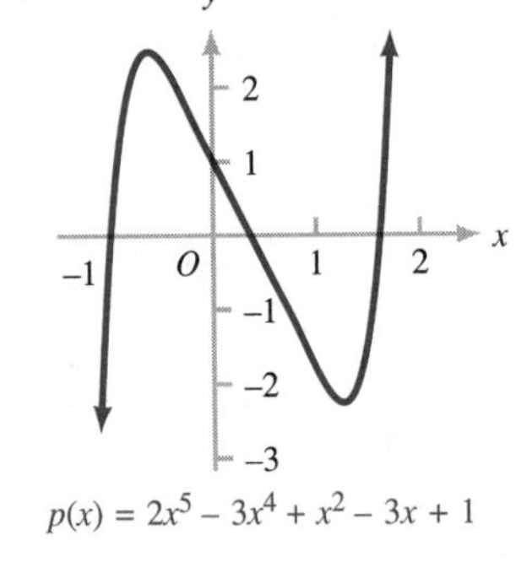

21. Starting from an interval of length one unit, how many times must the bisection method be used in succession so that the midpoint of the final interval gives an approximation to a zero of the function with an accuracy of three decimal places?

C **22.** As the computer-generated graph in Figure 8 shows, the polynomial function $f(x) = x^4 + 3x - 13$ has two real zeros. Use the bisection method to find these zeros with an accuracy of three decimal places.

Figure 8

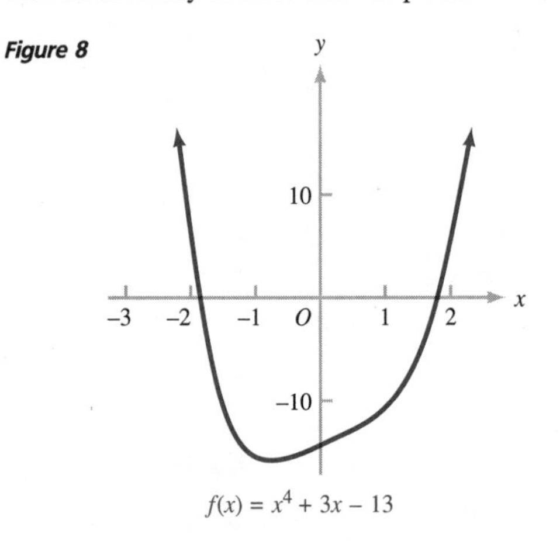

In Problems 23 and 24, the given polynomial function has exactly one real zero. Use the fixed-point method to find this zero to within the limits of accuracy of your calculator.

C **23.** $f(x) = x^3 + x + 1$

C **24.** $f(x) = x^3 + x^2 + x + 5$

C **25.** Use the fixed-point method to find the positive zero of $f(x) = x^4 + 3x - 13$ (Figure 8) to within the limits of accuracy of your calculator.

C **26.** Use the fixed-point method to find the negative zero of $f(x) = x^4 + 3x - 13$ (Figure 8) to within the limits of accuracy of your calculator.

C **27.** A *radome* is a structure designed to enclose a radar antenna and protect it from rain, wind, and snow. Radomes often have the shape of a sphere cut off at the bottom as shown in Figure 9. If r is the radius and h is the height of the radome, then the volume enclosed by the radome is given by $V = \pi r h^2 - \frac{\pi}{3} h^3$. Find the height h of a radome of radius $r = 7$ meters if it encloses a volume of $V = 1260$ cubic meters. Round off your answer to two decimal places.

Figure 9

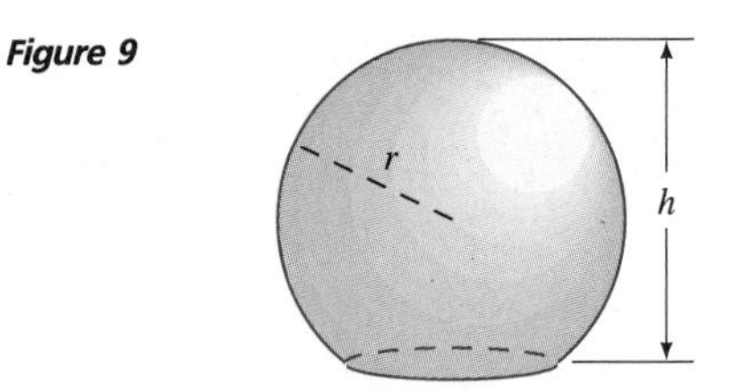

C **28.** In a right triangle with vertex angles of 20° and 70° and with a hypotenuse one unit long (Figure 10), it can be shown that the length x of the side opposite the 70° angle is the positive solution of the equation $8x^3 - 6x - 1 = 0$. (This fact is used in the proof that it is impossible to trisect an angle with a straightedge and a compass alone.) Find x to as many decimal places as possible on your calculator.

Figure 10

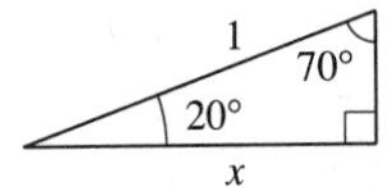

C **29.** A plot of land has the shape of an isosceles triangle with area A, perimeter p, and base b (Figure 11).

(a) Show that $2pb^3 - p^2b^2 + 16A^2 = 0$.

(b) Solve for b, rounded off to two decimal places, if $A = 7000$ square meters and $p = 400$ meters.

Figure 11

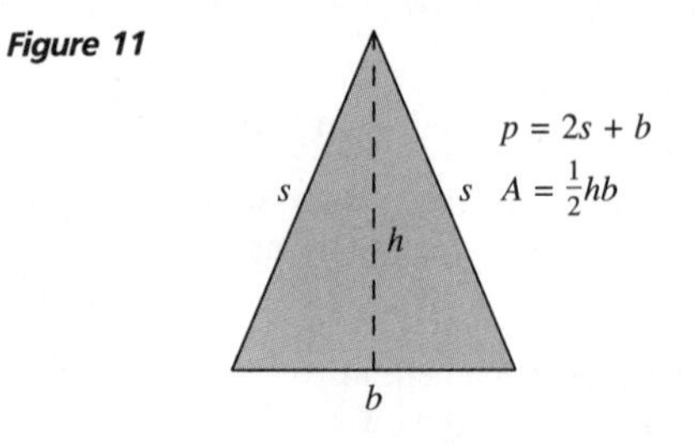

4.6 Complex Polynomials

If z represents a complex variable, then by a **complex polynomial in** z we mean an expression of the form

$$f(z) = a_n z^n + a_{n-1} z^{n-1} + \cdots + a_1 z + a_0,$$

where n is a nonnegative integer and the coefficients $a_n, a_{n-1}, \ldots, a_1, a_0$ are complex numbers. If $a_n \neq 0$, we say that the polynomial has degree n and we call a_n the leading coefficient. The zero polynomial (which has no nonzero coefficient) is not assigned a degree.

Federal Republic of Germany, five deutschemark coin honoring Gauss (1977)

Because the complex numbers have the same basic algebraic properties as the real numbers, much of our previous discussion of polynomials in Sections 1.4, 1.5, 2.1, 2.3, 4.3, and 4.4* extends almost verbatim to complex polynomials. In particular, the long-division procedure (page 242) works for complex polynomials, and it follows that the remainder theorem (page 245) and the factor theorem (page 246) hold as well for complex polynomials.

The following theorem, which was proved in 1799 by the great German mathematician Carl Friedrich Gauss (1777–1855), is so important that it is called the **fundamental theorem of algebra.**

Theorem 1 **The Fundamental Theorem of Algebra**

Every complex polynomial of degree $n \geq 1$ has a complex zero.

A proof of the fundamental theorem of algebra can be found in *A Modern Course on the Theory of Equations* by David Dobbs and Robert Hanks (Polygonal Publishing House, Passaic, N.J., 1980) on pages 111 to 112. By combining Theorem 1 with the factor theorem, we can obtain the following useful result.

Theorem 2 **The Complete Linear Factorization Theorem**

If $f(z)$ is a complex polynomial of degree $n \geq 1$, then there is a nonzero complex number a and there are complex numbers $c_1, c_2, \ldots, c_n$ such that

$$f(z) = a(z - c_1)(z - c_2) \cdots (z - c_n).$$

Proof By Theorem 1, there is a complex number c_1 such that

$$f(c_1) = 0.$$

Hence, by the factor thereom, $z - c_1$ is a factor of $f(z)$; that is,

$$f(z) = (z - c_1)Q_1(z),$$

* Of course, complex polynomials cannot be graphed in the Cartesian plane, and Descartes' rule of signs does not apply to such polynomials.

where $Q_1(z)$ is a complex polynomial of degree $n - 1$. If $n - 1 \geq 1$, we can again apply Theorem 1 and the factor theorem to $Q_1(z)$ to obtain

$$Q_1(z) = (z - c_2)Q_2(z),$$

for some complex number c_2, where $Q_2(z)$ is a complex polynomial of degree $n - 2$; therefore,

$$f(z) = (z - c_1)(z - c_2)Q_2(z).$$

If we continue this process, then after n steps, we arrive at

$$f(z) = (z - c_1)(z - c_2) \cdots (z - c_n)Q_n,$$

where Q_n has degree $n - n = 0$; that is, Q_n is a (complex) constant. Letting $a = Q_n$, we obtain

$$f(z) = a(z - c_1)(z - c_2) \cdots (z - c_n).$$

■

According to Theorem 2, every complex polynomial $f(z)$ of degree $n \geq 1$ can be factored completely into a nonzero complex constant a and exactly n linear (first-degree) factors

$$(z - c_1), (z - c_2), \ldots, (z - c_n).$$

Evidently, the complex number a is the leading coefficient of $f(z)$, and the complex numbers $c_1, c_2, \ldots, c_n$ are the zeros of $f(z)$. There can be no other zeros (Problem 38). The zeros $c_1, c_2, \ldots, c_n$ need not be distinct from one another, but *a complex polynomial $f(z)$ of degree n can have no more than n zeros.*

That the zeros $c_1, c_2, \ldots, c_n$ of $f(z)$ need not be distinct is illustrated by

$$f(z) = z^3 - 3z^2 + 4 = (z + 1)(z - 2)(z - 2),$$

which has the number 2 as a **double zero.** More generally, a complex zero c of a complex polynomial $f(z)$ is called a **zero of multiplicity** m if the linear factor $(z - c)$ appears exactly m times in the complete factorization of $f(z)$. Thus, if a zero of multiplicity m is counted m times, we can conclude from Theorem 2 that *a complex polynomial of degree $n \geq 1$ has exactly n zeros.* For instance,

$$f(z) = 7(z - 3)^4(z - i)^3(z + i)$$

is a complex polynomial of degree 8 with the following zeros:

$$\overbrace{3, 3, 3, 3}^{4},\ \overbrace{i, i, i}^{3},\ \overbrace{-i}^{1} \longleftarrow \text{multiplicity}$$

Example 1 Form a polynomial $f(z)$ that has zeros, 2, 2, $1 + i$. (Zeros are repeated to show multiplicity.)

Solution

$$\begin{aligned} f(z) &= (z - 2)(z - 2)[z - (1 + i)] = (z^2 - 4z + 4)[z - (1 + i)] \\ &= z^3 - (5 + i)z^2 + (8 + 4i)z - (4 + 4i) \end{aligned}$$

■

As we have seen in Section 2.3, the zeros of a quadratic polynomial with *real* coefficients and a negative discriminant are complex conjugates of each other. More generally, we have the following theorem.

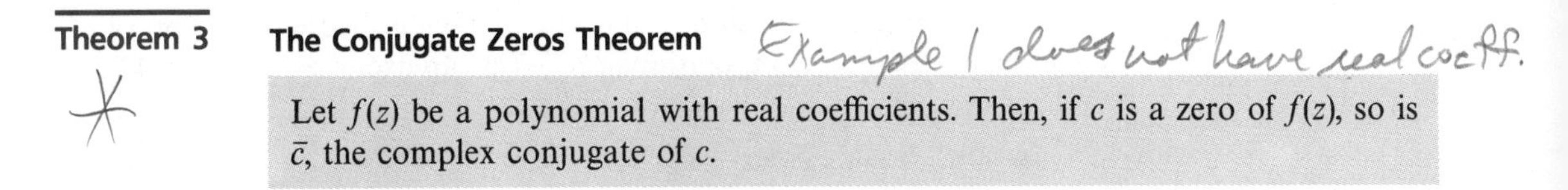

Theorem 3 **The Conjugate Zeros Theorem**

Let $f(z)$ be a polynomial with real coefficients. Then, if c is a zero of $f(z)$, so is $\bar{c}$, the complex conjugate of c.

Proof Suppose that

$$f(z) = a_n z^n + a_{n-1} z^{n-1} + \cdots + a_1 z + a_0,$$

where the coefficients $a_n, a_{n-1}, \ldots, a_1$, and a_0 are real numbers. If c is a zero of $f(z)$, then $f(c) = 0$, so

$$a_n c^n + a_{n-1} c^{n-1} + \cdots + a_1 c + a_0 = 0.$$

Using the fact that complex conjugation "preserves" addition and multiplication (see page 68), we take the complex conjugate of both sides of the last equation to obtain

$$\bar{a}_n(\bar{c})^n + \bar{a}_{n-1}(\bar{c})^{n-1} + \cdots + \bar{a}_1(\bar{c}) + \bar{a}_0 = \bar{0}.$$

Because all the coefficients are real numbers, they are equal to their own conjugates, and likewise $\bar{0} = 0$, so we have

$$a_n(\bar{c})^n + a_{n-1}(\bar{c})^{n-1} + \cdots + a_1(\bar{c}) + a_0 = 0.$$

In other words, $\bar{c}$ is also a complex zero of $f(z)$, and our proof is complete. ■

As a consequence of Theorem 3, we have the following result:

The nonreal zeros of a polynomial $f(z)$ with real coefficients must occur in conjugate pairs.

Therefore, *a polynomial with real coefficients must have an even number of nonreal zeros.* [*Note:* It may have 0 such nonreal zeros; but 0 is an even integer.] It follows that *a polynomial of odd degree with real coefficients must have at least one real zero* (Problem 39).

Example 2 Form a polynomial in z that has real coefficients, that has the smallest possible degree, and that has 2, -1, and $1 - i$ as zeros.

Solution Since we want a polynomial of the smallest possible degree, we assume that each zero has multiplicity 1. Because we want the coefficients to be real numbers, we must include $1 + i$, the complex conjugate of $1 - i$, as a zero. The polynomial is

$$\begin{aligned} f(z) &= (z - 2)[z - (-1)][z - (1 - i)][z - (1 + i)] \\ &= (z - 2)(z + 1)[z^2 - (1 + i)z - (1 - i)z + (1 - i)(1 + i)] \\ &= (z^2 - z - 2)(z^2 - 2z + 2) \\ &= z^4 - 3z^3 + 2z^2 + 2z - 4. \end{aligned}$$

■

If w is a complex number, then any complex number z such that $z^2 = w$ is called a **complex square root** of w. Since a complex square root of w is a zero of the polynomial $z^2 - w$, it follows that there are two such roots. These two complex square roots can be calculated by using the following theorem. The proof of this theorem, which is mainly computational, is left as an exercise (Problem 42).

Theorem 4 **Complex Square Roots**

Suppose that $w = p + qi$, where p and q are real numbers, and let $r = \sqrt{p^2 + q^2}$. Then $r + p \geq 0$ and $r - p \geq 0$. Furthermore, if

$$x = \sqrt{(r + p)/2} \qquad \text{and} \qquad y = \pm\sqrt{(r - p)/2},$$

where the $\pm$ sign is chosen so that y and q have the same algebraic sign, then

$$x + yi \qquad \text{and} \qquad -x - yi$$

are the two complex square roots of w.

Example 3 Find the two complex square roots of $w = 3 + 4i$.

Solution We use Theorem 4 with $p = 3$ and $q = 4$. Then $r = \sqrt{3^2 + 4^2} = 5$,

$$x = \sqrt{(5 + 3)/2} = 2 \qquad \text{and} \qquad y = \pm\sqrt{(5 - 3)/2} = \pm 1.$$

Since q is positive, we choose the plus sign, so that $y = 1$. Thus, the two complex square roots of w are

$$x + yi = 2 + i \qquad \text{and} \qquad -x - yi = -2 - i.$$

■

The quadratic formula (Section 2.3) is easily extended to complex quadratic polynomials, since the proof by completing the square still works. With the understanding that $\pm\sqrt{b^2 - 4ac}$ stands for two complex square roots of $b^2 - 4ac$, we can still write the quadratic formula

$$z = \frac{-b \pm \sqrt{b^2 - 4ac}}{2a}$$

for the roots of the complex quadratic equation

$$az^2 + bz + c = 0.$$

Example 4 Use the quadratic formula to find the complex roots of the quadratic equation

$$z^2 + (4 + 2i)z + (3 + 3i) = 0.$$

Solution Here $a = 1$, $b = 4 + 2i$, and $c = 3 + 3i$. Thus,

$$\begin{aligned} b^2 - 4ac &= (4 + 2i)^2 - 4(1)(3 + 3i) = 16 + 16i + 4i^2 - 12 - 12i \\ &= 16 + 16i - 4 - 12 - 12i = 4i. \end{aligned}$$

We are going to use Theorem 4 to find $\pm\sqrt{b^2 - 4ac} = \pm\sqrt{4i}$, that is, to find the two complex square roots of $w = 4i = 0 + 4i$. Putting $p = 0$ and $q = 4$ in Theorem 4, we find that $r = \sqrt{0^2 + 4^2} = 4$, $x = \sqrt{(4 + 0)/2} = \sqrt{2}$, and $y = \pm\sqrt{(4 - 0)/2} =$

$\pm\sqrt{2}$. Since q is positive, we have $y = \sqrt{2}$. Therefore, the two complex square roots of $4i$ are

$$x + yi = \sqrt{2} + \sqrt{2}i \qquad \text{and} \qquad -x - yi = -\sqrt{2} - \sqrt{2}i.$$

Hence,

$$z = \frac{-b \pm \sqrt{b^2 - 4ac}}{2a} = \frac{-(4 + 2i) \pm \sqrt{4i}}{2(1)} = \frac{-(4 + 2i) \pm (\sqrt{2} + \sqrt{2}i)}{2}.$$

In other words, the two roots are the complex numbers

$$z = \frac{-(4 + 2i) + (\sqrt{2} + \sqrt{2}i)}{2} = \frac{-4 + \sqrt{2}}{2} + \frac{-2 + \sqrt{2}}{2}i$$

and

$$z = \frac{-(4 + 2i) - (\sqrt{2} + \sqrt{2}i)}{2} = \frac{-4 - \sqrt{2}}{2} + \frac{-2 - \sqrt{2}}{2}i. \quad ■$$

Example 5 Find all complex zeros of $f(z) = z^3 - 2z^2 - 3z + 10$ and factor $f(z)$ completely into linear factors.

Solution Since $f(z)$ is a polynomial of degree 3 with real coefficients, we know that it has 3 zeros, at least one of which is a real number. By the rational-zeros theorem (Section 4.4), the only possible rational zeros are ± 1, ± 2, ± 5, and ± 10. Testing these possibilities one at a time (using synthetic division for efficiency), we find that -2 is, in fact, a zero:

$$\begin{array}{r|rrrr} -2 & 1 & -2 & -3 & 10 \\ & & -2 & 8 & -10 \\ \hline & 1 & -4 & 5 & 0. \end{array}$$

From this, we also see that the quotient polynomial is

$$z^2 - 4z + 5;$$

that is,

$$z^3 - 2z^2 - 3z + 10 = (z + 2)(z^2 - 4z + 5).$$

By the quadratic formula, the zeros of $z^2 - 4z + 5$ are given by

$$z = \frac{4 \pm \sqrt{16 - 20}}{2} = \frac{4 \pm \sqrt{-4}}{2} = \frac{4 \pm 2i}{2} = 2 \pm i.$$

Therefore, the zeros of $z^3 - 2z^2 - 3z + 10$ are

$$-2, \qquad 2 + i, \qquad \text{and} \qquad 2 - i,$$

and the polynomial factors completely into linear factors as follows:

$$\begin{aligned} z^3 - 2z^2 - 3z + 10 &= [z - (-2)][z - (2 + i)][z - (2 - i)] \\ &= (z + 2)(z - 2 - i)(z - 2 + i). \end{aligned} \quad ■$$

Problem Set 4.6

In Problems 1 to 6, form a polynomial $f(z)$ that has the indicated zeros. (Zeros are repeated to show multiplicity.)

1. $3, 3, i$

2. $3, 1 + i, 1 + i$

3. $1, -i, -i$

4. $1, 1, 1 + i, 1 + i$

5. $-2, -2, i, -i$

6. $-1, -1, 1 + i, 1 - i$

In Problems 7 to 12, form a polynomial $f(z)$ that has real coefficients, has the smallest possible degree, and has the indicated zeros. (Zeros are repeated to show multiplicity.)

7. $1, 2, i$

8. $2, 2, -1, 2 + i$

9. $1, 1, i, 1 + 3i$

10. $\frac{1}{2}, 2 + 3i$

11. $0, 0, 3, i$

12. $\frac{1}{2}, \frac{1}{3}, \frac{1}{4}, \frac{1}{4}, -i$

In Problems 13 to 18, find the two complex square roots of each complex number.

13. i

14. $5 + 12i$

15. $4 - 3i$

16. $1 + i$

17. $\sqrt{3} + i$

18. $2 - \sqrt{5}i$

In Problems 19 to 24, use the quadratic formula to find the complex roots of each quadratic equation.

19. $z^2 - 2z + 2 = 0$

20. $z^2 - 2z + 10 = 0$

21. $z^2 + iz - 1 = 0$

22. $iz^2 + 2z + i = 0$

23. $z^2 - (3 + i)z + (2 + 2i) = 0$

24. $4w^2 - 4w + 3i = 0$

In Problems 25 to 36, **(a)** find all of the complex zeros of each complex polynomial $f(z)$, **(b)** factor $f(z)$ completely into linear factors, and **(c)** determine the multiplicity of each zero of $f(z)$.

25. $f(z) = z^4 - 1$

26. $f(z) = z^4 - 81$

27. $f(z) = (z + 2)^2(z^2 + 4)$

28. $f(z) = (3z + 1)^5(z^2 - 5z - 6)$

29. $f(z) = (2z + 1)^3(z^2 - 2z - 3)$

30. $f(z) = (z^2 - 4)^2(z^2 + 4z + 13)$

31. $f(z) = z^3 - z^2 + z - 1$

32. $f(z) = z^4 + 4z^3 - 4z^2 - 40z - 33$

33. $f(z) = z^3 - 5z^2 + 8z - 6$

34. $f(z) = z^4 + 4z^3 + 5z^2 + 4z + 4$

35. $f(z) = z^4 - 3z^3 + z^2 + 4$

36. $f(z) = z^4 - 3z^3 + 3z^2 - 3z + 2$

37. If z_1 and z_2 are the complex roots of the quadratic equation $az^2 + bz + c = 0$, show that:

(a) $z_1 + z_2 = -b/a$ **(b)** $z_1 z_2 = c/a$

38. If $f(z) = a(z - c_1)(z - c_2) \cdots (z - c_n)$, where $a \neq 0$, and if c is a complex zero of $f(z)$, show that c must be one of the complex numbers $c_1, c_2, \ldots,$ or c_n.

39. Show that a polynomial with real coefficients and with odd degree must have at least one real zero.

40. Show that a nonconstant polynomial with real coefficients can be factored into linear and quadratic polynomials with real coefficients, where the quadratic factors have negative discriminants.

41. Given that $-\frac{1}{2} + \frac{\sqrt{3}}{2}i$ is a zero of the polynomial $f(z) = z^3 - 1$, find the remaining two zeros.

42. Prove Theorem 4.

43. If g is a polynomial function with odd degree $n \geq 3$ and with real coefficients, show that g has a fixed point. [*Hint:* Apply the result of Problem 39 to the polynomial function f defined by $f(x) = x - g(x)$.]

4.7 Rational Functions

Although the sum $f + g$, the difference $f - g$, and the product $f \cdot g$ of polynomial functions f and g are again polynomial functions, the quotient f/g can be a polynomial function only if g is a factor of f. A function of the form $R = f/g$, where f and g are polynomial functions, is called a **rational function.*** In other words, a rational function is a function such as

$$R(x) = \frac{x^3 - 2x^2 + 5x - 4}{7x^2 - 3x + 2},$$

whose values are given by a rational expression in the independent variable. *The domain of a rational function is the set of all real numbers except the zeros of the denominator function.*

Example 1 Determine the domain of each rational function.

(a) $F(x) = \dfrac{x + 2}{x}$ **(b)** $H(x) = \dfrac{x^3 - 3x^2 - 2x - 1}{4}$

(c) $h(x) = \dfrac{1}{(x + 2)(x - 3)}$ **(d)** $T(x) = \dfrac{3x - 2}{x^2 + 1}$

Solution **(a)** All real numbers except 0.

(b) All real numbers.

(c) All real numbers except -2 and 3.

(d) All real numbers. ■

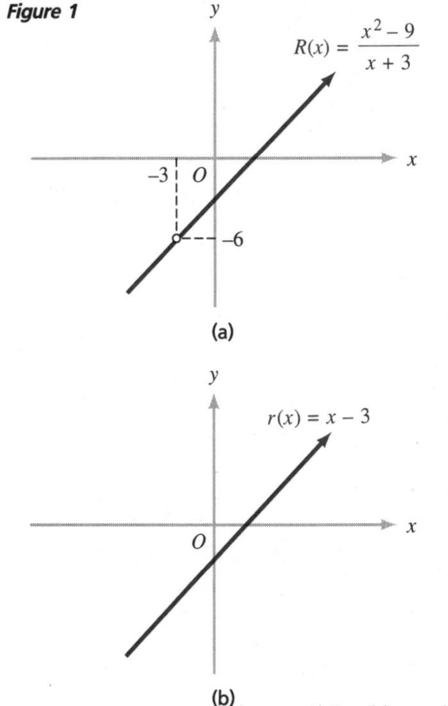

Recall that two rational expressions are said to be equivalent if one can be obtained from the other by canceling common factors or by multiplying numerator and denominator by the same nonzero polynomial. Strictly speaking, two rational functions defined by equivalent rational expressions may not be the same because their domains may be different. For instance, the rational function R defined by

$$R(x) = \frac{x^2 - 9}{x + 3} = \frac{(x - 3)(x + 3)}{x + 3}$$

isn't really the same as the rational function r defined by

$$r(x) = x - 3$$

because

$$R(x) = \frac{(x - 3)\cancel{(x + 3)}}{\cancel{x + 3}} = x - 3 = r(x)$$

only if $x \neq -3$. Notice that -3 does not belong to the domain of R, but it does belong to the domain of r. The graph of R (Figure 1a) is the same as the graph of r (Figure 1b), except that the point $(-3, -6)$ does not belong to the graph of R.

* In this section, we consider only rational functions for which the numerator and denominator polynomials have *real* coefficients.

Functions of the Form $R(x) = 1/x^n$

If n is a positive integer, the domain of the rational function $R(x) = 1/x^n$ consists of all real numbers except 0. Here we consider the graphs of such functions.

Example 2 Sketch the graph of each function.

(a) $F(x) = 1/x$ **(b)** $G(x) = 1/x^2$

Solution **(a)** The function F is odd,

$$F(-x) = \frac{1}{(-x)} = -\frac{1}{x} = -F(x),$$

so its graph is symmetric about the origin. Therefore, we begin by sketching the portion of the graph for $x > 0$, and then we obtain the complete graph by using the symmetry. For small positive values of x, the reciprocal $1/x$ will be large. As x increases, $1/x$ decreases, coming closer and closer to 0 as x gets larger. This variation is shown in Table 1. By using this information and plotting a few points, we obtain the graph of F (Figure 2a). Notice that although the graph comes closer and closer to the x and y axes, it never *touches* them; in other words, it has no x or y intercepts.

Table 1

x	$1/x$	$1/x^2$
$\frac{1}{100}$	100	10,000
$\frac{1}{10}$	10	100
$\frac{1}{3}$	3	9
$\frac{1}{2}$	2	4
1	1	1
2	$\frac{1}{2}$	$\frac{1}{4}$
3	$\frac{1}{3}$	$\frac{1}{9}$
10	$\frac{1}{10}$	$\frac{1}{100}$
100	$\frac{1}{100}$	$\frac{1}{10,000}$

(b) The graph of $G(x) = 1/x^2$ is obtained in much the same way as the graph of $F(x) = 1/x$ in part (a). The main differences are that $G(x)$ is always positive and that G is an even function,

$$G(-x) = \frac{1}{(-x)^2} = \frac{1}{x^2} = G(x),$$

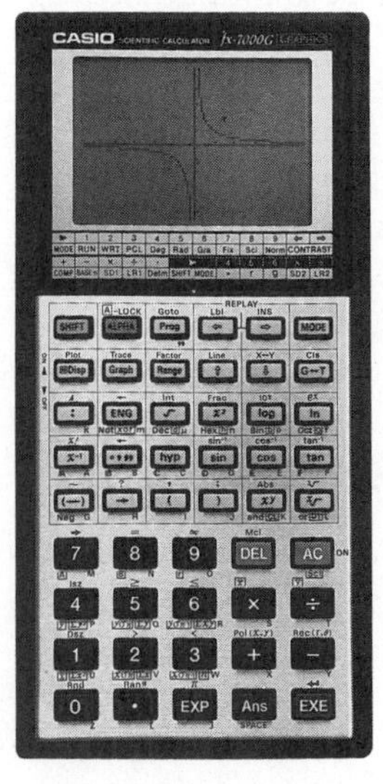

Casio fx-7000G showing the graph in Figure 2a.

Figure 2

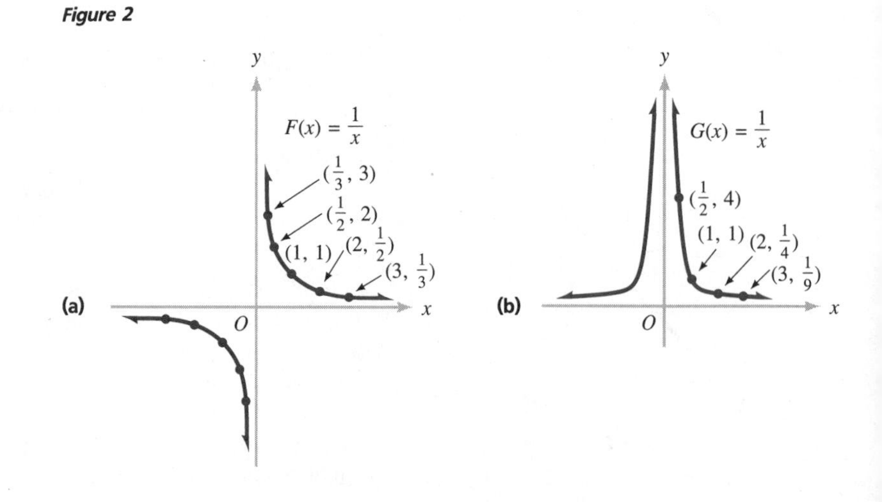

The graphs in Figure 2 are typical of the graphs of functions of the form $R(x) = 1/x^n$, where n is a positive integer. If n is odd, the graph resembles that of $F(x) = 1/x$ (Figure 2a), whereas if n is even, the graph resembles that of $G(x) = 1/x^2$ (Figure 2b). For larger values of n, the graphs become steeper near the origin and they approach the x axis more rapidly as one moves away from the origin.

For the graphs of the functions $R(x) = 1/x^n$, $n \geq 1$, the coordinate axes play a very special role. Indeed, they are **asymptotes** of these graphs in the following sense: If the values $f(x)$ of a function f approach a fixed number k as x (or $-x$) gets larger and larger without bound, we say that the line $y = k$ is a **horizontal asymptote** of the graph of f (Figure 3a). Similarly, if $|f(x)|$ gets larger and larger without bound as x approaches a fixed number h, the line $x = h$ is called a **vertical asymptote** of the graph of f (Figure 3b).

Figure 3

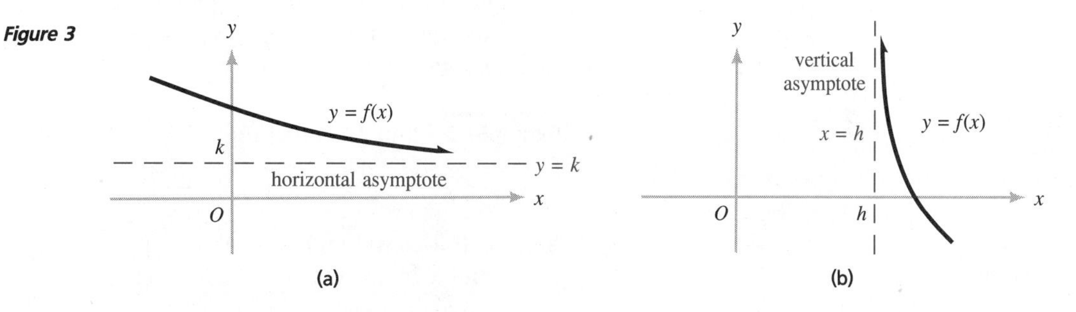

The general idea of an asymptote is illustrated in Figure 4, which shows a curve "approaching" a line L in the sense that, as the point P moves away from the origin along the curve, the distance between P and L becomes arbitrarily small. An asymptote that is neither horizontal nor vertical is called an **oblique** or **slant asymptote.** Although an asymptote isn't really part of the curve, it helps us to visualize how the curve behaves in distant regions of the xy plane.

Figure 4

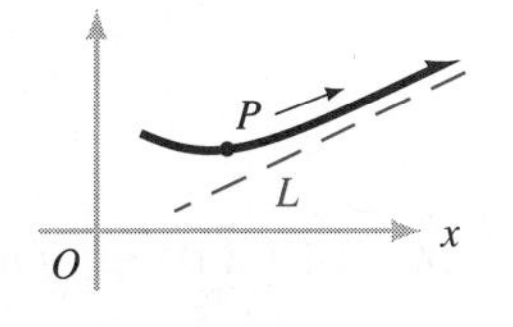

Figure 5

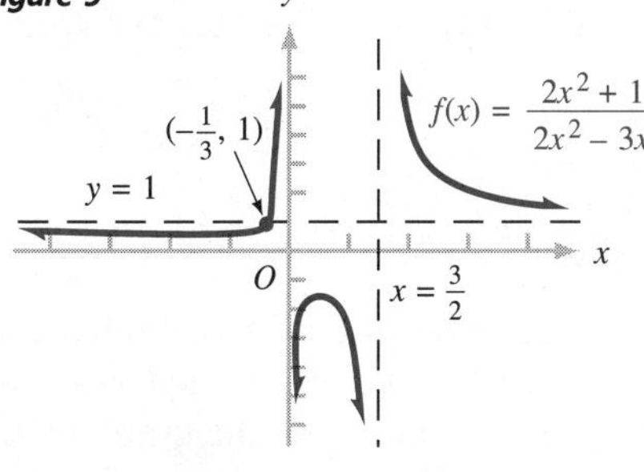

As we shall see, the graph of a rational function f may have several vertical asymptotes, but it can have at most one horizontal or oblique asymptote. Although the graph of f cannot intersect any of its own vertical asymptotes, it can intersect its own horizontal or oblique asymptote. For instance, consider the computer-generated graph of

$$f(x) = \frac{2x^2 + 1}{2x^2 - 3x}$$

in Figure 5. Since $f(-\frac{1}{3}) = 1$, the graph of f intersects its own horizontal asymptote $y = 1$ at the point $(-\frac{1}{3}, 1)$.

Shifting and Stretching

From our discussion in Section 3.5, if a, h, and k are constants and n is a positive integer, then the graph of

$$f(x) = \frac{a}{(x-h)^n} + k$$

is obtained by vertically stretching or flattening the graph of

$$R(x) = \frac{1}{x^n}$$

by a factor of $|a|$, reflecting it across the x axis if $a < 0$, shifting it horizontally by $|h|$ units, and shifting it vertically by $|k|$ units. After the vertical stretching, flattening, or reflecting, the x and y axes will still be asymptotes. However, *when you shift a graph, you shift its asymptotes as well; hence, $x = h$ is a vertical asymptote, and $y = k$ is a horizontal asymptote of the graph of f.*

Example 3 For the function

$$f(x) = \frac{3}{x-2} + 4,$$

find the horizontal and vertical asymptotes of the graph, determine any y and x intercepts, and sketch the graph.

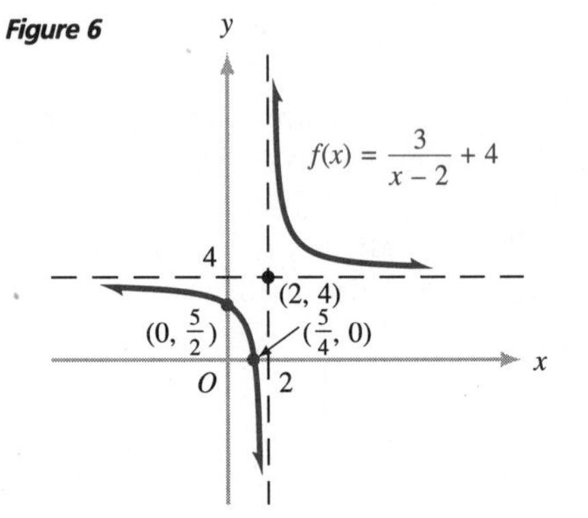

Figure 6

Solution The graph will have the same general shape as the graph of $F(x) = 1/x$ in Figure 2a, except that it will be stretched vertically by a factor of 3 and then shifted 2 units to the right and 4 units upward (Figure 6). The asymptotes are $x = 2$ and $y = 4$. The y intercept is given by $f(0) = 3/(0-2) + 4 = \frac{5}{2}$, and the x intercept is obtained by solving the equation $f(x) = 0$; that is,

$$\frac{3}{x-2} + 4 = 0 \quad \text{or} \quad 3 + 4(x-2) = 0.$$

Thus, $4x - 5 = 0$, so $x = \frac{5}{4}$. ■

Graphs of General Rational Functions

Although computers and calculators with graphing features may be used to generate graphs of rational functions, a little practice sketching a few graphs by hand will enhance your understanding of these functions. In many cases you can sketch an accurate graph of a rational function,

$$f(x) = \frac{p(x)}{q(x)},$$

by carrying out the following systematic procedure:

Step 1. Factor the numerator $p(x)$ and the denominator $q(x)$ and reduce the fraction $p(x)/q(x)$ by canceling all common factors. (Be careful though—as we saw in Figure 1, such common factors might produce "holes" in the graph of f.)

Step 2. Find the intercepts of the graph: If $q(0) \neq 0$, the y intercept is $f(0) = p(0)/q(0)$. Find the real zeros of the numerator p. These real zeros, which are the same as the real zeros of the function f, are the x intercepts of the graph.

Step 3. Find the real zeros of the denominator q. If $q(c) = 0$, then the line $x = c$ is a vertical asymptote of the graph of f. Note that the graph of f can have more than one vertical asymptote, but the graph cannot intersect any of these vertical asymptotes.

Step 4. Check whether the fraction is proper, that is, whether the degree of the numerator is smaller than the degree of the denominator. If the fraction is proper, then the x axis is the only horizontal asymptote of the graph. If the fraction is improper, perform a long division and use the quotient theorem (Section 4.3, Theorem 2) to write $f(x)$ as a sum of a polynomial and a proper fraction. If this polynomial has the form $ax + b$, then the line $y = ax + b$ is an asymptote of the graph of f; otherwise f has no asymptotes other than the vertical ones found in step 3.

Step 5. Check whether f has the form

$$f(x) = \frac{a}{(x-h)^n} + k.$$

If it does, proceed as in Example 3 and omit the following steps.

Step 6. Use the x intercepts and the vertical asymptotes to divide the x axis into intervals; then use convenient test numbers as in Section 2.7 to determine the algebraic sign of $f(x)$ on each of these intervals. This shows where the graph of f lies above or below the x axis.

Step 7. Check to see whether f is either an even or an odd function; if it is, then its graph is symmetric about the y axis or the origin, respectively.

Step 8. Sketch the graph of f by using the information obtained in the preceding steps together with the point-plotting method. Here a calculator may be useful.

[C] *In Examples 4 to 8, sketch the graph of each rational function.*

Example 4 $f(x) = \dfrac{-1}{x^2 + 4x + 4}$

Solution We follow the steps in the graph-sketching procedure for rational functions:

Step 1. Here the denominator is a perfect square, and we have

$$f(x) = \frac{-1}{(x+2)^2}.$$

Step 2. The y intercept is given by $f(0) = -\frac{1}{4}$. Because the equation $f(x) = 0$ has no solution, there is no x intercept.

Step 3. The denominator is zero when $x = -2$; hence, the line $x = -2$ is a vertical asymptote.

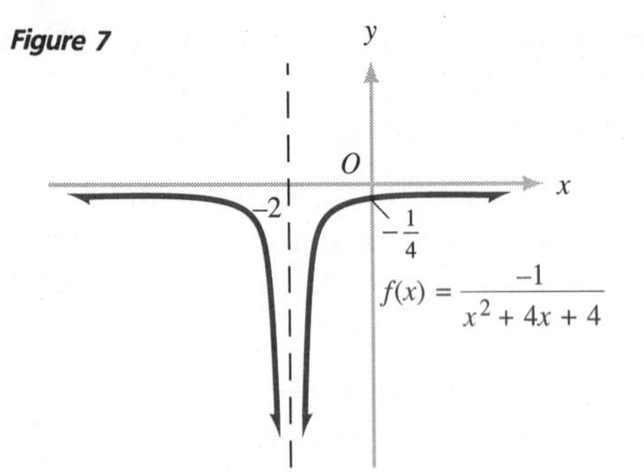

Figure 7

Step 4. The fraction is proper, so the x axis is a horizontal asymptote.

Step 5. The function f has the form

$$f(x) = \frac{a}{(x-h)^n} + k,$$

with $a = -1$, $h = -2$, $n = 2$, and $k = 0$; hence, the graph has the same shape as the graph of $G(x) = 1/x^2$ in Figure 2b, except that it is reflected across the x axis and then shifted 2 units to the left (Figure 7). ■

Example 5 $g(x) = \dfrac{2x - 1}{x + 1}$

Solution

Step 1. The fraction is in reduced form.

Step 2. The y intercept is given by $g(0) = -1$. The x intercept is $\frac{1}{2}$, since $2x - 1 = 0$ when $x = \frac{1}{2}$.

Step 3. The denominator is zero when $x = -1$; hence, the line $x = -1$ is a vertical asymptote.

Step 4. The fraction is improper because its numerator has the same degree as its denominator. We must divide and apply the quotient theorem:

$$\begin{array}{r|l} & 2 \\ x+1 & 2x - 1 \\ & 2x + 2 \\ \hline & -3 \end{array} \qquad \text{so} \qquad g(x) = 2 + \frac{-3}{x+1}.$$

Thus, $y = 2$ is a horizontal asymptote.

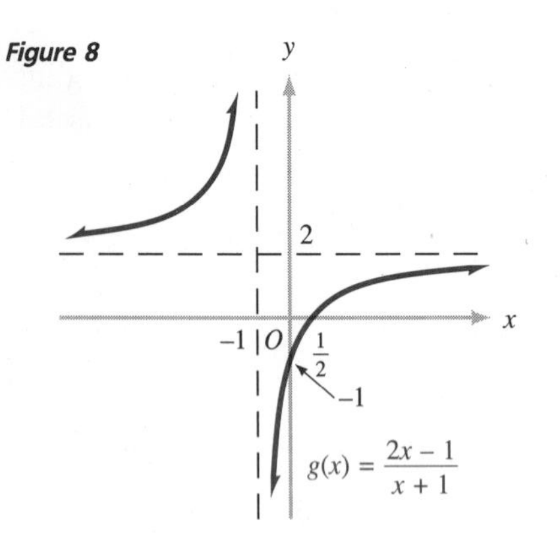

Figure 8

Step 5. The function g now has the form

$$g(x) = \frac{a}{(x-h)^n} + k,$$

with $a = -3$, $h = -1$, $n = 1$, and $k = 2$; hence, the graph has the same shape as the graph of $F(x) = 1/x$ in Figure 2a, except that it is stretched vertically by a factor of 3, reflected across the x axis, shifted 1 unit to the left, and shifted 2 units upward (Figure 8). ■

Example 6 $h(x) = \dfrac{x + 1}{x^2 + x - 2}$

Solution

Step 1. Factoring the denominator, we have $h(x) = \dfrac{x + 1}{(x + 2)(x - 1)}$.

Step 2. The y intercept is given by $h(0) = -\frac{1}{2}$. To find the x intercept, we set $x + 1 = 0$ to obtain $x = -1$.

Step 3. The denominator is zero when $x = -2$ and when $x = 1$; hence, the lines $x = -2$ and $x = 1$ are vertical asymptotes.

Step 4. The fraction is proper, so the x axis is a horizontal asymptote.

Step 5. The function does not have the form given in step 5 of the procedure.

Figure 9

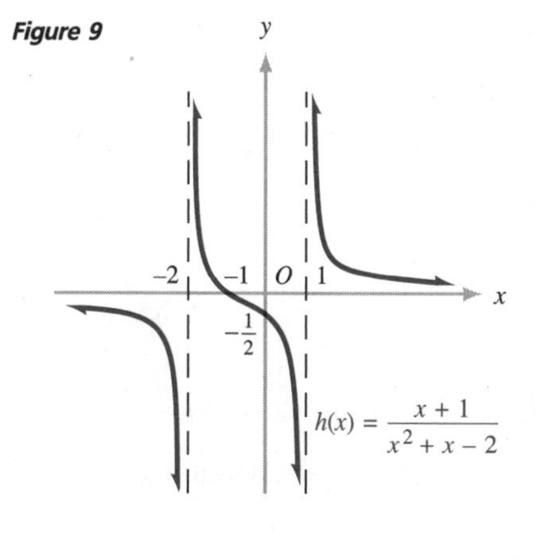

Step 6. The x intercept and the vertical asymptotes divide the x axis into the intervals $(-\infty, -2), (-2, -1), (-1, 1)$, and $(1, \infty)$, on which the function values $h(x)$ cannot change algebraic sign. Selecting, say, the test numbers $-3, -1.5, 0$, and 2 on these intervals and evaluating h at these numbers, we find that $h(-3) = -0.5$, $h(-1.5) = 0.4$, $h(0) = -0.5$, and $h(2) = 0.75$. Therefore, the graph of h is above the x axis on the intervals $(-2, -1)$ and $(1, \infty)$, and it is below the x axis on the intervals $(-\infty, -2)$ and $(-1, 1)$.

Step 7. The function h is neither even nor odd.

Step 8. Using the information above and plotting a few points, we obtain the graph shown in Figure 9. ■

Example 7 $F(x) = \dfrac{x^2 + 2x - 3}{x - 2}$

Solution

Step 1. Factoring the numerator, we have

$$F(x) = \frac{(x-1)(x+3)}{x-2}.$$

Step 2. The y intercept is given by $F(0) = \frac{3}{2}$. To find the x intercepts, we set $(x - 1)(x + 3) = 0$ and conclude that 1 and -3 are the x intercepts.

Step 3. The denominator is zero when $x = 2$; hence, the line $x = 2$ is a vertical asymptote.

Step 4. The fraction is improper because its numerator has degree greater than the degree of the denominator. We must divide and apply the quotient theorem:

$$\begin{array}{r|l} & x + 4 \\ \hline x - 2 & x^2 + 2x - 3 \\ & x^2 - 2x \\ \hline & \quad 4x - 3 \\ & \quad 4x - 8 \\ \hline & \qquad 5 \end{array} \qquad \text{so} \qquad F(x) = (x + 4) + \frac{5}{x - 2}.$$

Thus, $y = x + 4$ is an oblique asymptote.

Step 5. The function does not have the form given in step 5 of the procedure.

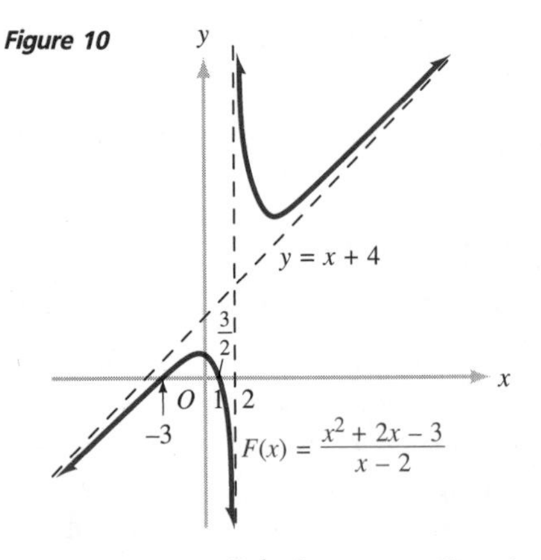

Figure 10

Step 6. The x intercepts and the vertical asymptote divide the x axis into the intervals $(-\infty, -3)$, $(-3, 1)$, $(1, 2)$, and $(2, \infty)$, on which the function values $F(x)$ cannot change algebraic sign. Using suitable test numbers, we find that the graph is above the x axis on $(-3, 1)$ and $(2, \infty)$, and below the x axis on $(-\infty, -3)$ and $(1, 2)$.

Step 7. The function F is neither even nor odd.

Step 8. Using the information above and plotting a few points, we obtain the graph shown in Figure 10. ■

Example 8 $G(x) = \dfrac{1}{x^2 + 1}$

Solution

Step 1. The denominator $x^2 + 1$ will not factor (over the real numbers), and the fraction is in reduced form.

Step 2. The y intercept is given by $G(0) = 1$. The equation $G(x) = 0$ has no real solution, so there are no x intercepts.

Step 3. The denominator has no real zeros, so there is no vertical asymptote.

Step 4. The fraction is proper, so the x axis is a horizontal asymptote.

Step 5. G does not have the form given in step 5 of the procedure.

Step 6. Since there are no x intercepts and no vertical asymptotes, there is only one interval to consider, namely $(-\infty, \infty)$. Evidently, $G(x)$ is always positive, so the graph is entirely above the x axis.

Step 7. G is an even function, so its graph is symmetric about the y axis.

Step 8. Using the information above and plotting a few points, we obtain the graph shown in Figure 11. ■

Figure 11

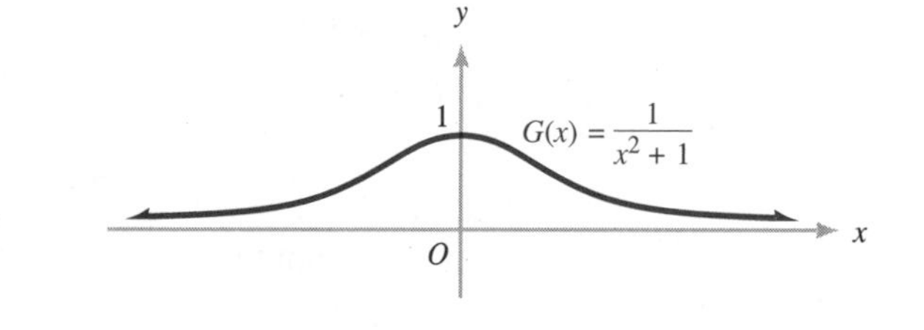

Problem Set 4.7

In Problems 1 to 6, determine the domain of each rational function.

1. $f(x) = \dfrac{-5}{x^4}$

2. $g(x) = \dfrac{-5}{x^2 + 5}$

3. $P(x) = \dfrac{x^4 - 16}{2x^2 - x + 3}$

4. $T(x) = \dfrac{x^4 + 1}{2x^2 + 9x - 3}$

5. $F(x) = \dfrac{x^2 + x + 1}{x^2 + 8x + 15}$

6. $p(x) = 1 + \dfrac{x^2 + x}{x^3 - x}$

In Problems 7 to 18, the graph of each rational function can be obtained by stretching, flattening, reflecting, and/or shifting the graph of a function of the form $R(x) = 1/x^n$, for $n \geq 1$. In each case, find the horizontal and vertical asymptotes of the graph, determine any y and x intercepts, and sketch the graph. [c] Use a calculator, if you wish, to determine more points on the graph and thus improve the accuracy of your sketch. [gc] Alternatively, use a graphing calculator.

7. $F(x) = -\dfrac{3}{x}$

8. $G(x) = -7x^{-2}$

9. $P(x) = \dfrac{-4}{x^4}$

10. $Q(x) = -7x^{-8}$

11. $f(x) = \dfrac{1}{x} + 1$

12. $F(x) = 2 - \dfrac{1}{x}$

13. $g(x) = \dfrac{1}{x^2} - 4$

14. $G(x) = -1 - \dfrac{1}{x^2}$

15. $Q(x) = \dfrac{-2}{x+1}$

16. $K(x) = \dfrac{-3}{(x+5)^2}$

17. $f(x) = \dfrac{3}{x-2} + 6$

18. $k(x) = \dfrac{-3}{(x+2)^3} - 4$

[c] In Problems 19 to 38, sketch the graph of each rational function by following the steps in the graph-sketching procedure. [gc] If you wish, you may use a graphing calculator to check your work.

19. $H(x) = \dfrac{2x+1}{x-4}$

20. $G(x) = \dfrac{4x^2 + 24x + 41}{x^2 + 6x + 9}$

21. $F(x) = \dfrac{x^2 - 2x + 1}{x - 1}$

22. $k(x) = \dfrac{3x^2 + 1}{x^2 - 1} + 1$

23. $K(x) = \dfrac{-2}{x^2 + 9}$

24. $G(x) = \dfrac{x^2 + 5}{x^2 - 4}$

25. $p(x) = \dfrac{x + 2}{x^2 + 2x - 15}$

26. $F(x) = \dfrac{3}{x^2 - 2x + 1}$

27. $f(x) = \dfrac{2x^2 + 4x + 7}{x^2 + 2x + 1}$

28. $P(x) = \dfrac{3x^2 - 18x + 28}{x^2 - 6x + 9}$

29. $g(x) = \dfrac{x^2 - x - 2}{x - 3}$

30. $H(x) = \dfrac{x^2 - 9}{x^2 - 1}$

31. $R(x) = \dfrac{x^2 - 4}{(x-3)^2}$

32. $q(x) = \dfrac{2x}{x^2 + 1}$

33. $g(x) = \dfrac{x^2 - 1}{x^2 + 5x + 4}$

34. $h(x) = \dfrac{x^2 + x - 6}{x^2 - x - 12}$

35. $Q(x) = \dfrac{x^3 + 5x^2 + 6x}{x + 3}$

36. $G(x) = \dfrac{3x^3 + 5x^2 - 26x + 8}{x^2 + 2x - 8}$

37. $r(x) = \dfrac{x^3 + x^2 + 6x}{x^2 + x - 2}$

38. $K(x) = \dfrac{x^3 - x^2 - 6x}{x^2 - 3x + 2}$

39. Find the point where the graph of $f(x) = \dfrac{2x^3 + x^2 + 3x}{x^2 + 1}$ intersects its own oblique asymptote.

40. Explain why Figure 12 cannot be the graph of a rational function.

Figure 12

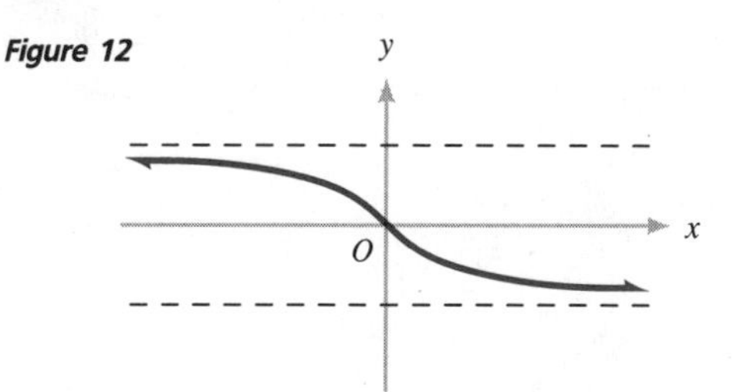

[c] 41. According to Boyle's law, the pressure P and the volume V of a sample of gas maintained at constant temperature are related by an equation of the form $P = K/V$, where K is a numerical constant. If $K = 3.5$, **(a)** sketch a graph of P as a function of V, [gc] using a graphing calculator if you wish; **(b)** determine the domain of this function; and **(c)** indicate whether P increases or decreases when V increases.

42. A function of the form $f(x) = (ax + b)/(cx + d)$ where a, b, c, and d are constants and c and d are not both zero, is called a **fractional linear function.** Discuss the graph of f with regard to **(a)** its x and y intercepts, **(b)** its horizontal and vertical asymptotes, and **(c)** its general shape.

[c] 43. The electrical resistance R per unit length of a wire depends on the diameter d of the wire according to the formula $R = K/d^2$, where K is a constant. If $K = 0.1$, **(a)** sketch a graph of R as a function of d, **(b)** determine the domain of this function, and **(c)** indicate whether

R increases or decreases when d decreases. gc Use a graphing calculator if you wish.

44. By adding x milliliters of a pure acid to a milliliters of a $p\%$ acid solution, a chemist obtains a mixture with a concentration of $c\%$ acid (Figure 13).

(a) Show that $c = f(x)$, where $f(x) = \dfrac{100x + pa}{x + a}$.

(b) Discuss the graph of f with regard to its domain, intercepts, asymptotes, and general shape.

(c) Solve the equation $c = f(x)$ for x in terms of c.

Figure 13

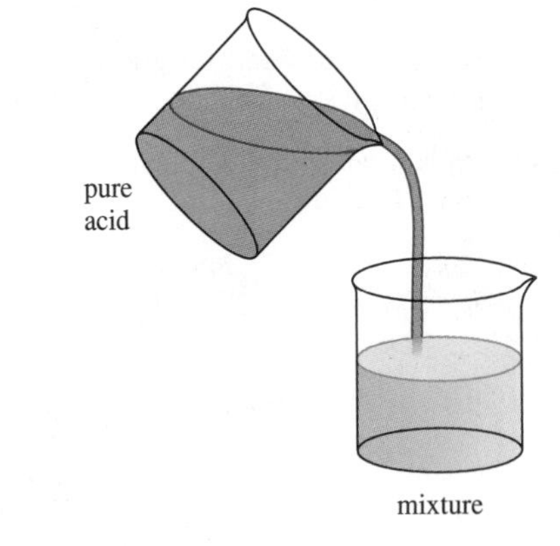

c **45.** A person learning to type has an achievement record given by $N(t) = 60[(t - 2)/t]$ for $3 \le t \le 10$, where $N(t)$ is the number of words per minute at the end of t weeks. Sketch the graph of the rational function N. gc Use a graphing calculator if you wish.

c **46.** The electrical power P in watts consumed by a certain electrical circuit is related to the load resistance R in ohms by the equation $P = 100R(0.5 + R)^{-2}$ for $R \ge 0$. Sketch a graph of P as a function of R and use it to estimate the value of R for which P is maximum. gc Use a graphing calculator if you wish.

47. Let k be a positive constant and let $f(x) = \dfrac{x^2}{x - k}$.

(a) Show that $y = x + k$ is an oblique asymptote of the graph of f.

(b) Show that $f(x) = 4k + \dfrac{(x - 2k)^2}{x - k}$.

(c) Use the result of part (b) to show that, for $x > k$, $f(x)$ takes on its smallest value when $x = 2k$.

(d) For $x < k$, show that $f(x)$ takes on its largest value when $x = 0$.

(e) If $k = \frac{1}{4}$, sketch the graph of f. gc Use a graphing calculator if you wish.

48. The logo for a world's fair is to consist of a transparent right circular cone of height h circumscribing a sphere of radius r (Figure 14). The total volume V enclosed by the cone is given by $V = \dfrac{\pi r^2}{3}\left(\dfrac{h^2}{h - 2r}\right)$.

(a) If r is fixed and V is regarded as a function of h, $V = f(h)$, what is the domain of f?

(b) Find the ratio h/r that produces the cone of smallest volume. [*Hint:* Use the results of Problem 47.]

Figure 14

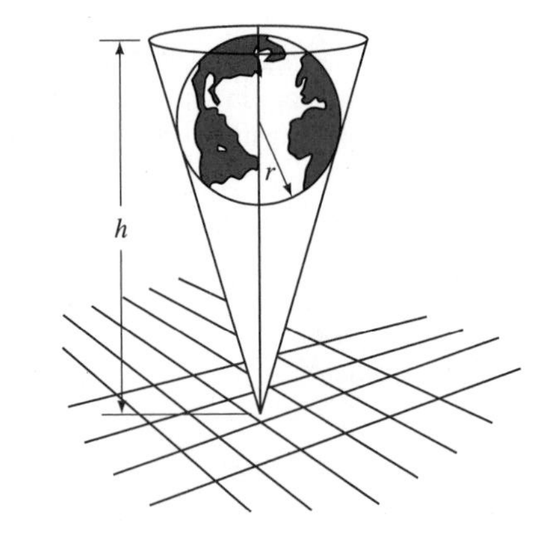

CHAPTER 4 Review Problem Set

In Problems 1 to 4, find the vertex and the y and x intercepts of the graph of the quadratic function, determine whether the graph opens upward or downward, sketch the graph, and find the domain and range of the function.

1. $f(x) = x^2 - 3x + 2$

2. $G(x) = -2x^2 + x + 10$

3. $F(x) = 10x - 25 - x^2$

4. $H(x) = 7x + 2x^2 - 39$

5. Use the graph obtained in Problem 1 to find the solution set of $x^2 - 3x + 2 \geq 0$.

6. Suppose that x_1 and x_2 are the zeros of the quadratic function f. Show that the x coordinate of the vertex of the graph of f is $\frac{1}{2}(x_1 + x_2)$.

7. Find two numbers whose sum is 21 and whose product is maximum.

8. Suppose that f is a quadratic function and that there is a number x_0 with $x_0 \neq 0$ and $f(x_0) = f(-x_0)$. Prove that f is an even function.

[c] **9.** The strength S of a new plastic is given by the mathematical model $S = 500 + 600T - 20T^2$, where T is the temperature in degrees Fahrenheit. At what temperature is the strength maximum?

10. Find all values of K such that the graph of $g(x) = 3x^2 + Kx - 4K$ has no x intercepts.

11. Sketch the graphs of the following functions f, g, F, and G on the same coordinate system: $f(x) = \frac{1}{3}x^3$; $g(x) = -\frac{1}{3}x^3$; $F(x) = \frac{1}{3}x^3 + 1$; $G(x) = \frac{1}{3}(x + 1)^3 + 4$.

12. Decide which functions are polynomial functions and which are not.

(a) $f(x) = \sqrt{3}x - 4x^4 + \frac{1}{2}$
(b) $g(x) = 2x^{-2} + 3x^{-1} + x + 1$
(c) $h(x) = 2\sqrt{x} + x - 4$
(d) $H(x) = (x - 1)(x - 2)(x - 3)$

In Problems 13 to 16, **(a)** find the x intercepts of the graph of the function, **(b)** sketch the graph, and **(c)** use the graph to solve the given inequality. [c] Use a calculator if you wish. [gc] Alternatively, use a graphing calculator.

13. $G(x) = (x + 1)(10x^2 - 3x - 18)$; $G(x) \geq 0$

14. $f(x) = -(x^2 - 4)(x^2 - 9)$; $f(x) < 0$

15. $h(x) = -(x + 3)(2 - x)^2$; $h(x) > 0$

16. $g(x) = x^2(x - 1)^2(x + 2)^2$; $g(x) \leq 0$

In Problems 17 to 20, use long division to find the quotient polynomial $q(x)$ and the remainder polynomial $r(x)$ when $f(x)$ is divided by $g(x)$. Check your work by verifying that $r(x)$ is either the zero polynomial or a polynomial of degree less than $g(x)$, and that $f(x) = g(x)q(x) + r(x)$.

17. $f(x) = x^2 + 5x + 2$; $g(x) = x + 3$

18. $f(x) = 6x^4 + 38x^3 + 44x^2 - 96x + 27$; $g(x) = 2x + 6$

19. $f(x) = x^5 - 32$; $g(x) = x^3 - 8$

20. $f(x) = 8x^3 - 5x^2 - 51x - 18$; $g(x) = 2x^3 + x^2 + 1$

In Problems 21 and 22, use the quotient theorem to rewrite each rational expression as the sum of a polynomial and a proper rational expression.

21. $\dfrac{4x^4 + x^3 - 2x + 3}{x + 1}$

22. $\dfrac{x^3 + 6x^2 + 10x}{x^2 + x + 2}$

In Problems 23 to 26, **(a)** use synthetic division to obtain the quotient polynomial $q(x)$ and the constant remainder R upon division of the given polynomial by $x - c$ for the indicated value of c, and [c] **(b)** verify by direct substitution that R is the value of the polynomial when $x = c$.

23. $x^3 - 2x^2 + 3x - 5$; $c = -2$

24. $5x^4 - 10x^3 - 12x - 7$; $c = -4$

25. $x^5 + 5x - 13$; $c = -1$

26. $3x^5 + 5x^4 - 2x^3 + x^2 - x + 1$; $c = 2$

[c] In Problems 27 and 28, use a calculator and Horner's method to find the indicated number rounded off to two decimal places.

27. $f(2.72)$ if $f(x) = 17.1x^4 + 33.3x^3 - 2.75x^2 + 11.1x + 21.8$

28. $g(3.14)$ if $g(x) = 13.5x^8 - 31.7x^6 + 22.1x^4 - 35.7x^2 + 21.2$

In Problems 29 to 32, [c] **(a)** use the remainder theorem to find the remainder R when each division is performed, and **(b)** verify your result using synthetic division.

29. $x^3 - 2x^2 + 3x - 5$ divided by $x + 2$

30. $5x^4 - 10x^3 - 12x - 7$ divided by $x - 4$

31. $x^5 + 5x - 13$ divided by $x - 1$

32. $3x^5 + 5x^4 - 2x^3 + x^2 - x + 1$ divided by $x - 2$

In Problems 33 and 34, use the factor theorem to determine (without actually dividing) whether or not the indicated binomial is a factor of the given polynomial.

33. $x^3 - 2x^2 + 3x + 4$; $x - 2$

34. $2x^4 + 3x - 26$; $x + 3$

In Problems 35 and 36, the indicated number is a zero of the polynomial function $f(x)$. Determine the multiplicity of this zero.

35. 1; $f(x) = x^4 - 4x^3 + 7x^2 - 6x + 2$

36. $\frac{2}{3}$; $f(x) = 27x^4 - 27x^3 - 18x^2 + 28x - 8$

In Problems 37 and 38, determine the number of variations in sign for $f(x)$ and $f(-x)$; use Descartes' rule of signs to determine the possible number of positive and negative zeros of $f(x)$; and find integers that are upper and lower bounds for the real zeros of $f(x)$.

37. **(a)** $f(x) = x^3 + 2x^2 - x - 2$
(b) $f(x) = 5x^4 - 2x^3 - 5x + 2$

38. **(a)** $f(x) = x^3 - x^2 - 5x - 3$
(b) $f(x) = 6x^4 - 7x^3 - 7x^2 + 6x + 1$

In Problems 39 to 42, find all rational zeros of each polynomial function.

39. $f(x) = x^3 - 8x^2 + 5x + 14$

40. $F(x) = x^4 - 4x^3 - 5x^2 + 36x - 36$

41. $h(x) = 4x^4 - 4x^3 - 7x^2 + 4x + 3$

42. $H(x) = 4x^4 - 2x^3 + 2x^2 + 10x + 3$

In Problems 43 and 44, factor each polynomial function completely into prime factors, and [c] sketch the graph of the function. [gc] Use a graphing calculator if you wish.

43. $f(x) = x^3 + 2x^2 - x - 2$

44. $F(x) = x^3 - x^2 - 14x + 24$

[c] In Problems 45 and 46, show that the given polynomial function has a zero in the indicated interval by using the change-of-sign property, and then find this zero, rounded off to two decimal places, by applying the bisection method seven times in succession.

45. $f(x) = x^3 - x^2 + 3x - 2$ in $[0, 1]$

46. $h(x) = 2x^4 - 3x^3 - 2x^2 - 4x + 1$ in $[2, 3]$

[c] **47.** Show that the function $f(x) = 2x^3 - 5x^2 + 4x + 1$ has a zero on the interval $[-1, 0]$ by using the change-of-sign property and then find this zero to within the limits of accuracy of your calculator by using the fixed-point method.

48. The function $g(x) = 4x^3 - 19x^2 + 31x - 15$ has exactly one fixed point. Find this fixed point.

In Problems 49 to 52, **(a)** find all complex zeros of each complex polynomial $f(z)$, **(b)** factor $f(z)$ completely into linear factors, and **(c)** determine the multiplicity of each zero of $f(z)$.

49. $z^4 + 5z^3 + 8z^2 + z - 15$

50. $z^4 - 8z^2 - 4z + 3$

51. $z^4 - 6z^3 + 3z^2 + 24z - 28$

52. $z^4 - 4z^3 + 5z^2 - 4z + 4$

In Problems 53 and 54, form a polynomial in z that has real coefficients, has the smallest possible degree, and has the indicated zeros. (Zeros are repeated to show multiplicity.)

53. 2, 2, 2, i

54. $3 + \sqrt{2}i$, $1 - \sqrt{2}i$

In Problems 55 to 62, indicate the domain of each rational function, find the y and x intercepts of its graph, determine all the asymptotes of the graph, and [c] sketch it. [gc] Use a graphing calculator if you wish.

55. $F(x) = \dfrac{2x}{x - 1}$

56. $H(x) = \dfrac{3x + 2}{2 - x}$

57. $p(x) = \dfrac{(x + 2)^2}{x^2 + 2x}$

58. $G(x) = \dfrac{4x^2}{x^2 - 4x}$

59. $T(x) = \dfrac{x^2 + 1}{x^2 - 3x}$

60. $f(x) = \dfrac{x - 1}{x^2 - 9}$

61. $h(x) = \dfrac{x^2 - 2x - 3}{x - 2}$

62. $F(x) = \dfrac{x^3 + x^2 - 2}{x^2 - 2x - 8}$

63. A manufacturer of sports trophies knows that the total cost C dollars of making x thousand trophies is given by $C = 600 + 60x$ and that the corresponding sales revenue R dollars is given by $R = 300x - 4x^2$. Find the number of trophies (in thousands) that will maximize the manufacturer's profit. [*Hint:* Profit = revenue − cost.]

64. A homeowner is planning a rectangular flower garden surrounded by an ornamental fence. The fencing for three sides of the garden costs \$20 per meter, but the fencing for the fourth side, which faces the house, costs \$30 per meter. If the homeowner has \$1200 to spend

on the fence, what dimensions of the garden will give it the maximum area?

65. In statistics, the third-degree **Chebyshev polynomial** is given by $C_3(x) = 4x^3 - 3x$. Sketch the graph of C_3. gc Use a graphing calculator if you wish.

66. An engineer is designing a new laser disk of fixed radius R centimeters to store information in a computer. The information, measured in units called *bytes*, will be stored optically along concentric circular tracks (Figure 1). For uniformity, each track will store the same number of bytes. Physical properties of the disk restrict the track density to c tracks per centimeter measured radially. Likewise, the information density is restricted to b bytes per centimeter along the innermost track, which will have a radius of r centimeters.

(a) Show that the total number of bytes that can be stored on the disk is given by $N = 2\pi bcr(R - r)$.

(b) Find the value of r that will maximize the information that can be stored on the disk.

Figure 1

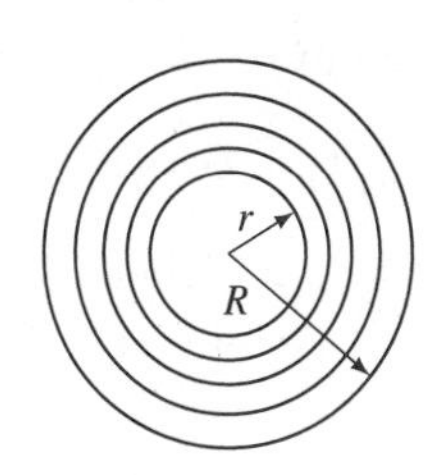

67. Find the original length of the edge of a cube if, after a slice 1 centimeter thick is cut from one side, the volume of the remaining solid is 448 cubic centimeters (Figure 2).

Figure 2

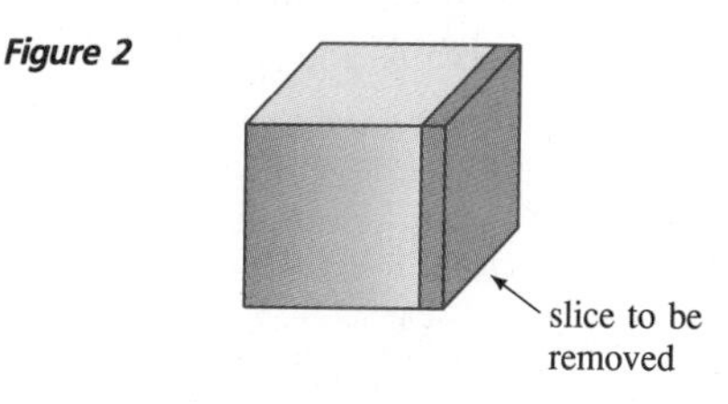

68. Let f be a polynomial function and suppose that a and b are constant real numbers with $a \neq b$. Show that the remainder upon division of $f(x)$ by $(x - a)(x - b)$ has the form

$$Ax + B,$$

where $A = \dfrac{f(b) - f(a)}{b - a}$ and $B = \dfrac{bf(a) - af(b)}{b - a}$.

69. Liquid helium in a cryogenic laboratory is stored in an insulated spherical tank of radius $r = 0.5$ meter (Figure 3). A sensor in the tank measures the depth d of the liquid helium. The volume V of liquid helium in the tank can then be computed using the formula $V = \dfrac{\pi}{3}(3rd^2 - d^3)$. Find d, rounded off to two decimal places, when $V = 0.13$ cubic meter.

Figure 3

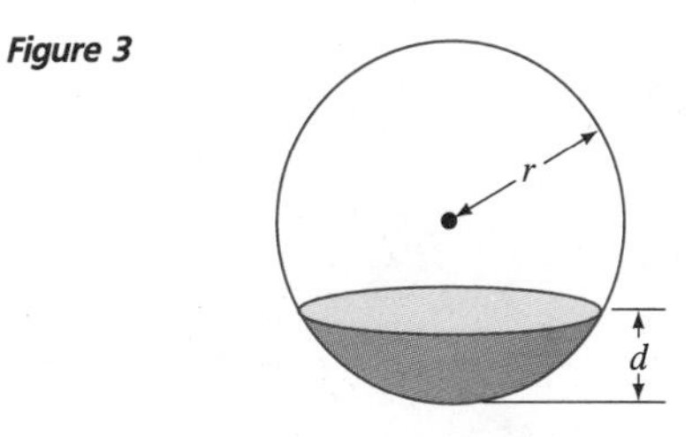

70. Corresponding to each polynomial function f is another polynomial function f' called its **derivative,** such that for each real number c, $f'(c) = q(c)$, where $q(x)$ is the quotient polynomial obtained by dividing $f(x)$ by $x - c$.

(a) Find the derivative of $f(x) = 3x^2 + 2x + 7$.

(b) Find the derivative of $f(x) = Ax^3 + Bx^2 + Cx + D$.

C **71.** If a resistor of resistance x ohms, $x \geq 0$, is connected in parallel with a resistor of resistance 1 ohm, the resulting net resistance y is given by $y = x/(1 + x)$ (Figure 4).

(a) Sketch a graph of y as a function of x. gc Use a graphing calculator if you wish.

(b) Determine the domain and the range of this function.

(c) Indicate whether y increases or decreases when x increases.

Figure 4

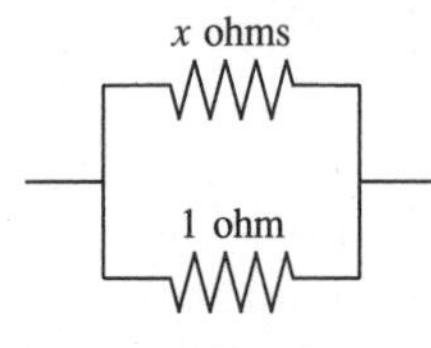

C **72.** The daredevil Mean MacDean plans to jump her motorcycle over a flaming vat of fuel oil (Figure 5). The ramp at the beginning of the jump has rise $a = 6$ feet and run $b = 10$ feet, and the speed of the motorcycle upon leaving the ramp will be $v = 60$ feet per second. If air resistance is neglected, the height h of the motorcycle x feet from the foot of the ramp will be given by

$$h = a + \frac{a}{b}x - \frac{16(a^2 + b^2)}{b^2v^2}x^2.$$

Round off your answers to the following questions to the nearest foot.

Figure 5

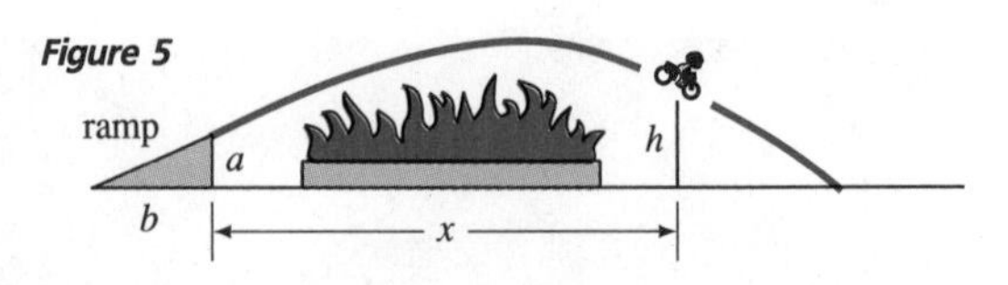

(a) Find the total length of MacDean's jump measured from the foot of the ramp.

(b) Find the value of x for which MacDean's height h is maximum.

(c) Find the maximum height of MacDean's jump.

C **73.** According to the Doppler effect in physics, if a source of sound of frequency n is moving away from an observer with speed u, the frequency N of the sound heard by the observer is given by $N = \dfrac{n}{1 + (u/v)}$, where v is the speed of sound in air. If $v = 768$ miles per hour and $n = 440$ hertz, sketch the graph of N as a function of u for $0 \leq u \leq 100$ miles per hour. gc Use a graphing calculator if you wish.

74. A large tank supplies water to both Mattoon Heights and East Mattoon. Every N days, when the tank is half full, fresh water is pumped in to refill it. Mattoon Heights uses water at a rate of half a tank every T days, and East Mattoon uses half a tank every $T + 2$ days.

(a) Find N as a function of T.

(b) Sketch the graph of the function obtained in part (a). gc Use a graphing calculator if you wish.

CHAPTER 4 Test

1. Let $f(x) = 3x^2 + 12x + 50$

(a) Write $f(x)$ in the form $a(x - h)^2 + k$.

(b) Find the vertex of the parabolic graph of f.

(c) Sketch the graph of f.

(d) Find the domain and range of f.

2. Sketch the graph of $f(x) = ax^4 + k$ for:

(a) $a = 2$; $k = -1$ **(b)** $a = -\frac{1}{2}$; $k = 2$

3. Let $f(x) = 4x^5 - 15x^4 - 20x^3 - 350x$.

(a) Find polynomial functions $q(x)$ and $r(x)$ such that $f(x) = q(x)(x - 6) + r(x)$.

(b) Use Horner's method to find $f(-2)$.

4. Let $f(x) = x^3 - 4x^2 + x + 6$.

(a) Find the zeros of f.

(b) Sketch a graph of f and label the x and y intercepts.

(c) Use the graph to solve the inequality $x^3 + x + 6 < 4x^2$.

5. Let $f(x) = x^3 + 3x^2 - 2x - 5$.

(a) Use the change-of-sign property to show that f has a zero in the interval $[1, 2]$.

(b) Apply the bisection method three times in succession to locate the zero in part (a) within an interval of length $\frac{1}{8}$. Show your work in a table with columns labeled as follows:

Number of Applications	Interval	Midpoint c	$f(c)$
0	$[1, 2]$	1.5	

(c) Use the midpoint of the final interval, rounded off to one decimal place, as an approximation to a zero of f.

6. Taking the number obtained in part (c) of Problem 5 as a first approximation, use the fixed-point method to find a zero of $f(x) = x^3 + 3x^2 - 2x - 5$ to within the limits of accuracy of your calculator.

7. (a) Let $f(x) = (4x^2 - 25)(3x + 1)^4(2x^2 - 5x)$. Find the zeros of f and indicate the multiplicity of each zero.

(b) Form a polynomial of degree 3 with real coefficients that has -2 and $3 + 2i$ as zeros.

8. Sketch each graph and label the intercepts and asymptotes, if any.

(a) $f(x) = 1 + \dfrac{2}{x + 2}$ **(b)** $g(x) = \dfrac{x^2 + x - 3}{x - 1}$

9. A rectangular field adjacent to a straight river must be fenced on three sides but not on the river bank. What is the largest area that can be enclosed if 50 meters of fencing is used?

10. Archie can paint a house in t days and Cecil can paint a house in $t + 1$ days. Find a rational function f such that $n = f(t)$ is the number of days required for Archie and Cecil, working together, to paint the house.

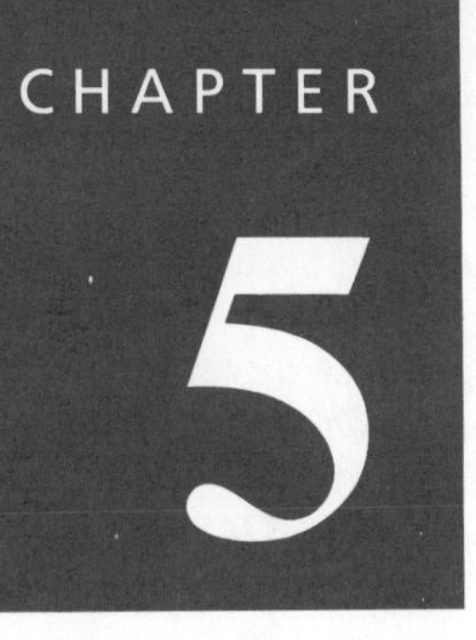

CHAPTER 5

Exponential and Logarithmic Functions

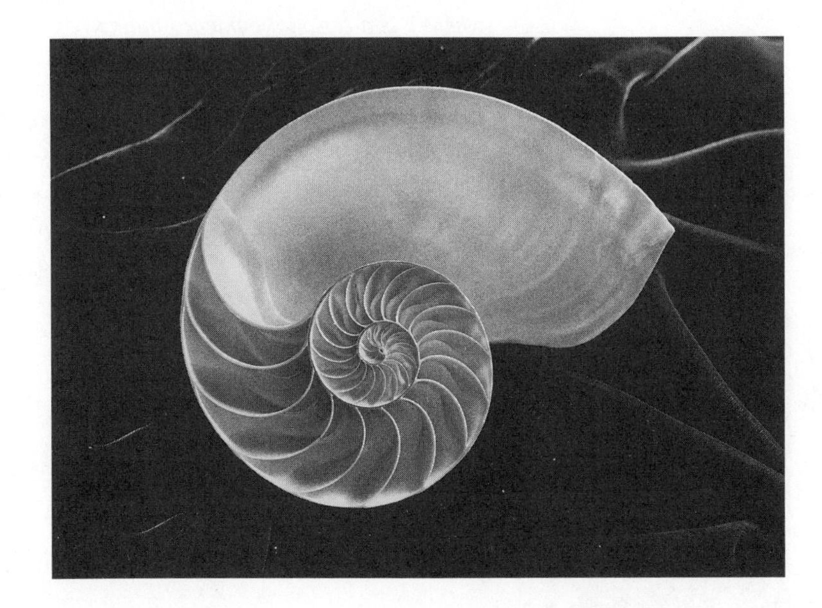

A split view of the shell of a chambered nautilus showing a logarithmic spiral.

Until now, we have considered only functions defined by equations involving algebraic expressions. Although such *algebraic functions* are useful and important, they are not sufficient for all the requirements of applied mathematics. In this chapter we begin our study of the *transcendental functions*—that is, functions that transcend (go beyond) purely algebraic methods. Here we consider the exponential and logarithmic functions and give some of their many applications to the life sciences, finance, earth sciences, engineering, electronics, and other fields.

5.1 Exponential Functions

In Section 4.2, we studied power functions

$$p(x) = x^n$$

in which the base is the variable x and the exponent n is constant. In this section we shall study functions of the form

$$f(x) = b^x$$

in which the exponent is the variable x and the base b is constant. Such a function is called an **exponential function.***

If b is a positive real number, then b^x is defined as in Section 1.8 for all *rational* values of x. It is possible to extend the definition so that *irrational* numbers can also be used as exponents. Although the technical details of the extended definition depend on methods studied in calculus, the basic idea is quite simple: *If $b > 0$ and x is an irrational number, then*

$$b^x \approx b^r,$$

where r is a rational number obtained by rounding off x to a finite number of decimal places. Better and better approximations to b^x are obtained by rounding off x to more and more decimal places. For instance,

$$b^\pi \approx b^{3.14},$$

and a better approximation is given by

$$b^\pi \approx b^{3.14159}.$$

Casio fx-7000G showing the graph in Figure 1b.

Figure 1

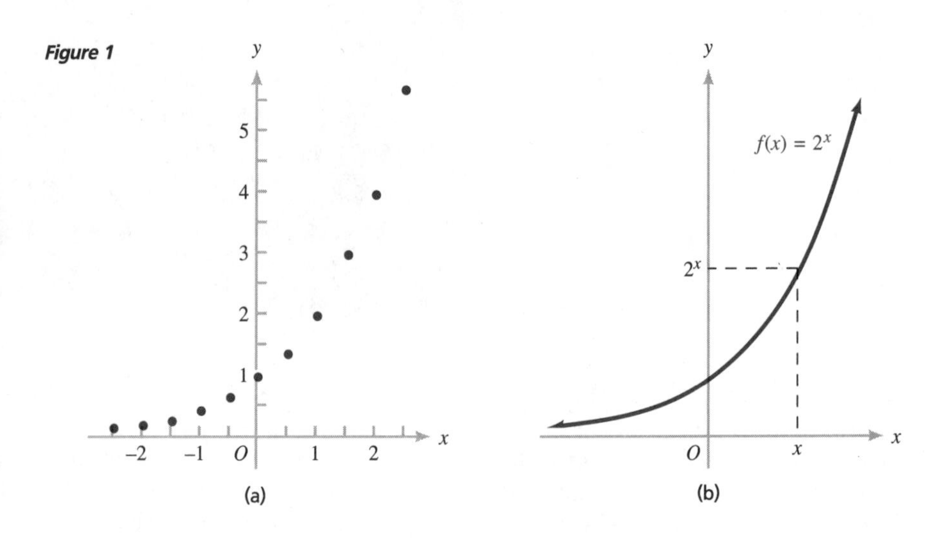

If b is a positive constant, and if you plot several points (x, b^x) for rational values of x, you will notice that these points seem to lie along a smooth curve. This curve is the graph of the exponential function $f(x) = b^x$ with base b. For instance, taking $b = 2$ and plotting several points $(x, 2^x)$ for rational values of x, we obtain Figure 1a. In Figure 1b, we have connected these points with a smooth curve to obtain the graph of $f(x) = 2^x$. The shape of the graph in Figure 1b can be confirmed by using a graphing calculator.

Example 1 Sketch the graphs of $f(x) = 3^x$ and $g(x) = (\frac{1}{3})^x$.

* In BASIC, the exponential function with base `B` is written `B X`.

Solution We begin by calculating values of $f(x) = 3^x$ and of $g(x) = (\frac{1}{3})^x$ for several integer values of x, as shown in the table in Figure 2. Then we plot the corresponding points and connect them by smooth curves to obtain the graphs of $f(x) = 3^x$ (Figure 2a) and $g(x) = (\frac{1}{3})^x$ (Figure 2b). Because

$$g(x) = \left(\frac{1}{3}\right)^x = \frac{1}{3^x} = 3^{-x} = f(-x),$$

these curves are reflections of each other across the y axis. The shapes of the graphs in Figure 2 can be confirmed by using a graphing calculator. ■

Figure 2

x	3^x	$(\frac{1}{3})^x$
-2	$\frac{1}{9}$	9
-1	$\frac{1}{3}$	3
0	1	1
1	3	$\frac{1}{3}$
2	9	$\frac{1}{9}$

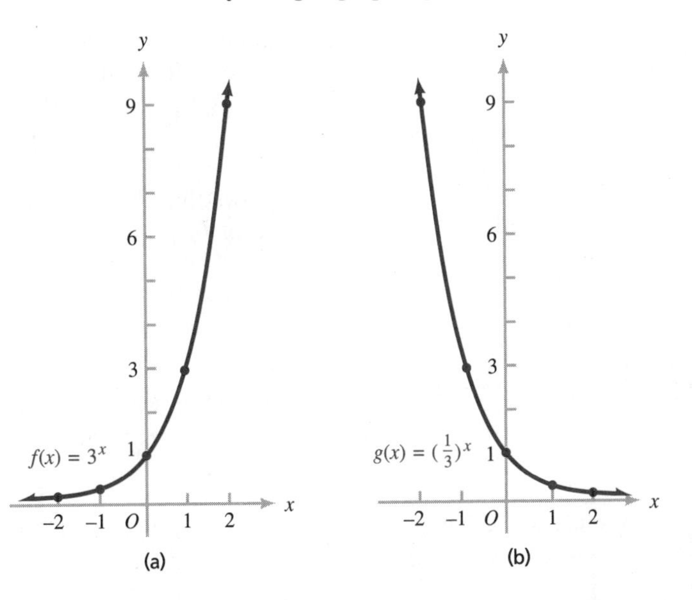

In general, graphs of exponential functions have shapes similar to the graphs in Figure 2. Thus, if $b > 1$, the graph of $f(x) = b^x$ is rising to the right (Figure 3a), while if $0 < b < 1$, the graph is falling to the right (Figure 3b). Of course, when $b = 1$, the graph is neither rising nor falling (Figure 3c). Notice that the graph of $f(x) = b^x$ always contains the point $(1, b)$ (because $b^1 = b$) and that its y intercept is always 1 (because $b^0 = 1$).

If $b > 0$, the domain of the exponential function $f(x) = b^x$ is $\mathbb{R}$. If $b > 1$ (Figure 3a), the graph of $f(x) = b^x$ comes as close to the x axis as we please if we move

Figure 3

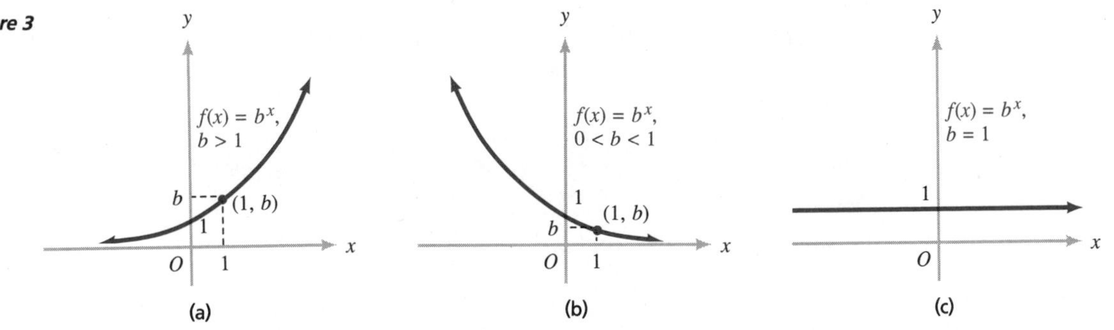

far enough to the left of the origin, but the curve never reaches the axis; in other words, the x axis is a horizontal asymptote. There is no vertical asymptote. As we move farther and farther to the right of the origin, the graph climbs higher and higher without bound; hence, the range of $f(x) = b^x$ is the interval $(0, \infty)$. Similar remarks apply to the graph of $f(x) = b^x$ for $0 < b < 1$ (Figure 3b).

Example 2 Sketch the graphs of $f(x) = 2^x$ and $F(x) = (\frac{1}{2})^x$ on the same coordinate system.

Solution The graph of $f(x) = 2^x$ has already been sketched in Figure 1b. Notice that this graph is similar to the graph of $g(x) = 3^x$ in Figure 2a, except that it does not rise as rapidly. Because

$$F(x) = (\tfrac{1}{2})^x = 2^{-x} = f(-x),$$

it follows that the graph of $F(x) = (\frac{1}{2})^x$ is obtained by reflecting the graph of $f(x) = 2^x$ across the y axis (Figure 4). In both cases, the x axis is a horizontal asymptote and the y intercept is 1. Accurate graphs can be sketched by using a calculator to plot more points or by using a graphing calculator. ■

Figure 4

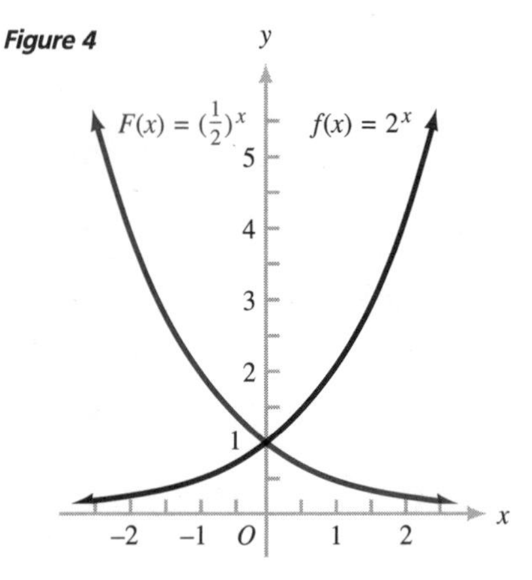

In dealing with exponential functions, a calculator with a y^x or x^y key is a most useful tool.

Example 3 Using a calculator with a y^x or x^y key, evaluate:

(a) $\sqrt{2}^{\sqrt{3}}$ **(b)** $\pi^{-\sqrt{2}}$

Solution On a 10-digit calculator, we obtain:

(a) $\sqrt{2}^{\sqrt{3}} \approx 1.414213562^{1.732050808} \approx 1.822634655$

In using a calculator, *it isn't necessary to write down the intermediate results*, such as 1.414213562 and 1.732050808. We do this only so you can check your calculator work.

(b) $\pi^{-\sqrt{2}} \approx 3.141592654^{-1.414213562} \approx 0.198117987$ ■

The properties of rational exponents (page 61) continue to hold for all real exponents, provided that all bases are positive. For instance, if a and b are positive, we have

$$a^x a^y = a^{x+y}, \qquad (a^x)^y = a^{xy}, \qquad \text{and} \qquad (ab)^x = a^x b^x$$

for all real values of x and y.

c Example 4 Using a calculator, verify that $\pi^{\sqrt{2}}\pi^{\sqrt{3}} = \pi^{\sqrt{2}+\sqrt{3}}$.

Solution On a 10-digit calculator, we obtain*

$$\pi^{\sqrt{2}}\pi^{\sqrt{3}} \approx (5.047497267)(7.262545035) \approx 36.65767622$$

and

$$\pi^{\sqrt{2}+\sqrt{3}} \approx 3.141592654^{3.146264370} \approx 36.65767622.$$

Therefore the equation $\pi^{\sqrt{2}}\pi^{\sqrt{3}} = \pi^{\sqrt{2}+\sqrt{3}}$ is confirmed to within the limits of accuracy of our calculator. ■

The techniques of graph sketching presented in Section 3.5 and illustrated throughout Chapter 4 can be applied to exponential functions.

In Examples 5 and 6, sketch the graph of the given function; determine its domain, its range, and any horizontal or vertical asymptotes; and indicate where the function is increasing or decreasing. c *Use a calculator if you wish to increase the accuracy of your sketch.* gc *Alternatively, use a graphing calculator.*

Example 5 $h(x) = 3^{x+2}$

Solution The graph of $g(x) = 3^x$ has already been sketched in Figure 2a. Since $g(x + 2) = 3^{x+2}$, we have

$$h(x) = g(x + 2).$$

Figure 5

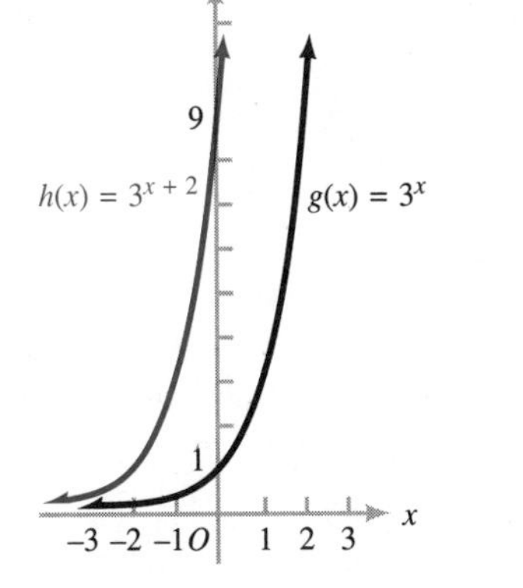

Therefore, the graph of h is obtained by shifting the graph of g two units to the left. Because

$$\begin{aligned} h(x) &= 3^{x+2} \\ &= 3^x \cdot 3^2 \\ &= 9 \cdot 3^x = 9g(x), \end{aligned}$$

the graph of h can also be obtained from the graph of g by multiplying each ordinate by 9. Using this information, and plotting a few points with the aid of a calculator, we obtain the graph shown in Figure 5. Evidently, the domain of h is $\mathbb{R}$, the range is the interval $(0, \infty)$, the x axis is a horizontal asymptote, there is no vertical asymptote, and h is increasing throughout its domain. ■

*Again, it isn't necessary to write down the intermediate results. If you are using your calculator efficiently, you should be able to make this calculation directly with only a few key strokes.

Example 6 $k(x) = -2^x + 1$

Solution The graph of $y = -2^x$ is obtained by reflecting the graph of $y = 2^x$ (Figure 1b) across the x axis (Figure 6), and the graph of $k(x) = -2^x + 1$ is obtained by shifting the graph of $y = -2^x$ one unit upward (Figure 6). Evidently, the domain of k is $\mathbb{R}$, the range is the interval $(-\infty, 1)$, the line $y = 1$ is a horizontal asymptote, there is no vertical asymptote, and k is decreasing throughout its domain. ■

Figure 6

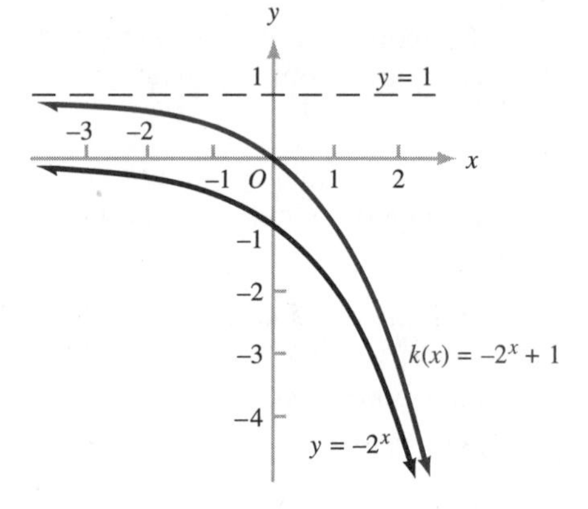

The Mathematics of Finance

The following examples illustrate the use of exponential functions in the world of business, investment, and finance. In such examples, we ordinarily round off our answers to the nearest cent.

C **Example 7** Bankers use the **compound interest** formula*

$$S = P\left(1 + \frac{r}{n}\right)^{nt}$$

for the **final value** S dollars of a **principal** P dollars invested for a **term** of t years at a **nominal annual interest rate** r **compounded** n **times per year.** If you invest $P = \$500$ at a nominal annual interest rate of 8% (that is, $r = 0.08$) compounded quarterly ($n = 4$), what is the final value S of your investment after a term of $t = 3$ years?

Solution Using a calculator, we find that

$$S = P\left(1 + \frac{r}{n}\right)^{nt} = 500\left(1 + \frac{0.08}{4}\right)^{4(3)} = 500(1.02)^{12} \approx \$634.12.$$ ■

* This formula is derived in a later section on mathematical induction.

Example 8 When a bank offers compound interest, it usually specifies not only the nominal annual interest rate r but also the **effective simple annual interest rate** R, that is, the rate of simple annual interest that would yield the same final value as the compound interest over a 1-year term. The formula

$$R = \left(1 + \frac{r}{n}\right)^n - 1$$

is used to calculate R in terms of r and n. Find the effective simple annual interest rate R corresponding to a nominal annual interest rate of 12% (that is, $r = 0.12$) compounded semiannually ($n = 2$).

Solution

$$R = \left(1 + \frac{r}{n}\right)^n - 1 = \left(1 + \frac{0.12}{2}\right)^2 - 1 = (1.06)^2 - 1 = 0.1236;$$

in other words, the effective simple annual interest rate is 12.36%. ■

Money that you will receive in the future is worth *less* to you than the same amount of money received now, because you miss out on the interest you could collect by investing the money now. For this reason, we use the idea of the *present value* of money to be received in the future. If you have an opportunity to invest P dollars at a nominal annual interest rate r compounded n times a year, this principal plus the interest it earns will amount to S dollars after t years, as given by the compound interest formula

$$S = P\left(1 + \frac{r}{n}\right)^{nt}.$$

Thus, P dollars in hand *right now* is worth S dollars to be received t years *in the future*. Solving the equation above for P in terms of S, we get an equation for the **present value** of an offer of S dollars to be received t years in the future:

$$P = S\left(1 + \frac{r}{n}\right)^{-nt}.$$

You could do just as well by investing P dollars now and collecting S dollars from your investment after t years.

The idea of present value helps people to make intelligent investment choices. It allows the investor to translate various complex arrangements for future payments into single figures that are easy to compare.

C Example 9 Find the present value of \$500 to be paid to you 2 years in the future, if investments during this period are earning a nominal annual interest rate of 10% compounded monthly.

Solution

Here $S = 500$, $r = 0.10$, $n = 12$, $t = 2$, and

$$P = S\left(1 + \frac{r}{n}\right)^{-nt} = 500\left(1 + \frac{0.10}{12}\right)^{-12(2)}$$

$$= 500\left(1 + \frac{0.10}{12}\right)^{-24} \approx \$409.70.$$

■

An interest-bearing debt, such as a car loan or a home mortgage, is said to be **amortized** if the principal P dollars and the interest I dollars are paid over a term of t years by regular successive payments of p dollars every $\frac{1}{n}$th of a year. For car loans and home mortgages, payments are ordinarily made monthly, so that $n = 12$. The **amortization formulas** are

$$p = \frac{Pr}{n\left[1 - \left(1 + \frac{r}{n}\right)^{-nt}\right]} \quad \text{and} \quad I = npt - P.$$

Example 10

Suppose that, after negotiating with a dealer, you agree to pay \$6749 for a used car. You will give the dealer a down payment of \$749, and the dealer will finance the remaining \$6000 at a nominal annual interest rate of 11.8% over a term of 3 years. Find **(a)** your monthly payment and **(b)** the total amount of interest that you will pay the dealer.

Solution

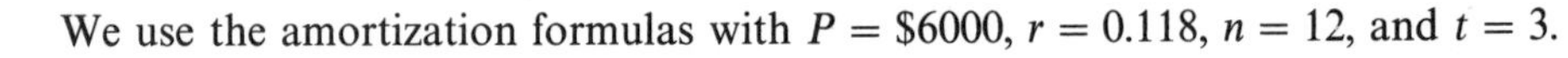

We use the amortization formulas with $P = \$6000$, $r = 0.118$, $n = 12$, and $t = 3$.

(a) Thus, your monthly payment will be

$$p = \frac{Pr}{n\left[1 - \left(1 + \frac{r}{n}\right)^{-nt}\right]} = \frac{(6000)(0.118)}{12\left[1 - \left(1 + \frac{0.118}{12}\right)^{-12(3)}\right]} \approx \$198.71.$$

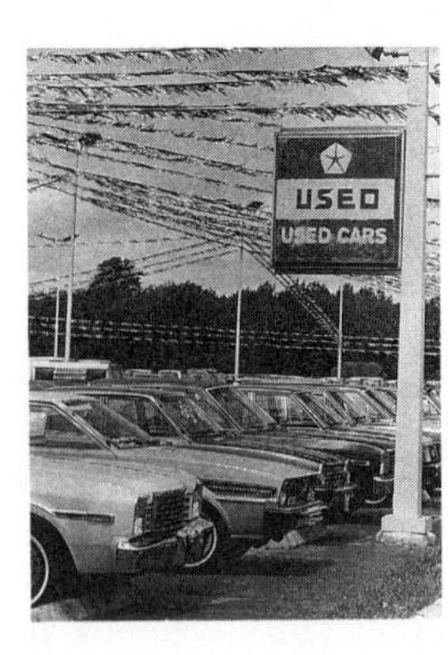

(b) The total interest that you will pay to the dealer will be

$$I = npt - P = 12(198.71)(3) - 6000 = \$1153.56.$$

■

The amortization formulas with $n = 12$ apply if you borrow P dollars from a bank for a home mortgage at a nominal annual interest rate r payable in successive equal monthly payments of p dollars each over a term of t years. The balance of the principal owed to the bank at the beginning of the kth month is customarily considered to be

$$P_k = P\frac{\left(1 + \frac{r}{12}\right)^{12t} - \left(1 + \frac{r}{12}\right)^{k-1}}{\left(1 + \frac{r}{12}\right)^{12t} - 1}.$$

[C] Example 11 Suppose that you purchase a condominium for \$65,000, paying \$15,000 down and taking out a 30-year mortgage on the remaining \$50,000 at a nominal annual interest rate of 11%.

(a) What is your monthly payment on the mortgage?

(b) After 15 years, how much of the original \$50,000 will you have paid off?

Solution **(a)** Your monthly payment will be

$$p = \frac{Pr}{n\left[1 - \left(1 + \frac{r}{n}\right)^{-nt}\right]} = \frac{(50{,}000)(0.11)}{12\left[1 - \left(1 + \frac{0.11}{12}\right)^{-12(30)}\right]} \approx \$476.16.$$

(b) After 15 years you will have made $15(12) = 180$ payments, and your next payment will be payment number $k = 181$. The balance of the principal owed to the bank at the beginning of the 181st month is

$$P_{181} = P\frac{\left(1 + \frac{r}{12}\right)^{12t} - \left(1 + \frac{r}{12}\right)^{k-1}}{\left(1 + \frac{r}{12}\right)^{12t} - 1}$$

$$= (50{,}000)\frac{\left(1 + \frac{0.11}{12}\right)^{12(30)} - \left(1 + \frac{0.11}{12}\right)^{181-1}}{\left(1 + \frac{0.11}{12}\right)^{12(30)} - 1}$$

$$\approx \$41{,}893.63.$$

Therefore, after 15 years you will have paid off

$$\$50{,}000 - \$41{,}893.63 = \$8106.37.$$

■

Problem Set 5.1

In each problem set, problems with colored numbers constitute a good representation of the main ideas of the section. Note that some of the even-numbered problems may be considerably more challenging than the odd-numbered ones.

1. By finding values of 4^x for $x = -2, -\frac{3}{2}, -1, -\frac{1}{2}, 0, \frac{1}{2}, 1, \frac{3}{2}$, and 2, plotting the resulting points $(x, 4^x)$, and drawing a smooth curve through these points, sketch the graph of $f(x) = 4^x$.

2. **(a)** Using the graph obtained in Problem 1 and approximating $\sqrt{2}$ as 1.4, find the approximate value of $4^{\sqrt{2}}$.

   [C] **(b)** Using a calculator with a y^x or x^y key, find the value of $4^{\sqrt{2}}$ to as many decimal places as you can.

3. Sketch the graph of $F(x) = (\frac{1}{4})^x$.

4. Sketch graphs of $g(x) = 5^x$ and $G(x) = (\frac{1}{5})^x$ on the same coordinate system.

C **5.** Use a calculator with a y^x or x^y key to find the value of each quantity to as many significant digits as you can.

(a) $2^{\sqrt{2}}$ **(b)** $2^{-\sqrt{2}}$ **(c)** 2^{π} **(d)** $2^{-\pi} - \pi^{-2}$
(e) $\sqrt{2}^{\sqrt{2}}$ **(f)** π^{π} **(g)** $\sqrt{3}^{-\sqrt{5}}$

6. Use the properties of exponents to simplify the expression $(2^x + 2^{-x})^2 - (2^x - 2^{-x})^2$.

C In Problems 7 and 8, use a calculator to verify each equation for the indicated values of the variables.

7. **(a)** $a^x a^y = a^{x+y}$ for $a = 3.074$, $x = 2.183$, $y = 1.075$
(b) $a^{x+y} = a^x a^y$ for $a = 2.471$, $x = 5.507$, $y = 0.012$
(c) $(a^x)^y = a^{xy}$ for $a = 1.777$, $x = -2.058$, $y = 3.333$

8. **(a)** $a^{x-y} = \dfrac{a^x}{a^y}$ for $a = \sqrt{2}$, $x = \sqrt{5}$, $y = \sqrt{3}$
(b) $(ab)^x = a^x b^x$ for $a = \sqrt{7}$, $b = \pi$, $x = \sqrt{\pi}$
(c) $\left(\dfrac{a}{b}\right)^x = \dfrac{a^x}{b^x}$ for $a = 2 + \pi$, $b = \sqrt{2} - 1$, $x = \sqrt{5} - \sqrt{3}$

In Problems 9 to 20, sketch the graph of the given function; determine its domain, its range, and any horizontal or vertical asymptotes; and indicate whether the function is increasing or decreasing. C Use a calculator if you wish. gc Alternatively, use a graphing calculator.

9. $f(x) = 2^x + 1$
10. $g(x) = (\frac{2}{3})^x$
11. $h(x) = 4^x - 1$
12. $F(x) = (0.2)^x$
13. $G(x) = 3 \cdot 2^x$
14. $H(x) = 3 \cdot 2^{-x}$
15. $f(x) = 2^{-x} - 3$
16. $g(x) = \frac{3}{2}(\frac{1}{3})^{-x}$
17. $g(x) = 2^{x-3} + 3$
18. $H(x) = (0.3)^x - 1$
19. $k(x) = (0.5)^x + 2$
20. $K(x) = 3^x - 2$

C In Problems 21 and 22, assume that you have invested a principal P dollars at a nominal annual interest rate r compounded n times per year for a term of t years. Calculate the final value S of your investment and the effective simple annual interest rate R. As usual, round off your answer to the nearest cent.

21. **(a)** $P = \$1000$, $r = 0.07$ (7%), $n = 1$, $t = 13$ years
(b) $P = \$1000$, $r = 0.07$ (7%), $n = 12$, $t = 13$ years
(c) $P = \$1000$, $r = 0.12$ (12%), $n = 1$, $t = 13$ years

22. **(a)** $P = \$1000$, $r = 0.12$ (12%), $n = 52$, $t = 13$ years
(b) $P = \$50{,}000$, $r = 0.135$ (13.5%), $n = 12$, $t = \frac{1}{2}$ year
(c) $P = \$25{,}000$, $r = 0.155$ (15.5%), $n = 52$, $t = \frac{1}{4}$ year

C **23.** Suppose that a bank offers to pay a nominal annual interest rate of 0.08 (8%) on money left on deposit for 2 years. Assume that a principal $P = \$1000$ is deposited. Find the final value S after the 2-year term if the interest is compounded **(a)** annually, **(b)** semiannually, **(c)** quarterly, **(d)** monthly, **(e)** weekly, **(f)** daily, and **(g)** hourly.

C **24.** Find out the nominal annual interest rate r offered by your local savings bank for regular savings accounts and the number of times n per year that the interest is compounded. Calculate the effective simple annual interest rate R.

C **25.** Find the present value of \$1000 five years in the future at a nominal annual rate of 12% ($r = 0.12$) compounded weekly.

C **26.** A fund compounds interest quarterly. If a principal $P = \$14{,}000$ yields a final value $S = \$45{,}510$ after a 5-year term, **(a)** find the nominal annual interest rate r, and **(b)** find the effective simple annual interest rate R.

C **27.** Suppose that someone owes you money and your local savings bank offers savings accounts at 8% nominal annual interest compounded monthly. Use this interest rate to determine the present value to you of money offered in the future. If your debtor offers to pay you \$100 six months from now, what is the present value to you of this offer?

C **28.** On Leroy's 16th birthday, his father promises to give him \$25,000 when he turns 21 to help set him up in business. Local banks are offering savings accounts at a nominal annual interest rate of 8% compounded quarterly. Leroy, who has studied the mathematics of finance, says, "Dad, I'll settle for ________ dollars right now!" Fill in the blank appropriately.

C In Problems 29 and 30, suppose that you are planning to borrow P dollars from your local bank to finance a new car. Assume that the bank charges a nominal annual interest rate r on new car loans with monthly payments for a term of $t = 3$ years. Find **(a)** the monthly payment and **(b)** the total interest that you will pay for the loan.

29. $P = \$9995.00$, $r = 0.118$ (11.8%)

30. $P = \$8400.00$, $r = 0.105$ (10.5%)

C In Problems 31 and 32, suppose that you are planning to obtain a home mortgage at a nominal annual interest rate r payable in successive equal monthly payments over a term of t years. Assume that the total cost of the home is C dollars and that you will pay a cash down payment of 20% of C. Find **(a)** the monthly payment, **(b)** the amount of the original mortgage you will have paid off after T years, and **(c)** the total amount of interest you will pay to the bank over the term of the mortgage.

31. $C = \$90{,}000$, $r = 0.11$ (11%), $t = 15$ years, $T = 5$ years

32. $C = \$120{,}000$, $r = 0.105$ (10.5%), $t = 30$ years, $T = 10$ years

C **33.** Many state lotteries pay the winner's prize in the form of an initial payment of q dollars followed by N annual payments of p dollars each for a total of $q + Np$ dollars. If investments are earning a nominal annual interest rate r compounded n times per year, then the present value of the prize to the winner upon receipt of the initial payment is actually $p = q + pb\dfrac{1 - b^N}{1 - b}$ dollars, where $b = \left(1 + \dfrac{r}{n}\right)^{-n}$. A certain state advertises a \$1,000,000 lottery prize in the form of an initial payment of \$50,000 followed by 19 more annual payments of \$50,000 each. Assuming interest rates of 9% compounded monthly, what is the actual present value of the alleged \$1,000,000 prize?

C **34.** In making used car loans, banks *depreciate* the value of a car according to its age. One popular depreciation formula is $B = C(S/C)^{t/n}$, for $0 \le t \le n$, where B dollars is the *book value* of the car when it is t years old, C dollars is the value of the car when it was new, n years is the *useful life* of the car, and S dollars is the *salvage value* of the car after n years. Suppose that a new car costing \$9575 will have a salvage value of \$500 at the end of its useful life of 8 years.

(a) Sketch a graph of B as a function of t.

(b) Find the book value of the car when it is 5 years old.

C **35.** Because of inflation, the value of a home often *increases* with time. State and local agencies frequently use the formula $S = C(1 + r)^n$ to assess the value S dollars of a home for property tax purposes, where C dollars is the cost of the home when new, n years is the age of the home, and r is the annual rate of inflation. If the inflation rate is 6%, use this formula to determine the assessed value of a 10-year-old home that originally cost \$55,000.

C **36.** In Example 11, what part of the 181st payment goes toward interest and what portion goes toward reduction of the balance of the principal owed?

C **37.** You have just won first prize in a lottery and you have your choice of the following: **(a)** \$30,000 will be placed in a savings account in your name, and the money will be compounded daily at a nominal annual rate of 10% or **(b)** one penny will be placed in a fund in your name, and the amount in the fund will be doubled every 6 months over the next 12 years. Which plan do you choose and why?

5.2 The Natural Exponential Function

If we increase the base b, the graph of the exponential function $f(x) = b^x$ rises more rapidly. This is illustrated in Figure 1 (page 302) for $b = 2$ and $b = 3$. Since the graphs of all exponential functions contain the point (0, 1), a good indication of how rapidly such a graph rises is its "steepness" at this point. The steepness of the graph of $f(x) = b^x$ at the point (0, 1) can be measured by the slope m of its **tangent line**—that is, the straight line that just grazes the curve at this point (Figure 2, page 302). Although the tangent line is easily sketched by eye, its precise determination requires the use of calculus.

As we increase the base b of the exponential function $f(x) = b^x$, the graph of f in Figure 2 becomes steeper at the point (0, 1) and the slope m of the tangent line increases. If you sketch accurate graphs for the values of b in the following table,

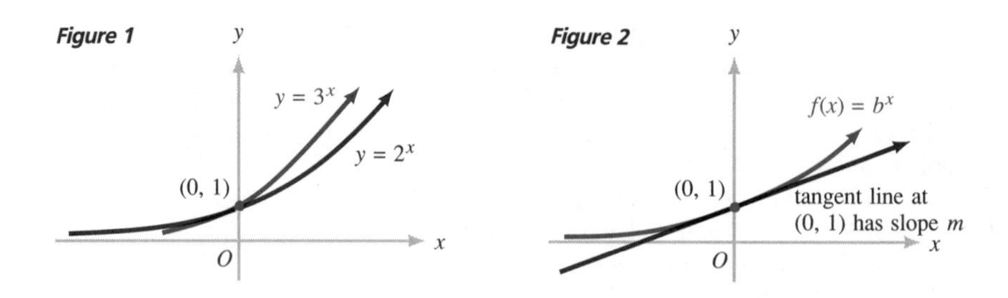

draw the tangent lines at (0, 1) by eye, and measure their slopes m, you will obtain approximately the values shown in the table below (Problem 1). Notice that m is less than 1 when $b = 2$, and that m is greater than 1 when $b = 3$.

Base b	0.5	1	2	3	4
Slope m	-0.7	0	0.7	1.1	1.4

The Swiss honor Leonhard Euler by picturing him on their 10-franc note.

As you might suspect, somewhere between 2 and 3 there is a value of the base b for which the slope m of the tangent line to the graph of $y = b^x$ at the point (0, 1) is exactly 1. This particular value of the base is denoted by e, in honor of the great Swiss mathematician Leonhard Euler (1707–1783)(pronounced "oiler"), who was one of the first to recognize its immense importance. Like π, the value of e is an irrational number. By using advanced mathematical methods and high-speed computers, the numerical value of e has been calculated to thousands of decimal places. Rounded off to three decimal places,

$$e \approx 2.718.$$

C Example 1 Sketch the graph of the exponential function

$$f(x) = e^x.$$

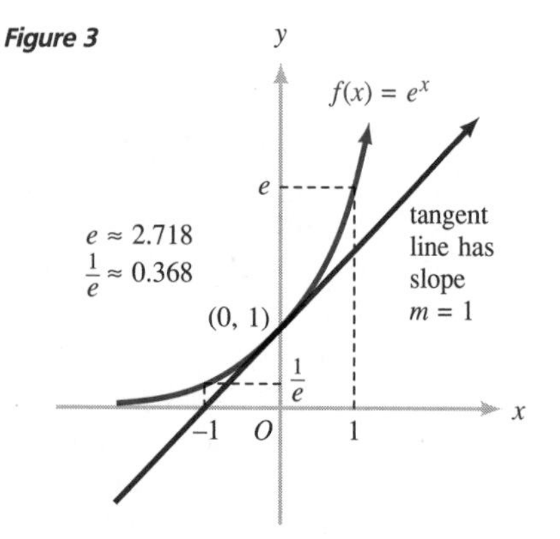

Solution The graph of the exponential function $f(x) = e^x$ rises at just the right rate so that the tangent line at (0, 1) has slope $m = 1$. Using this fact, recalling the general shape of graphs of exponential functions with bases greater than 1, and plotting points corresponding to

$$f(1) = e^1 = e \approx 2.718 \quad \text{and} \quad f(-1) = e^{-1} = \frac{1}{e} \approx \frac{1}{2.718} \approx 0.368,$$

we can sketch the graph of $f(x) = e^x$ (Figure 3). As usual, accuracy is enhanced if a calculator is used to plot additional points (Problem 2) or if a graphing calculator is used. ■

The exponential function with base e, $f(x) = e^x$, is called the **natural exponential function.** This function plays an important role in calculus, and it is essential in the applications of mathematics to many fields ranging from engineering to public health. You will find a variety of these applications in the remainder of this chapter. Indeed, the function $f(x) = e^x$ is used so often that people simply call it the **exponential function.** Whenever anyone uses this term without specifying the base, you can be certain that the function $f(x) = e^x$ is intended. On some scientific calculators,* and in many standard computer languages (such as BASIC), the natural exponential function is denoted by exp (or by EXP). Thus,

$$\exp(x) = e^x.$$

When you deal with algebraic expressions that involve the natural exponential function, keep in mind that e^x *is always positive* (Figure 3).

Example 2 Find the zeros of $f(x) = xe^x - e^x$.

Solution Factoring, we have

$$f(x) = (x - 1)e^x.$$

Since e^x cannot be zero, the only way that $f(x)$ can be equal to zero is to have

$$x - 1 = 0.$$

Therefore, $x = 1$ is the only zero of $f(x)$. ■

C **Example 3** Using a scientific calculator, evaluate:

(a) e^1

(b) $e^{\sqrt{2}}$

(c) $e^{-5.0321}$

Solution Using a 10-digit calculator, we find that:

(a) $e^1 = e \approx 2.718281828$

(b) $e^{\sqrt{2}} \approx e^{1.414213562} \approx 4.113250379$

(c) $e^{-5.0321} \approx 0.006525093476 = 6.525093476 \times 10^{-3}$ ■

Example 4 Sketch graphs of the given functions on the same coordinate system.

(a) $f(x) = e^{x-2}$ **(b)** $g(x) = -e^{x-2}$ **(c)** $h(x) = 1 - e^{x-2}$

* On certain calculators, you must press INV and then LN (inverse natural logarithm) for the exponential function. Also, on some calculators the EXP key is used to enter powers of 10. Check the instruction manual furnished with your particular calculator.

Figure 4

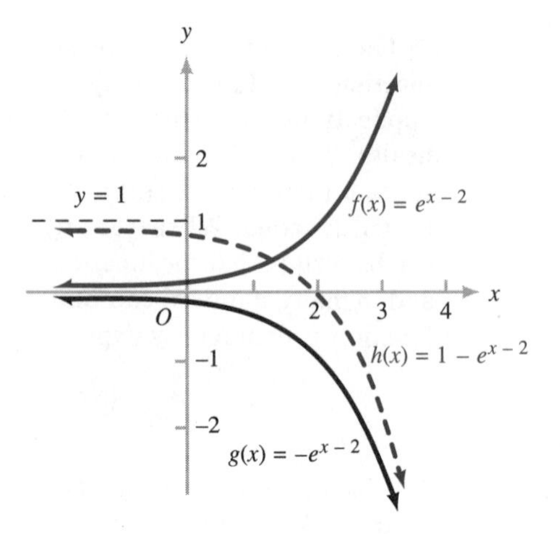

Solution

(a) The graph of $f(x) = e^{x-2}$ is obtained by shifting the graph of $y = e^x$ in Figure 3 exactly 2 units to the right (Figure 4).

(b) The graph of g is obtained by reflecting the graph of f in part (a) about the x axis (Figure 4).

(c) The graph of h is obtained by shifting the graph of g in part (b) exactly 1 unit upward (Figure 4). ■

Figure 5

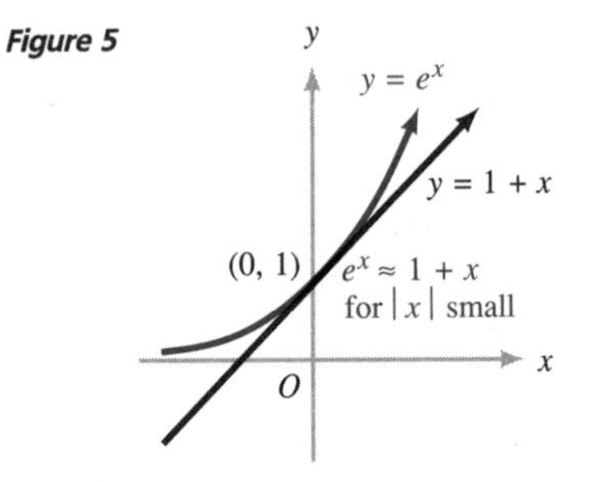

Because the tangent line to the graph of $y = e^x$ at the point (0, 1) has slope $m = 1$, the slope-intercept equation of the tangent line is $y = 1 + x$ (Figure 5). Obviously, the curve $y = e^x$ and the tangent line $y = 1 + x$ are very close together near the point (0, 1); hence, *for small values of* $|x|$, we have

$$e^x \approx 1 + x,$$

and this approximation becomes more and more accurate as $|x|$ gets smaller and smaller.

C Example 5 How accurate is the approximation $e^x \approx 1 + x$ if $x = 0.01$?

Solution

Using a calculator, we find that, rounded to nine decimal places,

$$e^{0.01} \approx 1.010050167.$$

Since

$$1 + 0.01 = 1.010000000,$$

the discrepancy in the approximation $e^{0.01} \approx 1 + 0.01$ first occurs in the fifth decimal place. ■

Some banks offer savings accounts with interest compounded not quarterly, not weekly, not daily, not hourly, but *continuously*. The formula for continuously compounded interest involves the exponential function. Although the derivation of this formula requires methods studied in calculus, we can derive it informally as follows. We begin with the formula

$$S = P\left(1 + \frac{r}{n}\right)^{nt}$$

for the final value S dollars of a principal P dollars invested for t years at a nominal annual interest rate r compounded n times a year. We're interested in what happens as n gets larger and larger. Let $x = r/n$ and notice that the larger n is, the smaller x is. Using the approximation $e^x \approx 1 + x$ for small x and the fact that $xn = r$, we have

$$S = P\left(1 + \frac{r}{n}\right)^{nt} = P(1 + x)^{nt} \approx P(e^x)^{nt} = Pe^{xnt} = Pe^{rt}.$$

As n becomes larger and larger, $x = r/n$ becomes smaller and smaller, and the approximation

$$S \approx Pe^{rt}$$

becomes more and more accurate. Therefore, for **continuously compounded** interest at a nominal annual rate r, bankers use the formula

$$S = Pe^{rt}$$

for the final value S dollars of a principal P dollars invested for a term of t years.

C Example 6 The New Mattoon Savings Bank offers a savings account with continuously compounded interest at a nominal annual rate of 7% (that is, $r = 0.07$).

(a) Sketch a graph showing the amount of money S dollars in such an account after t years, $0 \le t \le 20$, if a principal $P = \$100$ is deposited when $t = 0$.

(b) What is the final value S dollars of an investment of $P = \$100$ for a term of $t = 20$ years?

Figure 6

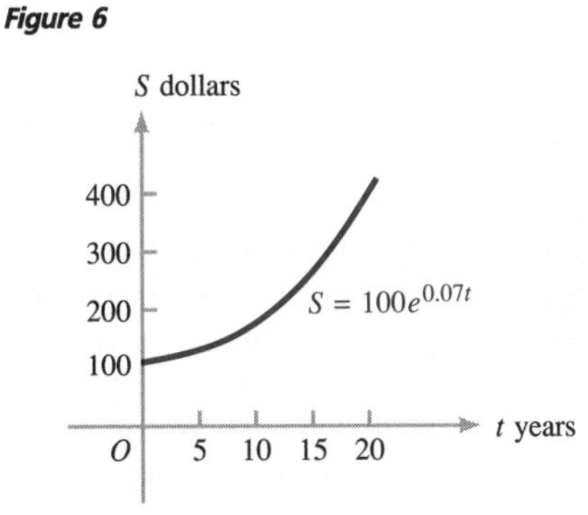

Solution

(a) Here $r = 0.07$, $P = 100$, and $S = Pe^{rt} = 100e^{0.07t}$. The graph is sketched in Figure 6.

(b) When $t = 20$ years,

$$S = 100e^{0.07(20)} = 100e^{1.4} \approx \$405.52.$$

■

As we have seen, for large values of n,

$$P\left(1 + \frac{r}{n}\right)^{nt} \approx Pe^{rt}.$$

Substituting $P = 1$, $r = 1$, and $t = 1$ in this formula, we find that

$$\left(1 + \frac{1}{n}\right)^n \approx e,$$

with *the approximation becoming better and better as n grows larger and larger*. This fact is so important in the study of calculus that it is sometimes taken as the very *definition* of the number e. The following table shows some values of $\left(1 + \frac{1}{n}\right)^n$ for successively larger values of n.

n	100	10,000	1,000,000	100,000,000
$\left(1 + \frac{1}{n}\right)^n$	2.704813829	2.718145927	2.718280469	2.718281815

The correct value of e, rounded off to 12 decimal places, is 2.718281828459; so, for $n = 100{,}000{,}000$, the first discrepancy in the approximation $\left(1 + \frac{1}{n}\right)^n \approx e$ occurs in the eighth decimal place.

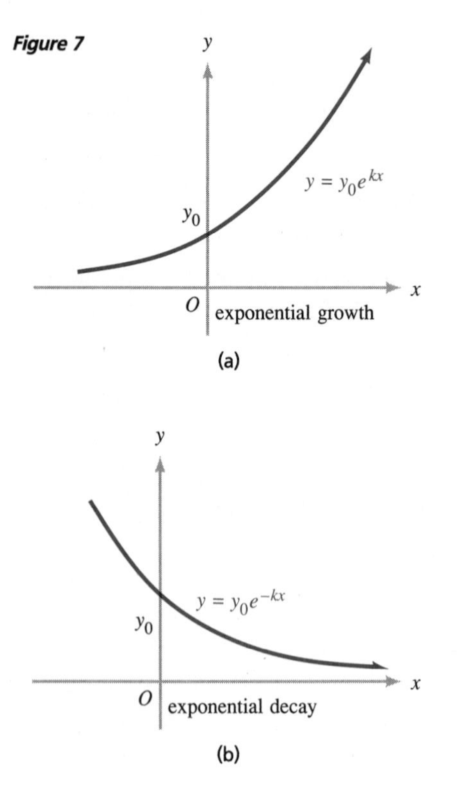

Figure 7

Exponential Growth and Decay

If x and y are variable quantities, we say that y **increases** or **grows exponentially** as a function of x if there are positive constants y_0 and k such that

$$y = y_0e^{kx}.$$

Similarly, if

$$y = y_0e^{-kx},$$

we say that y **decreases** or **decays exponentially** as a function of x. Graphs of y as a function of x for exponential growth and exponential decay are shown in Figure 7. Notice that y_0 is the y intercept in both of the graphs; that is, y_0 is the value of y when $x = 0$. (Why?) The constant k, which determines how rapidly the growth or decay takes place, is called the **growth constant** or the **decay constant.**

As we have seen, S dollars in a savings account with continuously compounded interest grows exponentially as a function of time. Since $S = Pe^{rt}$, the growth constant is equal to the nominal annual interest rate r. On the other hand, radioactive materials provide a good example of exponential decay.

C Example 7 Polonium, a radioactive element discovered by Marie Curie in 1898 and named after her native country Poland, decays exponentially. If y_0 grams of polonium are initially present, the number of grams y present after t days is given by

$$y = y_0 e^{-0.005t}.$$

(a) If $y_0 = 5$ grams, sketch a graph showing the amount y grams of polonium left after t days for $0 \le t \le 730$.

(b) Of a 5-gram sample of polonium, how much is left after 2 years (730 days)?

Solution **(a)** The graph of $y = 5e^{-0.005t}$ for $0 \le t \le 730$ is sketched in Figure 8.

(b) When $t = 730$, we have

$$y = 5e^{-(0.005)(730)} = 5e^{-3.65} \approx 0.13 \text{ gram.}$$ ■

Marie Curie is shown here receiving the Nobel Prize.

Figure 8

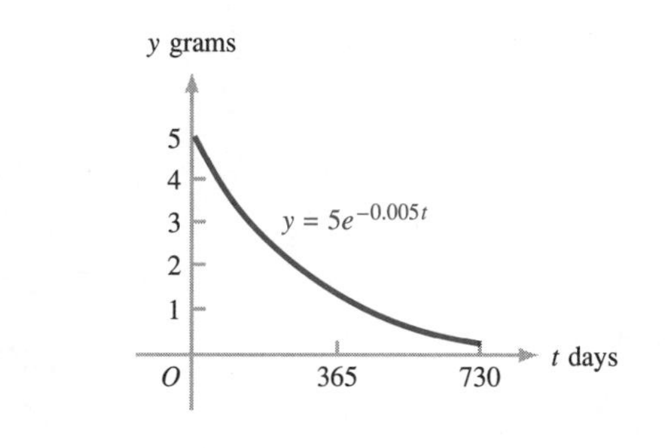

Problem Set 5.2

C **1.** Using a calculator, sketch accurate graphs of $y = b^x$ for the values of b in the table on page 302. Draw tangent lines to these graphs by eye at the point (0, 1), measure the slopes of the tangent lines, and compare your slopes to the entries in the table. gc Use a graphing calculator if you wish.

C **2.** Using a calculator and a sheet of graph paper (available at your college bookstore), sketch the graph of the exponential function $\exp(x) = e^x$ as accurately as you can for $-2 \le x \le 2$.

In Problems 3 to 10, use the graph of $y = e^x$ (Figure 3) and the techniques of shifting, stretching, and reflecting to sketch the graph of each function. C Use a calculator if you wish. gc Alternatively, use a graphing calculator.

3. $f(x) = e^x + 1$ **4.** $g(x) = e^{x+1}$

5. $h(x) = -e^x$ **6.** $F(x) = -3e^{x+1}$

7. $G(x) = e^{-x}$ **8.** $H(x) = e^{1-x}$

9. $f(x) = e^{-x} + 1$ **10.** $g(x) = 2e^{1-x} + 2$

C In Problems 11 and 12, use a calculator to verify each equation for the indicated values of the variables.

11. **(a)** $e^x e^y = e^{x+y}$ for $x = \sqrt{2}, y = \sqrt{3}$

(b) $e^{x+y} = e^x e^y$ for $x = \sqrt{5}, y = -\pi$

12. **(a)** $(e^x)^y = e^{xy}$ for $x = \frac{\pi}{2}$, $y = 1 - \sqrt{3}$

(b) $e^{x-y} = \frac{e^x}{e^y}$ for $x = 3.9$, $y = 2.5$

In Problems 13 and 14, find the zeros of each function.

13. $f(x) = 2xe^{-x} - x^2e^{-x}$ [*Hint:* e^{-x} is always positive.]

14. $h(x) = e^{2x} - 2e^x + 1$[*Hint:* $e^{2x} = (e^x)^2$.]

C 15. How accurate is the approximation $e^x \approx 1 + x$ if:

(a) $x = 0.05$ **(b)** $x = 0.1$
(c) $x = 0.5$ **(d)** $x = 1$

16. Simplify each expression.

(a) $\left(\frac{e^x + e^{-x}}{2}\right)^2 - \left(\frac{e^x - e^{-x}}{2}\right)^2$

(b) $\left(\frac{2}{e^x + e^{-x}}\right)^2 + \left(\frac{e^x - e^{-x}}{e^x + e^{-x}}\right)^2$

C 17. Sketch the graph of $h(x) = \frac{2}{e^x - e^{-x}}$. gc Use a graphing calculator if you wish. In more advanced mathematics, h is called the **hyperbolic cosecant** function.

C 18. The function

$$g(x) = \begin{cases} 0 & \text{if } x = 0 \\ e^{-1/x^2} & \text{if } x \neq 0 \end{cases}$$

is useful in advanced mathematics because its graph is very flat near the origin. Sketch the graph of g. gc Use a graphing calculator if you wish.

C 19. In calculus it is shown that, for small values of x, $e^x \approx 1 + x + (x^2/2)$ gives an even better approximation than does $e^x \approx 1 + x$. Sketch graphs of $y = e^x$ and $y = 1 + x + (x^2/2)$ on the same coordinate system. gc Use a graphing calculator if you wish.

20. If $f(x) = 1 + e^x$, show that $\frac{1}{f(x)} + \frac{1}{f(-x)} = 1$.

C 21. When a flexible cord or chain is suspended from its ends, it hangs in a curve called a **catenary** whose equation has the form $y = \frac{a}{2}(e^{x/a} + e^{-x/a})$. Sketch the graph of the catenary for $a = 2$. gc Use a graphing calculator if you wish.

22. The famous Gateway Arch to the West in St. Louis has the shape of an *inverted catenary* (see Problem 21). Find an equation of such an inverted catenary if it is formed by reflecting the catenary $y = \frac{a}{2}(e^{x/a} + e^{-x/a})$ across the x axis and then shifting it vertically so that its apex is on the y axis h units above the origin.

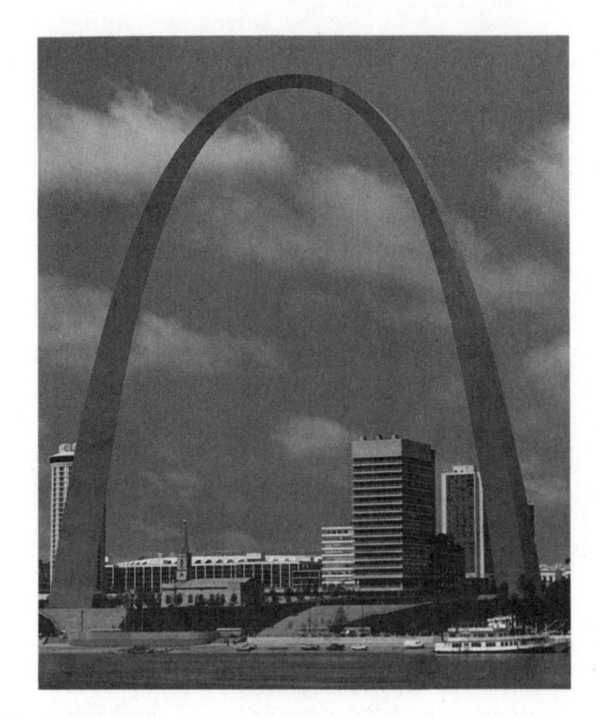

C 23. Suppose that you invest a principal of $P = \$1000$ at a nominal annual interest rate of 10% ($r = 0.1$) for a period of $t = 5$ years. Calculate the final value S of your investment if the interest is compounded **(a)** monthly and **(b)** continuously.

C 24. The concentration C of a drug in a person's circulatory system decreases as the drug is eliminated by the liver and kidneys or absorbed by other organs. Medical researchers often use the mathematical model $C = C_0e^{-kt}$ to predict the concentration C at a time t hours after the drug is administered, where C_0 is the initial concentration when $t = 0$ and k is a constant depending on the type of drug. If $C_0 = 3$ milligrams per liter and $k = 0.173$:

(a) Sketch a graph of C as a function of t for $0 \leq t \leq 4$ hours. gc Use a graphing calculator if you wish.

(b) Find C when $t = 4$ hours.

C **25.** Ecologists have determined that the approximate population N of bears in a certain protected forest area is given by the mathematical model $N = 225e^{0.02t}$, where t is the elapsed time in years since 1987.

(a) Sketch a graph showing the bear population N as a function of t for $0 \le t \le 10$ years. gc Use a graphing calculator if you wish.

(b) Estimate the number of bears that will inhabit the region in the year 2000.

26. If P dollars is invested for $t = 1$ year at a nominal annual interest rate r compounded continuously, the final value S dollars at the end of the year is given by $S = Pe^{r \cdot 1} = Pe^{r}$. Since P dollars invested for $t = 1$ year at a simple annual interest rate R yields a final value $S = P(1 + R)$ dollars, it follows that the effective simple annual interest rate R corresponding to the continuous nominal annual interest rate r satisfies the equation $P(1 + R) = Pe^{r}$.

(a) Solve for R in terms of r.

C **(b)** Find R if $r = 0.07$ (7%).

C **27.** Carbon 14, which is used in archaeology for radioactive dating, decays exponentially according to the equation $y = y_0e^{-0.0001212t}$ where y grams is the amount left after t years and y_0 grams is the initial amount.

(a) If $y_0 = 10$ grams, sketch a graph of y as a function of t for $0 \le t \le 10{,}000$ years. gc Use a graphing calculator if you wish.

(b) Of a 10-gram sample of carbon 14, how much will be left after 10,000 years?

C **28.** The electric current I amperes flowing in a series circuit having an inductance L henrys, a resistance R ohms, and a constant electromotive force E volts (Figure 9) satisfies the equation

$$I = \frac{E}{R} - \frac{E}{R}\exp\left(-\frac{Rt}{L}\right),$$

where t is the time in seconds after the current begins to flow. If $E = 12$ volts, $R = 5$ ohms, and $L = 0.03$ henry, sketch the graph of I as a function of t. gc Use a graphing calculator if you wish.

Figure 9

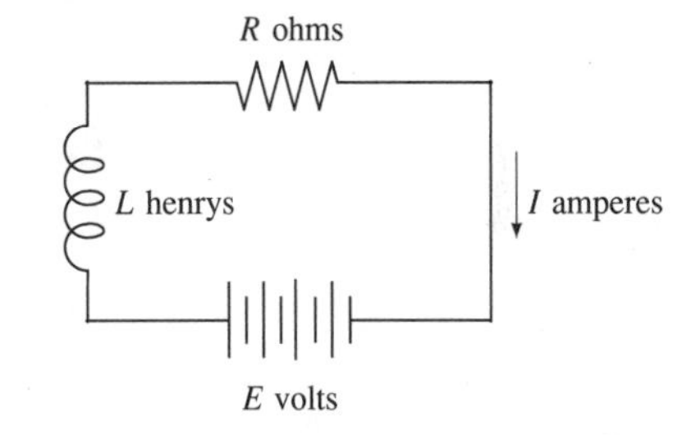

C **29.** The population N of a small country after t years is given by the mathematical model $N = 2{,}000{,}000e^{0.03t}$.

(a) What is the population when $t = 0$?

(b) What is the projected population when $t = 20$ years?

30. Find a formula for the present value P of S dollars t years in the future at a nominal annual interest rate r compounded continuously.

C **31.** A biologist finds that the number N of bacteria in a culture after t hours is given by the mathematical model $N = 2000e^{0.7t}$.

(a) How many bacteria were present when $t = 0$?

(b) How many bacteria will be present after $t = 12$ hours?

(c) Sketch a graph of N as a function of t for $0 \le t \le 6$ hours. gc Use a graphing calculator if you wish.

(d) Using the graph in part (c), estimate the time at which $N = 32{,}000$ bacteria.

C **32.** In statistics, the **normal probability density function** is defined by

$$f(x) = \frac{1}{\sigma\sqrt{2}}\exp\left[-\frac{(x-\mu)^2}{2\sigma^2}\right],$$

where σ and μ are certain constants called the **standard deviation** and the **mean,** respectively, and $\sigma > 0$. When students ask to be graded "on the curve," the curve in question is the graph of f. Sketch the graph of f for $\sigma = 1$ and $\mu = 0$. gc Use a graphing calculator if you wish.

c **33.** What is the percent error in estimating e by $\left(1 + \frac{1}{n}\right)^n$ for $n = 100{,}000$?

c **34.** In calculus it is shown that $\left(1 - \frac{1}{n}\right)^{-n} \approx e$, the approximation becoming better and better as n grows larger and larger. Make a table similar to the one on page 306 showing values of $\left(1 - \frac{1}{n}\right)^{-n}$ for the same values of n that appear in that table.

5.3 Logarithms and Their Properties

If you put \$100 in a savings account at 8% nominal annual interest compounded quarterly, the final value S dollars of your investment after t years is given by

$$S = 100\left(1 + \frac{0.08}{4}\right)^{4t} = 100(1.02)^{4t}.$$

It's natural to ask how long you'll have to wait until your money doubles. To find out, you have to solve the equation

$$200 = 100(1.02)^{4t} \qquad \text{or} \qquad 1.02^{4t} = 2$$

for t. We'll explain later (Example 2 in Section 5.5) how to solve this equation, but you can verify, using a calculator, that the solution is (approximately) $t = 8.75$. In other words, you'll have to wait 8 years and 9 months for your money to double. An equation such as $1.02^{4t} = 2$, in which an unknown appears in an exponent, is called an **exponential equation.**

The simple exponential equations in Example 1 can be solved by using the fact that *if* $b > 0$ *and* $b \neq 1$, *then*

$$b^x = b^y \quad \text{if and only if} \quad x = y$$

(Problem 18).

Example 1 Solve each exponential equation.

(a) $2^x = 64$ **(b)** $36^t = 216^{2t-1}$

Solution **(a)** We begin by expressing 64 as a power of 2 so that both sides of the equation will have the same base. Since $64 = 2^6$, we can rewrite $2^x = 64$ as $2^x = 2^6$, from which it follows that $x = 6$.

(b) Since $36 = 6^2$ and $216 = 6^3$, we can rewrite the given equation as

$$(6^2)^t = (6^3)^{2t-1} \qquad \text{or} \qquad 6^{2t} = 6^{3(2t-1)},$$

from which it follows that

$$2t = 3(2t - 1), \qquad 2t = 6t - 3, \qquad 4t = 3, \qquad \text{and} \qquad t = \tfrac{3}{4}.$$ ■

If $b > 0$, $b \neq 1$, and $c > 0$, the solution x of the exponential equation

$$b^x = c$$

is denoted by

$$x = \log_b c,$$

which is read "x equals the **logarithm to the base b of c**." In other words,

> $\log_b c$ is the power to which you must raise b to obtain c.

For instance, because $10^2 = 100$, it follows that $\log_{10} 100 = 2$. More generally, the fact that $\log_b c$ is the solution of $b^x = c$ is expressed by the equation

$$b^{\log_b c} = c.$$

As a useful memory device for this important identity, notice that *a logarithm is an exponent*.

Example 2 Find:

(a) $\log_2 16$ **(b)** $\log_3 27$ **(c)** $\log_{10} \frac{1}{10}$ **(d)** $\log_e e^5$

Solution **(a)** We ask ourselves, "To what power must we raise 2 to obtain 16?" Since $2^4 = 16$, the answer is 4. Therefore, $\log_2 16 = 4$.

(b) Since $3^3 = 27$, it follows that $\log_3 27 = 3$.

(c) Since $10^{-1} = \frac{1}{10}$, it follows that $\log_{10} \frac{1}{10} = -1$.

(d) We ask ourselves, "To what power must we raise e to obtain e^5?" The answer is 5, so $\log_e e^5 = 5$. ■

To generalize part (d) of Example 2, notice that, if you ask yourself, "To what power must I raise b in order to obtain b^y?", the obvious answer is y; hence,

$$\log_b b^y = y$$

holds for any real number y and any positive base $b \neq 1$. In particular, letting $y = 0$ and noting that $b^0 = 1$, you can see that

$$\log_b 1 = 0.$$

Similarly, letting $y = 1$ and noting that $b^1 = b$, you can see that

$$\log_b b = 1.$$

Using the fact that for $b > 0$, $b \neq 1$, and $c > 0$,

$$x = \log_b c \quad \text{if and only if} \quad b^x = c,$$

you can convert equations from logarithmic form to exponential form, and vice versa. For instance:

Logarithmic Form	Exponential Form
$2 = \log_2 4$	$2^2 = 4$
$\log_{10} 10{,}000 = 4$	$10{,}000 = 10^4$
$-\frac{1}{2} = \log_{64} \frac{1}{8}$	$64^{-1/2} = \frac{1}{8}$
$\log_b x = y$	$b^y = x \quad (b > 0, b \neq 1, x > 0)$
$k = \log_x d$	$x^k = d \quad (x > 0, x \neq 1, d > 0)$

Notice that whenever you write $\log_b c$, you must make sure that c is positive and that b is positive and not equal to 1.

Example 3 Solve each equation.

(a) $\log_3 x^2 = 4$ **(b)** $\log_x 25 = 2$ **(c)** $\log_t(6t + 7) = 2$

Solution **(a)** The equation $\log_3 x^2 = 4$ is equivalent to $3^4 = x^2$; that is, $x^2 = 81$. The solutions are $x = 9$ and $x = -9$.

(b) The equation $\log_x 25 = 2$ is equivalent to $x^2 = 25$, with the restriction that $x > 0$ and $x \neq 1$; hence, $x = 5$ is the solution.

(c) The equation $\log_t(6t + 7) = 2$ is equivalent to the equation $t^2 = 6t + 7$; that is, $t^2 - 6t - 7 = 0$, or $(t - 7)(t + 1) = 0$. Here again we have the restriction that $t > 0$ and $t \neq 1$, so $t = 7$ is the only solution ($t = -1$ is rejected). ■

Using the connection between logarithms and exponents, we can translate properties of exponents into properties of logarithms. Some of these properties are as follows.

Properties of Logarithms

Let M, N, and b be positive numbers, $b \neq 1$, and let y be any real number. Then:

(i) $\log_b(MN) = \log_b M + \log_b N$

(ii) $\log_b \frac{M}{N} = \log_b M - \log_b N$

(iii) $\log_b N^y = y \log_b N$

(iv) $\log_b \frac{1}{N} = -\log_b N$

We verify Properties (i) and (ii) here and leave it to you to check Properties (iii) and (iv) (Problems 36 and 38). Thus, let

$$x = \log_b M \quad \text{and} \quad y = \log_b N,$$

so that

$$b^x = M \quad \text{and} \quad b^y = N.$$

Then,

$$\log_b(MN) = \log_b(b^x b^y) = \log_b(b^{x+y}) = x + y = \log_b M + \log_b N$$

and

$$\log_b \frac{M}{N} = \log_b \frac{b^x}{b^y} = \log_b(b^{x-y}) = x - y = \log_b M - \log_b N.$$

■

Unfortunately, there is no useful formula for $\log_b(M + N)$ or for $\log_b(M - N)$. In general,

$\log_b(M + N)$ *is not the same as* $\log_b M + \log_b N$

and

$\log_b(M - N)$ *is not the same as* $\log_b M - \log_b N$.

Example 4 Use the properties of logarithms to rewrite each expression as a sum or difference of multiples of logarithms.

(a) $\log_3 2x$ **(b)** $\log_2 \frac{1}{4}$ **(c)** $\log_b \frac{z}{uv}$ **(d)** $\log_2 \frac{(x^2 + 5)(2x + 5)^{3/2}}{\sqrt[4]{3x + 1}}$

Solution **(a)** Assuming that x is positive, we have

$$\log_3 2x = \log_3 2 + \log_3 x. \qquad \text{[Property (i)]}$$

(b) $$\log_2 \tfrac{1}{4} = -\log_2 4 = -2 \qquad \text{[Property (iv)]}$$

(c) Assuming that z, u, and v are positive, we have

$$\begin{aligned} \log_b \frac{z}{uv} &= \log_b z - \log_b(uv) && \text{[Property (ii)]} \\ &= \log_b z - (\log_b u + \log_b v) && \text{[Property (i)]} \\ &= \log_b z - \log_b u - \log_b v. \end{aligned}$$

(d) Assuming that $2x + 5$ and $3x + 1$ are positive, we have

$$\log_2 \frac{(x^2+5)(2x+5)^{3/2}}{\sqrt[4]{3x+1}} = \log_2[(x^2+5)(2x+5)^{3/2}] - \log_2 \sqrt[4]{3x+1}$$

$$= \log_2(x^2+5) + \log_2(2x+5)^{3/2} - \log_2(3x+1)^{1/4}$$

$$= \log_2(x^2+5) + \tfrac{3}{2}\log_2(2x+5) - \tfrac{1}{4}\log_2(3x+1),$$

where we applied Property (iii) twice in the last step. ■

Example 5 Rewrite each expression as a single logarithm.

(a) $\log_5 7 + \log_5 x$ **(b)** $\log_4 t - \log_4 5$

(c) $2\log_{10} x + 3\log_{10}(x+1)$ **(d)** $\log_b\left(x + \frac{x}{y}\right) - \log_b\left(z + \frac{z}{y}\right)$

Solution Assuming that all quantities whose logarithms are taken are positive, we have the following:

(a) $\log_5 7 + \log_5 x = \log_5 7x$ [Property (i)]

(b) $\log_4 t - \log_4 5 = \log_4 \frac{t}{5}$ [Property (ii)]

(c) $2\log_{10} x + 3\log_{10}(x+1) = \log_{10} x^2 + \log_{10}(x+1)^3 = \log_{10}[x^2(x+1)^3]$

(d) $\log_b\left(x + \frac{x}{y}\right) - \log_b\left(z + \frac{z}{y}\right) = \log_b \frac{x + \frac{x}{y}}{z + \frac{z}{y}} = \log_b \frac{\left(1 + \frac{1}{y}\right)x}{\left(1 + \frac{1}{y}\right)z} = \log_b \frac{x}{z}$ ■

Example 6 Solve each equation.

(a) $\log_{10} x + \log_{10}(x+21) = 2$

(b) $\log_7(3t^2 - 5t - 2) - \log_7(t-2) = 1$

Solution **(a)** We begin by noticing that x must be positive for $\log_{10} x$ to be defined. If x is positive, so is $x + 21$, and $\log_{10}(x+21)$ is also defined. Applying Property (i), we rewrite

$$\log_{10} x + \log_{10}(x+21) = 2$$

as

$$\log_{10}[x(x+21)] = 2.$$

The last equation can be written in exponential form as

$$x(x+21) = 10^2;$$

that is,

$$x^2 + 21x - 100 = 0.$$

Factoring, we have

$$(x+25)(x-4) = 0,$$

so $x = -25$ or $x = 4$. Since x must be positive, we can eliminate $x = -25$ as an extraneous root. Therefore, the solution is $x = 4$.

(b) Applying Property (ii), we can rewrite the given equation

$$\log_7(3t^2 - 5t - 2) - \log_7(t - 2) = 1$$

as

$$\log_7 \frac{3t^2 - 5t - 2}{t - 2} = 1,$$

provided that both $3t^2 - 5t - 2$ and $t - 2$ are positive. The last equation can be simplified by reducing the fraction,

$$\frac{3t^2 - 5t - 2}{t - 2} = \frac{(3t + 1)\cancel{(t - 2)}}{\cancel{t - 2}} = 3t + 1,$$

so

$$\log_7(3t + 1) = 1;$$

that is,

$$3t + 1 = 7^1 = 7.$$

The solution of this equation is

$$t = (7 - 1)/3 = 2.$$

We must check this answer against the original restrictions on the variables. However, if $t = 2$, then $t - 2 = 0$ and $\log_7(t - 2)$ is undefined. Thus, $t = 2$ is an extraneous root, and the original equation has no solution. ■

It is often useful to rewrite a logarithm in terms of logarithms to other bases. This is accomplished by using the following formula, which holds for $a > 0$, $b > 0$, $a \neq 1$, $b \neq 1$, and $c > 0$.

The Base-Changing Formula

$$\log_a c = \frac{\log_b c}{\log_b a}$$

To prove the base-changing formula, let

$$x = \log_a c.$$

Then,

$$a^x = c,$$

and it follows that

$$\log_b a^x = \log_b c.$$

Using Property (iii), the last equation can be written as

$$x \log_b a = \log_b c,$$

or

$$x = \frac{\log_b c}{\log_b a}.$$

Therefore, since $x = \log_a c$,

$$\log_a c = \frac{\log_b c}{\log_b a}. \qquad \blacksquare$$

Example 7 Rewrite $\log_e 3$ in terms of logarithms to base 10.

Solution By the base-changing formula with $a = e$, $c = 3$, and $b = 10$, we have

$$\log_e 3 = \frac{\log_{10} 3}{\log_{10} e}. \qquad \blacksquare$$

In working with logarithms, you must be careful not to confuse the fact that

$$\log_b\left(\frac{c}{a}\right) = \log_b c - \log_b a$$

[Property (ii) on page 313] and the fact that

$$\frac{\log_b c}{\log_b a} = \log_a c,$$

which is a consequence of the base-changing formula.

Problem Set 5.3

In Problems 1 to 12, solve each exponential equation.

1. $2^x = 8$
2. $3^{x^2} = 81$
3. $25^x = 5$
4. $2^{x^3} = 256$
5. $3^{2x+1} = 27$
6. $(\frac{1}{10})^{4x} = 1000$
7. $3^{2-8x} = 9^{3x+1}$
8. $8^{3t} = 32^{4t-1}$
9. $5^{x^2+x} = 25$
10. $7^{x^2+x} = 1$
11. $3^{2t} - 3^t - 6 = 0$ [*Hint:* Let $x = 3^t$.]
12. $2^{2x+1} + 2^x = 10$

13. Find the *exact* value of each expression.
    (a) $\log_2 4$ (b) $\log_2 8$
    (c) $\log_3 81$ (d) $\log_9 9^5$
    (e) $\log_3 \frac{1}{9}$ (f) $\log_8 \frac{1}{64}$
    (g) $\log_{10} 100{,}000$ (h) $\log_{16} 0.125$

14. Find the *exact* value of each expression.
    (a) $\log_2 \frac{1}{4}$ (b) $\log_3 \sqrt{3}$
    (c) $\log_9 1$ (d) $\log_e e^\pi$
    (e) $\log_2 4^3$ (f) $\log_3 9^{-0.5}$
    (g) $\log_{10} \dfrac{1}{100{,}000}$ (h) $\log_{32} 0.125$

15. Rewrite each logarithmic equation as an equivalent exponential equation.
    (a) $\log_2 32 = 5$ (b) $\log_{16} 2 = \frac{1}{4}$
    (c) $\log_9 \frac{1}{3} = -\frac{1}{2}$ (d) $\log_e e = 1$
    (e) $\log_{\sqrt{3}} 9 = 4$ (f) $\log_{10} 10^n = n$
    (g) $\log_x x^5 = 5$

16. Give a geometric argument based on the graph of $f(x) = b^x$ to show that if $b > 0$, $b \neq 1$, and $c > 0$, then the exponential equation $b^x = c$ has exactly one solution.

17. Rewrite each exponential equation as an equivalent logarithmic equation.

(a) $8^0 = 1$ (b) $10^{-4} = 0.0001$
(c) $4^4 = 256$ (d) $27^{-1/3} = \frac{1}{3}$
(e) $8^{2/3} = 4$ (f) $a^c = y$

18. Using the result of Problem 16, show that if $b > 0$, $b \neq 1$, and x and y are real numbers such that $b^x = b^y$, it follows that $x = y$.

In Problems 19 to 34, solve each equation.

19. $x = \log_6 36$ **20.** $\log_5 x = 2$

21. $\log_x 125 = 3$ **22.** $x = \log_3 729$

23. $x = \log_5 \sqrt[4]{5}$ **24.** $x = \log_{3/4} \frac{4}{3}$

25. $\log_2(2x - 1) = 3$ **26.** $\log_5(2x - 3) = 2$

27. $\log_3(3x - 4) = 4$ **28.** $\log_7(2x - 7) = 0$

29. $\log_2(t^2 + 3t + 4) = 1$ **30.** $\log_5(y^2 - 4y) = 1$

31. $\log_4(9u^2 + 6u + 1) = 2$

32. $\log_3|3 - 2t| = 2$

33. $\log_x(10 - 3x) = 2$ **34.** $\log_x(1 - x + x^2) = 3$

35. Suppose that $\log_b 2 = 0.53$, $\log_b 3 = 0.83$, $\log_b 5 = 1.22$, and $\log_b 7 = 1.48$. Find:

(a) $\log_b 21$ (b) $\log_b 35$ (c) $\log_b \frac{2}{7}$
(d) $\log_b \frac{35}{3}$ (e) $\log_b \sqrt{7}$ (f) $\log_b \sqrt[3]{42}$

Round off all answers to two decimal places.

36. Verify Property (iii) on page 313.

37. Find $\log_2(\log_4 256)$.

38. Verify Property (iv) on page 313.

In Problems 39 to 46, rewrite each expression as a sum or difference of multiples of logarithms. (Make the necessary assumptions about the values of the variables.)

39. $\log_b[x(x + 1)]$ **40.** $\log_a(x^4\sqrt{y})$

41. $\log_{10}[x^2(x + 1)]$ **42.** $\log_c \sqrt{\dfrac{x}{x + 7}}$

43. $\log_3 \dfrac{x^3y^2}{z}$ **44.** $\log_e \dfrac{t(t + 1)}{(t + 2)^3}$

45. $\log_e \sqrt{x(x + 3)}$ **46.** $\log_b \sqrt[3]{(x + 1)^2\sqrt{x + 7}}$

In Problems 47 to 54, rewrite each expression as a single logarithm. (Make the necessary assumptions about the values of the variables.)

47. $2 \log_3 x + 7 \log_3 x$ **48.** $\log_{10} \dfrac{a^3}{b} + \log_{10} \dfrac{b^2}{5a}$

49. $\dfrac{1}{2}(\log_5 4 - \log_5 9)$ **50.** $\log_x \dfrac{y^5}{z^4} - \log_x \dfrac{y^3}{z^2}$

51. $\log_e \dfrac{x}{x - 1} + \log_e \dfrac{x^2 - 1}{x}$

52. $\log_b \dfrac{x + y}{z} - \log_b \dfrac{1}{x + y}$

53. $\log_3 \dfrac{x^2 + 14x - 15}{x^2 + 4x - 5} - \log_3 \dfrac{x^2 + 12x - 45}{x^2 + 6x - 27}$

54. $\log_e \dfrac{m^2 - 2m - 24}{m^2 - m - 30} + \log_e \dfrac{(m + 5)^2}{m^2 - 16}$

In Problems 55 to 62, solve each equation.

55. $\log_4 x + \log_4(x + 6) = 2$

56. $\log_{10} x + \log_{10}(x + 3) = 1$

57. $\log_7 x + \log_7(18x + 61) = 1$

58. $\log_2 x + \log_2(x - 2) = \log_2(9 - 2x)$

59. $\log_3(x^2 + x) - \log_3(x^2 - x) = 1$

60. $\log_5(4x^2 - 1) = 2 + \log_5(2x + 1)$

61. $\log_8(x^2 - 9) - \log_8(x + 3) = 2$

62. $2 \log_2 x - \log_2(x - 1) = 2$

63. Use the base-changing formula to rewrite $\log_{10} 5$ in terms of logarithms to base e.

64. Using the base-changing formula, evaluate $\dfrac{\log_7 9}{\log_7 3}$.

65. Suppose that $a > 0$, $b > 0$, $a \neq 1$, and $b \neq 1$. Using the base-changing formula and the fact that $\log_b b = 1$, show that

$$\log_a b = \frac{1}{\log_b a}.$$

66. Derive the following alternative base-changing formula. If $a > 0$, $b > 0$, $a \neq 1$, $b \neq 1$, and $c > 0$, then

$$\log_a c = (\log_a b)(\log_b c).$$

C 67. If a rocket with total mass M_0 moving in free space with speed V_0 is fired, its speed V after a mass M of propellent has been burned is given by

$$V = V_0 + V_1 \log_e \frac{M_0}{M},$$

where V_1 is the speed of the exhaust gases produced by the burn. Find V if $V_0 = 500$ meters per second, $V_1 = 300$ meters per second, and M is 20% of M_0. Round off the answer to the nearest meter per second.

68. Insect pests newly released into an environment (for instance, the Mediterranean fruit fly and the African killer bee) reproduce and spread out from the point of release. The mathematical model

$$D = Ke^{ct - kr}$$

may be used to predict the density D of insects per square kilometer t weeks after the initial infestation at a point r kilometers from the point of release. Here K, c, and k are constants depending on the intensity of the initial infestation, the rate at which the insects reproduce, and the rate at which they spread. Solve this equation for t in terms of the remaining quantities.

69. The equation $y = y_0 e^{-0.005t}$ governs the radioactive decay of polonium. (See Example 7 in Section 5.2.) Solve this equation for t in terms of the ratio y_0/y.

5.4 Logarithmic Functions

A function F of the form

$$F(x) = \log_b x,$$

where $b > 0$ and $b \neq 1$, is called a **logarithmic function with base *b*.** The domain of F is the interval $(0, \infty)$ of all positive real numbers. If we let

$$f(x) = b^x,$$

then, since

$$b^{\log_b x} = x \quad \text{for } x > 0, \qquad \text{and} \qquad \log_b b^x = x \quad \text{for } x \text{ in } \mathbb{R},$$

it follows that

$$f[F(x)] = x \quad \text{for } x > 0, \qquad \text{and} \qquad F[f(x)] = x \quad \text{for } x \text{ in } \mathbb{R}.$$

In other words, the function f and F are inverses of each other. (You may wish to review the idea of inverse functions in Section 3.7.) Therefore, *the graph of the logarithmic function $F(x) = \log_b x$ is the mirror image of the graph of the exponential function $f(x) = b^x$ across the line $y = x$.*

Example 1 Sketch the graphs of $F(x) = \log_3 x$ and $G(x) = \log_{1/3} x$.

Solution Graphs of the functions $f(x) = 3^x$ and $g(x) = (\frac{1}{3})^x$ were shown in Figure 2 of Section 5.1 (page 293). By reflecting these graphs across the line $y = x$, we obtain the graphs of $F(x) = \log_3 x$ (Figure 1a) and $G(x) = \log_{1/3} x$ (Figure 1b). Notice that the graph of G can be obtained by reflecting the graph of F across the x axis. (For the reason why, see Problem 36.) ■

Figure 1

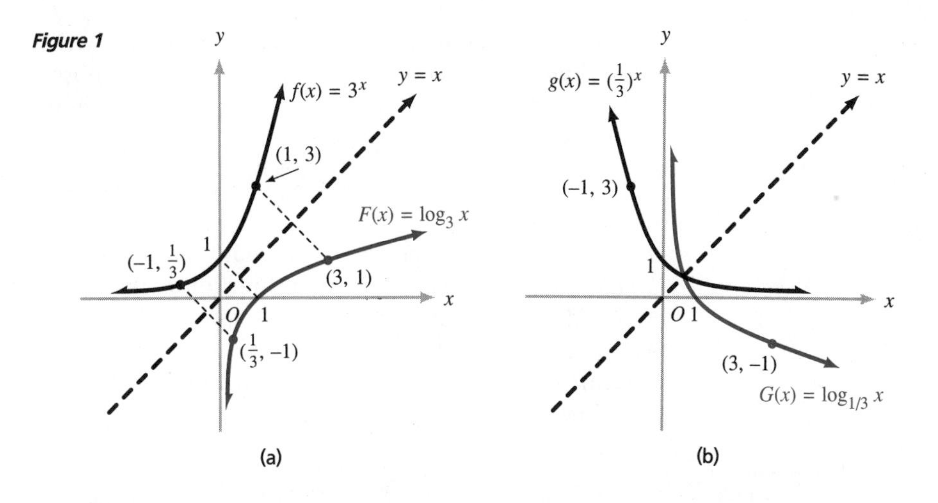

In general, graphs of logarithmic functions have the characteristic shapes shown in Figure 2. Thus, if $b > 1$, the graph of $F(x) = \log_b x$ rises to the right (Figure 2a), whereas if $0 < b < 1$, the graph falls to the right (Figure 2b).

Figure 2

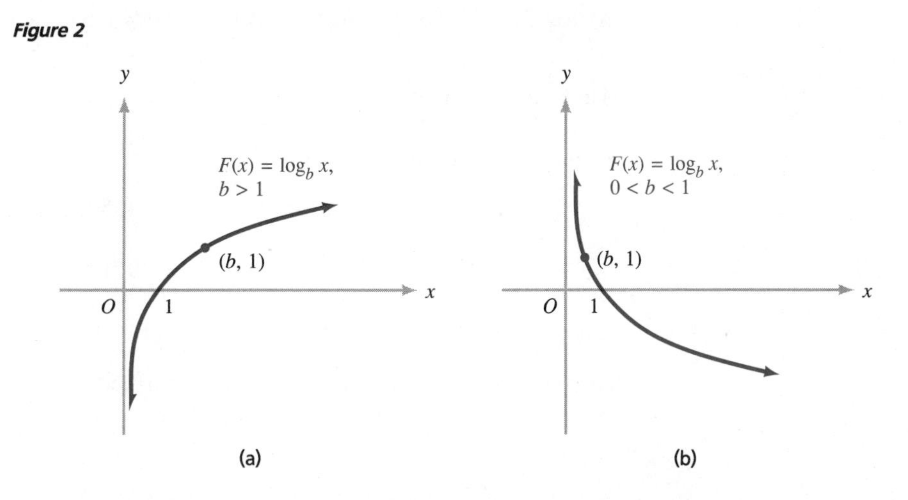

Notice that the graph of $F(x) = \log_b x$ always contains the point $(b, 1)$ (because $\log_b b = 1$) and that its x intercept is always 1 (because $\log_b 1 = 0$). There is no y intercept; in fact, the y axis is a vertical asymptote of the graph. The range of $F(x) = \log_b x$ is $\mathbb{R}$.

The Common and Natural Logarithm Functions

Before the development of electronic calculators and computers, logarithms were extensively used to facilitate numerical calculations. Because the usual positional system for writing numerals is based on 10, arithmetic calculation is easiest when logarithms with base 10 are used. Logarithms with base 10 are called **common logarithms,** and the symbol "log x"* (with no subscript) is often used as an abbreviation for $\log_{10} x$. Thus, the **common logarithm function** is defined by

$$\log x = \log_{10} x \qquad \text{for } x > 0.$$

Today, because of the wide availability of inexpensive and reliable electronic calculators, the common logarithm function is rarely used for purposes of numerical calculation. However, it still has many applications, ranging from the measurement of pH in chemistry to the measurement of sound pollution in the health sciences (see Section 5.5). For this reason, most scientific calculators have both a 10^x key and a log key.

C Example 2 Use a calculator with a log key to evaluate:

(a) log 2110 **(b)** log 0.004326

Solution On a 10-digit calculator, we obtain:

(a) $\log 2110 \approx 3.324282455$ **(b)** $\log 0.004326 \approx -2.363913485$ ■

C Example 3 Use a calculator to verify that $10^{\log x} = x$ for $x = \pi$.

Solution Rounded off to nine decimal places,

$$\pi \approx 3.141592654$$

and

$$\log \pi \approx 0.497149873.$$

Now,

$$10^{0.497149873} \approx 3.141592654 \approx \pi.$$

Therefore, the equation $10^{\log \pi} = \pi$ is confirmed to within the limits of accuracy of our calculator. ■

In advanced mathematics and its applications, many otherwise cumbersome formulas become much simpler if logarithmic and exponential functions with base $e \approx 2.718$ are used (see Section 5.2). Logarithms with base e are called **natural logarithms,** and the symbol "ln x," pronounced "el en x," is often used as an abbreviation for $\log_e x$. Thus, the **natural logarithm function** is defined by

$$\ln x = \log_e x \qquad \text{for } x > 0.$$

* In some textbooks, and in the BASIC computer language, log or LOG is used as an abbreviation for the logarithm to base e.

In other words, for $x > 0$,

$$y = \ln x \quad \text{if and only if} \quad e^y = x.$$

Of course, all scientific calculators have an ln key.

C Example 4 Use a calculator with an ln key to evaluate:

(a) $\ln 7124$ **(b)** $\ln 0.05319$

Solution On a 10-digit calculator, we obtain:

(a) $\ln 7124 \approx 8.871224644$ **(b)** $\ln 0.05319 \approx -2.933884870$ ■

C Example 5 Use a calculator to verify that $\ln e^x = x$ for $x = \sqrt{5}$.

Solution Rounded off to nine decimal places,

$$\sqrt{5} \approx 2.236067978$$

and $$e^{\sqrt{5}} \approx 9.356469017.$$

Now, $$\ln 9.356469017 \approx 2.236067978 \approx \sqrt{5}.$$

Therefore the equation $\ln e^{\sqrt{5}} = \sqrt{5}$ is confirmed to within the limits of accuracy of our calculator. ■

Earlier, on page 311, we established the formulas

$$b^{\log_b u} = u \quad \text{for } u > 0 \qquad \text{and} \qquad \log_b b^u = u \quad \text{for } u \text{ in } \mathbb{R}.$$

In particular, then, for $b = e$, we have

$$e^{\ln u} = u \quad \text{for } u > 0 \qquad \text{and} \qquad \ln e^u = u \quad \text{for } u \text{ in } \mathbb{R}.$$

Example 6 Simplify each expression.

(a) $\ln e^{x^2+3x}$ **(b)** $e^{4\ln(3s+5)}$ **(c)** $e^{5-3\ln t}$

Solution **(a)** $\ln e^{x^2+3x} = x^2 + 3x$.

(b) $e^{4\ln(3s+5)} = e^{\ln(3s+5)^4} = (3s + 5)^4$ provided that $3s + 5 > 0$; that is, $s > -\frac{5}{3}$.

(c) $e^{5-3\ln t} = e^5 e^{-3\ln t} = e^5 e^{\ln t^{-3}} = e^5 t^{-3} = \dfrac{e^5}{t^3}$, provided that $t > 0$. ■

Because *the natural logarithm function is the inverse of the natural exponential function*, the graph of

$$y = \ln x$$

Figure 3

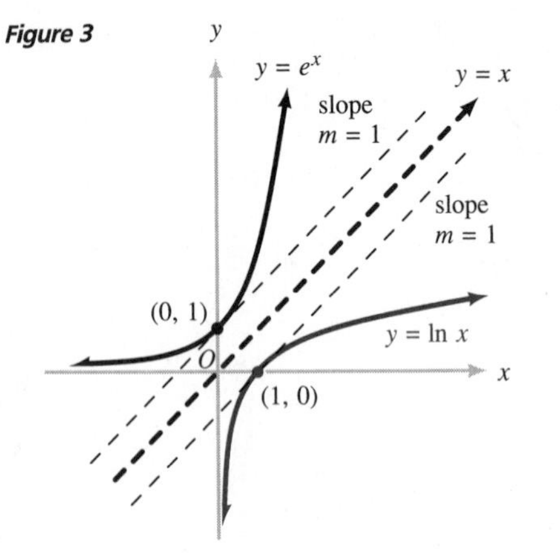

can be obtained by reflecting the graph of $y = e^x$ across the line $y = x$ (Figure 3). Recall that the tangent line to the graph of $y = e^x$ at (0, 1) has slope $m = 1$ (Section 5.2, Figure 3). It follows that the tangent line to the graph of $y = \ln x$ at (1, 0) also has slope $m = 1$. If you keep this fact in mind whenever you sketch the graph of the natural logarithm function, you will obtain a more accurate graph.

Although common and natural logarithm functions are sufficient for most purposes, there are situations in which logarithms with bases other than 10 and e are useful. For instance, in communications engineering and computer science, the bases 2, 8, and 16 are often used. Scientific calculators ordinarily have keys only for log and ln, you must use the *base-changing formula*

$$\log_a x = \frac{\log x}{\log a} \quad \text{or} \quad \log_a x = \frac{\ln x}{\ln a},$$

derived in Section 5.3, if you want to calculate logarithms with other bases.*

C Example 7 Use a calculator and the base-changing formula to find $\log_2 3$.

Solution

$$\log_2 3 = \frac{\log 3}{\log 2} \approx \frac{0.477121255}{0.301029996} \approx 1.584962501$$ ■

Graphs of Functions Involving Logarithms

In dealing with functions involving logarithms, you must keep in mind that logarithms of negative numbers and zero are undefined.

Example 8 Find the domain of each function:

(a) $h(x) = \log_2(x + 1)$ **(b)** $g(x) = \log x^2$

Solution **(a)** $\log_2(x + 1)$ is defined if and only if

$$x + 1 > 0; \qquad \text{that is, } x > -1.$$

Therefore, the domain of h is the interval $(-1, \infty)$.

(b) $\log x^2$ is defined if and only if

$$x^2 > 0; \qquad \text{that is, } x \neq 0.$$

Therefore, the domain of g is the set of all nonzero real numbers. ■

* In the BASIC computer language, where LOG refers to the natural logarithm, you must use the formula `LOG_A (X) = LOG (X)/LOG (A)` to obtain the logarithm to the base A of X.

Figure 4

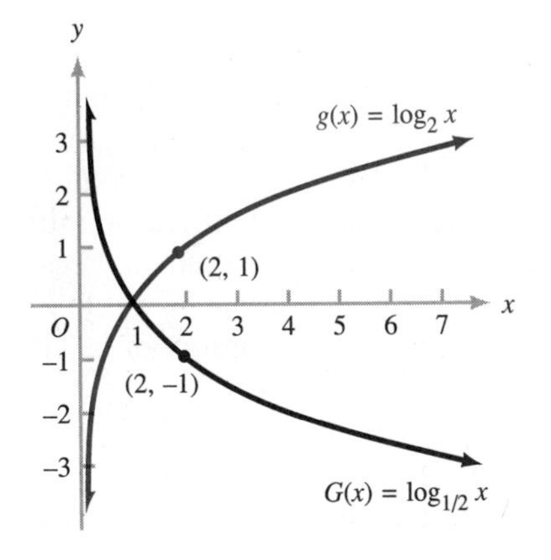

Example 9 Sketch the graphs of $g(x) = \log_2 x$ and $G(x) = \log_{1/2} x$ on the same coordinate system.

Solution The graph of $g(x) = \log_2 x$ has the characteristic shape shown in Figure 2a, contains the point (2, 1), and has x intercept 1 (Figure 4). The graph of $G(x) = \log_{1/2} x$ is obtained by reflecting the graph of $g(x) = \log_2 x$ across the x axis (Figure 4). ■

The techniques of graph sketching presented in Section 3.5 can be applied to functions involving logarithms. As usual, accuracy is enhanced by plotting more points, and you may use a calculator to find coordinates of such points quickly or you may use a graphing calculator.

In Examples 10 and 11, determine the domain of each function, find any y or x intercepts of its graph, use the techniques of Section 3.5 to sketch the graph, determine the range of the function, indicate where it is increasing or decreasing, and find any horizontal or vertical asymptotes of the graph. c *If you wish, use a calculator for greater accuracy.* gc *You may use a graphing calculator to check your work.*

Example 10 $H(x) = \log_3(-x)$

Solution The domain of H consists of all values of x for which $-x > 0$, that is, the interval $(-\infty, 0)$. Because $H(0)$ is undefined, there is no y intercept. The x intercept is the solution of the equation $H(x) = 0$; that is, $\log_3(-x) = 0$. Rewriting the last equation in exponential form, we obtain

$$-x = 3^0 = 1,$$

so the x intercept is $x = -1$. The graph of $H(x) = \log_3(-x)$ (Figure 5) is the mirror image of the graph of $F(x) = \log_3 x$ (Figure 1a) across the y axis. From the graph, we see that the range of H is the set $\mathbb{R}$ of all real numbers, and that the function H is decreasing over its entire domain. The y axis is a vertical asymptote, and there is no horizontal asymptote. ■

Figure 5

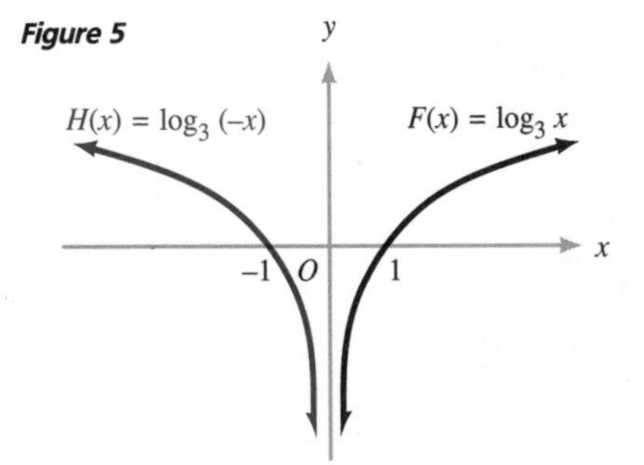

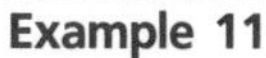
Example 11 $f(x) = 1 + \ln(x - 2)$

Solution The domain of f is the interval $(2, \infty)$; hence, $f(0)$ is undefined and there is no y intercept. The x intercept is the solution of the equation $f(x) = 0$; that is,

$$1 + \ln(x - 2) = 0 \qquad \text{or} \qquad \ln(x - 2) = -1.$$

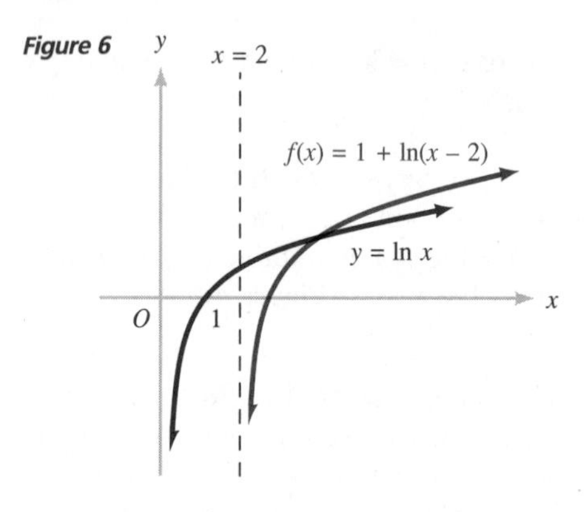

Figure 6

Rewriting the last equation in exponential form, we obtain

$$x - 2 = e^{-1},$$

so the x intercept is $x = 2 + e^{-1} \approx 2.37$. The graph of the function $f(x) = 1 + \ln(x - 2)$ (Figure 6) is obtained by shifting the graph of $y = \ln x$ (Figure 3) 1 unit upward and 2 units to the right. From the graph, we see that the range of f is the set $\mathbb{R}$ of all real numbers, and that the function f is increasing over its entire domain. Because the y axis is a vertical asymptote of the graph of $y = \ln x$, it follows that the line $x = 2$ is a vertical asymptote of the graph of f. There is no horizontal asymptote. ■

Problem Set 5.4

In Problems 1 and 2, sketch the graph of each logarithmic function by reflecting the graph of its inverse exponential function across the line $y = x$, as in Example 1. gc If you wish, you may use a graphing calculator to check your work.

1. **(a)** $f(x) = \log_4 x$ **(b)** $F(x) = \log_{1/4} x$

2. **(a)** $g(x) = \log_5 x$ **(b)** $G(x) = \log_{1/5} x$

c **3.** Use a calculator with a log key to evaluate:

(a) $\log 6.373$ **(b)** $\log 1230.4$
(c) $\log 0.03521$ **(d)** $\log(3.047 \times 10^{11})$
(e) $\log(6.562 \times 10^{-9})$

c **4.** Using a calculator, verify the property $\log(xy) = \log x + \log y$ for:

(a) $x = 31.27,\ y = 5.246$ **(b)** $x = \pi,\ y = \sqrt{2}$

c **5.** Use a calculator with an ln key to evaluate:

(a) $\ln 4126$ **(b)** $\ln 2.704$
(c) $\ln 0.040404$ **(d)** $\ln(7.321 \times 10^{8})$
(e) $\ln(1.732 \times 10^{-7})$

c **6.** Using a calculator, verify that:

(a) $\log 10^{\sqrt{7}} = \sqrt{7}$ **(b)** $10^{\log \sqrt{7}} = \sqrt{7}$

c **7.** Using a calculator, verify that:

(a) $\ln e^{\pi} = \pi$ **(b)** $e^{\ln \pi} = \pi$

c **8.** Using a calculator, verify that $\ln y^x = x \ln y$ for $x = 77.01$ and $y = 3.352$.

c **9.** Using a calculator and the base-changing formula, evaluate:

(a) $\log_2 25$ **(b)** $\log_3 2$ **(c)** $\log_8 e$
(d) $\log_{\pi} 5$ **(e)** $\log_{\sqrt{2}/2} 0.07301$

10. Show that, for $x > 0$, $\ln x = M \log x$, where $M = \ln 10$.

11. Find the domain of each function.

(a) $f(x) = \log(x - 2)$
(b) $g(x) = \log_2 \sqrt{x}$
(c) $h(x) = \log_8(4 - x)$
(d) $H(x) = \log |x|$
(e) $K(x) = \dfrac{1}{\ln x}$
(f) $Q(x) = \ln\left(\dfrac{1 - x}{1 + x}\right)$
(g) $G(x) = \log(x^2 - 5x + 7)$
(h) $k(x) = \ln(x^2 + 1)$

12. Find any y and x intercepts of the graph of each function in Problem 11.

13. The computer-generated graphs in Figures 7b, 7c, and 7d are obtained by reflecting and shifting the graph of $f(x) = \log_3 x$ in Figure 7a. In each case, write a formula for the function whose graph is shown.

Figure 7

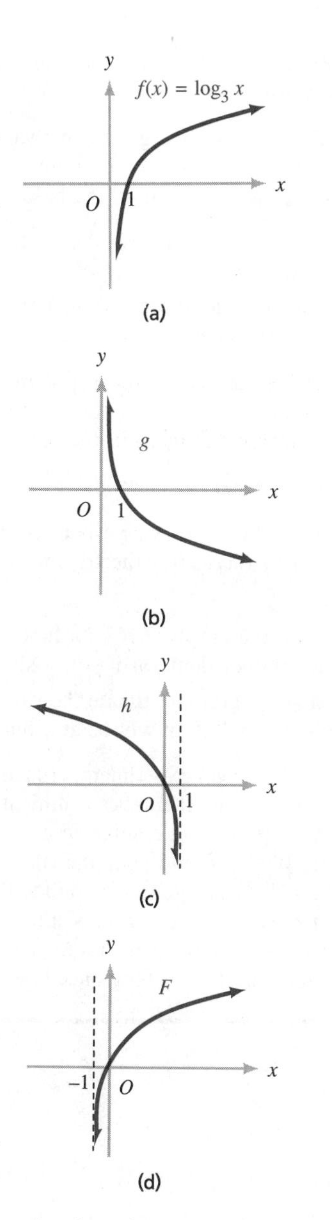

C **14.** Sketch the graphs of $y = \log_2 x$, $y = \ln x$, and $y = \log x$ on the same coordinate system and describe the relationships among these graphs. gc Use a graphing calculator if you wish.

In Problems 15 to 28, determine the domain of each function and find any y or x intercepts of its graph. Use the techniques of Section 3.5 to sketch the graph, determine the range of the function, indicate where it is increasing or decreasing, and find any horizontal or vertical asymptotes of the graph. C If you wish, use a calculator for greater accuracy. gc You may use a graphing calculator to check your work.

15. $f(x) = 1 + \log_2 x$

16. $g(x) = \log_5(-x)$

17. $h(x) = \log_2 x^{-2}$

18. $F(x) = \log|x|$

19. $G(x) = \log \sqrt{x}$

20. $H(x) = \log_5 \sqrt[3]{x}$

21. $f(x) = \log_2(x - 1)$

22. $g(x) = (\ln x) - 1$

23. $G(x) = -\log_3 x$

24. $H(x) = \ln(1 - x)$

25. $f(x) = 2 + \ln(x - 1)$

26. $g(x) = (\ln x)^{-1}$

27. $h(x) = \ln \dfrac{1}{x}$

28. $F(x) = 2 - \ln(1 - x)$

In Problems 29 and 30, justify each property of the natural logarithm function by using either its definition or the properties of logarithms given in Section 5.3.

29. **(a)** $\ln 1 = 0$ **(b)** $\ln \dfrac{1}{x} = -\ln x, \quad x > 0$

(c) $\ln e = 1$ **(d)** $\ln x = \dfrac{\log x}{\log e}, \quad x > 0$

30. **(a)** $\ln xy = \ln x + \ln y, \quad x > 0, y > 0$

(b) $\ln x = \dfrac{1}{\log_x e}, \quad x > 0, x \neq 1$

(c) $\ln \dfrac{x}{y} = \ln x - \ln y, \quad x > 0, y > 0$

(d) $\ln y^x = x \ln y, \quad y > 0, x \text{ in } \mathbb{R}$

31. Simplify each expression.

(a) $e^{\ln 7}$ **(b)** $e^{-3 \ln x}$ **(c)** $e^{6 + 5 \ln t}$

(d) $e^{\ln(1/x)}$ **(e)** $e^{-4 - 3 \ln p}$ **(f)** $\ln e^{x^2 - 9}$

(g) $e^{-3 + \ln r^2}$ **(h)** $e^{2 \ln x - 7 \ln y}$

32. Find the inverse of each function.

(a) $f(x) = e^{x+2}$ **(b)** $g(x) = e^{2x-1} + 1$

In Problems 33 and 34, write each expression as a sum or difference of multiples of natural logarithms.

33. $\ln \sqrt[3]{(x + 2)(3x - 7)(4x + 1)}$

34. $\ln \dfrac{x^2\sqrt{2x + 3}}{(3x + 4)^{3/2}}$

35. Explain why every real number y can be written in the form $y = \ln x$ for a suitable value of x.

36. If $b > 0$ and $b \neq 1$, show that $\log_{1/b} x = -\log_b x$ holds for $x > 0$.

37. The function $F(x) = \ln|x|$ is used quite often in calculus. Sketch a graph of this function.

38. Solve the equation $\log_x(2x)^{3x} = 4^{\log_4 4x}$.

39. Find the value of b if the graph of $y = \log_b x$ contains the point $(\frac{1}{2}, -1)$.

40. If $a > 0$, $b > 0$, and $b \neq 1$, find a base c such that $a \log_b x = \log_c x$ holds for all $x > 0$.

41. Using the fact that the tangent line to the graph of $y = \ln x$ at $(1, 0)$ has slope $m = 1$, explain why $\ln x \approx x - 1$ for values of x close to 1, with the approximation becoming more and more accurate as x comes closer and closer to 1.

42. Show that $\ln(1 + x) \approx x$ if $|x|$ is small and that the approximation becomes more and more accurate as $|x|$ gets smaller and smaller. [*Hint:* Use the result of Problem 41.]

43. The formula $y^x = e^{x \ln y}$ for $y > 0$ is used in calculus to write y^x in terms of the exponential and natural logarithm functions. Derive this formula. [*Hint:* $x \ln y = \ln y^x$.]

C **44.** Using a calculator with y^x, e^x, and ln keys, check the formula in Problem 43 for $x = 1.59$ and $y = 7.47$.

45. Criticize the following statement: Since $\ln x^2 = 2 \ln x$, the graph of $y = \ln x^2$ can be found by doubling all ordinates on the graph of $y = \ln x$.

46. The table on page 302 gives some values of the slope m of the tangent line to the graph of $y = b^x$ at the point $(0, 1)$ as a function of the base b.

(a) Using the data in this table, sketch a graph of m as a function of b.

(b) By looking at the graph in part (a), guess what function this is.

C **47.** A small lake is known to be polluted with coliform bacteria. Environmental biologists are planning to test a new bactericidal agent in the lake and have developed the mathematical model $N = 150\left(\frac{t}{14} - \ln \frac{t}{14}\right)$ for the number N of viable bacteria per milliliter of lake water t days after the treatment, for $1 \leq t \leq 30$ days.

(a) Sketch the graph of N as a function of t. gc Use a graphing calculator if you wish.

(b) Using the graph, estimate the value of t for which coliform pollution will be at a minimum.

C **48.** In a psychological experiment, students took equivalent forms of a mathematics examination at monthly intervals after having completed a semester of college algebra. It was found that the mathematical model $S = 100e^{-t/12} - 22$ gave a good fit to the average score S (in percent) of the students after t months. After approximately how many months had the students forgotten half of what they once knew?

5.5 Applications of Exponential and Logarithmic Functions

In this and the next section, we present a small sample of the many and varied applications of exponential and logarithmic functions.

Solving Exponential Equations

You can often solve an exponential equation by taking the logarithm of both sides and using the properties of logarithms (page 313) to simplify the resulting equation. For this purpose, you can use either common or natural logarithms.

C Example 1 Solve the exponential equation $7^{2x+1} = 3^{x-2}$.

Solution Taking the common logarithm of each side of the equation, we have

$$\log 7^{2x+1} = \log 3^{x-2}.$$

Thus, by Property (iii),

$$(2x + 1) \log 7 = (x - 2) \log 3.$$

Distributing, we have

$$\begin{aligned}
(2 \log 7)x + \log 7 &= (\log 3)x - 2 \log 3 \\
(2 \log 7 - \log 3)x &= -\log 7 - 2 \log 3 \\
(\log 7^2 - \log 3)x &= -(\log 7 + \log 3^2) \\
(\log 49 - \log 3)x &= -(\log 7 + \log 9) \\
(\log \tfrac{49}{3})x &= -\log 63 \\
x &= \frac{-\log 63}{\log \frac{49}{3}}.
\end{aligned}$$

This is the *exact* solution. Using a 10-digit calculator, we obtain an approximate solution rounded off to nine decimal places:

$$x = \frac{-\log 63}{\log \frac{49}{3}} \approx \frac{-1.799340549}{1.213074825} \approx -1.483289004.$$ ■

The following example shows how to answer the question that we raised in the introduction to Section 5.3.

C Example 2 If you put \$100 in a savings account at 8% nominal annual interest compounded quarterly, how long will it take for your money to double?

Solution The final value S dollars of your investment after t years is given by

$$S = 100\left(1 + \frac{0.08}{4}\right)^{4t} = 100(1.02)^{4t}.$$

If t is the time required to double your money, then

$$200 = 100(1.02)^{4t} \quad \text{or} \quad 1.02^{4t} = 2.$$

Taking the logarithm of both sides of the last equation, we get

$$4t \log 1.02 = \log 2$$

so that

$$t = \frac{\log 2}{4 \log 1.02} \approx 8.75 \text{ years.}$$ ■

A sum of P dollars earning interest compounded *continuously* at a nominal annual rate r will double in t years, where

$$Pe^{rt} = 2P \qquad \text{or} \qquad e^{rt} = 2.$$

Therefore,

$$rt = \ln 2$$

so that

$$t = \frac{\ln 2}{r} \approx \frac{0.69}{r} = \frac{69}{100r}.$$

In other words, the approximate doubling time for continuously compounded interest is obtained by dividing 69 by the interest rate expressed as a percent. However, money earning interest compounded *periodically* will take a bit longer to double, and this can be compensated for (roughly) by dividing 70 (rather than 69) by the interest rate expressed as a percent. Investors refer to this procedure as the **rule of 70.** For instance, applying the rule of 70 to Example 2, we again obtain

$$t \approx \frac{70}{8} = 8.75 \text{ years}$$

as the approximate doubling time.*

Applications in Chemistry, Earth Sciences, Psychophysics, and Physics

Measuring pH *in Chemistry* In chemistry, the **pH** of a substance is defined by

$$\text{pH} = -\log[\text{H}^+],$$

where $[\text{H}^+]$ is the concentration of hydrogen ions in the substance, measured in moles per liter. The pH of distilled water is 7. A substance with a pH of less than 7 is known as an *acid*, whereas a substance with a pH of greater than 7 is called a *base*.

Environmentalists constantly monitor the pH of rain and snow because of the destructive effects of "acid rain" caused largely by sulfur dioxide emissions from factories and coal-burning power plants. Because of dissolved carbon dioxide from the atmosphere, rain and snow have a natural concentration of $[\text{H}^+] = 2.5 \times 10^{-6}$ mole per liter.

C Example 3 Find the natural pH of rain and snow.

Solution

$$\text{pH} = -\log[\text{H}^+] = -\log(2.5 \times 10^{-6}) \approx -(-5.6) = 5.6$$

* For interest compounded only once a year, a **rule of 72** is used for a more accurate estimate of the doubling time.

Measuring Altitude: The Barometric Equation The **barometric equation**

$$h = (30T + 8000) \ln \frac{P_0}{P}$$

relates the height h in meters above sea level, the air temperature T in degrees Celsius, the atmospheric pressure P_0 in centimeters of mercury at sea level, and the atmospheric pressure P in centimeters of mercury at height h. The altimeters most commonly used in aircraft measure the atmospheric pressure P and display the altitude by means of a scale calibrated according to the barometric equation.

C Example 4 Atmospheric pressure at the summit of Pike's Peak in Colorado on a certain day measures 44.7 centimeters of mercury. If the average air temperature is 5°C and the atmospheric pressure at sea level is 76 centimeters of mercury, find the height of Pike's Peak. Round off the answer to the nearest 10 meters.

Solution We use the barometric equation with $T = 5°\text{C}$, $P_0 = 76$ centimeters of mercury, and $P = 44.7$ centimeters of mercury to obtain

$$h = [30(5) + 8000] \ln \frac{76}{44.7} = 8150 \ln \frac{76}{44.7}$$

$$\approx 4330 \text{ meters.}$$

■

Gustav Fechner

Measuring Sensation In 1860, the German physicist Gustav Fechner (1801–1887) published a psychophysical law relating the intensity S of a sensation to the intensity P of the physical stimulus causing it. This law, which was based on experiments originally reported in 1829 by the German physiologist Ernst Weber (1795–1878), states that the change in S caused by a small change in P is proportional not to the change in P as one might suppose, but rather to the *percentage* of change in P. Using calculus, it can be shown that, as a consequence, S varies linearly as the natural logarithm of P, so that

$$S = A + B \ln P,$$

where A and B are suitable constants.

Suppose that P_0 denotes the **threshold intensity** of the physical stimulus; that is, the largest value of the intensity P for which there is no sensation. Then, substituting 0 for S and P_0 for P in the equation above, we find that

$$0 = A + B \ln P_0 \quad \text{or} \quad A = -B \ln P_0.$$

It follows that

$$S = -B \ln P_0 + B \ln P = B(\ln P - \ln P_0) = B \ln \frac{P}{P_0}.$$

If you prefer to write the relationship between S and P in terms of the common logarithm, use the equation

$$\ln \frac{P}{P_0} = M \log \frac{P}{P_0}, \qquad \text{where } M = \ln 10$$

(see Problem 10 in Problem Set 5.4) to rewrite $S = B \ln \frac{P}{P_0}$ as

$$S = MB \log \frac{P}{P_0}.$$

Finally, letting $C = MB$, you will obtain the **Weber–Fechner** law in the form

$$S = C \log \frac{P}{P_0}.$$

The choice of the constant C determines the units in which S is measured.

Centennial model of Bell's telephone, 1976

Early in the development of the telephone, it became necessary to have a unit to measure the loudness of telephone signals at various points in the network. The proposed unit, called the *bel* in honor of Alexander Graham Bell, the inventor of the telephone, is obtained by taking the constant $C = 1$ in the Weber–Fechner law. (The power P of the electrical signal is measured in watts.) It turns out that $\frac{1}{10}$ of a bel, called a **decibel**, is roughly the smallest noticeable difference in the loudness of two sounds. For this reason, loudness is commonly measured in decibels rather than in bels. Thus, in electronics, an amplifier with an input of P_0 watts and an output of P watts is said to provide a *gain* of

$$S = 10 \log \frac{P}{P_0} \text{ decibels.}$$

Decibels are used to measure the loudness of sound produced by any source, not just a telephone receiver or an electronic amplifier. The intensity P of the air vibrations that produce the sensations of sound is commonly measured by the power (in watts) per square meter of wavefront. The threshold of human hearing is approximately

$$P_0 = 10^{-12} \text{ watt per square meter}$$

at the eardrum, and the formula

$$S = 10 \log \frac{P}{P_0}$$

thus gives the loudness in decibels produced by a sound wave of intensity P watts per square meter at the eardrum. The rustle of leaves in a gentle breeze corresponds to about 20 decibels, while loud thunder may reach 110 decibels or more. Sound in excess of 120 decibels can cause pain. In order to prevent "boilermaker's deafness," the American Academy of Ophthalmology and Otolaryngology recommends that no worker be exposed to a continuous sound level of 85 decibels or more for 5 hours or more per day without protective devices.

C Example 5 Suppose that an employee in a factory must work in the vicinity of a machine producing 60 decibels of sound during an 8-hour shift. A second machine producing 70 decibels is moved into the same vicinity. Does the worker now need ear protection?

Solution We can add sound intensities, but not decibels; thus, a combinaton of 60 decibels and 70 decibels does *not* amount to 130 decibels. In fact, suppose that P_1 and P_2 are the sound-wave intensities in watts per square meter produced by the two machines separately. Then the sound-wave intensity P produced by the machines operating together is given by $P = P_1 + P_2$. We know that

$$10 \log \frac{P_1}{P_0} = 60 \quad \text{and} \quad 10 \log \frac{P_2}{P_0} = 70,$$

so that

$$\log \frac{P_1}{P_0} = 6 \quad \text{and} \quad \log \frac{P_2}{P_0} = 7;$$

that is,

$$\frac{P_1}{P_0} = 10^6 \quad \text{and} \quad \frac{P_2}{P_0} = 10^7$$

or

$$P_1 = 10^6 P_0 \quad \text{and} \quad P_2 = 10^7 P_0.$$

It follows that

$$P = P_1 + P_2 = 10^6 P_0 + 10^7 P_0,$$

and the loudness of the sound produced by the machines operating together is

$$S = 10 \log \frac{P}{P_0} = 10 \log \frac{10^6 P_0 + 10^7 P_0}{P_0} = 10 \log(10^6 + 10^7)$$
$$= 10 \log 11{,}000{,}000 \approx 70.4 \text{ decibels},$$

well below the 85-decibel limit. Ear protection will not be needed. ■

Charles F. Richter

Measuring the Magnitude of Earthquakes In 1935, the U.S. seismologist Charles F. Richter (1900–1984) proposed the use of a logarithmic scale for measuring the magnitude of earthquakes. Since then, different versions of Richter's idea have been developed in which the magnitude of an earthquake is computed as a function of one variable or another, for instance, the degree of damage or the amount of actual ground motion. In 1956, Richter and B. Gutenberg proposed the formula

$$1.5M + 4.4 = \log E$$

to relate the magnitude M of an earthquake to the total amount of energy E in joules released by the earthquake. (One joule is approximately the amount of energy required to lift a one-pound weight a distance of 0.74 feet.)

Figure 1 shows a graph of the Richter–Gutenberg magnitude–energy equation and an indication of the magnitudes of several noteworthy earthquakes. The strongest earthquakes in modern times, one in Japan in 1933 and one on the Columbia–Ecuador border in 1906, both had a magnitude of 8.9 on the Richter

Figure 1

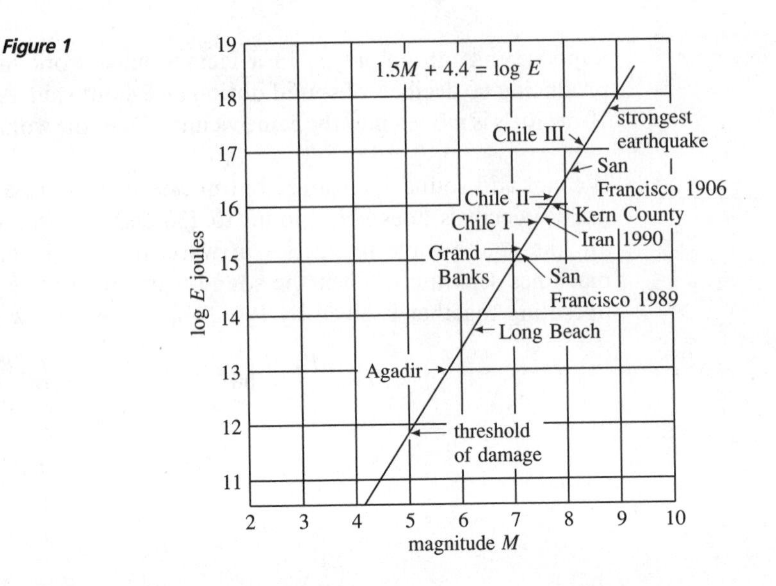

scale. A recent devastating earthquake in Iran on June 22, 1990, had a magnitude of 7.7. Magnitude 5 earthquakes, which are at the threshold of damage, may crack plaster or knock dishes from shelves. The following table indicates the degree of damage associated with earthquakes of various magnitudes.

Magnitude	Resulting Damage
8.9	Catastrophic damage
8.0	Serious damage to structures
7.5	Significant damage over a wide area
6.0	Hazardous, damage in a limited area
4.5	Felt by most people, little or no damage
2.0	Not perceptible without instruments

Although they can be detected only with seismographic instruments, numerous mild earthquakes (tremors) do occur with magnitudes ranging down to zero or even to negative values on the Richter scale.

Solving the magnitude–energy equation

$$1.5M + 4.4 = \log E$$

for the released energy E joules in terms of the magnitude M, we find that

$$E = 10^{\log E} = 10^{1.5M+4.4} = 10^{4.4} \cdot 10^{1.5M}.$$

In particular, the energy released by a tremor of magnitude 0, rounded off to two significant digits, is given by

$$E_0 = 10^{4.4} \approx 25{,}000 \text{ joules}.$$

Damage from the 1988 earthquake in Armenia.

C **Example 6** The 1988 Armenian earthquake had a magnitude of 6.8, and the 1985 earthquake in Mexico City had a magnitude of 7.8 on the Richter scale. Compare the energy release of the two earthquakes.

Solution Let E_1 denote the energy release of the magnitude 6.8 earthquake, and let E_2 denote the energy release of the magnitude 7.8 earthquake. Then,

$$\frac{E_2}{E_1} = \frac{10^{4.4} \cdot 10^{1.5(7.8)}}{10^{4.4} \cdot 10^{1.5(6.8)}} = 10^{1.5(7.8-6.8)} = 10^{1.5} \approx 32.$$

Thus, the earthquake in Mexico City released approximately 32 times as much energy as the earthquake in Armenia. ■

Measuring the Rate of Radioactive Decay As mentioned in Section 5.2, radioactive materials decay according to the formula

$$y = y_0 e^{-kt},$$

where y is the mass of the material at time t and y_0 is the original mass when $t = 0$.

Let's consider how much the mass y changes when t changes by one unit of time. The change is

$$y_0 e^{-kt} - y_0 e^{-k(t+1)} = y_0(e^{-kt} - e^{-kt-k}) = y_0(e^{-kt} - e^{-kt}e^{-k}) = y_0 e^{-kt}(1 - e^{-k}).$$

The percentage of the change in mass in one unit of time is therefore given by

$$\frac{\text{change in mass}}{\text{original mass}} \times 100\% = \frac{y_0 e^{-kt}(1 - e^{-k})}{y_0 e^{-kt}} \times 100\%$$
$$= (1 - e^{-k}) \times 100\%.$$

In other words, if the percentage of the change in mass in one unit of time is expressed as a decimal K, we have

$$K = 1 - e^{-k} \quad \text{or} \quad e^{-k} = 1 - K,$$

so that

$$-k = \ln(1 - K) \quad \text{or} \quad k = -\ln(1 - K).$$

For small values of K, it can be shown (Problem 26) that $k \approx K$.

The rate of decay of a radioactive material can be measured not only in terms of k or K, but in terms of its **half-life,** which is defined to be the period of time T required for it to decay to half of its original mass. Thus,

$$\tfrac{1}{2}y_0 = y_0 e^{-kT} \quad \text{or} \quad \tfrac{1}{2} = e^{-kT};$$

that is,

$$e^{kT} = 2 \quad \text{or} \quad kT = \ln 2.$$

It follows that

$$T = \frac{\ln 2}{k}.$$

Example 7 Potassium 42 is a radioactive element that is often used as a tracer in biological experiments. Its half-life is approximately 12.5 hours.

(a) If y_0 milligrams of potassium 42 are initially present, write a formula for the number of milligrams y present after t hours.

(b) What percent decrease in the amount of potassium 42 occurs in a sample during 1 hour?

(c) Of a 1-milligram sample of potassium 42, how much is left after 5 hours?

Solution

(a) $T = 12.5$; hence, from the equation $T = \dfrac{\ln 2}{k}$, we have (rounding off to three significant digits),

$$k = \frac{\ln 2}{T} = \frac{\ln 2}{12.5} \approx 0.0555.$$

Therefore,

$$y = y_0 e^{-0.0555t}.$$

(b) Here $k = 0.0555$, so

$$K = 1 - e^{-k} = 1 - e^{-0.0555} \approx 0.0540.$$

Thus the percentage of decrease per hour is about 5.4%.

(c) Putting $y_0 = 1$ milligram and $t = 5$ hours in the equation of part (a), we find that the amount left after 5 hours is

$$y = 1 \cdot e^{(-0.0555)(5)} = e^{-0.2775} \approx 0.758 \text{ milligram.}$$

■

Willard Libby is shown here receiving the Nobel Prize.

Measuring the Age of Fossils In 1960, the American scientist Willard Libby received the Nobel Prize in Physical Chemistry for his discovery of the technique of radiocarbon dating. When a plant or animal dies, it receives no more of the naturally occurring radioactive carbon (carbon 14, ^{14}C) from the atmosphere. Libby developed methods for determining the fraction F of the original ^{14}C that is left in a fossil and made an experimental determination of the half-life T of ^{14}C.

As we have seen,

$$T = \frac{\ln 2}{k}$$

so that

$$k = \frac{\ln 2}{T},$$

where k is the exponential decay constant for ^{14}C. If y_0 is the original amount of ^{14}C in the plant or animal when it died t years ago and y is the amount now present, then

$$y = y_0 e^{-kt}$$

and so

$$F = \frac{y}{y_0} = e^{-kt}.$$

Therefore,

$$-kt = \ln F,$$

and it follows that

$$t = -\frac{\ln F}{k} = -\frac{\ln F}{(\ln 2)/T} = -T\frac{\ln F}{\ln 2}.$$

C Example 8 In 1947, an Arab Bedouin herdsman of the Taamireh tribe climbed into a cave at Kirbext Qumran on the shores of the Dead Sea near Jericho to search for a stray goat. He came upon earthenware jars containing an incalculable treasure of ancient manuscripts called the *Dead Sea scrolls.* When the linen wrappings on the scrolls were analyzed, it was found that they contain 76% of their original ^{14}C. If the half-life of ^{14}C is $T = 5580$ years, find the age of the scrolls.

Solution Using the formula developed above, we find that

$$t = -T\frac{\ln F}{\ln 2} = -5580\frac{\ln 0.76}{\ln 2} \approx 2210 \text{ years.}$$

■

Problem Set 5.5

C In Problems 1 to 10, use logarithms to solve each exponential equation.

1. $4^x = 3$

2. $7.07^x = 2001$

3. $4^{2x+1} = 6^x$

4. $10^{\sqrt{x}} = 100$

5. $1000^{\sqrt[3]{x}} = 10$

6. $5^{3x-1} = 45^{-x}$

7. $3^{-3x+1} = e^{-x}$

8. $2^{2x+1} = 3^{2x+3}$

9. $(2.11)^{3x} = (1.77)^{x-1}$

10. $x^{\log x} = 1000x$

In Problems 11 and 12, solve for y in terms of x.

11. $3e^{4y+1} = x - 2$

12. $e^{2y} - 2xe^y - 1 = 0$

C **13.** If you invest $1000 in a savings account at 6% nominal annual interest compounded weekly, how long will it take for your money to double? Find **(a)** the exact number of weeks and **(b)** the approximate number of weeks using the rule of 70.

14. In calculus, an important function called the **inverse hyperbolic cosine** is obtained by solving the equation $x = \frac{1}{2}(e^y + e^{-y})$ for y in terms of x, where $x \geq 1$ and $y \geq 0$. Solve this equation.

C **15.** Find the pH of each substance (rounded off to two significant digits).

(a) eggs: $[H^+] = 1.6 \times 10^{-8}$ mole per liter

(b) tomatoes: $[H^+] = 6.3 \times 10^{-5}$ mole per liter

(c) milk: $[H^+] = 4 \times 10^{-7}$ mole per liter

(d) blood plasma: $[H^+] = 3.98 \times 10^{-8}$ mole per liter

C **16.** Find the hydrogen ion concentration $[H^+]$ in moles per liter of each substance (rounded off to two significant digits).

(a) vinegar: pH = 3.1

(b) beer: pH = 4.3

(c) lemon juice: pH = 2.3

(d) bile: pH = 7.9

C **17.** Atmospheric pressure at the summit of Mt. Everest on a certain day measures 25.1 centimeters of mercury, and the air temperature is 0°C. If the atmospheric pressure at sea level is 76 centimeters of mercury, use the barometric equation to find the approximate height of Mt. Everest.

C **18.** Suppose that the pilot of a light aircraft has neglected to recalibrate the altimeter, which is set to a sea-level pressure of 76 centimeters of mercury, when the actual sea-level pressure has dropped to 75 centimeters because of weather conditions. Using $T = 0°C$ in the barometric equation, find the amount of error in the pilot's altimeter reading.

C **19.** Use the barometric equation with $T = 0°C$ to find the elevation at which one-half of the atmosphere lies below and one-half lies above. [*Hint:* At this height, atmospheric pressure will have dropped to one-half its sea-level value.]

C **20.** Using the barometric equation with $T = 0°C$ and $P_0 = 76$ centimeters of mercury, sketch a graph of the height h as a function of atmospheric pressure P. gc Use a graphing calculator if you wish.

C **21.** At takeoff, a certain supersonic jet produces a sound wave of intensity 0.2 watt per square meter. Taking P_0 to be 10^{-12} watt per square meter, find the loudness in decibels of the takeoff.

22. Show that a combination of two sounds with loudnesses S_1 and S_2 decibels produces a sound of loudness $S = 10 \log(10^{S_1/10} + 10^{S_2/10})$ decibels.

C **23.** Find the ratio of the sound in watts per square meter at the threshold of pain (about 120 decibels) and at the threshold of hearing (0 decibels).

C **24.** Suppose that the smallest weight you can perceive is $W_0 = 0.5$ gram and that you can just barely notice the difference between 100 grams and 125 grams. Using the Weber–Fechner law, develop a scale of perceived heaviness. Call one unit on this scale a *heft* and write a formula for the number S of hefts corresponding to a weight of W grams.

C **25.** The brightness of a star perceived by the naked eye is measured in units called *magnitudes;* the brightest stars are of magnitude 1, and the dimmest are of magnitude 6. If I is the actual intensity of light from a star, and I_0 is the intensity of light from a just-visible star, the magnitude M corresponding to I is given by

$$M = 6 - 2.5 \log \frac{I}{I_0}.$$

Calculate the ratio of light intensities from a star of magnitude 1 and a star of magnitude 5.

26. If k is a small positive number, show that $k \approx 1 - e^{-k}$. [*Hint:* Use the approximation $e^x \approx 1 + x$ for small $|x|$.]

27. The earthquake in San Francisco on April 18, 1906, measured 8.2 on the Richter scale. The earthquake in the Santa Cruz mountains on October 18, 1989, 0004 Greenwich Mean Time, which affected the same city, had a magnitude of 7.1. Compare the energy release of the two earthquakes.

Earthquake, San Francisco, 1989

28. If E_0 is the energy released by an earthquake of magnitude 0 on the Richter scale, and E is the energy released by an earthquake of magnitude M, show that

$$M = \frac{2}{3} \log \frac{E}{E_0}.$$

C **29.** The half-life of radium is 1656 years.

(a) If y_0 grams of radium are initially present, write a formula for the number of grams y present after t years.

(b) What percent decrease occurs in the amount of radium in a sample over 1 year?

(c) How much of a 1-gram sample of radium is present after 20 years?

C **30.** The isotope strontium 90 with a half-life of 28 years was a by-product of the atmospheric nuclear tests conducted in the 1940s and 1950s. What fraction of the strontium 90 created by a nuclear test in 1950 will remain in the year 2000?

C **31.** A radioisotope supplies power to an automated space probe. As the radioisotope disintegrates, the power available to the space probe decreases according to the formula $p = 50e^{-0.001t}$, where t is the number of days since launch and p is the available power in watts.

(a) How much power will be available after 365 days?

(b) What is the half-life of the power supply?

C **32.** A manufacturing plant estimates that the value V dollars of a machine is decreasing exponentially according to the equation

$$V = 76{,}000e^{-0.14t},$$

where t is the number of years since the machine was placed in service.

(a) Find the value of the machine after 12 years.

(b) Find the "half-life" of the machine.

C **33.** Sociologists in a certain country estimate that of all couples married this year, 40% will be divorced within 15 years. Assuming that marriages fail exponentially, find the "half-life" of a marriage in that country.

C **34.** In oceanography, the *Beer–Lambert law* states that the intensity I of light below the surface of the ocean decreases exponentially with the depth d, the decay constant k depending on the amount of dissolved organic material, the pollution level, and other factors. In the surface layer of the ocean, called the *photic zone*, there is sufficient light for plant growth. Nearly all life in the ocean is dependent on microscopic plants called *phytoplankton*, which can live only in the photic zone. In coastal waters, the photic zone has a depth of as little as 5 meters, while in clear oceanic waters this zone may extend to a depth of 150 meters.

(a) Find the ratio of the decay constant k for the intensity of light in coastal waters to its value in clear oceanic waters.

(b) Assuming that the light intensity at the bottom of the photic zone is 1% of the surface light intensity, find the value of k for clear oceanic waters.

(c) Find the depth at which the light intensity in clear oceanic waters has dropped to half its value at the surface.

C **35.** Archaeologists find a scrap of bone that contains only 30% of its original amount of ^{14}C. Estimate the age of the bone.

C **36.** In engineering thermodynamics, it is shown that when n moles of a gas are compressed isothermally (that is, at a constant temperature T degrees Celsius) from volume V_0 to volume V, the resulting work W joules done on the gas is given by

$$W = n(8.314)(T + 273) \ln \frac{V_0}{V}.$$

Calculate the work done in compressing $n = 5$ moles of carbon dioxide isothermally at a temperature of $T = 100°C$ from an initial volume $V_0 = 0.5$ cubic meter to a final volume $V = 0.1$ cubic meter.

C **37.** In psychological tests it is often found that if a group of people memorize a list of nonsense words, the fraction F of these people who remember all the words t hours later is given by $F = 1 - k \ln(t + 1)$, where k is a constant depending on the length of the list of words and other factors. A certain group was given such a memory test, and after 3 hours only half of the group's members could remember all the words.

(a) Find the value of k for this experiment.

(b) Predict the approximate fraction of group members who will remember all the words after 5 hours.

5.6 Mathematical Models for Population Growth

In this section, we shall consider some of the mathematical models currently used in the life sciences to describe population growth and biological growth in general. The construction of realistic mathematical models nearly always requires the patient accumulation of experimental data, sometimes over a period of many years. In Figure 1, we have plotted points (t, N) showing the population N of the United States in the year $1790 + t$, according to the U.S. Census Bureau. These points seem to lie along a curve that is reminiscent of the graph of exponential growth (Figure 7a in

Figure 1

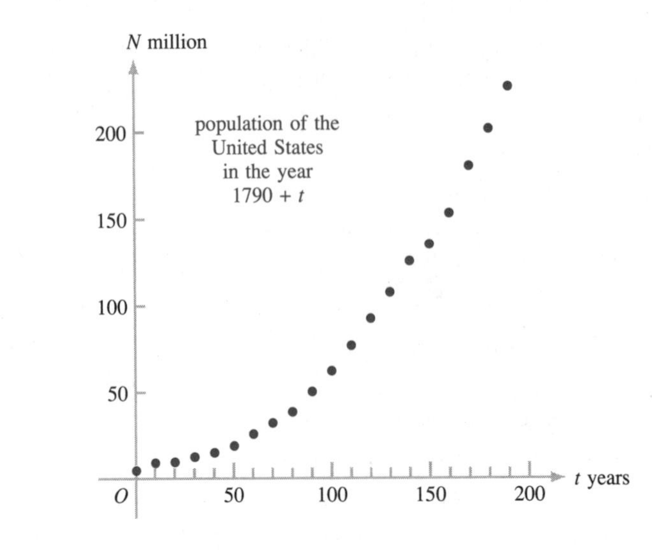

Section 5.2) and they therefore suggest the mathematical model $N = N_0 e^{kt}$ for the growth of the population of the United States.

In 1798, the English economist Thomas Malthus made similar observations about the world population in his *Essay on the Principle of Population.* Because Malthus also proposed a linear model for the expansion of food resources, he forecast that the exponentially growing population would eventually be unable to feed itself. This dire prediction had such an impact on economic thought that the exponential model for population growth came to be known as the **Malthusian model.**

Thomas Malthus (1766–1834)

Consider a population growing according to the Malthusian model

$$N = N_0 e^{kt}.$$

Of course, N_0 is the population when $t = 0$; but what is the meaning of the growth constant k? To find out, let's consider how the population changes in 1 year. At the beginning of the $(t + 1)$st year, the population is $N_0 e^{kt}$, and at the end of this year it is $N_0 e^{k(t+1)}$. During the year, the population increase is

$$N_0 e^{k(t+1)} - N_0 e^{kt} = N_0(e^{kt+k} - e^{kt}) = N_0(e^{kt}e^k - e^{kt}) = N_0 e^{kt}(e^k - 1).$$

The percentage of the increase in population during the year is therefore given by

$$\frac{\text{increase during the year}}{\text{population at the beginning of the year}} \times 100\% = \frac{N_0 e^{kt}(e^k - 1)}{N_0 e^{kt}} \times 100\%$$
$$= (e^k - 1) \times 100\%.$$

In other words, if the yearly percentage increase in population is expressed as a decimal K, we have

$$K = e^k - 1 \qquad \text{or} \qquad e^k = 1 + K;$$

so that,

$$k = \ln(1 + K).$$

Example 1 According to the U.S. Census Bureau, the population of the United States in 1980 was $N_0 = 226$ million. Suppose that the population grows according to the Malthusian model at 1.1% per year.

(a) Write an equation for the population N million of the United States t years after 1980.

(b) Predict N in the year 2000.

(c) Predict N in the year 2020.

Solution **(a)** 1.1% expressed as a decimal is $K = 0.011$. Therefore,

$$k = \ln(1 + K) = \ln 1.011 \approx 0.011,$$

and we have

$$N = N_0 e^{kt} = 226e^{0.011t}.$$

(b) In the year 2000, $t = 20$ and

$$N = 226e^{0.011(20)} = 226e^{0.22} \approx 282 \text{ million.}$$

(c) In the year 2020, $t = 40$ and

$$N = 226e^{0.011(40)} = 226e^{0.44} \approx 351 \text{ million.}$$

■

In Example 1, the growth constant k, rounded off to two significant digits, is the same as the yearly percentage of population increase expressed as a decimal K. This is no accident. Indeed, if K is small, then we can use the approximation $e^x \approx 1 + x$ (Section 5.2, page 304) with $x = K$ to get $e^K \approx 1 + K$, so that

$$k = \ln(1 + K) \approx \ln e^K = K.$$

If $K \leq 0.06$ (6% per year), the error in the approximation $k \approx K$ is less than 3%.

Whenever exponential growth is involved as in the Malthusian model, it is interesting to ask when the growing quantity doubles. If T is the **doubling time,** then

$$N_0 e^{k(t+T)} = 2N_0 e^{kt} \quad \text{or} \quad e^{kt+kT} = 2e^{kt};$$

that is

$$e^{kt}e^{kT} = 2e^{kt} \quad \text{or} \quad e^{kT} = 2.$$

Thus,

$$kT = \ln 2,$$

and we have the formula

$$T = \frac{\ln 2}{k} \approx \frac{\ln 2}{K}$$

for the doubling time.

Example 2 If the population of the United States grows according to the Malthusian model at 1.1% per year, in approximately how many years will it double?

Solution Here $K = 0.011$, so $T \approx \dfrac{\ln 2}{K} = \dfrac{\ln 2}{0.011} \approx 63$ years. ■

The Logistic Growth Model

When one-celled organisms reproduce by simple cell division in a culture containing an unlimited supply of nutrients, the Malthusian or exponential model $N = N_0 e^{kt}$ for the number N of organisms at time t is often quite accurate. In a natural environment, however, growth is often inhibited by various constraints that have a greater and greater effect as time goes on—depletion of the food supply, buildup of toxic wastes, physical crowding, and so on—and the Malthusian model may no longer apply.

If N_0 organisms are introduced into a habitat at time $t = 0$ and the population N of these organisms is plotted as a function of time t, the result is often an S-shaped curve (Figure 2). Typically, such a curve shows an increasing rate of growth up to a point P_I, called the **inflection point,** followed by a declining rate of growth as the population N levels off and approaches the maximum $N_{\max}$ that can be supported by the habitat. The horizontal line $N = N_{\max}$ is an asymptote of the graph. Notice that the graph is bending upward to the left of the inflection point and bending downward to the right of it.

Figure 2

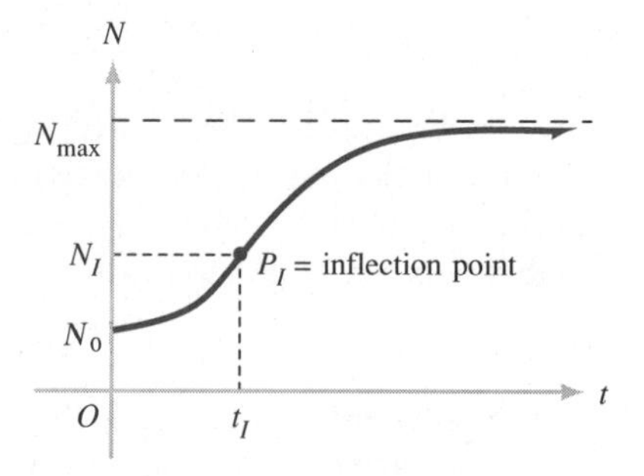

There are many equations that have graphs with the characteristic S-shape of Figure 2, and that are therefore used as mathematical models for population growth. Of these, one of the most popular is the **logistic model**

$$N = \frac{N_0 N_{\max}}{N_0 + (N_{\max} - N_0)e^{-kt}}$$

(Problem 8). Using calculus, it can be shown that the coordinates (t_I, N_I) of the inflection point P_I of the logistic model are

$$t_I = \frac{1}{k} \ln \frac{N_{\max} - N_0}{N_0}, \qquad N_I = \frac{N_{\max}}{2}.$$

Using logarithms, we can solve the logistic equation for t in terms of N; the result is

$$t = \frac{1}{k} \ln \frac{N(N_{\max} - N_0)}{N_0(N_{\max} - N)}$$

(Problem 10).

C Example 3 Suppose that the population N million of the United States grows according to the logistic model

$$N = \frac{N_0 N_{\max}}{N_0 + (N_{\max} - N_0)e^{-kt}},$$

where t is the time in years since 1780, and $k = 0.03$. If the population in 1780 was 3 million and the population in 1880 was 50 million:

(a) Find $N_{\max}$.

(b) Find N in the year 1980.

(c) Find N in the year 2000.

(d) Determine the year in which the inflection occurred.

Solution **(a)** From the data given, $N_0 = 3$ million in 1780 when $t = 0$, while in 1880 when $t = 100$, $N = 50$ million; that is,

$$50 = \frac{N_0 N_{\max}}{N_0 + (N_{\max} - N_0)e^{-kt}} = \frac{3N_{\max}}{3 + (N_{\max} - 3)e^{-0.03(100)}}.$$

It follows that

$$50[3 + (N_{\max} - 3)e^{-3}] = 3N_{\max}$$

or

$$150 + 50e^{-3}N_{\max} - 150e^{-3} = 3N_{\max}.$$

Thus,

$$(3 - 50e^{-3})N_{\max} = 150(1 - e^{-3}).$$

So, rounding off to two significant digits, we have

$$N_{\max} = \frac{150(1 - e^{-3})}{3 - 50e^{-3}} \approx 280 \text{ million.}$$

(b) In 1980, we have $t = 200$ years and

$$N = \frac{N_0 N_{\max}}{N_0 + (N_{\max} - N_0)e^{-kt}} = \frac{3(280)}{3 + (280 - 3)e^{-0.03(200)}} \approx 230 \text{ million.}$$

[*Note:* The correct value according to the 1980 census was 226 million, so our logistic model has predicted the correct value with an error of less than 2%.]

(c) In the year 2000, $t = 220$ years and

$$N = \frac{3(280)}{3 + (280 - 3)e^{-0.03(220)}} \approx 250 \text{ million.}$$

[Compare this with the prediction of 282 million according to the Malthusian model (Example 1). Notice that the "leveling off" built into the logistic model has apparently taken hold.]

(d) $t_I = \frac{1}{k} \ln \frac{N_{max} - N_0}{N_0} = \frac{1}{0.03} \ln \frac{280 - 3}{3} = \frac{1}{0.03} \ln \frac{277}{3} \approx 150$ years

Thus, according to the logistic model, the inflection would have taken place in the year $1780 + 150 = 1930$. ■

Problem Set 5.6

C **1.** The population of a small country was 10 million in 1980, and it is growing according to the Malthusian model at 3% per year.

(a) Write an equation for the population N million of the country t years after 1980.

(b) Predict N in the year 2000.

(c) Find the doubling time T for the population.

C **2.** The population of a certain city is growing according to the Malthusian model and it is expected to double in 35 years. Approximately what percent of growth will occur in this population over 1 year?

C **3.** The number of bacteria in an unrefrigerated chicken salad doubles in 3 hours. Assuming exponential growth, in how many hours will the number of bacteria be increased by a factor of 10?

4. Suppose that a population growing according to the Malthusian model increases by $100K\%$ per year. Write an *exact* formula (not an approximation) for the doubling time T in terms of K.

C **5.** The bacterium *Escherichia coli* is found in the human intestine. When *E. coli* is cultivated under ideal conditions in a biological laboratory, the population doubles in $T = 20$ minutes. Suppose that $N_0 = 10$ *E. coli* cells are placed in a nutrient broth medium extracted from yeast at time $t = 0$ minutes.

(a) Write an equation for the number N or *E. coli* bacteria in the colony t minutes later.

(b) Find N when $t = 60$ minutes.

6. Suppose that a population is growing according to the Malthusian model with doubling time T. If N_0 is the original size of the population when $t = 0$, show that the population t units of time later is given by the equation

$$N = N_0 2^{t/T}.$$

C **7.** The fruit fly *Drosophila melanogaster* is often used by biologists for genetic experiments because it breeds rapidly and has a short life cycle. Suppose a colony of *D. melanogaster* in a laboratory is observed to double in size in $T = 2$ days. Assuming a Malthusian model for growth of the colony, determine the approximate daily percentage increase in the size of the colony.

8. For the logistic model

$$N = \frac{N_0 N_{max}}{N_0 + (N_{max} - N_0)e^{-kt}},$$

show that:

(a) $N = N_0$ when $t = 0$.

(b) $N = N_{max}$ is a horizontal asymptote of the graph of N as a function of t. [*Hint:* As t gets larger and larger, e^{-kt} gets closer and closer to 0.]

C 9. Suppose that a herd of 300 deer, newly introduced into a game preserve, grows according to the logistic model (Problem 8) with $k = 0.1$ Assume that the herd has grown to 387 deer after 5 years.

(a) Find N_{max}, the maximum possible size of the herd in this habitat.

(b) Find the population of the herd after 7 years.

(c) When does the inflection occur in the population of the herd?

10. Solve the logistic equation (Problem 8) for t in terms of N.

C 11. Sketch an accurate graph of the population N of the deer herd in Problem 9 as a function of the time t in years since the herd was introduced into the habitat. gc Use a graphing calculator if you wish.

12. If a certain population N is thought to be growing according to the logistic model (Problem 8), the value of the constant k is often found experimentally as follows: First, an estimate is made for N_{max}, the maximum possible size of the population in the given habitat. Then values of N are measured corresponding to several different values of t and

$$y = \ln \frac{N}{N_{max} - N}$$

is calculated for each value of N. The points (t, y) are plotted on a graph (Figure 3) and a straight line L that "best fits" these points is drawn. The constant k

Figure 3

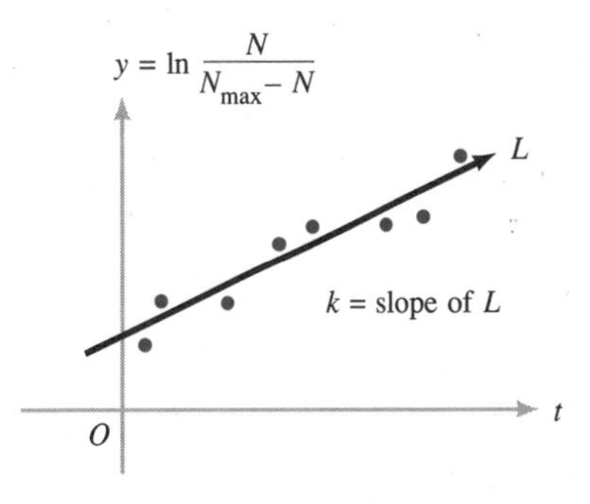

is taken to be the slope of the line L. Justify this procedure. [*Hint:* Use the result of Problem 10 to show that $kt = y - b$, where b is the value of y when $t = 0$.]

C 13. Consider the alternative growth model

$$N = N_{max}\left[1 - \left(1 - \frac{N_0}{N_{max}}\right)e^{-ct}\right]$$

for the deer herd in Problem 9, where $c = 0.05$, $N_0 = 300$, and $N_{max} = 700$. Sketch the graph of N as a function of t according to this model and compare it with the graph in Problem 11. gc Use a graphing calculator if you wish.

14. Outline a procedure for determining the constant c for the alternative growth model in Problem 13. The procedure should be similar to that in Problem 12, but with

$$y = \ln \frac{N_{max}}{N_{max} - N}.$$

C 15. The **Gompertz growth model** is $N = N_{max} \exp(-Be^{-Ct})$, where $B = \ln \dfrac{N_{max}}{N_0}$ and C is a positive constant. (Recall that $\exp x = e^x$.) Take $N_0 = 300$, $N_{max} = 700$, and $C = 0.07$; sketch the resulting graph of N as a function of t; and compare it with the graph in Problem 11. gc Use a graphing calculator if you wish.

16. For the Gompertz growth model (Problem 15), show that:

(a) N_0 is the value of N when $t = 0$.

(b) $N = N_{max}$ is a horizontal asymptote of the graph.

CHAPTER 5 Review Problem Set

1. Let $f(x) = 3^x$, $g(x) = (\frac{1}{5})^x$, $h(x) = 2^{-x}$, and $k(x) = \log_5 x$. Find:

(a) $f(3)$ **(b)** $g(0)$ **(c)** $g(-1)$ **(d)** $h(2)$
(e) $f[h(1)]$ **(f)** $g(\frac{1}{2})$ **(g)** $k(25)$ **(h)** $k(1)$
(i) $k(0.04)$ **(j)** $k(g(x))$

C **2.** Using a calculator, find the value of each quantity to as many significant digits as you can.

(a) $0.47308^{7.7703}$ **(b)** $32.273^{-0.35742}$
(c) $\ln 101$ **(d)** $\log(\log 14)$
(e) $e^{\log 14}$ **(f)** $\ln 0.00731$

In Problems 3 to 16, sketch the graph of the given function, determine its domain, its range, and any horizontal or vertical asymptotes, and indicate whether the function is increasing or decreasing. (Just make a rough sketch—do not use a calculator.)

3. $f(x) = 2^{3+x}$ **4.** $g(x) = (\frac{1}{3})^{2-x}$
5. $F(x) = (\frac{1}{10})^{x+1}$ **6.** $G(x) = 6^{-x} - 1$
7. $H(x) = 2^{-x} + 3$ **8.** $h(x) = 3 + (\frac{1}{3})^{x-3}$
9. $h(x) = 4e^{x-4}$ **10.** $F(x) = 3 - e^{2-x}$
11. $G(x) = 3e^{x-1} + 2$ **12.** $H(x) = 2 + \left(\frac{1}{e}\right)^x$
13. $f(x) = 1 + \ln(x - 1)$ **14.** $g(x) = 1 - \log x^3$
15. $h(x) = \log\sqrt{4 - x}$ **16.** $G(x) = \ln|2 - x|$

In Problems 17 and 18, solve each equation without using a calculator.

17. (a) $\log_3 81 = 5x$ **(b)** $x = \log_3 9\sqrt{3}$
(c) $\log_x \frac{1}{49} = 2$ **(d)** $\log_5|3x - 5| = 0$
(e) $\log_3 \frac{1}{81} = -2x$ **(f)** $\ln(x^2 - 4) = 0$
(g) $5^{2x} - 26(5^x) + 25 = 0$ **(h)** $2^x + 2^{-x} = 2$

18. (a) $1 + \log_7 49 = |x|$
(b) $\log|2x - 1| = 2$
(c) $\log_3(x + 1) + \log_3(x + 3) = 1$
(d) $\log_4(x + 3) - \log_4 x = 2$
(e) $7^{3x^2-2x} = 49(7^3)$
(f) $(2^{|x|+1} - 1)^{-1} = \frac{1}{3}$

C In Problems 19 and 20, *an error has been made* in each case. Show that the equation is false by evaluating both sides with the aid of a calculator.

19. (a) $\dfrac{\log \pi}{\log e} = \log \pi - \log e$? **(b)** $\dfrac{1}{\log \frac{2}{3}} = \log \frac{3}{2}$?
(c) $\log_2 3 = \log 8$?

20. (a) $\log 6 = (\log 2)(\log 3)$? **(b)** $-\log 5 = \log(-5)$?
(c) $\dfrac{1}{\ln 2} + \dfrac{1}{\ln 3} = \dfrac{1}{\ln 5}$?

21. Rewrite each expression as a single logarithm. (Make the necessary assumptions about the values of the variables.)

(a) $\ln \dfrac{x^2 - 4}{x^2 - 3x - 4} - \ln \dfrac{x^2 - 3x - 10}{2x^2 + x - 1}$

(b) $\ln \sqrt{\dfrac{x}{x + 2}} + \ln \dfrac{\sqrt{x^2 - 4}}{x^2}$

22. True or false: If $f(x) = b^x$, then $f(x^2) = [f(x)]^2$. Justify your answer.

23. Rewrite each expression as a sum or difference of multiples of logarithms. (Make the necessary assumptions about the values of the variables.)

(a) $\log \sqrt{\dfrac{4 - x}{4 + x}}$ **(b)** $\log \dfrac{y^2\sqrt{4 - y^2}}{16(3y + 7)^{3/2}y^4}$

24. Suppose that $\log_b 2 = 0.9345$, $\log_b 3 = 1.4812$, and $\log_b 7 = 2.6237$. Find each value.

(a) $\log_b 6$ **(b)** $\log_b \frac{7}{6}$ **(c)** $\log_b \frac{18}{7}$
(d) $\log_b 2^{64}$ **(e)** $\log_b 21$ **(f)** $\log_b 0.5$
(g) $\log_b 49$ **(h)** $\log_b \sqrt[5]{28}$ **(i)** $\log_b (36/\sqrt{56})$
(j) $\log_7 b$ C **(k)** b

25. Evaluate each expression without using a calculator or tables.

(a) $\log_2 16$ **(b)** $\log_{16} 2$ **(c)** $\log_5 \sqrt{5}$

(d) $\log_7 1$ (e) $\log_3 \frac{1}{27}$ (f) $\ln e^{33}$
(g) $\log 100^{-0.7}$ (h) $e^{2\ln 7}$ (i) $10^{-3\log 2}$

26. Simplify each expression.

(a) $\dfrac{10^{\log(x^2-3x+2)}}{x-2}$ (b) $e^{(1/2)\ln(x^2-2x+1)}$

(c) $\dfrac{\log y^x}{10^{\log x}}$

C **27.** Use a calculator and the base-changing formula to evaluate:

(a) $\log_2 11$ (b) $\log_8 10^{10}$
(c) $\log_{1/3} 0.707$ (d) $\log_{\sqrt{5}} \sqrt{2}$

C **28.** Use a calculator to verify each equation for the indicated values of the variables.

(a) $\log \sqrt{x} = \frac{1}{2}\log x$ for $x = 33.20477$
(b) $\log(xy) = \log x + \log y$ for $x = 72.1355$, $y = 0.774211$
(c) $\ln x = \dfrac{\log x}{\log e}$ for $x = 77.0809$
(d) $\log_a b = \dfrac{1}{\log_b a}$ for $a = 10$, $b = e$
(e) $\log y^x = x \log y$ for $y = 0.001507$, $x = 10.1333$
(f) $\log \dfrac{x}{y} = \log x - \log y$ for $x = 5.3204 \times 10^{-7}$, $y = 3.2211 \times 10^3$

C **29.** Use logarithms and a calculator to solve each exponential equation.

(a) $10^x = 20$ (b) $e^{x+3} = 5^{-2}$
(c) $5^{2x} = 4(3^x)$ (d) $3^{2x-1} = 4^{x+2}$

30. Find the domain and the x and y intercepts of the graph for each function.

(a) $f(x) = \log_2(4x-3)$ (b) $g(x) = \ln|x+1|$
(c) $h(x) = \ln(x^2-4x-4)$ (d) $F(x) = \log e^x$

C **31.** A savings account earns a nominal annual interest of 6% compounded quarterly.

(a) If there is \$5000 in the account now, how much money will it contain after 5 years?
(b) Use the rule of 70 to approximate the time required for the money to double.

C **32.** A certain bank advertises interest at a nominal annual rate of 6.5% compounded hourly. Find the effective simple annual interest rate.

C **33.** Suppose that your local savings bank offers certificates of deposit paying a nominal annual interest of 8% compounded semiannually. Use this interest rate to determine the present value to you of money in the future. If a debtor offers to pay you \$1000 eighteen months from now, what is the present value to you of this offer?

C **34.** A savings bank with \$28,000,000 in regular savings accounts is paying interest at a nominal rate of 5.5% compounded quarterly. The bank is contemplating offering the same rate of interest, but compounding continuously rather than quarterly. How much more interest will the bank have to pay out per year if the new plan is adopted?

C **35.** Consider an \$80,000 mortgage on a home for a term of 30 years at a nominal annual interest rate of 10%.

(a) What is the monthly payment on the mortgage?
(b) After 14 years, how much of the original \$80,000 will have been paid off?

C **36.** Suppose a bank offers savings accounts at a nominal annual rate r compounded n times per year. By regularly depositing a fixed sum of p dollars in such an account every $\frac{1}{n}$th of a year, money can be accumulated for future needs. Such a plan is called a *sinking fund.* If it is required that such a sinking fund yield an amount S dollars after a term of t years, the periodic payment p dollars must be

$$p = \frac{Sr}{n\left[\left(1+\dfrac{r}{n}\right)^{nt} - 1\right]}.$$

If a restaurant anticipates a capital expenditure of \$80,000 for expanding in 5 years, how much money should be deposited quarterly in a sinking fund, at a nominal annual interest rate of 10% compounded quarterly, in order to provide the required amount for the expansion?

C **37.** Consider an \$8000 car loan to be paid over a term of 36 months at a nominal annual rate of 11.5%.

(a) What is the monthly payment on the loan?

(b) At the end of the 36 months, what is the total amount of interest that will have been paid?

C **38.** An automobile dealer's advertisement reads: "No money down, $400 per month for 36 months, 12 percent annual interest rate compounded monthly on the unpaid balance, puts you in the driver's seat of a new Wildebeest." How much of the $14,400 to be paid goes toward the car and how much is interest?

39. In Problem 37, suppose that, just after the fourteenth payment is made, the car is totaled in an accident. If the owner wants to use money from the insurance settlement to pay off the remainder of the car loan in a lump sum, and if the lender agrees and assesses no charge for the early payment, how much will the lump sum be?

40. A couple wishes to invest their life savings, P dollars, but finds that inflation is reducing the value of a dollar by $100K\%$ per year. They also find that they will have to pay state and local income taxes amounting to $100Q\%$ of their interest at the end of each year. Write a formula for the nominal annual interest rate r, compounded continuously, that the couple must receive on their investment just to break even at the end of the year.

C **41.** Assume that y grows or decays exponentially as a function of x, that y_0 is the value of y when $x = 0$, and that k is the growth or decay constant. Find the indicated quantity from the information given.

(a) Growth, $k = 5$, $y_0 = 3$. Find y when $x = 2$.

(b) Decay, $k = 0.12$, $y_0 = 5000$. Find y when $x = 10$.

(c) Growth, $y = 5$ when $x = 1$, $y = 7$ when $x = 3$. Find y_0.

(d) Decay, $k = 4$. Find the percent change in y when x changes from 10 to 11.

42. Infusion of a glucose solution into the bloodstream is a standard medical technique. Medical technicians use the formula $y = A + (B - A)e^{-kt}$, where A, B, and k are positive constants, to determine the concentration y of glucose in the blood t minutes after the beginning of the infusion.

(a) Interpret the meaning of the constant B.

(b) Interpret the meaning of the constant A.

(c) The constant k has to do with the rate at which glucose is converted and removed from the bloodstream. If glucose is being removed from the bloodstream rapidly, does this suggest that k is relatively large or small?

C **43.** Suppose that the population N of cancer cells in an experimental culture grows exponentially as a function of time. If $N_0 = 3000$ cells are present when $t = 0$ days, and if the number of cells is increasing at a rate of 0.35% per day:

(a) How many cells are present in the culture after 100 days?

(b) What is the approximate doubling time?

44. Suppose that y is growing or decaying exponentially as a function of x and that k is the growth or decay constant. Assume that $y = y_1$ when $x = x_1$ and that $y = y_2$ when $x = x_2$. If $x_1 \neq x_2$, prove that

$$k = \frac{1}{x_2 - x_1} \ln \frac{y_2}{y_1}.$$

C **45.** In 1921, President Warren G. Harding presented Marie Curie a gift of 1 gram of radium on behalf of the women of the United States. Using the fact that the half-life of radium is 1656 years, determine how much of the 1-gram gift was left in 1990.

C **46.** As part of a psychological experiment, a group of students took an exam on material they had just studied in a physics course. At monthly intervals thereafter they were given equivalent exams. After t months, the average score S of the group was found to be given by $S = 78 - 55\log(t + 1)$. What was the average score **(a)** when they took the original exam and **(b)** 5 months later? **(c)** When will the group have forgotten everything they learned?

C **47.** Suppose that a supersonic plane on takeoff produces 120 decibels of sound. How many decibels of sound would be produced by two such planes taking off side by side?

C **48.** An alternative method for measuring the loudness of sound is the *sone* scale. Whereas loudness S in decibels is given by $S = 10\log(P/P_0)$, loudness s in sones is given by $s = 10^{2.4}P^{0.3}$, where P is the intensity of the sound wave in watts per square meter and $P_0 = 10^{-12}$ watt per square meter. Show that a 10-decibel increase in loudness doubles the loudness on the sone scale.

49. An alternative version of the formula for measuring the magnitude M of an earthquake is $M = \log\left(\frac{a}{T}\right) + B$, where a is the amplitude (in micrometers) of the vertical ground motion at a seismographic station, T is the period (in seconds) of the seismic wave, and B is a constant that accounts for the weakening of the seismic wave as it travels from the epicenter of the earthquake to the station. Find the magnitude M of an earthquake if $a = 250$ micrometers, $T = 20$ seconds, and $B = 4.2$. (Round off your answer to one decimal place.)

50. A heart pacemaker consists of a battery, a capacitor, and an electronically actuated switch (Figure 1). When the switch is at position A, the battery charges the capacitor, and when the switch is at position B, the capacitor discharges, sending an electrical stimulus to the heart. Here the heart acts as an electrical resistor. If E_0 is the voltage supplied by the battery, R is the resistance of the heart in ohms, and C is the capacitance of the capacitor in farads, then the voltage drop across the heart t seconds after the switch moves to position B is given by $E = E_0\exp[-t/(RC)]$. Solve this equation for t in terms of E, E_0, R, and C.

Figure 1

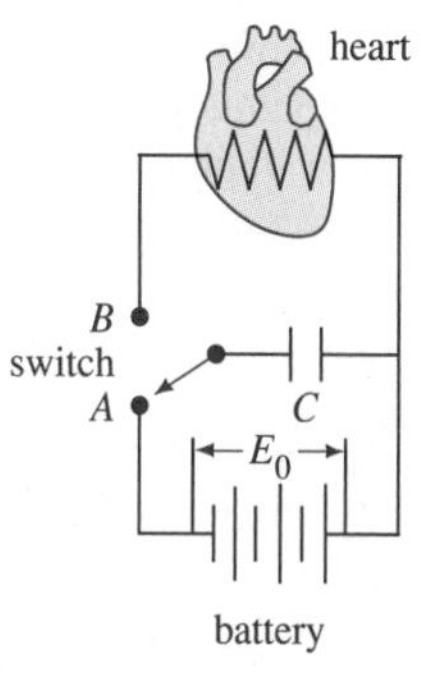

51. A population of rabbits is growing exponentially at a rate of 11.2% every 2 months. How many months will be required for the rabbit population to double?

C **52.** According to **Newton's law of cooling,** if a hot object with initial temperature T_0 is placed at time $t = 0$ in a surrounding medium with a lower temperature T_1, the object cools in such a way that its temperature T at time t is given by

$$T = T_1 + (T_0 - T_1)e^{-kt},$$

where the constant k depends on the materials involved. Suppose an iron ball is heated to 300°F and allowed to cool in a room where the air temperature is 80°F. If after 10 minutes the temperature of the ball has dropped to 250°F, what is its temperature after 20 minutes?

C **53.** A new species of plant is introduced on a small island. Assume that 500 plants were introduced initially, and that the number N of plants increases according to the logistic model (Section 5.6, page 340) with $k = 0.62$. Suppose that there are 1700 plants on the island 2 years after they were introduced.

(a) Find N_{max}, the maximum possible number of plants the island will support.

(b) Find N after 10 years.

(c) When does the inflection occur?

(d) Sketch a graph of N as a function of time t in years. gc Use a graphing calculator if you wish.

C **54.** A rumor started by a single individual is spreading among students at a university. Assume that the number N of students who have heard the rumor t days after it was started grows according to the logistic model, with $N_0 = 1$ and $N_{max} = 20{,}000$ (the entire student body). If 10,000 students have heard the rumor after 2 days:

(a) Find the value of k.

(b) Determine how long it takes until 15,000 students have heard the rumor.

CHAPTER 5 Test

1. Sketch the graph of each function, determine the domain and range, indicate whether the function is increasing or decreasing, and identify any horizontal or vertical asymptotes.

(a) $f(x) = 2^x - 1$ **(b)** $g(x) = 2e^{-x} + 3$

(c) $h(x) = \ln(x + 1)$ **(d)** $G(x) = \log_2|x + 2|$

(e) $F(x) = (\frac{1}{3})^x$

2. Find the zeros of $f(x) = 3x^2e^{3x} + 7xe^{3x} + 4e^{3x}$.

3. Simplify each expression.

(a) $\ln e^{2x^2+x}$ **(b)** $e^{8\ln(5x+1)}$ **(c)** $e^{5\ln x - 3\ln y}$

4. Solve each equation.

(a) $3^{x^2-2x} = 27$ **(b)** $\log_4(3x - 2) = 1$

(c) $\log(x + 1) - \log x = 1$

5. Write each expression as a sum or difference of multiples of logarithms. (Make the necessary assumptions about the values of the variables.)

(a) $\ln \sqrt[3]{\dfrac{3x+4}{x^5}}$ **(b)** $\log_3 \sqrt{x\sqrt{xy}}$

6. Solve the equation $5^{3x-1} = 7$ with the aid of a calculator. Round off your answer to five decimal places.

7. With the aid of a calculator, evaluate each quantity to five decimal places.

(a) $3^{-\sqrt{7}}$ **(b)** $\ln 0.0183$

(c) $\log_5 3$ **(d)** $\log 3172$

8. Bank A offers continuous compounding on one-year certificates at 8% nominal annual interest. Bank B offers the same interest but compounds daily. Find the size of a \$10,000 initial investment at each bank after one year.

9. The equation $1.5M + 4.4 = \log E$ relates the energy E in joules released by an earthquake to the magnitude M of the earthquake on the Richter scale. The Alaskan earthquake of 1964 measured 8.4 on the Richter scale. How much energy did this earthquake release?

10. The half-life of radioactive carbon 14 is 5580 years. Of 100 milligrams of carbon 14, how much will be left after 2000 years?

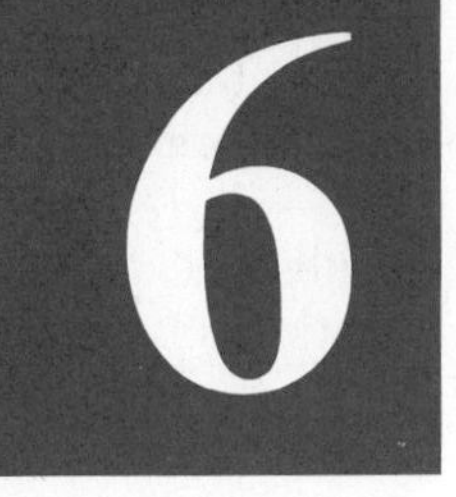

Systems of Equations and Inequalities

Intersecting planes illustrate solutions of systems of linear equations.

In Chapter 2, we studied methods for solving equations and inequalities in *one unknown.* Now we turn our attention to equations and inequalities containing *two or more unknowns.* Because applications of mathematics frequently involve many unknown quantities, it is very important to be able to deal with more than one unknown. In this chapter, we study systems of equations, determinants, systems of linear inequalities, and linear programming.

6.1 Systems of Linear Equations

A collection of two or more equations is called a **system** of equations. The equations in such a system are customarily written in a column enclosed by a brace on the left; for instance,

$$\begin{cases} 2x + y = -4 \\ x + 2y = 1 \end{cases}$$

is a system of two linear equations in the two unknowns x and y. Such a system is usually written as shown, with first-degree polynomials on the left and constants on the right of the equal signs.

Figure 1

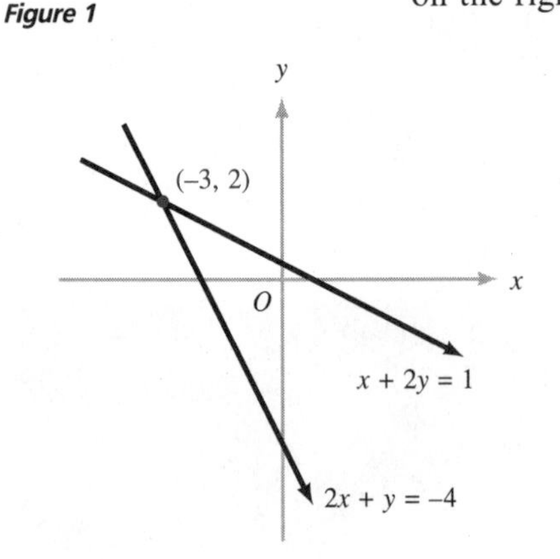

If *every* equation in a system is true when we substitute particular numbers for the unknowns, we say that the substitution is a **solution** of the system. For instance, as you can quickly verify, the substitution $x = -3$, $y = 2$ is a solution of the system

$$\begin{cases} 2x + y = -4 \\ x + 2y = 1. \end{cases}$$

Sometimes we write such a solution as an ordered pair $(-3, 2)$, with the value of x first and the value of y second. Then the solution $(-3, 2)$ can be interpreted geometrically as the point where the graph of $2x + y = -4$ intersects the graph of $x + 2y = 1$ (Figure 1).

In this section, we concentrate on solving systems of linear equations—later (in Section 6.6) we study the more general case in which nonlinear equations may be involved. The most general system of two linear equations in x and y can be written as

$$\begin{cases} ax + by = h \\ cx + dy = k, \end{cases}$$

where a, b, c, d, h, and k are constants. Unless both a and b, or both c and d, are zero, the graphs of the equations in this system are straight lines. If you draw these lines on the same coordinate system, one of the following cases will occur.

Case 1. *The two lines intersect at exactly one point and there is exactly one solution (corresponding to this point). In this case, we say that the system is* **well-determined.***

Case 2. *The two lines are parallel and therefore do not intersect. In this case, there is no solution, and we say that the equations in the system are* **inconsistent,** *or that the system is* **overdetermined.**

Case 3. *The two lines coincide. In this case, every point on the common line corresponds to a solution, and we say that the equations in the system are* **dependent,** *or that the system is* **underdetermined.**

If a system of equations has at least one solution, as in cases 1 and 3, it is said to be **consistent.**

* In applied mathematics, a problem is said to be **well-posed** or **well-determined** if it has a unique solution. If too many conditions are specified, so that the problem has no solution, it is said to be **overdetermined;** if too few conditions are specified, so that the problem has more than one solution, it is said to be **underdetermined.**

Example 1 Use graphs to decide whether each system is well-determined, inconsistent, or dependent and indicate the solutions (if any) on the graphs.

(a) $\begin{cases} x - 2y = 2 \\ x + y = 5 \end{cases}$ **(b)** $\begin{cases} x + 2y = 4 \\ 2x + 4y = -3 \end{cases}$ **(c)** $\begin{cases} 2x + 4y = 6 \\ x + 2y = 3 \end{cases}$

Solution The graphs of the equations in systems (a), (b), and (c) are shown in Figure 2.

Figure 2

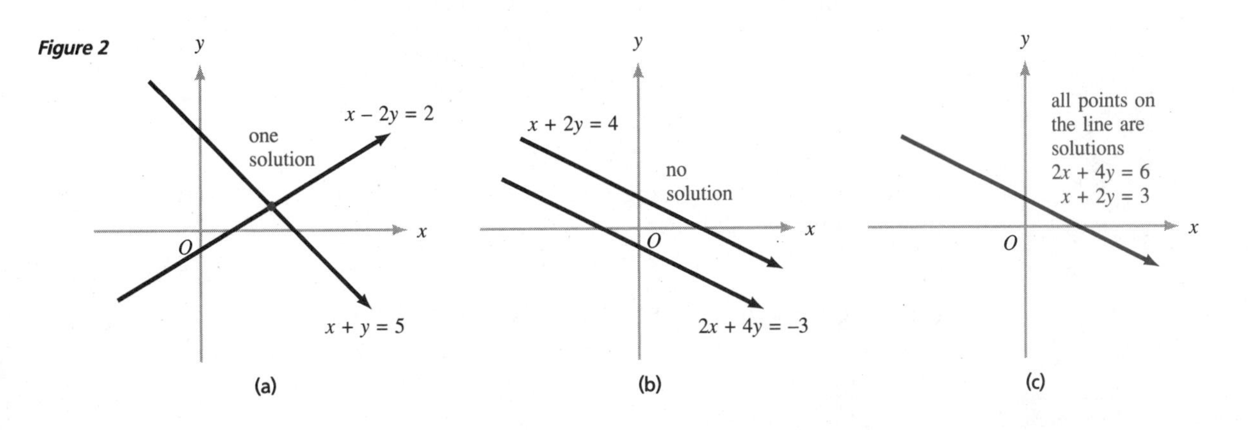

(a) In Figure 2a, the graphs intersect at a point, so there is one solution, and the system is well-determined (and therefore consistent).

(b) In Figure 2b, the two lines have the same slope ($m = -\frac{1}{2}$), so they are parallel, there is no solution, and the system is inconsistent.

(c) In Figure 2c, the two equations have the same graph, so there are infinitely many solutions (one for each point on the graph), and the system is dependent (and therefore consistent). ■

The Substitution Method

There are various algebraic methods for solving systems of equations. The **substitution method** works as follows:

> Choose one of the equations and solve it for *one* of the unknowns in terms of the remaining ones. Then substitute this solution into the remaining equation or equations. You will then have a system involving one fewer equation and one fewer unknown.

If necessary, you can repeat this procedure until the resulting system becomes simple enough to be solved.

In Examples 2 and 3, use the substitution method to solve each system of equations.

Example 2

$$\begin{cases} 2x + 3y = 1 \\ 3x - y = 7 \end{cases}$$

Solution

The second equation is easily solved for y in terms of x to obtain

$$y = 3x - 7.$$

We now substitute $3x - 7$ for y in the first equation $2x + 3y = 1$ to get

$$2x + 3(3x - 7) = 1$$

or

$$2x + 9x - 21 = 1;$$

that is,

$$11x = 22 \qquad \text{or} \qquad x = 2.$$

To obtain the corresponding value of y, we go back to the equation $y = 3x - 7$ and substitute $x = 2$; hence,

$$y = 3(2) - 7 = -1.$$

Therefore, the solution is $x = 2$ and $y = -1$, or $(2, -1)$. ■

Example 3

$$\begin{cases} 2x + y - z = 3 \\ 2x - 2y + 8z = -24 \\ x + 3y + 5z = -2 \end{cases}$$

Solution

We begin by solving the first equation for z in terms of x and y:

$$z = 2x + y - 3.$$

Now, we substitute $2x + y - 3$ for z in the second and third equations to obtain

$$\begin{cases} 2x - 2y + 8(2x + y - 3) = -24 \\ x + 3y + 5(2x + y - 3) = -2. \end{cases}$$

Simplifying these two equations, we have

$$\begin{cases} 18x + 6y = 0 \\ 11x + 8y = 13. \end{cases}$$

This simpler system can now be solved by another use of the substitution procedure. Solving the equation $18x + 6y = 0$ for y in terms of x, we get

$$y = -3x.$$

Substituting $y = -3x$ in the equation $11x + 8y = 13$, we have

$$11x + 8(-3x) = 13 \qquad \text{or} \qquad -13x = 13;$$

hence,

$$x = -1.$$

Now, we substitute $x = -1$ in the previous equation $y = -3x$ to get

$$y = -3(-1) = 3.$$

Having found that $x = -1$ and $y = 3$, we need only substitute these values back into the equation $z = 2x + y - 3$ to find that

$$z = 2(-1) + 3 - 3 = -2.$$

Hence, our solution is $x = -1$, $y = 3$, $z = -2$. This solution can also be written as an ordered *triple* $(-1, 3, -2)$. ■

In solving systems of equations by any method, it is always good practice to *check* the solutions by substitution into the original equations. We suggest that you do this in Examples 2 and 3 above, as well as in all examples and problems that follow.

The Elimination Method

Although the method of substitution can be quite efficient for solving a system of two or three equations, it tends to become cumbersome when more than three equations are involved. A more serious drawback of the substitution method is that it does not lend itself to being programmed on a computer. An alternative method, called the **method of elimination,** is similar in spirit to the method of substitution but is much more systematic. It leads directly to **matrix methods** of solution, which are easily performed by computers.

The method of elimination is based on the idea of *equivalent* systems of equations. Two systems of equations (linear or not) are said to be **equivalent** if they have exactly the same solutions. You can change a system of equations into an equivalent system by any of the following three **elementary operations:**

1. Interchange the position of two equations in the system.
2. Multiply or divide an equation by a nonzero constant.
3. Add a constant multiple of one equation to another equation.

To solve a system of linear equations by the method of elimination, use a sequence of elementary operations to reduce the given system to a simple equivalent system whose solution is obvious. Do this by using the elementary operations to eliminate unknowns from the equations (which accounts for the name of the method). The following examples illustrate the method of elimination.

Example 4 Use the method of elimination to solve the system of linear equations

$$\begin{cases} \frac{1}{3}x + y = 3 \\ -2x + 5y = 4. \end{cases}$$

Solution We begin by multiplying the first equation by 3 in order to remove the fraction $\frac{1}{3}$. The result is the equivalent system

$$\begin{cases} x + 3y = 9 \\ -2x + 5y = 4. \end{cases}$$

Next we multiply the first equation by 2 and add it to the second equation in order to eliminate x from the latter. (The actual addition of

$$2x + 6y = 18 \qquad \text{to} \qquad -2x + 5y = 4 \qquad \text{to obtain} \qquad 11y = 22$$

is best done separately to avoid messing up the system with arithmetic calculations.) The resulting equivalent system is

$$\begin{cases} x + 3y = 9 \\ \quad\;\; 11y = 22. \end{cases}$$

(Note that the first equation was *not changed* by this operation—it was used to eliminate x from the second equation, but *only* the second equation was changed.) Now we divide the second equation by 11 to obtain the equivalent system

$$\begin{cases} x + 3y = 9 \\ \qquad\; y = 2. \end{cases}$$

The resulting system is so simple that its solution is at hand. We just substitute $y = 2$ from the second equation back into the first equation to get

$$x + 3(2) = 9 \qquad \text{or} \qquad x = 3.$$

Thus, the solution is

$$x = 3 \text{ and } y = 2 \qquad \text{or} \qquad (3, 2).$$

■

In the solution above, the final step in which the value of y is substituted into a previous equation is called **back substitution.**

Example 5 Solve the system of linear equations

$$\begin{cases} x + 2y = 3 \\ 3x + 6y = 10. \end{cases}$$

Solution By adding -3 times the first equation to the second, we obtain

$$\begin{cases} x + 2y = 3 \\ \qquad\; 0 = 1. \end{cases}$$

The last equation cannot be true, so the system has *no* solution. Thus, the system is inconsistent. ■

Applications of Systems of Linear Equations

Word problems that lead to a system of linear equations can be solved by using a slight variation of the procedure given in Section 2.2 (page 85). We only need to modify step 2 of that procedure by introducing as many letters as may be necessary to represent all of the unknown quantities in the problem. Step 3 will then produce a *system* of equations to be solved for these unknowns.

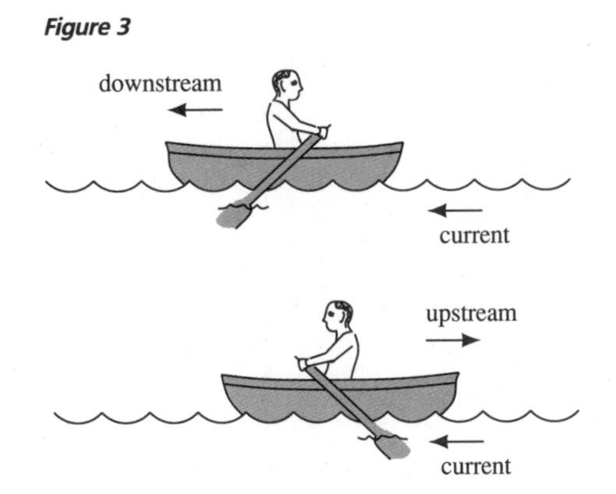

Figure 3

Example 6 A person can row a boat 9 kilometers downstream in 2 hours. Rowing back upstream, it takes 3 hours to return to the starting point (Figure 3). Find the speed with which the boat is rowed through the water and the speed of the current, assuming that both of these are constant.

Solution Let x denote the speed with which the boat is rowed through the water and let y denote the speed of the current. Then, the speed of the boat downstream is $x + y$ kilometers per hour, and the speed upstream is $x - y$ kilometers per hour. Using the formula

$$\text{distance} = (\text{rate of speed}) \cdot (\text{time}),$$

we have, for the trip downstream,

$$9 = (x + y)(2),$$

and for the trip upstream,

$$9 = (x - y)(3).$$

The second equation can be rewritten as $3 = x - y$, and so we have to solve the following system of linear equations:

$$\begin{cases} 2x + 2y = 9 \\ x - y = 3. \end{cases}$$

Solving this system using either the substitution or the elimination method (Problem 10), we find that

$$x = 3.75 \qquad \text{and} \qquad y = 0.75.$$

Thus, the boat is rowed through the water at a speed of 3.75 kilometers per hour, and the speed of the current is 0.75 kilometer per hour. ■

Problem Set 6.1

In each problem set, problems with colored numbers constitute a good representation of the main ideas of the section. Note that some of the even-numbered problems may be considerably more challenging than the odd-numbered ones.

In Problems 1 to 8, use graphs to determine whether each system is well-determined, inconsistent, or dependent and indicate the solutions (if any) on the graphs.

1. $\begin{cases} 4x + y = 5 \\ 3x - y = 2 \end{cases}$

2. $\begin{cases} y = 5x - 2 \\ y = 2x + 1 \end{cases}$

3. $\begin{cases} 3x - y = 4 \\ -6x + 2y = -8 \end{cases}$

4. $\begin{cases} x - y = 4 \\ -3x - 3y = -12 \end{cases}$

5. $\begin{cases} 2x + y = 3 \\ 4x + 2y = 7 \end{cases}$

6. $\begin{cases} x = 2 \\ y = x \end{cases}$

7. $\begin{cases} \frac{1}{6}x + \frac{1}{4}y = \frac{1}{6} \\ \frac{1}{4}x - \frac{1}{2}y = 2 \end{cases}$

8. $\begin{cases} 0.5x - 1.2y = 0.3 \\ 0.7x + 1.5y = 3.6 \end{cases}$

9. Solve the systems in Problems 1, 3, and 5 by the substitution method.

10. Solve the system of equations in Example 6.

In Problems 11 to 18, use the substitution method to solve each system of linear equations.

11. $\begin{cases} x - y = -2 \\ 2x - 3y = -7 \end{cases}$

12. $\begin{cases} x = 3 - y \\ 5y = 12 - 2x \end{cases}$

13. $\begin{cases} 2u + 3v = 5 \\ u - 2v = 6 \end{cases}$

14. $\begin{cases} 6s + 5t = 7 \\ 3s - 7t = 13 \end{cases}$

15. $\begin{cases} \frac{1}{2}x - \frac{3}{4}y = 1 \\ 3x + y = 1 \end{cases}$

16. $\begin{cases} 13x + 11y = 21 \\ 7x + 6y = -3 \end{cases}$

17. $\begin{cases} x - 2y + 3z = -3 \\ 2x - 3y - z = 7 \\ 3x + y - 2z = 6 \end{cases}$

18. $\begin{cases} s - 5t + 4u = 8 \\ 3s + t - 2u = 7 \\ -9s - 3t + 6u = 5 \end{cases}$

In Problems 19 to 26, use the method of elimination to solve each system of linear equations.

19. $\begin{cases} 2x + y = 10 \\ 3x - y = 5 \end{cases}$

20. $\begin{cases} 5u + v = 14 \\ 2u + v = 5 \end{cases}$

21. $\begin{cases} 2x + 3y = 18 \\ -7x + 9y = 15 \end{cases}$

22. $\begin{cases} x + \frac{1}{2}y = 2 \\ 3x - y = 1 \end{cases}$

23. $\begin{cases} x + y + z = 6 \\ 2x - y + z = 3 \\ x + 2y - 3z = -4 \end{cases}$

24. $\begin{cases} x - 5y + 4z = 8 \\ 3x + y - 2z = 7 \\ 9x + 3y - 6z = -5 \end{cases}$

25. $\begin{cases} 2x + y + z = 20 \\ x + 2y + 2z = 16 \\ x + y + 2z = 12 \end{cases}$

26. $\begin{cases} 2x_1 + 3x_2 - 2x_3 = 3 \\ 8x_1 + x_2 + x_3 = 2 \\ 2x_1 + 2x_2 + x_3 = 1 \end{cases}$

27. Solve the system

$$\begin{cases} \frac{1}{x} + \frac{3}{y} = -1 \\ \frac{2}{x} - \frac{1}{y} = 5 \end{cases}$$

for x and y. $\left[\textit{Hint: } \text{Let } u = \frac{1}{x} \text{ and let } v = \frac{1}{y}.\right]$

28. Solve the system

$$\begin{cases} \frac{1}{s} + \frac{4}{t} - \frac{3}{u} = 4 \\ \frac{2}{s} - \frac{3}{t} + \frac{1}{u} = 1 \\ -\frac{3}{s} + \frac{2}{t} + \frac{2}{u} = -3 \end{cases}$$

for s, t, and u. $\left[\textit{Hint: } \text{Let } x = \frac{1}{s}, y = \frac{1}{t}, \text{ and } z = \frac{1}{u}.\right]$

29. An appliance store sells dryers for $380 each and washing machines for $515 each. On a certain day, the store sells a total of 39 washers and dryers, and its total receipts for them are $16,575. Let x denote the number of dryers sold on this day and let y denote the number of washing machines sold.

(a) Write two linear equations that must be satisfied by x and y.

(b) Solve the resulting system.

30. A truck enters an expressway traveling 60 miles per hour. Meanwhile, the state police are tipped off that the truck is carrying an illegal cargo. Fifteen minutes after the truck entered the expressway, a police cruiser enters the expressway at the same point and pursues the truck at a speed of 80 miles per hour (Figure 4). How long does it take for the cruiser to overtake the truck?

Figure 4

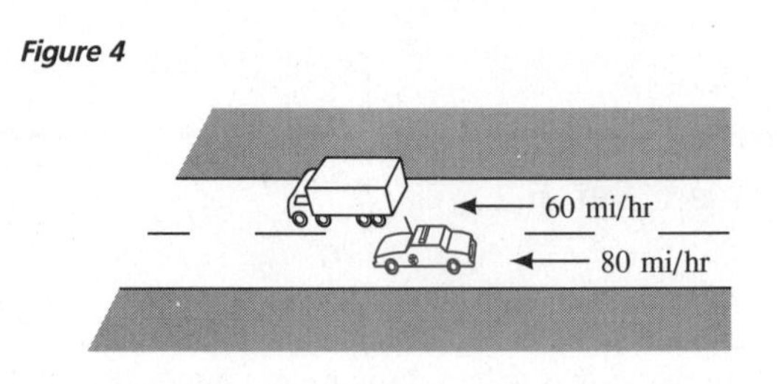

31. A motor boat traveled 60 kilometers upstream in 6 hours. Traveling back downstream, it took 4 hours to reach the starting point. Find the speed of the boat through the water and the speed of the current, assuming that both of these are constant.

32. A roof truss has the shape of an isosceles triangle (Figure 5). The outer perimeter of the truss is 20 meters, and twice the length b of the span is 3 meters more than three times the length r of the rafters. Find b and r.

Figure 5

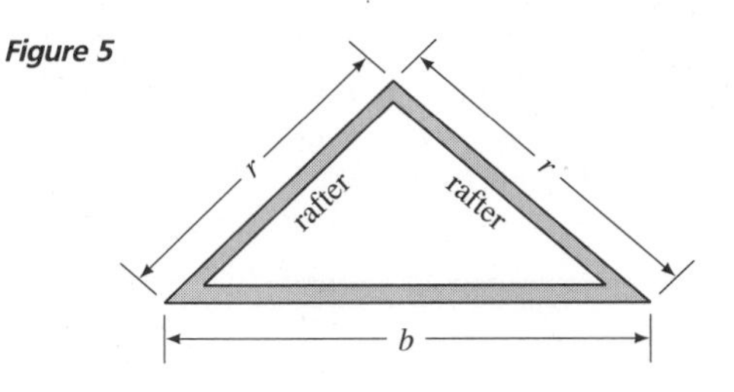

33. The price of admission for a sporting event was \$2 for adults and \$1 for children. Altogether, 925 tickets were sold, and the resulting revenue was \$1150. How many adults and how many children attended the game?

34. A veterinarian has put certain animals on a diet. Each animal receives, among other things, exactly 25 grams of protein and 9.5 grams of fat for each feeding. If the veterinarian buys two food mixes, the first containing 10% protein and 8% fat, and the second containing 20% protein and 2% fat, how many grams of each should be combined to provide the right diet for a single feeding of 10 animals?

35. The price of admission to a play was \$2.00 for adults, \$1.00 for senior citizens, and \$0.50 for children. Altogether, 270 tickets were sold and the total revenue was \$360. Twice as many children as senior citizens attended the play. How many adults, how many children, and how many senior citizens attended the play?

36. A certain three-digit number is 56 times the sum of its digits. The unit's digit is 4 more than the ten's digit. If the unit's digit and the hundred's digit were interchanged, the resulting number would be 99 less than the original number. Find the number.

37. One angle x of a triangle is 10° greater than a second angle y, and 40° less than the third angle z. Find x, y, and z.

38. A collection of dimes and quarters amounts to \$2.70. If the total number of coins is 15, how many coins of each type are in the collection?

39. Suppose that the demand and supply equations for coal in a certain marketing area are $q = -2p + 150$ and $q = 3p$, respectively, where p is the price per ton in dollars and q is the quantity of coal in thousands of tons. In economics, *market equilibrium* is said to occur when these equations hold simultaneously. Solve the system for market equilibrium.

40. A chemist has two solutions, the first containing 20% acid and the second containing 50% acid. The two solutions will be mixed to obtain 9 liters of a 30% acid solution. How many liters of each solution should be used?

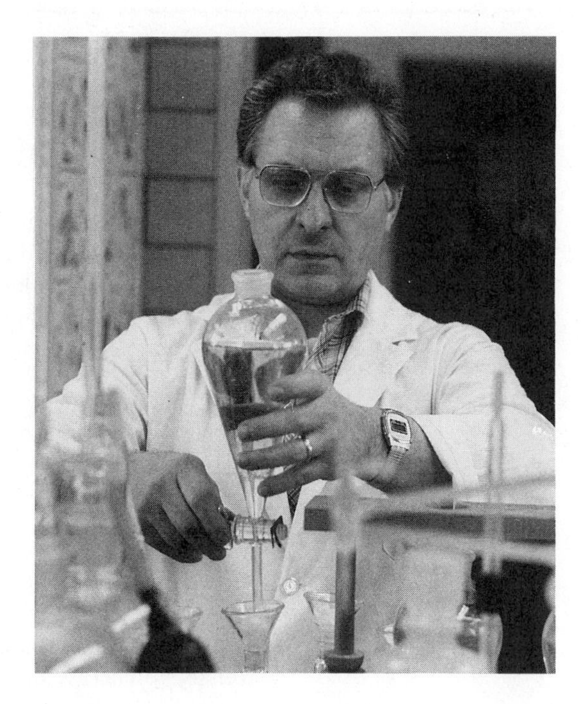

41. Suppose that x dollars is invested at a simple annual interest rate of 8.5%, and that y dollars is invested at 9.5%. If the total amount invested is \$17,000 and the total interest from the two investments at the end of the year is \$1535, find x and y.

42. A company has two hydraulic presses, an old one and an improved model. With both presses working together, a certain job is done in 2 hours and 24 minutes. On another job of the same kind, the old press is operated alone for 3 hours, and then the new press is also put into operation and the two presses finish the job together in an additional 1 hour and 12 minutes. How long would it take each press operating alone to do this job?

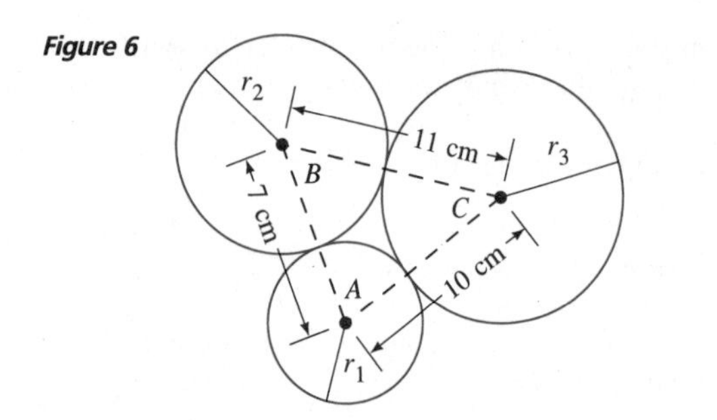

Figure 6

43. Figure 6 shows three circles of radii r_1, r_2, and r_3 that are mutually tangent to each other. The line segments between the centers A, B, and C have lengths $|\overline{AB}| =$ 7 centimeters, $|\overline{BC}| = 11$ centimeters, and $|\overline{AC}| =$ 10 centimeters. Find r_1, r_2, and r_3.

44. Determine A, B, and C so that the graph of the equation $y = Ax^2 + Bx + C$ contains the points $P_1 = (1, 3)$, $P_2 = (3, -5)$, and $P_3 = (5, 3)$.

6.2 The Elimination Method Using Matrices

When you solve a system of linear equations by the elimination method, your arithmetic involves only the numerical coefficients and constants in the equations—the unknowns just "go along for the ride." This suggests abbreviating a system of linear equations by writing only the coefficients and constants. For instance, to abbreviate the system

$$\begin{cases} \frac{1}{3}x + y = 3 \\ -2x + 5y = 4, \end{cases}$$

we write only the coefficients on the left and the constants on the right:

$$\begin{array}{rr|r} \frac{1}{3} & 1 & 3 \\ -2 & 5 & 4. \end{array}$$

(The dashed line separating the coefficients and the constants is optional.) It is customary to enclose the resulting array of numbers in square brackets,

$$\left[\begin{array}{rr|r} \frac{1}{3} & 1 & 3 \\ -2 & 5 & 4 \end{array}\right],$$

and to refer to it as the **matrix*** of the system.

Example 1 Write the matrix of the system

$$\begin{cases} y - 2x + \frac{2}{3}z = 7 \\ \phantom{y - {}} z + 5x = 3 \\ -y - z = 4. \end{cases}$$

* Some authors call this the **augmented** matrix of the system.

Solution

We begin by rewriting the equations so that the unknowns on the left appear in the order x, y, z. Missing unknowns are written with zero coefficients:

$$\begin{cases} -2x + y + \frac{2}{3}z = 7 \\ 5x + 0y + \phantom{\frac{2}{3}}z = 3 \\ 0x - y - \phantom{\frac{2}{3}}z = 4. \end{cases}$$

The corresponding matrix is

$$\left[\begin{array}{ccc|c} -2 & 1 & \frac{2}{3} & 7 \\ 5 & 0 & 1 & 3 \\ 0 & -1 & -1 & 4 \end{array}\right].$$

■

Example 2

Write the system of linear equations in x, y, and z, corresponding to the matrix

$$\left[\begin{array}{ccc|c} 0 & 5 & -1 & \frac{2}{3} \\ 1 & 2 & 0 & 0 \end{array}\right].$$

Solution

Since there are only two horizontal rows in the matrix, there are only two equations in the corresponding system:

$$\begin{cases} 0x + 5y - z = \frac{2}{3} \\ x + 2y + 0z = 0 \end{cases} \quad \text{or} \quad \begin{cases} 5y - z = \frac{2}{3} \\ x + 2y = 0. \end{cases}$$

■

The horizontal rows in a matrix are called simply the **rows,** and the vertical columns are called simply the **columns.** The numbers that appear in the matrix are called the **entries** or **elements** of the matrix. To specify a particular entry in a matrix, you can give its "address" by indicating the row and column to which it belongs. For instance, in the matrix

$$\begin{bmatrix} 0 & 8 & -5 & \frac{2}{3} \\ -5 & \frac{3}{7} & 4 & 11 \end{bmatrix},$$

the entry in the first row and third column is -5; the entry in the second row and second column is $\frac{3}{7}$; and so forth.

The three elementary operations on the equations of a system (Section 6.1) are represented by the following **elementary row operations** on the corresponding matrix:

1. Interchange two rows.
2. Multiply or divide all elements of a row by a nonzero constant.
3. Add a constant multiple of the elements of one row to the corresponding elements of another row.

We denote an interchange of the ith and jth rows of a matrix by

$$R_i \rightleftarrows R_j.$$

The operation of replacing the ith row by a nonzero constant multiple of the ith row is symbolized by

$$cR_i \to R_i,$$

where c is the constant multiplier. Similarly, the operation of replacing the jth row by the result of adding c times the ith row to the jth row is denoted by

$$cR_i + R_j \to R_j.$$

For instance, $2R_1 + R_2 \to R_2$ is read as "2 times R_1 added to R_2 replaces R_2." By performing these elementary row operations on a matrix, you can solve the corresponding system of linear equations by the elimination method. When an unknown is eliminated from a particular equation, a zero will appear in the corresponding row and column of the matrix.

For instance, the solution of Example 4 in Section 6.1 is abbreviated as follows:

$$\left[\begin{array}{rr|r} \frac{1}{3} & 1 & 3 \\ -2 & 5 & 4 \end{array}\right] \xrightarrow{3R_1 \to R_1} \left[\begin{array}{rr|r} 1 & 3 & 9 \\ -2 & 5 & 4 \end{array}\right] \xrightarrow{2R_1 + R_2 \to R_2} \left[\begin{array}{rr|r} 1 & 3 & 9 \\ 0 & 11 & 22 \end{array}\right]$$

$$\xrightarrow{\frac{1}{11}R_2 \to R_2} \left[\begin{array}{rr|r} 1 & 3 & 9 \\ 0 & 1 & 2 \end{array}\right];$$

that is,

$$\begin{cases} x + 3y = 9 \\ \quad\;\; y = 2, \end{cases}$$

from which the solution $x = 3$, $y = 2$ is found, as before, by back substitution. In this calculation, the last matrix is in **echelon form** in the sense that the following conditions are satisfied:

1. The **leading entry**—that is, the first nonzero entry, reading from left to right—in each row is 1.
2. The leading entry in each row after the first is to the right of the leading entry in the previous row.
3. If there are any rows with no leading entry—that is, rows consisting entirely of zeros—they are placed at the bottom of the matrix.

By a sequence of elementary row operations, the matrix corresponding to any system of linear equations can be brought into echelon form, and then the solution can be obtained by back substitution.

Example 3 Solve the system of equations

$$\begin{cases} 2x + y - z = 5 \\ 2x - 2y + 8z = -10 \\ \quad\;\; 4y + z = 7. \end{cases}$$

Solution We write the matrix of the system and then reduce it to echelon form by a sequence of elementary row operations as follows:

$$\left[\begin{array}{rrr|r} 2 & 1 & -1 & 5 \\ 2 & -2 & 8 & -10 \\ 0 & 4 & 1 & 7 \end{array}\right] \xrightarrow{(-1)R_1 + R_2 \to R_2} \left[\begin{array}{rrr|r} 2 & 1 & -1 & 5 \\ 0 & -3 & 9 & -15 \\ 0 & 4 & 1 & 7 \end{array}\right] \xrightarrow{(-\frac{1}{3})R_2 \to R_2}$$

$$\left[\begin{array}{rrr|r} 2 & 1 & -1 & 5 \\ 0 & 1 & -3 & 5 \\ 0 & 4 & 1 & 7 \end{array}\right] \xrightarrow{(-4)R_2 + R_3 \to R_3} \left[\begin{array}{rrr|r} 2 & 1 & -1 & 5 \\ 0 & 1 & -3 & 5 \\ 0 & 0 & 13 & -13 \end{array}\right] \xrightarrow{\frac{1}{13}R_3 \to R_3}$$

$$\left[\begin{array}{rrr|r} 2 & 1 & -1 & 5 \\ 0 & 1 & -3 & 5 \\ 0 & 0 & 1 & -1 \end{array}\right] \xrightarrow{\frac{1}{2}R_1 \to R_1} \left[\begin{array}{rrr|r} 1 & \frac{1}{2} & -\frac{1}{2} & \frac{5}{2} \\ 0 & 1 & -3 & 5 \\ 0 & 0 & 1 & -1 \end{array}\right].$$

The last matrix is in echelon form and corresponds to the system

$$\begin{cases} x + \frac{1}{2}y - \frac{1}{2}z = \frac{5}{2} \\ \quad\quad y - 3z = 5 \\ \quad\quad\quad\quad z = -1. \end{cases}$$

Now we back substitute $z = -1$ from the third equation into the second equation $y - 3z = 5$ to obtain

$$y - 3(-1) = 5 \quad \text{or} \quad y = 2.$$

Finally, we back substitute $z = -1$ and $y = 2$ into the first equation $x + \frac{1}{2}y - \frac{1}{2}z = \frac{5}{2}$ to obtain

$$x + \tfrac{1}{2}(2) - \tfrac{1}{2}(-1) = \tfrac{5}{2} \quad \text{or} \quad x = 1.$$

Therefore, the solution is

$$x = 1, \quad y = 2, \quad \text{and} \quad z = -1 \quad \text{or} \quad (1, 2, -1).$$ ■

For the relatively simple systems of linear equations considered here and in the problems at the end of this section, you can find, by trial and error, a suitable sequence of elementary row operations that reduces the matrix to echelon form. More advanced textbooks on linear algebra describe step-by-step procedures (eliminating all guesswork) for doing this. In these textbooks, you can also find a proof that, for any system of linear equations, there are just three possibilities:

Case 1. *There is exactly one solution, in which case we say that the system is* **well-determined.**

Case 2. *There is no solution, in which case we say that the equations in the system are* **inconsistent,** *or that the system is* **overdetermined.**

Case 3. *There is an infinite number of solutions, in which case we say that the equations in the system are* **dependent,** *or that the system is* **underdetermined.**

If a system of equations has at least one solution, as in cases 1 and 3, it is said to be **consistent.**

In working with systems of linear equations, you can use whatever letters you please to denote the unknowns. Sometimes it is convenient to use just one letter with different subscripts (for instance, $x_1, x_2, x_3, \ldots, x_m$) to represent the different unknowns. In any case, before you can form the corresponding matrix, you must decide in what order the unknowns are to be written in the equations. When subscripts are used, the unknowns are usually written in the numerical order of the subscripts.

In solving a system of linear equations by the matrix method, it is often convenient to bring the matrix of the system into **row reduced echelon form;** that is, to continue applying the elementary row operations until the entries *above* each leading entry (as well as below) are zero.

Example 4 Solve the system $\begin{cases} x_1 + 3x_2 - 5x_3 = 1 \\ 2x_1 + 5x_2 - 2x_3 = 4. \end{cases}$

Solution We begin by writing the matrix of the system and bringing it into echelon form:

$$\left[\begin{array}{ccc|c} 1 & 3 & -5 & 1 \\ 2 & 5 & -2 & 4 \end{array}\right] \xrightarrow{(-2)R_1 + R_2 \to R_2} \left[\begin{array}{ccc|c} 1 & 3 & -5 & 1 \\ 0 & -1 & 8 & 2 \end{array}\right]$$

$$\xrightarrow{(-1)R_2 \to R_2} \left[\begin{array}{ccc|c} 1 & 3 & -5 & 1 \\ 0 & 1 & -8 & -2 \end{array}\right].$$

The matrix is now in echelon form, but it is not in row reduced echelon form because the entry above the leading entry in the second row is not zero. To bring it into row reduced echelon form, we continue as follows:

$$\left[\begin{array}{ccc|c} 1 & 3 & -5 & 1 \\ 0 & 1 & -8 & -2 \end{array}\right] \xrightarrow{(-3)R_2 + R_1 \to R_1} \left[\begin{array}{ccc|c} 1 & 0 & 19 & 7 \\ 0 & 1 & -8 & -2 \end{array}\right].$$

The last matrix is the matrix of the system

$$\begin{cases} x_1 + 19x_3 = 7 \\ x_2 - 8x_3 = -2, \end{cases}$$

or equivalently,

$$\begin{cases} x_1 = -19x_3 + 7 \\ x_2 = 8x_3 - 2. \end{cases}$$

By assigning x_3 any value, say $x_3 = t$, we express x_1, x_2, and x_3 in terms of t:

$$x_1 = -19t + 7$$
$$x_2 = 8t - 2$$
$$x_3 = t.$$

Thus, the system is dependent, since there are infinitely many solutions, namely, $(-19t + 7, 8t - 2, t)$. For instance, $t = 0$ gives the solution $(7, -2, 0)$, $t = \frac{1}{2}$ gives the solution $(-\frac{5}{2}, 2, \frac{1}{2})$, and $t = \pi$ gives the solution $(-19\pi + 7, 8\pi - 2, \pi)$. ■

Example 5 A single storage tank supplies water through three separate outlet pipes A, B, and C to the nearby communities of Lower Mattoon, Central Mattoon, and Mattoon Heights (Figure 1). When all three pipes are delivering water, they remove 90 gallons per minute from the tank. If only pipes A and B are delivering water, they remove 60 gallons per minute, and if only pipes B and C are delivering water, they remove 70 gallons per minute from the tank. Find the rate at which each of the pipes removes water from the tank when all three are delivering water.

Figure 1

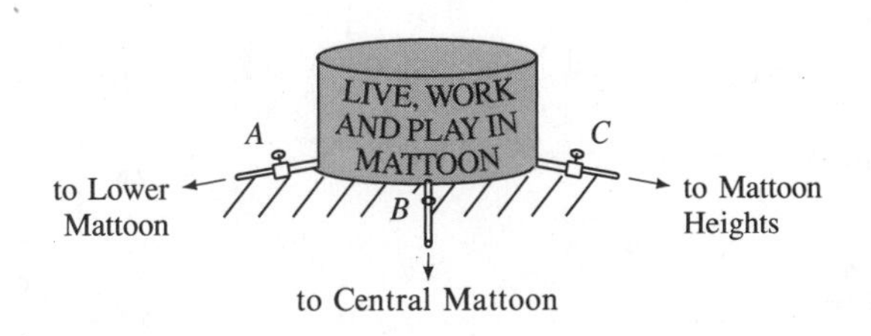

Solution Let r_1, r_2, and r_3 represent the removal rates, in gallons per minute, for pipes A, B, and C, respectively. The information given in the statement of the problem leads to the following system of linear equations:

$$\begin{cases} r_1 + r_2 + r_3 = 90 \\ r_1 + r_2 \qquad\ = 60 \\ \qquad r_2 + r_3 = 70. \end{cases}$$

The matrix of this system is

$$\left[\begin{array}{ccc|c} 1 & 1 & 1 & 90 \\ 1 & 1 & 0 & 60 \\ 0 & 1 & 1 & 70 \end{array}\right].$$

Bringing the matrix into row reduced echelon form, we obtain

$$\left[\begin{array}{ccc|c} 1 & 0 & 0 & 20 \\ 0 & 1 & 0 & 40 \\ 0 & 0 & 1 & 30 \end{array}\right],$$

which is the matrix of the system

$$\begin{cases} r_1 = 20 \\ r_2 = 40 \\ r_3 = 30. \end{cases}$$

Thus, pipes A, B, and C are removing water from the tank at rates of 20, 40, and 30 gallons per minute, respectively. ■

Problem Set 6.2

In Problems 1 and 2, write the system of linear equations in x and y or in x, y, and z corresponding to each matrix.

1. (a) $\left[\begin{array}{cc|c} 1 & 3 & 0 \\ 2 & -4 & 1 \end{array}\right]$ **(b)** $\left[\begin{array}{ccc|c} 2 & 5 & 3 & 1 \\ -3 & 7 & \frac{1}{2} & \frac{3}{4} \\ 0 & \frac{2}{3} & 0 & -\frac{4}{5} \end{array}\right]$

2. (a) $\left[\begin{array}{cc|c} 0.1 & 3.2 & -1.7 \\ -4.4 & 0 & 0 \end{array}\right]$ **(b)** $\left[\begin{array}{cc|c} 3 & 2 & 3 \\ 1 & \frac{2}{3} & 1 \\ 5 & 0 & -2 \end{array}\right]$

In Problems 3 and 4, write the matrix of each system of linear equations.

3. (a) $\begin{cases} \frac{3}{4}x - \frac{2}{3}y = \frac{1}{7} \\ -x + 5y = 6 \end{cases}$ **(b)** $\begin{cases} 40x - z + 22y = -17 \\ y + z = 0 \\ 17y - 13x + 12z = 5 \end{cases}$

4. (a) $\begin{cases} 0.5x_1 + 3.2x_2 = 7.1 \\ 5.3x_1 - 3.0x_2 = -6.5 \end{cases}$ **(b)** $\begin{cases} 3x_3 - 2x_2 = x_1 \\ 2x_1 - 5x_3 = -x_2 \\ x_2 + x_3 = 6 \end{cases}$

In Problems 5 to 24, write the matrix of the system of linear equations and solve the system by the elimination method using matrices.

5. $\begin{cases} x + y = 4 \\ x - 4y = 8 \end{cases}$

6. $\begin{cases} x + 6y = 7 \\ 11x - 7y = -10 \end{cases}$

7. $\begin{cases} 3x + y = 15 \\ 3x - 7y = 15 \end{cases}$

8. $\begin{cases} -6x_1 + 2x_2 = 3 \\ 2x_1 + 5x_2 = 3 \end{cases}$

9. $\begin{cases} 4s + 3t = 17 \\ 2s + 3t = 13 \end{cases}$

10. $\begin{cases} \frac{1}{2}x + \frac{2}{3}y = 6 \\ -\frac{3}{2}x + \frac{1}{2}y = -3 \end{cases}$

11. $\begin{cases} x + y = 1 \\ 2x + 2y = 0 \end{cases}$

12. $\begin{cases} 5x - 2y = y - 1 \\ 4x - 5y = 3 - 2x \end{cases}$

13. $\begin{cases} \frac{1}{3}x_1 + \frac{1}{6}x_2 = 1 \\ x_1 - x_2 = 3 \end{cases}$

14. $\begin{cases} x + by = 2 \\ bx + y = 3 \end{cases}$

15. $\begin{cases} x + 5y - z = -7 \\ 3x + 4y - 2z = 2 \\ 2x - 3y + 5z = 19 \end{cases}$

16. $\begin{cases} 3x + 2y + 5z = 7 \\ 2x - 3y - 2z = -3 \\ x + 2y + 3z = 5 \end{cases}$

17. $\begin{cases} x + 3y + z = 0 \\ 2y + 4z = 1 \\ -x + 3z = 2 \end{cases}$

18. $\begin{cases} x + 2y - z = 0 \\ 2x - y + 3z = 1 \\ 3x - 2y = -1 \end{cases}$

19. $\begin{cases} 2x + 3z = 1 \\ x - y + 2z = 0 \\ 3x - y + 5z = -1 \end{cases}$

20. $\begin{cases} 3x_1 + 5x_2 - x_3 = 2 \\ x_2 + x_3 = 3 \\ 2x_1 + 3x_2 - x_3 = 4 \end{cases}$

21. $\begin{cases} 3x + 2y + z = 6 \\ x - 3y + 5z = 3 \end{cases}$

22. $\begin{cases} 2x_1 - x_2 - x_3 = 4 \\ x_1 + 2x_2 + 3x_3 = 8 \end{cases}$

23. $\begin{cases} x_1 + 2x_2 - x_3 = 0 \\ 2x_1 - 2x_2 + x_3 = 0 \\ 6x_1 + 4x_2 + 3x_3 = 0 \end{cases}$

24. $\begin{cases} \frac{2}{3}x_1 + \frac{1}{4}x_2 - \frac{1}{3}x_3 = 3 \\ -\frac{3}{2}x_1 + \frac{1}{8}x_2 + x_3 = 1 \\ \frac{1}{2}x_1 - x_2 + x_3 = 4 \end{cases}$

In Problems 25 to 34, write a system of linear equations corresponding to the information given in each problem, write the matrix for the system of equations, and solve the problem by bringing the matrix into echelon or row reduced echelon form.

25. A sociologist is planning an opinion survey on school busing by making a total of 600 telephone contacts and 210 home contacts. A first survey team can make 40 telephone contacts and 10 home contacts in one hour, and a second survey team can make 30 telephone contacts and 15 home contacts in one hour. How many hours should be scheduled for each survey team?

26. A plumber charges a fixed charge plus an hourly rate for service on a house call. The plumber charged \$70 to repair a water tank that required 2 hours of labor and \$100 to repair a water tank that took 3.5 hours. Find the plumber's fixed charge and hourly rate.

27. A department store has sold 80 men's suits of three different types at a discount. If the suits had been sold at their original prices—type I suits for \$80, type II suits for \$90, and type III suits for \$95—the total receipts would have been \$6825. However, the suits were sold for \$75, \$80, and \$85, respectively, and the total receipts amounted to \$6250. Determine the number of suits of each type sold during the sale.

28. Find an equation of a circle having the form $x^2 + y^2 + Ax + By + C = 0$ that contains the three points $P_1 = (-2, 5)$, $P_2 = (1, 4)$, and $P_3 = (-3, 6)$.

29. On a certain date, 3 pounds of coffee, 4 quarts of milk, and 2 cans of tuna fish cost \$13.20. One can of tuna fish cost twice as much as 1 quart of milk. Six months later, because of inflation, the price of coffee had increased by 15%, the price of milk by 5%, and the price of tuna fish by 10%; so the same grocery order costs \$14.82. Find the original prices of a pound of coffee, a quart of milk, and a can of tuna fish.

30. Figure 2 shows the intersection of two one-way streets, one running from west to east and the other running from north to south. Traffic engineers have found that, on the average, a total of 200 cars enter the intersection (from the north and from the west) and 90 cars leave the intersection headed east between 4:00 and 5:00 P.M. each business day. On the average, of all cars entering the intersection, 25% make a turn and 75% go straight. On the average, how many cars enter the intersection from the north and how many cars that leave the intersection headed south actually entered from the west during this period of time?

Figure 2

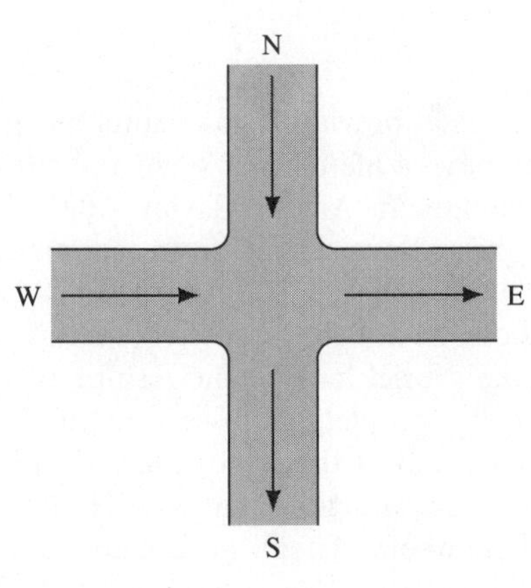

31. A petrochemical company produces three industrial-grade organic solvents, A, B, and C. In each case, the manufacturing process emits carbon monoxide (CO), hydrogen sulfide (H_2S), and oxides of nitrogen (NO_x). The following table shows the number of cubic meters of each gas emitted in producing one metric ton of each solvent, and it shows the maximum total amount of each gas that can be emitted per day according to pollution standards.

	CO	H_2S	NO_x
A	19	10	12
B	18	9	7
C	15	4	8
Max	200	80	100

How many metric tons of each solvent can be produced per day at the maximum allowable pollution standards?

C **32.** For demographic purposes, the population of a certain industrial country is divided into residents of three regions, C = city residents, S = residents of suburbs, and R = residents of rural areas. The following table shows the percent of the population that moves from region to region over a period of one year:

		To	
	City	Suburb	Rural
From City	—	15%	5%
From Suburb	4%	—	6%
From Rural	2%	3%	—

At the end of a one-year period, $C = 5.00$ million, $S = 4.56$ million, and $R = 2.24$ million. Find C, S, and R at the beginning of this one-year period. Round off your answers to three significant digits.

33. Three small colleges, A, B, and C, located in different parts of the country, agree to a student exchange program. The total enrollment of the three colleges, taken together, is 8300 students. Under the exchange program, 5% of students at college A go to college B, 7% of students at college B go to college C, and 4% of students at college C go to college A. The bursars at the three colleges are surprised when it turns out that in spite of the exchange program, student enrollment remains unchanged at each of the individual colleges. Find the student enrollment at colleges A, B, and C.

34. Professor Grumbles, who teaches morning and afternoon statistics classes of equal size, is accused of male chauvinism because in the two classes taken together, 80% of the male students passed, but only 20% of the female students did. However, the professor contends that the accusation is false because in the morning class, 10% of the men and 10% of the women passed; whereas in the afternoon class, 90% of the men and 90% of the women passed. Furthermore, the total number of men in the two classes is the same as the total number of women. Is this possible, and if so, how?*

35. According to Kirchhoff's laws for an electrical network, such as that shown in Figure 3:

(i) The sum of the voltage increases around each closed loop must equal the sum of the voltage decreases around that loop (*loop law*).

(ii) At any junction point (such as points P and Q in Figure 3), the sum of the currents into the junction must equal the sum of the currents out of the junction (*junction law*).

The voltage across a battery (such as E_1 volts and E_2 volts in Figure 3) increases in the direction from negative $(-)$ to positive $(+)$. The voltage decreases across a resistor R ohms in the direction of the current flow I amperes through the resistor, and the amount of the decrease is IR volts (*Ohm's law*). Write a system of equations that describes the network in Figure 3 according to Kirchhoff's laws.

Figure 3

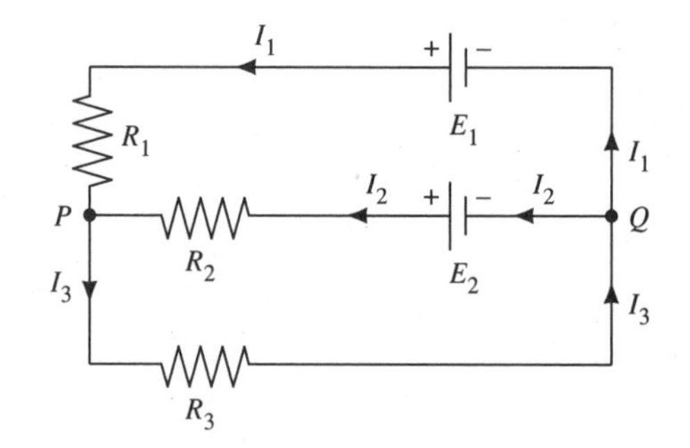

36. For the electrical network in Figure 3, find the currents I_1, I_2, and I_3 amperes given that $R_1 = 1$ ohm, $R_2 = 2$ ohms, $R_3 = 2$ ohms, $E_1 = 12$ volts, and $E_2 = 16$ volts. Use the equations obtained in Problem 35.

* This problem illustrates a situation, called *Simpson's paradox*, that appears to go against one's numerical intuition. For more information on this curious paradox, see C.H. Wagner, "Simpson's Paradox in Real Life," *The American Statistician*, Vol. 36, No. 1, 1982, pp. 46–48.

Arthur Cayley

6.3 Matrix Algebra

In Section 6.2, we used matrices merely as abbreviations for simultaneous systems of linear equations. Matrices, however, have a life of their own, and since their invention in 1858 by the English mathematician Arthur Cayley (1821–1895), they have played an ever increasing role in applications ranging from economics to quantum mechanics.

Under suitable conditions, matrices can be added, subtracted, multiplied, and (in a sense) divided. In this section, we take a brief look at the resulting "algebra" of matrices. Here we make no attempt to be complete, and we recommend that you look into a textbook on linear algebra or matrix theory for more details.

We use capital letters A, B, C, and so on to denote matrices. Here we consider only matrices whose entries are real numbers. These entries are arranged in a rectangular pattern of horizontal rows and vertical columns; for instance, the matrix

$$A = \begin{bmatrix} -1 & 0 & 5 & 7 & 2 \\ 3 & -2 & \frac{2}{3} & 0 & \frac{1}{2} \\ \frac{5}{4} & 0 & -4 & 5 & 1 \end{bmatrix}$$

has 3 rows and 5 columns. If a matrix has n rows and m columns, we call it an **n by m matrix.** Thus, A is a 3 by 5 matrix. By a **square matrix,** we mean a matrix with the same number of rows as columns. For instance, the matrix

$$\begin{bmatrix} 1 & -5 \\ -3 & 2 \end{bmatrix}$$

is a square 2 by 2 matrix. Notice that an n by m matrix has nm elements.

If A and B are n by m matrices and if each element of A is equal to the corresponding element of B, we say that the matrices A and B are **equal** and we write

$$A = B.$$

Such a **matrix equation** represents nm ordinary equations in a highly compact form.

If C and D are two n by m matrices, then their **sum** $C + D$ is defined to be the n by m matrix obtained by adding the corresponding elements of C and D. Likewise, the **difference** $C - D$ is defined to be the n by m matrix obtained by subtracting the elements of D from the corresponding elements of C.

Example 1 Let $C = \begin{bmatrix} 2 & 1 & 4 \\ 3 & -5 & 8 \end{bmatrix}$ and $D = \begin{bmatrix} -3 & 2 & -1 \\ 4 & -1 & 5 \end{bmatrix}$. Find:

(a) $C + D$ **(b)** $C - D$

Solution

(a)
$$C + D = \begin{bmatrix} 2 & 1 & 4 \\ 3 & -5 & 8 \end{bmatrix} + \begin{bmatrix} -3 & 2 & -1 \\ 4 & -1 & 5 \end{bmatrix}$$
$$= \begin{bmatrix} 2 + (-3) & 1 + 2 & 4 + (-1) \\ 3 + 4 & -5 + (-1) & 8 + 5 \end{bmatrix}$$
$$= \begin{bmatrix} -1 & 3 & 3 \\ 7 & -6 & 13 \end{bmatrix}$$

(b)
$$C - D = \begin{bmatrix} 2 & 1 & 4 \\ 3 & -5 & 8 \end{bmatrix} - \begin{bmatrix} -3 & 2 & -1 \\ 4 & -1 & 5 \end{bmatrix}$$
$$= \begin{bmatrix} 2 - (-3) & 1 - 2 & 4 - (-1) \\ 3 - 4 & -5 - (-1) & 8 - 5 \end{bmatrix} = \begin{bmatrix} 5 & -1 & 5 \\ -1 & -4 & 3 \end{bmatrix}$$ ■

Notice that you can add or subtract matrices only if they have the same shape—that is, only if they have the same number of rows and the same number of columns. Because matrices are added by adding their corresponding elements, it follows from the commutative and associative properties of real numbers (Section 1.1) that matrix addition is also commutative and associative. Thus, if A, B, and C are matrices of the same shape, we have

$$A + B = B + A \qquad \text{(commutative property of addition)}$$

and $\quad A + (B + C) = (A + B) + C \qquad$ (associative property of addition).

A matrix all of whose elements are zero is called a **zero matrix.** The zero matrix with n rows and m columns is denoted by $0_{n,m}$. For instance,

$$0_{1,3} = [0 \quad 0 \quad 0], \qquad 0_{2,2} = \begin{bmatrix} 0 & 0 \\ 0 & 0 \end{bmatrix}, \qquad 0_{4,3} = \begin{bmatrix} 0 & 0 & 0 \\ 0 & 0 & 0 \\ 0 & 0 & 0 \\ 0 & 0 & 0 \end{bmatrix},$$

and so forth. The subscripts indicating the shape of a zero matrix are often omitted because you can tell from the context how many rows and how many columns are involved. For instance, we have the property

$$A + 0 = A \qquad \text{(additive identity property)},$$

where it is understood that 0 denotes the zero matrix with the same shape as the matrix A.

If A is a matrix, we define $-A$ to be the matrix obtained by multiplying each element of A by -1. For instance,

$$-\begin{bmatrix} 2 & -\frac{1}{2} & 4 & 0 \\ -3 & 2 & 0 & -5 \end{bmatrix} = \begin{bmatrix} -2 & \frac{1}{2} & -4 & 0 \\ 3 & -2 & 0 & 5 \end{bmatrix}.$$

More generally, if k is any real number, we define kA to be the matrix obtained by multiplying each element of A by k. In particular,

$$(-1)A = -A.$$

Notice that $-A$ is the **additive inverse** of A in the sense that

$$A + (-A) = 0 \qquad \text{(additive inverse property)}.$$

Example 2 If $A = \begin{bmatrix} 3 & -1 & 2 \\ 0 & 2 & -4 \end{bmatrix}$, find **(a)** $-A$ and **(b)** $\frac{1}{2}A$.

Solution

(a) $-A = -\begin{bmatrix} 3 & -1 & 2 \\ 0 & 2 & -4 \end{bmatrix} = \begin{bmatrix} -3 & 1 & -2 \\ 0 & -2 & 4 \end{bmatrix}$

(b) $\frac{1}{2}A = \frac{1}{2}\begin{bmatrix} 3 & -1 & 2 \\ 0 & 2 & -4 \end{bmatrix} = \begin{bmatrix} \frac{3}{2} & -\frac{1}{2} & 1 \\ 0 & 1 & -2 \end{bmatrix}$ ■

The product of matrices is defined in a somewhat unexpected way. In order to introduce matrix multiplication, we begin by considering special matrices having only one row or column. A 1 by n matrix,

$$R = [x_1 \quad x_2 \quad x_3 \quad \cdots \quad x_n],$$

is called a **row vector,** and an n by 1 matrix,

$$C = \begin{bmatrix} y_1 \\ y_2 \\ y_3 \\ \vdots \\ y_n \end{bmatrix},$$

is called a **column vector.**

If the number of elements in R is the same as the number of elements in C, we define the **product** RC of R and C to be the number obtained by pairing each element of R with the corresponding element of C, multiplying these pairs, and adding the resulting products. Thus,

$$RC = x_1y_1 + x_2y_2 + x_3y_3 + \cdots + x_ny_n.$$

Example 3 Find RC if $R = [2 \quad 8 \quad 3]$ and $C = \begin{bmatrix} 60 \\ 20 \\ 300 \end{bmatrix}$.

Solution

$$RC = [2 \quad 8 \quad 3]\begin{bmatrix} 60 \\ 20 \\ 300 \end{bmatrix} = 2(60) + 8(20) + 3(300) = 1180$$

■

Although the "row by column" product may seem contrived at first, it has many practical uses. For instance, if a furniture store sells 2 tables, 8 chairs, and 3 sofas for \$60 per table, \$20 per chair, and \$300 per sofa, the product of

the **demand vector** $[2 \quad 8 \quad 3]$

and the **revenue vector** $\begin{bmatrix} 60 \\ 20 \\ 300 \end{bmatrix}$

gives the **total revenue**

$$[2 \quad 8 \quad 3]\begin{bmatrix} 60 \\ 20 \\ 300 \end{bmatrix} = 2(60) + 8(20) + 3(300) = 1180 \text{ dollars.}$$

As another example, notice that the linear equation

$$a_1x_1 + a_2x_2 + a_3x_3 + \cdots + a_nx_n = k$$

can be written in matrix form as

$$[a_1 \quad a_2 \quad a_3 \quad \cdots \quad a_n]\begin{bmatrix} x_1 \\ x_2 \\ x_3 \\ \vdots \\ x_n \end{bmatrix} = k.$$

The rows of an n by m matrix can be regarded as row vectors, and its columns as column vectors. This permits us to give the following definition.

Definition 1 **The Product of Matrices**

Let A be an n by m matrix and let B be an m by p matrix. We define the **product** AB to be the n by p matrix determined by the following procedure: To find the element in the ith row and jth column of AB, we multiply the ith row of A by the jth column of B.

Example 4 Find AB if $A = \begin{bmatrix} 3 & -2 & 4 \\ 5 & 1 & -3 \end{bmatrix}$ and $B = \begin{bmatrix} 3 & 6 \\ -4 & 5 \\ 2 & -2 \end{bmatrix}$.

Solution Here A is a 2 by 3 matrix and B is a 3 by 2 matrix, so AB will be a 2 by 2 matrix.

$$AB = \begin{bmatrix} 3 & -2 & 4 \\ 5 & 1 & -3 \end{bmatrix}\begin{bmatrix} 3 & 6 \\ -4 & 5 \\ 2 & -2 \end{bmatrix} = \begin{bmatrix} \text{1st row of } A \text{ times} & \text{1st row of } A \text{ times} \\ \text{1st column of } B & \text{2nd column of } B \\ \text{2nd row of } A \text{ times} & \text{2nd row of } A \text{ times} \\ \text{1st column of } B & \text{2nd column of } B \end{bmatrix}$$

$$= \begin{bmatrix} 3(3) + (-2)(-4) + 4(2) & 3(6) + (-2)(5) + 4(-2) \\ 5(3) + 1(-4) + (-3)(2) & 5(6) + 1(5) + (-3)(-2) \end{bmatrix} = \begin{bmatrix} 25 & 0 \\ 5 & 41 \end{bmatrix}.$$ ■

Notice that two matrices can be multiplied only if they fit together in the sense that the first matrix has as many columns as the second matrix has rows. If they do fit together in this way, the product has as many rows as the first matrix and as many columns as the second. If A is an n by m matrix, B is an m by p matrix, and C is a p by q matrix, it can be shown that

$$A(BC) = (AB)C \qquad \text{(associative property of multiplication).}$$

Distributive properties can also be proved, so that for matrices of the appropriate shapes,

$$A(B + C) = AB + AC \qquad \text{(distributive property)}$$

and

$$(D + E)F = DF + EF \qquad \text{(distributive property).}$$

If A and B are square matrices of the same size, we can form the product AB and also the product BA; however, in general, $AB \neq BA$, so:

The commutative property of multiplication fails for matrices.

Example 5 Let $A = \begin{bmatrix} 1 & -1 \\ 2 & 3 \end{bmatrix}$ and $B = \begin{bmatrix} 5 & -2 \\ 4 & 1 \end{bmatrix}$. Find:

(a) AB **(b)** BA

Solution

(a) $AB = \begin{bmatrix} 1 & -1 \\ 2 & 3 \end{bmatrix}\begin{bmatrix} 5 & -2 \\ 4 & 1 \end{bmatrix} = \begin{bmatrix} 1(5) + (-1)4 & 1(-2) + (-1)1 \\ 2(5) + 3(4) & 2(-2) + 3(1) \end{bmatrix} = \begin{bmatrix} 1 & -3 \\ 22 & -1 \end{bmatrix}$

(b) $BA = \begin{bmatrix} 5 & -2 \\ 4 & 1 \end{bmatrix}\begin{bmatrix} 1 & -1 \\ 2 & 3 \end{bmatrix} = \begin{bmatrix} 5(1) + (-2)2 & 5(-1) + (-2)3 \\ 4(1) + 1(2) & 4(-1) + 1(3) \end{bmatrix} = \begin{bmatrix} 1 & -11 \\ 6 & -1 \end{bmatrix}$

Notice that $AB \neq BA$. ■

Using the idea of matrix multiplication, you can write a system of linear equations such as

$$\begin{cases} a_1x_1 + a_2x_2 + a_3x_3 = k_1 \\ b_1x_1 + b_2x_2 + b_3x_3 = k_2 \\ c_1x_1 + c_2x_2 + c_3x_3 = k_3 \end{cases}$$

in the matrix form

$$\begin{bmatrix} a_1 & a_2 & a_3 \\ b_1 & b_2 & b_3 \\ c_1 & c_2 & c_3 \end{bmatrix}\begin{bmatrix} x_1 \\ x_2 \\ x_3 \end{bmatrix} = \begin{bmatrix} k_1 \\ k_2 \\ k_3 \end{bmatrix}.$$

Thus,

$$AX = K,$$

where A represents the **matrix of coefficients**

$$A = \begin{bmatrix} a_1 & a_2 & a_3 \\ b_1 & b_2 & b_3 \\ c_1 & c_2 & c_3 \end{bmatrix},$$

X represents the column vector of unknowns, and K represents the column vector of constants:

$$X = \begin{bmatrix} x_1 \\ x_2 \\ x_3 \end{bmatrix}, \qquad K = \begin{bmatrix} k_1 \\ k_2 \\ k_3 \end{bmatrix}.$$

If $AX = K$ were an ordinary equation, you could multiply both sides by the reciprocal of A to obtain the solution

$$X = A^{-1}K.$$

As we shall see, it is often possible to solve the matrix equation in much the same way by using the **multiplicative inverse** A^{-1} of the matrix A.

A square n by n matrix with 1 in each position on the diagonal running from upper left to lower right, and zeros elsewhere, is called the ***n* by *n* identity matrix** and is denoted by I_n. For instance.

$$I_2 = \begin{bmatrix} 1 & 0 \\ 0 & 1 \end{bmatrix} \quad \text{and} \quad I_3 = \begin{bmatrix} 1 & 0 & 0 \\ 0 & 1 & 0 \\ 0 & 0 & 1 \end{bmatrix}.$$

The subscript indicating the size of the identity matrix is often omitted because you can tell from the context how many rows and columns are involved. An identity matrix I plays a role in matrix algebra similar to the role played by 1 in ordinary algebra. In particular, for matrices C and D of the appropriate shape,

$$IC = C \quad \text{and} \quad DI = D \qquad \text{(multiplicative identity property).*}$$

By analogy with the reciprocal a^{-1} of a nonzero number in ordinary algebra, we have the following definition in matrix algebra.

Definition 2 **The Inverse of a Square Matrix**

Let A be a square n by n matrix. We say that A is **nonsingular** if there exists an n by n matrix A^{-1} such that

$$AA^{-1} = A^{-1}A = I.$$

If A^{-1} exists, it is called the **inverse** of the matrix A.

It can be shown that a nonsingular matrix A has a unique inverse A^{-1}, which you can find by carrying out the following procedure:

Form the n by $2n$ matrix

$$[A \mid I]$$

in which the first n columns are the columns of A and the last n columns are the columns of the identity matrix I_n. Then, using elementary row operations (Section 6.2), reduce this matrix to the form

$$[I \mid B],$$

so that the first n columns are the columns of I_n. Then the matrix B formed by the last n columns is the inverse of A; that is,

$$B = A^{-1}.$$

Example 6 Find A^{-1} if $A = \begin{bmatrix} 1 & -1 & 1 \\ 0 & 2 & -1 \\ 2 & 3 & 0 \end{bmatrix}$.

Solution We start by forming the matrix $[A \mid I]$:

$$\left[\begin{array}{ccc|ccc} 1 & -1 & 1 & 1 & 0 & 0 \\ 0 & 2 & -1 & 0 & 1 & 0 \\ 2 & 3 & 0 & 0 & 0 & 1 \end{array}\right].$$

* See Problems 17, 19, and 27.

Now, we execute elementary row operations on the entire matrix until the left half is transformed into the identity matrix:

$$\left[\begin{array}{rrr|rrr} 1 & -1 & 1 & 1 & 0 & 0 \\ 0 & 2 & -1 & 0 & 1 & 0 \\ 2 & 3 & 0 & 0 & 0 & 1 \end{array}\right] \xrightarrow{-2R_1 + R_3 \to R_3} \left[\begin{array}{rrr|rrr} 1 & -1 & 1 & 1 & 0 & 0 \\ 0 & 2 & -1 & 0 & 1 & 0 \\ 0 & 5 & -2 & -2 & 0 & 1 \end{array}\right]$$

$$\xrightarrow{\frac{1}{2}R_2 \to R_2} \left[\begin{array}{rrr|rrr} 1 & -1 & 1 & 1 & 0 & 0 \\ 0 & 1 & -\frac{1}{2} & 0 & \frac{1}{2} & 0 \\ 0 & 5 & -2 & -2 & 0 & 1 \end{array}\right]$$

$$\xrightarrow{R_2 + R_1 \to R_1} \left[\begin{array}{rrr|rrr} 1 & 0 & \frac{1}{2} & 1 & \frac{1}{2} & 0 \\ 0 & 1 & -\frac{1}{2} & 0 & \frac{1}{2} & 0 \\ 0 & 5 & -2 & -2 & 0 & 1 \end{array}\right]$$

$$\xrightarrow{-5R_2 + R_3 \to R_3} \left[\begin{array}{rrr|rrr} 1 & 0 & \frac{1}{2} & 1 & \frac{1}{2} & 0 \\ 0 & 1 & -\frac{1}{2} & 0 & \frac{1}{2} & 0 \\ 0 & 0 & \frac{1}{2} & -2 & -\frac{5}{2} & 1 \end{array}\right]$$

$$\xrightarrow{R_3 + R_2 \to R_2} \left[\begin{array}{rrr|rrr} 1 & 0 & \frac{1}{2} & 1 & \frac{1}{2} & 0 \\ 0 & 1 & 0 & -2 & -2 & 1 \\ 0 & 0 & \frac{1}{2} & -2 & -\frac{5}{2} & 1 \end{array}\right]$$

$$\xrightarrow{-1R_3 + R_1 \to R_1} \left[\begin{array}{rrr|rrr} 1 & 0 & 0 & 3 & 3 & -1 \\ 0 & 1 & 0 & -2 & -2 & 1 \\ 0 & 0 & \frac{1}{2} & -2 & -\frac{5}{2} & 1 \end{array}\right]$$

$$\xrightarrow{2R_3 \to R_3} \left[\begin{array}{rrr|rrr} 1 & 0 & 0 & 3 & 3 & -1 \\ 0 & 1 & 0 & -2 & -2 & 1 \\ 0 & 0 & 1 & -4 & -5 & 2 \end{array}\right].$$

Therefore,

$$A^{-1} = \begin{bmatrix} 3 & 3 & -1 \\ -2 & -2 & 1 \\ -4 & -5 & 2 \end{bmatrix}.$$

By using matrix multiplication, you can check that $AA^{-1} = I$ and $A^{-1}A = I$ (Problem 43). ■

The following example illustrates the use of the inverse of a matrix to solve a system of linear equations.

Example 7 Use the inverse of the matrix of coefficients to solve the system of linear equations

$$\begin{cases} x - y + z = 8 \\ 2y - z = -7 \\ 2x + 3y = 1. \end{cases}$$

Solution The matrix of coefficients is

$$A = \begin{bmatrix} 1 & -1 & 1 \\ 0 & 2 & -1 \\ 2 & 3 & 0 \end{bmatrix}.$$

If we let

$$X = \begin{bmatrix} x \\ y \\ z \end{bmatrix} \quad \text{and} \quad K = \begin{bmatrix} 8 \\ -7 \\ 1 \end{bmatrix},$$

we can write the system of linear equations as the matrix equation

$$AX = K.$$

In the previous example, we showed that A is nonsingular with

$$A^{-1} = \begin{bmatrix} 3 & 3 & -1 \\ -2 & -2 & 1 \\ -4 & -5 & 2 \end{bmatrix}.$$

If we multiply both sides of $AX = K$ on the left by A^{-1}, we obtain

$$\begin{aligned} A^{-1}(AX) &= A^{-1}K \\ (A^{-1}A)X &= A^{-1}K \\ IX &= A^{-1}K \\ X &= A^{-1}K. \end{aligned}$$

Therefore,

$$\begin{aligned} \begin{bmatrix} x \\ y \\ z \end{bmatrix} &= X = A^{-1}K \\ &= \begin{bmatrix} 3 & 3 & -1 \\ -2 & -2 & 1 \\ -4 & -5 & 2 \end{bmatrix} \begin{bmatrix} 8 \\ -7 \\ 1 \end{bmatrix} \\ &= \begin{bmatrix} 3(8) + \quad 3(-7) + (-1)1 \\ (-2)8 + (-2)(-7) + 1(1) \\ (-4)8 + (-5)(-7) + 2(1) \end{bmatrix} = \begin{bmatrix} 2 \\ -1 \\ 5 \end{bmatrix}. \end{aligned}$$

It follows that $x = 2$, $y = -1$, and $z = 5$. ■

Problem Set 6.3

In Problems 1 to 8, find **(a)** $A + B$, **(b)** $A - B$, **(c)** $-3A$, and **(d)** $-3A + 2B$.

1. $A = \begin{bmatrix} 2 & -3 \\ 5 & 1 \end{bmatrix}$, $B = \begin{bmatrix} 6 & 4 \\ 3 & -2 \end{bmatrix}$

2. $A = \begin{bmatrix} 3 & 2 & -4 & 1 \\ 0 & 3 & -5 & 6 \end{bmatrix}$,
$B = \begin{bmatrix} -7 & 3 & 0 & 4 \\ 1 & 0 & -1 & -3 \end{bmatrix}$

3. $A = \begin{bmatrix} 3 & 2 \\ -2 & 5 \\ 2 & 1 \\ -4 & 4 \end{bmatrix}$, $B = \begin{bmatrix} -2 & 3 \\ 3 & 1 \\ 4 & -2 \\ 1 & 0 \end{bmatrix}$

4. $A = \begin{bmatrix} 1 & \frac{1}{3} & 3 \\ 2 & 0 & -\frac{4}{3} \\ 1 & \sqrt{3} & -2 \end{bmatrix}$, $B = \begin{bmatrix} 1 & \frac{1}{2} & -1 \\ 3 & -1 & 0 \\ 2 & 0 & -\frac{3}{2} \end{bmatrix}$

5. $A = \begin{bmatrix} 2 & -3 & 2 & -3 \\ -3 & 2 & 1 & 1 \\ 4 & 1 & -3 & 4 \end{bmatrix}$,
$B = \begin{bmatrix} 2 & -3 & 0 & 2 \\ 3 & 2 & -1 & 5 \\ 0 & -2 & 1 & 0 \end{bmatrix}$

6. $A = \begin{bmatrix} 0 \\ 1 \\ 2 \\ 3 \\ 4 \end{bmatrix}$, $B = \begin{bmatrix} -1 \\ 3 \\ -4 \\ 2 \\ 0 \end{bmatrix}$

7. $A = \begin{bmatrix} 1 & \frac{1}{6} & 0 \\ \frac{4}{3} & \pi & -2 \\ 1 & 0 & \frac{5}{3} \end{bmatrix}$, $B = \begin{bmatrix} 1 & -\frac{5}{6} & \sqrt{2} \\ \frac{3}{2} & 1 & 0 \\ 3 & \frac{5}{2} & 0 \end{bmatrix}$

8. $A = \begin{bmatrix} 3.1 & 2.5 \\ 6.8 & 1.1 \\ 4.7 & -8.2 \end{bmatrix}$, $B = \begin{bmatrix} 1.9 & 0 \\ 7.4 & 1 \\ -1 & 2 \end{bmatrix}$

In Problems 9 to 20, let

$$A = \begin{bmatrix} a_1 & a_2 \\ a_3 & a_4 \end{bmatrix}, \quad B = \begin{bmatrix} b_1 & b_2 \\ b_3 & b_4 \end{bmatrix}, \quad \text{and} \quad C = \begin{bmatrix} c_1 & c_2 \\ c_3 & c_4 \end{bmatrix},$$

and let p and q denote arbitrary numbers. Verify each equation by direct calculation.

9. $A + (B + C) = (A + B) + C$

10. $p(B + C) = pB + pC$

11. $(p + q)A = pA + qA$

12. $(pq)A = p(qA)$

13. $A + (-A) = 0$

14. $A(BC) = (AB)C$

15. $A + 0 = A$

16. $A(B + C) = AB + AC$

17. $AI = A$

18. $(A + B)C = AC + BC$

19. $IA = A$

20. $0A = 0$

In Problems 21 to 42, let

$$A = \begin{bmatrix} 1 & -1 & 3 \\ 2 & 0 & 4 \\ 2 & -3 & 6 \end{bmatrix}, \quad B = \begin{bmatrix} -1 & 2 & -1 \\ -3 & 4 & 3 \\ 0 & -1 & 2 \end{bmatrix},$$

$$C = \begin{bmatrix} 2 & -1 \\ 0 & 2 \\ -3 & 1 \end{bmatrix}, \quad D = \begin{bmatrix} 1 & 3 & 2 \\ 4 & -1 & -2 \end{bmatrix},$$

$$E = \begin{bmatrix} 1 & 3 \\ -1 & 2 \end{bmatrix}, \quad F = \begin{bmatrix} 2 & 0 \\ 4 & -1 \end{bmatrix}.$$

Find each product, if it exists.

21. EF
22. CE
23. FE
24. EC
25. EE
26. BC
27. AI
28. CB
29. AB
30. BA
31. CD
32. AC
33. AA
34. ABC
35. EFE
36. ACB
37. $(A + B)C$
38. $A(A + B)$
39. $E(F - I)$
40. $EDCF$
41. DE
42. $(A - B)(A + B)$

43. Check the solution to Example 6 by verifying that $AA^{-1} = I$ and that $A^{-1}A = I$.

44. Let

$$A = \begin{bmatrix} a & b \\ c & d \end{bmatrix}$$

and suppose that $ad - bc \neq 0$. Prove that A is nonsingular with

$$A^{-1} = (ad - bc)^{-1} \begin{bmatrix} d & -b \\ -c & a \end{bmatrix}.$$

In Problems 45 to 58, find the inverse of each matrix if it exists. Check your answers (see Problem 43).

45. $\begin{bmatrix} 1 & 1 \\ 1 & -4 \end{bmatrix}$

46. $\begin{bmatrix} 1 & 6 \\ 11 & -7 \end{bmatrix}$

47. $\begin{bmatrix} -6 & 2 \\ 2 & 5 \end{bmatrix}$

48. $\begin{bmatrix} 3 & 6 \\ 1 & 2 \end{bmatrix}$

49. $\begin{bmatrix} 1 & -1 \\ 9 & 3 \end{bmatrix}$

50. $\begin{bmatrix} 1 & b \\ b & 1 \end{bmatrix}$

51. $\begin{bmatrix} 1 & 1 \\ 1 & 1 \end{bmatrix}$

52. $\begin{bmatrix} 0 & 1 \\ 1 & 0 \end{bmatrix}$

53. $\begin{bmatrix} 1 & 2 & -1 \\ 2 & -1 & 3 \\ 3 & -2 & 3 \end{bmatrix}$

54. $\begin{bmatrix} 1 & 5 & -1 \\ 2 & 1 & 1 \\ 1 & -1 & 2 \end{bmatrix}$

55. $\begin{bmatrix} 1 & 2 & -1 \\ 2 & -2 & 1 \\ 6 & 4 & 3 \end{bmatrix}$

56. $\begin{bmatrix} 1 & 2 & -1 \\ 1 & 3 & 2 \\ 2 & 5 & 1 \end{bmatrix}$

57. $\begin{bmatrix} 1 & 3 & 1 \\ 0 & 2 & 4 \\ -1 & 0 & 3 \end{bmatrix}$

58. $\begin{bmatrix} 0 & 0 & 1 \\ 0 & 1 & 0 \\ 1 & 0 & 0 \end{bmatrix}$

In Problems 59 to 68, use the inverse of the matrix of coefficients to solve each system of linear equations.

59. $\begin{cases} x + y = 4 \\ x - 4y = 8 \end{cases}$ (See Problem 45.)

60. $\begin{cases} x + 6y = 7 \\ 11x - 7y = -10 \end{cases}$ (See Problem 46.)

61. $\begin{cases} -6x + 2y = 3 \\ 2x + 5y = 3 \end{cases}$ (See Problem 47.)

62. $\begin{cases} x_1 + 6x_2 = a \\ 11x_1 - 7x_2 = b \end{cases}$ (See Problem 46.)

63. $\begin{cases} r - s = 6 \\ 9r + 3s = 14 \end{cases}$ (See Problem 49.)

64. $\begin{cases} x + by = 2 \\ bx + y = 3 \end{cases}$ (See Problem 50.)

65. $\begin{cases} x + 2y - z = 6 \\ 2x - y + 3z = -13 \\ 3x - 2y + 3z = -16 \end{cases}$ (See Problem 53.)

66. $\begin{cases} u + 5v - w = 2 \\ 2u + v + w = 7 \\ u - v + 2w = 11 \end{cases}$ (See Problem 54.)

67. $\begin{cases} x + 3y + z = 0 \\ 2y + 4z = 1 \\ -x + 3z = 2 \end{cases}$ (See Problem 57.)

68. $\begin{cases} x + 2y - z = 0 \\ 2x - 2y + z = 0 \\ 6x + 4y + 3z = 0 \end{cases}$ (See Problem 55.)

69. In economics, a square matrix in which the element in the ith row and jth column indicates the number of units of commodity number i used to produce one unit of commodity number j is called a **technology matrix.** Let $T = \begin{bmatrix} a & b \\ c & d \end{bmatrix}$ be the technology matrix for a simple economic model involving only two commodities. The column vector $X = \begin{bmatrix} x_1 \\ x_2 \end{bmatrix}$ in which x_1 (respectively, x_2) represents the number of units of commodity number 1 (respectively, commodity number 2) produced in unit time is called the **intensity vector.**

(a) Give an economic interpretation of the product TX.

(b) Give an economic interpretation of $X - TX$.

70. In Problem 69, let d_1 (respectively, d_2) denote the surplus number of units of commodity number 1 (respectively, commodity number 2) required per unit time for export. If $I - T$ is a nonsingular matrix and $D = \begin{bmatrix} d_1 \\ d_2 \end{bmatrix}$, give an economic interpretation of $(I - T)^{-1}D$.

71. ACME Electronics manufactures both scientific and graphing calculators. Each scientific calculator requires 2 minutes to assemble from standard components and 3 minutes for testing and packaging. Each graphing calculator requires 3 minutes to assemble and 4 minutes for testing and packaging.

(a) Write a system of two linear equations for the total number of minutes a required for assembly and the total number of minutes t required for testing and packaging if s scientific and g graphing calculators are manufactured.

(b) If A is the coefficient matrix for the system in part (a), find A^{-1}.

(c) Write the system in part (a) in matrix form, using the coefficient matrix A, the *production vector* $\begin{bmatrix} s \\ g \end{bmatrix}$, and the *time vector* $\begin{bmatrix} a \\ t \end{bmatrix}$.

(d) Use A^{-1} to solve the matrix equation in part (c) for the production vector in terms of the time vector.

6.4 Determinants and Cramer's Rule

In this section, we consider an alternative method, called *Cramer's rule*, for solving systems of linear equations. Although Cramer's rule applies to systems of n linear equations in n unknowns for any positive integer n, its practical use is usually limited to the cases $n = 2$ and $n = 3$. For larger values of n, the elimination method using matrices (Section 6.2) is usually more efficient.

Cramer's rule is based on the idea of a *determinant*. If a, b, c, and d are any four numbers, the symbol

$$\begin{vmatrix} a & b \\ c & d \end{vmatrix}$$

is called a 2 by 2 **determinant** with *entries* or *elements* a, b, c, and d. Its **value** is defined to be the number $ad - cb$; that is,

$$\begin{vmatrix} a & b \\ c & d \end{vmatrix} = ad - cb.$$

The memory aid

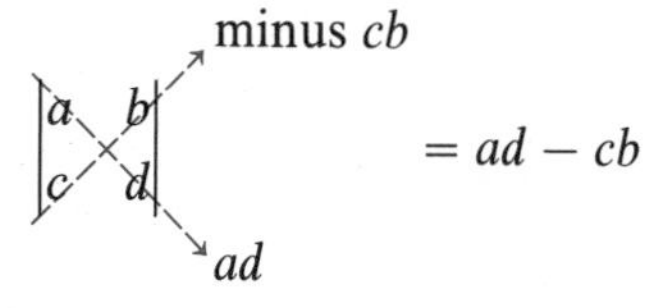

is often helpful.

Example 1 Evaluate the determinant $\begin{vmatrix} 4 & -3 \\ 2 & 1 \end{vmatrix}$.

Solution $\begin{vmatrix} 4 & -3 \\ 2 & 1 \end{vmatrix} = 4(1) - 2(-3) = 10$ ∎

Now, let's see how determinants can be used to solve systems of linear equations. Consider the system

$$\begin{cases} ax + by = h \\ cx + dy = k. \end{cases}$$

If we multiply the first equation by d and the second equation by b, we obtain

$$\begin{cases} adx + bdy = hd \\ bcx + bdy = bk. \end{cases}$$

So, subtracting the second equation from the first, we eliminate the terms involving y and get

$$adx - bcx = hd - bk$$

or

$$(ad - cb)x = hd - kb.$$

Using determinants, we can rewrite the last equation as

$$\begin{vmatrix} a & b \\ c & d \end{vmatrix} x = \begin{vmatrix} h & b \\ k & d \end{vmatrix}.$$

Therefore,

$$x = \frac{\begin{vmatrix} h & b \\ k & d \end{vmatrix}}{\begin{vmatrix} a & b \\ c & d \end{vmatrix}},$$

provided that $\begin{vmatrix} a & b \\ c & d \end{vmatrix} \neq 0$. A similar calculation (Problem 51) yields

$$y = \frac{\begin{vmatrix} a & h \\ c & k \end{vmatrix}}{\begin{vmatrix} a & b \\ c & d \end{vmatrix}}.$$

Thus, if

$$D = \begin{vmatrix} a & b \\ c & d \end{vmatrix}, \quad D_x = \begin{vmatrix} h & b \\ k & d \end{vmatrix}, \quad D_y = \begin{vmatrix} a & h \\ c & k \end{vmatrix},$$

and $D \neq 0$, then the system

$$\begin{cases} ax + by = h \\ cx + dy = k \end{cases}$$

has one and only one solution:

$$x = \frac{D_x}{D}, \qquad y = \frac{D_y}{D}.$$

This is **Cramer's rule** for two linear equations in two unknowns.

In Cramer's rule, D is called the **coefficient determinant** because its entries are the coefficients of the unknowns in the system:

$$\begin{cases} ax + by = h \\ cx + dy = k, \end{cases} \qquad D = \begin{vmatrix} a & b \\ c & d \end{vmatrix}.$$

Notice that D_x is obtained by replacing the *first* column of D (the coefficients of x), and that D_y is obtained by replacing the *second* column of D (the coefficients of y) by the constants on the right in the system of equations:

$$\begin{cases} ax + by = h \\ cx + dy = k, \end{cases} \qquad D_x = \begin{vmatrix} h & b \\ k & d \end{vmatrix}, \qquad D_y = \begin{vmatrix} a & h \\ c & k \end{vmatrix}.$$

Keep in mind that Cramer's rule can be applied only when $D \neq 0$. If $D = 0$, it can be shown (Problem 53) that the system is either inconsistent or dependent.

Example 2 Use Cramer's rule (if applicable) to solve each system.

(a) $\begin{cases} 2x - y = 7 \\ x + 3y = 14 \end{cases}$ **(b)** $\begin{cases} 2x - y = 7 \\ 4x - 2y = 3 \end{cases}$

Solution **(a)** $D = \begin{vmatrix} 2 & -1 \\ 1 & 3 \end{vmatrix} = 2(3) - 1(-1) = 7, \quad D_x = \begin{vmatrix} 7 & -1 \\ 14 & 3 \end{vmatrix} = 7(3) - 14(-1) = 35,$

$D_y = \begin{vmatrix} 2 & 7 \\ 1 & 14 \end{vmatrix} = 2(14) - 1(7) = 21;$

hence, by Cramer's rule,

$$x = \frac{D_x}{D} = \frac{35}{7} = 5 \quad \text{and} \quad y = \frac{D_y}{D} = \frac{21}{7} = 3.$$

(b) Here

$$D = \begin{vmatrix} 2 & -1 \\ 4 & -2 \end{vmatrix} = 2(-2) - 4(-1) = 0,$$

so Cramer's rule does not apply. (Actually, the system of equations is inconsistent—it has no solution.) ■

To extend Cramer's rule to systems of three linear equations in three unknowns, we begin by extending the definition of a determinant. The **value** of a 3 by 3 determinant is defined in terms of 2 by 2 determinants as follows:

$$\begin{vmatrix} a_1 & a_2 & a_3 \\ b_1 & b_2 & b_3 \\ c_1 & c_2 & c_3 \end{vmatrix} = a_1 \begin{vmatrix} b_2 & b_3 \\ c_2 & c_3 \end{vmatrix} - a_2 \begin{vmatrix} b_1 & b_3 \\ c_1 & c_3 \end{vmatrix} + a_3 \begin{vmatrix} b_1 & b_2 \\ c_1 & c_2 \end{vmatrix}.$$

We refer to this as the **expansion formula** for 3 by 3 determinants. Notice the *negative sign* on the middle term.

In the expansion formula, notice that each entry in the first row is multiplied by the 2 by 2 determinant that remains when the row and column containing the multiplier are (mentally) crossed out. Thus,

$$a_1 \text{ is multiplied by } \begin{vmatrix} \not{a_1} & a_2 & a_3 \\ \not{b_1} & b_2 & b_3 \\ \not{c_1} & c_2 & c_3 \end{vmatrix} = \begin{vmatrix} b_2 & b_3 \\ c_2 & c_3 \end{vmatrix},$$

$$a_2 \text{ is multiplied by } \begin{vmatrix} a_1 & \not{a_2} & a_3 \\ b_1 & \not{b_2} & b_3 \\ c_1 & \not{c_2} & c_3 \end{vmatrix} = \begin{vmatrix} b_1 & b_3 \\ c_1 & c_3 \end{vmatrix},$$

and

$$a_3 \text{ is multiplied by } \begin{vmatrix} a_1 & a_2 & \not{a_3} \\ b_1 & b_2 & \not{b_3} \\ c_1 & c_2 & \not{c_3} \end{vmatrix} = \begin{vmatrix} b_1 & b_2 \\ c_1 & c_2 \end{vmatrix}.$$

To **expand** a 3 by 3 determinant means to find its value by using the expansion formula. Again, we emphasize: When expanding a 3 by 3 determinant, *don't forget the negative sign on the middle term.*

Example 3 Expand the determinant $\begin{vmatrix} 3 & 1 & 2 \\ -4 & 2 & 4 \\ 1 & 0 & 5 \end{vmatrix}$.

Solution

$$\begin{vmatrix} 3 & 1 & 2 \\ -4 & 2 & 4 \\ 1 & 0 & 5 \end{vmatrix} = 3\begin{vmatrix} 2 & 4 \\ 0 & 5 \end{vmatrix} - 1\begin{vmatrix} -4 & 4 \\ 1 & 5 \end{vmatrix} + 2\begin{vmatrix} -4 & 2 \\ 1 & 0 \end{vmatrix}$$
$$= 3[2(5) - 0(4)] - 1[(-4)5 - 1(4)] + 2[(-4)0 - 1(2)]$$
$$= 3(10) - 1(-24) + 2(-2) = 50$$

Now we can state **Cramer's rule** for solving a system

$$\begin{cases} a_1x + a_2y + a_3z = k_1 \\ b_1x + b_2y + b_3z = k_2 \\ c_1x + c_2y + c_3z = k_3 \end{cases}$$

of three linear equations in three unknowns: Form the **coefficient determinant**

$$D = \begin{vmatrix} a_1 & a_2 & a_3 \\ b_1 & b_2 & b_3 \\ c_1 & c_2 & c_3 \end{vmatrix}.$$

If $D \neq 0$, form the determinants

$$D_x = \begin{vmatrix} k_1 & a_2 & a_3 \\ k_2 & b_2 & b_3 \\ k_3 & c_2 & c_3 \end{vmatrix}, \quad D_y = \begin{vmatrix} a_1 & k_1 & a_3 \\ b_1 & k_2 & b_3 \\ c_1 & k_3 & c_3 \end{vmatrix}, \quad D_z = \begin{vmatrix} a_1 & a_2 & k_1 \\ b_1 & b_2 & k_2 \\ c_1 & c_2 & k_3 \end{vmatrix}.$$

Then the solution of the system of linear equations is

$$x = \frac{D_x}{D}, \qquad y = \frac{D_y}{D}, \qquad z = \frac{D_z}{D}.$$

If $D \neq 0$, this is the *only* solution of the system; that is, the system is *well-determined* (see page 361). If $D = 0$, the system is either *inconsistent* or *dependent*, and Cramer's rule is not applicable. You can find a proof of Cramer's rule in a textbook on linear algebra.

Example 4 Use Cramer's rule to solve the system

$$\begin{cases} 3x - 2y + z = -9 \\ x + 2y - z = 5 \\ 2x - y + 3z = -10. \end{cases}$$

Solution

$$D = \begin{vmatrix} 3 & -2 & 1 \\ 1 & 2 & -1 \\ 2 & -1 & 3 \end{vmatrix} = 3\begin{vmatrix} 2 & -1 \\ -1 & 3 \end{vmatrix} - (-2)\begin{vmatrix} 1 & -1 \\ 2 & 3 \end{vmatrix} + 1\begin{vmatrix} 1 & 2 \\ 2 & -1 \end{vmatrix}$$

$$= 3[2(3) - (-1)(-1)] + 2[1(3) - 2(-1)] + 1[1(-1) - 2(2)]$$

$$= 3(5) + 2(5) + 1(-5) = 20$$

Because $D \neq 0$, the system is well-determined and we can solve it by applying Cramer's rule:

$$D_x = \begin{bmatrix} -9 & -2 & 1 \\ 5 & 2 & -1 \\ -10 & -1 & 3 \end{bmatrix} = -9\begin{vmatrix} 2 & -1 \\ -1 & 3 \end{vmatrix} - (-2)\begin{vmatrix} 5 & -1 \\ -10 & 3 \end{vmatrix} + 1\begin{vmatrix} 5 & 2 \\ -10 & -1 \end{vmatrix} = -20,$$

$$D_y = \begin{vmatrix} 3 & -9 & 1 \\ 1 & 5 & -1 \\ 2 & -10 & 3 \end{vmatrix} = 3\begin{vmatrix} 5 & -1 \\ -10 & 3 \end{vmatrix} - (-9)\begin{vmatrix} 1 & -1 \\ 2 & 3 \end{vmatrix} + 1\begin{vmatrix} 1 & 5 \\ 2 & -10 \end{vmatrix} = 40,$$

$$D_z = \begin{vmatrix} 3 & -2 & -9 \\ 1 & 2 & 5 \\ 2 & -1 & -10 \end{vmatrix} = 3\begin{vmatrix} 2 & 5 \\ -1 & -10 \end{vmatrix} - (-2)\begin{vmatrix} 1 & 5 \\ 2 & -10 \end{vmatrix} + (-9)\begin{vmatrix} 1 & 2 \\ 2 & -1 \end{vmatrix} = -40;$$

hence,

$$x = \frac{D_x}{D} = \frac{-20}{20} = -1, \qquad y = \frac{D_y}{D} = \frac{40}{20} = 2, \qquad z = \frac{D_z}{D} = \frac{-40}{20} = -2.$$ ■

Properties of Determinants

Determinants have a number of useful properties, some of which we now state (without proof).

Property 1. If you interchange any two rows or any two columns of a determinant, you change its algebraic sign.

For instance,

$$\begin{vmatrix} 2 & 3 \\ 5 & 6 \end{vmatrix} = -\begin{vmatrix} 5 & 6 \\ 2 & 3 \end{vmatrix}.$$

(Check this yourself.)

Property 2. If you multiply every entry in one row or one column of a determinant by a constant k, the effect is to multiply the value of the determinant by k.

For instance,

$$\begin{vmatrix} 2k & 3 \\ 5k & 6 \end{vmatrix} = k\begin{vmatrix} 2 & 3 \\ 5 & 6 \end{vmatrix}.$$

(Check this yourself.)

Property 2 allows you to "factor out" a common factor of all the elements of a single row or column of a determinant. For instance,

$$\begin{vmatrix} 8 & 28 & 3 \\ 3 & -14 & 2 \\ 5 & 35 & 4 \end{vmatrix} = 7\begin{vmatrix} 8 & 4 & 3 \\ 3 & -2 & 2 \\ 5 & 5 & 4 \end{vmatrix}.$$

Property 3. If you add a constant multiple of the entries in any one row of a determinant to the corresponding entries in any other row, the value of the determinant will not change. Likewise for columns.

For instance, in the determinant

$$\begin{vmatrix} -3 & 5 \\ 6 & -4 \end{vmatrix},$$

if you add two times the first row to the second row, the value of the determinant won't change; that is,

$$\begin{vmatrix} -3 & 5 \\ 6 & -4 \end{vmatrix} = \begin{vmatrix} -3 & 5 \\ 6 + 2(-3) & -4 + 2(5) \end{vmatrix} = \begin{vmatrix} -3 & 5 \\ 0 & 6 \end{vmatrix}.$$

(Check this yourself.)

Property 4. If any two rows or any two columns of a determinant are the same, its value is zero. More generally, if the corresponding entries in any two rows or in any two columns are proportional, the value of the determinant is zero.

For instance,

$$\begin{vmatrix} 1 & 5 & -7 \\ 2 & 3 & 1 \\ 1 & 5 & -7 \end{vmatrix} = 0 \qquad \text{and} \qquad \begin{vmatrix} 2 & -1 & 10 \\ 3 & 7 & 15 \\ 4 & 2 & 20 \end{vmatrix} = 0.$$

(Check this yourself.)

The **main diagonal** of a determinant is the diagonal running from upper left to lower right. For instance, in the determinant

$$\begin{vmatrix} a & b & c \\ u & v & w \\ x & y & z \end{vmatrix}$$

the entries on the main diagonal are a, v, and z. A determinant is said to be in **triangular form** if all entries below the main diagonal are zero. For instance,

$$\begin{vmatrix} 3 & 5 & -7 \\ 0 & 2 & 4 \\ 0 & 0 & 6 \end{vmatrix}$$

is in triangular form.

Property 5. If a determinant is in triangular form, its value is the product of the entries on its main diagonal.

For instance,

$$\begin{vmatrix} 3 & 5 & -7 \\ 0 & 2 & 4 \\ 0 & 0 & 6 \end{vmatrix} = 3(2)(6) = 36.$$

(Check this yourself.)

By using Properties 1 to 5, you can often evaluate a determinant more easily than by applying the expansion formula. The usual idea is to use Properties 1 to 3 to bring the determinant into triangular form and then to apply Property 5.

Example 5 Use the properties of determinants to evaluate:

(a) $\begin{vmatrix} 3 & -1 & 2 \\ 6 & -2 & 4 \\ 7 & 0 & 3 \end{vmatrix}$ **(b)** $\begin{vmatrix} 4 & 3 & 3 \\ 1 & 0 & 2 \\ 6 & 6 & 7 \end{vmatrix}$

Solution **(a)** The second row is proportional to the first, so by Property 4,

$$\begin{vmatrix} 3 & -1 & 2 \\ 6 & -2 & 4 \\ 7 & 0 & 3 \end{vmatrix} = 0.$$

(b) $\begin{vmatrix} 4 & 3 & 3 \\ 1 & 0 & 2 \\ 6 & 6 & 7 \end{vmatrix} = 3\begin{vmatrix} 4 & 1 & 3 \\ 1 & 0 & 2 \\ 6 & 2 & 7 \end{vmatrix}$ (We factored out 3 from the second column, using Property 2.)

$= -3\begin{vmatrix} 1 & 4 & 3 \\ 0 & 1 & 2 \\ 2 & 6 & 7 \end{vmatrix}$ (We interchanged the first and second columns, using Property 1.)

$= -3\begin{vmatrix} 1 & 4 & 3 \\ 0 & 1 & 2 \\ 0 & -2 & 1 \end{vmatrix}$ (We added -2 times the first row to the third row, using Property 3.)

$= -3\begin{vmatrix} 1 & 4 & 3 \\ 0 & 1 & 2 \\ 0 & 0 & 5 \end{vmatrix}$ (We added 2 times the second row to the third row, using Property 3.)

$= -3(1)(1)(5)$ (We used Property 5.)

$= -15.$ ■

It should come as no surprise to you to learn that 4 by 4, 5 by 5, and, in general, n by n determinants can be defined, and Properties 1 to 5 continue to hold for them as well. You can learn more about determinants from a textbook on linear algebra. Determinants have a wide variety of uses (other than for solving systems of linear equations) in linear algebra, calculus, and other branches of mathematics.

Problem Set 6.4

In Problems 1 to 8, evaluate each determinant.

1. $\begin{vmatrix} 2 & 3 \\ 1 & 4 \end{vmatrix}$

2. $\begin{vmatrix} e & \pi \\ \sqrt{3} & \sqrt{2} \end{vmatrix}$

3. $\begin{vmatrix} 6 & -4 \\ 3 & 7 \end{vmatrix}$

4. $\begin{vmatrix} x & -y \\ y & x \end{vmatrix}$

5. $\begin{vmatrix} \sqrt{6} & -2\sqrt{5} \\ 3\sqrt{5} & 4\sqrt{6} \end{vmatrix}$

6. $\begin{vmatrix} x+y & x+y \\ x-y & x+y \end{vmatrix}$

7. $\begin{vmatrix} \log 100 & 2 \\ \log 10 & 3 \end{vmatrix}$

8. $\begin{vmatrix} x & -x \\ y & -y \end{vmatrix}$

In Problems 9 to 14, use Cramer's rule (when applicable) to solve each system of linear equations.

9. $\begin{cases} 5x + 7y = -2 \\ 3x + 4y = -1 \end{cases}$

10. $\begin{cases} \frac{1}{2}x - \frac{2}{3}y = \frac{3}{4} \\ \frac{1}{3}x + 2y = \frac{5}{6} \end{cases}$

11. $\begin{cases} 2x_1 + x_2 = 5 \\ x_1 - 2x_2 = 0 \end{cases}$

12. $\begin{cases} 3u - 4v = 1 \\ -4u + \frac{16}{3}v = 2 \end{cases}$

13. $\begin{cases} 8u + 3v = 9 \\ 4u - 6v = 7 \end{cases}$

14. $\begin{cases} ax + y = 0 \\ x + ay = 0 \end{cases}$

In Problems 15 to 22, expand each determinant.

15. $\begin{vmatrix} 2 & 3 & -1 \\ 5 & 7 & 0 \\ 2 & -3 & 1 \end{vmatrix}$

16. $\begin{vmatrix} 1 & 5 & -7 \\ 3 & 0 & 2 \\ -1 & 4 & 1 \end{vmatrix}$

17. $\begin{vmatrix} 2 & 0 & 4 \\ 1 & 5 & 0 \\ 0 & 7 & 1 \end{vmatrix}$

18. $\begin{vmatrix} a & b & c \\ x & y & z \\ 1 & 1 & 1 \end{vmatrix}$

19. $\begin{vmatrix} 2 & -3 & 1 \\ 1 & 2 & 3 \\ 0 & 1 & 2 \end{vmatrix}$

20. $\begin{vmatrix} e & \sqrt{2} & \sqrt{3} \\ \pi & 0 & 1 \\ -1 & 2 & 0 \end{vmatrix}$

21. $\begin{vmatrix} \frac{1}{2} & 1 & -\frac{2}{3} \\ \frac{5}{2} & -\frac{4}{3} & 1 \\ \frac{3}{2} & 0 & \frac{3}{4} \end{vmatrix}$ **22.** $\begin{vmatrix} 0 & a & b \\ a & 0 & c \\ b & c & 0 \end{vmatrix}$

In Problems 23 to 28, use Cramer's rule (when applicable) to solve each system of linear equations.

23. $\begin{cases} x + 2y - z = -3 \\ 2x - y + z = 5 \\ 3x + 2y - 2z = -3 \end{cases}$ **24.** $\begin{cases} -u + 2v + w = -1 \\ 4u - 2v - w = 3 \\ 4u + 2v - w = 5 \end{cases}$

25. $\begin{cases} -3x + 4y + 6z = 30 \\ x + 2z = 6 \\ -x - 2y + 3z = 8 \end{cases}$ **26.** $\begin{cases} 2y - 3x = 1 \\ 3z - 2y = 5 \\ x + z = 4 \end{cases}$

27. $\begin{cases} 2x + 5y - z = 3 \\ -3x - 2y + 7z = 4 \\ -x + 3y + 6z = 0 \end{cases}$ **28.** $\begin{cases} 3x + y - z = 14 \\ x + 3y - z = 16 \\ x + y - 3z = -10 \end{cases}$

In Problems 29 to 36, use Properties 1 to 5 to evaluate each determinant.

29. $\begin{vmatrix} 1 & -2 & 3 \\ 2 & 3 & -2 \\ 3 & 1 & -1 \end{vmatrix}$ **30.** $\begin{vmatrix} 5 & 2 & 3 \\ 4 & -5 & -6 \\ 7 & -8 & -9 \end{vmatrix}$

31. $\begin{vmatrix} 1 & -5 & 2 \\ -4 & -1 & 5 \\ 3 & -4 & 3 \end{vmatrix}$ **32.** $\begin{vmatrix} 2 & 4 & 3 \\ -6 & 0 & 4 \\ -1 & 1 & 3 \end{vmatrix}$

33. $\begin{vmatrix} -1 & 1 & 1 \\ 4 & 2 & 3 \\ 1 & 3 & 0 \end{vmatrix}$ **34.** $\begin{vmatrix} 4 & 5 & 6 \\ 2 & 2 & 2 \\ 7 & 2 & -7 \end{vmatrix}$

35. $\begin{vmatrix} 3 & 1 & 4 \\ 1 & 7 & 3 \\ 5 & -10 & 5 \end{vmatrix}$ **36.** $\begin{vmatrix} 9 & 3 & 3 \\ -2 & 0 & 6 \\ 2 & 1 & 1 \end{vmatrix}$

C In Problems 37 and 38, evaluate each determinant with the aid of a calculator.

37. $\begin{vmatrix} 2.03 & -7.07 & 1.55 \\ 3.71 & 2.22 & 5.77 \\ 6.65 & -8.56 & 3.65 \end{vmatrix}$

38. $\begin{vmatrix} 0.071 & 0.029 & -0.095 \\ 0.101 & 0.210 & 0.055 \\ 0.077 & -0.101 & 0.039 \end{vmatrix}$

In Problems 39 to 44, evaluate each determinant mentally. Indicate which property or properties you use.

39. $\begin{vmatrix} 3 & \sqrt{2} & 19 \\ 0 & 1 & \frac{5}{2} \\ 0 & 0 & -4 \end{vmatrix}$ **40.** $\begin{vmatrix} 3 & 1 & 4 \\ -6 & -2 & -8 \\ 1 & 5 & -7 \end{vmatrix}$

41. $\begin{vmatrix} 1 & 1 & 1 \\ -1 & -1 & -1 \\ a & b & c \end{vmatrix}$ **42.** $\begin{vmatrix} a & b & c \\ 0 & d & h \\ 0 & 0 & \sqrt{5} \end{vmatrix}$

43. $\begin{vmatrix} 0 & 1 & 0 \\ 1 & 0 & 0 \\ 0 & 0 & 1 \end{vmatrix}$ **44.** $\begin{vmatrix} 1 & 2 & 3 \\ 3 & 2 & 1 \\ 4 & 4 & 4 \end{vmatrix}$

In Problems 45 to 50, assume that $\begin{vmatrix} a & b & c \\ u & v & w \\ x & y & z \end{vmatrix} = -3.$

Find the value of each determinant.

45. $\begin{vmatrix} u & v & w \\ a & b & c \\ x & y & z \end{vmatrix}$ **46.** $\begin{vmatrix} u & v & w \\ x & y & z \\ a & b & c \end{vmatrix}$

47. $\begin{vmatrix} c & a & b \\ w & u & v \\ z & x & y \end{vmatrix}$ **48.** $\begin{vmatrix} -a & -b & -c \\ 3u & 3v & 3w \\ 4x & 4y & 4z \end{vmatrix}$

49. $\begin{vmatrix} a+u & b+v & c+w \\ u & v & w \\ x+u & y+v & z+w \end{vmatrix}$

50. $\begin{vmatrix} a & b & c \\ u-2a & v-2b & w-2c \\ 3x & 3y & 3z \end{vmatrix}$

51. Complete the proof of Cramer's rule for two linear equations in two unknowns (page 378) by showing that if $D \neq 0$, then $y = D_y/D$. [*Hint:* Multiply the first equation by c and the second equation by a; then subtract the first equation from the second.]

52. Show that Property 4 can be derived from Property 2 and Property 3.

53. If $\begin{vmatrix} a & b \\ c & d \end{vmatrix} = 0$, show that the system $\begin{cases} ax + by = h \\ cx + dy = k \end{cases}$ is either inconsistent or dependent. [*Hint:* The equations $Dx = D_x$ and $Dy = D_y$ are true even if $D = 0$.]

54. Solve each equation for x.

(a) $\begin{vmatrix} 4 & -1 \\ x & 3 \end{vmatrix} = 2$ **(b)** $\begin{vmatrix} -1 & 5 & -2 \\ 2 & -2 & x \\ 3 & 1 & 0 \end{vmatrix} = -3.$

55. Show that in the xy plane, an equation of the line that contains the two points (a, b) and (c, d) is

$$\begin{vmatrix} x & y & 1 \\ a & b & 1 \\ c & d & 1 \end{vmatrix} = 0.$$

56. Show that

$$\begin{vmatrix} a_1 + a_2 & b_1 + b_2 & c_1 + c_2 \\ u & v & w \\ x & y & z \end{vmatrix} = \begin{vmatrix} a_1 & b_1 & c_1 \\ u & v & w \\ x & y & z \end{vmatrix} + \begin{vmatrix} a_2 & b_2 & c_2 \\ u & v & w \\ x & y & z \end{vmatrix}.$$

57. Solve the equation

$$\begin{vmatrix} 1 & x-2 & 2 \\ -2 & x-1 & 3 \\ 1 & 2 & x \end{vmatrix} = 0.$$

58. Show that $x - a$ and $x - b$ are factors of

$$\begin{vmatrix} 1 & 1 & 1 \\ x & a & b \\ x^2 & a^2 & b^2 \end{vmatrix}.$$

59. By considering the areas A, A_1, A_2, and A_3 of the triangles in Figure 1, show that

$$ad = \tfrac{1}{2}ab + \tfrac{1}{2}cd + \tfrac{1}{2}(a-c)(d-b) + A;$$

then solve this equation for A to show that $A = \frac{1}{2}\begin{vmatrix} a & b \\ c & d \end{vmatrix}$.

Figure 1

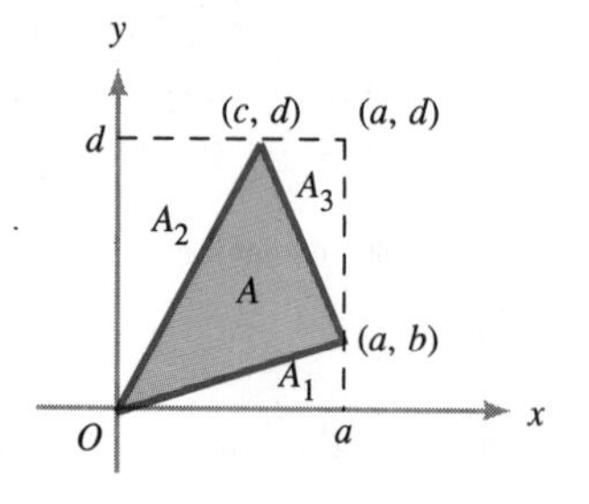

60. Let A be the area of the triangle PQR in Figure 2. If $P = (p_1, p_2)$, $Q = (q_1, q_2)$, and $R = (r_1, r_2)$, show that $A = \frac{1}{2}\begin{vmatrix} p_1 & p_2 & 1 \\ q_1 & q_2 & 1 \\ r_1 & r_2 & 1 \end{vmatrix}$. [*Hint:* Consider triangles OQP, OQR, and OPR and use the result of Problem 59.]

Figure 2

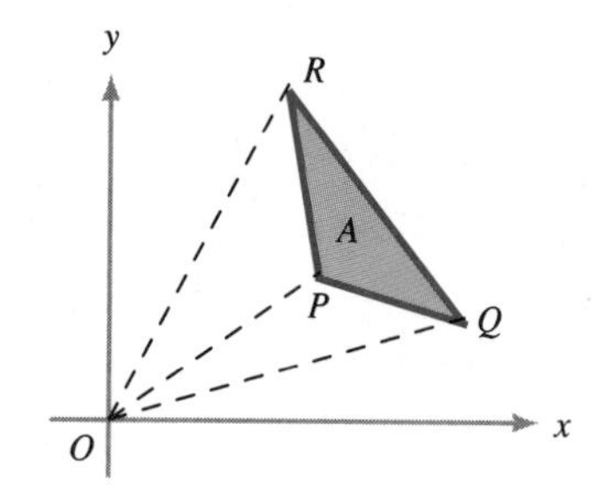

61. Show that the determinant $\begin{vmatrix} a & b \\ c & d \end{vmatrix}$ can be interpreted as the area A of the parallelogram having the vectors $\mathbf{V} = \langle a, b \rangle$ and $\mathbf{W} = \langle c, d \rangle$ as adjacent edges, provided that the angle from $\mathbf{V}$ to $\mathbf{W}$ turns in a counterclockwise direction (Figure 3). [*Hint:* Use the result of Problem 59.]

Figure 3

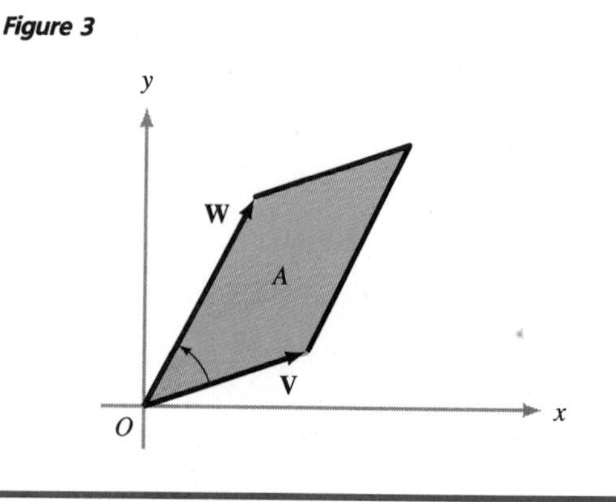

6.5 Partial Fractions

In applied mathematics, it is often necessary to rewrite a fraction as a sum of simpler fractions. In many cases, this can be accomplished by setting up and solving a system of linear equations.

If you add the two fractions $\frac{2}{x-2}$ and $\frac{3}{x+1}$, you obtain

$$\frac{2}{x-2} + \frac{3}{x+1} = \frac{2(x+1) + 3(x-2)}{(x-2)(x+1)} = \frac{5x-4}{(x-2)(x+1)}.$$

The reverse process of "taking the fraction $\frac{5x-4}{(x-2)(x+1)}$ apart" into the sum of simpler fractions,

$$\frac{5x-4}{(x-2)(x+1)} = \frac{2}{x-2} + \frac{3}{x+1},$$

is called **decomposing** $\frac{5x-4}{(x-2)(x+1)}$ into the **partial fractions** $\frac{2}{x-2}$ and $\frac{3}{x+1}$. The process of decomposing fractions into simpler partial fractions is used routinely in calculus and other branches of mathematics. Here we shall give you a brief introduction to the subject—you can find further details in calculus textbooks.

It is easiest to decompose a rational expression into partial fractions if the degree of the numerator is less than the degree of the denominator, and if the denominator is (or can be) factored into linear factors* that are all different from one another. In this case, you simply provide a partial fraction of the form

$$\frac{\text{constant}}{\text{linear factor}}$$

for each linear factor in the denominator.

Example 1 Decompose $\frac{6x^2 + 2x + 2}{x(x-2)(2x+1)}$ into partial fractions.

Solution The linear factors in the denominator are x, $x - 2$, and $2x + 1$. Since these factors are different from each other, we must provide partial fractions

$$\frac{\text{constant}}{x}, \quad \frac{\text{constant}}{x-2}, \quad \text{and} \quad \frac{\text{constant}}{2x+1}.$$

We denote the three constants by A, B, and C, so that

$$\frac{6x^2 + 2x + 2}{x(x-2)(2x+1)} = \frac{A}{x} + \frac{B}{x-2} + \frac{C}{2x+1}.$$

* A linear factor has the form $ax + b$, where a and b are constants and $a \neq 0$.

To determine the values of A, B, and C, we begin by multiplying both sides of this equation by $x(x-2)(2x+1)$ to clear the fractions. Thus, we have

$$6x^2 + 2x + 2 = A(x-2)(2x+1) + Bx(2x+1) + Cx(x-2)$$

or

$$6x^2 + 2x + 2 = A(2x^2 - 3x - 2) + B(2x^2 + x) + C(x^2 - 2x).$$

Collecting like powers on the right, we obtain

$$6x^2 + 2x + 2 = (2A + 2B + C)x^2 + (-3A + B - 2C)x - 2A.$$

Now we equate the coefficients of like powers of x on both sides of the equation:

$$6 = 2A + 2B + C, \qquad 2 = -3A + B - 2C, \qquad \text{and} \qquad 2 = -2A.$$

In other words, the constants A, B, and C satisfy the system of linear equations

$$\begin{cases} 2A + 2B + C = 6 \\ -3A + B - 2C = 2 \\ -2A = 2. \end{cases}$$

Solving this system by one of the methods given in Sections 6.1, 6.2, and 6.4, we find that

$$A = -1, \qquad B = 3, \qquad \text{and} \qquad C = 2.$$

Therefore,

$$\frac{6x^2 + 2x + 2}{x(x-2)(2x+1)} = \frac{-1}{x} + \frac{3}{x-2} + \frac{2}{2x+1}.$$

■

A factor of the form $(ax + b)^2$ in the denominator requires two partial fractions:

$$\frac{B}{ax+b} + \frac{C}{(ax+b)^2}.$$

Example 2 Decompose $\dfrac{2x^3 + 7x^2 + 6x + 2}{x^3 + 2x^2 + x}$ into partial fractions.

Solution When the degree of the numerator is greater than or equal to the degree of the denominator, we first perform a long division and then decompose into partial fractions. By long division, we have

$$\frac{2x^3 + 7x^2 + 6x + 2}{x^3 + 2x^2 + x} = 2 + \frac{3x^2 + 4x + 2}{x^3 + 2x^2 + x} = 2 + \frac{3x^2 + 4x + 2}{x(x+1)^2}.$$

Notice that we have factored the denominator in preparation for the method of partial fractions. Now we write

$$\frac{3x^2 + 4x + 2}{x(x+1)^2} = \frac{A}{x} + \frac{B}{x+1} + \frac{C}{(x+1)^2}.$$

Multiplying both sides of the equation by $x(x + 1)^2$, we have

$$\begin{aligned} 3x^2 + 4x + 2 &= A(x + 1)^2 + Bx(x + 1) + Cx \\ &= A(x^2 + 2x + 1) + B(x^2 + x) + Cx \\ &= (A + B)x^2 + (2A + B + C)x + A. \end{aligned}$$

Equating coefficients of like powers of x on both sides of the last equation, we obtain the system

$$\begin{cases} A + B = 3 \\ 2A + B + C = 4 \\ A = 2. \end{cases}$$

Solving this system, we find that $A = 2$, $B = 1$, and $C = -1$. Hence,

$$\frac{3x^2 + 4x + 2}{x(x + 1)^2} = \frac{2}{x} + \frac{1}{x + 1} + \frac{-1}{(x + 1)^2}.$$

Therefore,

$$\frac{2x^3 + 7x^2 + 6x + 2}{x^3 + 2x^2 + x} = 2 + \frac{2}{x} + \frac{1}{x + 1} + \frac{-1}{(x + 1)^2}.$$

■

A prime quadratic factor of the form $ax^2 + bx + c$ in the denominator requires a partial fraction

$$\frac{Bx + C}{ax^2 + bx + c}.$$

Example 3 Decompose $\dfrac{8x^2 + 3x + 20}{(x + 1)(x^2 + 4)}$ into partial fractions.

Solution We begin by writing

$$\frac{8x^2 + 3x + 20}{(x + 1)(x^2 + 4)} = \frac{A}{x + 1} + \frac{Bx + C}{x^2 + 4}.$$

Multiplying both sides by $(x + 1)(x^2 + 4)$, we obtain

$$\begin{aligned} 8x^2 + 3x + 20 &= A(x^2 + 4) + (Bx + C)(x + 1) \\ &= Ax^2 + 4A + Bx^2 + Bx + Cx + C \\ &= (A + B)x^2 + (B + C)x + 4A + C. \end{aligned}$$

Equating coefficients, we get the system

$$\begin{cases} A + B = 8 \\ B + C = 3 \\ 4A + C = 20. \end{cases}$$

The solution of this system is $A = 5$, $B = 3$, and $C = 0$; hence,

$$\frac{8x^2 + 3x + 20}{(x + 1)(x^2 + 4)} = \frac{5}{x + 1} + \frac{3x}{x^2 + 4}.$$

■

Problem Set 6.5

In Problems 1 to 26, decompose each fraction into partial fractions.

1. $\dfrac{3}{(x-3)(x-2)}$

2. $\dfrac{x}{(x-1)(x-4)}$

3. $\dfrac{x+2}{(x+5)(x-1)}$

4. $\dfrac{3x+7}{x^2-2x-3}$

5. $\dfrac{x^2-5x-3}{x(x-2)(x+2)}$

6. $\dfrac{x+12}{x^3-x^2-6x}$

7. $\dfrac{8x+2}{x^3-x}$

8. $\dfrac{x^2-16x-12}{x^3-3x^2-4x}$

9. $\dfrac{3x^3+4x^2-17x-1}{x^3+x^2-6x}$

10. $\dfrac{2x^2+5x-4}{x^3+x^2-2x}$

11. $\dfrac{-2x^2+x-1}{(x-3)(x-1)^2}$

12. $\dfrac{13x-12}{x^2(x-3)}$

13. $\dfrac{x^3-x^2+1}{x^2(x-1)}$

14. $\dfrac{3y+4}{(y+2)^2(y-6)}$

15. $\dfrac{3x^2+18x+15}{(x-1)(x+2)^2}$

16. $\dfrac{s+4}{(s+1)^2(s-1)^2}$

17. $\dfrac{4x^2-7x+10}{(x+2)(3x-2)^2}$

18. $\dfrac{1}{x^4-2x^3+x^2}$

19. $\dfrac{t+10}{(t+1)(t^2+1)}$

20. $\dfrac{x^4-x^2-2x+3}{(x-2)(x^2+2x+2)}$

21. $\dfrac{x^5+9x^3+1}{x^3+9x}$

22. $\dfrac{4x+3}{(x^2+1)(x^2+2)}$

23. $\dfrac{t+3}{t(t^2+1)}$

24. $\dfrac{x}{(x+1)^2(x^2+1)}$

25. $\dfrac{u}{u^4-1}$

26. $\dfrac{3s+1}{s^2(s^2+1)}$

In Problems 27 to 30, find the values of the constants in each decomposition into partial fractions.

27. $\dfrac{3x^2-2x-4}{x^3(x+2)} = \dfrac{A}{x} + \dfrac{B}{x^2} + \dfrac{C}{x^3} + \dfrac{D}{x+2}$

28. $\dfrac{x^3+3x^2+1}{(x^2+1)^2} = \dfrac{Ax+B}{x^2+1} + \dfrac{Cx+D}{(x^2+1)^2}$

29. $\dfrac{t^3-t^2}{(t^2+3)^2} = \dfrac{At+B}{t^2+3} + \dfrac{Ct+D}{(t^2+3)^2}$

30. $\dfrac{x^5-2x^4+2x^3+x-2}{x^2(x^2+1)^2}$
$$= \dfrac{A}{x} + \dfrac{B}{x^2} + \dfrac{Cx+D}{x^2+1} + \dfrac{Ex+G}{(x^2+1)^2}$$

6.6 Systems Containing Nonlinear Equations

In this section, we consider the solution of systems containing nonlinear equations. To avoid technical difficulties, we shall study only the relatively simple case of two equations in two unknowns. The solution of such a system can be found (at least approximately) by sketching graphs of the two equations on the same coordinate system and determining the points where the two graphs intersect.

The substitution and elimination methods introduced for systems of linear equations in Section 6.1 can often be used (with minor modifications) for systems containing nonlinear equations. Even then, graphs can be sketched, or generated on a computer or graphing calculator, to determine the number of solutions and as a rough check on the calculations.

In Examples 1 and 2, sketch graphs to determine the number of solutions of each system and then solve the system.

Example 1

$$\begin{cases} x^2 - 2y = 0 \\ x + 2y = 6 \end{cases}$$

Solution The graph of $x^2 - 2y = 0$, or $y = \frac{1}{2}x^2$, is a parabola opening upward with vertex at the origin; the graph of $x + 2y = 6$ is a line that intersects the parabola at two points (Figure 1). Thus, there are two solutions of the system. To find these solutions algebraically, we use the method of elimination. Adding the second equation to the first (to eliminate y), we obtain the equivalent system

$$\begin{cases} x^2 + x = 6 \\ x + 2y = 6. \end{cases}$$

Figure 1

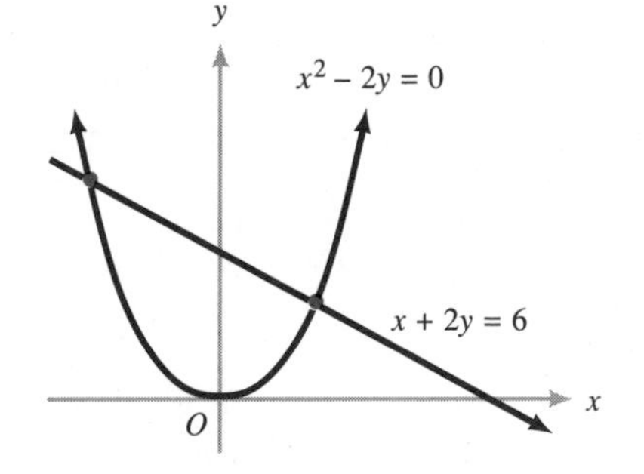

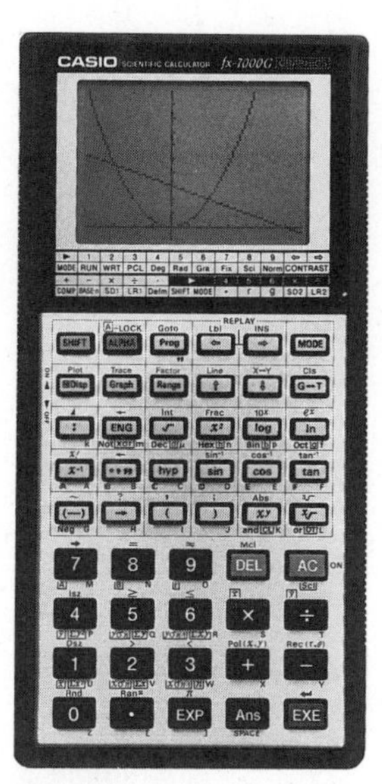

Casio fx-7000G showing the graph in Figure 1.

Now the first equation is quadratic in x and can be solved by factoring:

$$x^2 + x - 6 = 0 \quad \text{or} \quad (x + 3)(x - 2) = 0,$$

so $x = -3$ or $x = 2$. Substituting these values, one at a time, into the second equation $x + 2y = 6$, we find that

$$y = \tfrac{9}{2} \text{ when } x = -3 \quad \text{and} \quad y = 2 \text{ when } x = 2.$$

Hence, the two solutions are $(-3, \frac{9}{2})$ and $(2, 2)$. ■

Example 2 $\begin{cases} x^2 + y^2 = 25 \\ x^2 + y = 13 \end{cases}$

Figure 2

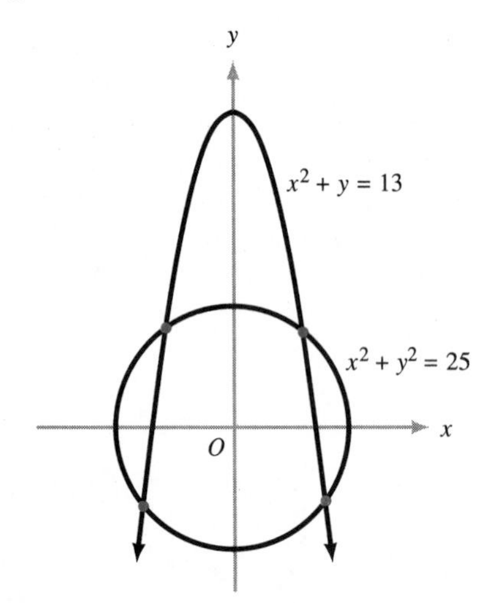

Solution Sketching the graphs of the two equations (a circle and a parabola) on the same coordinate system, we see four points of intersection (Figure 2). Thus, there are four solutions of the system. Again, we can use the method of elimination. Subtracting the second equation from the first (to eliminate x^2), we obtain the equivalent system

$$\begin{cases} y^2 - y = 12 \\ x^2 + y = 13. \end{cases}$$

Now the first equation is quadratic in y and can be solved by factoring:

$$y^2 - y - 12 = 0 \qquad \text{or} \qquad (y + 3)(y - 4) = 0,$$

so $y = -3$ or $y = 4$. Substituting $y = -3$ into the second equation, $x^2 + y = 13$, we obtain

$$x^2 - 3 = 13 \qquad \text{or} \qquad x^2 = 16;$$

hence, $x = 4$ or $x = -4$. Similarly, substituting $y = 4$ into the second equation, we obtain

$$x^2 + 4 = 13 \qquad \text{or} \qquad x^2 = 9;$$

hence, $x = 3$ or $x = -3$. Therefore, the solutions are $(-4, -3)$, $(4, -3)$, $(-3, 4)$, and $(3, 4)$. ■

When the elimination method can be used, as in the examples above, it is usually the most efficient way to solve the system. Otherwise, try the substitution method.

Example 3 Solve the system $\begin{cases} x^2 + 2y^2 = 18 \\ xy = 4. \end{cases}$

Solution Here there doesn't seem to be a simple way to eliminate variables, so we try the substitution method. Because the second equation is easily solved for y in terms of x, we begin there. If $xy = 4$, then $x \neq 0$ and

$$y = \frac{4}{x}.$$

Substituting $4/x$ for y in the first equation, we obtain

$$x^2 + 2\left(\frac{4}{x}\right)^2 = 18 \qquad \text{or} \qquad x^2 + \frac{32}{x^2} = 18.$$

Multiplying both sides of the last equation by x^2, we get

$$x^4 + 32 = 18x^2 \qquad \text{or} \qquad x^4 - 18x^2 + 32 = 0.$$

Factoring, we have

$$(x^2 - 16)(x^2 - 2) = 0 \qquad \text{or} \qquad (x - 4)(x + 4)(x - \sqrt{2})(x + \sqrt{2}) = 0.$$

Setting each factor equal to zero gives us

$$x = 4, \qquad x = -4, \qquad x = \sqrt{2}, \qquad \text{or} \qquad x = -\sqrt{2}.$$

For each of these values of x, there is a corresponding value of y given by $y = 4/x$. Therefore, the solutions are

$$(4, 1), \qquad (-4, -1), \qquad (\sqrt{2}, 2\sqrt{2}), \qquad \text{and} \qquad (-\sqrt{2}, -2\sqrt{2}).$$

■

If a system of equations contains exponential or logarithmic functions, the properties of these functions may be used to help find the solutions.

Example 4 Solve the system $\begin{cases} y - \log_4(6x + 10) = 1 \\ y + \log_4 x = 2. \end{cases}$

Solution We use the method of elimination. Subtracting the first equation from the second, we obtain

$$\log_4 x + \log_4(6x + 10) = 1.$$

Now we recall a basic property of logarithms [Property (i), page 313] and the definition of logarithms to rewrite the last equation as

$$\log_4[x(6x + 10)] = 1 \qquad \text{or} \qquad x(6x + 10) = 4^1;$$

that is,

$$6x^2 + 10x - 4 = 0 \qquad \text{or} \qquad 3x^2 + 5x - 2 = 0.$$

Factoring, we get

$$(3x - 1)(x + 2) = 0;$$

hence,

$$x = \tfrac{1}{3} \qquad \text{or} \qquad x = -2.$$

Because $\log_4 x$ is undefined when x is negative, $x = -2$ is an extraneous root, and we must reject it. Substituting $x = \frac{1}{3}$ into the equation $y = 2 - \log_4 x$, we find that

$$y = 2 - \log_4 \tfrac{1}{3} = 2 - \log_4 3^{-1} = 2 - (-1) \log_4 3 = 2 + \log_4 3.$$

Hence, the solution is $(\frac{1}{3}, 2 + \log_4 3)$. ■

We close this section with a brief indication of the way in which systems of equations, not all of which need be linear, arise in practical situations. We choose economics as our area of application. If p denotes the price per unit of a commodity, and q denotes the number of units of the commodity demanded in the marketplace

Figure 3

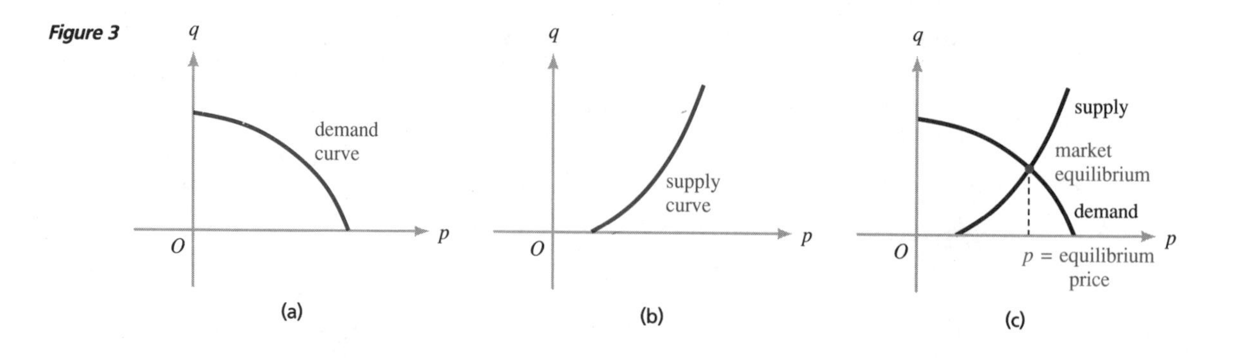

at price p, a graph of q as a function of p produces a **demand curve** (Figure 3a). Note that q will ordinarily be a decreasing function of p. (Why?) On the other hand, if p represents the price per unit of a commodity in the marketplace, and q denotes the number of units that manufacturers are willing to supply at that price, a graph of q as a function of p produces a **supply curve** (Figure 3b). In this case, q will ordinarily be an increasing function of p. (Why?) If the supply and demand curves are plotted on the same coordinate system, the point where they intersect is called the **market equilibrium point** (Figure 3c). At the equilibrium price (the p coordinate of the market equilibrium point), the quantity supplied will be equal to the quantity demanded. Under the usual interpretations of price, supply, and demand, only the portions of the supply and demand curves that fall in the first quadrant are economically meaningful.

Example 5 Suppose that the weekly demand q million gallons of synthetic fuel made from oil shale is related to the price p in dollars per gallon by the demand equation $8p^2 + 5q = 100$. If the market price is p dollars per gallon, assume that producers are willing to supply q million gallons of the synthetic fuel per week according to the supply equation $6p^2 - p - 3q = 5$. Find:

(a) The market equilibrium point. **(b)** The equilibrium price.

Solution **(a)** The market equilibrium point is found by solving the system

$$\begin{cases} 8p^2 + 5q = 100 \\ 6p^2 - p - 3q = 5. \end{cases}$$

Multiplying the second equation by 5 and adding 3 times the first equation to the result, we obtain the equivalent system

$$\begin{cases} 8p^2 + 5q = 100 \\ 54p^2 - 5p = 325. \end{cases}$$

The second equation can be solved by factoring:

$$54p^2 - 5p - 325 = 0 \quad \text{or} \quad (2p - 5)(27p + 65) = 0;$$

hence, $p = \frac{5}{2}$ or $p = -\frac{65}{27}$. Because of the economic interpretation, p cannot be negative, so we reject the second solution and retain only the solution $p = \frac{5}{2}$. Substituting $p = \frac{5}{2}$ into the first equation $8p^2 + 5q = 100$, we obtain

$$8(\tfrac{5}{2})^2 + 5q = 100 \quad \text{or} \quad 5q = 50,$$

so

$$q = 10.$$

Therefore, the market equilibrium point is $(p, q) = (\frac{5}{2}, 10)$.

(b) The equilibrium price is

$$p = \tfrac{5}{2} = \$2.50 \text{ per gallon.}$$

■

Problem Set 6.6

In Problems 1 to 6, sketch graphs to determine the number of solutions of each system of equations and then solve the system. gc Use a graphing calculator if you wish.

1. $\begin{cases} x^2 - 2y = 0 \\ 3x + 2y = 10 \end{cases}$

2. $\begin{cases} x^2 - 2y = 3 \\ x - y = 1 \end{cases}$

3. $\begin{cases} y^2 - 3x = 0 \\ 2x - y = 3 \end{cases}$

4. $\begin{cases} x^2 + y^2 = 25 \\ x^2 + y = 19 \end{cases}$

5. $\begin{cases} x^2 + y^2 = 4 \\ x - 2y = 4 \end{cases}$

6. $\begin{cases} 2x^2 + y = 9 \\ y - x^2 - 5x = 1 \end{cases}$

In Problems 7 to 26, solve each system of equations by the elimination or substitution method.

7. $\begin{cases} 2x^2 + y = 4 \\ 2x - y = 1 \end{cases}$

8. $\begin{cases} x^2 + y^2 = 1 \\ x^2 - y = 3 \end{cases}$

9. $\begin{cases} x^2 - y^2 = 3 \\ -2x + y = 1 \end{cases}$

10. $\begin{cases} x^2 + y^2 = 1 \\ -x + 2y = -2 \end{cases}$

11. $\begin{cases} x^2 + 2y^2 = 22 \\ 2x^2 + y^2 = 17 \end{cases}$

12. $\begin{cases} x^2 + y^2 = 625 \\ x + y = 35 \end{cases}$

13. $\begin{cases} 2s^2 - 4t^2 = 8 \\ s^2 + 2t^2 = 10 \end{cases}$

14. $\begin{cases} 4r^2 + 7s^2 = 23 \\ -3r^2 + 11s^2 = -1 \end{cases}$

15. $\begin{cases} 4h^2 + 7k^2 = 32 \\ -3h^2 + 11k^2 = 41 \end{cases}$

16. $\begin{cases} 2a^2 - 5b^2 + 8 = 0 \\ a^2 - 7b^2 + 4 = 0 \end{cases}$

17. $\begin{cases} x^2 - y = 2 \\ 2x^2 + y = 6x + 7 \end{cases}$

18. $\begin{cases} x^2 + y^2 - 8y = -7 \\ y - x^2 = 1 \end{cases}$

19. $\begin{cases} x^2 - xy + 2y^2 = 8 \\ xy = 4 \end{cases}$

20. $\begin{cases} 4x^2 - 6xy + 9y^2 = 63 \\ 2x - 3y + 3 = 0 \end{cases}$

21. $\begin{cases} x - y = 21 \\ \sqrt{x} + \sqrt{y} = 7 \end{cases}$

22. $\begin{cases} \dfrac{3}{x^2} - \dfrac{2}{y^2} = 1 \\ \dfrac{7}{x^2} - \dfrac{6}{y^2} = 2 \end{cases}$

23. $\begin{cases} y - \sqrt[4]{x} = 0 \\ y^2 - \sqrt[4]{x} = 2 \end{cases}$

24. $\begin{cases} y = \sqrt[4]{x - 2} \\ y^2 = \sqrt[4]{x - 2} + 12 \end{cases}$

25. $\begin{cases} (x - y)^2 + (x + y)^2 = 17 \\ (x - y)^2 - 4(x + y)^2 = 12 \end{cases}$

26. $\begin{cases} \dfrac{2}{(x + 1)^2} - \dfrac{5}{(y - 1)^2} = 3 \\ \dfrac{1}{(x + 1)^2} + \dfrac{3}{(y - 1)^2} = 7 \end{cases}$

In Problems 27 to 32, use appropriate properties of the exponential and logarithmic functions to solve each system.

27. $\begin{cases} y - \log_6(x + 3) = 1 \\ y + \log_6(x + 4) = 2 \end{cases}$

28. $\begin{cases} y + \log_3(x + 1) = 2 \\ y - \log_3(x + 3) = 1 \end{cases}$

29. $\begin{cases} x - 5^y = 0 \\ x - 25^y = -20 \end{cases}$

30. $\begin{cases} y + 15 = 10^x \\ y - 10^{2-x} = 0 \end{cases}$

31. $\begin{cases} y = 5 + \log_{10} x \\ y - \log_{10}(x + 6) = 7 \end{cases}$

32. $\begin{cases} \log_4(x^2 + y^2) = 2 \\ 2y - x = 4 \end{cases}$

33. A commercial artist is using a triangular template whose altitude exceeds its base by 4 inches and whose area is 30 square inches. Find the altitude and base of the template.

C **34.** A certain northern state stores road salt, used to melt ice and snow in winter, in buildings made of galvanized sheet iron in the shape of right circular cones (Figure 4). Each building has a height h equal to 1.5 times the radius r of its base, and each building has a volume of 12,000 cubic feet. Find h and r rounded off to two decimal places.

Figure 4

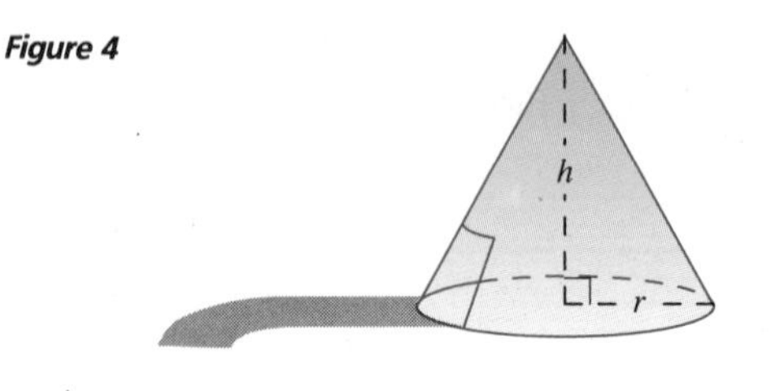

35. A farmer uses 30 meters of fence to enclose a rectangular pen next to a barn, with the barn wall forming one side of the pen (Figure 5). What are the dimensions of the pen if it encloses an area of 108 square meters?

Figure 5

36. A sporting goods store sells two circular targets for archery. The radius of the larger target is 10 centimeters more than the radius of the smaller target, and the difference between the areas of the two targets is 2300 square centimeters. Find the radii of the targets.

37. An open-ended cylindrical pipe used in an irrigation system has an outside surface area of 2700π square centimeters and a volume of 4050π cubic centimeters (Figure 6). Find the radius r and the length l of the pipe.

Figure 6

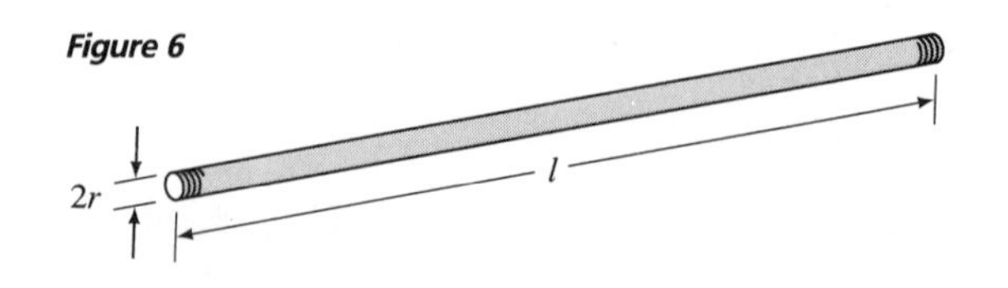

38. Find formulas for the length l and the width w of a rectangle in terms of its area A and its perimeter P.

39. The sum of the squares of the digits of a certain two-digit number is 61. The product of the number and the number with the digits reversed is 3640. What is the number?

40. A child's sandbox is made of galvanized sheet iron with a square base (Figure 7). If the volume of the sandbox is 50 cubic feet, and the combined area of the sides and the bottom is 65 square feet, find the dimensions of the box.

Figure 7

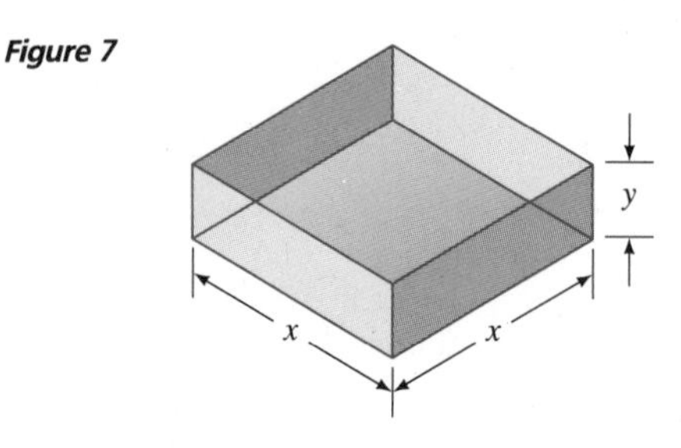

C **41.** An outfitting company manufactures backpacking tents with ends in the shape of equilateral triangles (Figure 8). Find the dimensions s and l of the tent if

Figure 8

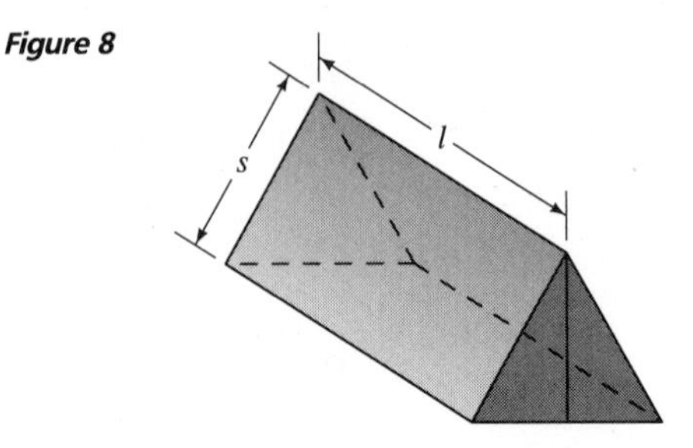

its volume is 6 cubic meters and if three times more material is used for the two sides and the bottom than for the two ends.

42. The sum of the radii of two circular pulleys is 25 centimeters, and the sum of the areas of the pulleys is 325π square centimeters (Figure 9). Find the radii of the two pulleys.

Figure 9

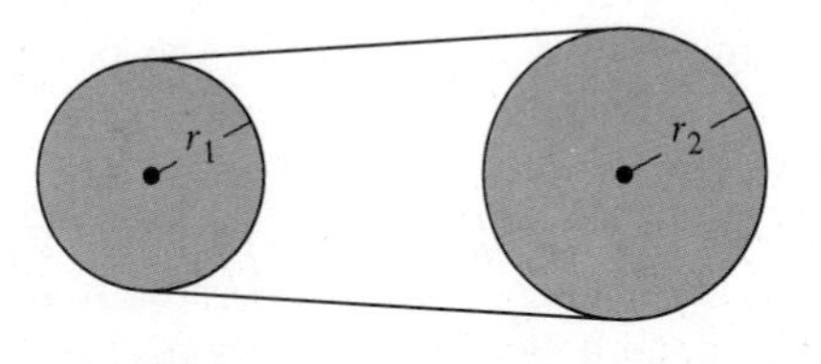

43. A company determines that its total monthly production cost C in thousands of dollars satisfies the equation $C^2 = 8x + 4$ and that its monthly revenue R in thousands of dollars satisfies the equation $8R - 3x^2 = 0$, where x is the number of thousands of units of its product manufactured and sold per month. When the cost of manufacturing the product equals the revenue obtained from selling it, the company *breaks even.* How many units must the company produce in order to break even?

44. A commercial jet was delayed for 15 minutes before takeoff because of a problem with baggage loading. To make up for the delay, the pilot increased the jet's air speed and took advantage of a favorable tail wind. This increased the jet's average speed by 36 miles per hour, with the result that it landed on time, 3 hours and 45 minutes after takeoff. Find the usual average ground speed and the distance between the two airports.

45. Suppose that the annual demand, q million cars, for a front-wheel-drive economy car is related to its sticker price, p thousand dollars, by the demand equation $2q^2 = 9 - p$. At a sticker price of p thousand dollars, the manufacturers are willing to build q million cars a year according to the supply equation $q^2 + 5q = p - 1$.

(a) Find the market equilibrium point.

(b) Find the equilibrium price of a car.

6.7 Systems of Inequalities and Linear Programming

In this section, we consider problems involving systems of inequalities in *two* unknowns. The solution of such a system can be represented by a set of points in the xy plane, and the problem can then be solved geometrically. Many real-world applications of mathematics—especially those pertaining to the allocation of limited resources—give rise to systems of *linear* inequalities and thus to problems in *linear programming*. Therefore, before considering more general types of inequalities, we begin by considering linear inequalities.

The Graph of an Inequality

By a **linear inequality** in the two unknowns x and y, we mean an inequality having one of the forms

$$ax + by + c > 0, \quad ax + by + c < 0, \quad ax + by + c \geq 0, \quad \text{or} \quad ax + by + c \leq 0,$$

where a, b, and c are constants and a and b are not both zero. The first two inequalities are called **strict;** the second two, **nonstrict.** The **graph** of such an inequality is defined to be the set of all points (x, y) in the xy plane whose coordinates satisfy the inequality.

Figure 1

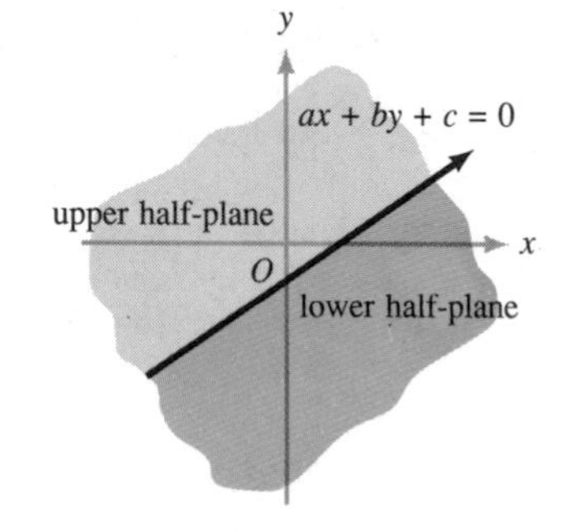

In order to study the graph of a linear inequality in x and y, we begin by considering the graph of the linear *equation*

$$ax + by + c = 0$$

obtained by (temporarily) replacing the inequality sign with an equal sign. This graph is a straight line which divides the xy plane into two regions called **half-planes,** one on each side of the line (Figure 1). A half-plane is called **closed** if the points on the boundary line belong to it; a half-plane is said to be **open** if the points on the boundary line do not belong to it.

In Figure 1, the expression $ax + by + c$ is zero only for points on the line separating the two half-planes, so it is either positive or negative for points (x, y) in the open half-planes. Actually, $ax+by+c$ is positive on one of the open half-planes and negative on the other (Problems 52 and 53). Thus, we have the following procedure.

Procedure for Sketching the Graph of a Linear Inequality

Step 1. Draw the graph of the linear equation obtained by (temporarily) replacing the inequality sign with an equal sign. If the inequality is strict ($>$ or $<$), draw a dashed line; if the inequality is nonstrict ($\geq$ or $\leq$), draw a solid line.

Step 2. Select any convenient test point in one of the two open half-planes determined by the line in step 1. If the coordinates of the test point satisfy the inequality, shade the half-plane that contains it; otherwise, shade the other half-plane.

The shaded half-plane is the graph of the inequality. A dashed boundary line indicates an *open* half-plane, and a solid boundary line indicates a *closed* one.

Figure 2

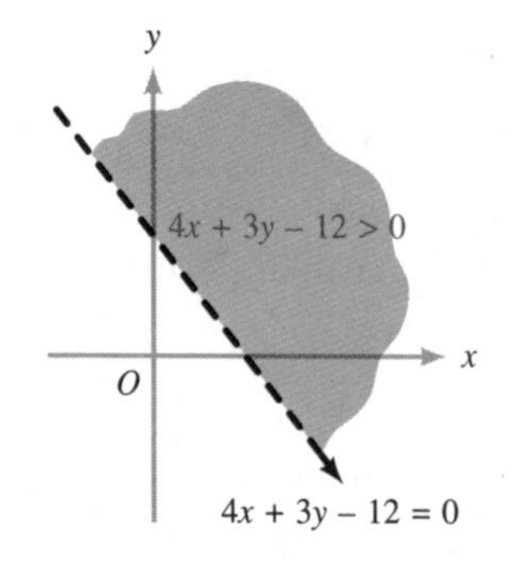

In Examples 1 and 2, sketch the graph of each linear inequality.

Example 1 $4x + 3y - 12 > 0$

Solution We follow the procedure above.

1. Since the inequality is strict, we draw the graph of the equation $4x + 3y - 12 = 0$ as a dashed line (Figure 2).

2. We test the inequality at the origin $O = (0, 0)$ by substituting $x = 0$ and $y = 0$ to obtain $-12 > 0$, which is *false*. Therefore, we shade the open half-plane *not* containing the origin (Figure 2). ■

Example 2 $-2x + y \leq 0$

Solution Again, we follow the procedure.

1. Since the inequality is nonstrict, we draw the graph of the equation $-2x + y = 0$ as a solid line (Figure 3).

2. We test the inequality at the point $(1, 0)$ by substituting $x = 1$ and $y = 0$ to obtain $-2 \le 0$, which is *true*. Therefore, we shade the closed half-plane containing the point $(1, 0)$ (Figure 3). ■

Figure 3

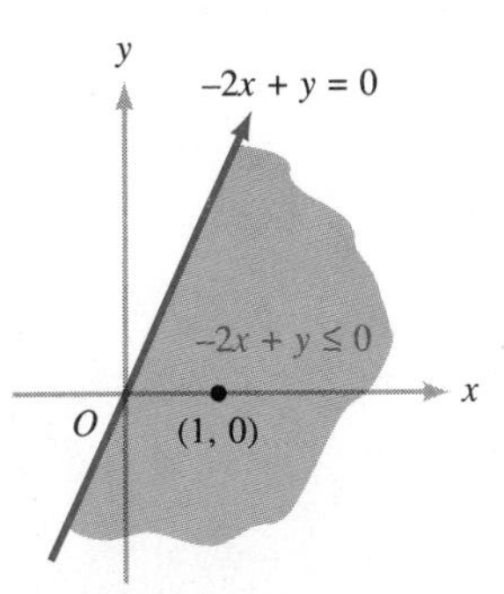

The method for sketching the graph of a nonlinear inequality is similar to the procedure illustrated above, except that the graph of the equation obtained in step 1 might divide the plane into more than two regions. (In special cases, it might not divide the plane at all.) In carrying out step 2 for nonlinear inequalities, it is necessary to select a test point in *each* of these regions and to shade only those regions (if any) for which the chosen test point satisfies the given inequality.

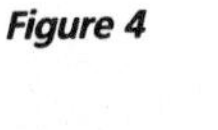

Figure 4

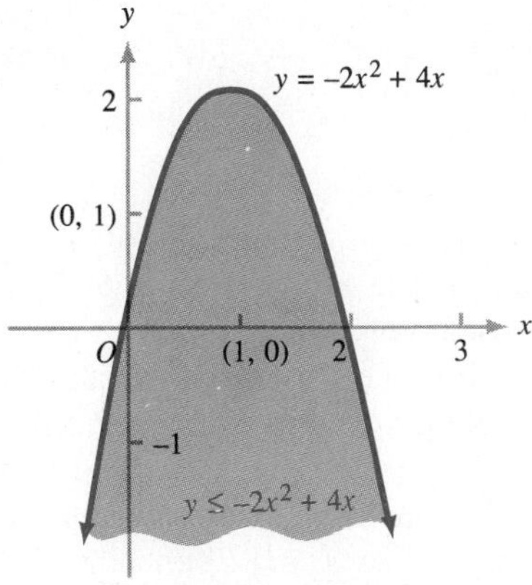

Example 3 Sketch the graph of the inequality $y \le -2x^2 + 4x$.

Solution Since the inequality is nonstrict, we begin by sketching the graph of $y = -2x^2 + 4x$ as a solid curve (Figure 4). The result is a parabola that divides the plane into two regions, one region above the parabola and one region below it. We select (say) the test point $(0, 1)$ in the upper region and the test point $(1, 0)$ in the lower one. Substituting $x = 0$ and $y = 1$ in the inequality $y \le -2x^2 + 4x$, we obtain the *false* statement $1 \le 0$, so we do not shade the upper region. Substituting $x = 1$ and $y = 0$, we obtain the *true* statement $0 \le 2$, so we shade the region below the parabola. ■

The Graph of a System of Inequalities

The **graph** of a *system* of inequalities is defined to be the set of all points (x, y) in the xy plane whose coordinates satisfy every inequality in the system. Such a graph is obtained by sketching the graphs of all the inequalities on the same coordinate system. The region where all of these graphs overlap is the graph of the system of inequalities.

In Examples 4 and 5, sketch the graph of each system of linear inequalities.

Example 4

$$\begin{cases} x + y \le 2 \\ -x + 3y \ge 4 \end{cases}$$

Solution Using the procedure given above, we sketch the graphs of $x + y \le 2$ and $-x + 3y \ge 4$ on the same coordinate system (Figure 5). The graph of $x + y \le 2$ is the closed

Figure 5

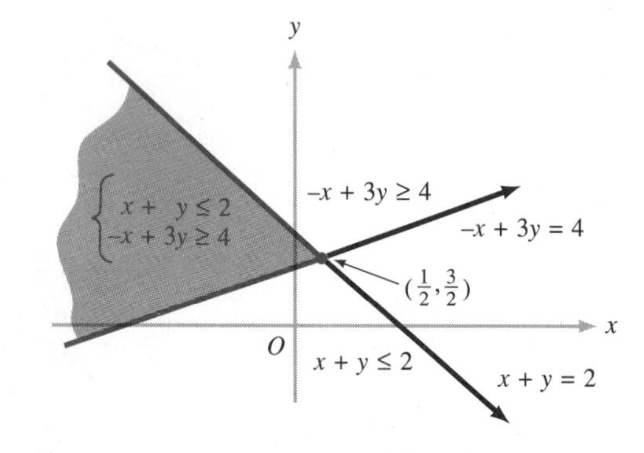

half-plane below the line $x + y = 2$, and the graph of $-x + 3y \ge 4$ is the closed half-plane above the line $-x + 3y = 4$. These two half-planes overlap in the region shaded in Figure 5; hence, this shaded region is the graph of the system of inequalities. ■

Example 5

$$\begin{cases} 2x + y \ge 2 \\ x - 2y \le 3 \\ x \ge 0 \\ y \ge 0 \end{cases}$$

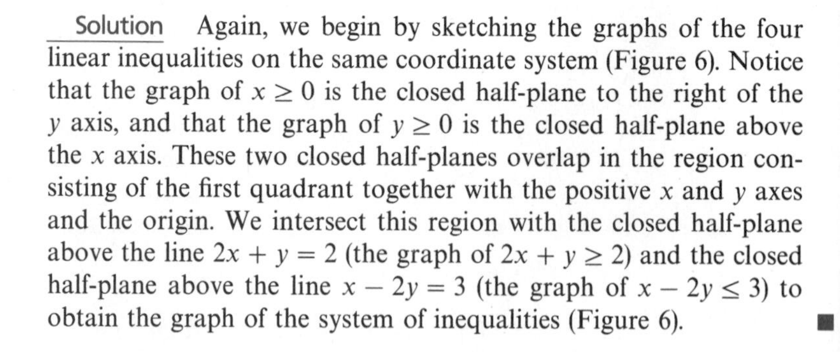

Solution Again, we begin by sketching the graphs of the four linear inequalities on the same coordinate system (Figure 6). Notice that the graph of $x \ge 0$ is the closed half-plane to the right of the y axis, and that the graph of $y \ge 0$ is the closed half-plane above the x axis. These two closed half-planes overlap in the region consisting of the first quadrant together with the positive x and y axes and the origin. We intersect this region with the closed half-plane above the line $2x + y = 2$ (the graph of $2x + y \ge 2$) and the closed half-plane above the line $x - 2y = 3$ (the graph of $x - 2y \le 3$) to obtain the graph of the system of inequalities (Figure 6). ■

Figure 6

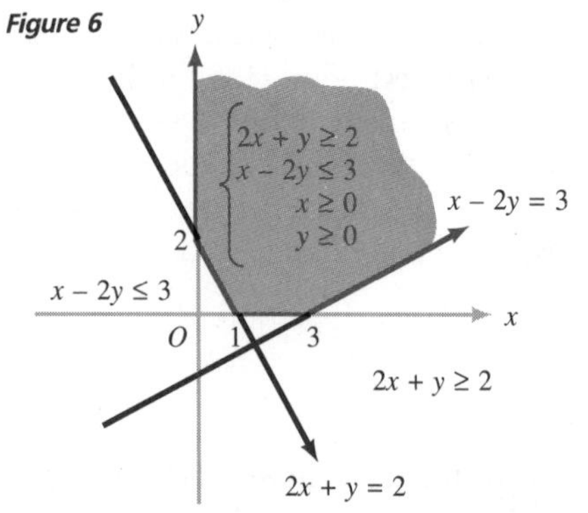

If the graph of a system of linear inequalities in x and y is a nonempty set of points, it will be bounded by straight-line segments or rays meeting at "corner points" called **vertices.** After you have sketched the graph, you can see these vertices and you can find their coordinates by solving appropriate pairs of linear equations.

Example 6 Find the vertices of the graphs in:

(a) Figure 5.

(b) Figure 6.

Solution **(a)** In Figure 5, the vertex, found by solving the system

$$\begin{cases} x + y = 2 \\ -x + 3y = 4, \end{cases}$$

is $(\frac{1}{2}, \frac{3}{2})$.

(b) The graph in Figure 6 has three vertices, (0, 2), (1, 0), and (3, 0). ■

Because the graphs in Figures 5 and 6 extend indefinitely in certain directions, we say that they are **unbounded.** The graph in the next example is **bounded** in the sense that it is cut off by boundary curves in every direction.

Example 7 Sketch the graph of the system of inequalities

$$\begin{cases} -x^2 + y \geq 1 \\ x + y < 3. \end{cases}$$

Solution We begin by sketching the graph of the first inequality. Using test points above and below the graph of the parabola $-x^2 + y = 1$, that is, $y = x^2 + 1$, we find that the inequality $-x^2 + y \geq 1$ is satisfied for points on or above the parabola (Figure 7a). Next, we sketch the graph of the second inequality and find that it consists of all points below the line $x + y = 3$ (Figure 7b). Finally, we sketch both the parabola and the line on the same coordinate system (Figure 7c) and shade the region above (or on) the parabola *and* below the line. ■

Figure 7

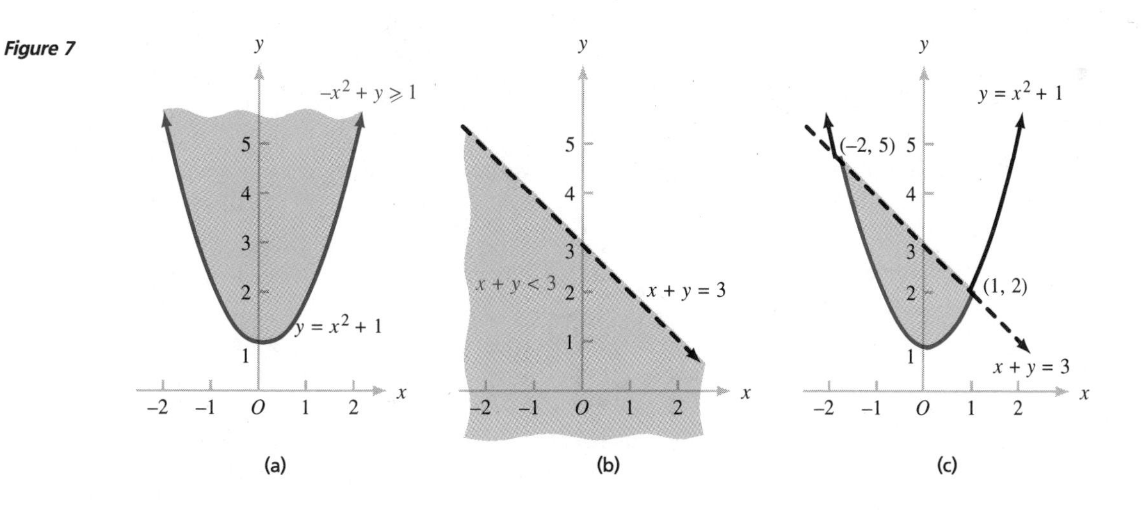

Linear Programming

Suppose that an oil company's profit depends on what portion of a limited allocation of crude oil the company refines into gasoline and what portion it converts into heating oil. Assume that there is a governmental restriction requiring that at least a certain fraction of the crude oil be converted into heating oil. In order to achieve the greatest possible profit under the governmental constraint, the company will have to plan or "program" its activities.

A problem in which it is necessary to find the maximum (largest) or minimum (smallest) value of a certain quantity (such as profit, cost, revenue, distance, or time) under a given set of constraints (restrictions) is sometimes called a **programming** problem. Here the word "programming" is intended to suggest planning. The quantity whose maximum or minimum value is desired is called the **objective function.** The objective function depends on one or more variables, and the constraints are expressed as conditions on the possible values of these variables. When the objective function is a linear (first-degree) polynomial in two or more variables and the constraints are expressed as a system of nonstrict linear inequalities, we have a **linear programming** problem.

Here we consider only linear programming problems in which the objective function F depends on two variables x and y, and the graph G of the system of linear inequalities that express the constraints is a *bounded* region in the xy plane. Thus, F has the form

$$F = ax + by + c,$$

where a, b, and c are constants and G is bounded by a finite number of line segments meeting at a finite number of vertices. Under these circumstances, it can be shown that F *has a maximum and a minimum value on* G, *and these values occur at certain vertices of* G. Thus:

> To find the maximum and minimum values, you merely list all the vertices of G and calculate the value of F at each one.

The largest and smallest of these values are the maximum and minimum values of F on the region G.

Example 8 Find the maximum and minimum values of the objective function

$$F = 3x + 4y + 1,$$

subject to the constraints

$$\begin{cases} x + 2y \le 8 \\ x + y \le 5 \\ x \ge 0 \\ y \ge 0. \end{cases}$$

Figure 8

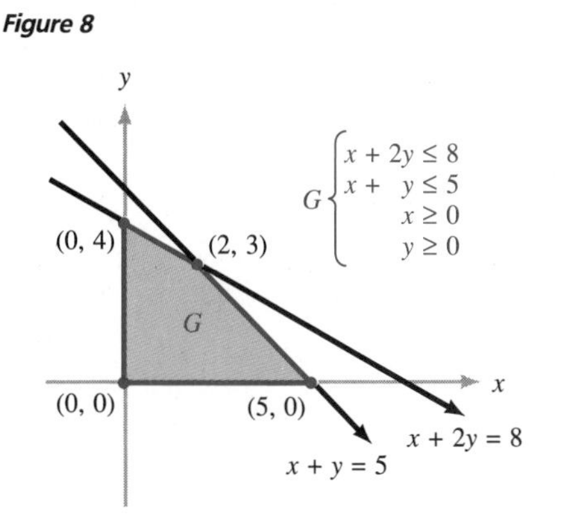

Solution We begin by using the method described above to sketch the graph G of the system representing the constraints (Figure 8). By solving appropriate pairs of linear equations, we find the coordinates of the vertices of G. These coordinates are listed in Table 1 along with the corresponding values of $F = 3x + 4y + 1$ at each vertex. From Table 1, the minimum value of F, which is 1, occurs at (0, 0); and the maximum value of F, which is 19, occurs at (2, 3).

Table 1

Vertex	Value of $F = 3x + 4y + 1$	
(0, 0)	1	minimum
(5, 0)	16	
(2, 3)	19	maximum
(0, 4)	17	

■

The following example illustrates a typical application of linear programming.

Example 9 A large school system wants to design a lunch menu containing two food items X and Y. Each ounce of X supplies 1 unit of protein, 2 units of carbohydrates, and 1 unit of fat. Each ounce of Y supplies 1 unit of protein, 1 unit of carbohydrates, and 1 unit of fat. The two items together must provide at least 7 units of protein, at least 10 units of carbohydrates, and no more than 8 units of fat per serving. If each ounce of X costs 12 cents and each ounce of Y costs 8 cents, how many ounces of each item should each serving contain to meet the dietary requirements at the lowest cost?

Solution Let x and y denote the number of ounces per serving of X and Y, respectively. Since each ounce of X costs 12 cents and each ounce of Y costs 8 cents, the cost per serving is

$$F = 12x + 8y \text{ cents.}$$

Since each ounce of X or of Y supplies 1 unit of protein, each serving will supply $x + y$ units of proteins. To meet the protein requirement, we must have

$$x + y \ge 7.$$

Similarly, to meet the carbohydrate requirements, we must have

$$2x + y \ge 10,$$

and to meet the fat requirement,

$$x + y \le 8.$$

Because the number of ounces per serving cannot be negative, we also have the conditions $x \geq 0$ and $y \geq 0$. Therefore, the problem is to minimize the objective function

$$F = 12x + 8y,$$

subject to the constraints

$$\begin{cases} x + y \geq 7 \\ 2x + y \geq 10 \\ x + y \leq 8 \\ x \geq 0 \\ y \geq 0. \end{cases}$$

Figure 9

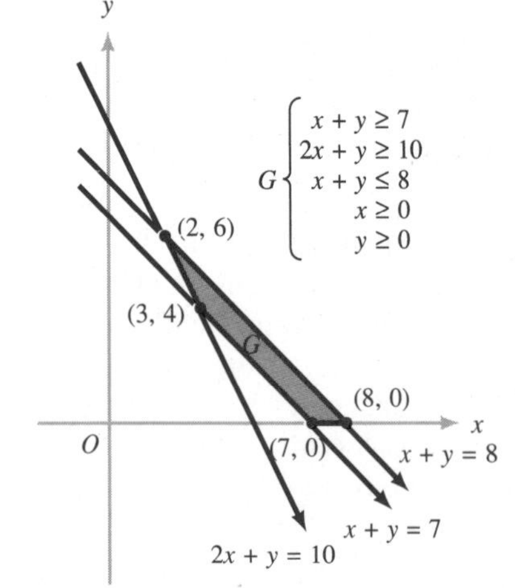

The graph G of the system representing these constraints is sketched in Figure 9, from which we find the vertices listed in Table 2. Thus, if each serving contains 3 ounces of item X and 4 ounces of item Y, all dietary requirements will be met at the minimum cost of 68 cents per serving.

Table 2

Vertex	Value of $F = 12x + 8y$	
(7, 0)	84	
(8, 0)	96	
(2, 6)	72	
(3, 4)	68	minimum

■

Problem Set 6.7

In Problems 1 to 10, sketch the graph of each linear inequality.

1. $x \geq 0$

2. $y < 1$

3. $2x + y - 3 \leq 0$

4. $5x - 2y \leq 2$

5. $3x + 2y \geq 6$

6. $2x \geq 2 - y$

7. $3x + 5y < 15$

8. $4x > 3 - 2y$

9. $x < y$

10. $1 < x - y \leq 5$

In Problems 11 to 22, **(a)** sketch the graph of the system of linear inequalities, **(b)** determine whether the graph is bounded or unbounded, and **(c)** find all vertices of the graph.

11. $\begin{cases} 2x + 3y \leq 7 \\ 3x - y \leq 5 \end{cases}$

12. $\begin{cases} 2x + y < 3 \\ x + 3y > 4 \end{cases}$

13. $\begin{cases} x + y \leq 4 \\ y > 2x - 4 \end{cases}$

14. $\begin{cases} x + y < 1 \\ -x + 2y \geq 4 \\ y > 0 \end{cases}$

15. $\begin{cases} y + 1 < 3x \\ y - x > 3 \end{cases}$

16. $\begin{cases} 3x + y < 4 \\ y - 2x \geq -1 \\ x \leq 0 \end{cases}$

17. $\begin{cases} 4x + 7y \leq 28 \\ 2x - 3y \geq -6 \\ y \geq -2 \end{cases}$

18. $\begin{cases} x + 3y \leq 6 \\ 3x - 2y \leq 4 \\ y \geq 0 \end{cases}$

19. $\begin{cases} -2x + y \leq 2 \\ x + 2y \leq 8 \\ x \geq 0 \\ y \geq 0 \end{cases}$

20. $\begin{cases} 2x - 3y \geq 2 \\ y \geq 6 - x \\ x \geq 4 \end{cases}$

21. $\begin{cases} x + 2y \leq 6 \\ 3x + y \leq 9 \\ x \geq 0 \\ y \geq 0 \end{cases}$

22. $\begin{cases} \frac{x}{2} - 1 \leq y \leq 3 + 3x \\ 0 \leq 4x \leq 12 - 3y \end{cases}$

In Problems 23 to 30, sketch the graph of each system of inequalities.

23. $\begin{cases} -x^2 + y \geq 0 \\ x + y \leq 2 \end{cases}$

24. $\begin{cases} -x^2 + y \geq 0 \\ x^2 + y^2 < 1 \end{cases}$

25. $\begin{cases} x^2 + y^2 \leq 1 \\ (x + 1)^2 + y^2 \leq 1 \end{cases}$

26. $\begin{cases} x^2 + y^2 \leq 4 \\ (x - 2)^2 + (y + 2)^2 \geq 4 \end{cases}$

27. $\begin{cases} \sqrt{x} - y > 0 \\ x - 4y \leq 0 \end{cases}$

28. $\begin{cases} x^2 + y \leq 13 \\ x^2 + y^2 \geq 25 \\ x \geq 0 \end{cases}$

29. $\begin{cases} |x + y| \geq 1 \\ x^2 + y^2 \leq 1 \end{cases}$

30. $\begin{cases} |x| + |y| \leq 1 \\ 2x^2 + 2y^2 \geq 1 \\ xy \geq 0 \end{cases}$

31. Rewrite the system $\begin{cases} 0 \leq x \leq 7 - 2y \\ 0 \leq 8y \leq 5x + 3 \end{cases}$ as an equivalent system of four inequalities and sketch the graph of the resulting system.

32. Sketch the graph of the system

$$\begin{cases} |x - 2| < 2 \\ |y + 1| < 3. \end{cases}$$

In Problems 33 to 42, find the maximum and minimum values of the objective function F subject to the indicated constraints.

33. $F = 2x - y + 3$

$\begin{cases} x + y \leq 1 \\ x \geq 0 \\ y \geq 0 \end{cases}$

34. $F = 3x + 2y - 1$

$\begin{cases} x + 2y \leq 7 \\ 5x - 8y \leq -3 \\ x \geq 0 \\ y \geq 0 \end{cases}$

35. $F = 5x + 4y$

$\begin{cases} x + 2y \geq 3 \\ x + 2y \leq 5 \\ x \geq 0 \\ y \geq 0 \end{cases}$

36. $F = 4x - y + 7$

$\begin{cases} 3x + 8y \leq 120 \\ 3x + y \leq 36 \\ x \geq 0 \\ y \geq 0 \end{cases}$

37. $F = 3x - 5y + 2$

$\begin{cases} x + y \leq 10 \\ x - 3y \geq -18 \\ x \geq 0 \\ y \geq 0 \end{cases}$

38. $F = 10x + 12y$

$\begin{cases} 0.2x + 0.4y \leq 30 \\ 0.2x + 0.2y \leq 20 \\ x \geq 0 \\ y \geq 0 \end{cases}$

39. $F = \frac{3}{2}x + y$

$\begin{cases} \frac{1}{2}x + y \geq 2 \\ \frac{1}{2}x - y \geq 1 \\ x \leq 6 \\ y \geq 0 \end{cases}$

40. $F = -10x + 5y + 3$

$\begin{cases} x + \frac{3}{2}y \geq -60 \\ x + y \geq -50 \\ x \leq 0 \\ y \leq 0 \end{cases}$

41. $F = \frac{1}{2}x + \frac{3}{4}y$

$\begin{cases} \frac{1}{3}x + y \leq 30 \\ x + \frac{1}{2}y \leq 40 \\ x + y \geq 10 \end{cases}$

42. $F = 0.15x + 0.1y$

$\begin{cases} 4x + 5y \leq 2000 \\ 12x + 5y \leq 3000 \\ x + y \geq 100 \\ x \geq 0 \\ y \geq 0 \end{cases}$

43. A hardware store has display space for at most 40 spray cans of rustproofing paint, x cans of red and y cans of gray. If there are to be at least 10 cans of each type in the display, **(a)** sketch a graph showing the possible numbers of red and gray cans in the display, and **(b)** find all vertices of the graph.

44. A gymnasium will be built on a rectangular plot of land surrounded by an outdoor jogging track (Figure 10). The track is to consist of two line segments of length x along two opposite sides of the rectangular plot connected by two semicircles with diameter y

along the other two sides of the plot. The jogging track is to have a total length of at least 1000 meters, the plot of land is to have an area of no more than 40,000 square meters, and neither dimension of the plot is to exceed 300 meters. Write a system of inequalities in x and y describing these constraints and sketch a graph of the system.

Figure 10

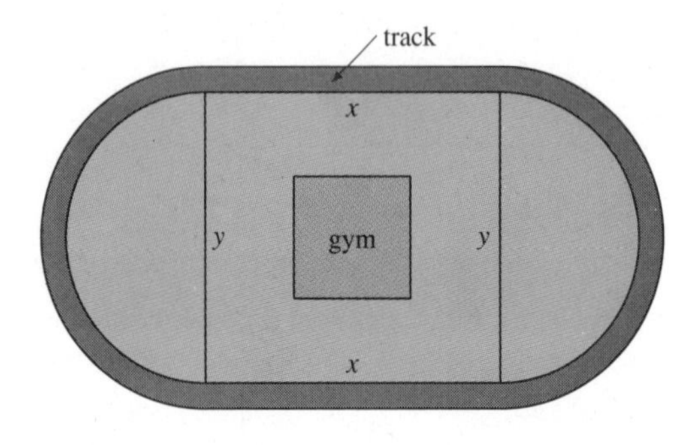

45. An electronics company manufactures two models of household smoke detectors. Model A requires 1 unit of labor and 4 units of parts; model B requires 1 unit of labor and 3 units of parts. If 90 units of labor and 320 units of parts are available, and if the company makes a profit of \$5 on each model A detector and \$4 on each model B detector, how many of each model should it manufacture to maximize its profit?

C 46. A family owns and operates a 312-acre farm on which it grows cotton and peanuts. The task of planting, picking, ginning, and baling the cotton requires 35 person-hours of labor per acre; the task of planting, harvesting, and bagging the peanuts requires 27 person-hours of labor per acre. The family is able to devote 9500 person-hours to these activities. If the profit for each acre of cotton grown is \$173 and the profit for each acre of peanuts grown is \$152, how many acres should be planted in cotton and how many in peanuts to maximize the family's profit?

47. In Problem 45, suppose that the company raises its prices so that its profit on each model A detector is \$7 and on each model B detector is \$5. Now how many detectors of each type should it manufacture to maximize its profit?

48. A supplier has 105 pounds of leftover beef which must be sold before it spoils. Of this beef, 15 pounds is prime grade, 40 pounds is grade A, and the rest is utility grade. A local restaurant will buy ground beef consisting of 20% prime grade, 40% grade A, and 40% utility grade for 75 cents per pound. A hamburger stand will buy ground beef consisting of 40% grade A and 60% utility grade for 55 cents per pound. How much ground beef should the supplier sell to the restaurant and how much to the hamburger stand in order to maximize its revenue?

49. A town operates two recycling centers. Each day that center I is open, 300 pounds of glass, 200 pounds of paper, and 100 pounds of aluminum are deposited there. Each day that center II is open, 100 pounds of glass, 600 pounds of paper, and 100 pounds of aluminum are deposited there. The town has contracted to supply at least 1200 pounds of glass, 2400 pounds of paper, and 800 pounds of aluminum per week to a salvage company. Supervision and maintenance at center I cost the town \$40.00 each day it is open, and center II costs \$50.00 each day it is open. How many days a week should each center remain open so as to minimize the total weekly cost for supervision and maintenance, yet allow the town to fulfill its contract with the salvage company?

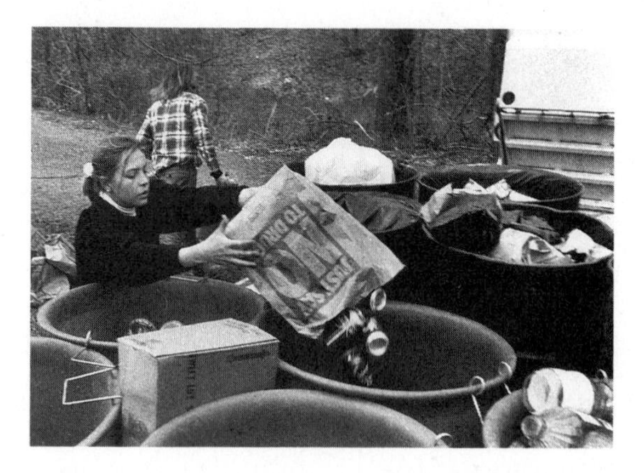

50. In Problem 48, suppose that the owner of the restaurant learns that the meat is in danger of spoiling and decides to pay only \$0.50 per pound for it. If the hamburger stand will still pay \$0.55 per pound, how much ground beef should the supplier now sell to the restaurant and how much to the hamburger stand?

51. Find a condition on the positive constants a and b such that the objective function

$$F = ax + by,$$

subject to the constraints $2x + 3y \leq 9$, $x - y \leq 2$, $x \geq 0$, and $y \geq 0$, will take on the *same* maximum value at the vertex $(0, 3)$ and the vertex $(3, 1)$.

52. Suppose that a, b, and c are constants and $b \neq 0$. Let L be the line

$$ax + by + c = 0.$$

(a) If $b > 0$, show that $ax + by + c > 0$ for (x, y) above L, and that $ax + by + c < 0$ for (x, y) below L.

(b) If $b < 0$, show that $ax + by + c < 0$ for (x, y) above L, and that $ax + by + c > 0$ for (x, y) below L.

53. Suppose that a and c are constants and $a \neq 0$. Let L be the vertical line

$$ax + c = 0.$$

(a) If $a > 0$, show that $ax + c > 0$ for (x, y) to the right of L, and that $ax + c < 0$ for (x, y) to the left of L.

(b) If $a < 0$, show that $ax + c < 0$ for (x, y) to the right of L, and that $ax + c > 0$ for (x, y) to the left of L.

CHAPTER 6 Review Problem Set

In Problems 1 and 2, use graphs to decide whether each system is well-determined, inconsistent, or dependent and indicate the solutions (if any) on the graphs.

1. (a) $\begin{cases} y = 2x + 7 \\ 2x - y = 5 \end{cases}$ **(b)** $\begin{cases} 4x - y = 3 \\ -2x + 3y = 1 \end{cases}$

(c) $\begin{cases} 5x + 5y = 10 \\ -3x - 3y = -6 \end{cases}$

2. (a) $\begin{cases} 4x + 2y = 3 \\ 2x + y = 5 \end{cases}$ **(b)** $\begin{cases} -2x + 6y = -8 \\ x - 3y = 4 \end{cases}$

(c) $\begin{cases} x - 3y = 4 \\ 2x + y = 15 \end{cases}$

In Problems 3 to 6, use the substitution method to solve each system of linear equations.

3. $\begin{cases} 3u + 2v = 7 \\ -u + 4v = 3 \end{cases}$ **4.** $\begin{cases} 6x_1 = 15 + 9x_2 \\ x_1 = 7 - \frac{3}{2}x_2 \end{cases}$

5. $\begin{cases} 2x - y - z = 0 \\ 2x + 3y = 1 \\ 8x - 3z = 4 \end{cases}$ **6.** $\begin{cases} x_1 + x_2 - x_3 = 0 \\ x_1 - x_2 + x_3 = 2 \\ 2x_1 + x_2 - 4x_3 = -8 \end{cases}$

In Problems 7 to 10, use the method of elimination (without matrices) to find the solution of each system of linear equations.

7. $\begin{cases} 6x - 9y = -3 \\ 2x + y = 3 \end{cases}$ **8.** $\begin{cases} -4x_1 + 5x_2 = 1 \\ x_1 - 2x_2 = -1 \end{cases}$

9. $\begin{cases} 2x - z = 12 \\ x + y = 7 \\ 5x + 4z = -9 \end{cases}$ **10.** $\begin{cases} x_1 + x_2 + x_3 = 6 \\ 2x_1 - x_2 + x_3 = 3 \\ 3x_1 + x_2 - x_3 = 2 \end{cases}$

In Problems 11 to 16, solve the system of linear equations by the elimination method using matrices.

11. $\begin{cases} 5u - v = 19 \\ -2u + 3v = 8 \end{cases}$ **12.** $\begin{cases} s + 26 = -4t \\ 3s + 8 = 2t \end{cases}$

13. $\begin{cases} 2x + y - z = -3 \\ 3x - 2y + 2z = 13 \\ x + 3y + 4z = 0 \end{cases}$ **14.** $\begin{cases} 2x + y = 11 \\ x + 2y - 4z = 17 \\ 3x - y + 3z = 1 \end{cases}$

15. $\begin{cases} 2x_1 - x_2 + 2x_3 = -8 \\ 3x_1 - x_2 - 4x_3 = 3 \\ x_1 + 2x_2 - 3x_3 = 9 \end{cases}$ **16.** $\begin{cases} \frac{1}{2}x + \frac{1}{3}y - \frac{1}{4}z = 2 \\ \frac{1}{3}x + \frac{1}{4}y - \frac{1}{2}z = \frac{1}{6} \\ -\frac{3}{2}x + 3y + 2z = 23 \end{cases}$

17. Let $A = \begin{bmatrix} 2 & -1 & 3 \\ 1 & 2 & 4 \end{bmatrix}$, $B = \begin{bmatrix} -3 & 2 & 1 \\ 5 & -2 & 4 \end{bmatrix}$, and $C = \begin{bmatrix} 2 & 1 & -1 \\ 3 & 1 & 2 \end{bmatrix}$. Find:

(a) $A + 3B$ **(b)** $2A - B$

(c) $\frac{1}{2}A - 2B + C$ **(d)** $2C + B - 3A$

18. Let $A = \begin{bmatrix} 1 & -1 \\ 2 & 3 \end{bmatrix}$ and $B = \begin{bmatrix} 1 & 3 \\ -2 & 1 \end{bmatrix}$. Find:

(a) A^2 **(b)** B^2 **(c)** A^2B^2

19. Let $A = \begin{bmatrix} 2 & 1 & -1 \\ 3 & 1 & 2 \end{bmatrix}$ and $B = \begin{bmatrix} -1 & 2 \\ 2 & 4 \\ 0 & 5 \end{bmatrix}$. Find:

(a) AB **(b)** BA

20. Find a value of x for which $AB = BA$ if $A = \begin{bmatrix} 1 & x \\ 0 & -1 \end{bmatrix}$ and $B = \begin{bmatrix} 2 & 3 \\ 0 & 1 \end{bmatrix}$.

21. Let $A = \begin{bmatrix} 2 & -4 \\ 3 & 7 \end{bmatrix}$. Find:

(a) A^{-1} **(b)** AA^{-1} **(c)** $A^{-1}A$

22. Let $A = \begin{bmatrix} 1 & 2 \\ 3 & -1 \end{bmatrix}$ and $B = \begin{bmatrix} -2 & 4 \\ 1 & 3 \end{bmatrix}$. Find:

(a) A^{-1} **(b)** B^{-1} **(c)** $(AB)^{-1}$

(d) $B^{-1}A^{-1}$ **(e)** $(AB)^{-1} - A^{-1}B^{-1}$

23. Find A^{-1} if $A = \begin{bmatrix} 2 & -1 & 0 \\ 1 & 0 & 1 \\ 1 & -2 & 0 \end{bmatrix}$.

24. If $A = \begin{bmatrix} a & b \\ c & d \end{bmatrix}$, the **determinant** of A, in symbols, det A, is defined by $\det A = \begin{vmatrix} a & b \\ c & d \end{vmatrix}$. Let A and B be 2 by 2 matrices.

(a) Show that $\det(AB) = (\det A)(\det B)$.

(b) If I is the 2 by 2 identity matrix, show that $\det I = 1$.

(c) If A is nonsingular, show that $\det A \neq 0$ and $\det(A^{-1}) = (\det A)^{-1}$.

(d) Let $d = \det A$. If $d \neq 0$, show that A is nonsingular.

In Problems 25 and 26, write the system of linear equations in the form $AX = K$ for suitable matrices A, X, and K and solve each system with the aid of A^{-1}.

25. (a) $\begin{cases} 3x + 2y = 11 \\ 4x - 3y = 9 \end{cases}$ **(b)** $\begin{cases} 3x + 2y + z = 7 \\ 2x + 3z = 10 \\ 5x - y = -8 \end{cases}$

26. (a) $\begin{cases} 7x_1 + 10x_2 = -3 \\ 5x_1 + 2x_2 = 3 \end{cases}$

C **(b)** $\begin{cases} 7x - 4z = k_1 \\ 5y + 3z = k_2 \\ -3x + 7y = k_3 \end{cases}$

In Problems 27 to 30, evaluate each determinant.

27. (a) $\begin{vmatrix} \frac{1}{2} & \frac{4}{3} \\ -\frac{5}{2} & \frac{2}{3} \end{vmatrix}$ **(b)** $\begin{vmatrix} 7 & 4 & 5 \\ 6 & -5 & 1 \\ 3 & 2 & 0 \end{vmatrix}$

C **28. (a)** $\begin{vmatrix} 5.007 & 13.142 \\ -3.733 & 2.501 \end{vmatrix}$

(b) $\begin{vmatrix} 4.21 & 3.72 & -2.02 \\ -1.59 & 7.07 & -8.83 \\ 2.22 & 3.14 & 2.03 \end{vmatrix}$

29. (a) $\begin{vmatrix} 1 & -1 & 2 \\ 2 & -2 & 4 \\ \sqrt{2} & -\sqrt{2} & 2\sqrt{2} \end{vmatrix}$

(b) $\begin{vmatrix} 1 & -1 & 2 \\ 2 & -2 & 1 \\ \sqrt{2} & -\sqrt{2} & 2\sqrt{2} \end{vmatrix}$

30. (a) $\begin{vmatrix} 4 & \sqrt{2} & \pi \\ 0 & 5 & e \\ 0 & 0 & -1 \end{vmatrix}$

(b) $\begin{vmatrix} 1 & 1 & 1 \\ x & a & b \\ x^2 & a^2 & b^2 \end{vmatrix}$

In Problems 31 and 32, assume that

$$\begin{vmatrix} a & b & c \\ u & v & w \\ x & y & z \end{vmatrix} = 2.$$

Find the value of each determinant.

31. (a) $\begin{vmatrix} c & a & b \\ w & u & v \\ z & x & y \end{vmatrix}$ **(b)** $\begin{vmatrix} 2a & 2b & 2c \\ 2u & 2v & 2w \\ 2x & 2y & 2z \end{vmatrix}$

(c) $\begin{vmatrix} a+u & b+v & c+w \\ u+x & v+y & w+z \\ -x & -y & -z \end{vmatrix}$

32. (a) $\begin{vmatrix} -a & -b & -c \\ -u & -v & -w \\ -x & -y & -z \end{vmatrix}$ **(b)** $\begin{vmatrix} a & b+a & 2c \\ u & v+u & 2w \\ x & y+x & 2z \end{vmatrix}$

(c) $\begin{vmatrix} a & u & x \\ b & v & y \\ c & w & z \end{vmatrix}$

In Problems 33 to 36, use Cramer's rule to solve each system.

33. $\begin{cases} 3x - 4y = -2 \\ 4x + y = 7 \end{cases}$

34. $\begin{cases} 3x_1 + 10x_2 = 3 \\ 6x_1 - 5x_2 = 16 \end{cases}$

35. $\begin{cases} x + y + 2z = 10 \\ 5x + 3y - z = 1 \\ 3x - y - 3z = -3 \end{cases}$

36. $\begin{cases} 3r + 6s - 5t = -1 \\ r - 2s + 4t = 4 \\ 5r + 6s - 7t = -5 \end{cases}$

37. If $P = (p_1, p_2)$, $Q = (q_1, q_2)$, and $R = (r_1, r_2)$ are three points in the Cartesian plane, then the area A of triangle PQR is given by one-half the absolute value of $\begin{vmatrix} p_1 & p_2 & 1 \\ q_1 & q_2 & 1 \\ r_1 & r_2 & 1 \end{vmatrix}$. (See Problem 60 in Problem Set 6.4.) Use this formula to find the area of each triangle.

(a) $P = (4, -3)$, $Q = (-2, 5)$, $R = (5, 1)$

(b) $P = (9, 4)$, $Q = (2, 9)$, $R = (-3, -2)$

38. If $P = (a, b)$ and $Q = (c, d)$ are points in the xy plane and $P \neq (0, 0)$, show that the perpendicular distance from Q to the line through $O = (0, 0)$ and P is the absolute value of

$$\begin{vmatrix} a & b \\ c & d \end{vmatrix}(a^2 + b^2)^{-1/2}.$$

In Problems 39 to 44, decompose each fraction into partial fractions.

39. $\dfrac{2x}{(x - 1)(x + 1)}$

40. $\dfrac{4x + 7}{x^2 + 5x + 4}$

41. $\dfrac{4x^2 + 13x - 9}{x(x + 3)(x - 1)}$

42. $\dfrac{-x^2 + 13x - 26}{(x + 1)^2(x - 4)}$

43. $\dfrac{x^2 + 4}{x(x^2 + 1)}$

44. $\dfrac{6x^3 + 5x^2 + 21x + 12}{x(x + 1)(x^2 + 4)}$

In Problems 45 to 50, solve each system of equations.

45. $\begin{cases} y^2 - x^2 = 3 \\ xy = 2 \end{cases}$

46. $\begin{cases} 9x^2 - 4y^2 = 36 \\ x^2 + y^2 = 43 \end{cases}$

47. $\begin{cases} \log_4(x^2 - 4y^2) = 1 \\ x - y = 1 \end{cases}$

48. $\begin{cases} x - y = 2 \\ 2^{x^2 - y^2} = 256 \end{cases}$

49. $\begin{cases} \log_2(2x^2 + y) = 2 \\ \log_2(2x - y) = 0 \end{cases}$

50. $\begin{cases} \dfrac{3}{x^2} - \dfrac{2}{y^2} = \dfrac{1}{2} \\ \dfrac{6}{x^2} + \dfrac{1}{y} = \dfrac{5}{2} \end{cases}$

In Problems 51 to 54, **(a)** sketch the graph of each system of linear inequalities, **(b)** determine whether the graph is bounded or unbounded, and **(c)** find all of its vertices.

51. $\begin{cases} 3y + x \leq 2 \\ y > x + 1 \end{cases}$

52. $\begin{cases} y \leq x + 1 \\ x + y \leq 4 \\ x \geq 0 \end{cases}$

53. $\begin{cases} y - 3x \leq 2 \\ y + 2x \leq 4 \\ y \geq 0 \end{cases}$

54. $\begin{cases} x + y \leq 500 \\ 3y \leq x \\ x \leq 400 \\ y \geq 60 \end{cases}$

In Problems 55 and 56, sketch the graph of each system of inequalities.

55. $\begin{cases} x^2 + 2y \geq 0 \\ x + 2y \leq 6 \end{cases}$

56. $\begin{cases} x^2 - y \leq 2 \\ 2x^2 + y < 6x + 7 \end{cases}$

In Problems 57 and 58, find the maximum and minimum values of the objective function F subject to the indicated constraints.

57. $F = 4x + 7y + 1$

$\begin{cases} y - x + 1 \geq 0 \\ 2y + x - 10 \geq 0 \\ x \geq 0 \\ y \leq 5 \end{cases}$

58. $F = 3x + 5y$

$\begin{cases} y \leq 4 - x \\ y \geq \frac{1}{2}x - 2 \\ x \geq 0 \\ y \geq 0 \end{cases}$

59. Two wind turbines supply a total of 60 kilowatts of electrical power. If the larger of the two turbines supplies twice as much power as the smaller, find the amount of power supplied by each turbine.

60. A patient's daily intake of vitamins must include 50 milligrams of vitamin A and 120 milligrams of vitamin C. The local drugstore sells brand X and brand Y vitamin capsules. One capsule of brand X contains 10 milligrams of vitamin A and 20 milligrams of vitamin C. One capsule of brand Y contains 10 milligrams of vitamin A and 30 milligrams of vitamin C. How many capsules of each brand should the patient take each day?

61. The ratio of two positive numbers is 3 to 8, and their product is 864. Find the numbers.

62. An airline that flies from New York to Kansas City with a stopover in Chicago charges a fare of \$178 to Chicago and a fare of \$208 to Kansas City. A total of

160 passengers boarded in New York, and fares totaled $30,370. How many passengers got off the plane in Chicago?

63. If the supermarket price of a certain cut of beef is p dollars per pound, then q million pounds will be sold according to the demand equation $p + q = 4$. When the supermarket price is p dollars per pound, a large packing company will supply q million pounds of this meat according to the supply equation $p - 4q + 2 = 0$. Find the point of market equilibrium.

64. Two cars start from points 400 kilometers apart and travel toward each other at constant speeds. If the first car travels 20 kilometers per hour faster than the second, and if they both meet after 4 hours, find the speeds of both vehicles.

C **65.** A person invested a total of $40,000, part of it in a conservative investment at 8.5% simple annual interest, and the rest in a riskier investment at 11.2% simple annual interest. At the end of the year, both investments paid off at the stipulated rates for a total interest of $4129. How much was invested at each rate of interest?

66. An industrial chemist, in preparing a batch of fertilizer, mixes 10 tons of nitrogen compounds with 12 tons of phosphates. The total cost of the ingredients is $5400. A second batch of fertilizer, prepared for special soil conditions, is mixed from 15 tons of nitrogen compounds and 8 tons of phosphates, for a total cost of $6100. Find the cost per ton of the nitrogen compounds and the cost per ton of the phosphates.

67. Byron, Jason, and Adrian receive a total weekly allowance of $12, which is split three ways. If Jason's allowance plus twice the sum of Byron's and Adrian's is $20, and if Adrian's allowance plus twice the sum of Byron's and Jason's is $22, find each boy's weekly allowance.

68. Joe is 4 years younger than twice Gus's age, and Jamal is 3 years older than Gus. Six years from now, Joe will be $\frac{4}{3}$ times as old as Gus will be. Find their present ages.

69. The length of the hypotenuse of a right triangle is 25 centimeters, and the perimeter is 56 centimeters. Find the lengths of the legs.

70. A manufacturer of drafting supplies makes right triangles from plastic sheets. The perimeters of the triangles are 60 centimeters, and each triangle has a hypotenuse of 25 centimeters. Find the lengths of the sides of the triangles.

71. A town parking lot is a rectangle with an area of 7500 square meters. The town engineer suggests enlarging the lot by increasing both its length and its width by 10 meters and points out that this will increase the area of the lot by 1850 square meters. Find the present dimensions of the lot.

72. In three years, Joshua will be twice as old as Miriam. At present, twice his age is the same as the product of Rebecca's and Miriam's ages. If Rebecca is now twice as old as Miriam, find the ages of all three children.

73. A psychology professor is planning a 50-minute class. The professor will devote x minutes to a review of old material and y minutes to a lecture on new material. The remaining time will be used for classroom discussion of the new material. At least 20 minutes must be reserved for the lecture on new material; at least 10 minutes will be needed for the classroom discussion; and no more than 15 minutes will be required for the review of old material. Sketch a graph in the xy plane illustrating the professor's options and find all vertices of the graph.

C **74.** A woman has $20,000 to invest and two investment opportunites, one conservative and one somewhat riskier. The conservative investment pays 8.5% simple annual interest, and the risky investment pays 9.5% simple annual interest. She wants to invest at most three times as much in the risky investment as in the conservative one. Find her maximum possible dividend after 1 year if she decides to invest no more than $16,000 in the conservative investment and at least $2400 in the risky investment.

75. A firm manufactures two products, A and B. For each product, two different machines, M_1 and M_2, are used. Each machine can operate up to 20 hours per day. Product A requires 1 hour of time on machine M_1 and 3 hours of time on machine M_2 per 100 units. Product B requires 2 hours of time on machine M_1 and 1 hour of time on machine M_2 per 100 units. The firm makes a profit of $20 per unit on product A and $10 per unit on product B. Determine how many units of each product should be manufactured per day in order to maximize the profit.

CHAPTER 6 Test

1. Use graphs to determine whether each system is well-determined, inconsistent, or dependent.

   (a) $\begin{cases} x + \frac{1}{3}y = 2 \\ 3x + y = -2 \end{cases}$ **(b)** $\begin{cases} 2x + 3y = 2 \\ x - 2y = 8 \end{cases}$

   (c) $\begin{cases} 2x - 3y = -7 \\ 4x - 6y = -14 \end{cases}$

2. Consider the system $\begin{cases} 13x + 11y = 21 \\ 7x + 6y = -3. \end{cases}$

   (a) Solve the system using the substitution method.

   (b) Solve the system using matrices and the elimination method.

   (c) Write the system in the form $AX = K$ for suitable matrices A, X, and K.

   (d) Solve the system with the aid of A^{-1}.

3. Let $A = \begin{bmatrix} 2 & 4 \\ 5 & -2 \\ -4 & 3 \end{bmatrix}$, $B = \begin{bmatrix} -1 & 0 \\ 3 & 4 \\ 0 & -5 \end{bmatrix}$, and $C = \begin{bmatrix} 6 & 2 \\ -4 & 1 \end{bmatrix}$. Find:

   (a) $4A + 2B$ **(b)** $BC - 2A$ **(c)** C^2

4. **(a)** Evaluate $\begin{vmatrix} 3 & 10 \\ 6 & -5 \end{vmatrix}$.

   (b) Use Cramer's rule to solve the system $\begin{cases} 3x_1 + 10x_2 = 3 \\ 6x_1 - 5x_2 = 16. \end{cases}$

5. If $\begin{vmatrix} a & b & c \\ u & v & w \\ x & y & z \end{vmatrix} = 2$, find the value of $\begin{vmatrix} a & b + 2a & 3c \\ u & v + 2u & 3w \\ x & y + 2x & 3z \end{vmatrix}$.

6. Find the partial-fraction decomposition of
$$\frac{x^2 - x - 21}{(2x - 1)(x^2 + 4)}.$$

7. Solve the system $\begin{cases} x^2 - y = 0 \\ x^2 + 4y^2 = 4. \end{cases}$

8. Consider the system of linear inequalities
$$\begin{cases} x + y \leq 4 \\ 4x + y \leq 7 \\ x \geq 0 \\ y \geq 0. \end{cases}$$

   (a) Sketch the graph of the system.

   (b) Determine whether the graph is bounded or unbounded.

   (c) Find all the vertices of the graph.

   (d) Find the maximum and minimum values of the objective function $F = x + 5y$ subject to the inequalities in the system.

9. A farmer is planning to fence three adjacent rectangular fields as shown in Figure 1. If the length l of the large rectangle formed by the three fields taken together is 100 meters more than the width w, and if a total of 2600 meters of fencing is required, find l and w.

Figure 1

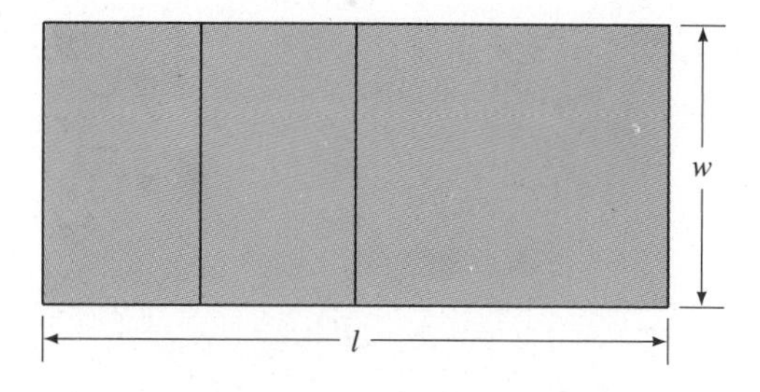

10. A candy company is to supply its distributors with a total of at least 60,000 units of bubble gum and candy bars, of which at least 40,000 units must be candy bars. If the company's costs are 1¢ for each unit of bubble gum and 2¢ for each unit of candy bars, how many units of each kind should it produce in order to minimize its production costs?

CHAPTER 7

Additional Topics in Algebra

A roulette wheel numbered 00 to 36 illustrates the idea of probability in games of chance.

In this chapter, we give a brief introduction to several topics in algebra that extend, supplement, and round out the ideas presented in earlier chapters. These topics include mathematical induction, the binomial theorem, sequences, series, permutations and combinations, probability, and conic sections. Here we can only give you the "flavor" of these topics—we urge you to consult more advanced textbooks for a more detailed and systematic presentation.

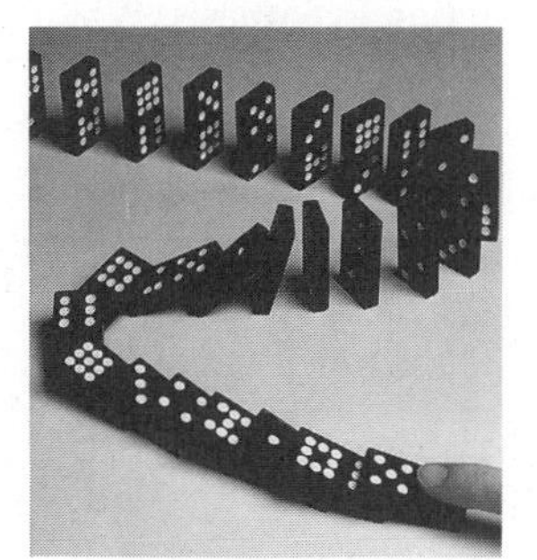

7.1 Mathematical Induction

The principle of *mathematical induction* is often illustrated by a row of oblong wooden blocks, called dominoes, set on end as in Figure 1 on page 414. If we are guaranteed two conditions,

(i) that the first domino is toppled over, and

(ii) that if any domino topples over, it will hit the next one and topple it over,

then we can be certain that *all* the dominoes will topple over.

Figure 1

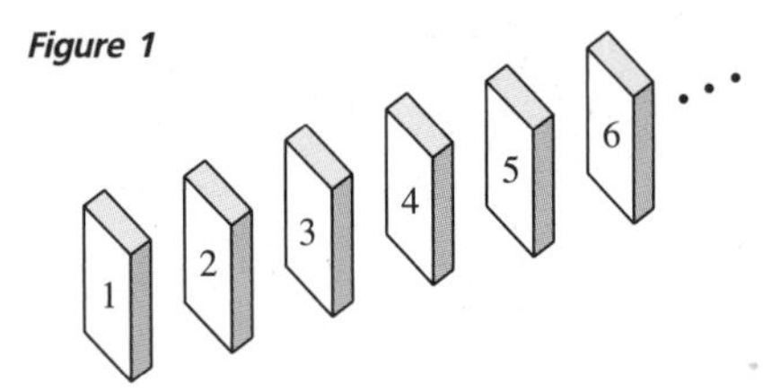

The principle of mathematical induction concerns a sequence of statements

$$S_1, S_2, S_3, S_4, S_5, S_6, \ldots.$$

Let's think of this sequence as corresponding to a row of dominoes; the first statement S_1 corresponds to the first domino, the second statement S_2 corresponds to the second domino, and so on. Each statement can be either true or false. If a statement proves to be true, let's think of the corresponding domino toppling over. The two conditions given above for the row of dominoes can then be interpreted as follows.

(i) The first statement S_1 is true.

(ii) If any statement S_k is true, then the next statement S_{k+1} is also true.

If these two conditions hold, then, by analogy with the toppling over of *all* the dominoes, we conclude that *all* statements $S_1, S_2, S_3, \ldots$ are true. That this argument is valid is the **principle of mathematical induction.**

Although the principle of mathematical induction can be established formally, the required argument is beyond the scope of this book. We ask you to accept this principle on the intuitive basis of the domino analogy. Thus, we have the following.

Procedure for Making a Proof by Mathematical Induction

Begin by clearly identifying the sequence $S_1, S_2, S_3, \ldots$ of statements to be proved. This is usually done by specifying the meaning of S_n, where n denotes an arbitrary positive integer, $n = 1, 2, 3, \ldots$. Then carry out the following two steps:

Step 1. Show that S_1 is true.

Step 2. Let k denote an arbitrary positive integer. Assume that S_k is true and show, on the basis of this assumption, that S_{k+1} is also true.

If both steps can be carried out, conclude that S_n is true for all positive integer values of n.

The assumption in step 2 that S_k is true is called the **induction hypothesis.** When you make the induction hypothesis, you're not saying that S_k is *in fact* true—you're just *supposing* that it is true to see if the truth of S_{k+1} follows from this supposition. It's as if you are checking to make sure that if the kth domino *were* to topple over, it *would* knock over the next domino.

In Examples 1 and 2, use mathematical induction to prove each result.

Example 1 For every positive integer n, $1 + 2 + 3 + \cdots + n = \dfrac{n(n+1)}{2}$.

Solution

Let S_n be the statement

$$S_n\colon\quad 1 + 2 + 3 + \cdots + n = \frac{n(n+1)}{2}.$$

In other words, S_n asserts that the sum of the first n positive integers is $n(n+1)/2$. For instance,

$$S_2 \text{ says that} \qquad 1 + 2 = \frac{2(2+1)}{2} = 3$$

and

$$S_{20} \text{ says that} \qquad 1 + 2 + 3 + \cdots + 20 = \frac{20(20+1)}{2} = 210.$$

Step 1. The statement

$$S_1\colon\quad 1 = \frac{1(1+1)}{2}$$

is clearly true.

Step 2. Here we must deal with the statements

$$S_k\colon\quad 1 + 2 + 3 + \cdots + k = \frac{k(k+1)}{2}$$

and

$$S_{k+1}\colon\quad 1 + 2 + 3 + \cdots + (k+1) = \frac{(k+1)[(k+1)+1]}{2}$$

obtained by replacing n in S_n by k and then by $k + 1$. The statement S_k is our induction hypothesis, and we assume, for the sake of argument, that it is true. Our goal is to prove that, on the basis of this assumption, S_{k+1} is true. To this end, we add $k + 1$ to both sides of the equation expressing S_k to obtain

$$\begin{aligned} 1 + 2 + 3 + \cdots + k + (k+1) &= \frac{k(k+1)}{2} + (k+1) = \frac{k(k+1) + 2(k+1)}{2} \\ &= \frac{(k+1)(k+2)}{2} = \frac{(k+1)[(k+1)+1]}{2}, \end{aligned}$$

which is the equation expressing S_{k+1}. This completes the proof by mathematical induction. ■

Example 2

If a principal of P dollars is invested at a compound interest rate R per conversion period, the final value F dollars of the investment at the end of n conversion periods is given by $F = P(1 + R)^n$. (For *compound interest*, the interest is periodically calculated and added to the principal. The time interval between successive conversions of interest into principal is called the *conversion period*.)

Solution Let S_n be the statement

$$S_n:\quad F = P(1 + R)^n \text{ at the end of } n \text{ conversion periods.}$$

Step 1. The statement

$$S_1:\quad F = P(1 + R) \text{ at the end of the first conversion period}$$

is true, because the interest on P dollars for one conversion period is PR dollars, so the final value of the investment at the end of the first conversion period is

$$P + PR = P(1 + R) \text{ dollars.}$$

Step 2. Assume that the statement

$$S_k:\quad F = P(1 + R)^k \text{ at the end of } k \text{ conversion periods}$$

is true. It follows that, over the $(k + 1)$st conversion period, interest at rate R is paid on $P(1 + R)^k$ dollars. Adding this interest, $P(1 + R)^k R$ dollars, to the value of the investment at the beginning of the $(k + 1)$st conversion period, we find that the value of the investment at the end of the $(k + 1)$st conversion period is given by

$$\begin{aligned} F &= P(1 + R)^k + P(1 + R)^k R \\ &= P(1 + R)^k(1 + R) \\ &= P(1 + R)^{k+1} \text{ dollars.} \end{aligned}$$

Hence, S_{k+1} is true, and the proof by mathematical induction is complete. ■

Problem Set 7.1

In each problem set, problems with colored numbers constitute a good representation of the main ideas of the section. Note that some of the even-numbered problems are considerably more challenging than the odd-numbered ones.

In Problems 1 to 14, use mathematical induction to prove that the assertion is true for all positive integers n.

1. $1 + 3 + 5 + \cdots + (2n - 1) = n^2$; in other words, the sum of the first n odd positive integers is n^2.

2. $2 + 4 + 6 + \cdots + 2n = n(n + 1)$; in other words, the sum of the first n even positive integers is $n(n + 1)$.

3. $1^2 + 2^2 + 3^2 + \cdots + n^2 = \frac{1}{6}n(n + 1)(2n + 1)$; that is, the sum of the first n perfect squares is given by $\frac{1}{6}n(n + 1)(2n + 1)$.

4. $1 + 5 + 9 + \cdots + (4n - 3) = n(2n - 1)$

5. $1^2 + 3^2 + 5^2 + \cdots + (2n - 1)^2 = \frac{1}{3}n(2n - 1)(2n + 1)$

6. $1^3 + 2^3 + 3^3 + \cdots + n^3 = \left[\frac{n(n + 1)}{2}\right]^2$

7. $1 \cdot 2 + 2 \cdot 3 + 3 \cdot 4 + \cdots + n(n + 1) = \frac{1}{3}n(n + 1)(n + 2)$

8. $\frac{1}{1 \cdot 2} + \frac{1}{2 \cdot 3} + \frac{1}{3 \cdot 4} + \cdots + \frac{1}{n(n + 1)} = \frac{n}{n + 1}$

9. $(ab)^n = a^n b^n$

10. If $h \geq 0$, then $1 + nh \leq (1 + h)^n$.

11. $2 + 2^2 + 2^3 + \cdots + 2^n = 2(2^n - 1)$

12. $r + r^2 + r^3 + \cdots + r^n = \dfrac{r(r^n - 1)}{r - 1}$

13. $n < 2^n$

14. $\left(1 + \dfrac{1}{1}\right)\left(1 + \dfrac{1}{2}\right)\left(1 + \dfrac{1}{3}\right) \cdots \left(1 + \dfrac{1}{n}\right) = n + 1$

c **15.** Using a calculator, verify the identities in odd Problems 1 to 7 for $n = 15$ by directly calculating both sides.

16. The *domino theory* was a tenet of U.S. foreign policy subscribed to by the administrations of Presidents Eisenhower, Kennedy, Johnson, and Nixon. Explain the connection, if any, between the domino theory and mathematical induction.

17. A grocer stacks cans of pineapple on special sale in the form of a triangle as shown in Figure 2. If there are n cans in the bottom row of the triangle, develop a formula for the total number of cans in the triangular stack.

Figure 2

18. Recall that a *diagonal* of a polygon is a line segment between two nonadjacent vertices (Problem 70 in Problem Set 2.3). Figure 3 shows all of the diagonals of a polygon with 6 vertices.

(a) For a polygon with n vertices, prove that the number of diagonals that meet at any one vertex is $n - 3$.

Figure 3

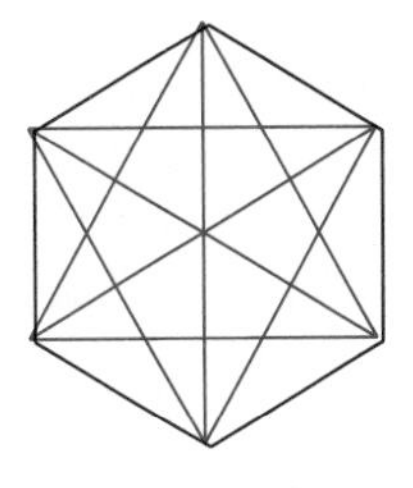

(b) Using the result of part (a), show that the total number of diagonals of a polygon with n vertices is given by $\frac{1}{2}n(n - 3)$. [*Hint:* If you multiply $n - 3$ by n, you will count each diagonal twice.]

19. Consider a network in the plane consisting of points called *vertices* and line segments called *edges* subject to the following three conditions:

(i) Each vertex is an endpoint of at least one edge.

(ii) Both endpoints of each edge are vertices.

(iii) If two edges intersect, they do so only at a common vertex.

Such a network is called a **planar graph.*** A region of the plane that is entirely bounded by edges is called a *face* of the planar graph. Figure 4 shows a planar graph with 9 vertices, 12 edges, and 4 faces. For an arbitrary planar graph with V vertices, E edges, and F faces, prove that $V - E + F = 1$. [*Hint:* Start with the simplest planar graph (which would consist of 1 edge, 2 vertices, and 0 faces) and proceed inductively, considering what happens each time a new edge is added to an already existing planar graph.]

Figure 4

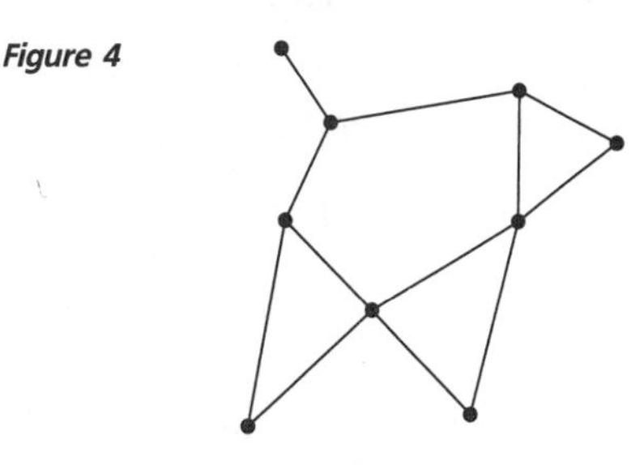

* This alternative use of the word "graph" should not be confused with the graph of a function.

7.2 The Binomial Theorem

The principle of mathematical induction can be used to establish a general formula for the expansion of $(a + b)^n$, where the exponent n is an arbitrary positive integer. The expression $a + b$ is a binomial, so the formula is called the **binomial theorem.**

Of course, the expansion of $(a + b)^n$ for small values of n can be obtained by direct calculation. For instance,

$$(a + b)^1 = a + b$$

$$(a + b)^2 = a^2 + 2ab + b^2$$

$$(a + b)^3 = a^3 + 3a^2b + 3ab^2 + b^3$$

$$(a + b)^4 = a^4 + 4a^3b + 6a^2b^2 + 4ab^3 + b^4$$

$$(a + b)^5 = a^5 + 5a^4b + 10a^3b^2 + 10a^2b^3 + 5ab^4 + b^5,$$

and so forth.

Notice that we have written the expansions in descending powers of a and ascending powers of b. A certain pattern is already apparent. Indeed, in the expansion of $(a + b)^n$:

1. There are $n + 1$ terms, beginning with a^n and ending with b^n.

2. As we move from each term to the next, powers of a decrease by 1 and powers of b increase by 1. Therefore, in each term, the exponents of a and b add up to n.

3. The successive terms can be written in the form

$$(\text{numerical coefficient})a^{n-j}b^j$$

for $j = 0, 1, 2, \ldots, n$.

That statements 1 to 3 are true for every positive integer n is part of the binomial theorem. The remaining part concerns the values of the numerical coefficients, which are called the **binomial coefficients.** In order to obtain a useful formula for the binomial coefficients, we begin by introducing *factorial notation.*

Definition 1 **Factorial Notation**

If n is a positive integer, we define $n!$, read ***n* factorial,** by

$$n! = n(n - 1)(n - 2) \cdots 3 \cdot 2 \cdot 1.$$

We also define

$$0! = 1.$$

In words, *the factorial of a positive integer n is the product of n and all smaller positive integers.* For instance,

$$7! = 7 \cdot 6 \cdot 5 \cdot 4 \cdot 3 \cdot 2 \cdot 1 = 5040.$$

Table 1

$0! = 1$
$1! = 1$
$2! = 2$
$3! = 6$
$4! = 24$
$5! = 120$
$6! = 720$
$7! = 5040$

The special definition $0! = 1$ extends the usefulness of the factorial notation. As a direct consequence of Definition 1, we have the important **recursion formula** for factorials

$$(n + 1)! = (n + 1)n!,$$

which holds for every integer $n \geq 0$ (Problem 35). Using the recursion formula, you can make a table of values of $n!$ (Table 1). The table can be continued indefinitely; for instance, to obtain the next entry, use the recursion formula as follows:

$$8! = 8(7!) = 8(5040) = 40{,}320.$$

The factorials increase very rapidly. Indeed, 10! is over 3 million, and 70! is beyond the range of many calculators.

Using factorial notation, we now introduce special symbols for what, as we shall soon see, are the binomial coefficients.

Definition 2 **The Binomial Coefficients**

If n and j are integers with $n \geq j \geq 0$, the symbol $\binom{n}{j}$ is defined as follows:

$$\binom{n}{j} = \frac{n!}{(n-j)!j!}.$$

Example 1 Evaluate

(a) $\binom{5}{2}$ **(b)** $\binom{7}{6}$

(c) $\binom{n}{0}$ **(d)** $\binom{n}{n}$

if n is a nonnegative integer.

Solution Using Definition 2, we have:

(a) $\binom{5}{2} = \frac{5!}{(5-2)!2!} = \frac{5!}{3!2!} = \frac{5 \cdot 4 \cdot \cancel{3 \cdot 2 \cdot 1}}{(\cancel{3 \cdot 2 \cdot 1})(2 \cdot 1)} = \frac{5 \cdot 4}{2} = 10$

(b) $\binom{7}{6} = \frac{7!}{(7-6)!6!} = \frac{7!}{1!6!} = \frac{7!}{6!} = \frac{7(\cancel{6!})}{\cancel{6!}} = 7$

(c) $\binom{n}{0} = \frac{n!}{(n-0)!0!} = \frac{\cancel{n!}}{\cancel{n!}0!} = \frac{1}{0!} = \frac{1}{1} = 1$

(d) $\binom{n}{n} = \frac{\cancel{n!}}{(n-n)!\cancel{n!}} = \frac{1}{0!} = \frac{1}{1} = 1$ ■

Example 2 If k and j are integers with $k \geq j > 0$, show that

$$\binom{k}{j-1} + \binom{k}{j} = \binom{k+1}{j}.$$

Solution Using Definition 2 and the recursion formula for factorials, we have

$$\begin{aligned}\binom{k}{j-1} + \binom{k}{j} &= \frac{k!}{[k-(j-1)]!(j-1)!} + \frac{k!}{(k-j)!j!} \\ &= \frac{jk!}{(k-j+1)!j(j-1)!} + \frac{(k-j+1)k!}{(k-j+1)(k-j)!j!} \\ &= \frac{jk!}{(k-j+1)!j!} + \frac{(k+1)k! - jk!}{(k-j+1)!j!} \\ &= \frac{(k+1)k!}{(k-j+1)!j!} = \frac{(k+1)!}{[(k+1)-j]!j!} \\ &= \binom{k+1}{j}.\end{aligned}$$

■

We can now state and prove the binomial theorem.

Theorem 1 **The Binomial Theorem**

If n is a positive integer and if a and b are any two numbers, then

$$(a+b)^n = \binom{n}{0}a^n + \binom{n}{1}a^{n-1}b + \cdots + \binom{n}{j}a^{n-j}b^j + \cdots + \binom{n}{n}b^n.$$

Proof The proof is by mathematical induction. Thus, for each positive integer n, let S_n be the statement

$$S_n\colon\ (a+b)^n = \binom{n}{0}a^n + \binom{n}{1}a^{n-1}b + \cdots + \binom{n}{j}a^{n-j}b^j + \cdots + \binom{n}{n}b^n.$$

Step 1. The statement

$$S_1\colon\ (a+b)^1 = \binom{1}{0}a^1 + \binom{1}{1}b^1$$

is true because $\binom{1}{0} = 1$ and $\binom{1}{1} = 1$.

Step 2. Assume the induction hypothesis

$$S_k\colon\ (a+b)^k = \binom{k}{0}a^k + \binom{k}{1}a^{k-1}b + \cdots + \binom{k}{j}a^{k-j}b^j + \cdots + \binom{k}{k}b^k,$$

where k is a positive integer. Notice that the term immediately preceding $\binom{k}{j}a^{k-j}b^j$ in this equation is $\binom{k}{j-1}a^{k-(j-1)}b^{j-1}$, that is, $\binom{k}{j-1}a^{k-j+1}b^{j-1}$. We use this fact in the following computation:

$$
\begin{aligned}
&(a+b)^{k+1}\\
&\quad = (a+b)(a+b)^k = a(a+b)^k + b(a+b)^k\\
&\quad = a\left[\binom{k}{0}a^k + \binom{k}{1}a^{k-1}b + \cdots + \binom{k}{j}a^{k-j}b^j + \cdots + \binom{k}{k}b^k\right]\\
&\qquad + b\left[\binom{k}{0}a^k + \binom{k}{1}a^{k-1}b + \cdots + \binom{k}{j-1}a^{k-j+1}b^{j-1} + \cdots + \binom{k}{k}b^k\right]\\
&\quad = \binom{k}{0}a^{k+1} + \binom{k}{1}a^kb + \cdots + \binom{k}{j}a^{k-j+1}b^j + \cdots + \binom{k}{k}ab^k\\
&\qquad + \binom{k}{0}a^kb + \cdots + \binom{k}{j-1}a^{k-j+1}b^j + \cdots + \binom{k}{k-1}ab^k + \binom{k}{k}b^{k+1}\\
&\quad = \binom{k}{0}a^{k+1} + \left[\binom{k}{1} + \binom{k}{0}\right]a^kb + \cdots + \left[\binom{k}{j} + \binom{k}{j-1}\right]a^{k+1-j}b^j\\
&\qquad + \cdots + \left[\binom{k}{k} + \binom{k}{k-1}\right]ab^k + \binom{k}{k}b^{k+1}.
\end{aligned}
$$

This expression of $(a+b)^{k+1}$ will fulfill the requirements of statement S_{k+1}, provided that the following three conditions hold:

(i) $\binom{k}{0} = \binom{k+1}{0}$, **(ii)** $\binom{k}{j} + \binom{k}{j-1} = \binom{k+1}{j}$, **(iii)** $\binom{k}{k} = \binom{k+1}{k+1}$.

That (i) and (iii) hold is left as an exercise (Problem 36). We just proved (ii) in Example 2 above. Hence S_{k+1} is established, and the proof by induction is complete. ■

Blaise Pascal

By arranging the binomial coefficients in a triangle as shown in Figure 1a on page 422, we obtain **Pascal's triangle,** named in honor of its discoverer, the French mathematician Blaise Pascal (1623–1662). The numerical values of the entries in the first six horizontal rows of Pascal's triangle are shown in Figure 1b on page 422. Notice that:

> Each number in Pascal's triangle (other than those on the border, which are all equal to 1) can be obtained by adding the two numbers diagonally above it.

This is called the **additive property** of Pascal's triangle, and it is a consequence of the identity established in Example 2.

Figure 1

$$\binom{0}{0}$$
$$\binom{1}{0} \quad \binom{1}{1}$$
$$\binom{2}{0} \quad \binom{2}{1} \quad \binom{2}{2}$$
$$\binom{3}{0} \quad \binom{3}{1} \quad \binom{3}{2} \quad \binom{3}{3}$$
$$\binom{4}{0} \quad \binom{4}{1} \quad \binom{4}{2} \quad \binom{4}{3} \quad \binom{4}{4}$$
$$\binom{5}{0} \quad \binom{5}{1} \quad \binom{5}{2} \quad \binom{5}{3} \quad \binom{5}{4} \quad \binom{5}{5}$$
$$\vdots \quad \vdots \quad \vdots \quad \vdots \quad \vdots \quad \vdots$$

(a)

1
1 1
1 2 1
1 3 3 1
1 4 6 4 1
1 5 10 10 5 1
⋮ ⋮ ⋮ ⋮ ⋮ ⋮

(b)

Example 3 Find the numerical entries in the seventh horizontal row of Pascal's triangle.

Solution The seventh horizontal row is obtained from the sixth horizontal row in Figure 1b by using the additive property:

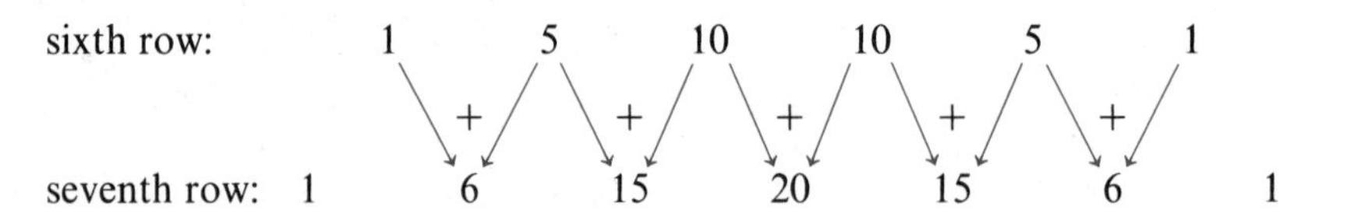

■

According to the binomial theorem:

> The binomial coefficients in the expansion of $(a + b)^n$ appear in row number $n + 1$ of Pascal's triangle.

For instance, using the result of Example 3, we have

$$(a + b)^6 = 1a^6 + 6a^5b + 15a^4b^2 + 20a^3b^3 + 15a^2b^4 + 6ab^5 + 1b^6.$$

In Examples 4 and 5, use the binomial theorem and Pascal's triangle to expand each of the given expressions.

Example 4 $(2x + y)^4$

Solution In the binomial theorem (Theorem 1), let $a = 2x$, $b = y$, and $n = 4$. The appropriate binomial coefficients (1, 4, 6, 4, 1) are found in the *fifth* horizontal row of Pascal's triangle (Figure 1b), since $n + 1 = 4 + 1 = 5$. Therefore,

$$\begin{aligned}(2x + y)^4 &= 1(2x)^4 + 4(2x)^3 y + 6(2x)^2 y^2 + 4(2x)y^3 + 1y^4 \\ &= 16x^4 + 32x^3 y + 24x^2 y^2 + 8xy^3 + y^4.\end{aligned}$$

■

Example 5 $(x - y)^5$

Solution In the binomial theorem, let $a = x$, $b = -y$, and $n = 5$. The appropriate binomial coefficients (1, 5, 10, 10, 5, 1) are found in the *sixth* horizontal row of Pascal's triangle. Thus,

$$\begin{aligned}(x - y)^5 &= 1x^5 + 5x^4(-y) + 10x^3(-y)^2 + 10x^2(-y)^3 + 5x(-y)^4 + 1(-y)^5 \\ &= x^5 - 5x^4 y + 10x^3 y^2 - 10x^2 y^3 + 5xy^4 - y^5.\end{aligned}$$

■

According to the binomial theorem (Theorem 1), the $(j + 1)$st term in the expansion of $(a + b)^n$ is given by the formula

$$(j + 1)\text{st term} = \binom{n}{j} a^{n-j} b^j \qquad \text{for } 0 \leq j \leq n.$$

You can use this formula to find a specified term in a binomial expansion.

Example 6 Find the tenth term in the expansion of $(x - \frac{1}{2}y)^{12}$.

Solution The tenth term is obtained by putting $j = 9$, $a = x$, $b = -\frac{1}{2}y$, and $n = 12$ in the formula above. Thus,

$$\begin{aligned}10\text{th term} &= \binom{12}{9} a^{12-9} b^9 \\ &= \binom{12}{9} x^3(-\tfrac{1}{2}y)^9 \\ &= -\binom{12}{9} \frac{x^3 y^9}{2^9}.\end{aligned}$$

Now,

$$\binom{12}{9} = \frac{12!}{(12 - 9)!9!} = \frac{12!}{3!9!} = \frac{12 \cdot 11 \cdot 10(9!)}{3!9!} = \frac{12 \cdot 11 \cdot 10}{6} = 220,$$

and

$$2^9 = 512,$$

so we have

$$10\text{th term} = -220 \frac{x^3 y^9}{512} = -\frac{55}{128} x^3 y^9.$$

■

Example 7 Find the term involving y^4 in the expansion of $(x + y)^8$.

Solution With $a = x$, $b = y$, and $n = 8$, we have

$$(j + 1)\text{st term} = \binom{n}{j} a^{n-j} b^j = \binom{8}{j} x^{8-j} y^j.$$

Hence, the term involving y^4, obtained by putting $j = 4$, is the

$$5\text{th term} = \binom{8}{4} x^{8-4} y^4 = \binom{8}{4} x^4 y^4.$$

Now,

$$\binom{8}{4} = \frac{8!}{(8-4)!4!} = \frac{8!}{4!4!} = \frac{8 \cdot 7 \cdot 6 \cdot 5(4!)}{4!4!} = \frac{8 \cdot 7 \cdot 6 \cdot 5}{4 \cdot 3 \cdot 2 \cdot 1} = 2 \cdot 7 \cdot 5 = 70.$$

Therefore, the term involving y^4 is the

$$5\text{th term} = 70x^4 y^4.$$

■

Problem Set 7.2

In Problems 1 to 10, find the numerical value of each expression.

1. $9!$
2. $10!$
3. $\dfrac{7!}{5!3!}$
4. $\dfrac{8!6!}{5!3!}$
5. $\dbinom{7}{3}$
6. $\dbinom{12}{10}$
7. $\dbinom{5}{5}$
8. $\dbinom{12}{2}$
9. $\dbinom{6}{3}$
10. $\dbinom{50}{49}$

11. Using the additive property of Pascal's triangle and the fact that the numerical entries in the seventh horizontal row are (1, 6, 15, 20, 15, 6, 1), find the numerical entries in the eighth row.

12. Construct Pascal's triangle showing the numerical entries in the first 10 horizontal rows.

In Problems 13 to 24, use the binomial theorem and Pascal's triangle to expand each expression.

13. $(a + 3x)^4$
14. $(3a - 2b)^5$
15. $(x - y)^5$
16. $(x + y^2)^5$
17. $(c + 2)^6$
18. $(r - 3s)^5$
19. $(1 - c^3)^7$
20. $(b^2 + 10)^4$
21. $(\sqrt{x} - \sqrt{y})^6$
22. $\left(\dfrac{1}{x} + \dfrac{1}{y}\right)^6$
23. $\left(\dfrac{x}{2} - 2y\right)^6$
24. $(1 - 1)^6$

In Problems 25 to 32, find and simplify the specified term in the binomial expansion of the indicated expression.

25. The third term of $(s - t)^5$.
26. The fourth term of $(a^2 - b)^6$.
27. The fifth term of $\left(2x^2 + \dfrac{y^3}{4}\right)^{10}$.
28. The sixth term of $(\pi - \sqrt{2})^9$.
29. The term involving y^4 in $(x + y)^5$.
30. The term involving b^6 in $(a - 3b^3)^4$.
31. The term involving c^6 in $(c + 2d)^{10}$.
32. The term involving x^3 in $(\sqrt{x} + \sqrt{y})^8$.

33. Use mathematical induction to show that $2^{n+3} < (n + 3)!$ for all positive integers n.

C 34. Using the recursion formula and a calculator, extend Table 1 to show values of $n!$ for $n = 0, 1, 2, \ldots, 15$.

35. Verify the recursion formula for factorials: $(n + 1)! = (n + 1)n!$ for every nonnegative integer n.

36. If k is a nonnegative integer, show that $\binom{k}{0} = \binom{k+1}{0}$ and that $\binom{k}{k} = \binom{k+1}{k+1}$.

37. Show that Pascal's triangle is symmetric about a vertical line through its apex by showing that for integers n and j with $n \geq j \geq 0$, $\binom{n}{j} = \binom{n}{n-j}$.

38. If n and j are integers with $n \geq j > 0$, show that

$$\binom{n}{j} = \frac{\overbrace{n(n-1)\cdots(n-j+1)}^{j \text{ factors}}}{\underbrace{j(j-1)\cdots 1}_{j \text{ factors}}}.$$

39. If n is a positive integer, show that

$$(1 + x)^n = 1 + nx + \binom{n}{2}x^2 + \cdots + \binom{n}{j}x^j + \cdots + x^n.$$

40. Show that the sum of all the entries in row number $n + 1$ of Pascal's triangle is 2^n; that is,

$$\binom{n}{0} + \binom{n}{1} + \binom{n}{2} + \cdots + \binom{n}{n} = 2^n.$$

[*Hint:* Use the binomial theorem to expand $(1 + 1)^n$.]

41. If, in a term of the binomial expansion of $(a + b)^n$, we multiply the coefficient by the exponent of a and divide by the number of the term, show that we obtain the coefficient of the next term.

C 42. **Stirling's approximation,**

$$n! \approx \sqrt{2n\pi}\left(\frac{n}{e}\right)^n\left(1 + \frac{1}{12n - 1}\right),$$

is often used to approximate $n!$ for large values of n.

(a) Use Stirling's approximation to estimate 15! and compare the result with the true value.

(b) Use Stirling's approximation to estimate 50!.

7.3 Sequences

A **sequence** is a function

$$n \longmapsto a_n$$

whose domain is the set of positive integers.* Following tradition, we write the function values as a_n rather than as $a(n)$ and call them the **terms** of the sequence. Thus, a_1 is the **first term,** a_2 is the **second term,** and, in general, a_n is the ***n*th term** of the sequence $n \longmapsto a_n$. The symbolism

$$a_1, a_2, a_3, \ldots, a_n, \ldots$$

is often used to denote the sequence $n \longmapsto a_n$; it suggests a never-ending list in which the terms appear in order, with the nth term in the nth position for each positive integer n. In this notation, a set of three dots is read "and so on." The more compact notation $\{a_n\}$, in which the general term is enclosed in braces, is also used to denote the sequence.

* In some cases the domain of a sequence is taken to be all nonnegative integers or, more generally, all integers greater than or equal to a fixed integer.

To specify a particular sequence, we give a rule by which the nth term a_n is determined. This is often done by means of a formula.

Example 1 Find the first five terms of each sequence.

(a) $a_n = 2n - 1$ **(b)** $b_n = \dfrac{(-1)^n}{n(n+1)}$

Solution **(a)** The first five terms of the sequence are found by substituting the positive integers 1, 2, 3, 4, and 5 for n in the formula for the general term. Thus,

$$a_1 = 2(1) - 1 = 1, \quad a_2 = 2(2) - 1 = 3, \quad a_3 = 2(3) - 1 = 5,$$
$$a_4 = 2(4) - 1 = 7, \quad a_5 = 2(5) - 1 = 9;$$

so the first five terms of the sequence $\{a_n\}$ are

$$1, 3, 5, 7, 9, \ldots .$$

(b) We have

$$b_1 = \frac{(-1)^1}{1(1+1)} = -\frac{1}{2}, \quad b_2 = \frac{(-1)^2}{2(2+1)} = \frac{1}{6}, \quad b_3 = \frac{(-1)^3}{3(3+1)} = -\frac{1}{12},$$
$$b_4 = \frac{(-1)^4}{4(4+1)} = \frac{1}{20}, \quad b_5 = \frac{(-1)^5}{5(5+1)} = -\frac{1}{30};$$

so the first five terms of the sequence $\{b_n\}$ are

$$-\tfrac{1}{2}, \tfrac{1}{6}, -\tfrac{1}{12}, \tfrac{1}{20}, -\tfrac{1}{30}, \ldots .$$

■

A formula that relates the general term a_n of a sequence to one or more of the terms that come before it is called a **recursion formula.** A sequence that is specified by giving the first term (or the first few terms) together with a recursion formula is said to be **defined recursively.**

Example 2 Find the first five terms of the sequence $\{a_n\}$ defined recursively by $a_1 = 1$ and $a_n = na_{n-1}$.

Solution We are given $a_1 = 1$. By the recursion formula $a_n = na_{n-1}$, we have

$$a_2 = 2a_{2-1} = 2a_1 = 2(1) = 2.$$

Now, using the fact that $a_2 = 2$, we find that

$$a_3 = 3a_{3-1} = 3a_2 = 3(2) = 6.$$

Continuing in this way, we have

$$a_4 = 4a_{4-1} = 4a_3 = 4(6) = 24$$

and

$$a_5 = 5a_{5-1} = 5a_4 = 5(24) = 120.$$

Thus, the first five terms of the sequence $\{a_n\}$ are

$$1, 2, 6, 24, 120, \ldots .$$

The pattern here is clear: $a_n = n!$. This can be proved by mathematical induction (Problem 20). ■

Arithmetic and **geometric** sequences (also called *progressions*) satisfy the following simple recursion formulas:

arithmetic sequence:	$a_n = d + a_{n-1}$,	where d is a fixed number called the **common difference**
geometric sequence:	$a_n = ra_{n-1}$,	where r is a fixed number known as the **common ratio**

In an arithmetic sequence, each term (after the first) is obtained by adding the fixed number d to the term just before it. Therefore, d is the common difference between successive terms; that is, $d = a_n - a_{n-1}$ for $n = 2, 3, 4, \ldots$. For instance, the sequence

$$5, 8, 11, \ldots, 3n + 2, \ldots$$

with general term $a_n = 3n + 2$ is an arithmetic sequence, and the common difference is

$$\begin{aligned} d = a_n - a_{n-1} &= (3n + 2) - [3(n - 1) + 2] \\ &= 3n + 2 - 3n + 3 - 2 = 3. \end{aligned}$$

If $\{a_n\}$ is an arithmetic sequence with common difference d, then by the recursion formula $a_n = d + a_{n-1}$ we have

$$a_2 = a_1 + d,$$

$$a_3 = a_2 + d = (a_1 + d) + d = a_1 + 2d,$$

$$a_4 = a_3 + d = (a_1 + 2d) + d = a_1 + 3d,$$

and so on.

Evidently, for every positive integer n,

$$a_n = a_1 + (n - 1)d.$$

This formula for the nth term of an arithmetic sequence can be proved by mathematical induction (Problem 37).

Example 3 Find the seventeenth term of the arithmetic sequence $2, 5, 8, 11, \ldots$.

Solution Here, the common difference of successive terms is $d = 3$, and the first term is $a_1 = 2$. Using the formula $a_n = a_1 + (n - 1)d$ with $n = 17$, we obtain

$$a_{17} = 2 + (17 - 1)(3) = 50.$$ ■

In a geometric sequence, each term (after the first) is obtained by multiplying the term just before it by r. Therefore, if the terms are nonzero, *r is the common ratio of successive terms;* that is, $r = a_n/a_{n-1}$ for $n = 2, 3, 4, \ldots$.

For instance, the sequence

$$6, 12, 24, \ldots, 3(2^n), \ldots$$

with general term $a_n = 3(2^n)$ is a geometric sequence, and the common ratio is

$$r = \frac{3(2^n)}{3(2^{n-1})} = 2^{n-(n-1)} = 2.$$

If $\{a_n\}$ is a geometric sequence with common ratio r, then by the recursion formula $a_n = ra_{n-1}$ we have

$$a_2 = ra_1,$$
$$a_3 = ra_2 = r(ra_1) = r^2a_1,$$
$$a_4 = ra_3 = r(r^2a_1) = r^3a_1,$$

and so on. Evidently, for every positive integer n,

$$a_n = r^{n-1}a_1.$$

This formula for the nth term of a geometric sequence can be proved by mathematical induction (Problem 38).

Example 4 Find the sixth term of the geometric sequence 7, 21, 63, ...

Solution Here, the common ratio of successive terms is $r = 3$, and the first term is $a_1 = 7$. Using the formula $a_n = r^{n-1}a_1$ with $n = 6$, we obtain

$$a_6 = 3^{6-1}(7) = 3^5(7) = 1701.$$

Problem Set 7.3

In Problems 1 to 10, find the first five terms of the sequence with the specified general term.

1. $a_n = \dfrac{1}{n+1}$

2. $b_n = \dfrac{(-1)^n}{n^2}$

3. $b_n = (-1)^n n$

4. $c_n = \dfrac{1+(-1)^n}{1+3n}$

5. $a_n = \dfrac{2n-1}{2n+1}$

6. $b_n = \left(-\dfrac{3}{2}\right)^{n-1}$

7. $c_n = \left(\dfrac{1}{3}\right)^n$

8. $b_n = \dfrac{n^2-1}{n^2+1}$

9. $a_n = (n+1)^2$

10. $b_n = \dfrac{1}{n(1+n)}$

In Problems 11 to 19, use the formula in the second box on page 427 to find the indicated term in each arithmetic sequence.

11. The sixth term of 3, 9, 15, . . .

12. The eighth term of 12, 24, 36, . . .

13. The twentieth term of 5, 10, 15, . . .

14. The twenty-fifth term of -20, -16, -12, . . .

15. The tenth term of -0.8, 0, 0.8, . . .

16. The fifteenth term of $5\sqrt{2}$, $7\sqrt{2}$, $9\sqrt{2}$, . . .

17. The eighth term of $m + r$, $m + 2r$, $m + 3r$, . . .

18. The tenth term of $2 + b$, $2 + 4b$, $2 + 7b$, . . .

19. The thirteenth term of $\sqrt{5}$, 0, $-\sqrt{5}$, . . .

20. If the sequence $\{a_n\}$ has the first term $a_1 = 1$ and satisfies the recursion formula $a_n = na_{n-1}$, use mathematical induction to prove that $a_n = n!$ for all positive integers n.

In Problems 21 to 30, use the boxed formula on page 428 to find the indicated term in each geometric sequence.

21. The eighth term of 2, 1, $\frac{1}{2}$, . . .

22. The fourth term of 8, 4, 2, . . .

23. The sixth term of $\frac{2}{3}$, 2, 6, . . .

24. The tenth term of $\frac{1}{4}$, $\frac{1}{8}$, $\frac{1}{16}$, . . .

25. The fifth term of -90, -9, -0.9, . . .

26. The seventh term of $\sqrt{2}/2$, -1, $\sqrt{2}$, . . .

27. The fifteenth term of $2x$, $2x^2$, $2x^3$, . . .

28. The tenth term of c^{-4}, $-c^{-2}$, 1, . . .

29. The sixth term of 0.3, 0.03, 0.003, . . .

30. The fifth term of 0.12, 0.012, 0.0012, . . .

In Problems 31 to 36, find the first five terms of the recursively defined sequence $\{a_n\}$.

31. $a_1 = 2$ and $a_n = -3a_{n-1}$

C **32.** $a_1 = 2$ and $a_n = (a_{n-1})^{n-1}$

33. $a_1 = 5$ and $a_n = 7 - 2a_{n-1}$

34. $a_1 = 1$, $a_2 = 2$, and $a_n = a_{n-1}a_{n-2}$ for $n \geq 3$

35. $a_1 = 1$ and $a_n = \frac{1}{n} a_{n-1}$

36. $a_1 = 0$, $a_2 = 1$, and $a_n = a_{n-1} - a_{n-2}$ for $n \geq 3$

37. Use mathematical induction to prove that for every positive integer n, the nth term of an arithmetic sequence $\{a_n\}$ with first term a_1 and common difference d is given by $a_n = a_1 + (n - 1)d$.

38. Use mathematical induction to prove that for every positive integer n, the nth term of a geometric sequence $\{a_n\}$ with first term a_1 and common ratio r is given by $a_n = r^{n-1} a_1$.

39. The **Fibonacci sequence** $\{F_n\}$, which was introduced by Leonardo Fibonacci in 1202, is defined recursively as follows:

$$F_1 = 1,\ F_2 = 1,\ \text{and}\ F_n = F_{n-1} + F_{n-2}\ \text{for}\ n \geq 3.$$

Find the first 10 terms of the Fibonacci sequence.

40. An elastic ball is dropped on a hard surface from an initial height of h meters. Each time the ball hits the surface, it bounces back to a height r times that from which it fell (Figure 2). Let d_n denote the height from which the ball falls just before it hits the surface for the nth time.

(a) Show that $\{d_n\}$ is a geometric sequence.

(b) Write a formula for d_n in terms of n, h, and r.

Figure 2

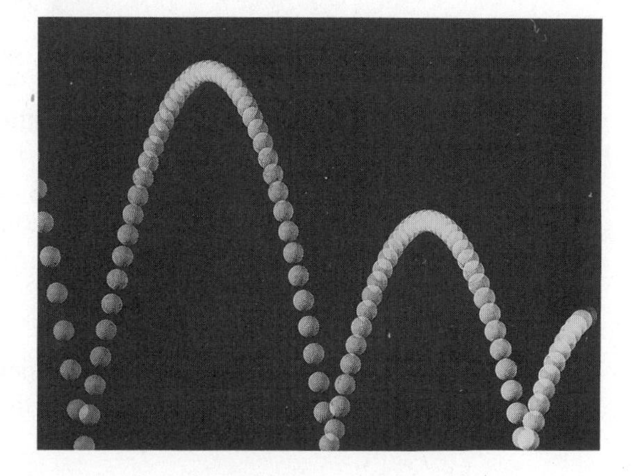

41. For the Fibonacci sequence $F_1, F_2, F_3, \ldots$, in Problem 39, define

$$R_n = \frac{F_n}{F_{n-1}} \quad \text{for } n \geq 2.$$

(a) For $n \geq 3$, show that

$$R_n = 1 + \frac{1}{R_{n+1}}.$$

(b) It can be shown that as n becomes larger and larger, R_n comes closer and closer to a number R that satisfies the equation $R = 1 + (1/R)$. Show that $1/R$ is the golden ratio (Problem 73 in Problem Set 2.3).

[C] **(c)** For $n = 10$, find the percent error in the approximation $R_n \approx R$.

42. At each stroke, a vacuum pump exhausts $100r$ percent of the air from a container. Find a formula for the percent of air that remains in the container after the nth stroke of the pump.

43. Suppose that the population of a certain city increases at the rate of 5% per year and that the present population is 300,000.

(a) If a_n denotes the population of the city n years from now, show that $\{a_n\}$ is a geometric sequence.

[C] **(b)** Find the population of the city in 5 years.

7.4 Series

In many applications of mathematics, we have to find the sum of the terms of a sequence. Such a sum is called a **series.** Although the sum of the first n terms of a sequence $\{a_n\}$ can be written as

$$a_1 + a_2 + a_3 + \cdots + a_n,$$

a more compact notation is sometimes required. The capital Greek letter Σ (*sigma*), which corresponds to the letter S, is used for this purpose, and we write the sum in **sigma notation** as

$$\sum_{k=1}^{n} a_k = a_1 + a_2 + a_3 + \cdots + a_n.$$

Here, Σ indicates a sum and k is called the **index of summation.** That the summation begins with $k = 1$ and ends with $k = n$ is indicated by writing $k = 1$ below Σ and n above it. For some purposes, it is useful to allow sequences that begin with a "zeroth term" a_0, and to write

$$\sum_{k=0}^{n} a_k = a_0 + a_1 + a_2 + \cdots + a_n.$$

In any case, it is understood that the summation index runs through all *integer* values starting with the value shown below Σ and ending with the value shown above it.

In Examples 1 and 2, evaluate each sum.

Example 1 $\sum_{k=1}^{4} (2k+1)$

Solution

$$\sum_{k=1}^{4} (2k+1) = [2(1)+1] + [2(2)+1] + [2(3)+1] + [2(4)+1]$$
$$= 3 + 5 + 7 + 9$$
$$= 24$$

■

Example 2 $\sum_{k=0}^{3} \frac{2^k}{4k-5}$

Solution

$$\sum_{k=0}^{3} \frac{2^k}{4k-5} = \frac{2^0}{4(0)-5} + \frac{2^1}{4(1)-5} + \frac{2^2}{4(2)-5} + \frac{2^3}{4(3)-5}$$
$$= \frac{1}{-5} + \frac{2}{-1} + \frac{4}{3} + \frac{8}{7}$$
$$= -\frac{21}{105} - \frac{210}{105} + \frac{140}{105} + \frac{120}{105}$$
$$= \frac{29}{105}$$

■

If C is a constant, the notation $\sum_{k=1}^{n} C$ is understood to mean the sum of the first n terms of the sequence $C, C, C, \ldots, C, \ldots$. Thus,

$$\sum_{k=1}^{n} C = \overbrace{C + C + C + \cdots + C}^{n \text{ terms}} = nC.$$

Example 3 Evaluate $\sum_{k=1}^{7} 5$.

Solution

$$\sum_{k=1}^{7} 5 = 7(5) = 35$$

■

In calculations involving sums in sigma notation, the following properties may be used.

Properties of Summation

If $\{a_n\}$ and $\{b_n\}$ are sequences and C is a constant, then:

(i) $\sum_{k=1}^{n} (a_k + b_k) = \sum_{k=1}^{n} a_k + \sum_{k=1}^{n} b_k$

(ii) $\sum_{k=1}^{n} (a_k - b_k) = \sum_{k=1}^{n} a_k - \sum_{k=1}^{n} b_k$

(iii) $\sum_{k=1}^{n} Ca_k = C \sum_{k=1}^{n} a_k$

(iv) $\sum_{k=1}^{n} C = nC$

Properties (i), (ii), and (iii) can be verified by expanding both sides of the expression or by using mathematical induction. We have already discussed Property (iv). Of course, Properties (i), (ii), and (iii) continue to hold if all summations begin with $k = 0$. In Property (iv), if we begin with $k = 0$, we obtain

$$\sum_{k=0}^{n} C = (n + 1)C$$

(Problem 41).

In addition to Properties (i) through (iv), we have:

(v) $\sum_{k=1}^{n} k = \dfrac{n(n + 1)}{2}$

(vi) $\sum_{k=1}^{n} r^{k-1} = \dfrac{1 - r^n}{1 - r}$ if $r \neq 0, \quad r \neq 1$

Property (v), which is a formula for the sum of the first n positive integers, was verified by mathematical induction in Example 1 of Section 7.1. Property (vi) can also be confirmed by mathematical induction, but the following argument is more interesting: Let $r \neq 0$, $r \neq 1$, and let $x = \sum_{k=1}^{n} r^{k-1}$. Then

$$x = 1 + r + r^2 + \cdots + r^{n-2} + r^{n-1},$$

so

$$rx = r + r^2 + r^3 + \cdots + r^{n-1} + r^n.$$

If we subtract the second equation from the first, we find that

$$x - rx = 1 - r^n \qquad \text{or} \qquad (1 - r)x = 1 - r^n.$$

Since $r \neq 1$, the last equation can be rewritten as $x = \dfrac{1 - r^n}{1 - r}$, which is Property (vi).

Using Properties (i) to (vi), we can derive formulas for the sum of the first n terms of an arithmetic or geometric sequence. We begin with the arithmetic case.

Theorem 1 **Sum of the First n Terms of an Arithmetic Sequence**

Let $\{a_n\}$ be an arithmetic sequence with first term a_1 and common difference d. Then, for every positive integer n,

$$\sum_{k=1}^{n} a_k = na_1 + \frac{n}{2}(n-1)d = \frac{n}{2}(a_1 + a_n).$$

Proof We have already obtained the formula $a_k = a_1 + (k-1)d$ for the kth term of $\{a_n\}$ (page 427). Thus,

$$\begin{aligned}
\sum_{k=1}^{n} a_k &= \sum_{k=1}^{n} [a_1 + (k-1)d] = \sum_{k=1}^{n} a_1 + \sum_{k=1}^{n} (k-1)d && \text{[by Property (i)]} \\
&= na_1 + d\sum_{k=1}^{n} (k-1) && \text{[by Properties (iv) and (iii)]} \\
&= na_1 + d[0 + 1 + 2 + \cdots + (n-1)] \\
&= na_1 + d\left[\frac{(n-1)n}{2}\right] && \text{[by Property (v)]} \\
&= na_1 + \frac{n}{2}(n-1)d,
\end{aligned}$$

which proves the first part of the formula. To obtain the remaining part, we use the fact that $a_n = a_1 + (n-1)d$. Thus,

$$\begin{aligned}
na_1 + \frac{n}{2}(n-1)d &= \left(\frac{n}{2}a_1 + \frac{n}{2}a_1\right) + \frac{n}{2}(n-1)d \\
&= \frac{n}{2}[a_1 + a_1 + (n-1)d] \\
&= \frac{n}{2}(a_1 + a_n).
\end{aligned}$$

■

Example 4 Find the sum of the first six terms of the arithmetic sequence 4, 8, 12, 16, 20, 24, ...

Solution We use the formula

$$\sum_{k=1}^{n} a_k = \frac{n}{2}(a_1 + a_n)$$

of Theorem 1. Here, $n = 6$, $a_1 = 4$, and $a_6 = 24$, so

$$\sum_{k=1}^{6} a_k = \frac{6}{2}(4 + 24) = 84.$$

■

Example 5 Find the sum of the first twelve terms of the arithmetic sequence 5, 1, −3, −7, ...

Solution Here $d = -4$, $a_1 = 5$, and $n = 12$. Using the formula

$$\sum_{k=1}^{n} a_k = na_1 + \frac{n}{2}(n-1)d$$

of Theorem 1, we have

$$\sum_{k=1}^{12} a_k = 12(5) + \frac{12}{2}(12-1)(-4) = 60 - 264 = -204.$$

■

Example 6 A paperboy receives \$30 from a newspaper route for the first month. During each succeeding month he earns \$2 more than he did the month before. If this pattern continues, how much will he earn over a period of 2 years?

Solution The paperboy's monthly income forms an arithmetic sequence with $a_1 = 30$ dollars and $d = 2$ dollars. His total income over a 2-year period is given by the formula

$$\sum_{k=1}^{n} a_k = na_1 + \frac{n}{2}(n-1)d$$

with $n = 24$. We have

$$\sum_{k=1}^{24} a_k = 24(30) + \frac{24}{2}(24-1)(2) = \$1272.$$

■

Now, we turn our attention to the geometric case.

Theorem 2 **Sum of the First *n* Terms of a Geometric Sequence**

Let $\{a_n\}$ be a geometric sequence with first term a_1 and common ratio r, where $r \neq 0$ and $r \neq 1$. Then, for every positive integer n,

$$\sum_{k=1}^{n} a_k = \sum_{k=1}^{n} a_1 r^{k-1} = a_1 \frac{1-r^n}{1-r}.$$

Proof We have already obtained the formula $a_k = r^{k-1}a_1 = a_1 r^{k-1}$ for the kth term of a geometric sequence with common ratio r (page 428). Therefore,

$$\sum_{k=1}^{n} a_k = \sum_{k=1}^{n} a_1 r^{k-1} = a_1 \sum_{k=1}^{n} r^{k-1} \qquad \text{[by Property (iii)]}$$

$$= a_1 \frac{1-r^n}{1-r} \qquad \text{[by Property (vi)]}$$

and the proof is complete. ■

Example 7 Find the sum of the first eight terms of the geometric sequence 3, -6, 12, $-24, \ldots$.

Solution We use the formula of Theorem 2, with $a_1 = 3$, $r = -2$, and $n = 8$. Thus,

$$\sum_{k=1}^{n} a_k = a_1 \frac{1 - r^n}{1 - r} = (3)\frac{1 - (-2)^8}{1 - (-2)} = (3)\frac{1 - 256}{3} = -255. \quad \blacksquare$$

C Example 8 For doing a certain job, you are offered 1¢ the first day, 3¢ the second day, 9¢ the third day, and so forth, so that your daily earnings form a geometric sequence with first term $a_1 = 1$ cent and common ratio $r = 3$. How much will you earn in 14 days of work?

Solution By Theorem 2,

$$\sum_{k=1}^{n} a_1 r^{k-1} = \sum_{k=1}^{14} 3^{k-1} = \frac{1 - 3^{14}}{1 - 3}$$
$$= \frac{1 - 3^{14}}{-2} = \frac{3^{14} - 1}{2}.$$

Using a calculator, we find that the total earnings amount to 2,391,484¢ or $23,914.84. (Not bad for two weeks work!) ■

Infinite Geometric Series

An indicated sum of all the terms of a sequence $\{a_n\}$, such as

$$a_1 + a_2 + a_3 + \cdots + a_n + \cdots$$

is called an **infinite series** or simply a **series.** Using the sigma notation, we can write such a series more compactly as

$$\sum_{k=1}^{\infty} a_k.$$

Although we cannot literally add an infinite number of terms, it is sometimes useful to assign a numerical value as the "sum" of an infinite series. This is accomplished by using the idea of a "limit," which is studied in calculus.

Although the general idea of the sum of an infinite series is beyond the scope of this textbook, you can easily get a feeling for this concept by considering an infinite geometric series of the form

$$\sum_{k=1}^{\infty} a_1 r^{k-1} = a_1 + a_1 r + a_1 r^2 + \cdots + a_1 r^{n-1} + \cdots$$

for the case in which $0 < |r| < 1$. By Theorem 2, the sum of the *first n terms* of this series is

$$\sum_{k=1}^{n} a_1 r^{k-1} = a_1 + a_1 r + \cdots + a_1 r^{n-1} = a_1 \frac{1 - r^n}{1 - r}.$$

Because $0 < |r| < 1$, it follows that r^n gets smaller and smaller as n gets larger and larger. (Try it on a calculator, say for $r = 0.75$ with $n = 10$, $n = 50$, $n = 100$, and so forth.) Therefore, as n gets larger and larger, r^n gets closer and closer to 0 and

$$a_1 \frac{1 - r^n}{1 - r} \quad \text{comes closer and closer to} \quad a_1 \frac{1 - 0}{1 - r} = \frac{a_1}{1 - r}.$$

In other words, as you add more and more terms of the geometric series, the sum comes closer and closer to $a_1/(1 - r)$. This suggests the following definition.

Definition 1 **Sum of an Infinite Geometric Series**

If $|r| < 1$, we define

$$\sum_{k=1}^{\infty} a_1 r^{k-1} = a_1 + a_1 r + a_1 r^2 + \cdots + a_1 r^{n-1} + \cdots = \frac{a_1}{1 - r}.$$

In Examples 9 and 10, find the sum of each infinite geometric series.

Example 9 $\frac{1}{2} + \frac{1}{4} + \frac{1}{8} + \cdots$

Solution Here $a_1 = \frac{1}{2}$ and $r = \frac{1}{2}$, so by Definition 1,

$$\frac{1}{2} + \frac{1}{4} + \frac{1}{8} + \cdots = \frac{a_1}{1 - r} = \frac{\frac{1}{2}}{1 - \frac{1}{2}} = \frac{\frac{1}{2}}{\frac{1}{2}} = 1.$$ ■

Example 10 $\sum_{k=1}^{\infty} 4\left(-\frac{1}{3}\right)^{k-1}$

Solution Here $a_1 = 4$ and $r = -\frac{1}{3}$, so by Definition 1,

$$\sum_{k=1}^{\infty} 4\left(-\frac{1}{3}\right)^{k-1} = \frac{a_1}{1 - r}$$

$$= \frac{4}{1 - (-\frac{1}{3})} = \frac{4}{1 + \frac{1}{3}} = \frac{4}{\frac{4}{3}} = 3.$$ ■

When we write an infinite decimal

$$x = 0.d_1 d_2 d_3 d_4 \ldots d_n \ldots,$$

we are actually forming the sum of an infinite series

$$x = \frac{d_1}{10} + \frac{d_2}{100} + \frac{d_3}{1000} + \frac{d_4}{10{,}000} + \cdots + \frac{d_n}{10^n} + \cdots.$$

Thus, you can use the formula in Definition 1 to convert a repeating decimal to a quotient of integers.

Example 11 Rewrite the repeating decimal $0.\overline{31}$ as a quotient of integers.

Solution

$$0.\overline{31} = 0.31313131\ldots = 0.31 + 0.0031 + 0.000031 + 0.00000031 + \cdots,$$

which is an infinite geometric series with first term $a_1 = 0.31$ and common ratio $r = 0.01$. Hence, by Definition 1,

$$0.\overline{31} = \frac{a_1}{1-r} = \frac{0.31}{1-0.01} = \frac{0.31}{0.99} = \frac{31}{99}.$$

(Note that the same answer is obtained by using the method in Example 6 of Section 2.1.) ■

Problem Set 7.4

In Problems 1 to 6, write out and evaluate each sum.

1. $\sum_{k=1}^{4} (3k-2)$
2. $\sum_{k=1}^{5} [2+(-1)^k]$
3. $\sum_{k=1}^{5} \left(\frac{1}{4}k+3\right)$
4. $\sum_{k=2}^{6} (-1)^k 3^k$
5. [C] $\sum_{k=0}^{5} \frac{2^k}{3k+1}$
6. $\sum_{j=0}^{4} \frac{5^j}{2j-1}$

In Problems 7 to 12, use Properties (i) to (vi) of summation to evaluate each expression.

7. $\sum_{k=1}^{40} 3k$
8. $\sum_{k=1}^{30} (3-4k)$
9. $\sum_{k=1}^{24} (2^k+1)$
10. $\sum_{k=0}^{19} 2^{-k}$
11. $\sum_{k=1}^{100} \frac{1}{10^k}$
12. $\sum_{j=0}^{100} (5^{j+1}-5^j)$

In Problems 13 to 20, find the sum of the first n terms of each arithmetic sequence for the given value of n.

13. $1, 2, 3, \ldots,$ for $n = 10$
14. $5, 9, 13, \ldots,$ for $n = 13$
15. $36, 48, 60, \ldots,$ for $n = 8$
16. $27, 33, 39, \ldots,$ for $n = 30$
17. $2, -2, -6, \ldots,$ for $n = 15$
18. $0.6, 0.4, 0.2, \ldots,$ for $n = 16$
19. $x, 0, -x, \ldots,$ for $n = 10$
20. $-2y, 0, 2y, \ldots,$ for $n = 10$

In Problems 21 to 28, find the sum of the first n terms of each geometric sequence for the given value of n.

21. $4, 40, 400, \ldots,$ for $n = 5$
22. $-3, 15, -75, \ldots,$ for $n = 6$
23. $-\frac{1}{3}, -\frac{1}{9}, -\frac{1}{27}, \ldots,$ for $n = 10$
24. $-\frac{1}{16}, -\frac{1}{8}, -\frac{1}{4}, \ldots,$ for $n = 8$
25. $5^{10}, 5^8, 5^6, \ldots,$ for $n = 9$
26. $1, -1, 1, \ldots,$ for $n = 10$
27. $0.3, 0.03, 0.003, \ldots,$ for $n = 6$
28. $c^4, c^6, c^8, \ldots,$ for $n = 7$

In Problems 29 to 34, find the sum of each infinite geometric series.

29. $\sum_{k=1}^{\infty} \left(\frac{2}{3}\right)^{k-1}$
30. $\sum_{k=1}^{\infty} \left(-\frac{1}{3}\right)^{k-1}$
31. $\sum_{k=1}^{\infty} \left(-\frac{2}{5}\right)^{k-1}$
32. $\sum_{k=1}^{\infty} \left(\frac{3}{7}\right)^{k-1}$
33. $\sum_{k=1}^{\infty} 5\left(-\frac{3}{5}\right)^{k-1}$
34. $\sum_{k=0}^{\infty} ar^k$ for $|r| < 1$

In Problems 35 and 36, use the formula for the sum of an infinite geometric series to rewrite each infinite repeating decimal as a quotient of integers.

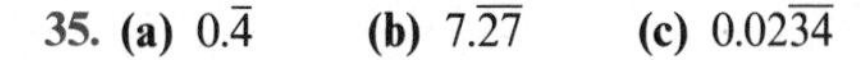

35. **(a)** $0.\overline{4}$ **(b)** $7.\overline{27}$ **(c)** $0.0\overline{234}$

36. **(a)** $4.\overline{53}$ **(b)** $0.\overline{279}$ **(c)** $-1.\overline{981}$

37. True or false: $0.\overline{9} < 1$. Explain your answer.

38. In the discussion leading up to Definition 1, it was assumed that $0 < |r| < 1$; however, in Definition 1, we only require that $|r| < 1$. Explain.

39. If a and b are positive constants, find a formula for the sum of the first n terms of the geometric sequence $\frac{a}{b}, -1, \frac{b}{a}, \ldots$.

40. If $\{a_n\}$ is a geometric sequence with common ratio r, show that $\sum_{k=1}^{n} a_k = \frac{a_1 - a_n r}{1 - r}$.

41. Justify the formula $\sum_{k=0}^{n} C = (n+1)C$. [*Hint:* Use $\sum_{k=0}^{n} a_k$, where $a_k = C$ for all values of k.]

42. If $\{a_n\}$ is an arithmetic sequence and if p and q are positive integers, show that

$$\sum_{k=p}^{p+q} a_k = \frac{q+1}{2}(a_p + a_{p+q}).$$

43. Use summation notation to rewrite the results of Problems 1, 3, and 6 in Problem Set 11.1 in more compact form.

44. The **arithmetic mean,** or **average,** of the numbers $a_1, a_2, a_3, \ldots, a_n$ is defined to be the number $m = \frac{1}{n}\sum_{k=1}^{n} a_k$. Find a formula in terms of n, a, and d for the arithmetic mean of the first n terms of an arithmetic sequence with first term a and common difference d.

45. A company had sales of \$200,000 during its first year of operation, and sales increased by \$30,000 per year during each successive year. Find the total sales of the company during its first 8 years.

46. A freely falling object dropped from rest falls $\frac{1}{2}g(2k - 1)$ units of distance during the kth second of its fall, where g is the acceleration of gravity. Find a formula for the total distance through which the body falls in n seconds.

47. If points are arranged in a triangular pattern as shown in Figure 1, with n points on the bottom row, $n - 1$ points on the next row up, and so on, then the total number of points forming the pattern is called the nth **triangular number** T_n.

(a) Show that the sequence $\{T_n\}$ of triangular numbers satisfies the recursive relationship $T_{n+1} = n + 1 + T_n$ for all integers $n \geq 1$.

(b) Find a formula for T_n.

(c) Find T_{10}.

Figure 1

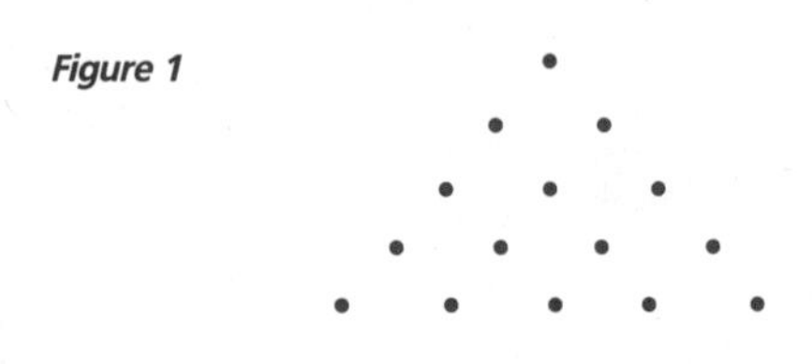

48. The rungs of a ladder decrease uniformly (that is, linearly) in length from a inches at the bottom to b inches at the top (Figure 2). If there are n rungs in the ladder, find a formula in terms of a, b, and n for the total length of the wood in all the rungs.

Figure 2

C **49.** Suppose that the distance traveled by a point on a pendulum in any swing is 5% less than in the previous swing. If the point travels 6 centimeters on the first swing, find the total distance traveled by the point in seven swings.

C **50.** In Problem 49, find the total distance covered by the point before the pendulum comes to rest.

51. A ball is dropped from a height of 256 centimeters, and on each rebound it rises to $\frac{1}{2}$ the height from which it last fell. Find a formula for the total distance traveled by the ball when it hits the floor for the nth time.

52. The **Cantor set,** named after the German mathematician George Cantor, is constructed as follows: Begin by removing the open interval $(\frac{1}{3}, \frac{2}{3})$ from the closed unit interval $[0, 1]$ (Figure 3a). This procedure is called "removing the middle third." Now remove the middle thirds from the remaining two intervals (Figure 3b). Then remove the middle thirds from the remaining four intervals (Figure 3c). Continue in this way forever. The Cantor set consists of all points of $[0, 1]$ that are not removed by this procedure. Find the sum of the lengths of all the open intervals that are removed.

Figure 3

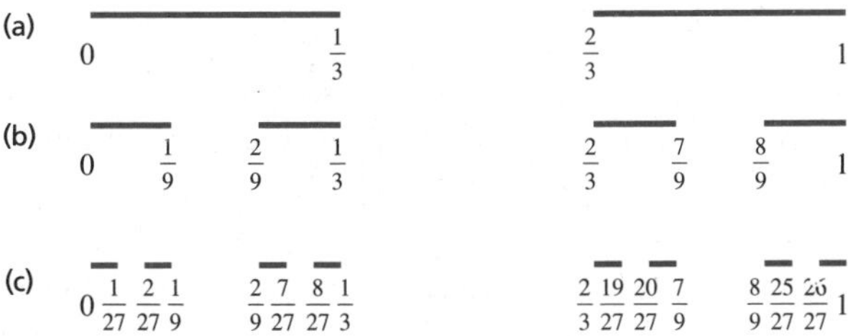

(a) $0 \quad \frac{1}{3} \qquad \frac{2}{3} \quad 1$

(b) $0 \quad \frac{1}{9} \quad \frac{2}{9} \quad \frac{1}{3} \qquad \frac{2}{3} \quad \frac{7}{9} \quad \frac{8}{9} \quad 1$

(c) $0 \; \frac{1}{27} \; \frac{2}{27} \; \frac{1}{9} \quad \frac{2}{9} \; \frac{7}{27} \; \frac{8}{27} \; \frac{1}{3} \qquad \frac{2}{3} \; \frac{19}{27} \; \frac{20}{27} \; \frac{7}{9} \quad \frac{8}{9} \; \frac{25}{27} \; \frac{26}{27} \; 1$

53. In Problem 51, find the total distance the ball travels before it comes to rest.

54. An equilateral triangle has sides of length 1 unit. A second equilateral triangle is constructed within the first by placing its vertices at the midpoints of the sides of the first triangle. A third equilateral triangle is constructed within the second in the same way, and the process is continued forever (Figure 4). Find the sum of the perimeters of all of the triangles.

Figure 4

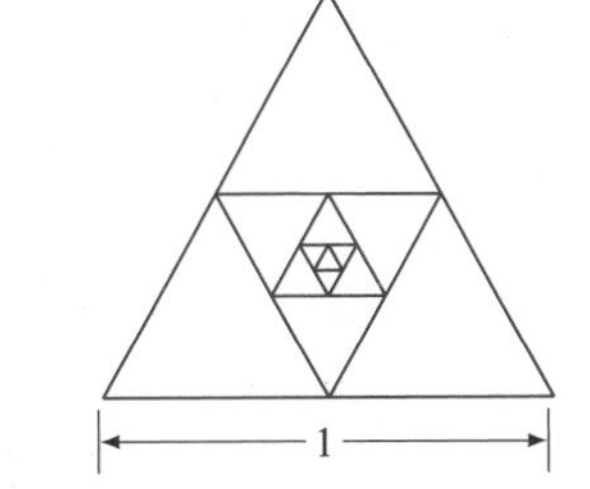

55. A famous mathematical problem concerns a fly that shuttles back and forth between the front wheels of two bicycles that are traveling directly toward each other, each moving at a constant speed of 10 miles per hour. The fly travels at a constant speed of 15 miles per hour, and the two bicycles start at a distance of 20 miles apart. The question is, how far does the fly travel until the two bicycles meet? What makes the problem interesting is that, although it can be solved by setting up and summing an infinite series, it can also be solved very quickly by using the right idea. Solve the problem by both methods.

7.5 Permutations and Combinations

It is often necessary to calculate the number of different ways in which something can be done or can happen. For instance, in order to find the odds of being dealt a full house in a game of five-card poker, it is necessary to calculate the total number of different possible poker hands. In this section we discuss some of the methods used in making such calculations.

We begin by stating the additive and multiplicative principles of counting, which are the basic tools for the calculations made in this section.

The Additive Principle of Counting

Let A and B be two events that cannot occur simultaneously. Then, if A can occur in a ways and B can occur in b ways, it follows that the number of ways in which either A or B can occur is $a + b$.

The additive principle of counting can be extended to three or more events in an obvious way.

Example 1 Suppose you have six signal flags—two red, two green, and two blue. By displaying a single flag, you can send three different signals. By displaying two flags, one above the other, you can send nine different signals. How many different signals can you send using one or two flags?

Solution Let A be the event that a one-flag signal is sent, so that A can occur in $a = 3$ ways. Let B be the event that a two-flag signal is sent, so that B can occur in $b = 9$ ways. Since a one-flag signal and a two-flag signal can't be sent at the same time, the additive principle of counting applies, and $a + b = 3 + 9 = 12$ different signals can be sent using either one or two flags. ■

The Multiplicative Principle of Counting

If a first event A can occur in a ways and, independently, a second event B can occur in b ways, then the number of ways in which both A and B can occur is ab.

The multiplicative principle of counting can be extended to three or more events in an obvious way.

Example 2 A restaurant offers a meal consisting of one of three entrees and one of four desserts at a special price. How many different meals can be ordered at this price?

Solution By the multiplicative principle of counting, $3 \cdot 4 = 12$ different meals can be ordered. ■

The idea of a *permutation*, introduced in the following definition, is useful in many counting problems.

Definition 1 **Permutations**

A **permutation** is an *ordered* arrangement of distinct objects in a row. Let S be a set of n distinct elements. An ordered arrangement of r of these elements in a row is called a *permutation of size r chosen from the n elements of S.* The symbol ${}_nP_r$ denotes the number of different permutations of size r chosen from a set of n elements.*

Example 3 Let $S = \{a, b, c, d\}$. List all permutations of size 2 chosen from the 4 elements of S, and count these permutations to find ${}_4P_2$.

* The symbol ${}_nP_r$ is read as "the number of permutations of n things taken r at a time."

Solution

We list the permutations according to which element is in the first position:

a first	b first	c first	d first
ab	ba	ca	da
ac	bc	cb	db
ad	bd	cd	dc

Thus, ${}_4P_2 = 12$. ■

Theorem 1

A Formula for ${}_nP_r$

If n and r are positive integers with $n \geq r$, then

$$
\begin{aligned}
{}_nP_r &= \overbrace{n(n-1)(n-2)\cdots(n-r+1)}^{r\text{ factors}} \\
&= \frac{n!}{(n-r)!}.
\end{aligned}
$$

Proof

In forming a permutation of r elements chosen from a set of n elements, we have n choices for the element to be placed in the first position in the row. Once this choice is made, there remain $n - 1$ elements to be placed in the remaining $r - 1$ positions. Therefore, we have $n - 1$ choices for the element to be placed in the second position. By the multiplicative principle of counting, there are $n(n - 1)$ ways to choose distinct elements to place in the first two positions. Similarly, there are $n(n - 1)(n - 2)$ ways to choose distinct elements to place in the first three positions. Continuing in this way, we see that all r positions in the row can be filled with distinct elements in $n(n - 1)(n - 2) \cdots [n - (r - 1)]$ ways. Therefore,

$$
\begin{aligned}
{}_nP_r &= \overbrace{n(n-1)(n-2)\cdots(n-r+1)}^{r\text{ factors}} \\
&= \frac{n(n-1)(n-2)\cdots(n-r+1)(n-r)!}{(n-r)!} = \frac{n!}{(n-r)!}.
\end{aligned}
$$

■

Example 4

Evaluate:

(a) ${}_7P_1$ **(b)** ${}_5P_3$ **(c)** ${}_4P_4$ **(d)** ${}_9P_4$

Solution

Using the formula ${}_nP_r = \overbrace{n(n-1)(n-2)\cdots(n-r+1)}^{r\text{ factors}}$, we have:

(a) ${}_7P_1 = 7$ **(b)** ${}_5P_3 = 5 \cdot 4 \cdot 3 = 60$

(c) ${}_4P_4 = 4 \cdot 3 \cdot 2 \cdot 1 = 24$ **(d)** ${}_9P_4 = 9 \cdot 8 \cdot 7 \cdot 6 = 3024$ ■

Example 5 A portfolio manager knows of 13 stocks that meet her investment criteria. She will choose three of these and rank them first, second, and third in order of preference. In how many ways can she do this?

Solution $_{13}P_3 = 13 \cdot 12 \cdot 11 = 1716$ ways

Suppose that we have n distinct objects that are to be arranged in a definite order. By Theorem 1, this can be done in

$$_nP_n = n(n-1)(n-2)\cdots 1 = n!$$

different ways.

C **Example 6** In how many ways can 12 different books be arranged on a shelf?

Solution Using a calculator,* we find that there are

$$12! = 479{,}001{,}600$$

different arrangements.

The value of a poker hand doesn't depend on the order in which the cards are arranged in the hand, but only on which cards are present. Thus, the idea of a permutation, which involves *order*, isn't the right tool for counting poker hands. Such situations, in which order isn't important, suggest the following definition.

Definition 2 **Combinations**

Let S be a set consisting of n distinct elements. A subset consisting of r of these elements is called a **combination of size r chosen from the n elements in S.** No importance is attached to the order of the r chosen elements. The symbol $_nC_r$ denotes the number of different combinations of size r chosen from a set of n elements.†

Example 7 Let $S = \{a, b, c, d\}$. List all combinations of size 2 chosen from the 4 elements of S, and count these combinations to find $_4C_2$.

Solution The subsets of S that consist of two elements are

$$\{a, b\}, \quad \{a, c\} \quad \{a, d\}, \quad \{b, c\}, \quad \{b, d\}, \quad \text{and} \quad \{c, d\}.$$

* Some calculators have special keys for evaluating $_nP_r$.

† The symbol $_nC_r$ is read as "the number of combinations of n things taken r at a time," or as "n choose r."

(Because order isn't important, the combination $\{a, b\}$ is the same as the combination $\{b, a\}$, the combination $\{a, c\}$ is the same as the combination $\{c, a\}$, and so on.) Thus, ${}_4C_2 = 6$. ■

There is a simple relationship between permutations and combinations. To obtain this relationship, we consider a set S consisting of n distinct elements. To specify a permutation of size r chosen from these n elements, we can first select the r elements that will appear in the permutation, and then we can give the order in which the selected elements are to be arranged. The first step, the selection of a combination of r elements from S, can be done in ${}_nC_r$ ways; the second step, the ordering of these r elements, can be accomplished in $r!$ ways. Therefore, by the multiplicative principle of counting,

$$ {}_nP_r = ({}_nC_r)r!. $$

Using this relationship, we can prove the following theorem.

Theorem 2 **A Formula for ${}_nC_r$**

If n and r are integers with $n \geq r \geq 0$, then

$$ {}_nC_r = \frac{\overbrace{n(n-1)(n-2)\cdots(n-r+1)}^{r \text{ factors}}}{\underbrace{r(r-1)(r-2)\cdots 1}_{r \text{ factors}}} = \frac{n!}{(n-r)!r!}. $$

Proof By the relationship established above and Theorem 1, we have

$$ {}_nC_r = \frac{{}_nP_r}{r!} = \frac{n(n-1)(n-2)\cdots(n-r+1)}{r(r-1)(r-2)\cdots 1}. $$

Since ${}_nP_r = \dfrac{n!}{(n-r)!}$, we can also write

$$ {}_nC_r = \frac{{}_nP_r}{r!} = \frac{n!}{(n-r)!r!}. $$ ■

Notice that ${}_nC_r$, the number of combinations of size r chosen from a set of n elements, is the same as the binomial coefficient $\binom{n}{r}$ (Definition 2, page 419).

Example 8 Evaluate:

(a) ${}_7C_3$ **(b)** ${}_7C_7$ **(c)** ${}_7C_1$

Solution Using the formula

$$_nC_r = \frac{n(n-1)(n-2)\cdots(n-r+1)}{r(r-1)(r-2)\cdots 1}$$

in Theorem 2, we have:

$$\textbf{(a)}\ _7C_3 = \frac{7\cdot 6\cdot 5}{3\cdot 2\cdot 1} = 35 \qquad \textbf{(b)}\ _7C_7 = \frac{7!}{7!} = 1 \qquad \textbf{(c)}\ _7C_1 = \frac{7}{1} = 7$$

■

Example 9 Find the number of different committees of 4 students each that can be formed from a group of 36 students.

Solution Here the set S consists of the 36 students, so $n = 36$. A committee of 4 of these students is a subset of S consisting of $r = 4$ elements. The order of students within a committee is not important, so each committee is a combination of size $r = 4$ chosen from the $n = 36$ elements of S. The number of such committees is

$$_{36}C_4 = \frac{36\cdot 35\cdot 34\cdot 33}{4\cdot 3\cdot 2\cdot 1} = \frac{\overset{3}{\cancel{36}}\cdot 35\cdot 34\cdot 33}{\cancel{4}\cdot\cancel{3}\cdot 2} = \frac{3\cdot 35\cdot \overset{17}{\cancel{34}}\cdot 33}{\cancel{2}}$$
$$= 3\cdot 35\cdot 17\cdot 33 = 58{,}905.^*$$

■

Example 10 Find the number of different hands in five-card poker.

Solution There are $n = 52$ cards in the deck, and the order of the cards in a hand isn't important. Thus, the number of different poker hands is given by

$$_{52}C_5 = \frac{52\cdot 51\cdot \overset{10}{\cancel{50}}\cdot 49\cdot 48}{\cancel{5}\cdot 4\cdot 3\cdot 2\cdot 1} = \frac{\overset{13}{\cancel{52}}\cdot 51\cdot 10\cdot 49\cdot 48}{\cancel{4}\cdot 3\cdot 2} = \frac{13\cdot 51\cdot 10\cdot 49\cdot \overset{8}{\cancel{48}}}{\cancel{3}\cdot\cancel{2}}$$
$$= 13\cdot 51\cdot 10\cdot 49\cdot 8 = 2{,}598{,}960.$$

■

Suppose that we have r red balls, w white balls, and b black balls, and that balls within a color group are indistinguishable from one another. Altogether, we have $n = r + w + b$ balls. In how many *distinguishable ways* can we arrange these n balls in an ordered row; that is, how many **distinguishable permutations** are there of the n colored balls?

To specify a distinguishable arrangement of the balls in an ordered row, we first decide which of the n positions in the row will be occupied by the red balls. This amounts to specifying a subset of r positions among the n available positions, and it can be done in $_nC_r$ ways. Having made this decision, we place the red balls in the specified positions. There remain $n - r$ positions to be filled with the remaining white and black balls. We choose w of these positions for the white balls. This choice can be made in $_{n-r}C_w$ ways. The remaining $n - r - w$ positions will have to be filled with black balls—here we have no choice. Therefore, by the multiplicative principle of counting, we can specify $(_nC_r)(_{n-r}C_w)$ distinguishable ordered arrangements or permutations of the colored balls.

* Some calculators have special keys for evaluating $_nC_r$.

Using the formula in Theorem 2 and the fact that $n = r + w + b$, so that $n - r - w = b$, we have

$$({}_nC_r)({}_{n-r}C_w) = \frac{n!}{(n-r)!r!} \cdot \frac{(n-r)!}{[(n-r)-w]!w!}$$
$$= \frac{n!(n-r)!}{(n-r)!r!b!w!} = \frac{n!}{r!w!b!}.$$

This formula for the number of distinguishable permutations of the colored balls can be generalized as follows.

Theorem 3 **Distinguishable Permutations**

Consider a set of n objects of which n_1 are alike of one kind, n_2 are alike of another kind, . . . , and n_k are alike of a last kind. Then, the number of distinguishable permutations of the objects is

$$\frac{n!}{n_1!n_2! \cdots n_k!}.$$

Example 11 How many distinguishable permutations are there of the letters in the word CINCINNATI?

Solution There are $n = 10$ letters altogether. Of these, there are $n_1 = 2$ C's, $n_2 = 3$ I's, $n_3 = 3$ N's, $n_4 = 1$ A, and $n_5 = 1$ T. By Theorem 3, the number of distinguishable permutations of these letters is given by

$$\frac{n!}{n_1!n_2!n_3!n_4!n_5!} = \frac{10!}{2!3!3!1!1!} = 50{,}400.$$ ■

Problem Set 7.5

1. There are six ways of rolling a 7 with two dice, and there are two ways of rolling an 11 with two dice. In how many ways can you roll 7 or 11?

2. There are three different roads connecting Newberry with New Mattoon, and four different roads connecting New Mattoon with Gainesburg. In how many different ways can a person drive from Newberry to Gainesburg, passing through New Mattoon on the way?

3. How many different 2-letter "words" can be formed using the 26 letters of the alphabet if repeated letters are allowed? (The "words" need not be in a dictionary.)

4. How many different 3-letter "words" can be formed using the 26 letters of the alphabet if repeated letters are allowed but the middle letter must be a, e, i, o, or u?

5. How many positive three-digit integers less than 500 can be formed using only the digits 1, 3, 5, and 7 if repetitions are allowed?

6. A combination lock for a bicycle has three levers, each of which can be set in nine positions. A thief attempting to steal the bicycle begins trying all possible arrangements of the levers. If the thief tries two

arrangements every second, what is the maximum time required to open the lock?

7. List and count all permutations of size 2 chosen from the elements of $S = \{a, b, c, d, e\}$ and thus determine ${}_5P_2$.

8. List and count all permutations of size 3 chosen from the elements of the set $S = \{a, b, c, d\}$ and thus determine ${}_4P_3$.

In Problems 9 to 16, find the value of the expression by using Theorem 1.

9. ${}_4P_3$

10. ${}_8P_6$

C 11. ${}_9P_9$

12. ${}_{12}P_1$

13. ${}_5P_2$

C 14. ${}_{13}P_{13}$

15. ${}_{11}P_3$

16. ${}_8P_2$

C 17. In how many ways can 5 of 10 books be chosen and arranged next to each other on a shelf?

18. In how many ways can a left end, a right end, and a center for a football team be picked from among Dean, Dolores, Carlos, Carmine, Gus, and Olga?

C 19. In how many ways can seven people line up at a ticket window?

20. How many signals can be sent by using four distinguishable flags, one above the other, on a flag pole, if the flags can be used one, two, three, or four at a time?

C 21. In how many ways can a baseball team of nine players be arranged in batting order if tradition is followed and the pitcher must bat last, but there are no other restrictions?

22. A poll consists of five questions. In how many different orders can these questions be asked?

23. If 10 runners are entered in a race for which first, second, and third prizes will be awarded, in how many different ways can the prizes be distributed? (Assume that there are no ties.)

24. How many integers that do not contain repeated digits are there between 1 and 1000, inclusive?

25. List and count all combinations of size 3 chosen from the elements of the set $S = \{a, b, c, d\}$ and thus determine ${}_4C_3$.

26. List and count all combinations of size 2 chosen from the elements of $S = \{a, b, c, d, e\}$ and thus determine ${}_5C_2$.

In Problems 27 to 35, find the value of the expression by using Theorem 2.

27. ${}_4C_4$

28. ${}_{48}C_3$

29. ${}_5C_2$

C 30. ${}_{52}C_{13}$

31. ${}_6C_3$

32. ${}_{10}C_{10}$

33. ${}_{10}C_9$

34. ${}_{10}C_1$

35. ${}_{10}C_0$

36. Show that ${}_nC_0 = 1$ holds for every integer $n \geq 0$ and give an interpretation of this fact.

37. A total of how many games will be played by eight teams in a league if each team plays the other teams just once?

38. A dealer has 15 different models of television sets and wishes to display 3 models in the store window. Disregarding the arrangement of the 3 sets in the window, in how many ways can this be done?

39. An election is to be held to choose 4 delegates from among 10 nominees to represent a district at a political convention. In how many ways can the delegation be formed?

40. A department store plans to fill 10 positions with 4 men and 6 women. In how many ways can these positions be filled if there are 9 men and 11 women applicants?

41. "Taxicab geometry" is based on the idea that a taxicab, in going from point A to point B in a city, must follow a network of streets rather than going on a straight line from A to B. The diagram in Figure 1a shows part of a rectangular network of streets in a certain city and one of the paths that a taxicab might take to go from point A to point B. In Figure 1b, each street corner is labeled with a number showing how many different paths a taxicab might follow, starting at the top of the diagram, to arrive at that corner without backtracking.

(a) Explain why each number in this diagram (other than those that are equal to 1 and lie on the border) is obtained by adding the two numbers diagonally above it.

(b) Account for the fact that the numbers in Figure 1b correspond to the numbers in Pascal's triangle (Section 7.2).

Figure 1

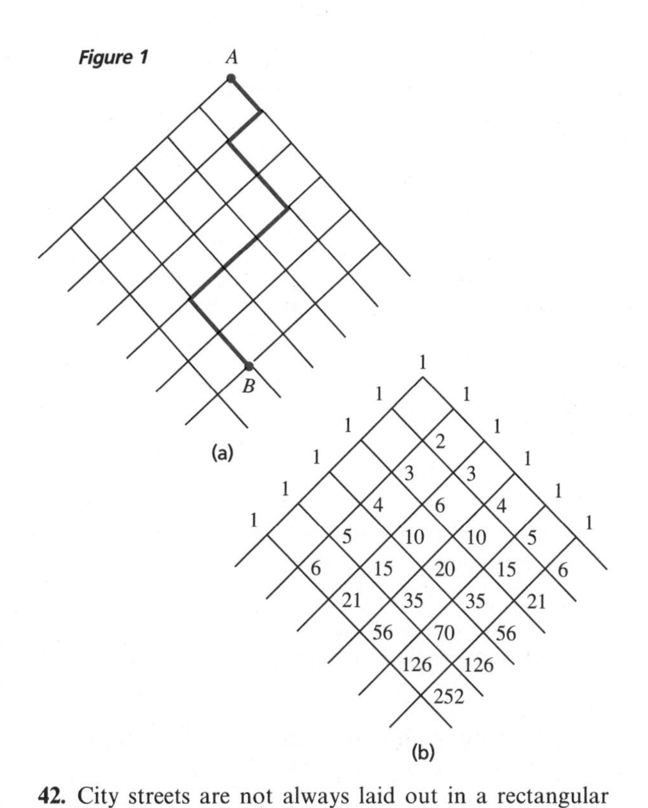

Figure 2

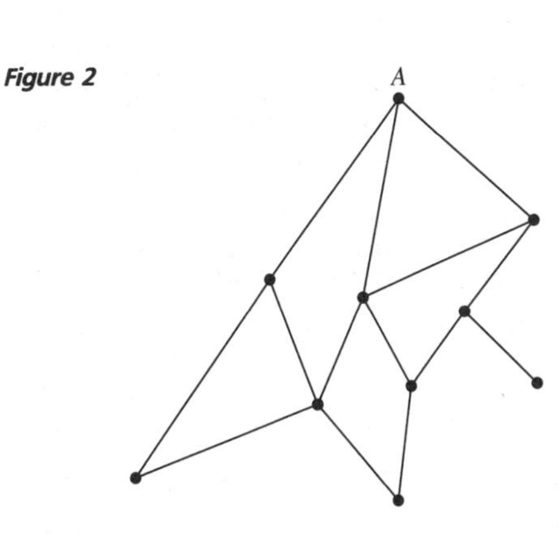

42. City streets are not always laid out in a rectangular pattern as in Figure 1. For the pattern shown in Figure 2, label each street corner with a number showing how many different paths a taxicab might follow, starting at point A, to arrive at that corner without backtracking.

C In Problems 43 to 48, determine the number of distinguishable permutaions of the letters in each word.

43. OHIO

44. MASSACHUSETTS

45. REARRANGEMENT

46. MISSISSIPPI

47. TENNESSEE

48. COMMITTEE

49. In how many distinguishable ways can eight people be arranged in a police lineup if two of them are identical twins and three are identical triplets?

50. There are three copies of a chemistry book, two copies of a biology book, and four copies of a sociology book to be placed on a shelf. In how many distinguishable ways can this be done?

7.6 Probability

The mathematical theory of probability has applications in nearly every area of human activity. These applications range from medical statistics to actuarial science, from statistical physics to law, and from decision making in the business world to gambling in Las Vegas. Here we can only touch the subject in the most superficial way, and we can give only the simplest examples, many of which involve playing cards or dice.

Consider an experiment that has a finite number of **equally likely outcomes.** For instance, if you roll a balanced die (a small cube with different numbers of spots on its faces), the possible outcomes are 1, 2, 3, 4, 5, or 6, and they are equally likely. An **event** is something that may or may not occur, depending on which outcome is obtained. For instance, if you roll a die, the event "rolling an odd number" occurs if the outcome is 1, 3, or 5 and fails to occur if the outcome is 2, 4, or 6. The outcomes for which an event occurs are said to be **favorable** to the event. For instance, the outcomes 1, 3, and 5 are favorable to the event "rolling an odd number."

Definition 1 **The Probability of an Event**

Let A be an event associated with an experiment with N equally likely outcomes. If $n(A)$ is the number of outcomes favorable to A, the **probability** of A, in symbols $P(A)$, is defined by

$$P(A) = \frac{n(A)}{N}.$$

The number $P(A)$ in Definition 1 is the fraction of all possible outcomes that are favorable to A. Hence, $P(A)$ can be regarded as a numerical measure of the likelihood, on a scale from 0 to 1, that A will occur if the experiment is performed.

Example 1 Find the probability of rolling an odd number with a balanced die (Figure 1).

Solution In Definition 1, let A denote the event "rolling an odd number." When a balanced die is rolled, there are $N = 6$ equally likely outcomes; $n(A) = 3$ of these outcomes are favorable to the event A, so

Figure 1

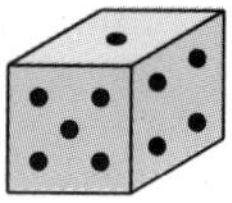

$$P(A) = \frac{n(A)}{N} = \frac{3}{6} = \frac{1}{2}.$$

We can also express $P(A)$ in decimal form as $P(A) = 0.5$, or as $P(A) = 50\%$. ■

Example 2 Find the probability of drawing a face card (jack, queen, or king) from a well-shuffled deck of 52 cards.

Solution Here there are $N = 52$ possible outcomes of the experiment of drawing one card from the deck. Since the deck is well shuffled, these outcomes are all equally likely. Of the 52 cards, there are 4 jacks, 4 queens, and 4 kings; hence, the number of face cards is $4 + 4 + 4 = 12$. Thus,

$$P(\text{"drawing a face card"}) = \frac{12}{52} = \frac{3}{13}.$$ ■

Example 3 From a group of nine people, five male and four female, a committee of three is to be selected by chance. What is the probability that no females will be on the committee?

Solution

Here the experiment is the selection of a committee, and the outcome is the selected committee. There are

$$N = {}_9C_3 = 84$$

ways of selecting a committee of 3 from 9 people. Since the committee is to be selected by chance, all 84 of these outcomes are equally likely. The event in question, "no females on the committee," occurs if and only if the committee consists only of males. There are

$$n = {}_5C_3 = 10$$

ways of selecting a committee consisting entirely of males from among the 5 available. It follows that

$$P(\text{"no females on committee"}) = \frac{n}{N} = \frac{10}{84} \approx 11.9\%.$$

■

Example 4

What is the probability of being dealt a full house (three cards of one denomination—say, three 6's or three kings—and two cards of another denomination—say, two aces or two 9's) from a well-shuffled deck in five-card poker?

Solution

Here an outcome is a 5-card hand. There are $N = {}_{52}C_5$ such hands, all of which are equally likely. Of these, we must calculate the number n of full houses. This is best done by asking ourselves in how many ways we could form a 5-card hand with 3 cards of one denomination and 2 cards of another. The common denomination of the 3 cards can be chosen in 13 ways, and the 3 cards themselves can be chosen from among the 4 cards of this denomination in ${}_4C_3$ ways. The common denomination of the other 2 cards can then be chosen in any of the remaining 12 ways, and the 2 cards themselves can be chosen from among the 4 cards of this denomination in ${}_4C_2$ ways. By the multiplicative principle of counting, we can therefore form full houses in

$$n = (13)({}_4C_3)(12)({}_4C_2)$$

different ways. It follows that

$$\begin{aligned}
P(\text{full house}) = \frac{n}{N} &= \frac{(13)({}_4C_3)(12)({}_4C_2)}{{}_{52}C_5} \\
&= \frac{(13)\left(\dfrac{4 \cdot 3 \cdot 2}{3 \cdot 2 \cdot 1}\right)(12)\left(\dfrac{4 \cdot 3}{2 \cdot 1}\right)}{\dfrac{52 \cdot 51 \cdot 50 \cdot 49 \cdot 48}{5 \cdot 4 \cdot 3 \cdot 2 \cdot 1}} \\
&= \frac{13 \cdot 4 \cdot 12 \cdot 2 \cdot 3}{52 \cdot 51 \cdot 10 \cdot 49 \cdot 2} \\
&= \frac{6}{4165}.
\end{aligned}$$

Example 5 What is the probability of rolling a 7 with two dice?

Solution Call one of the dice the *first die* and the other the *second die*. An outcome of a toss of the two dice can be denoted by an ordered pair (x, y), where x is the number on the first die and y the number on the second. Here x can have any of 6 possible values and y can have any of 6 possible values, so by the multiplicative principle of counting, there are $N = 6 \cdot 6 = 36$ possible outcomes. Of these, the outcomes favorable to the event "rolling a 7" are

$$(1, 6), \quad (2, 5), \quad (3, 4), \quad (4, 3), \quad (5, 2), \quad \text{and} \quad (6, 1).$$

Thus, there are $n = 6$ outcomes favorable to this event. It follows that

$$P(\text{"rolling a 7"}) = \frac{n}{N} = \frac{6}{36} = \frac{1}{6}.$$

■

Expectation and Odds

Suppose that you play a gambling game in which the probability of winning is p; the payoff, if you win, is w dollars; and it costs you c dollars for each play of the game. If you play the game a very large number of times N, then you will win approximately Np of the games and you will lose approximately $N - Np = (1 - p)N$ of the games. To play the N games will cost you cN dollars, and your total winnings will be $w(Np)$ dollars. Your net gain G dollars from playing the N games is the amount that you win minus the cost to play, so that

$$G = w(Np) - cN = (wp - c)N \text{ dollars.}$$

Thus, your average gain per play of the game is given by

$$\frac{G}{N} = (wp - c) \text{ dollars.}$$

In the theory of probability, the quantity $wp - c$ is called the **expected value,** or the **expectation,** of the game described above and is denoted by

$$E = wp - c.$$

Thus, the expectation E can be considered to be the long-run average gain per play of the game. Gambling games offered by casinos have a negative expectation E, indicating that in the long run your net loss will average approximately $|E|$ dollars on each play of the game. This is what produces the profit for the casino and is precisely why it is in business.

Example 6 In the game of roulette offered by American casinos, there are 38 possible outcomes. Of these, 18 are black, 18 are red, and 2 are green. If you bet one dollar on black, the dollar is returned to you along with an additional dollar if the outcome is black; otherwise, you lose your dollar. Find the expected value of this game.

Solution

Here we can consider that the cost to play the game is $c = 1$ dollar and that the payoff if you win is $w = 2$ dollars. The probability of winning is $p = \frac{18}{38} = \frac{9}{19}$, and so

$$E = wp - c = 2(\tfrac{9}{19}) - 1 = -\tfrac{1}{19}.$$

Thus, in the long run, you can expect to lose an average of about one-nineteenth of a dollar (just over a nickel) every time you play this game. ■

A gambling game is said to be **fair** if its expected value is zero, that is, if in the long run you can expect neither to win nor to lose in playing it. Using the notation established above, $E = wp - c$, so the game is fair if and only if

$$c = wp.$$

Example 7 (C)

In a certain state lottery, a prize of \$1,000,000 is offered for guessing which of six distinct numbers will be drawn at random* from among the numbers 1 through 36. The state charges one dollar to play this game. What would be the fair price to pay for playing the game?

Solution

Here, $w = \$1{,}000{,}000$ and $p = 1/{}_{36}C_6$, so if the game were fair (which it isn't), the cost to play would be

$$c = wp = \frac{1{,}000{,}000}{{}_{36}C_6}.$$

Using a calculator, we find that

$$c \approx \$0.51.$$ ■

Gamblers usually measure the likelihood of winning a game in terms of *odds*, rather than probability. By definition, to say that the **odds** on a certain event are a to b means that the probability of the event is $a/(a + b)$. Conversely, if the probability of the event, expressed as a fraction, is a/s, then the odds on the event are a to $s - a$.

Example 8

If you bet on black in an American game of roulette, what are your odds on winning?

Solution

In Example 6, we found that the probability of winning is $p = \frac{9}{19}$, so the odds on winning are 9 to $19 - 9$, that is, 9 to 10. This means that in the long run, out of every 19 times you play this game, you can expect to win 9 times and lose 10 times. ■

* Things are said to be drawn, or chosen, **at random** if any one of the things has the same probability of being chosen as any other.

Problem Set 7.6

In Problems 1 to 14, find the probability of the event.

1. Obtaining heads when tossing a coin.
2. Drawing a king from a deck of 52 well-shuffled cards.
3. Drawing a green ball, blindfolded, from a hat containing six red balls and eight green balls all of the same size.
4. Failing to draw one of the kings from a deck of 52 well-shuffled cards.
5. Rolling a 5 with a single die.
6. Rolling less than a 3 with a single die.

C 7. Being dealt four of a kind (four cards of the same denomination) in a game of five-card poker.

C 8. Being dealt a flush (all cards of the same suit) in a game of five-card poker.

C 9. Being dealt a royal flush (10, jack, queen, king, and ace, all of the same suit) in a game of five-card poker.

C 10. Being dealt a straight (five cards whose denominations form a sequence such as 7, 8, 9, 10, jack) in a game of five-card poker. (The ace can count either as a 1 or as the denomination just above the king.)

11. Rolling a 6 with a pair of dice.
12. Rolling a 2 (snake eyes) with a pair of dice.
13. Rolling at least one 6 in two rolls of a single die.
14. Two people, chosen at random, having birthdays on the same day of the year. (Disregard leap year complications.)
15. A committee of 5 people is to be chosen at random from among 10 men and 3 women. What is the probability that there will be 3 men and 2 women on the committee?
16. A committee of seven people is to be chosen at random from among five skilled workers, three unskilled workers, and four supervisory personnel. What is the probability that all four supervisory personnel and no unskilled workers will be on the committee?
17. Two coins are tossed, and you are told that at least one coin fell heads. What is the probability that the other coin fell heads? (Be careful—consider the possible outcomes in the face of the given information and how many of these are favorable to the event in question.)
18. Four Eastern states have daily lotteries in which numbers from 0 to 999 are drawn. What is the probability that two (or more) of these states will draw the same lucky number on the same day? [*Hint:* To count the number of outcomes favorable to the event, begin by counting the number of outcomes *unfavorable* to the event.]
19. Cards are drawn one at a time from a well-shuffled deck. They are not replaced after being drawn. What is the probability that the last ace in the deck will be drawn before the last king in the deck?
20. Two cards are drawn from a well-shuffled deck. The first card is *not* replaced before the second card is drawn. If the second card is a face card, what is the probability that the first card was a face card?
21. The probability p of "making the point 6" in a game of craps—that is, rolling a 6 before rolling a 7 in a sequence of rolls with a pair of dice—is the sum of the infinite geometric series $\frac{5}{36} + \frac{5}{36}(\frac{25}{36}) + \frac{5}{36}(\frac{25}{36})^2 + \cdots$.
    - **(a)** Find p.
    - **(b)** Find the odds on making the point 6.

C 22. The probability of winning a "pass line bet" in a game of casino craps is $p = 0.49\overline{29}$. At a "$10 table," it costs $10 to make this bet, and if you win, you get your $10 back plus another $10. Of course, if you lose, you lose your $10. Find the expectation E for a $10 pass line bet.

C 23. In the European version of roulette, there are 37 possible outcomes, 18 black, 18 red, and 1 green. Find the expected value of a 100-franc bet on black in a European casino.

24. A *wager*, or *bet*, on an event is conducted as follows: You place a dollars in the pot, and your opponent places b dollars in the pot, for a total *stake* of $s = a + b$ dollars. If the event occurs, you get the total stake; otherwise, your opponent gets it. Show that this game is fair if and only if the odds on the event are a to b.

C 25. The state lottery in Example 7 is conducted once a week. If there is no winner in a given week, the prize for the next lottery is increased to $2,000,000. If that prize is not won, the prize for the subsequent week is increased to $3,000,000, and so on. The price of a lottery ticket is always $1, regardless of the amount of the prize. How large would the prize have to be to make the game fair?

26. A box contains three cards. One of the cards is red on both sides, one is black on both sides, and one is red on one side and black on the other. One of the cards is drawn at random and placed on a table. The card is red on its top face. Your opponent remarks, "I've got $5 that says the other side of the card is also red." Assuming that you accept this bet, how much money should you be willing to bet against your opponent's $5?

27. Consider a gambling game in which the probability of winning is p; the payoff, if you win, is w dollars; and it costs you c dollars for each play of the game. If you win one play of the game, your net gain is $(w - c)$ dollars, the amount you win minus the cost to play. If you lose one play of the game, your net gain is $-c$ dollars. (A loss is counted as a negative gain.) Show that the expected value of the game is the probability of winning times the net gain if you win plus the probability of losing times the net gain if you lose.

28. If taxes do not go up next year, your net gain from a certain business venture will be $100,000; but if taxes do go up, your net gain will be only $30,000. If there is a 75% probability that taxes will go up next year, what is the expected value of your business venture? (See Problem 27.)

29. Suppose that $A_1, A_2, A_3, \ldots, A_n$ are events such that, on each play of a certain game, one and only one of the events will occur. Let $p_k = P(A_k)$ for $k = 1, 2, 3, \ldots, n$. For $k = 1, 2, 3, \ldots, n$, let g_k denote your *net* gain (what you win minus what it costs to play) for one play of the game if event A_k occurs. (Note that g_k can be negative.) Suppose that you play this game a very large number of times N.

(a) Explain why $p_1 + p_2 + p_3 + \cdots + p_n = 1$.

(b) Write a formula for the approximate number of times out of the N plays of the game that the event A_k will occur for $k = 1, 2, 3, \ldots, n$.

(c) Using the result of part (b), show that your net gain from playing the N games is approximately $G = g_1p_1N + g_2p_2N + g_3p_3N + \cdots + g_np_nN$.

(d) Conclude from part (c) that your expected average net gain per play of the game is $E = g_1p_1 + g_2p_2 + g_3p_3 + \cdots + g_np_n$.

30. The formula derived in Problem 29 is extended to an infinite sequence of mutually exclusive and exhaustive events $A_1, A_2, A_3, \ldots$, with probabilities $p_k = P(A_k)$ for $k = 1, 2, 3, \ldots$, and with corresponding net gains $g_1, g_2, g_3, \ldots$, by defining $E = \sum_{k=1}^{\infty} g_kp_k$. If a coin is tossed repeatedly until tails appears, the probability that tails appears first on the kth toss is given by $p_k = 1/2^k$ for $k = 1, 2, 3, \ldots$.

(a) Show that $\sum_{k=1}^{\infty} p_k = 1$.

(b) Suppose that a prize of 2^k cents will be awarded if tails appears first on the kth toss of the coin. What would be a fair price to pay for the privilege of playing this game?*

* This is the famous **St. Petersburg paradox.**

7.7 The Ellipse

In the next three sections we study the **conic sections** (or **conics,** for short). These graceful curves were well known to the ancients; however, their study is immensely enhanced by the use of analytic geometry and calculus. They are obtained by sectioning, or cutting, a right circular cone of two nappes with a suitable plane (Figure 1).

Figure 1

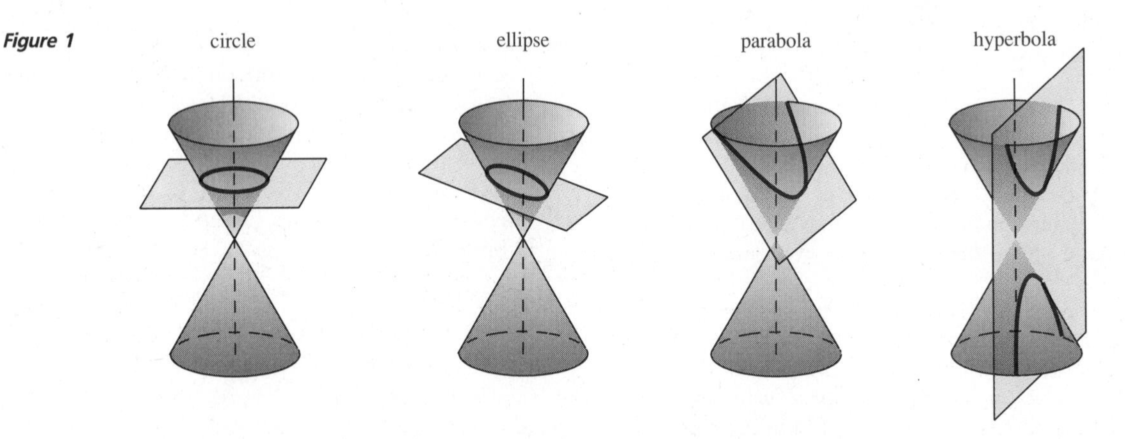

By shining a flashlight onto a white wall, you can see examples of the conic sections. If the axis of the flashlight is perpendicular to the wall, then the illuminated region is *circular* (Figure 2a); if the flashlight is tilted slightly upward, the illuminated region elongates and its boundary becomes an *ellipse* (Figure 2b). As the flashlight is tilted further, the ellipse becomes more and more elongated until it changes into a *parabola* (Figure 2c). Finally, if the flashlight is tilted still further, the edges of the parabola become straighter and it changes into a portion of a *hyperbola* (Figure 2d).

Figure 2

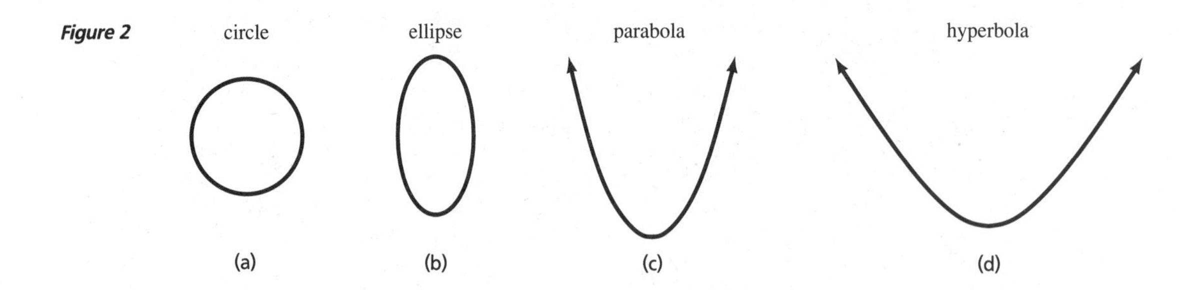

In this section, we begin by obtaining equations for ellipses in Cartesian form. Ellipses are of practical importance in fields ranging from art to astronomy. For instance, a circular object viewed in perspective forms an ellipse, and an orbiting satellite (natural or artificial) moves in an elliptical path. The geometric definition of an ellipse is as follows.

Definition 1 **Ellipse**

An **ellipse** is the set of all points P in the plane such that the sum of the distances from P to two fixed points F_1 and F_2 is constant. Here F_1 and F_2 are called the **focal points,** or the **foci,** of the ellipse. The midpoint C of the line segment $\overline{F_1F_2}$ is called the **center** of the ellipse.

Figure 2a shows two fixed pins F_1 and F_2 and a loop of string of length l stretched tightly about them to the point P. Here we have

$$|\overline{PF_1}| + |\overline{PF_2}| + |\overline{F_1F_2}| = l \qquad \text{or} \qquad |\overline{PF_1}| + |\overline{PF_2}| = l - |\overline{F_1F_2}|.$$

Hence, as P is moved about, $|\overline{PF_1}| + |\overline{PF_2}|$ always has the constant value $l - |\overline{F_1F_2}|$. Thus, if a pencil point P is inserted into the loop of string and moved so as to keep the string tight, it traces out an ellipse (Figure 2b).

Figure 2

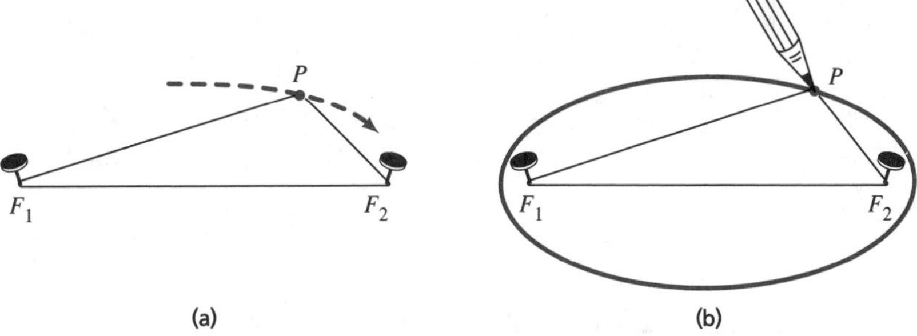

(a) (b)

Figure 3

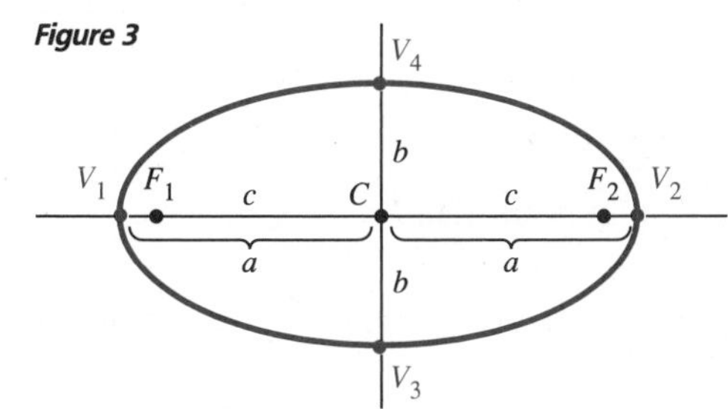

Consider the ellipse in Figure 3 with foci F_1 and F_2 and center C. We denote by c the distance between the center C and either focus F_1 or F_2. Notice that the ellipse is symmetric about the line through F_1 and F_2. Let V_1 and V_2 be the points where this line intersects the ellipse. The center C bisects the line segment $\overline{V_1V_2}$, and the ellipse is symmetric about the line through C perpendicular to $\overline{V_1V_2}$. Let V_3 and V_4 be the points where this perpendicular intersects the ellipse. The four points V_1, V_2, V_3, and V_4 are called the **vertices** of the ellipse. The line segment $\overline{V_1V_2}$ is called the **major axis,** and the line segment $\overline{V_3V_4}$ is called the **minor axis** of the ellipse. Let $2a$ denote the length of the major axis, and let $2b$ denote the length of the minor axis (Figure 3). The numbers a and b are called the **semimajor axis** and the **semiminor axis** of the ellipse.

Figure 4

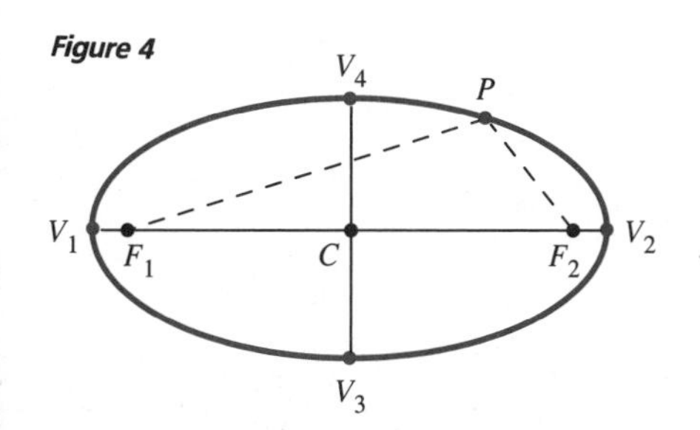

If a point P moves along the ellipse in Figure 4, then, by definition, the sum

$$|\overline{PF_1}| + |\overline{PF_2}|$$

does not change. Therefore, its value when P reaches V_1 is the same as its value when P reaches V_4; that is,

$$|\overline{V_1F_1}| + |\overline{V_1F_2}| = |\overline{V_4F_1}| + |\overline{V_4F_2}|.$$

By symmetry,

$$|\overline{V_4F_1}| = |\overline{V_4F_2}|,$$

so the previous equation can be rewritten

$$|\overline{V_1F_1}| + |\overline{V_1F_2}| = 2|\overline{V_4F_2}|.$$

But, by symmetry again,

$$|\overline{V_1F_2}| = |\overline{V_2F_1}|.$$

Hence, $\quad 2|\overline{V_4F_2}| = |\overline{V_1F_1}| + |\overline{V_1F_2}| = |\overline{V_1F_1}| + |\overline{V_2F_1}| = |\overline{V_1V_2}| = 2a,$

from which it follows that

$$|\overline{V_4F_2}| = a.$$

Figure 5

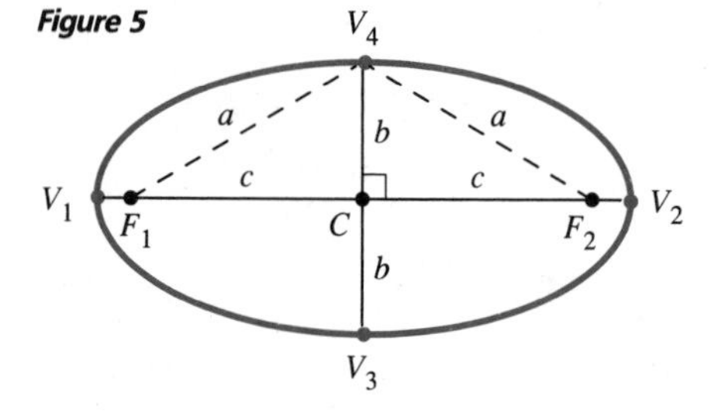

Therefore, applying the Pythagorean theorem to the right triangle V_4CF_2 (Figure 5), we find that

$$a^2 = b^2 + c^2.$$

If we place the ellipse of Figure 5 in the xy plane so that its center C is at the origin O and the foci F_1 and F_2 lie on the negative and positive portions of the x axis, respectively, then we can derive an equation of the ellipse as follows.

Theorem 1 **Ellipse Equation in Cartesian Form**

An equation of the ellipse with foci at $F_1 = (-c, 0)$ and $F_2 = (c, 0)$ is

$$\frac{x^2}{a^2} + \frac{y^2}{b^2} = 1,$$

where a is the semimajor axis, b is the semiminor axis, and $a^2 = b^2 + c^2$.

Proof Let $P = (x, y)$ be any point on the ellipse (Figure 6). As Figure 5 shows, when $P = (0, b)$, we have

$$|\overline{PF_1}| + |\overline{PF_2}| = 2a.$$

Figure 6

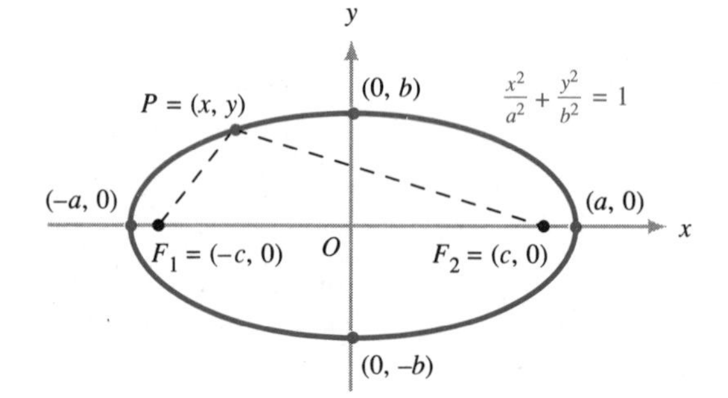

Therefore, by Definition 1, the equation

$$|\overline{PF_1}| + |\overline{PF_2}| = 2a$$

holds for every point $P = (x, y)$ on the ellipse. Using the distance formula, we can rewrite this as

$$\sqrt{(x + c)^2 + y^2} + \sqrt{(x - c)^2 + y^2} = 2a$$

or

$$\sqrt{(x + c)^2 + y^2} = 2a - \sqrt{(x - c)^2 + y^2}.$$

Squaring both sides of the last equation, we have

$$x^2 + 2cx + c^2 + y^2 = 4a^2 - 4a\sqrt{(x - c)^2 + y^2} + x^2 - 2cx + c^2 + y^2,$$

so that

$$4cx - 4a^2 = -4a\sqrt{(x - c)^2 + y^2} \quad \text{or} \quad cx - a^2 = -a\sqrt{(x - c)^2 + y^2}.$$

Squaring both sides of the last equation, we obtain

$$c^2x^2 - 2a^2cx + a^4 = a^2(x^2 - 2cx + c^2 + y^2),$$

so that

$$a^4 - a^2c^2 = (a^2 - c^2)x^2 + a^2y^2 \quad \text{or} \quad a^2(a^2 - c^2) = (a^2 - c^2)x^2 + a^2y^2.$$

Since $a^2 = b^2 + c^2$, we have $a^2 - c^2 = b^2$, and the equation above can be rewritten as

$$a^2b^2 = b^2x^2 + a^2y^2.$$

If both sides of the last equation are divided by a^2b^2, the result is

$$1 = \frac{x^2}{a^2} + \frac{y^2}{b^2}$$

as desired. By reversing the argument above, it can be shown that, conversely, if the equation $(x^2/a^2) + (y^2/b^2) = 1$ holds, then the point $P = (x, y)$ is on the ellipse (Problem 22). ■

Figure 7

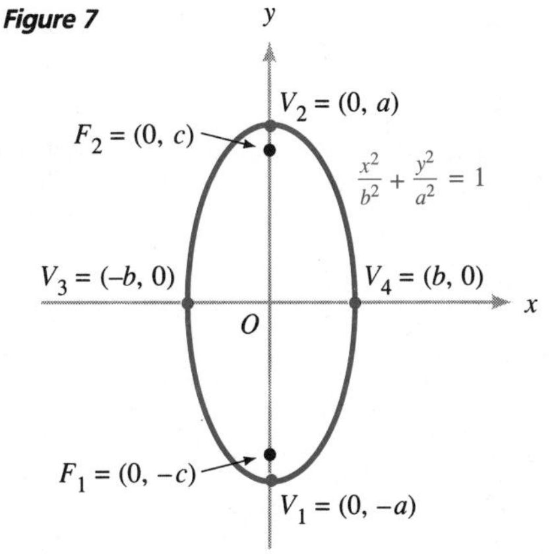

If a and b are positive constants and $a > b$, the Cartesian equation

$$\frac{x^2}{a^2} + \frac{y^2}{b^2} = 1$$

is called the **standard form** for the equation of an ellipse with *center at the origin O* and with *a horizontal major axis.*

It is not difficult to derive an equation of an ellipse with center at the origin and with a vertical major axis (Figure 7). In this case, the ellipse has foci $F_1 = (0, -c)$ and $F_2 = (0, c)$ on the y axis and vertices $V_1 = (0, -a)$, $V_2 = (0, a)$, $V_3 = (-b, 0)$, and $V_4 = (b, 0)$. The semimajor axis is a, and the semiminor axis is b. The equation can be derived as in Theorem 1, the argument being word for word the same except that the variables x and y interchange their roles. Therefore, the equation is

$$\frac{x^2}{b^2} + \frac{y^2}{a^2} = 1, \qquad \text{where } a > b.$$

This equation is also called the **standard form** for the equation of an ellipse with *center at the origin O* and with a *vertical major axis.*

The results obtained above are summarized in Table 1. In this table, we assume that $a > b$ and $c^2 = a^2 - b^2$

Table 1 Ellipses

Major Axis	Center	Foci	Vertices		Standard Form Equation $a > b$ and $c^2 = a^2 - b^2$
			On Major Axis	On Minor Axis	
Horizontal (on the x axis)	(0, 0)	$(-c, 0), (c, 0)$	$(-a, 0), (a, 0)$	$(0, -b), (0, b)$	$\frac{x^2}{a^2} + \frac{y^2}{b^2} = 1$
Vertical (on the y axis)	(0, 0)	$(0, -c), (0, c)$	$(0, -a), (0, a)$	$(-b, 0), (b, 0)$	$\frac{x^2}{b^2} + \frac{y^2}{a^2} = 1$

Example 1 Find the coordinates of the four vertices and the two foci of each ellipse and sketch the graph.

(a) $4x^2 + 9y^2 = 36$ **(b)** $4x^2 + y^2 = 4$

Solution **(a)** We divide both sides of the equation by 36 to obtain

$$\frac{x^2}{9} + \frac{y^2}{4} = 1;$$

that is,

$$\frac{x^2}{a^2} + \frac{y^2}{b^2} = 1,$$

with $a = 3$ and $b = 2$. By Table 1, this is the equation of an ellipse, centered at the origin, with major axis on the x axis, and with vertices $(-3, 0)$, $(3, 0)$, $(0, -2)$, and $(0, 2)$ (Figure 8a). Also, the foci are $(-c, 0)$ and $(c, 0)$, where

$$c^2 = a^2 - b^2 = 9 - 4 = 5;$$

that is, $c = \sqrt{5}$. Thus, $F_1 = (-\sqrt{5}, 0)$ and $F_2 = (\sqrt{5}, 0)$.

(b) We divide both sides of the equation by 4 to obtain

$$\frac{x^2}{1} + \frac{y^2}{4} = 1; \qquad \text{that is,} \qquad \frac{x^2}{b^2} + \frac{y^2}{a^2} = 1$$

Figure 8

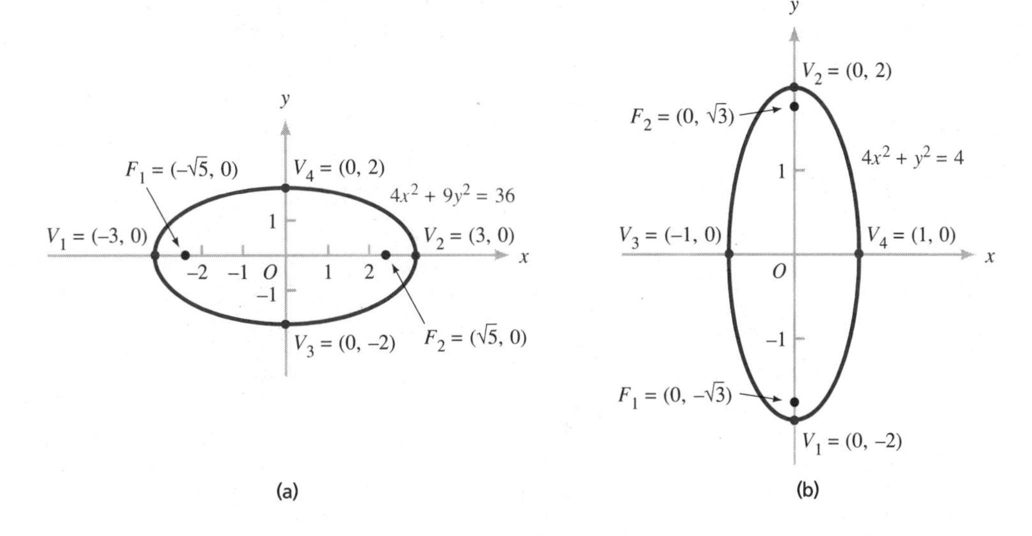

with $b = 1$ and $a = 2$. By Table 1, this is the equation of an ellipse, centered at the origin, with major axis on the y axis, and with vertices $(0, -2)$, $(0, 2)$, $(-1, 0)$, and $(1, 0)$ (Figure 8b). Also, the foci are $(0, -c)$ and $(0, c)$, where

$$c^2 = a^2 - b^2 = 4 - 1 = 3;$$

that is, $c = \sqrt{3}$. Thus, $F_1 = (0, -\sqrt{3})$ and $F_2 = (0, \sqrt{3})$. ■

Example 2 Find the equation in standard form of the ellipse with foci $F_1 = (-\sqrt{3}, 0)$ and $F_2 = (\sqrt{3}, 0)$ and vertices $V_1 = (-2, 0)$ and $V_2 = (2, 0)$. Also, find the coordinates of the remaining two vertices, V_3 and V_4, and sketch the graph.

Solution Here $c = \sqrt{3}$, $a = 2$; hence, $b = \sqrt{a^2 - c^2} = \sqrt{4 - 3} = 1$, and the equation is

$$\frac{x^2}{4} + \frac{y^2}{1} = 1.$$

Also, $V_3 = (0, -1)$ and $V_4 = (0, 1)$. The graph appears in Figure 9. ■

Figure 9

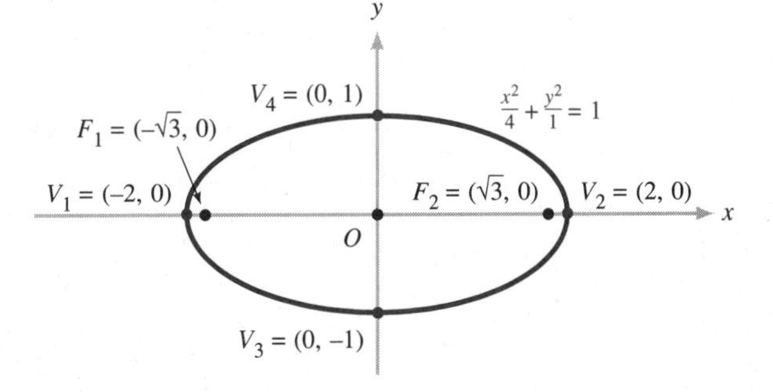

By reasoning as in Section 3.5, you can see that if an ellipse with center at the origin, semimajor axis a, and semiminor axis b is *shifted* so that its center is at the point $C = (h, k)$, then the equation of the shifted ellipse will have one of the following **standard forms.**

Standard Forms for Ellipses Centered at (h, k)

(i) $\dfrac{(x-h)^2}{a^2} + \dfrac{(y-k)^2}{b^2} = 1$, if $a > b$, so that the major axis is horizontal (Figure 10a).

(ii) $\dfrac{(x-h)^2}{b^2} + \dfrac{(y-k)^2}{a^2} = 1$, if $b < a$, so that the major axis is vertical (Figure 10b).

Figure 10

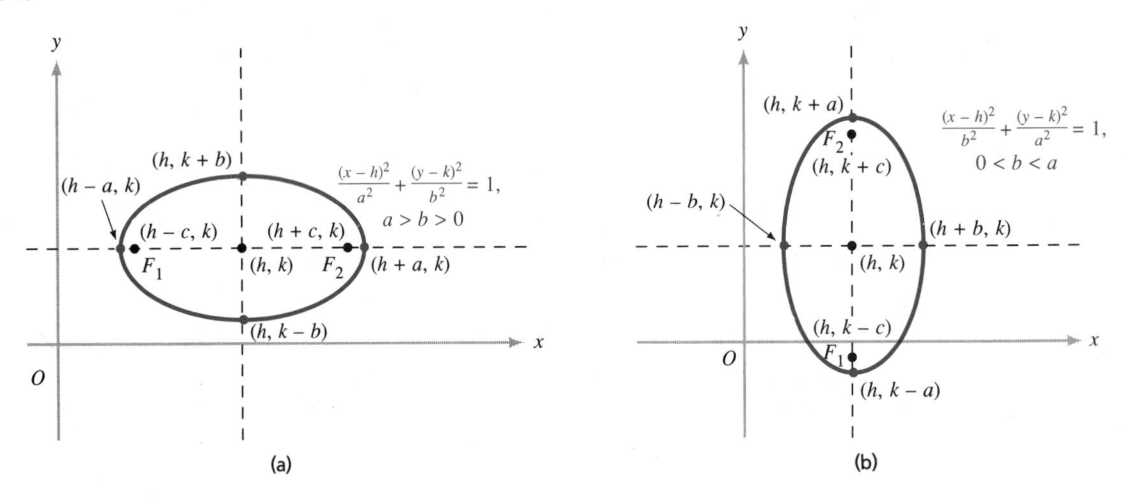

Example 3 Show that the graph of $25x^2 + 9y^2 - 100x - 54y = 44$ is an ellipse. Find the coordinates of the center, the vertices, and the foci and sketch the graph.

Solution Here we have

$$25(x^2 - 4x \qquad) + 9(y^2 - 6y \qquad) = 44.$$

Completing the squares in the last equation, we obtain

$$25(x^2 - 4x + 4) + 9(y^2 - 6y + 9) = 44 + 25(4) + 9(9)$$

or

$$25(x - 2)^2 + 9(y - 3)^2 = 225.$$

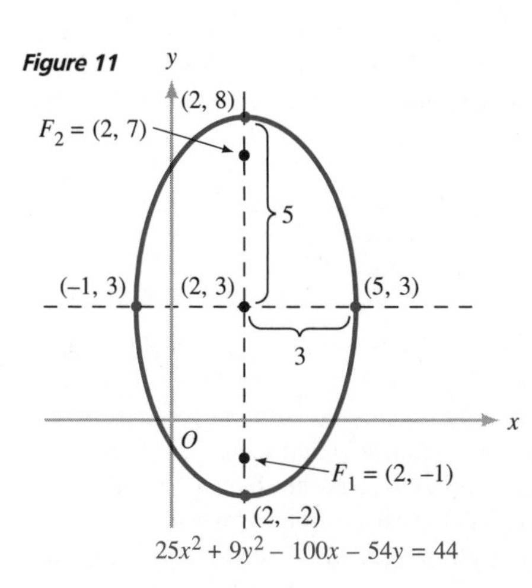

Figure 11 $25x^2 + 9y^2 - 100x - 54y = 44$

Dividing the last equation by 225, we have

$$\frac{(x-2)^2}{9} + \frac{(y-3)^2}{25} = 1;$$

that is,

$$\frac{(x-h)^2}{b^2} + \frac{(y-k)^2}{a^2} = 1,$$

with $h = 2$, $b = 3$, $k = 3$, and $a = 5$. This is the standard form for an ellipse with a vertical major axis with center $C = (h, k) = (2, 3)$. The distance c between the center and the foci is given by

$$c = \sqrt{a^2 - b^2} = \sqrt{25 - 9} = 4 \text{ units.}$$

Thus, the foci F_1 and F_2 have xy coordinates

$$F_1 = (2, 3 - 4) = (2, -1) \quad \text{and} \quad F_2 = (2, 3 + 4) = (2, 7),$$

and the xy coordinates of the four vertices are

$$(2, -2), \quad (2, 8), \quad (-1, 3), \quad \text{and} \quad (5, 3).$$

The graph is sketched in Figure 11. ■

Elliptically shaped surfaces have the interesting property that light or sound produced at one focus will be reflected to the other focus. This **reflecting property** of the ellipse is useful in the design of optical apparatus, and it accounts for the strange **"whispering gallery"** effect sometimes noticed in buildings with elliptical ceilings. In such a building, words whispered softly at one focus can be heard clearly at the other focus. Among the more famous whispering galleries are the Mormon Tabernacle in Salt Lake City, St. Paul's Cathedral in London, and the National Statuary Hall of the Capitol in Washington, D.C.

Problem Set 7.7

In each problem set, problems with colored numbers constitute a good representation of the main ideas of the section. Note that some of the even-numbered problems are considerably more challenging than the odd-numbered ones.

In Problems 1 to 8, find the coordinates of the vertices and foci of each ellipse and sketch its graph.

1. $\frac{x^2}{16} + \frac{y^2}{4} = 1$

2. $\frac{x^2}{9} + y^2 = 1$

3. $4x^2 + y^2 = 16$

4. $36x^2 + 9y^2 = 144$

5. $x^2 + 16y^2 = 16$

6. $16x^2 + 25y^2 = 400$

7. $9x^2 + 36y^2 = 4$

8. $x^2 + 4y^2 = 1$

In Problems 9 to 12, find the equation in standard form of the ellipse that satisfies the conditions given.

9. Foci $F_1 = (-4, 0)$ and $F_2 = (4, 0)$; vertices $V_3 = (0, -3)$ and $V_4 = (0, 3)$

10. Vertices $V_1 = (-5, 0)$ and $V_2 = (5, 0)$; horizontal major axis; $c = 3$ units

11. Foci $F_1 = (0, -12)$ and $F_2 = (0, 12)$; vertices $V_1 = (0, -13)$ and $V_2 = (0, 13)$

12. Foci $F_1 = (0, -8)$ and $F_2 = (0, 8)$; semiminor axis $b = 6$ units

In Problems 13 to 20, find the coordinates of the center, the vertices, and the foci of each ellipse and sketch its graph.

13. $\dfrac{(x-1)^2}{9} + \dfrac{(y+2)^2}{4} = 1$

14. $\dfrac{(x+2)^2}{16} + \dfrac{(y-1)^2}{4} = 1$

15. $4(x+3)^2 + y^2 = 36$

16. $25(x+1)^2 + 16(y-2)^2 = 400$

17. $x^2 + 2y^2 + 6x + 7 = 0$

18. $4x^2 + y^2 - 8x + 4y - 8 = 0$

19. $2x^2 + 5y^2 + 20x - 30y + 75 = 0$

20. $9x^2 + 4y^2 + 18x - 16y - 11 = 0$

21. The segment cut by an ellipse from a line containing a focus and perpendicular to the major axis is called a **focal chord** of the ellipse (Figure 12).

(a) Show that $2b^2/a$ is the length of a focal chord of the ellipse $(x^2/a^2) + (y^2/b^2) = 1$.

(b) Find the length of a focal chord of the ellipse $9x^2 + 16y^2 = 144$.

Figure 12

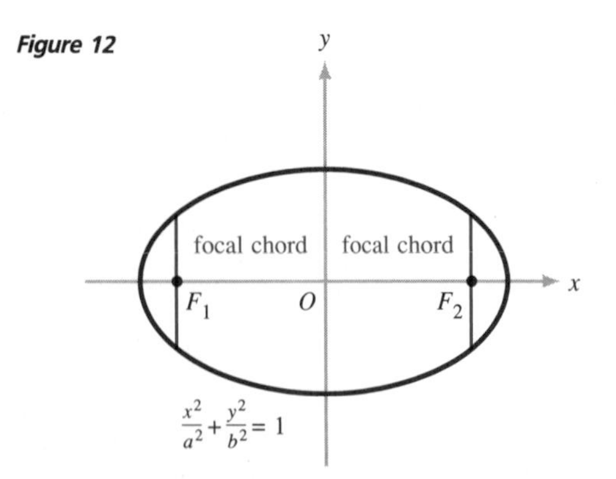

22. Suppose that $a > b > 0$, where a and b are constants and let $c = \sqrt{a^2 - b^2}$. Assume that the numbers x and y satisfy the equation $(x^2/a^2) + (y^2/b^2) = 1$ and let $P = (x, y)$, $F_1 = (-c, 0)$, and $F_2 = (c, 0)$. Prove the following without reference to geometric diagrams:

(a) $c|x| < a^2$ **(b)** $\sqrt{(x-c)^2 + y^2} < 2a$

(c) $|\overline{PF_1}| + |\overline{PF_2}| = 2a$

23. How long a loop of rope should be used to lay out an elliptical flower bed 7 meters wide and 20 meters long? How far apart should the two stakes (foci) be? (See Figure 2.)

24. **(a)** In Figure 2b, show that the semiminor axis b of the ellipse is given by

$$b = \tfrac{1}{2}\sqrt{l^2 - 4lc}, \qquad \text{where } c = \tfrac{1}{2}|\overline{F_1F_2}|.$$

(b) Find a formula for the semimajor axis of the ellipse in Figure 2b in terms of l and $c = \frac{1}{2}|\overline{F_1F_2}|$.

25. The Mormon Tabernacle in Salt Lake City, Utah, built between 1863 and 1867, has the shape of a whispering gallery. A vertical cross section forms the upper half of an ellipse with a horizontal major axis 76 meters long and a semiminor axis (measured vertically from floor to ceiling) of 24 meters. Find the distance between the two foci of this ellipse.

Mormon Tabernacle in Salt Lake City, Utah

26. An arch in the shape of the upper half of an ellipse with a horizontal major axis is to support a bridge over a river 100 meters wide (Figure 13). The center of the arch is to be 25 meters above the surface of the river. Find the equation in standard form for the ellipse.

Figure 13

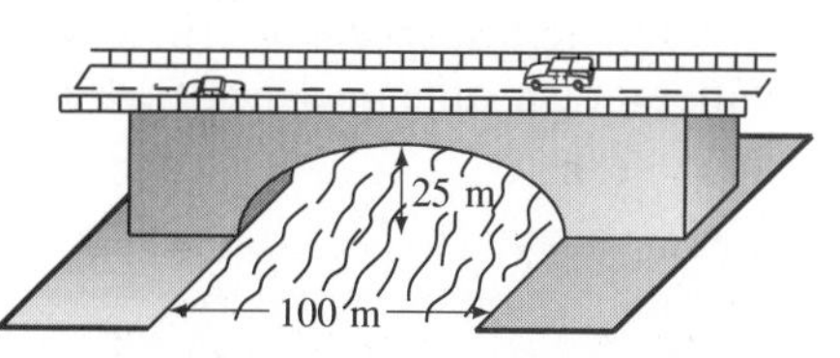

27. The Ellipse in Washington, D.C., is a park located between the White House and the Washington Monument. It is bounded by an elliptical path with a semimajor axis of 229 meters and a semiminor axis of 195 meters. What is the distance between the foci of this ellipse?

The Ellipse, Washington, D.C.

28. Except for minor perturbations, a satellite orbiting the earth moves in an ellipse with the center of the earth at one focus. Suppose that a satellite at perigee (nearest point to center of earth) is 400 kilometers from the surface of the earth and at apogee (farthest point to center of earth) is 600 kilometers from the surface of the earth. Assume that the earth is a sphere of radius 6371 kilometers. Find the semiminor axis b of the elliptical orbit.

29. The earth moves in an elliptical orbit with the sun at one focus. In early July, the earth is farthest from the sun (aphelion), a distance of 94,448,000 miles; in early January, it is closest to the sun (perihelion), a distance of 91,341,000 miles. Find the distance between the sun and the other focus of the earth's orbit.

30. A mathematician has accepted a position at a new university situated 6 kilometers from the straight shoreline of a large lake (Figure 14). The professor wishes to build a home that is half as far from the university as it is from the shore of the lake. The possible homesites satisfying this condition lie along a curve. Describe this curve, and find its equation with respect to a coordinate system having the shoreline as the x axis and the university at the point (0, 6) on the y axis.

Figure 14

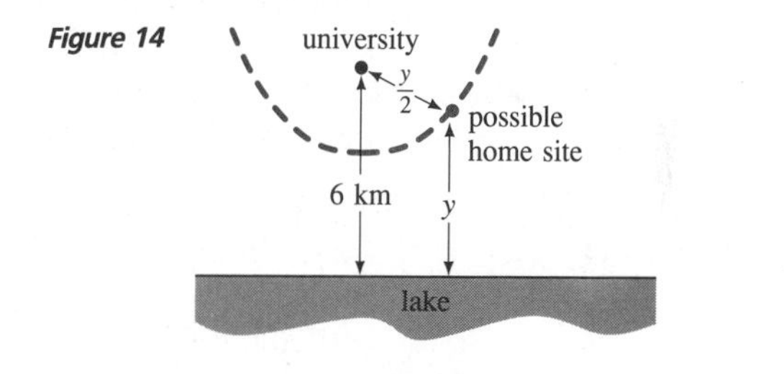

7.8 The Parabola

In Section 4.1, we mentioned that the graphs of equations of the form $y = ax^2$ are "examples of curves called *parabolas*." In this section we give a proof of this fact based on the following geometric definition.

Definition 1 **Parabola**

A **parabola** is the set of all points P in the plane such that the distance from P to a fixed point F (the **focus**) is equal to the distance from P to a fixed line D (the **directrix**).

Parabolas appear often in the real world. A ball thrown up at an angle travels along a parabolic arc (Figure 1a), a main cable in a suspension bridge forms an arc of a parabola (Figure 1b), and the familiar dish antennas have parabolic cross sections (Figure 1c).

Figure 1

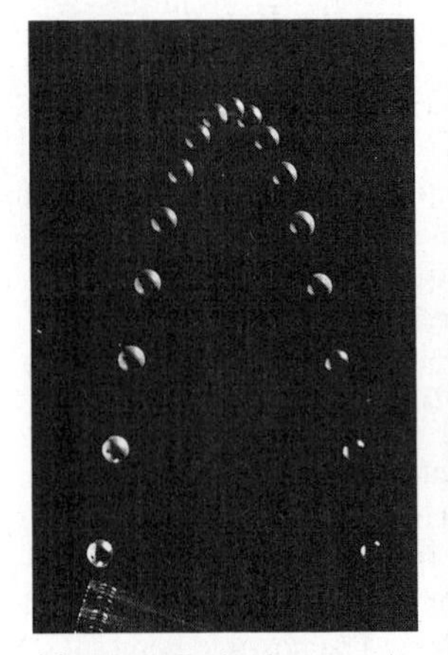

(a) *Multiflash photo showing the parabolic path of a ball thrown into the air*

(b) *Bay Bridge, San Francisco, showing parabolic main cables*

(c) *Parabolic dish antenna*

If the focus F of a parabola is placed on the positive y axis at the point $(0, p)$ and if the directrix D is placed parallel to the x axis and p units below it, the resulting parabola appears as in Figure 2. Its Cartesian equation is derived in the following theorem.

Theorem 1 **Parabola Equation in Cartesian Form**

An equation of the parabola with focus $F = (0, p)$ and with directrix $y = -p$ is $x^2 = 4py$; that is, $y = \frac{1}{4p}x^2$.

Figure 2

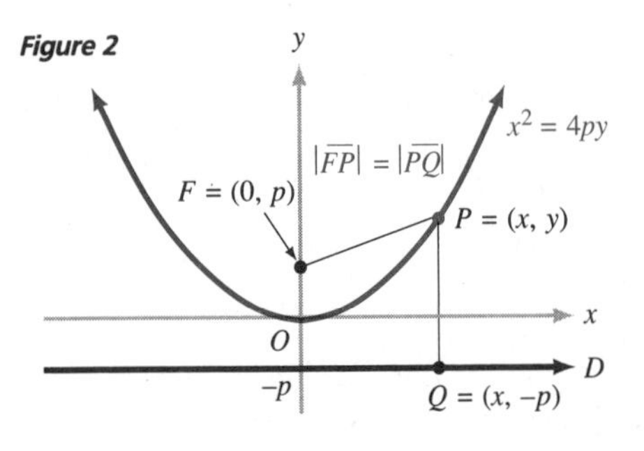

Proof Let $P = (x, y)$ be any point and let $Q = (x, -p)$ be the point at the foot of the perpendicular from P to the directrix D (Figure 2). The requirement for P to be on the parabola is $|\overline{FP}| = |\overline{PQ}|$; that is,

$$\sqrt{x^2 + (y - p)^2} = \sqrt{(y + p)^2}.$$

The last equation is equivalent to

$$x^2 + (y - p)^2 = (y + p)^2;$$

that is,

$$x^2 + y^2 - 2py + p^2 = y^2 + 2py + p^2 \qquad \text{or} \qquad x^2 = 4py.$$

■

Obvious modifications of the argument in Theorem 1 provide Cartesian equations for parabolas opening to the *right* (Figure 3a), to the *left* (Figure 3b), and *downward* (Figure 3c). These cases are summarized in Table 1. We assume that $p > 0$ for each situation.

Figure 3

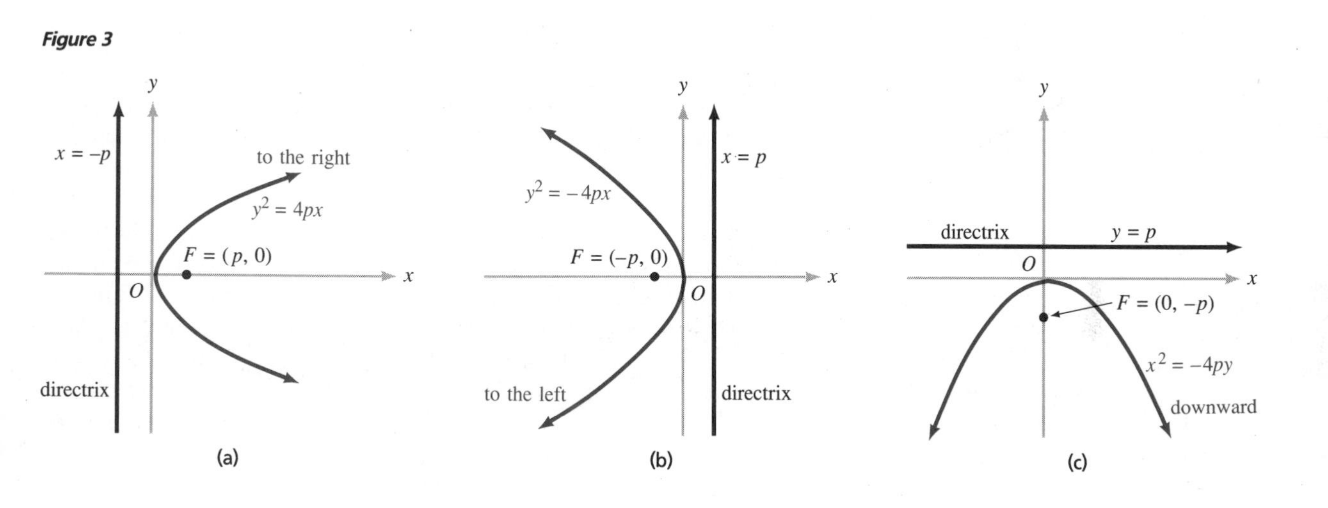

Table 1 **Standard Forms for Parabolas with Vertex at the Origin**

Parabola Opens	Axis of Symmetry	Vertex	Focus	Directrix	Standard Form Equation
Right (Figure 3a)	x axis	(0, 0)	$(p, 0)$	$x = -p$	$y^2 = 4px$
Left (Figure 3b)	x axis	(0, 0)	$(-p, 0)$	$x = p$	$y^2 = -4px$
Upward (Figure 2)	y axis	(0, 0)	$(0, p)$	$y = -p$	$x^2 = 4py$
Downward (Figure 3c)	y axis	(0, 0)	$(0, -p)$	$y = p$	$x^2 = -4py$

Figure 4

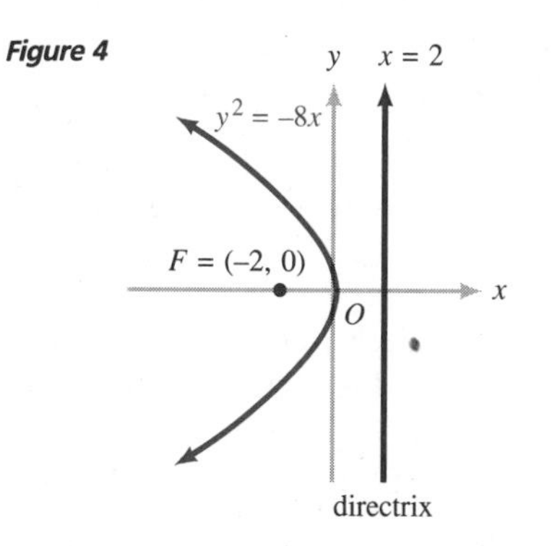

Example 1 Find the coordinates of the focus and an equation of the directrix of the parabola $y^2 = -8x$. Also determine its direction of opening and sketch the graph.

Solution The equation has the form $y^2 = -4px$ with $p = 2$; hence, it corresponds to Figure 3b. Therefore, the graph is a parabola opening to the left with focus given by

$$F = (-p, 0) = (-2, 0)$$

and directrix $x = p$; that is, $x = 2$ (Figure 4). ■

Figure 5

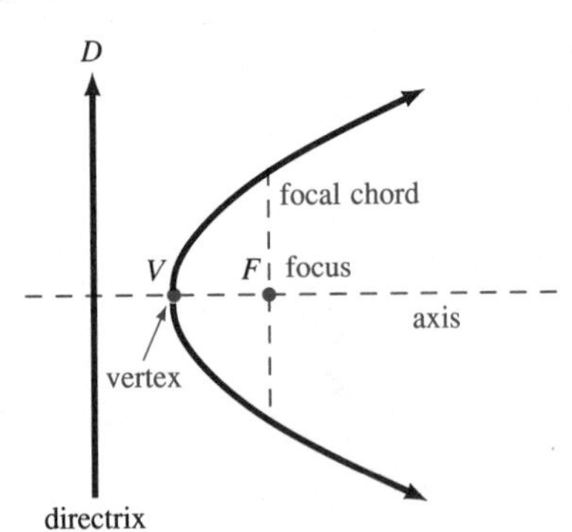

A parabola with focus F and directrix D is evidently symmetric about the line through F perpendicular to D. This line is called the **axis** of the parabola (Figure 5). The axis intersects the parabola at the **vertex** V, which is located midway between the focus and the directrix. The segment cut by the parabola on the line through the focus and perpendicular to its axis is called the **focal chord** of the parabola (Figure 5). The following example shows how to find the length of the focal chord.

Example 2 A parabola opens to the right, has its vertex at the origin, and contains the point (3, 6). Find its equation, sketch the parabola, and find the length of its focal chord.

Figure 6

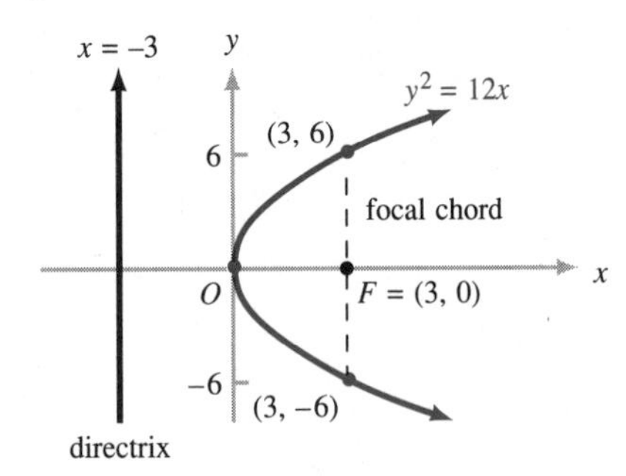

Solution The equation must have the form $y^2 = 4px$. Since the point (3, 6) belongs to the graph, we put $x = 3$ and $y = 6$ in the equation to obtain $36 = 12p$, and we conclude that $p = 3$. Thus, an equation of the parabola is $y^2 = 12x$ (Figure 6). The focus is given by $F = (p, 0) = (3, 0)$. The focal chord lies along the line $x = 3$. Putting $x = 3$ in the equation $y^2 = 12x$ and solving for y, we obtain $y^2 = 36$, $y = \pm 6$. Thus, the points (3, 6) and (3, -6) are the endpoints of the focal chord, and therefore its length is 12 units. ■

A parabola that opens upward, to the right, downward, or to the left is said to be **in standard position.** If a parabola is in standard position, then its axis of symmetry is either horizontal or vertical. The parabolas in Table 1 are in standard position with vertices at the *origin.* By shifting these parabolas horizontally and vertically (as in Section 4.1), you can obtain all parabolas in standard position with vertices at a point (h, k). The results are summarized in Table 2. Again we assume that $p > 0$.

Table 2 Standard Forms for Parabolas with Vertex (h, k)

Vertex	Opening	Equation
(h, k)	To the right	$(y - k)^2 = 4p(x - h)$
(h, k)	To the left	$(y - k)^2 = -4p(x - h)$
(h, k)	Upward*	$(x - h)^2 = 4p(y - k)$
(h, k)	Downward*	$(x - h)^2 = -4p(y - k)$

Example 3 Find the coordinates of the vertex V and the focus F of each parabola, determine the direction in which it opens, find an equation for its directrix, determine the length of the focal chord, and sketch the graph.

* Note that a parabola with a vertical axis has an equation of the form $y = a(x - h)^2 + k$, where $a = \pm 1/(4p)$, in conformity with Theorem 1 in Section 4.1.

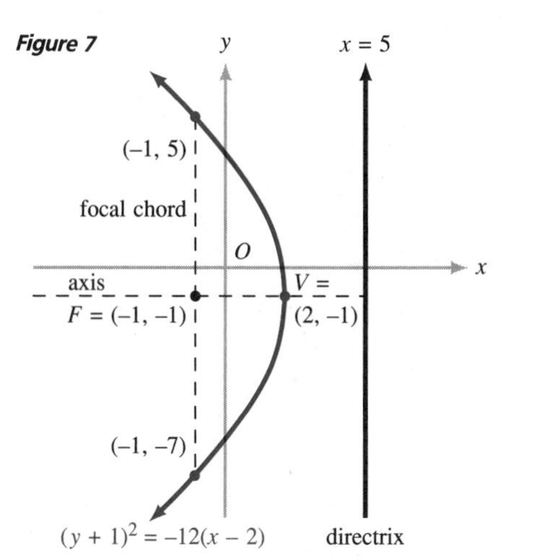

Figure 7

(a) $(y+1)^2 = -12(x-2)$ **(b)** $x^2 + 4x - 10y + 34 = 0$

Solution **(a)** The equation can be written as

$$(y-k)^2 = -4p(x-h)$$

with $p = 3$, $h = 2$, and $k = -1$. By Table 2, the parabola opens to the left, $V = (2, -1)$, $F = (2-3, -1) = (-1, -1)$, and the directrix is $x = 5$. Since the x coordinate of the focus is -1, the focal chord lies along the vertical line $x = -1$. Putting $x = -1$ in the equation of the parabola, we obtain $(y+1)^2 = 36$, so that $y + 1 = \pm 6$; that is, $y = 5$ or $y = -7$. Therefore, the endpoints of the focal chord are $(-1, 5)$ and $(-1, -7)$, and its length is 12 units (Figure 7).

(b) Completing the square, we obtain

$$x^2 + 4x + 4 - 10y + 34 = 4$$

or

$$(x+2)^2 = 10y - 30;$$

that is, $(x+2)^2 = 10(y-3)$. Thus, $p = \frac{10}{4} = \frac{5}{2}$, and the graph is a parabola opening upward with vertex $V = (-2, 3)$, focus $F = (-2, \frac{11}{2})$, and directrix $y = \frac{1}{2}$ (Figure 8). Here, the focal chord lies along the horizontal line $y = \frac{11}{2}$. Putting $y = \frac{11}{2}$ in the equation of the parabola, we obtain $(x+2)^2 = 10(\frac{11}{2}) - 30 = 25$, so $x = 3$ or $x = -7$. Therefore, the endpoints of the focal chord are $(-7, \frac{11}{2})$ and $(3, \frac{11}{2})$, and its length is 10 units. ■

Figure 8

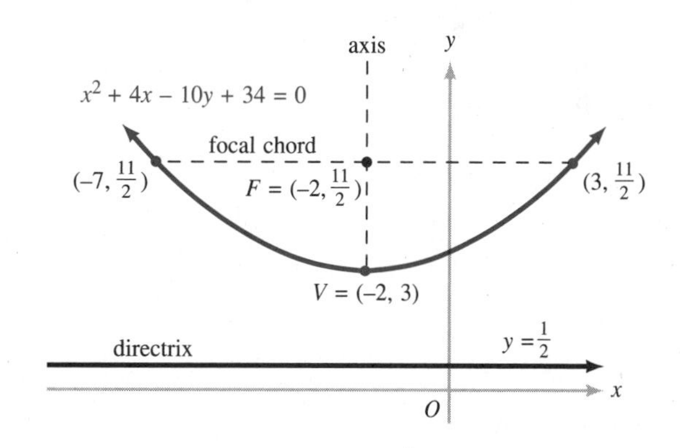

Figure 9

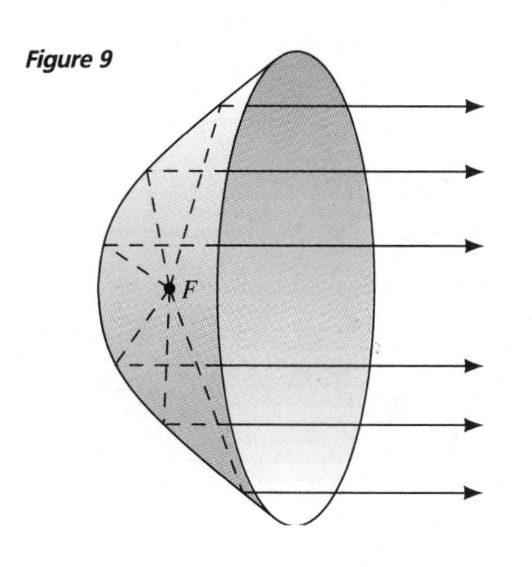

Parabolas have a **reflecting property** analogous to the reflecting property of an ellipse. Sound, light, or electromagnetic radiation emanating from the focus of a parabolic reflector are always reflected parallel to the axis (Figure 9). Thus, if an intense source of light such as a carbon arc or an incandescent filament is placed at the focus of a parabolic mirror, the light is reflected and projected in a parallel beam. The same principle is used in reverse in a reflecting telescope—parallel rays of light from a distant object are brought together at the focus of a parabolic mirror.

Figure 10

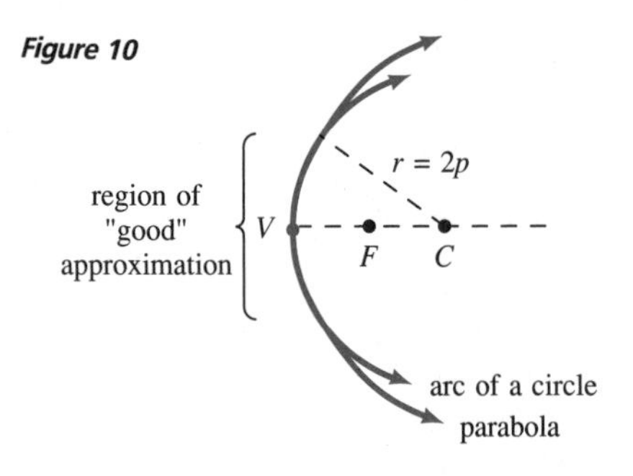

In practice, it is very difficult to manufacture large parabolic mirrors, so it is often necessary to make do with mirrors whose cross section is a portion of a circle approximating the appropriate parabola (Figure 10). It can be shown that the circle that "best approximates" the parabola near its vertex V has its center C located on the axis of the parabola twice as far from the vertex V as the focus F of the parabola and so has radius $r = 2|\overline{VF}| = 2p$.

Problem Set 7.8

In Problems 1 to 6, find the coordinates of the vertex and the focus of the parabola. Also find an equation of the directrix and the length of the focal chord. Sketch the graph.

1. $y^2 = 4x$
2. $y^2 = -9x$
3. $x^2 - y = 0$
4. $x^2 - 4y = 0$
5. $x^2 + 9y = 0$
6. $3x^2 - 4y = 0$

7. Find an equation of the parabola whose focus is the point $(0, 3)$ and whose directrix is the line $y = -3$.

8. Find the vertex of the parabola $y = Ax^2 + Bx + C$, where A, B, and C are constants and $A \neq 0$.

In Problems 9 to 16, find the coordinates of the vertex and the focus of the parabola. Also, find an equation of the directrix and the length of the focal chord. Sketch the graph.

9. $(y - 2)^2 = 8(x + 3)$
10. $(y + 1)^2 = -4(x - 1)$
11. $(x - 4)^2 = 12(y + 7)$
12. $(x + 1)^2 = -8y$
13. $y^2 - 8y - 6x - 2 = 0$
14. $2x^2 + 8x - 3y + 4 = 0$
15. $x^2 - 6x - 8y + 1 = 0$
16. $y^2 + 10y - x + 21 = 0$

In Problems 17 to 22, find an equation of the parabola that satisfies the conditions given.

17. Focus at $(4, 2)$ and directrix $x = 6$.
18. Focus at $(3, -1)$ and directrix $y = 5$.
19. Vertex at $(-6, -5)$ and focus at $(2, -5)$.
20. Vertex at $(2, -3)$ and directrix $x = -8$.
21. Axis is parallel to the x axis, vertex $(-\frac{1}{2}, -1)$, and contains the point $(\frac{5}{8}, 2)$.
22. Axis coincides with the y axis and the parabola contains the points $(2, 3)$ and $(-1, -2)$.
23. The Golden Gate Bridge in San Francisco has a main span of 1280 meters between the two towers that support the parabolic main cables, and these towers extend 160 meters above the road surface. Assuming that a main cable is tangent to the road at the center of the bridge and that the road is horizontal, find the equation of a main cable with respect to an xy coordinate system with a vertical y axis, with the x axis running along the road surface, and with the origin at the center of the bridge.

Golden Gate Bridge, San Francisco

24. Show that, if p is the distance between the vertex and the focus of a parabola, then the focal chord of the parabola has length $4p$.

25. A roadway 400 meters long is held up by a parabolic main cable. The main cable is 100 meters above the roadway at the ends and 4 meters above the roadway at the center (Figure 11). Vertical supporting cables run at 50-meter intervals along the roadway. Find the lengths of these vertical cables. [*Hint:* Set up an xy coordinate system with vertical y axis and having the vertex of the parabola 4 units above the origin.]

Figure 11

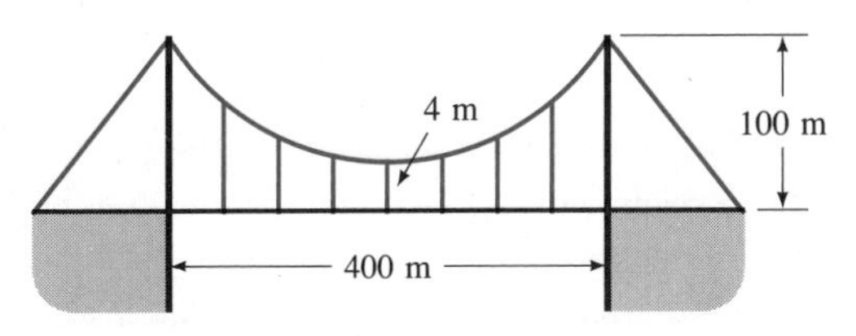

26. Let A, B, and C be constants with $A > 0$. Show that $y = Ax^2 + Bx + C$ is an equation of a parabola with a vertical axis of symmetry, opening upward. Find the coordinates of the vertex V and the focus F. Find p and the length of the focal chord. Find conditions for the graph to intersect the x axis.

27. Figure 12 shows a cross section of a parabolic dish antenna. Show that the focus F of the antenna is p units above the vertex V, where $p = a^2/(16b)$.

Figure 12

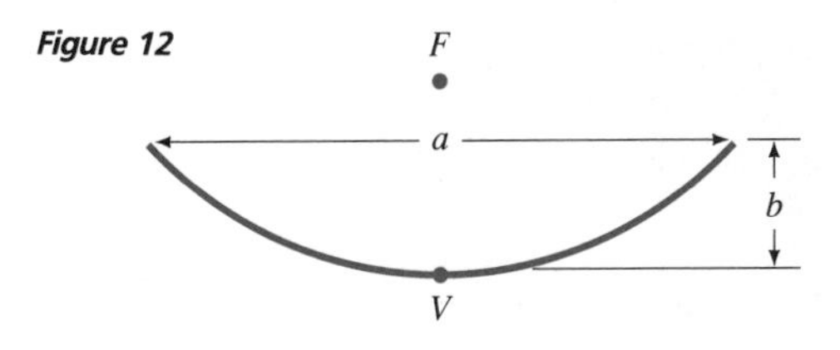

28. The surface of a roadway over a stone bridge follows a parabolic curve with the vertex in the middle of the bridge (Figure 13). The span of the bridge is 60 meters, and the road surface is 1 meter higher in the middle than at the ends. How much higher than the ends is a point on the roadway 15 meters from an end?

Figure 13

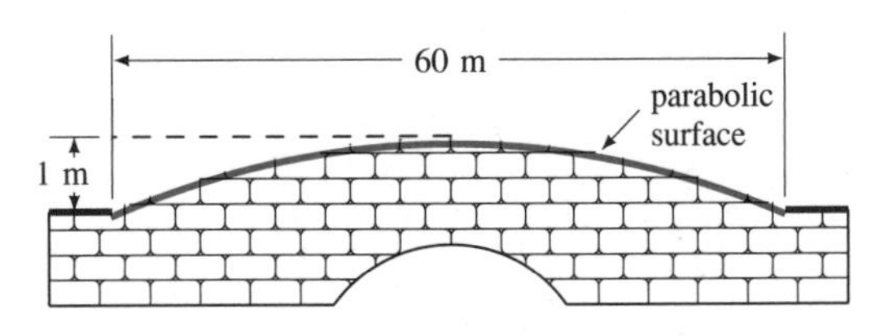

29. Figure 14b shows the parabolic cross section of the reflector of a small spotlight (Figure 14a). The reflector is 16 centimeters across and 6 centimeters deep. The light source is placed at the focus F of the parabola. Find the distance from F to the vertex V of the parabola.

Figure 14

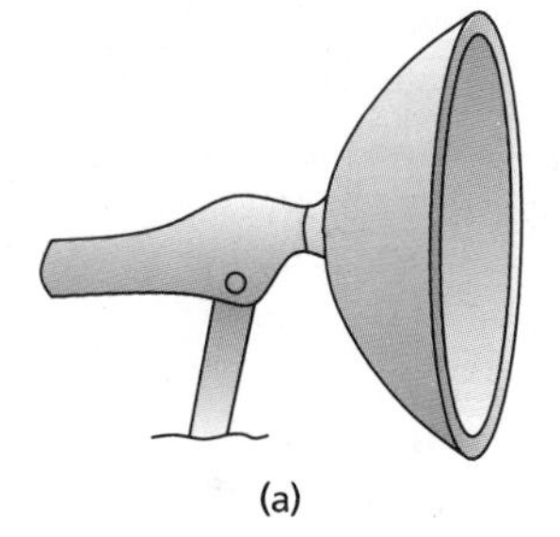

(a)

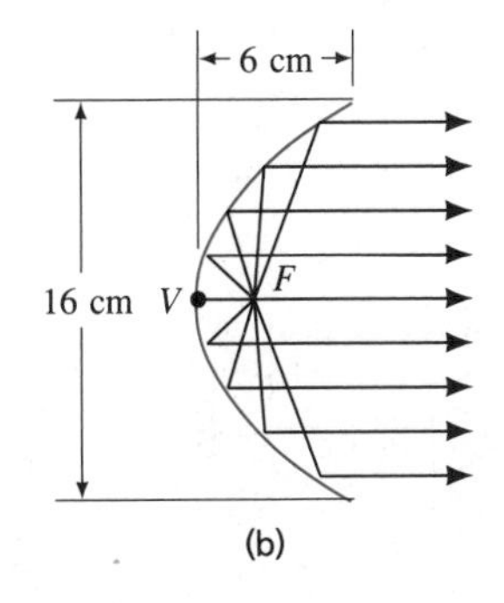

(b)

30. A parabola may be thought of as an enormous ellipse with one vertex infinitely far away. To see this, consider the ellipse in Figure 15 on page 470. Show that if we hold the lower vertex fixed at O and hold the lower focus fixed at $(0, p)$, but allow the upper vertex to approach $+\infty$ along the y axis, then the ellipse approaches the parabola $y = (1/4p)x^2$ as a limiting curve.

Figure 15

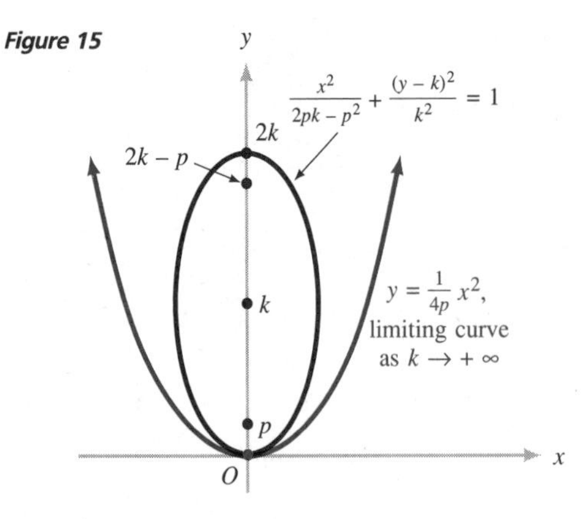

Figure 16

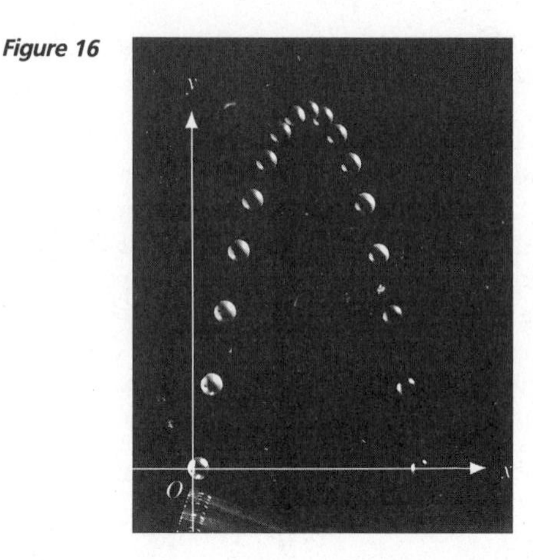

31. The parabolic path of a golf ball is 30 meters wide at the base and 50 meters high at the vertex (Figure 16). Taking the origin at the point of departure of the golf ball, write an equation that represents its path.

32. A student says, "If you've seen one parabola, you've seen them all."

(a) Explain why this is essentially true.

(b) Explain why a similar statement cannot be made for ellipses.

7.9 The Hyperbola

Hyperbolas are of practical importance in fields ranging from engineering to navigation. The natural-draft evaporative cooling towers used at large electric power stations have hyperbolic cross sections (Figure 1a); a comet or other object moving with more than enough kinetic energy to escape the sun's gravitational pull traces out one branch of a hyperbola (Figure 1b); and the long-range radio navigation system known as LORAN locates a ship or plane at the intersection of two hyperbolas (Figure 1c).

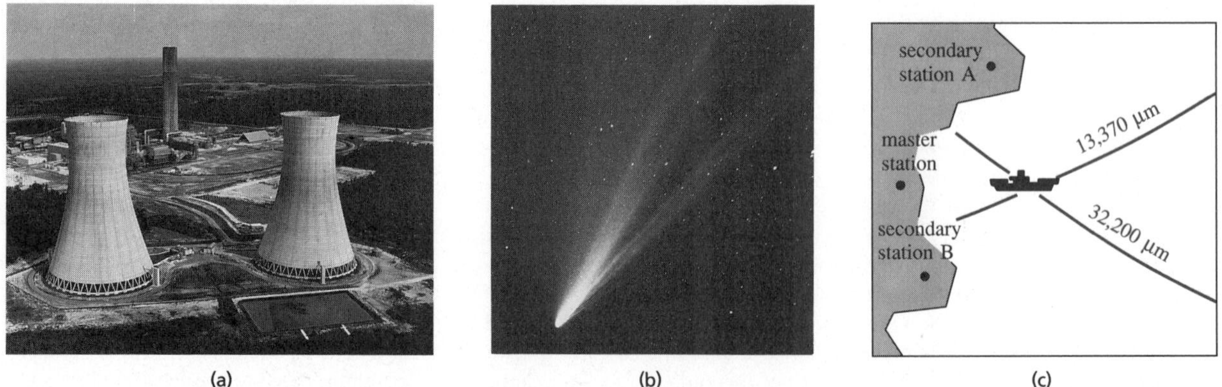

Figure 1
(a) *Hyperbolic cooling towers for an electric power station.* (b) *Comet photographed through a telescope.* (c) *Synchronized pulses transmitted from three stations locate a ship at the intersection of two hyperbolas plotted on a LORAN chart.*

The geometric definition of a hyperbola is as follows.

Definition 1 **Hyperbola**

A **hyperbola** is the set of all points P in the plane such that the absolute value of the difference of the distances from P to two points F_1 and F_2 is a constant positive number. Here F_1 and F_2 are called the **focal points,** or the **foci,** of the hyperbola. The midpoint C of the line segment $\overline{F_1F_2}$ is called the **center** of the hyperbola.

Figure 2 shows a hyperbola with foci F_1 and F_2. Notice that the line through the two foci is an axis of symmetry for the hyperbola, and so is the perpendicular bisector of the line segment $\overline{F_1F_2}$. The two points V_1 and V_2 where the two branches of the hyperbola intersect the line through F_1 and F_2 are called the **vertices,** and the line segment $\overline{V_1V_2}$ is called the **transverse axis** of the hyperbola. The distance from the center C to either focus is denoted by c, and the distance from the center C to either vertex is denoted by a. Thus, the length of the transverse axis is $2a$, and the distance between the two foci is $2c$.

As the point P in Figure 2 moves along the right-hand branch toward V_2, the difference

$$|\overline{PF_1}| - |\overline{PF_2}|$$

maintains a constant value (Definition 1). When P reaches V_2,

$$\begin{aligned}|\overline{PF_1}| - |\overline{PF_2}| &= |\overline{V_2F_1}| - |\overline{V_2F_2}| \\ &= (c + a) - (c - a) = 2a.\end{aligned}$$

Therefore, for any point P on the hyperbola,

$$\left||\overline{PF_1}| - |\overline{PF_2}|\right| = 2a$$

by Definition 1.

If we place the hyperbola in Figure 2 in the xy plane so that its center C is at the origin O and the foci F_1 and F_2 lie on the x axis (Figure 3), we can derive its Cartesian equation as in the following theorem.

Figure 2

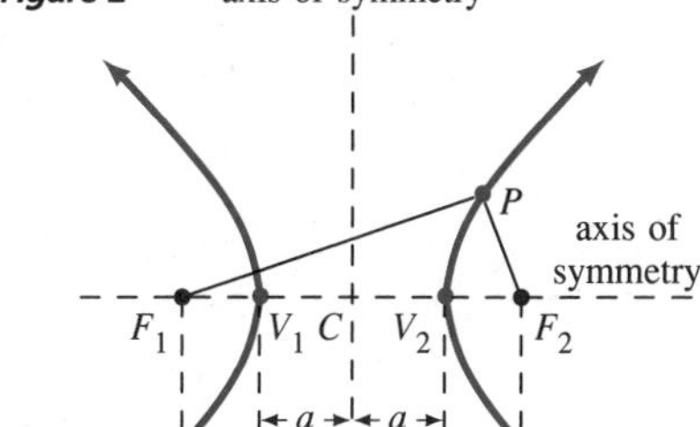

Figure 3

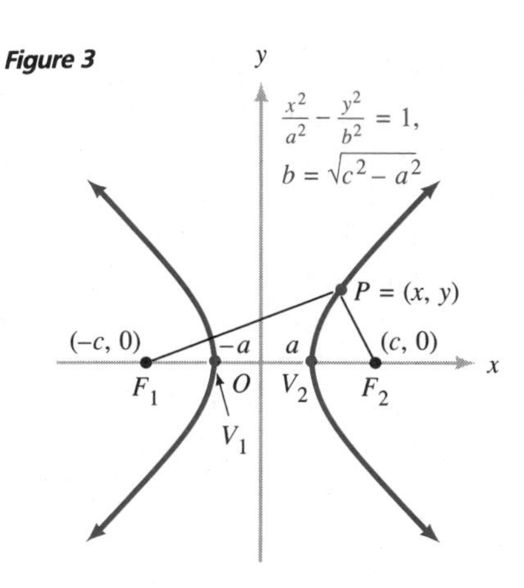

Theorem 1 **Hyperbola Equation in Cartesian Form**

An equation of the hyperbola with foci $F_1 = (-c, 0)$, $F_2 = (c, 0)$ and vertices $V_1 = (-a, 0)$, $V_2 = (a, 0)$ is

$$\frac{x^2}{a^2} - \frac{y^2}{b^2} = 1, \qquad \text{where } b = \sqrt{c^2 - a^2}.$$

The proof of Theorem 1 is quite similar to the proof of Theorem 1 in Section 7.7; hence, it is left as an exercise (Problem 26). The equation

$$\frac{x^2}{a^2} - \frac{y^2}{b^2} = 1$$

is called the **standard form** for the equation of a hyperbola (Figure 3). We solve this equation for y in terms of x as follows:

$$\frac{y^2}{b^2} = \frac{x^2}{a^2} - 1,$$

so

$$y^2 = b^2\left(\frac{x^2}{a^2} - 1\right) = \left(\frac{b^2x^2}{a^2}\right)\left(1 - \frac{a^2}{x^2}\right).$$

Hence,

$$y = \pm\left(\frac{bx}{a}\right)\sqrt{1 - \frac{a^2}{x^2}},$$

provided that $x \neq 0$. Note that, as $|x|$ gets very large, a^2/x^2 will get very small and the expression under the square root will come closer and closer to 1. In other words, for large values of $|x|$, points on the hyperbola come very close to the lines

$$y = \frac{b}{a}x \qquad \text{and} \qquad y = -\frac{b}{a}x.$$

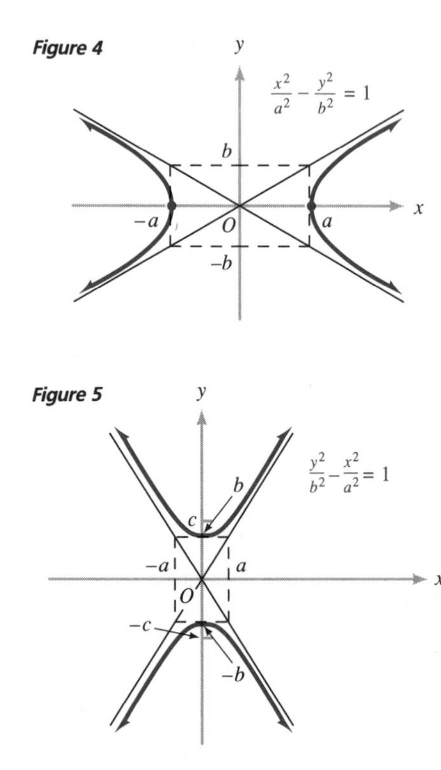

Figure 4

Figure 5

Thus, these lines are **asymptotes** of the hyperbola (see Section 4.7). They are good approximations to the hyperbola itself at large distances from the origin.

Although the asymptotes of the hyperbola are *not* part of the hyperbola itself, they are useful in sketching it. For instance, to sketch the graph of the equation $(x^2/a^2) - (y^2/b^2) = 1$, begin by sketching the rectangle with height $2b$ and horizontal base $2a$ whose center is at the origin (Figure 4). The asymptotes are then drawn through the two diagonals of this rectangle. If you keep in mind that the vertices of the hyperbola are located at the midpoints of the left and right sides of the rectangle, and that the hyperbola approaches the asymptotes as it moves out away from the vertices, then it becomes an easy matter to sketch the graph (Figure 4).

Suppose that we wish to find an equation of the hyperbola in Figure 5, which has a *vertical* transverse axis, center at the origin, vertices $V_1 = (0, -b)$ and $V_2 = (0, b)$, and foci $F_1 = (0, -c)$ and $F_2 = (0, c)$. Using Theorem 1 but interchanging x with y and interchanging a with b, we obtain the equation

$$\frac{y^2}{b^2} - \frac{x^2}{a^2} = 1, \qquad \text{where } a = \sqrt{c^2 - b^2}.$$

This equation is also called the **standard form** for the equation of a hyperbola. The asymptotes are still given by

$$y = \frac{b}{a}x \quad \text{and} \quad y = -\frac{b}{a}x.$$

(Why?)

Unlike the ellipse equation $(x^2/a^2) + (y^2/b^2) = 1$ and the equation $(x^2/b^2) + (y^2/a^2) = 1$, *there is no requirement that* $a > b$ *in the hyperbola equation* $(x^2/a^2) - (y^2/b^2) = 1$ *or in the hyperbola equation* $(y^2/b^2) - (x^2/a^2) = 1$.

Example 1 Find the coordinates of the foci and the vertices and find equations of the asymptotes of each hyperbola. Also, sketch the graph.

(a) $\dfrac{x^2}{4} - \dfrac{y^2}{1} = 1$ **(b)** $\dfrac{y^2}{16} - \dfrac{x^2}{9} = 1$

Solution **(a)** The equation has the form

$$\frac{x^2}{a^2} - \frac{y^2}{b^2} = 1,$$

with $a = 2$, $b = 1$. Hence, $c = \sqrt{a^2 + b^2} = \sqrt{5}$. Thus, the foci are $F_1 = (-\sqrt{5}, 0)$, $F_2 = (\sqrt{5}, 0)$, and the vertices are $V_1 = (-2, 0)$, $V_2 = (2, 0)$. The asymptotes are given by $y = \frac{1}{2}x$ and $y = -\frac{1}{2}x$ (Figure 6a).

(b) The equation has the form

$$\frac{y^2}{b^2} - \frac{x^2}{a^2} = 1,$$

with $b = 4$, $a = 3$. Hence $c = \sqrt{a^2 + b^2} = \sqrt{25} = 5$. Thus, the foci are $F_1 = (0, -5)$, $F_2 = (0, 5)$, and the vertices are $V_1 = (0, -4)$, $V_2 = (0, 4)$. The asymptotes are given by $y = \frac{4}{3}x$ and $y = -\frac{4}{3}x$ (Figure 6b). ■

Figure 6

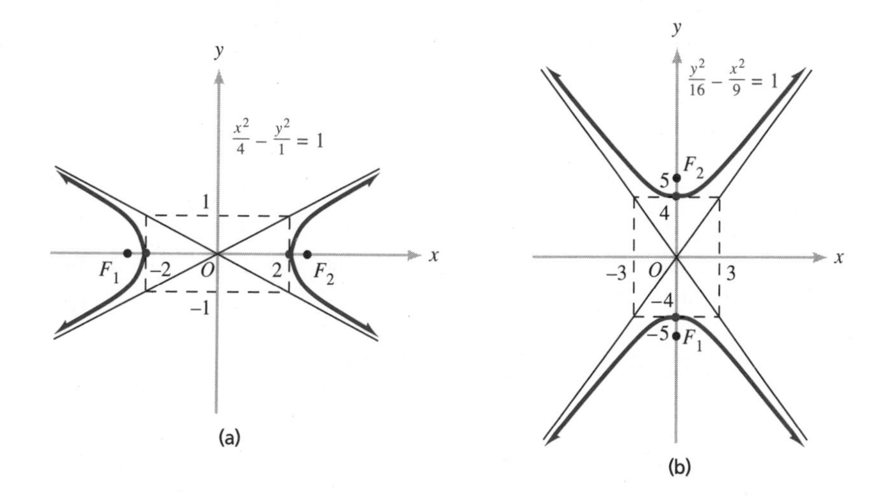

As with the ellipse in Section 7.7, if a hyperbola with either horizontal or vertical transverse axis is shifted so that its center is at the point $C = (h, k)$, then the equation of the shifted hyperbola will have one of the following standard forms.

Standard Forms for Hyperbolas Centered at (*h*, *k*)

(i) $\dfrac{(x-h)^2}{a^2} - \dfrac{(y-k)^2}{b^2} = 1$ if the transverse axis is horizontal (Figure 7a).

(ii) $\dfrac{(y-k)^2}{b^2} - \dfrac{(x-h)^2}{a^2} = 1$ if the transverse axis is vertical (Figure 7b).

In either case, the asymptotes have the equations

$$y - k = \frac{b}{a}(x-h) \quad \text{and} \quad y - k = -\frac{b}{a}(x-h),$$

and the distance from the center (h, k) to either focus is given by

$$c = \sqrt{a^2 + b^2}.$$

Figure 7

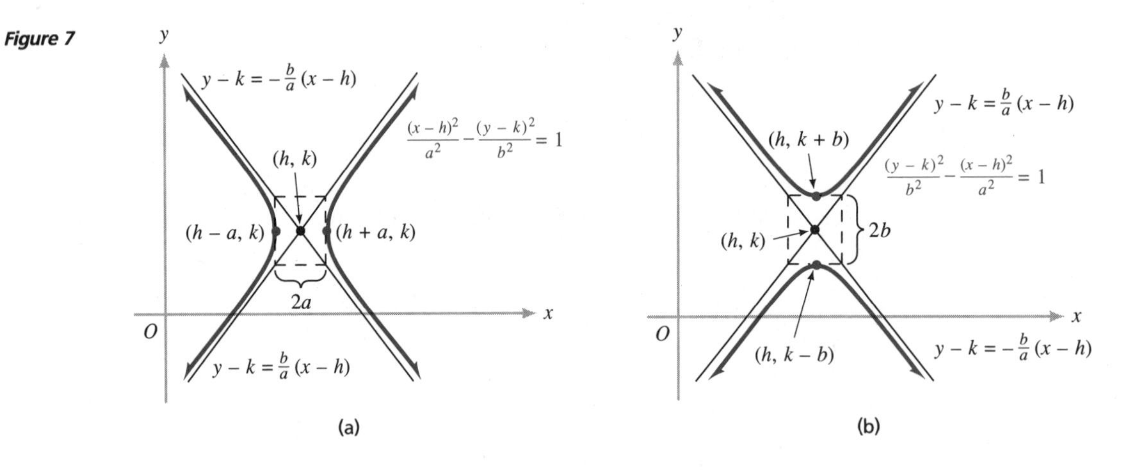

Example 2 Find the coordinates of the center, the foci, and the vertices of the hyperbola $y^2 - 4x^2 - 8x - 4y - 4 = 0$. Also, find equations of its asymptotes and sketch its graph.

Solution Completing the squares, we have

$$y^2 - 4y + 4 - 4(x^2 + 2x + 1) = 4$$

or

$$(y-2)^2 - 4(x+1)^2 = 4.$$

Dividing by 4, we obtain

$$\frac{(y-2)^2}{4} - \frac{(x+1)^2}{1} = 1,$$

the equation of a hyperbola with center $(-1, 2)$ and with a vertical transverse axis. Since $a = 1$ and $b = 2$, the equations of the asymptotes are

$$y - 2 = \frac{2}{1}(x+1) \quad \text{and} \quad y - 2 = \frac{-2}{1}(x+1);$$

that is,

$$y = 2x + 4 \quad \text{and} \quad y = -2x.$$

Also, $c = \sqrt{a^2 + b^2} = \sqrt{5}$, so $F_1 = (-1, 2 - \sqrt{5})$, $F_2 = (-1, 2 + \sqrt{5})$, $V_1 = (-1, 0)$, and $V_2 = (-1, 4)$ (Figure 8). ■

Figure 8

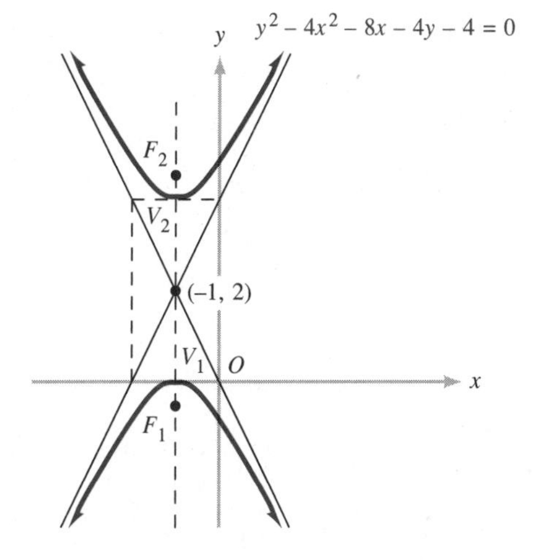

Figure 9

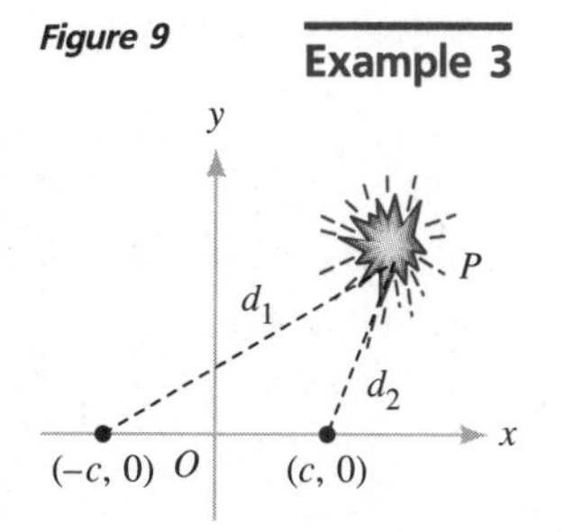

Example 3

Two microphones are located at the points $(-c, 0)$ and $(c, 0)$ on the x axis (Figure 9). An explosion occurs at an unknown point P to the right of the y axis. The sound of the explosion is detected by the microphone at $(c, 0)$ exactly T seconds before it is detected by the microphone at $(-c, 0)$. Assuming that sound travels in air at a constant speed of v feet per second, show that point P must be located on the right-hand branch of the hyperbola whose equation is $(x^2/a^2) - (y^2/b^2) = 1$, where

$$a = \frac{vT}{2} \quad \text{and} \quad b = \frac{\sqrt{4c^2 - v^2T^2}}{2}.$$

Solution Let d_1 and d_2 denote the distances from P to $(-c, 0)$ and $(c, 0)$, respectively. The sound of the explosion reaches $(-c, 0)$ in d_1/v seconds, and it reaches $(c, 0)$ in d_2/v seconds; hence, $(d_1/v) - (d_2/v) = T$, so $d_1 - d_2 = vT$. Putting $a = vT/2$, we notice that the condition $d_1 - d_2 = 2a$ requires that P belong to a hyperbola with foci $F_1 = (-c, 0)$ and $F_2 = (c, 0)$. By Theorem 1 we can write an equation of this hyperbola as $(x^2/a^2) - (y^2/b^2) = 1$, where

$$a = \frac{vT}{2} \quad \text{and} \quad b = \sqrt{c^2 - a^2} = \sqrt{c^2 - \left(\frac{vT}{2}\right)^2} = \frac{\sqrt{4c^2 - v^2T^2}}{2}.$$

Problem Set 7.9

In Problems 1 to 8, find the coordinates of the vertices and the foci of each hyperbola. Also, find equations of the asymptotes and sketch the graph.

1. $\dfrac{x^2}{9} - \dfrac{y^2}{4} = 1$

2. $\dfrac{x^2}{1} - \dfrac{y^2}{9} = 1$

3. $\dfrac{y^2}{16} - \dfrac{x^2}{4} = 1$

4. $\dfrac{y^2}{4} - \dfrac{x^2}{1} = 1$

5. $4x^2 - 16y^2 = 64$

6. $49x^2 - 16y^2 = 196$

7. $36y^2 - 10x^2 = 360$

8. $y^2 - 4x^2 = 1$

In Problems 9 to 11, find an equation of the hyperbola that satisfies the conditions given.

9. Vertices at $(-4, 0)$ and $(4, 0)$, foci at $(-6, 0)$ and $(6, 0)$.

10. Vertices at $(0, -\frac{1}{2})$ and $(0, \frac{1}{2})$, foci at $(0, -1)$ and $(0, 1)$.

11. Vertices at $(-4, 0)$ and $(4, 0)$, the equations of the asymptotes are $y = -\frac{5}{4}x$ and $y = \frac{5}{4}x$.

12. Determine the values of a^2 and b^2 so that the graph of the equation $b^2x^2 - a^2y^2 = a^2b^2$ contains the pair of points **(a)** $(2, 5)$ and $(3, -10)$ and **(b)** $(4, 3)$ and $(-7, 6)$.

In Problems 13 to 20, find the coordinates of the center, the vertices, and the foci of each hyperbola. Also, find equations of the asymptotes and sketch the graph.

13. $\dfrac{(x-1)^2}{9} - \dfrac{(y+2)^2}{4} = 1$

14. $\dfrac{(x+3)^2}{1} - \dfrac{(y-1)^2}{9} = 1$

15. $\dfrac{(y+1)^2}{16} - \dfrac{(x+2)^2}{25} = 1$

16. $4x^2 - y^2 - 8x + 2y + 7 = 0$

17. $x^2 - 4y^2 - 4x - 8y - 4 = 0$

18. $16x^2 - 9y^2 + 180y = 612$

19. $9x^2 - 25y^2 + 72x - 100y + 269 = 0$

20. $9x^2 - 16y^2 - 90x - 256y = 223$

21. Find an equation of the hyperbola that satisfies the conditions given.

    (a) Foci at $(1, -1)$ and $(7, -1)$; length of transverse axis is 2.

    (b) Vertices at $(-4, 3)$ and $(0, 3)$, foci at $(-\frac{9}{2}, 3)$ and $(\frac{1}{2}, 3)$.

    (c) Center at $(2, 3)$, one vertex at $(2, 8)$, and one focus at $(2, -3)$.

22. The segment cut by a hyperbola from a line containing a focus and perpendicular to the transverse axis is called a **focal chord** of the hyperbola (Figure 10).

    (a) Show that the length of a focal chord of the hyperbola $(x^2/a^2) - (y^2/b^2) = 1$ is $2b^2/a$.

    (b) Find the length of a focal chord of the hyperbola $x^2 - 8y^2 = 16$.

Figure 10

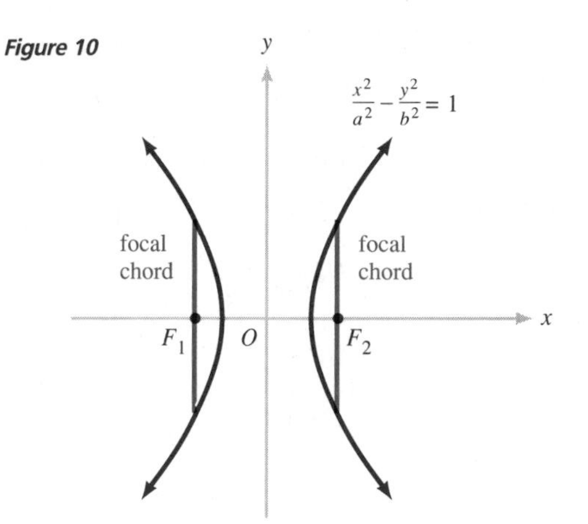

23. A hyperbola is said to be **equilateral** if its two asymptotes are perpendicular. Find an equation of an equilateral hyperbola with horizontal transverse axis and center at the origin. (Denote the distance from the center to a vertex by a.)

24. Sketch the graph of the hyperbola

$$\frac{(y+b)^2}{b^2} - \frac{x^2}{2bp+p^2} = 1.$$

Show that as $b \to +\infty$, the upper branch of this hyperbola approaches the parabola $y = (1/4p)x^2$.

25. In Figure 11, hold the center C and the asymptotes fixed but allow the foci F_1 and F_2 to move in toward the center. What happens to the hyperbola?

Figure 11

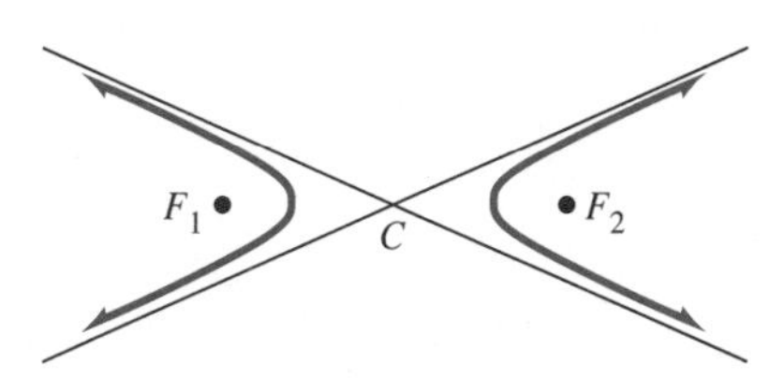

26. Give a proof of Theorem 1.

27. Sound travels with speed s in air, and a bullet travels with speed b from a gun at $(-h, 0)$ to a target at $(h, 0)$ in the xy plane. At what points (x, y) can the boom of the gun and the ping of the bullet hitting the target be heard simultaneously?

28. Space vehicles are sent to great distances with a minimal expenditure of fuel by using the *slingshot effect*, which occurs when the vehicle approaches, but does not collide with a planet. With an xy coordinate system in the plane of motion of the vehicle, and with the planet at the origin, the path of the vehicle will form one branch of a hyperbola with the origin as one focus (Figure 12). If the vehicle approaches the planet along a path with the line $y = m(x + c)$ as an asymptote and comes within r units of the planet, show that an equation of the path is $y^2 = m^2[(x + c)^2 - (c - r)^2]$.

Figure 12

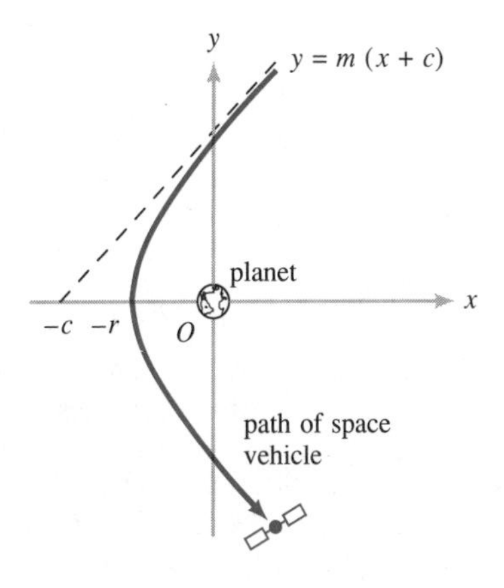

29. Two LORAN stations are positioned at points A and B along an east–west coastline, with point A 100 nautical miles due west of point B (Figure 13). Signals sent out from stations A and B travel at the rate of 0.16 nautical mile per microsecond (millionth of a second). Suppose that the signal from station A is received by a ship 500 microseconds after the same signal is received from station B. The ship is known to be in the first quadrant of an xy coordinate system with origin at the midpoint of $\overline{AB}$ and with the y axis pointing due north. Find an equation for the branch of the hyperbola on which the ship must be located.

Figure 13

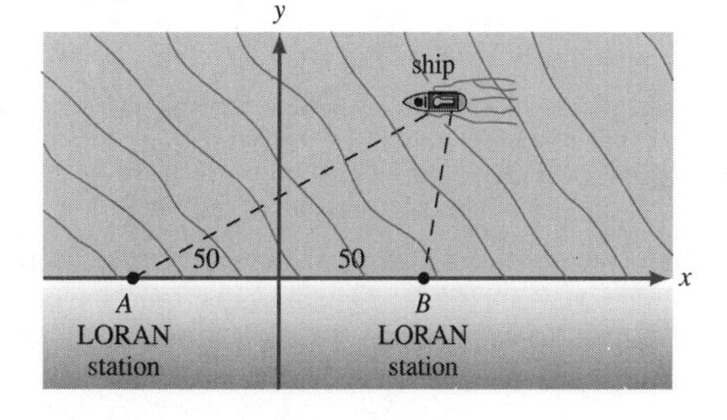

30. The **reflection property** for hyperbolas is illustrated in Figure 14. State this property in words.

Figure 14

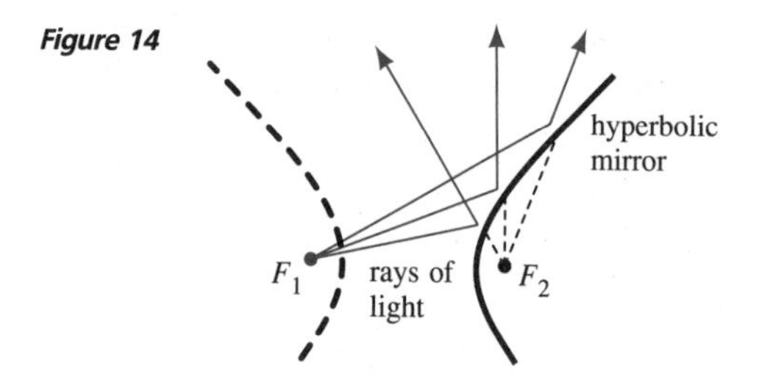

31. *Scattering experiments*, in which moving particles are deflected by various forces, are fundamental in modern physics. In a typical scattering experiment, a particle P, originally moving in the direction of a ray l_1, enters a region in which it is acted on by a force. The force deflects the particle, so that it emerges from the region in the direction of a ray l_2 (Figure 15). Assume that l_1 and l_2 have slopes of m and $-m$, respectively. One of the first scattering experiments, conducted about 1911 by Ernest Rutherford, led to the concept of the nucleus of an atom. In this experiment, alpha particles were deflected along hyperbolic paths by the nuclei of gold atoms. If V is the vertex of such a path, show that an equation of the path in standard form is $\frac{x^2}{a^2} - \frac{y^2}{(ma)^2} = 1$, where $a = |\overline{OV}|$.

Figure 15

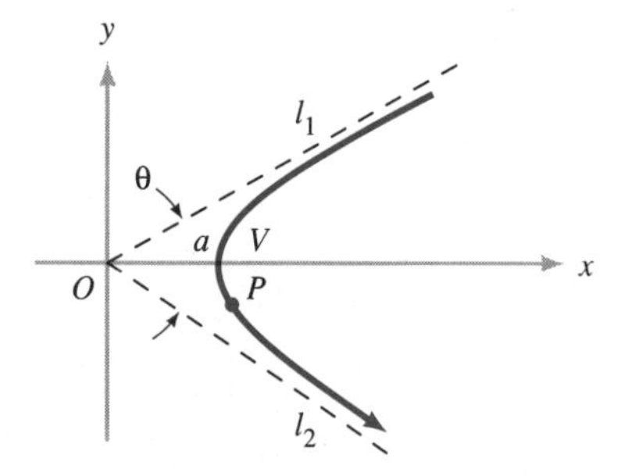

32. A mirror forms the right-hand branch of the hyperbola $x^2 - y^2 = 16$. If a ray of light from the left-hand focus F_1 of the hyperbola strikes the mirror at the point (5, 3), find an equation for the reflected ray (Figure 16). [*Hint:* See Problem 30.]

Figure 16

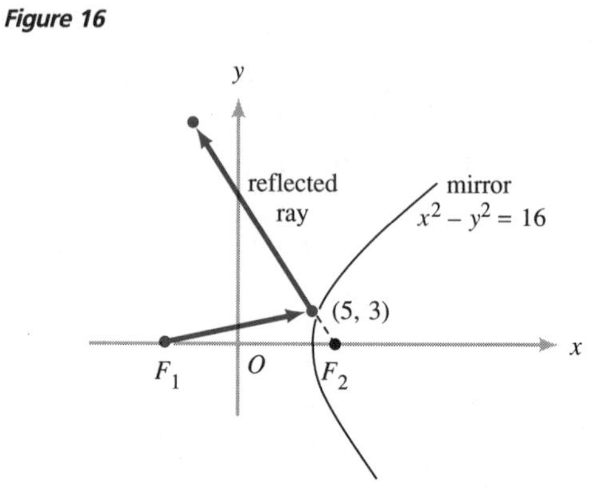

33. In the Rutherford scattering experiment described in Problem 31, the path of the alpha particle forms only the right-hand branch of a hyperbola. The nucleus of the gold atom causing the deflection of the alpha particle is actually located at the focus F_1 of the left-hand branch of the hyperbola. Find the coordinates of F_1 in terms of m and a.

34. It can be shown that the graph of the equation $xy = 1$ is a hyperbola with center at the origin and with the x and y axes as asymptotes (Figure 17). Find the coordinates of the foci of this hyperbola. [*Hint:* The transverse axis makes a 45° angle with the x axis.]

Figure 17

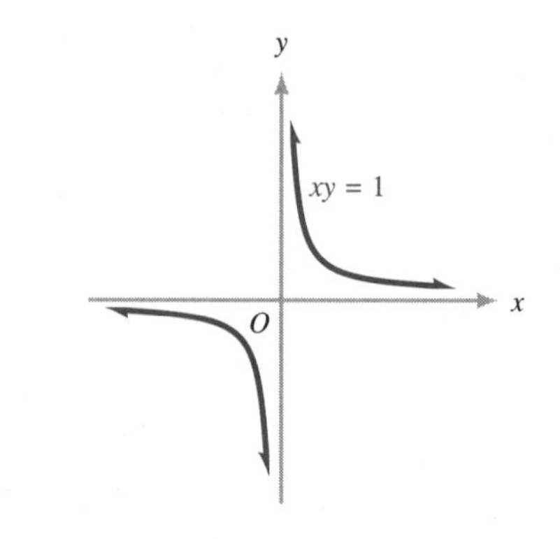

35. Let a be a fixed positive number and let $Q = (-a, 0)$, $R = (a, 0)$ (Figure 18). If the point $P = (x, y)$ moves in such a way that, for $P \neq Q$ and $P \neq R$, the product of the slopes of the segments $\overline{PQ}$ and $\overline{PR}$ is equal to a positive constant k, show that P moves along a branch of a hyperbola and that the asymptotes of this hyperbola have slopes $\sqrt{k}$ and $-\sqrt{k}$.

Figure 18

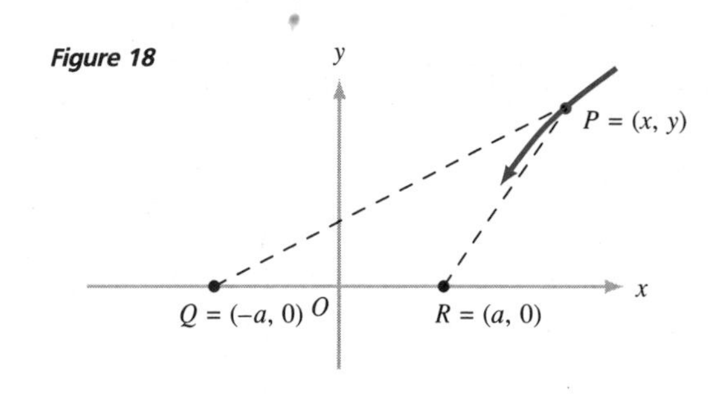

CHAPTER 7 Review Problem Set

In Problems 1 to 4, use mathematical induction to prove that the assertion is true for all positive integers n.

1. $n^2 + 3n$ is an even integer

2. 3 is an exact integer divisor of $n^3 + 6n^2 + 11n$

3. $1 \cdot 2 \cdot 3 + 2 \cdot 3 \cdot 4 + 3 \cdot 4 \cdot 5 + \cdots + n(n+1)(n+2) = \frac{1}{4}n(n+1)(n+2)(n+3)$

4. $2 \cdot 4 + 4 \cdot 6 + 6 \cdot 8 + \cdots + 2n(2n+2) = \frac{n}{3}(2n+2)(2n+4)$

In Problems 5 to 8, use the binomial theorem and Pascal's triangle to expand each expression.

5. $(2 + x)^5$

6. $(3x - 4y)^4$

C **7.** $(3x^2 - 2y^2)^7$

8. $(2a + 3b)^6$

In Problems 9 and 10, find and simplify the specified term in the binomial expansion of the expression.

9. **(a)** The fourth term of $(2x + y)^9$

(b) The term containing x^5 in $(3x + y)^{10}$

10. **(a)** The fifth term of $(x - 2y)^7$

(b) The term containing x^{10} in $(2x^2 - 3)^{11}$

In Problems 11 and 12, find the first five terms of the sequence with the given general term.

11. $a_n = \dfrac{(-1)^n}{n+1}$

12. $b_n = \dfrac{(-1)^{n+1}}{n^2 + 4n + 1}$

In Problems 13 and 14, find the first six terms of the recursively defined sequence $\{a_n\}$.

13. $a_1 = 0$, $a_2 = \frac{1}{2}$, and $a_n = \frac{1}{n} a_{n-2}$

14. $a_1 = 1$, $a_2 = 0$, and $(n+1)(n+2)a_{n+2} = -na_n$

In Problems 15 to 20, evaluate each sum.

15. $\displaystyle\sum_{k=0}^{3} \frac{7^k}{1 + 2^k}$

16. $\displaystyle\sum_{k=1}^{50} [3 + (-1)^k]$

17. $\displaystyle\sum_{k=1}^{20} \left(\frac{k}{3} + 2\right)$

18. $\displaystyle\sum_{k=0}^{15} \frac{3}{2^k}$

19. $\displaystyle\sum_{k=0}^{2} \frac{(-5)^k}{1 + 3^k}$

20. $\displaystyle\sum_{k=0}^{25} (4^{k+1} - 4^k)$

In Problems 21 to 24, find **(a)** the nth term, and **(b)** the sum of the first n terms of each arithmetic sequence for the given value of n.

21. 2, 6, 10, 14, . . . , for $n = 15$

22. 12, 13.5, 15, 16.5, . . . , for $n = 20$

23. 3, $\frac{8}{3}$, $\frac{7}{3}$, 2, . . . , for $n = 11$

24. $3x - 2$, $-x + 1$, . . . , for $n = 10$

In Problems 25 to 28, find **(a)** the nth term, and **(b)** the sum of the first n terms of each geometric sequence for the given value of n.

25. 48, 96, 192, . . . , for $n = 8$

26. -81, -27, -9, . . . , for $n = 12$

27. $\frac{3}{4}$, 3, 12, . . . , for $n = 10$

28. 0.2, 0.002, 0.00002, . . . , for $n = 5$

In Problems 29 to 32, find the sum of each infinite geometric series.

29. $\sum_{k=1}^{\infty} 3\left(\frac{2}{3}\right)^{k-1}$ **30.** $\sum_{k=1}^{\infty} 3\left(-\frac{2}{3}\right)^{k-1}$

31. $\sum_{k=1}^{\infty} \left(\sqrt{\frac{5}{7}}\right)^{k-1}$ **32.** $\sum_{k=0}^{\infty} (-1)^k\left(\frac{3}{4}\right)^{k}$

In Problems 33 and 34, rewrite each infinite repeating decimal as a quotient of integers.

33. **(a)** $0.\overline{83}$ **(b)** $0.\overline{91}$

34. **(a)** $4.65\overline{223}$ **(b)** $3.1\overline{9}$

In Problems 35 and 36, find the value of each expression.

35. **(a)** ${}_6P_4$ **(b)** ${}_{11}P_2$ **(c)** ${}_{16}C_2$
(d) ${}_9C_3$ **(e)** ${}_8C_8$

36. **(a)** ${}_7P_4$ **(b)** ${}_8P_3$ C **(c)** ${}_{16}C_7$
(d) ${}_7C_5$ **(e)** ${}_{19}C_0$

37. A mathematics club raffles a calculator by selling 100 sealed tickets numbered in order 1, 2, 3, ... The tickets are drawn at random by purchasers who pay the number of cents indicated by the number on the ticket. How much money does the club receive if all tickets are sold?

38. A clock strikes on the hour. How many times does it strike between 12:00 noon on one day and 12:00 noon on the next day, inclusive?

39. Gus saved \$200 the first year he was employed. Each year thereafter, he saved \$50 more than the year before. How much had he saved at the end of 8 years?

40. A car costs \$7670 and depreciates in value by 31% during the first year, by 26% during the second year, by 21% during the third year, and so on. What is the value of the car at the end of 5 years?

C **41.** Suppose that on each separate swing, a pendulum describes an arc whose length is 0.98 of the length of the preceding arc. If the length of the first arc is 24 centimeters, what total distance is covered by the pendulum before it comes to rest?

C **42.** In a lottery, the first ticket drawn will pay the ticket holder \$1, and each succeeding ticket will pay twice as much as the preceding one. If 15 tickets are drawn, what is the total amount paid in prize money?

43. A pyramid of cannonballs in an armaments museum stands on a square base having n cannonballs on a side. How many cannonballs are there in the pile?

44. Work Problem 43 if the base has the form of an equilateral triangle.

45. A woman has a choice of four airlines between Chicago and New York, and three airlines between New York and London. In how many ways can she fly from Chicago to London if she stops in New York?

C **46.** **(a)** How many different 4-letter "words" can be formed using the 26 letters of the alphabet if repeated letters are allowed?

(b) In how many ways can the letters of the word WORTH be arranged?

47. How many different four-course meals can be ordered in a restaurant if there is a choice of three soups, four entrees, two salads, and five desserts?

C **48.** In how many ways can 7 of 12 books be chosen and arranged next to each other on a shelf?

49. How many different signals can be sent by arranging up to five distinguishable flags, one above the other, on a flag pole, with the condition that the flags can be used one, two, three, four, or five at a time?

50. How many straight lines can be drawn through pairs of 10 points if no 3 points are in a straight line?

51. Twelve people meet and shake hands. If everyone shakes hands with everyone else, how many handshakes are exchanged?

C **52.** A bridge hand consists of 13 cards dealt from a deck of 52. How many different bridge hands are there?

C **53.** A corporation owns 20 motels. If there are 25 people eligible to be managers of these motels, in how many ways can the motel managers be appointed?

54. In how many distinguishable ways can the letters of the word MINIMUM be arranged?

55. In how many distinguishable ways can three identical racquet balls, four identical golf balls, and five identical tennis balls be placed in a row?

56. If three dimes are tossed in the air, find the probability that all three fall heads.

57. You meet a married couple and learn that they have two children. During the course of the conversation, it becomes clear that at least one of their children is a girl. What is the probability that the other child is a girl?

58. What is the probability of rolling an 11 with two dice?

59. Four married couples draw lots to decide who will be partners in a card game. If partners must be of opposite sexes, what is the probability that no woman will be paired with her own husband?

60. A bag contains four white balls, six black balls, three red balls, and eight green balls. If two balls are selected blindly, what is the probability that they are of the same color?

62. Three coins will be tossed, and a prize of \$5 will be paid if they all fall heads or if they all fall tails. What is a fair price to pay for the privilege of playing this game?

61. Professor Grumbles has four suits of clothes, each consisting of a vest, trousers, and a jacket. If he dresses at random, what is the probability that his clothes will match?

63. Three dice will be rolled, and a prize of \$12.00 will be awarded if all three dice show the same number on top. It costs \$0.50 to play this game.

(a) What are the odds on winning?

(b) What is the expected value of the game?

64. The probability of "making the point 9" in a game of craps—that is, rolling a 9 before rolling a 7 in a sequence of rolls with a pair of dice—is $\frac{2}{5}$. An "odds bet" on making the point 9 is one of the few fair bets offered by gambling casinos. (However, for the privilege of making this bet, you must have already made an unfair pass line bet.) If you make an odds bet of \$10 on making the point 9 and if you win this bet, how much money will you win?

[C] In Problems 65 to 84, rewrite each equation in standard form and identify the conic. Find the center (when applicable), the vertex (or vertices), the focus (foci), the asymptotes (when applicable), and the directrix (when applicable). Also, sketch the graph. [gc] Use a graphing calculator if you wish.

65. $3x^2 + 2y^2 = 24$

66. $3x^2 + 2y^2 = 1$

67. $y^2 = 4x$

68. $-x^2 = 5y$

69. $x^2 - 9y^2 = 72$

70. $y^2 - 9x^2 = 54$

71. $9x^2 + 25y^2 + 18x - 50y = 191$

72. $3x^2 + 4y^2 - 28x - 16y + 48 = 0$

73. $9x^2 + 4y^2 + 72x - 48y + 144 = 0$

74. $9x^2 + 4y^2 + 36x - 24y = 252$

75. $x^2 = -4(y - 1)$

76. $x + 8 - y^2 + 2y = 0$

77. $x^2 - 4y^2 + 4x + 24y - 48 = 0$

78. $16x^2 - 9y^2 = 96x$

79. $x^2 = 8(y + 1)$

80. $x + 4 + 2y = y^2$

81. $4y^2 - x^2 - 24y + 2x + 34 = 0$

82. $11y^2 - 66y - 25x^2 + 100x = 276$

83. $x^2 + y^2 - 4x + 2y = 4$

84. $4x^2 + 4y^2 + 4x - 4y + 1 = 0$

In Problems 85 and 86, find the equation in standard form for the ellipse that satisfies the indicated conditions.

85. Vertices $(-3, 1)$, $(5, 1)$, $(1, -4)$, and $(1, 6)$.

86. Center $(0, 0)$; major axis horizontal; ellipse contains the points $(4, 3)$ and $(6, 2)$.

In Problems 87 and 88, find the equation in standard form for the parabola that satisfies the indicated conditions.

87. Vertex at $(0, -6)$; horizontal axis; parabola contains the point $(-9, -3)$.

88. Vertical axis; parabola contains the points $(9, -1)$, $(3, -4)$, and $(-9, 8)$.

In Problems 89 and 90, find the equation in standard form for the hyperbola that satisfies the indicated conditions.

89. Asymptotes $y = -2x$ and $y = 2x$; hyperbola contains the point $(1, 1)$.

90. Vertices $(2, -8)$ and $(2, 2)$; asymptotes $25x - 9y = 77$ and $25x + 9y = 23$.

91. For a main cable in a suspension bridge, the dimensions H and L shown in Figure 1 are called the *sag* and the *span*, respectively. If the weight of the suspended roadbed (together with that of the cables) is uniformly distributed horizontally, the main cables form parabolas. Taking the origin of an xy coordinate system at the lowest point of a parabolic main cable with sag H and span L, find an equation of the cable.

Figure 1

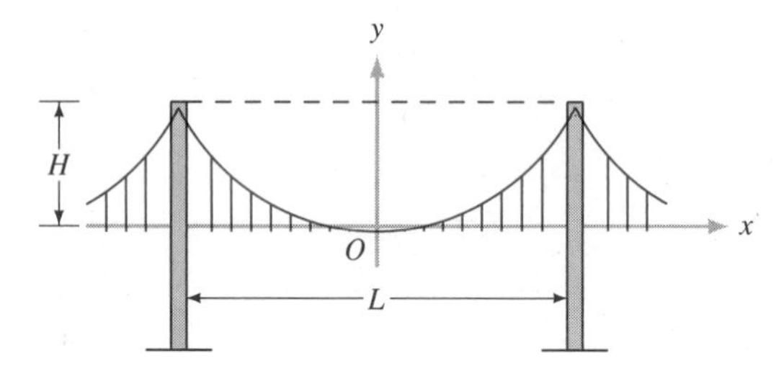

92. A point $P = (x, y)$ moves so that the product of the slopes of the line segments $\overline{PQ}$ and $\overline{PR}$ is -6, where $Q = (3, -2)$ and $R = (-2, 1)$. Find an equation of the curve traced out by P, identify the curve, and sketch it.

93. Suppose that the towers of a suspension bridge are 240 meters apart, the tops of the towers are 60 meters above the roadway, and the lowest point of the parabolic main cables is 20 meters above the roadway (Figure 2). Find the vertical distance from the roadway to the cable at intervals of 20 meters.

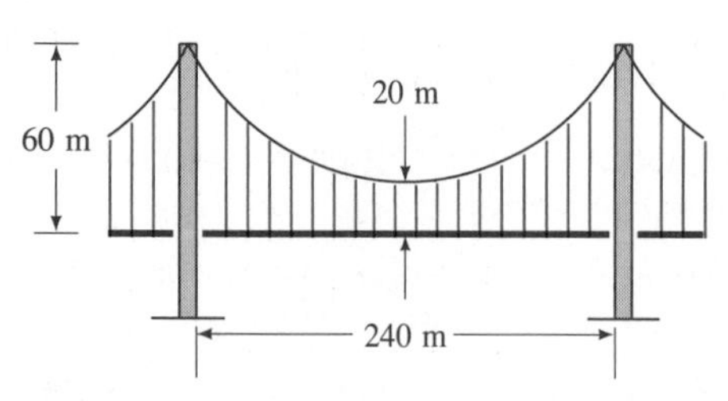

94. Two points are 800 meters apart. At one of these points the report of a cannon is heard 1 second later than at the other. Show that the cannon is somewhere on a certain hyperbola and write an equation for the hyperbola after making a suitable choice of axes. (Consider the velocity of sound to be 332 meters per second.)

c **95.** The Yale Bowl, completed in 1914, is a football stadium modeled after the Colosseum in Rome. The concrete stands surround a central region in the shape of an ellipse with a major axis 148 meters long and a minor axis 84 meters long. Find an equation in standard form of this ellipse.

Yale Bowl

c **96.** An indoor track is constructed in the shape of an ellipse with a major axis 70 meters long and a minor axis 40 meters long (Figure 3).

Figure 3

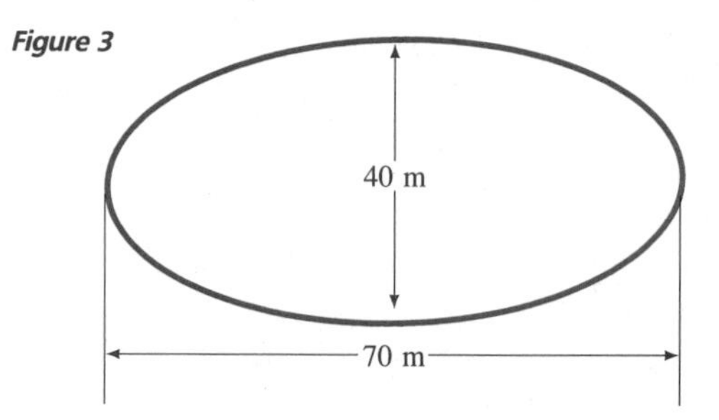

(a) Write an equation for the ellipse with respect to an xy coordinate system with origin at the center of the ellipse and with the x axis along the major axis of the ellipse.

(b) Find the focal chord of the ellipse.

c **97.** The Ellipse in front of the White House in Washington, D.C., has a semimajor axis of 229 meters and a semiminor axis of 195 meters. Find an equation of this ellipse in standard form.

c **98.** It is shown in calculus that the ellipsoid generated by revolving an ellipse with semimajor axis a and semiminor axis b about its major axis encloses a volume given by $V = 4\pi ab^2/3$ (Figure 4). A fruit market sells two varieties of watermelon, both of which have approximately the shape of such an ellipsoid. For the first type, $a = 6$ inches and $b = 5$ inches; for the second type, $a = 7$ inches and $b = 6$ inches. Both watermelons are of equal quality, and both have rinds that are 1 inch thick. The market charges \$3.65 for a watermelon of the first type, and \$5.95 for one of the second type. Which type of watermelon is the better buy?

Figure 4

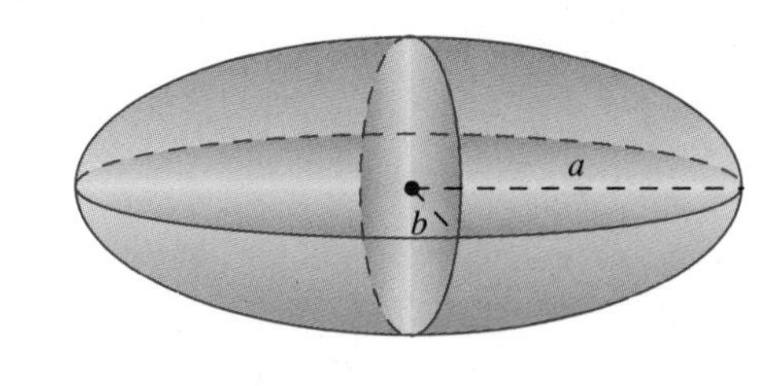

CHAPTER 7 Test

1. Use mathematical induction to prove that
$$3 + 9 + 15 + \cdots + (6n - 3) = 3n^2$$
is true for all positive integers n.

2. Use the binomial theorem to expand $(3x - 2y^2)^5$.

3. Find and simplify the specified term in the binomial expansion of each expression.

   (a) The fifth term of $(3a - 2b)^7$

   (b) The term containing x^4 in $(2x + 3y)^9$

4. Find the first five terms of the sequence $\left\{\dfrac{(-1)^{n+1}}{n^2 + 4n + 1}\right\}$.

5. Find the ninth term and the sum of the first nine terms of the geometric sequence $12, 6, 3, \frac{3}{2}, \ldots$.

6. Evaluate each sum.

   (a) $\displaystyle\sum_{k=0}^{3} \frac{7^k}{1 + 2^k}$ **(b)** $\displaystyle\sum_{k=1}^{\infty} \frac{1}{2^{k-1}}$

7. Evaluate each expression.

   (a) ${}_9C_3$ **(b)** ${}_5P_3$

8. An investor receives a sequence of monthly payments from a certain financial venture. The first payment is \$400, and the payments increase by \$25 each month. How much will the investor receive over the first 24 months of payments?

9. A certain culture initially contains 1000 bacteria. If the number of bacteria in the culture increases by 15% every hour, how many bacteria will be in the culture at the end of 8 hours?

10. If eight tennis players are in a tournament, find the number of different ways that the first, second, and third places can be decided, assuming that ties are not allowed.

11. A committee of 3 women and 2 men is to be chosen from among 7 women and 5 men. In how many different ways can this be done?

12. A certain state offers a gambling game in which you choose an integer between 0 and 999. It costs \$0.50 to play, and if your number is drawn, you win \$250.

    (a) What is the probability of winning?

    (b) What are the odds on winning?

    (c) What is the expected value E of this game?

In Problems 13 to 18, find the center (when applicable), the vertex (vertices), the focus (foci), and the asymptotes (when applicable) and sketch the graph of each conic.

13. $y^2 = 4x$

14. $x = 2y^2 + 8y + 3$

15. $3x^2 + 2y^2 = 4$

16. $16x^2 + y^2 - 32x + 4y - 44 = 0$

17. $9y^2 - 16x^2 = 144$

18. $4y^2 - x^2 - 8y + 2x + 7 = 0$

In Problems 19 to 21, find an equation of the conic that satisfies the given conditions.

19. The ellipse with vertices $(0, 0)$, $(6, 0)$, $(3, -5)$, and $(3, 5)$.

20. The parabola with focus at $(0, 0)$ and directrix $x = 2$.

21. The hyperbola with vertices at $(-3, 0)$ and $(3, 0)$ and with asymptotes $y = -4x/3$ and $y = 4x/3$.

22. A point P moves so that the product of the slopes of the line segments that join it to the points $(-5, 0)$ and $(5, 0)$ is equal to 9. Identify the curve traced out by the point.

APPENDIX

I Review of Plane Geometry

This appendix is intended as a concise review of the plane geometry that you will need in the textbook. To facilitate your review, the material is presented in condensed outline form with little or no attempt to justify the basic results.

Lines and Line Segments

The line segment with endpoints A and B is denoted by $\overline{AB}$, and its length by $|\overline{AB}|$. The point M on the segment $\overline{AB}$ for which $|\overline{AM}| = |\overline{MB}|$ is called the **midpoint** of $\overline{AB}$. The line L that is perpendicular to $\overline{AB}$ and intersects $\overline{AB}$ at its midpoint M is called the **perpendicular bisector** of $\overline{AB}$ (Figure 1).

If P is any point on the perpendicular bisector L of $\overline{AB}$, then P is equidistant from A and B; that is,

$$|\overline{PA}| = |\overline{PB}|.$$

Figure 1

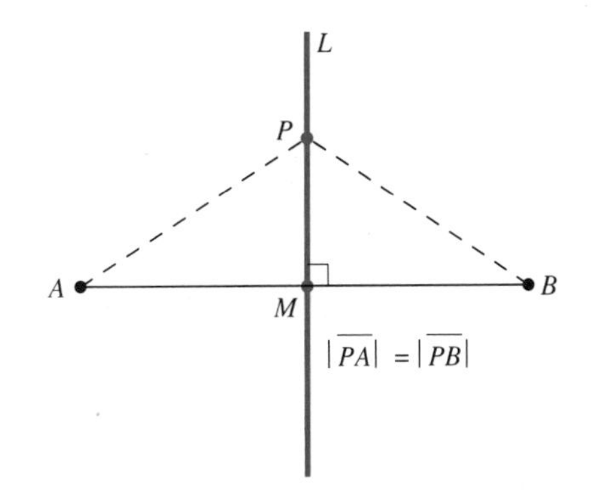

Angles

Two lines L_1 and L_2 that intersect at a point V will form four angles α, β, γ, and δ, as shown in Figure 2. Angles greater than 0° and less than 90°, such as angles α

and γ, are called **acute angles.** Angles greater than 90° and less than 180°, such as angles β and δ, are referred to as **obtuse angles.**

Two angles, such as α and β or γ and δ in Figure 2, that have a common vertex V and a common side between them are said to be **adjacent angles.** Nonadjacent angles formed by two intersecting lines, such as angles α and γ or angles β and δ, are called **vertical angles** or **opposite angles.**

Vertical angles formed by intersecting lines are equal.

Thus, in Figure 2, $\alpha = \gamma$ and $\beta = \delta$.

Figure 2

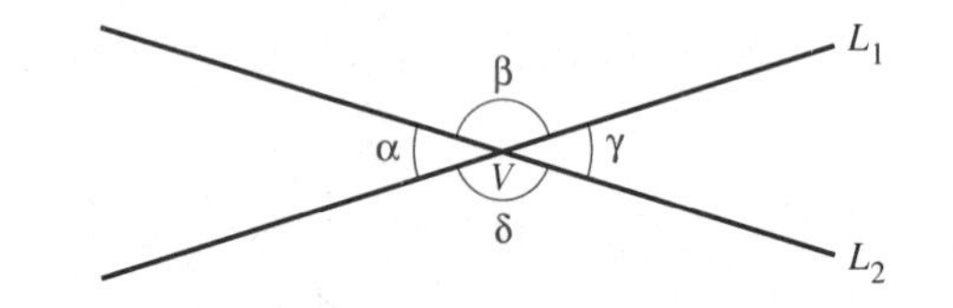

A line T that intersects two parallel lines L_1 and L_2 as in Figure 3 is called a **transversal.** Notice that T forms four angles with L_1 and four more with L_2. Of these, the four angles that lie between L_1 and L_2 (β, γ, α', and δ' in Figure 3) are said to be **interior angles.** Pairs of interior angles that lie on opposite sides of the transversal, such as β and δ' or γ and α', are called **alternate interior angles.**

Alternate interior angles are equal.

Thus, in Figure 3, $\beta = \delta'$ and $\gamma = \alpha'$.

Figure 3

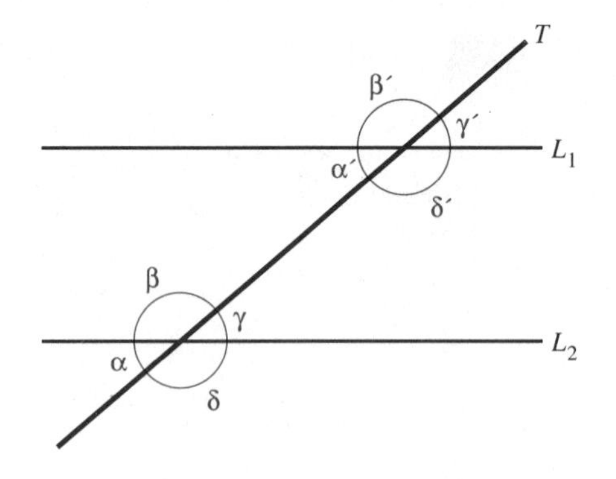

Two angles that lie on the same side of the transversal and on the same relative side of the parallel lines are called **corresponding angles.** In Figure 3, α and α', β and β', γ and γ', and δ and δ' are pairs of corresponding angles.

Corresponding angles are equal.

Figure 4

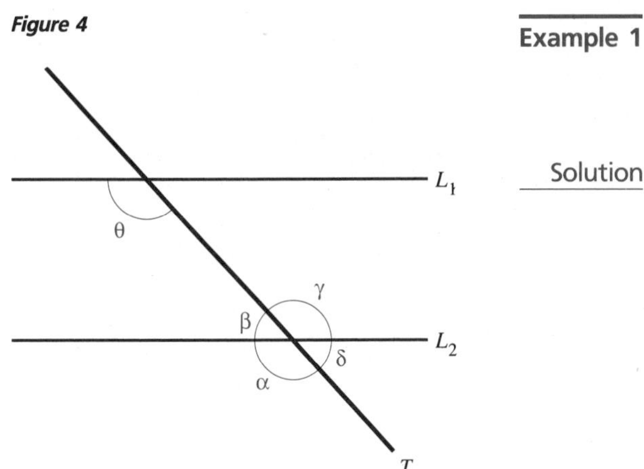

Example 1 In Figure 4, L_1 and L_2 are parallel lines and the measure of angle θ is 130°. Find the measures of angles α, β, δ, and γ.

Solution Because θ and γ are alternate interior angles, they are equal, so

$$\gamma = \theta = 130°.$$

Also, $$\beta + \gamma = 180°;$$

hence, $\beta = 180° - \gamma = 180° - 130° = 50°$.

Since α and γ are vertical angles formed by straight lines, we have

$$\alpha = \gamma = 130°$$

and, likewise,

$$\delta = \beta = 50°.$$ ■

Triangles

A triangle with two sides of equal length is said to be **isosceles** (Figure 5a).

> Angles opposite the equal sides of an isosceles triangle are equal.

Thus, in Figure 5a, $\alpha = \beta$.

A triangle is called **equilateral** if all three of its sides are equal in length (Figure 5b).

> In an equilateral triangle, all three angles have a measure of 60° ($\pi/3$ radians).

A **right triangle** (Figure 5c) is a triangle in which one of the three angles is a right angle, that is, an angle with a measure of 90° ($\pi/2$ radians). In a right triangle, the side opposite the right angle is called the **hypotenuse.** The **Pythagorean theorem** states that:

> In a right triangle, the square of the length of the hypotenuse is the sum of the squares of the lengths of the other two sides.

Figure 5

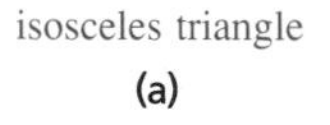

isosceles triangle
(a)

equilateral triangle
(b)

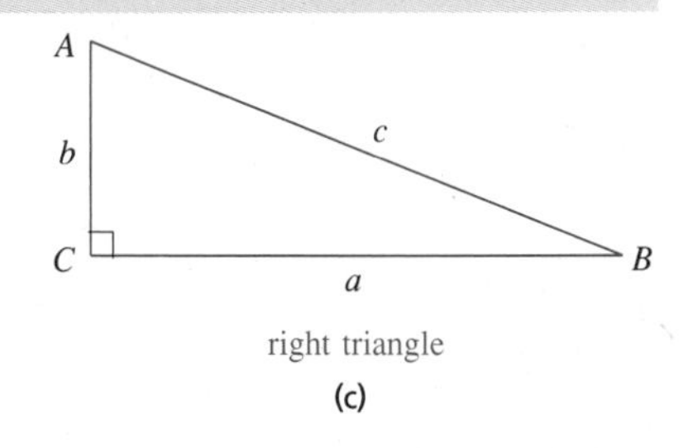

right triangle
(c)

Thus, in Figure 5c, we have $c^2 = a^2 + b^2$. The converse of the Pythagorean theorem also holds:

> If the lengths a, b, and c of the three sides of a triangle satisfy $a^2 + b^2 = c^2$, then the angle opposite the side of length c is a right angle.

Example 2 In right triangle ACB (Figure 5c), suppose $a = 7$ and $c = 25$. Find b.

Solution By the Pythagorean theorem,

$$b^2 = c^2 - a^2 = 25^2 - 7^2 = 625 - 49 = 576,$$

so

$$b = \sqrt{576} = 24.$$

■

If the three vertex angles in a first triangle ABC are equal to corresponding vertex angles in a second triangle DEF as in Figure 6, we say that the two triangles are **similar.**

Figure 6

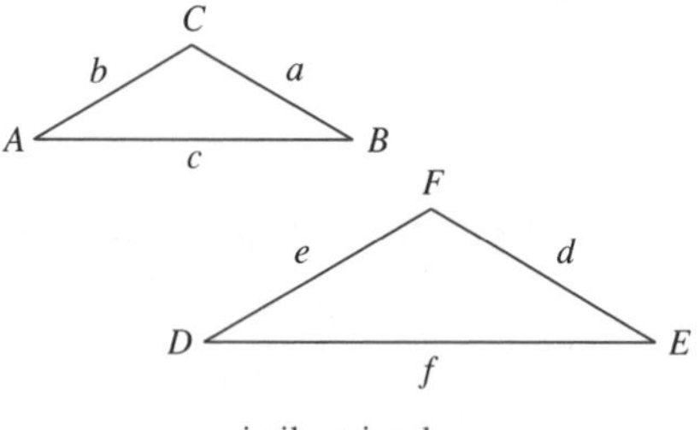

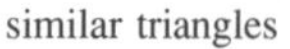
similar triangles

> Corresponding side lengths of two similar triangles are proportional.

Thus, in Figure 6, we have

$$\frac{a}{d} = \frac{b}{e} = \frac{c}{f},$$

from which it also follows that

$$\frac{a}{b} = \frac{d}{e}, \quad \frac{a}{c} = \frac{d}{f}, \quad \text{and} \quad \frac{b}{c} = \frac{e}{f}.$$

> The sum of the three vertex angles of a triangle is always 180° (π radians).

Therefore,

> Two vertex angles of a triangle determine the third one.

In particular,

> If two vertex angles of one triangle are equal to two vertex angles of another triangle, then the two triangles are similar.

If one triangle can be obtained from another by vertical shifting, horizontal shifting, rotation, or reflection across a line, the triangles are said to be **congruent.** Congruent triangles have exactly the same size and shape, their corresponding sides have the same length, and their corresponding vertex angles are the same (Figure 7).

Figure 7

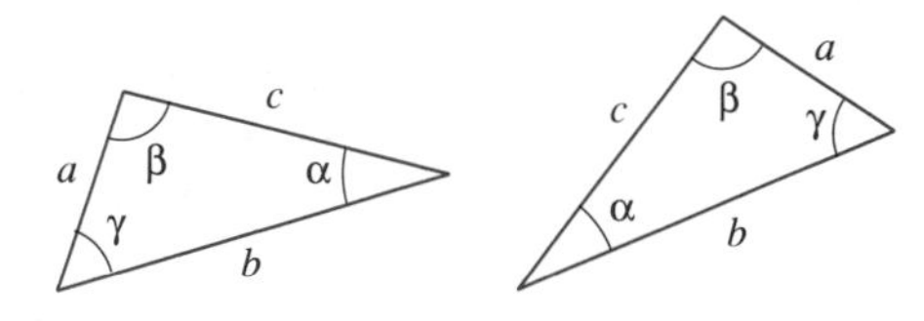

congruent triangles

> If two triangles satisfy any of the following three conditions, then they are congruent:
>
> **1.** (SSS) Three sides of one triangle are equal, respectively, to three sides of the other.
>
> **2.** (SAS) Two sides and the included angle of one triangle are equal, respectively, to two sides and the included angle of the other.
>
> **3.** (ASA) Two angles and the included side of one triangle are equal, respectively, to two angles and the included side of the other.

Circles

Figure 8

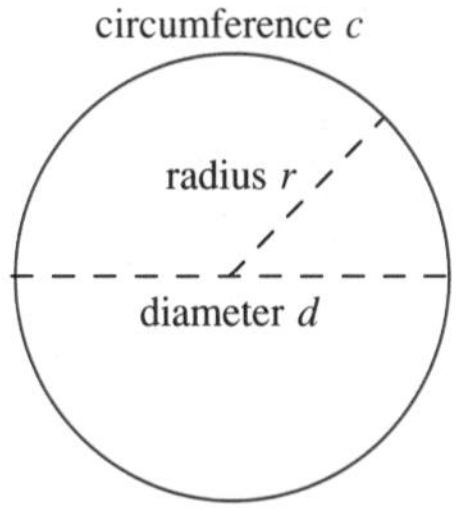

circumference c

radius r

diameter d

The ratio of the **circumference** c to the **diameter** d of a circle (Figure 8) is the constant π (the Greek letter pi). The number π is irrational; hence, written as a decimal, it is nonrepeating and nonterminating,

$$\pi = 3.14159265358979323\ldots.$$

Because $\pi = c/d$, it follows that

$$c = \pi d = 2\pi r.$$

Example 3 The radius of the earth is approximately 3959 miles. Find the circumference of the earth at the equator, rounded off to four significant digits.

Solution

$$c = 2\pi r = 2\pi(3959) \approx 24{,}880 \text{ miles.}$$

■

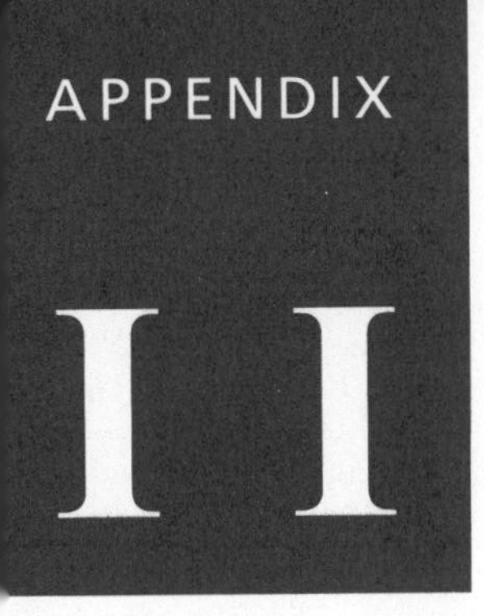

Formulas for Areas and Volumes

In this appendix, we summarize the formulas for areas of plane regions and for surface areas and volumes of solids.

Areas of Plane Regions

1. Triangle

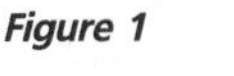
Figure 1

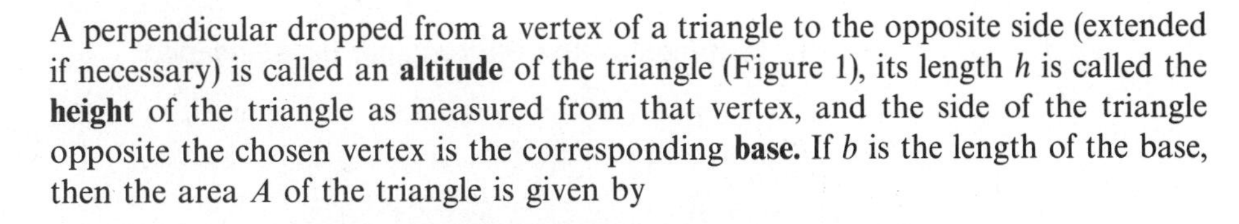

A perpendicular dropped from a vertex of a triangle to the opposite side (extended if necessary) is called an **altitude** of the triangle (Figure 1), its length h is called the **height** of the triangle as measured from that vertex, and the side of the triangle opposite the chosen vertex is the corresponding **base.** If b is the length of the base, then the area A of the triangle is given by

$$A = \tfrac{1}{2}hb.$$

2. Polygon

Figure 2

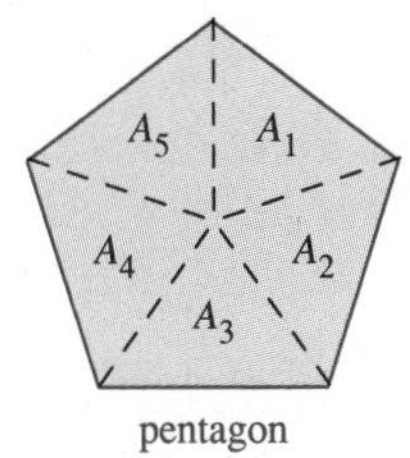

A region in the plane that is bounded by line segments is called a **polygon.** Figure 2 shows a polygon with five sides (called a **pentagon**). The area of a polygon can be found by breaking it up into triangles and adding the areas of the triangles. For instance, the area A of the pentagon in Figure 2 is given by

$$A = A_1 + A_2 + A_3 + A_4 + A_5.$$

3. Parallelogram

Figure 3

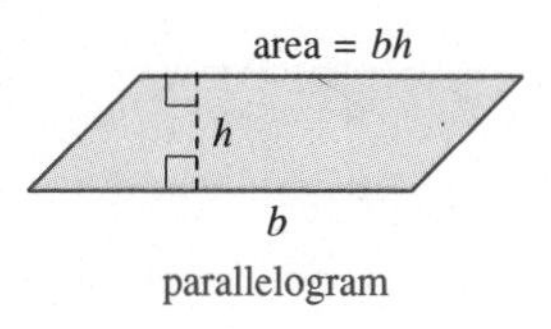

A polygon with four sides is called a **quadrilateral.** A quadrilateral in which opposite sides are parallel and have equal lengths is called a **parallelogram** (Figure 3). If one of the sides of a parallelogram is chosen and called the **base,** then the perpendicular distance between that side and the opposite side is the **height** h of the parallelogram with respect to that base. If b is the length of the base, then the area A of the parallelogram is given by

$$A = bh.$$

Figure 4

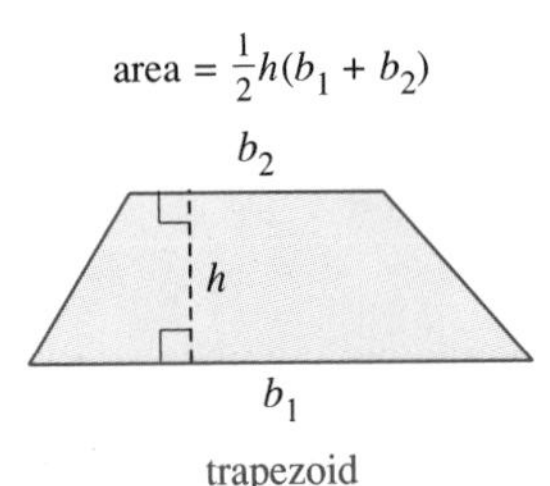

trapezoid

4. Trapezoid

A quadrilateral with two parallel sides is called a **trapezoid** (Figure 4). The two parallel sides are called the **bases,** and the perpendicular distance between the two bases is the **height** h of the trapezoid. If b_1 and b_2 are the lengths of the two bases, then the area A of the trapezoid is given by

$$A = \tfrac{1}{2}h(b_1 + b_2).$$

Figure 5

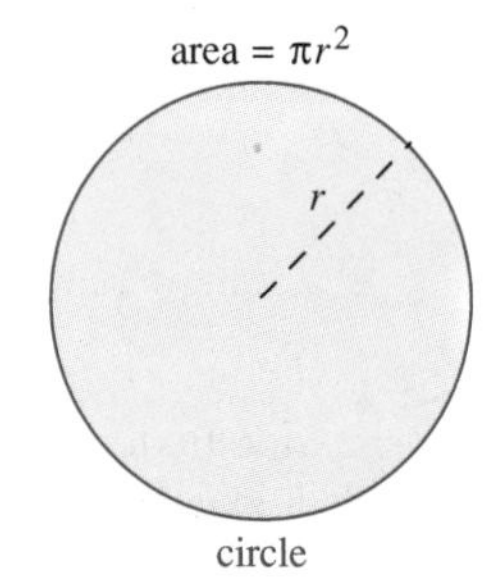

circle

5. Circle

The area A of a circle of radius r (Figure 5) is given by

$$A = \pi r^2.$$

Volume and Surface Area of Solids

1. Rectangular Box (Rectangular Parallelepiped)

Figure 6

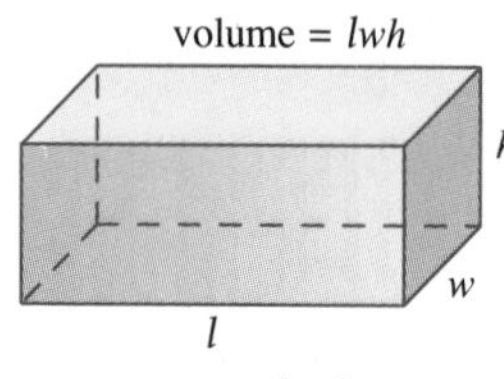

rectangular box

The volume V of a rectangular box (Figure 6) is the product of its **length** l, its **width** w, and its **height** h:

$$V = lwh.$$

The total surface area A of such a box is the sum of the areas of its six rectangular faces. Thus,

$$A = 2(lw + lh + wh).$$

If the box is hollow, but not closed, its total surface area is the sum of the areas of the rectangular faces that are present. For instance, if the box has no top, then

$$A = lw + 2(lh + wh).$$

2. Right Circular Cylinder

Figure 7

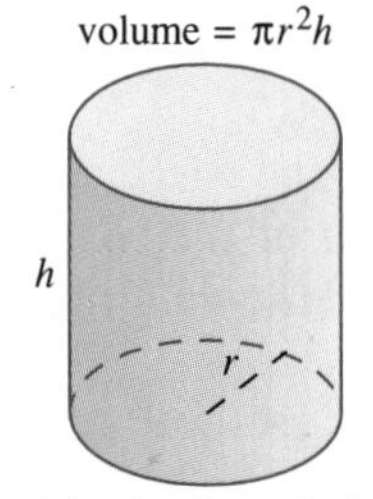

right circular cylinder

The volume V of a right circular cylinder (Figure 7) of **radius** r is the product of its **height** h and the **area** πr^2 of its circular base:

$$V = \pi r^2 h.$$

The lateral surface area S_L of the cylinder is the product of its **height** h and the **circumference** $2\pi r$ *of its circular base:*

$$S_L = 2\pi rh.$$

The total surface area S of the cylinder is the sum of the lateral surface area and the areas of its two circular bases:

$$S = 2\pi r^2 + 2\pi rh.$$

Figure 8

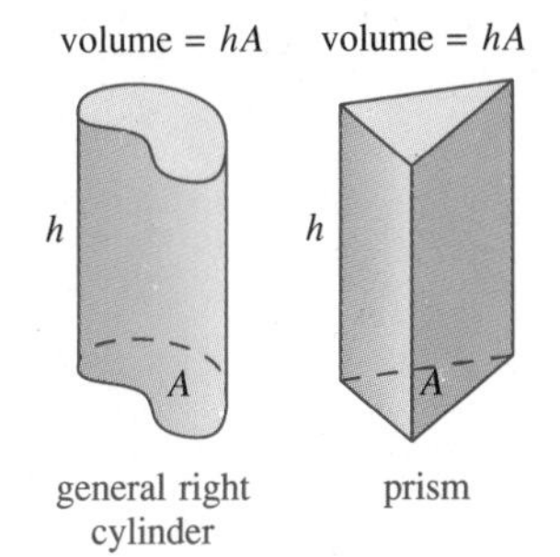

3. General Right Cylinder

The volume V of a general right cylinder (Figure 8a) is the product of its **height** h and the **area** A of one of its bases:

$$V = hA.$$

In particular, this holds for a **prism,** that is, a right cylinder with a base in the shape of a triangle or more general polygon (Figure 8b).

Figure 9

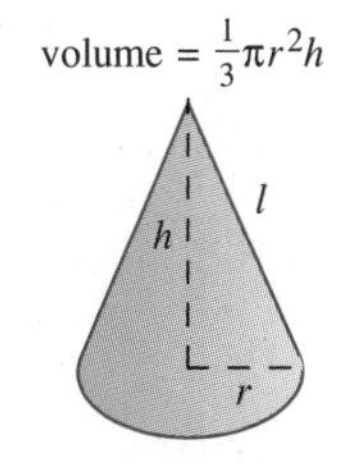

4. Right Circular Cone

The volume V of a right circular cone (Figure 9) with **base radius** r is one-third of the product of its **height** h and the **area** πr^2 of its base:

$$V = \tfrac{1}{3}\pi r^2 h.$$

If l is the **slant height** of the cone, then the lateral surface area S_L of the cone is given by

$$S_L = \pi rl,$$

and the total surface area S is the sum of the lateral surface area and the area of the circular base:

$$S = \pi rl + \pi r^2.$$

Figure 10

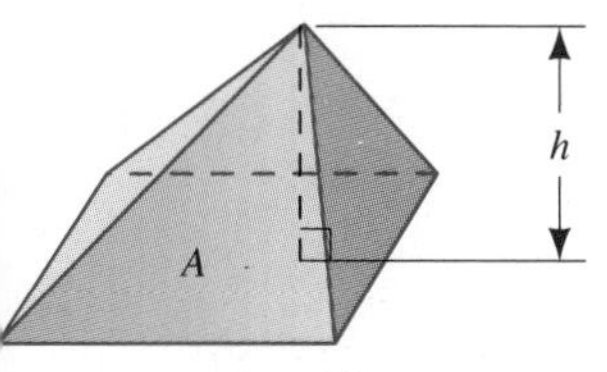

5. Pyramid

The volume V of a pyramid with a base in the shape of a triangle, rectangle, or other polygon (Figure 10) is one-third of the product of its **height** h and the **area** A of its base:

$$V = \tfrac{1}{3}Ah.$$

Figure 11

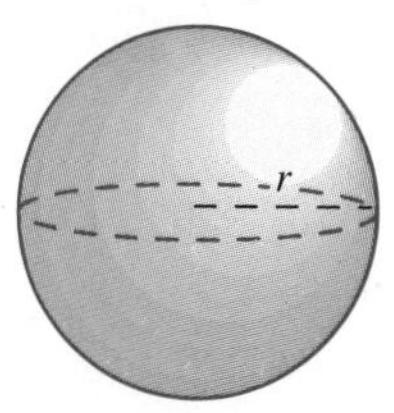

6. Sphere

The volume V and the surface area S of a sphere of **radius** r (Figure 11) are given by the formulas

$$V = \tfrac{4}{3}\pi r^3 \quad \text{and} \quad S = 4\pi r^2.$$

APPENDIX

III Computer Program for Locating Zeros by the Bisection Method

The following computer program, written in BASIC by Robert J. Weaver, uses the bisection method (Section 4.5, page 263) to locate zeros. For demonstration purposes, the function $F(X) = X^3 + X - 1$ is entered in line 100 of the program (see Example 3 in Section 4.5, page 264). To use the program for other functions, edit line 100 and replace X^3 + X − 1 by the function of your choice (written in BASIC). The program is especially useful for working the exercises in Problem Set 4.5, page 266.

```
100 DEF FN F(X) = X^3 + X - 1
110 PRINT "========== ENTER ENDPOINTS =========="
120 PRINT
130 PRINT "WHAT IS YOUR INITIAL LEFT ENDPOINT";
140 INPUT A
150 PRINT "WHAT IS YOUR INITIAL RIGHT ENDPOINT";
160 INPUT B
170 PRINT
180 IF A < B THEN GOTO 230
190 PRINT "THE LEFT ENDPOINT MUST BE LESS THAN"
200 PRINT "THE RIGHT ENDPOINT.  TRY AGAIN"
210 PRINT
220 GOTO 110
230 PRINT "F(";A;") = "; FN F(A)
240 PRINT "F(";B;") = "; FN F(B)
250 PRINT
260 REM - VARIABLE 'S' DETERMINES A CHANGE OF SIGN
270 LET S = FN F(A)*FN F(B)
280 IF S < 0 THEN GOTO 410
290 REM - ENDPOINTS NOT APPROPRIATE
300 PRINT "=============== NOTICE ==============="
310 PRINT
320 PRINT "F DOES NOT HAVE DIFFERENT ALGEBRAIC"
330 PRINT "SIGN AT THE ENDPOINTS YOU HAVE CHOSEN."
340 PRINT
350 PRINT "DO YOU WISH TO TRY AGAIN?"
360 PRINT "'Y' OR 'N'";
370 INPUT A$
380 IF A$ = "N" THEN GOTO 920
390 PRINT
400 GOTO 110
410 PRINT "HOW MANY TIMES DO YOU WISH TO BISECT"
420 PRINT "THE INTERVAL FROM ";A;" TO ";B;
430 INPUT N
440 IF N = INT(N) AND N > 0 THEN GOTO 490
```

```
450 PRINT
460 PRINT "ENTER A POSITIVE INTEGER, PLEASE!"
470 PRINT
480 GOTO 410
490 PRINT
500 PRINT "DO YOU WISH TO PAUSE AFTER EACH"
510 PRINT "BISECTION? 'Y' OR 'N'";
520 INPUT P$
530 PRINT
540 PRINT "=============== BEGIN ================"
550 PRINT
560 LET L = A
570 LET R = B
580 FOR I = 1 TO N
590 LET C = (L + R)/2
600 LET F1 = FN F(L)
610 LET F2 = FN F(C)
620 LET S = F1 * F2
630 IF S = 0 THEN GOTO 870
640 IF S > 0 THEN GOTO 680
650 REM - F CHANGES SIGN ON THE LEFT-HAND INTERVAL
660 LET R = C
670 GOTO 700
680 REM - F CHANGES SIGN ON THE RIGHT-HAND INTERVAL
690 LET L = C
700 PRINT "AFTER BISECTION NUMBER ";I
710 PRINT "F(";L;") = "; FN F(L)
720 PRINT "F(";R;") = "; FN F(R)
730 IF P$ <> "Y" THEN GOTO 770
740 PRINT "CONTINUE - 'Y' OR 'N'";
750 INPUT Q$
760 IF Q$ = "N" THEN GOTO 790
770 PRINT "------------------------------------"
780 NEXT I
790 REM - CALCULATE THE MIDPOINT OF THE LAST INTERVAL
800 LET C = (L + R)/2
810 PRINT
820 PRINT "THERE IS A ZERO IN THE INTERVAL FROM"
830 PRINT L;" TO ";R;"."
840 PRINT "THE MIDPOINT OF THIS INTERVAL IS"
850 PRINT C
860 GOTO 920
870 REM - COME HERE WHEN THERE IS A ZERO FOUND
880 REM - DIRECTLY FROM CALCULATION AT A MIDPOINT
890 PRINT "THERE IS A ZERO AT X = ";C
900 PRINT "THIS WAS FOUND AT A MIDPOINT AFTER"
910 PRINT I;" BISECTION(S)."
920 END
```

Answers to Selected Problems

CHAPTER 1

Problem Set 1.1, page 9

1. **(a)** 5^8; **(b)** $(-7)^4$; **(c)** $3^4 \cdot 4^3$; **(d)** $8^2 + (-6)^3$
3. $z = 2(x + y)$ **5.** $A = \pi r^2$ **7.** $x = 0.05n$
9. $A = \frac{1}{2}bh$ **11.** $A = p + prt$ **13.** $V = lw^2$
15. $A = \$6077.53$ **17.** **(a)** Commutative for $+$; **(b)** commutative for $\times$; **(c)** commutative for $\times$; **(d)** distributive; **(e)** identity for $\times$; **(f)** associative for $\times$; **(g)** associative for $+$; **(h)** inverse for $\times$
19. **(a)** Negation (i); **(b)** negation (iii); **(c)** negation (ii); **(d)** zero factor (i); **(e)** cancellation (i)
21. **(a)** $(3 + 5)^2 = 64$; **(b)** $1/(3 + 5) = 1/8$
23. $A = \{4, 6, 8, 10\}$

0 1 2 3 4 5 6 7 8 9 10 11

25. **(a)** F; **(b)** T; **(c)** T; **(d)** F; **(e)** T **27.** **(a)** 0.8; **(b)** -0.375; **(c)** -0.26; **(d)** -0.085; **(e)** -0.625; **(f)** $-1.\bar{6}$; **(g)** $2.\overline{142857}$ **29.** **(a)** $\frac{41}{100}$; **(b)** $\frac{-4}{125}$; **(c)** $\frac{-581}{1000}$; **(d)** $\frac{456}{5}$; **(e)** $\frac{10,451}{10,000}$ **31.** **(a)** 0.11; **(b)** 0.0103; **(c)** $4.3\overline{2}$; **(d)** 0.000006 **33.** **(a)** 50%; **(b)** 2.4%; **(c)** $66.\bar{6}$%; **(d)** $28.\overline{571428}$% **35.** **(a)** 2.1; **(b)** 15%
37. 20% increase **39.** $16.\bar{6}$% decrease
41. 25% decrease

Problem Set 1.2, page 16

1. 1.55×10^4 **3.** 5.8761×10^7 **5.** 1.86×10^{11}
7. 9.01×10^{-7} **9.** 33,300 **11.** 0.00004102
13. 10,010,000 **15.** 6.2×10^{-2} s
17. 9.29×10^7 mi; 1.92×10^{13} mi **19.** 2.51×10^{-10} s
21. 4.8×10^8 **23.** 7.00×10^{31} **25.** 6.56×10^{-4}
27. 2 **29.** 5 **31.** 5300 **33.** 0.015
35. 110,000 **37.** 2.14×10^{-13} **39.** 2.3102×10^2
41. 4.867×10^{-4} **43.** 1.51 **45.** 1.06×10^{27}
47. 3.47×10^5 **49.** **(a)** $\$2.1208 \times 10^{12}$; **(b)** \$19,220

Problem Set 1.3, page 23

1. $Q = kV$ **3.** $d = kt^2$ **5.** $P = kv^3$
7. $E = kAT/d$ **9.** $F = kQ_1Q_2/d^2$ **11.** $r = kp(1 - p)$
13. $R = k(T - S)$ **15.** **(a)** $y = \frac{1}{2}x$; **(b)** 6
17. **(a)** $w = 12/x$; **(b)** $\frac{3}{2}$ **19.** **(a)** $u = 200/v^2$; **(b)** $\frac{200}{9}$
21. **(a)** $V = 5.049T/P$; **(b)** 74.43 **23.** 40 gal
25. $\frac{15}{8}$ days **27.** **(a)** $d = F/15$; **(b)** 0.6 m
29. $\approx 2.36 \times 10^{-3}$ N **31.** $\approx 8.6 \times 10^{-7}$ N

Problem Set 1.4, page 31

1. **(a)** 18; **(b)** 400 **3.** **(a)** monomial, polynomial in x, degree 2, coefficients: -4, 0, 0; **(b)** trinomial, multinomial in x and y; **(c)** multinomial, polynomial in x and z, degree 3, coefficients: 3, $-\frac{6}{11}$, -1, -1, 2; **(d)** multinomial, polynomial in x, degree 5, coefficients: $\sqrt{2}, \pi, -\sqrt{\pi}, \frac{12}{13}, -5$; **(e)** multinomial, polynomial in x, y, z, and w, degree 3, coefficients: 1, 1, 1, 1 **5.** **(a)** $2x + 12$; **(b)** $z^2 + z + 2$; **(c)** $18x^2 + x + 5$; **(d)** $5s^2 + \pi r^2$ **7.** **(a)** $-x + 9$; **(b)** $4z^2 + 10z + 3$; **(c)** $-7t^2 + 14t - 9$; **(d)** $x^2 - 2xy - 1$
9. **(a)** $6x^7$; **(b)** 2^{10n}; **(c)** $(3x + y)^6$; **(d)** t^{48}
11. **(a)** u^{5n^2}; **(b)** $324v^6$; **(c)** $675a^7b^8$; **(d)** $864(x + 3y)^{22}$
13. $15x^4 - 20x^4y^2 + 20x^3yz$ **15.** $-12x^2 + 7xy + 10y^2$
17. $x^3 + 8y^3$ **19.** $625c^4 - d^4$
21. $3x^4 + 20x^3y - 9x^2y^2 - 2x^3 - 15x^2y + 2x^2 + 8xy^3 - 5xy^2 + 14xy - y^4 + y^3 - 2y^2$
23. $4p^4q^2 - 11p^3q^3 - 7p^3q^2 + 6p^2q^4 - p^2q^3 - 15p^2q^2$
25. $16 + 24x + 9x^2$
27. $16t^4 + 8t^2 + 1 - 8st^2 - 2s + s^2$ **29.** $9r^2 - 4s^2$
31. $4a^2 - 12ab + 9b^2 - 16c^2$ **33.** $8 + t^2$
35. $8x^6 + 36x^4y + 54x^2y^2 + 27y^3$
37. $4x^2 - 4xy + y^2 + 4xz - 2yz + z^2$
39. $t^6 - 4t^5 + 14t^4 - 16t^3 + 17t^2 + 20t + 4$
45. **(a)** $14x - 12$; **(b)** $10x^2 - 3x - 27$
47. $64x^3 + 144x^2 + 108x + 27$
49. $x^4 + 8x^3y + 24x^2y^2 + 32xy^3 + 16y^4$
51. $32x^5 - 80x^4y^2 + 80x^3y^4 - 40x^2y^6 + 10xy^8 - y^{10}$

Problem Set 1.5, page 38

1. $a^2b(a + 2 + b)$ **3.** $2r(x - y)(r + h)$
5. $(8x - 6y)(8x + 6y)$ **7.** $(3y - z)(3y + z)(9y^2 + z^2)$
9. $(x + y - 7z)(x + y + 7z)$
11. $8(x + 5)(x^2 - 5x + 25)$
13. $(4x - y^2)(16x^2 + 4xy^2 + y^4)$
15. $(x^2 + 8y^2)(x^4 - 8x^2y^2 + 64y^4)$
17. $(4x + 7)(2x - 1)$ **19.** $(2v - 5)(v + 4)$
21. $(2r - 3s)^2$ **23.** $(3x^2 + 2)(2x^2 + 3)$ **25.** Prime
27. Factorable, $(4r - 1)(3r - 2)$ **29.** Prime
31. $(3 + y)(x - 2)$ **33.** $(a - 1)(a + 1)(x + d)$
35. $(2x + 3 - 3y)(2x + 3 + 3y)$
37. $3(5x - 4y)(5x + 4y)$ **39.** $7x(y - z)(1 + 2x)$
41. $(11w - 10z)(3w + 4z)$
43. $(4x^2 + p - q)(16x^4 - 4x^2p + 4x^2q + p^2 - 2pq + q^2)$
45. $(x + 1)(x^2 - x + 1)(x - 2)(x + 2)$
47. $(2u - 3)(2u + 3)(3v - 1)$
49. $(p - q)^2(p^2 + pq + q^2)$
51. $(x^2 + y^2 + xy)(x^2 + y^2 - xy)$ **53.** $(t + 2)^2(t - 2)^2$
55. $(2s + 1)(a - b)(a^2 + ab + b^2)$
57. $(x^4 + 16y^4)(x^2 + 4y^2)(x + 2y)(x - 2y)$
59. $y^3z(5x + y)(2x - 3y)$ **61.** $(x - 3y)(x + 3y + 1)$
63. $(9a - 6b + 7)(3a - 2b - 2)$
65. $4t(5t^2 + 1)(3t^4 + 2)^3(15t^4 + 12t^2 + 10)$
67. $2\pi r(r + h)$

Problem Set 1.6, page 47

1. $\dfrac{by}{3}$ **3.** $\dfrac{c}{2}$ **5.** $\dfrac{-1}{5 + t}$ **7.** $\dfrac{3y + 1}{y + 5}$ **9.** 1
11. $\dfrac{3c + 1}{2c - 1}$ **13.** $\dfrac{3r + 3t + 2}{2r + 2t - 3}$ **15.** $c + 2$
17. $\dfrac{3(x^2 + 5)(x + 1)}{(x^2 + 6x + 15)(x - 1)}$ **19.** $\dfrac{1}{a + 1}$ **21.** $(x - y)^2$
23. $\dfrac{2u - 1}{u + 2}$ **25.** $\dfrac{2x + 3}{x + 3}$ **27.** $\dfrac{t + 5}{t - 6}$ **29.** $x - 1$
31. $\dfrac{36u^2 - 40u - 13}{4(2u + 1)(2u - 1)}$ **33.** $\dfrac{2y^2 + 2y - 1}{(y - 2)(y + 1)}$ **35.** $\dfrac{4}{t - 2}$
37. $\dfrac{-u(3u^3 - u^2 + 8u + 2)}{(u^2 + 3)(u - 1)^2(u + 2)}$ **39.** $\dfrac{5x^2 - 4x + 1}{x^3(2x - 1)^2}$
41. $\dfrac{1}{cd(c + d)}$ **43.** $(a + b)ab$ **45.** $\dfrac{pq(q + p)}{q^2 + qp + p^2}$
47. -1 **49.** $-\dfrac{2x + h}{x^2(x + h)^2}$ **51.** $\dfrac{2(27x^2 - 14x + 2)}{(7x + 2)^4}$
53. **(a)** $\dfrac{4 - x}{4} = 1 - \dfrac{x}{4} \neq 1 - x$; **(b)** cannot cancel x since it is not a factor of both the entire numerator and the entire denominator; **(c)** cannot cancel c
55. y is getting closer and closer to 8 **59.** **(a)** $\frac{3}{4}$; **(b)** $\frac{1}{30}$; **(c)** $\frac{9}{5}$; **(d)** $\frac{3}{200}$; **(e)** 5 **61.** 30,000 N
63. ≈ 0.32 lb **65.** 2500

Problem Set 1.7, page 55

1. $\frac{3}{4}$ **3.** Not rational **5.** $\frac{2}{5}$ **7.** $\frac{1}{2}$ **9.** $\frac{3}{4}$
11. $3x\sqrt{3x}$ **13.** $5x^2y^4\sqrt{3y}$ **15.** $2ab\sqrt[3]{3ab^2}$
17. $(x - 1)^2$ **19.** $y/(2x^4)$ **21.** $\sqrt{15y}/(6y)$
23. $(-3x/y^7)\sqrt[3]{x}$ **25.** -66 **27.** $5t$
29. $x^2y^4\sqrt[4]{y^3}$ **31.** 1 **33.** $[3/(2y)]\sqrt[3]{xz}$
35. $\frac{5}{2}\sqrt[3]{x^2y}$ **37.** 98 **39.** $13\sqrt[3]{3}$ **41.** $19t^3\sqrt{5t}$
43. $\frac{1}{6}\sqrt[4]{6}$ **45.** $6(\sqrt{2} - \sqrt{3})$ **47.** $-1 - \sqrt{3}$
49. -13 **51.** 4 **53.** $4 + 2\sqrt{6}$ **55.** $8\sqrt{11}/11$
57. $\sqrt{10}/3$ **59.** $\dfrac{\sqrt{2(x + 2y)}}{3(x + 2y)}$ **61.** $4\sqrt[3]{25}$
63. $\dfrac{5\sqrt{3} - 3}{22}$ **65.** $\dfrac{7 + 2\sqrt{10}}{3}$
67. $\dfrac{2a + 5 + 2\sqrt{a^2 + 5a}}{5}$ **69.** $\dfrac{5(4 + 2\sqrt[3]{x} + \sqrt[3]{x^2})}{8 - x}$
71. $\dfrac{a - b - 2\sqrt[3]{a^2b} + 2\sqrt[3]{ab^2}}{a + b}$ **73.** $\dfrac{1}{\sqrt{x + h} + \sqrt{x}}$
75. $\dfrac{2x + h}{\sqrt{(x + h)^2 + 1} + \sqrt{x^2 + 1}}$ **79.** 2.42×10^9 Hz
83. $(90\sqrt{3} + 20\sqrt{147})$ m **85.** $\approx 36{,}064$

Problem Set 1.8, page 63

1. 27 **3.** $\frac{1}{64}$ **5.** $y^4/(x^2z)$ **7.** -1 **9.** $-28c^2$
11. $y^2/(2x^5)$ **13.** $9t/(9 + t)$ **15.** $1/x^n$ **17.** $\dfrac{p^2 - 3}{p^2 - 1}$
19. $\dfrac{(t - 2)^2}{t + 2}$ **21.** $(a/b)^{3n}$ **23.** $-1/(2q)$
25. **(a)** 27; **(b)** -32; **(c)** 8; **(d)** not defined; **(e)** 4; **(f)** 30
27. **(a)** ≈ 2.374647381; **(b)** ≈ 79.702251523; **(c)** ≈ -1.912931183; **(d)** ≈ -0.522757959
29. a^2 **31.** $16p^{12}$ **33.** $u^3/8$ **35.** $x^3/8$
37. $1/x^{11}$ **39.** $2r^3s^2/3$ **41.** y^4/x^2
43. $\dfrac{3x + 2y}{4r + 3t}$ **45.** $\dfrac{1}{(m^2 - n^2)^{1/3}}$ **47.** $\dfrac{(x + y)^2}{\sqrt{x}}$
49. $\dfrac{-(p + 3)}{(p - 1)^3}$ **51.** $\dfrac{2(8t + 1)}{(4t - 1)^2(2t + 1)^2}$
53. $2(3x + x^{-1})(6x - 1)^4(63x + 9x^{-1} + x^{-2} - 3)$
55. $2(7y + 3)^{-3}(2y - 1)^3(14y + 19)$

57. **(a)** Exponent not reduced to lowest terms; **(b)** $\sqrt[m]{a^n} = (\sqrt[m]{a})^n$ only when $\sqrt[m]{a}$ is defined; **(c)** to apply rule, all expressions must be defined; **(d)** exponent not reduced to lowest terms; **(e)** can't apply rule when number is negative **59.** ≈ 13.86, so shoe size is 14 **61.** ≈ 689 days **63.** $\approx 4.69 \times 10^{22}$

Problem Set 1.9, page 69

1. **(a)** $\sqrt{2}i$; **(b)** $3\sqrt{3}i$ **3.** $3 + 4i$ **5.** $0 + 30\sqrt{2}i$ **7.** $9 + i$ **9.** $10 - i$ **11.** $19 - 4i$ **13.** $-2 - 2i$ **15.** $-11 - 4i$ **17.** $-1 - 14i$ **19.** $-3 + 11i$ **21.** $15 - 23i$ **23.** $64 - 40i$ **25.** $-20 + 0i$ **27.** $\frac{13}{36} + 0i$ **29.** $0 + i$ **31.** $0 + i$ **33.** $12 + 16i$ **35.** $-\frac{1}{2} + (\sqrt{3}/2)i$ **37.** $\frac{2}{13} - \frac{3}{13}i$ **39.** $\frac{21}{53} - \frac{6}{53}i$ **41.** $\frac{1}{5} - \frac{11}{10}i$ **43.** $\frac{11}{34} + \frac{41}{34}i$ **45.** $-\frac{13}{10} - \frac{19}{10}i$ **47.** $\frac{63}{290} - \frac{201}{290}i$ **49.** **(a)** $2 - i$; **(b)** 4; **(c)** $2i$; **(d)** 5 **51.** **(a)** i; **(b)** 0; **(c)** $-2i$; **(d)** 1 **53.** **(a)** $-i$; **(b)** 0; **(c)** $2i$; **(d)** 1 **55.** **(a)** $-\frac{1}{5} + \frac{11}{10}i$; **(b)** $-\frac{2}{5}$; **(c)** $-\frac{11}{5}i$; **(d)** $\frac{5}{4}$ **61.** 0 **63.** $14.1 - 2.1i$ ohms

Review Problem Set

CHAPTER 1, page 70

1. $w = 3xy/z$ **3.** $s = \frac{1}{2}(a + b + c)$ **5.** $P = 1000(3^n)$ **7.** 315 J **9.** **(a)** Commutative for $\times$; **(b)** distributive; **(c)** associative for $+$; **(d)** identity for $+$; **(e)** commutative for $\times$; **(f)** identity for $+$; **(g)** identity for $\times$; **(h)** associative for $\times$; **(i)** inverse for $\times$; **(j)** inverse for $+$ **11.** **(a)** No common denominator; **(b)** unless both a and b are zero, $\sqrt{a + b} \neq \sqrt{a} + \sqrt{b}$ **13.** **(a)** 0.22 or 22%; **(b)** -0.003 or -0.3%; **(c)** -0.085 or -8.5% **(d)** -0.875 or -87.5%; **(e)** 3.25 or 325%; **(f)** $-2.\bar{3}$ or $-233.\bar{3}\%$ **15.** 6.25% **17.** **(a)** 5.712×10^{10}; **(b)** 7.14×10^{-7} **19.** **(a)** 17,320,000; **(b)** 0.0000000312 **21.** **(a)** 5×10^{-6} g; **(b)** 6.6×10^{-17} g **23.** **(a)** 14.78976; **(b)** 61.8003×10^{31} **25.** **(a)** 3; **(b)** 1 **27.** **(a)** 17,000; **(b)** 0.003 **29.** 3.6×10^4 **31.** 2.65×10^{33} **33.** 1.8×10^{-5} **35.** **(a)** $F = kAv^2$; **(b)** $L = kAN^2/l$; **(c)** $r = kN(P - N)$; **(d)** $I_C = k\beta I_B$ **37.** **(a)** $y = 4x^2/(z + 3)$; **(b)** 4 **39.** **(a)** $w = 7x/(3y)$; **(b)** $\frac{28}{3}$ **41.** **(a)** $V = \frac{4.19}{8}d^3$; **(b)** ≈ 524 m^3 **43.** **(a)** Monomial, polynomial, degree 2, coefficients: $-2, 0, 0$; **(b)** rational expression, monomial; **(c)** trinomial, multinomial, polynomial, degree 2, coefficients: $3, -5, -1$; **(d)** binomial, multinomial, radical expression; **(e)** binomial, multinomial, polynomial, degree 3, coefficients: $\sqrt{2}, 0, 0, -\sqrt[5]{7}$; **(f)** binomial, multinomial, equivalent to rational expression $(\sqrt{\pi}\, y + x)/y$ **45.** **(a)** $20y^5$; **(b)** $-x^6y^4$; **(c)** p^8; **(d)** $-x^{14}y^7$; **(e)** $a^{1+n}b^{2n}$ **47.** $3x^3 - 3x^2 + 2x + 6$ **49.** $2x^2y^4 + 7xy^2 + 3$ **51.** $2p^3 - 5p^2 - 10p + 3$ **53.** $2x^3 - 3x^2 - 5x + 6$ **55.** $4x^4 - 20x^2yz + 25y^2z^2$ **57.** $4x^2 + y^2 + 9z^2 - 4xy + 12xz - 6yz$ **59.** $27x^9 - 54x^7y + 36x^5y^2 - 8x^3y^3$ **61.** $3xy^2(3x - 4y^2)$ **63.** $(a + b)c^2(a + b - c^2)$ **65.** $(x - y + z)(x - y - z)$ **67.** $x^2(5 + 7y)(5 - 7y)$ **69.** $(2p + 3q)(4p^2 - 6pq + 9q^2)$ **71.** $(x + 6)(x - 4)$ **73.** $(3x - 2y)(2x + 3y)$ **75.** $(4uv + 1)(uv - 2)$ **77.** $(4 + 3x)(5 - 2x)$ **79.** $(p + 3q + 2)(p + 3q - 2)$ **81.** $\dfrac{t + 3}{t + 2}$ **83.** $\dfrac{c - 3}{c - 2}$ **85.** $\dfrac{2(x + 1)}{x + 2}$ **87.** $2y$ **89.** $1/t$ **91.** $3/(c + 3)$ **93.** $1/(a - 3)$ **95.** $-1/(a + 5)$ **97.** **(a)** $\frac{240}{13}$; **(b)** $\frac{4000}{1}$; **(c)** $\frac{1}{40}$; **(d)** $\frac{264}{5}$; **(e)** $\frac{25}{16}$; **(f)** $\frac{8}{5}$ **99.** $x = \frac{5}{2}$ **101.** 4 **103.** 5000 **105.** $\sqrt[5]{0.6}$ is irrational **107.** $\sqrt[4]{-81}$ is not real **109.** $2x^2$ **111.** $(a + b)^3/(3a)$ **113.** $a^2\sqrt[3]{5a}$ **115.** $13\sqrt[3]{2p}$ **117.** $5\sqrt{6}$ **119.** $2a - 3$ **121.** $2a + b - 2\sqrt{a^2 + ab}$ **123.** $\dfrac{3\sqrt{2x}}{x}$ **125.** $\dfrac{a + \sqrt{ab}}{a - b}$ **127.** $\dfrac{\sqrt{a - 1} - a + 1}{2 - a}$ **129.** $5(\sqrt[3]{4} + \sqrt[3]{2} + 1)$ **131.** $\dfrac{9x - y}{5(3\sqrt{x} - \sqrt{y})}$ **133.** $\dfrac{x + y}{x^2y^2}$ **135.** $\dfrac{x^2 - 1}{x^4}$ **137.** $\dfrac{3}{a^6}$ **139.** $\dfrac{x^{13}}{y^6}$ **141.** $1/(25p^{10})$ **143.** **(a)** $\frac{4}{9}$; **(b)** $\frac{1}{512}$ **145.** y **147.** a^3 **149.** $y^{3/2}$ **151.** $2^{14}b^{13}c^{2/3}$ **153.** $x^2y^{-12}(x + y)^{-4}$ **155.** $(y + 2)^{-2/3}(y + 1)^{-1/3}(3y + 5)$ **157.** $10 + 5i$ **159.** $13 - 8i$ **161.** $47 - 45i$ **163.** $-24 - 2i$ **165.** $\frac{16}{13} + \frac{11}{13}i$ **167.** $0 - i$ **169.** $0 + \frac{1}{27}i$ **171.** $12 - 5i$

Chapter 1 Test, page 76

1. **(a)** Distributive; **(b)** associative for $\times$; **(c)** identity for $\times$; **(d)** inverse for $+$; **(e)** commutative for $\times$; **(f)** inverse for $\times$ **2.** **(a)** F; **(b)** F; **(c)** F; **(d)** T; **(e)** T; **(f)** F; **(g)** F; **(h)** T; **(i)** T **3.** **(a)** $(3y + 1)/(y + 2)$; **(b)** $-(14a + 3)$; **(c)** $(a - b)^2$; **(d)** $a^2 + b^2$; **(e)** $-x^5\sqrt{x}$; **(f)** $xy\sqrt[5]{9x^4y}$; **(g)** $\frac{6}{5}$; **(h)** $27x^3 - 8y^3$ **4.** **(a)** $(x + 2)(x^2 - 2x + 4)(x - 1)(x^2 + x + 1)$; **(b)** $(3x + 5)^{-2/3}(4x + 1)(28x + 41)$ **5.** **(a)** $\dfrac{x + 4\sqrt{x} + 4}{x - 4}$; **(b)** $\dfrac{-1}{\sqrt{1 - x} + 1}$ **6.** **(a)** $\dfrac{2x + 3}{x^2 - 9}$;

(b) $\dfrac{x}{(x^2+1)(x-1)}$; **(c)** $-(x+2)$ **7. (a)** $V = 10{,}000/p$; **(b)** 2500 sales/wk **8.** $\approx 2.81 \times 10^{-21}$ m^3 **9.** $\approx 2 \times 10^{10}$ yr **10. (a)** $\frac{7}{2} - 3i$; **(b)** $8 + i$; **(c)** $\frac{21}{10} + \frac{7}{10}i$; **(d)** i

CHAPTER 2

Problem Set 2.1, page 83

1. (a) -2; **(b)** $\frac{4}{3}$; **(c)** $\frac{9}{2}$ **3.** -1.538 **5.** 9 **7.** 3 **9.** $-\frac{19}{2}$ **11.** 3 **13.** 12 **15.** 7 **17.** $-\frac{5}{6}$ **19.** 5 **21.** 4 **23.** $\frac{1}{3}$ **25.** $-\frac{5}{2}$ **27.** No solution **29.** ≈ -0.287 **31. (a)** $\frac{2}{9}$; **(b)** $\frac{563}{165}$; **(c)** $\frac{13}{330}$; **(d)** $-\frac{9917}{9900}$; **(e)** $\frac{7}{900}$ **33.** $\dfrac{5a+c}{b-10}$ **35.** $\dfrac{4(cd-b)}{4a-1}$ **37.** $\dfrac{a+dm}{2}$ **39.** $\dfrac{mn-1}{bc+d}$ **41.** $-b$ **43.** $\dfrac{V}{\pi r^2}$ **45.** $\dfrac{A-P}{Pr}$ **47.** $\dfrac{9C+160}{5}$ **49.** $\dfrac{S-a}{S-l}$ **51.** $\dfrac{IR}{(E-Ir)}$ **53. (a)** Identity; **(b)** identity; **(c)** conditional **55.** No; -6 is a root of the second equation, but not the first **57.** Yes; they have exactly the same (real) roots **59. (a)** $S - \dfrac{(A + B\sqrt{V} - CV)(S-T)}{A + 2B - 4C}$; **(b)** ≈ -3.26°F

Problem Set 2.2, page 92

1. 8, 20 **3.** 81 **5. (a)** -40°C $= -40$°F; **(b)** $-7.\overline{27}$° **7.** Antique car, 43 yr; replica car, 3 yr **9.** \$12,250 **11.** \$65,000 at 6% and \$75,000 at 6.5% **13.** \$75,000 **15.** \$750 **17.** 75 kg **19.** 200,000 gal of 9%, 100,000 gal of 12% **21.** 30 mL **23.** ≈ 1.8 L **25.** 31 five-dollar bills, 5 ten-dollar bills, and 37 twenty-dollar bills **27.** 42 ft **29.** 22 km/hr **31.** 61 mi/hr **33.** 3390 km (to three significant digits) **35.** Length is 26 m and width is 14 m **37.** 2.8 m **39.** 27 hr **41.** $4\frac{2}{7}$ hr **43.** 4.5 hr

Problem Set 2.3, page 104

1. 0, 7 **3.** -3, 1 **5.** -3, 7 **7.** $-\frac{3}{2}$, 5 **9.** $-\frac{1}{3}, \frac{5}{2}$ **11. (a)** $x^2 + 6x + 9 = (x+3)^2$; **(b)** $x^2 - 5x + \frac{25}{4} = (x - \frac{5}{2})^2$; **(c)** $x^2 + \frac{3}{4}x + \frac{9}{64} = (x + \frac{3}{8})^2$ **13. (a)** $-2 \pm \sqrt{19}$; **(b)** $-1 \pm (\sqrt{21}/3)$ **15.** $-\frac{3}{5}$, 2 **17.** $(-5 \pm \sqrt{13})/6$ **19.** $(1 \pm \sqrt{17})/4$ **21. (a)** Real, unequal; **(b)** $(5 \pm \sqrt{53})/2$ **23. (a)** Rational, double root; **(b)** $\frac{1}{3}, \frac{1}{3}$ **25. (a)** Complex conjugates; **(b)** $(-1 \pm \sqrt{71}\,i)/12$ **27. (a)** Real, unequal; **(b)** $-\frac{1}{2}, \frac{1}{3}$ **29. (a)** Complex conjugates; **(b)** $(2 \pm \sqrt{6}\,i)/5$ **31. (a)** Complex conjugates; **(b)** $(1 \pm \sqrt{3}\,i)/2$ **33.** $\frac{1}{5}, \frac{4}{3}$ **35.** $(1 \pm \sqrt{7})/4$ **37.** $-3, -3$ **39.** $-2 \pm \sqrt{3}$ **41.** $\frac{1}{3}, \frac{1}{3}$ **43.** -4, 2 **45.** $-\frac{27}{8}$, 1 **47.** $\frac{4}{5}, \frac{4}{5}$ **49.** -5, 21 **51.** $-\frac{5}{2}$, 1 **53.** $-\frac{5}{2}, \frac{7}{2}$ **55.** $(5 \pm \sqrt{13})/3$ **57.** ≈ -1.85, ≈ 0.38 **59. (a)** $\dfrac{-v_0 \pm \sqrt{{v_0}^2 + 2gs}}{g}$; **(b)** $\dfrac{-R \pm \sqrt{R^2 - (4L/C)}}{2L}$ **61.** 10 days **63.** 6 hr **65.** 100 s **67.** 25 poles **69.** 30 ft and 40 ft **71.** 12% **73. (a)** $(-1 + \sqrt{5})/2$; **(b)** no

Problem Set 2.4, page 112

1. $\pm\frac{1}{2}$ **3.** No real solutions **5.** $-\sqrt[3]{4}$ **7.** -2 **9.** 8 **11.** $\sqrt[5]{9}$ **13.** $\frac{31}{64}, \frac{33}{64}$ **15.** -1, 8, $(7 \pm \sqrt{17})/2$ **17.** 7 **19.** -2 **21.** 3 **23.** 8 **25.** 27 **27.** 2 **29.** 2 **31.** $\pm\sqrt{3}, \pm 2$ **33.** 1, 729 **35.** $\pm\frac{1}{4}, \pm 3$ **37.** $-1 \pm \sqrt{13}$ **39.** -19, 61 **41.** ± 1 **43.** ± 1 **45.** 1 **47.** 5 m by 12 m **49.** 2.25 km **51.** $l = \dfrac{d^2 + d\sqrt{d^2 + a^2}}{a}$

53. $V = \left[\dfrac{B \pm \sqrt{B^2 - 4C\left(\dfrac{H}{S-T} - A\right)}}{2C}\right]^2$

Problem Set 2.5, page 121

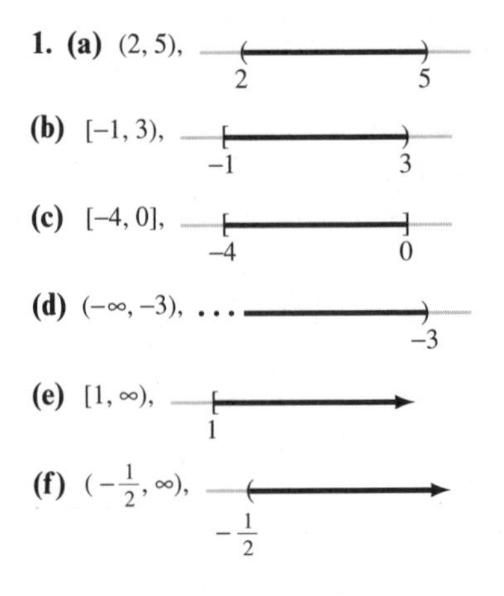

(g) $[-\frac{3}{2}, \frac{5}{2}]$,

(h) $(-4, 0]$,

3.

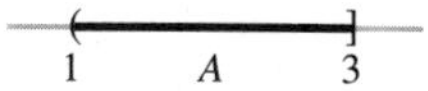

5.

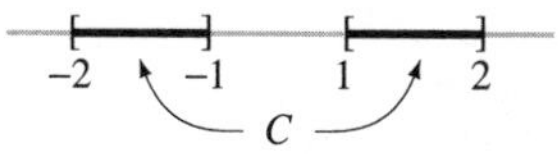

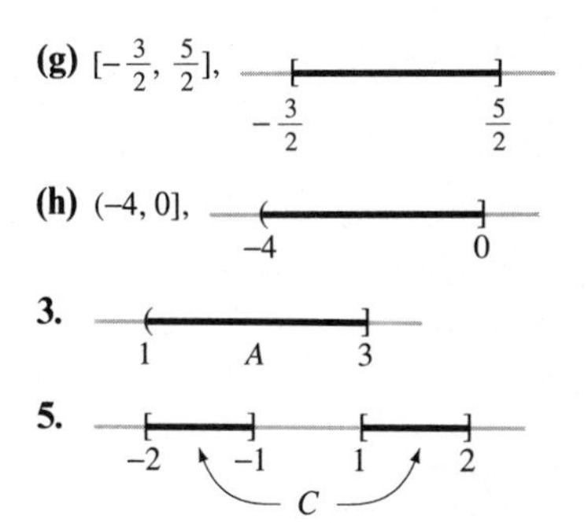

7. **(a)** Addition property; **(b)** transitivity property; **(c)** multiplication property; **(d)** trichotomy property
9. **(a)** $x + 4 > 1$; **(b)** $x - 4 > -7$; **(c)** $5x > -15$; **(d)** $-5x < 15$

11. $(-\infty, 5)$, . . .

13. $(-\infty, 5]$, . . .

15. $[-\frac{22}{3}, \infty)$,

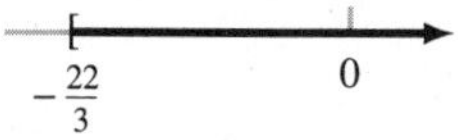

17. $(-\infty, \frac{24}{7})$, . . .

19. $[-13, \infty)$,

21. $[2, \infty)$,

23. $[1, 5]$,

25. $[-3, 2)$,

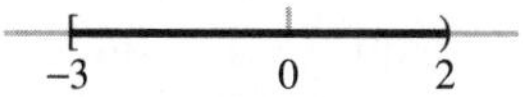

27. $[-2, 1]$,

29. $(-4, 4]$,

31. $[\frac{10}{9}, 10]$,

33. $[\frac{1}{3}, 1]$,

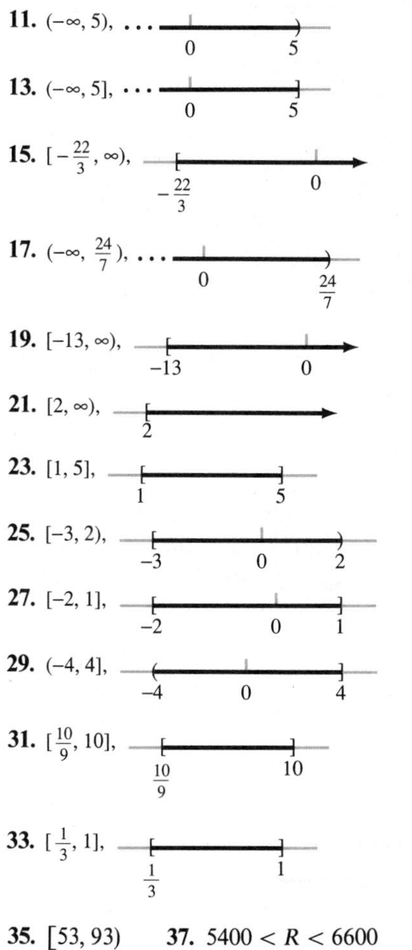

35. $[53, 93)$ **37.** $5400 < R < 6600$ **39.** $\frac{5}{2} \le p \le 4$
41. $(\frac{10}{3}, \frac{35}{6})$ **43.** $(11.7, 15.6)$ **45.** **(a)** $(-\infty, \frac{2}{3}]$;
(b) $(-\frac{5}{3}, \infty)$ **47.** $-2 < 1$, but $(-2)^2 \not< 1^2$

Problem Set 2.6, page 131

1. **(a)** 8; **(b)** $\frac{1}{8}$; **(c)** 6.043; **(d)** ≈ 3.146264370
3. **(a)** 8; **(b)** 32; **(c)** 32; **(d)** 8; **(e)** 8; **(f)** $\frac{5}{3}$; **(g)** 0; **(h)** 0
5. $-3, 3$ **7.** 2, 6 **9.** $-4, 5$ **11.** $p \ge 0$
13. No solution **15.** $\frac{4}{3}, \frac{5}{3}$ **17.** $-\frac{1}{5}, \frac{1}{3}$
19. $-4, -\frac{2}{3}$ **21.** 1, 3 **23.** **(a)** $|x - 3| \le 0.05$;
(b) $|x - a| < \delta$; **(c)** $|x - 2| \ge 5$

25. $[-2, 2]$,

27. $(1, 5)$,

29. $(-\infty, -\frac{3}{2}) \cup (3, \infty)$, . . .

31. $[\frac{6}{5}, 2]$,

33. $(-\infty, -\frac{13}{8}] \cup [-\frac{11}{8}, \infty)$, . . .

35. $[\frac{4}{7}, \frac{8}{7}]$,

37. $(-\infty, 1] \cup [\frac{5}{2}, \infty)$, . . .

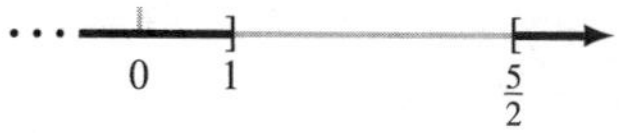

39. $(-\infty, 2) \cup (\frac{7}{2}, \infty)$, . . .

41. $\mathbb{R}$, . . .

43. **(a)** $(2 - \delta, 2 + \delta)$; **(b)** $(2 - \delta, 2 + \delta)$, $x \ne 2$; **(c)** in part (a), $x = 2$ is a solution; in part (b), $x = 2$ is not a solution **45.** **(b)** $|3x + 9| = \begin{cases} 3x + 9, & x \ge -3 \\ -3x - 9, & x < -3 \end{cases}$
47. If $x < 0$, then $0 < -x = |x|$ and $-|x| = -(-x) = x < 0$; if $x \ge 0$, then $-|x| = -x \le 0 \le x = |x|$.
49. **(b)** $\left|a - \dfrac{a+b}{2}\right| = \left|\dfrac{a-b}{2}\right|$ and $\left|\dfrac{a+b}{2} - b\right| = \left|\dfrac{a-b}{2}\right|$
51. **(a)** If $x \ge y$, then $\frac{1}{2}(x + y + |x - y|) = \frac{1}{2}(x + y + x - y) = \frac{1}{2}(2x) = x$; if $x < y$, then $\frac{1}{2}(x + y + |x - y|) = \frac{1}{2}[x + y - (x - y)] = \frac{1}{2}(2y) = y$; **(b)** minimum $(x, y) = \frac{1}{2}(x + y - |x - y|)$, since if $x > y$, we get $\frac{1}{2}(x + y - x + y) = y$; if $x < y$, we get $\frac{1}{2}[x + y + x - y] = x$; and if $x = y$, we get $\frac{1}{2}(x + x - 0) = x$. **53.** $(5.95, 7.65)$

55. $[5 \times 10^{-3}, 15 \times 10^{-3}]$ **57.** $A = 450, B = 50$
59. $A = 1, B = 0.02$ **61.** $\approx 0.6\%$ **63.** 10 iterations

Problem Set 2.7, page 139

1. **(a)** $x^2 - 6x + 8 < 0$; **(b)** $3x^2 + 3x + 5 > 0$; **(c)** $x^2 - 10x + 25 \geq 0$; **(d)** $x^3 - 3x^2 + 2x \leq 0$ **3.** $(2, 4)$
5. $\{7\}$ **7.** $[\frac{1}{3}, \frac{1}{2}]$ **9.** $\mathbb{R}$ **11.** $[-1, \frac{1}{2}]$
13. $[-2, -1] \cup [1, \infty)$ **15.** $(-1, 0) \cup (1, \infty)$
17. $(-\infty, -\frac{1}{2}) \cup (\frac{1}{2}, 2)$ **19.** $\mathbb{R}$ **21.** **(a)** $\dfrac{2x - 1}{x + 3} > 0$;
(b) $\dfrac{(x + 2)(x - 3)}{(x + 3)(x + 2)} \leq 0$ **23.** All real numbers except -2
25. $(-3, \frac{5}{3}]$ **27.** $(-\infty, -3) \cup (-1, 2) \cup (3, \infty)$
29. $(-\infty, 2)$ **31.** $(-\infty, -2) \cup (-1, 1) \cup (2, \infty)$
33. $(-\infty, 3) \cup [-2, 1) \cup [1, 2) \cup [3, \infty)$
35. **(a)** $(-\infty, -5] \cup [-\frac{7}{3}, \infty)$; **(b)** $(-\infty, -3] \cup (-\frac{5}{2}, 1) \cup [5, \infty)$
37. **(a)** $(-\infty, 0.203] \cup [0.684, \infty)$; **(b)** $(-\infty, -4.379) \cup (0.141, 0.794)$ **39.** $n \leq 34$
41. $(10, 20)$ **43.** $(0, \frac{20}{19})$ **45.** At least 5
47. $[0.38, 10.6]$ hr

Review Problem Set
CHAPTER 2, page 140

1. 5 **3.** -3 **5.** 6 **7.** $-\frac{35}{2}$ **9.** -2
11. No solution unless $m = 0$, in which case, x is any nonzero number **13.** -5 to one significant digit
15. 4 L **17.** \$6500 at 7%, \$3500 at 8% **19.** 30 m
21. **(a)** $x^2 + x + \frac{1}{4}$; **(b)** $x^2 - 24x + 144$; **(c)** $x^2 - 9x + \frac{81}{4}$; **(d)** $x^2 + \sqrt{3}x + \frac{3}{4}$ **23.** $-\frac{1}{2}, \frac{3}{5}$ **25.** $(7 \pm \sqrt{89})/10$
27. $\approx -2.867, \approx 0.135$ **29.** $(3 \pm i\sqrt{39})/8$
31. $(-3 \pm \sqrt{13})/4$ **33.** 450 knots **35.** 16 in. by 12 in. **37.** $(15 - \sqrt{33})/2$ **39.** $-\frac{10}{3}$ **41.** 3
43. $\pm 1, \pm 2$ **45.** 25 **47.** $-8, 1, 3 \pm \sqrt{17}$
49. $-\frac{1}{3}, \frac{1}{2}$ **51.** -1 **53.** $-\sqrt{2}, \sqrt{2}, 1$
55. $-1; \pm\dfrac{\sqrt{2}}{8}$ **57.** 2 **59.** $\frac{4}{5}$ **61.** $L^3F/(4d^3wY)$
63. $100\left(\dfrac{a + 100}{100}\right)^{1/12} - 100$ **65.** **(a)** 7; **(b)** 5; **(c)** $\frac{19}{6}$; **(d)** 7.3
67. $(3, \infty)$,

69. $(\frac{24}{13}, \infty)$,

71. $[73, 98)$ **73.** **(a)** $[10.5, 12.2]$; **(b)** $[8.75, 15.25]$
75. $\frac{1}{4}$
77. $(-\infty, -\frac{1}{2}] \cup [2, \infty)$, ...

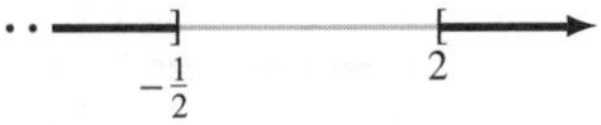

79. $(\frac{3}{2}, \frac{11}{6})$,

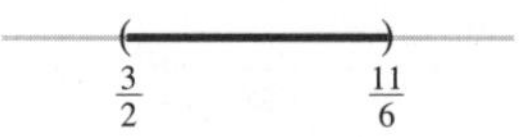

81. $(\frac{3}{2}, \frac{7}{4})$,

83. $(-\frac{2}{3}, \frac{5}{2})$,

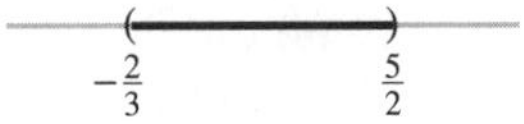

85. $[-\frac{3}{2}, \frac{2}{3}]$,

87. $(-\infty, \frac{1}{3}) \cup (\frac{5}{2}, \infty)$, ...

89. $(\frac{7}{5}, \frac{13}{7})$,

91. $(-\infty, -3) \cup (-2, -1) \cup (0, 2)$,

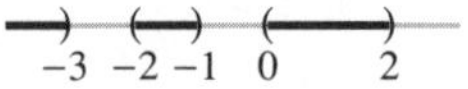

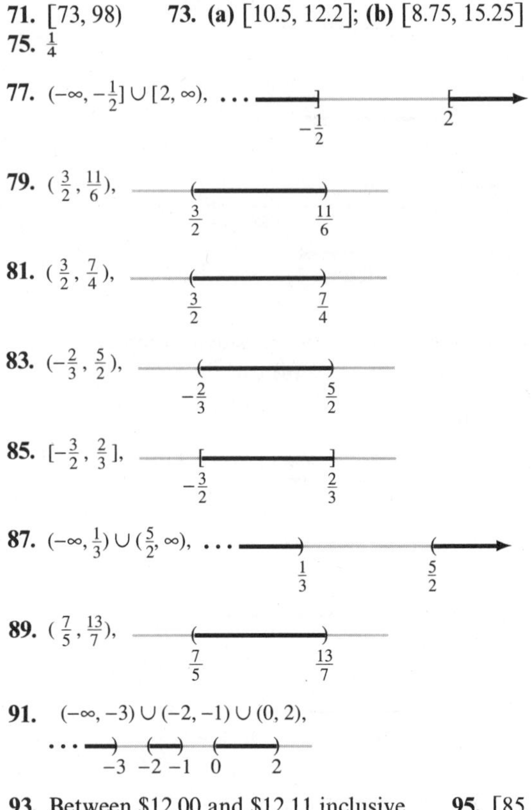

93. Between \$12.00 and \$12.11 inclusive **95.** $[85, 105]$
97. $k = 10$

Chapter 2 Test, page 145

1. **(a)** $-3, 9$; **(b)** $8t/(5 - 6t)$; **(c)** $-\frac{1}{5}, 1$; **(d)** $2 \pm \sqrt{2}i$; **(e)** $(-1 \pm \sqrt{13})/3$; **(f)** 79; **(g)** $-8, 125$; **(h)** $-2, -\frac{1}{3}$; **(i)** 0
2. **(a)** Order-reversing property; **(b)** addition property; **(c)** transitive property; **(d)** multiplication property
3. **(a)** $[6, \infty)$,

(b) $[-4, 5]$,

(c) $[-2, 3]$,

(d) $\mathbb{R}$, ...

(e) $(-\infty, -2) \cup (1, \infty)$, ...

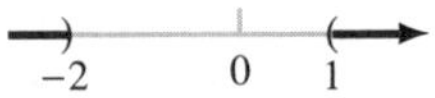

(f) $(-\infty, -3] \cup (2, \infty)$, ...

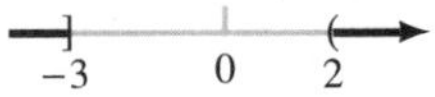

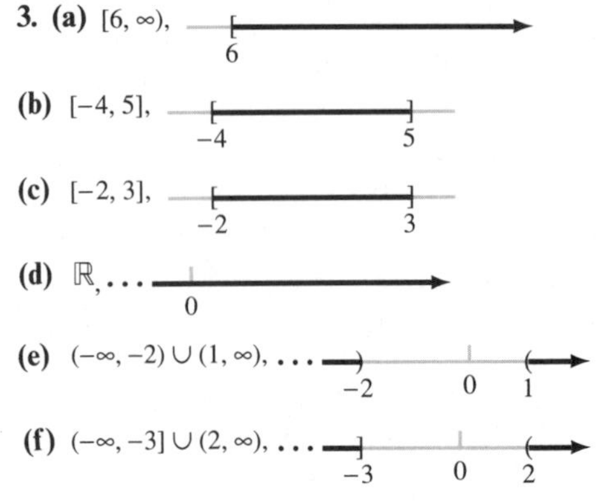

4. $\frac{271}{45}$ **5.** 40 nickels, 80 dimes, and 90 quarters
6. 9.55% **7.** 30 hr

CHAPTER 3

Problem Set 3.1, page 154

1. **(a)** I; **(b)** II; **(c)** IV; **(d)** III; **(e)** y axis; **(f)** x axis; **(g)** y axis; **(h)** IV; **(i)** I
3. **(a)** $Q = (3, -2)$, $R = (-3, 2)$, $S = (-3, -2)$; **(b)** $Q = (-1, -3)$, $R = (1, 3)$, $S = (1, -3)$
5. **(a)** (2, 3); **(b)** $2\sqrt{2}$ **7.** **(a)** (3, 4); **(b)** $2\sqrt{26}$
9. **(a)** $(-4, -3)$; **(b)** 10 **11.** **(a)** $\left(\frac{2a+1}{2}, \frac{2b+1}{2}\right)$; **(b)** $\sqrt{2}$ **13.** **(a)** (0.2105, 6.047); **(b)** ≈ 6.224
15. $\sqrt{145}/2$ **17.** Area = 12 square units
21. Isosceles **23.** -17 or 7 **25.** -2 or 3
27. **(a)** Yes; **(b)** no **29.** (3, 4) **31.** $x^2 + y^2 = 16$
33. $(x + 1)^2 + (y - 3)^2 = 4$
35. $(x + 4)^2 + (y - 6)^2 = 41$
37. $(x - 1)^2 + (y - 3)^2 = 9$
39. $(x - \frac{1}{2})^2 + (y - \frac{3}{2})^2 = \frac{13}{2}$
41. $C = (0, 0)$, $r = 6$

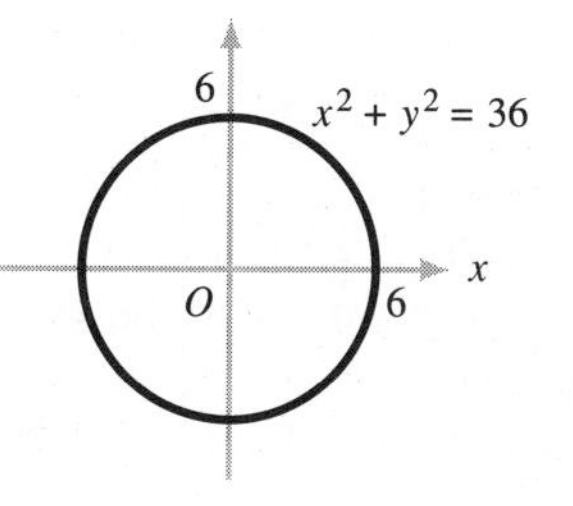

43. $C = (3, 5)$, $r = 7$ **45.** $C = (2, 5)$, $r = 5$

$(x - 3)^2 + (y - 5)^2 = 49$ $(x - 2)^2 + (y - 5)^2 = 25$

47. $C = (-1, -2)$, $r = 1$ **49.** 13 nautical miles

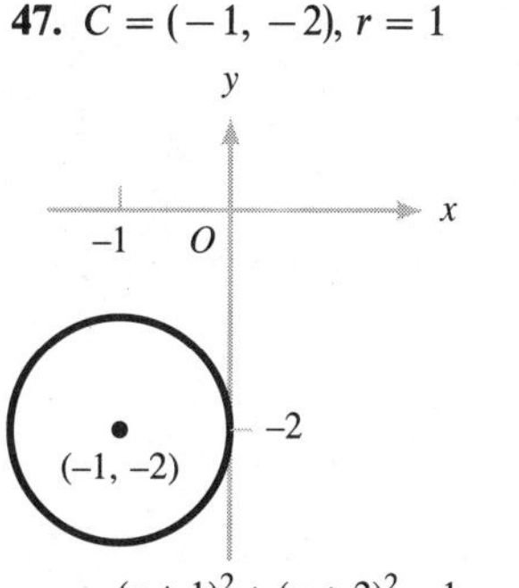
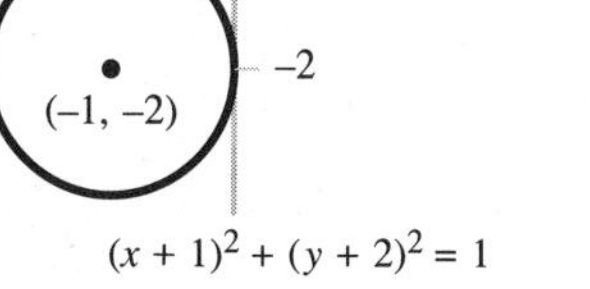

Problem Set 3.2, page 164

1. **(a)** -1; **(b)** -1; **(c)** $\frac{1}{4}$; **(d)** $\frac{1}{2}$
3. **(a)** **(b)** **(c)**

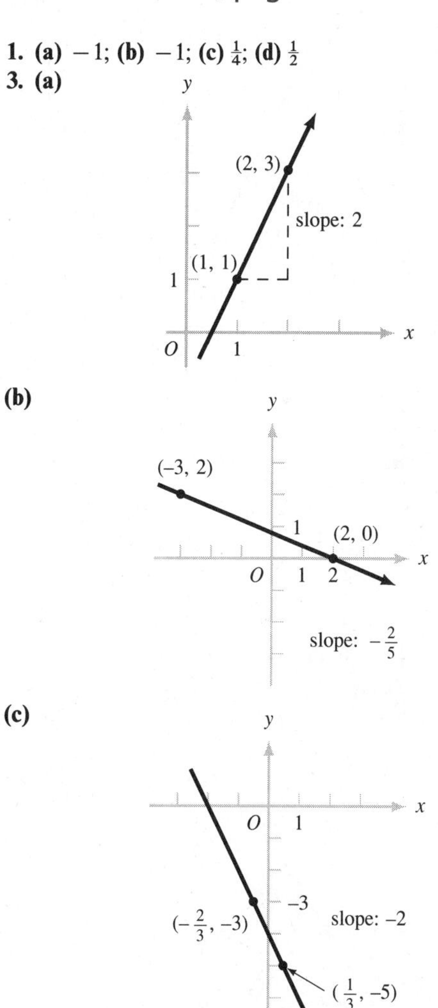

(d)

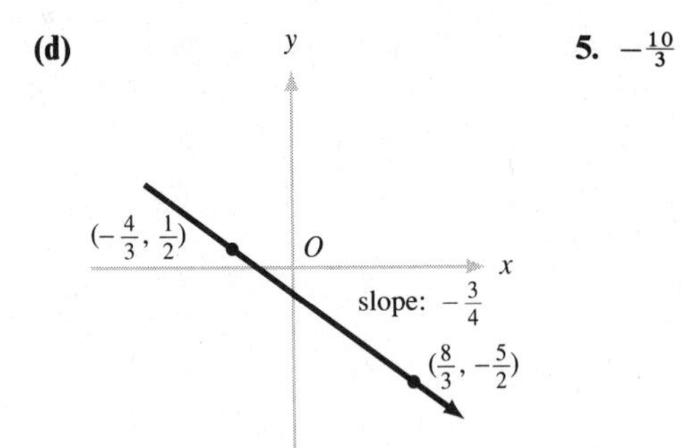

5. $-\frac{10}{3}$

7. (a) Collinear; **(b)** noncollinear; **(c)** collinear

9. (a)

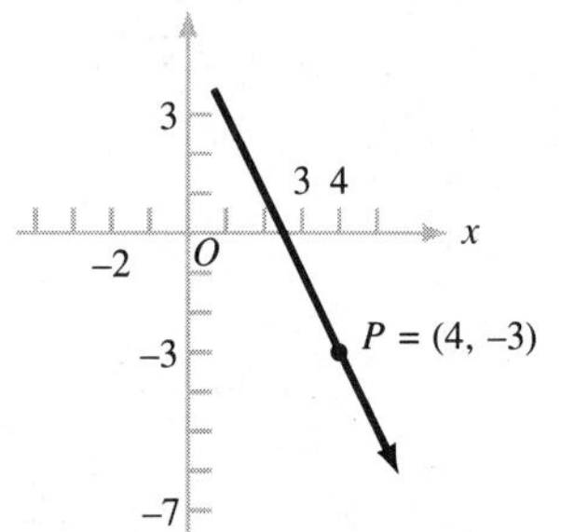

(b)

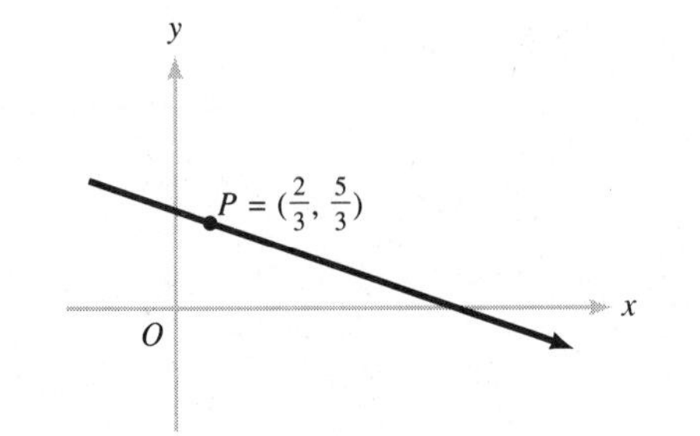

13. (a) $y - 2 = \frac{3}{4}(x - 3)$; **(b)** $y - 1 = -\frac{1}{7}(x - 4)$ or $y - 2 = -\frac{1}{7}(x + 3)$; **(c)** $y - 5 = 1(x + 3)$

15. $y = \frac{3}{2}x - 3$, $m = \frac{3}{2}$, $b = -3$

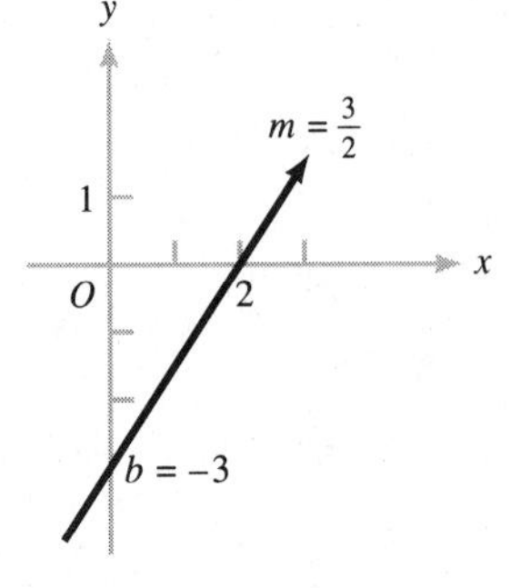

17. $y = 3x + 1$, $m = 3$, $b = 1$

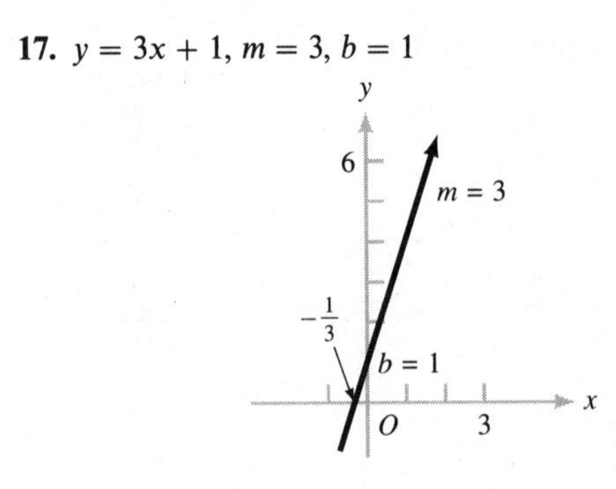

19. $y = -\frac{1}{3}x + 3$, $m = -\frac{1}{3}$, $b = 3$

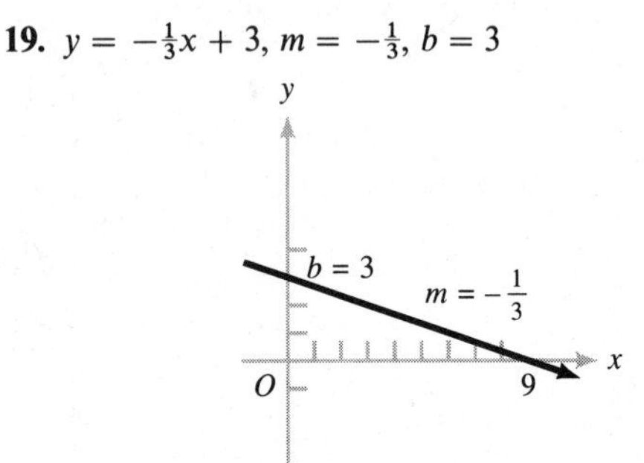

21. (a) $y - 2 = 4(x + 5)$; **(b)** $y = 4x + 22$; **(c)** $4x - y + 22 = 0$ **23. (a)** $y - 5 = -3(x - 0)$; **(b)** $y = -3x + 5$; **(c)** $-3x - y + 5 = 0$
25. (a) $y - 5 = -\frac{5}{3}(x - 0)$; **(b)** $y = -\frac{5}{3}x + 5$; **(c)** $5x + 3y - 15 = 0$ **27. (a)** $y + 4 = 0(x - 3)$; **(b)** $y = 0 \cdot x - 4$; **(c)** $y + 4 = 0$
29. (a) $y + 4 = \frac{2}{5}(x - 4)$; **(b)** $y = \frac{2}{5}x - \frac{28}{5}$; **(c)** $2x - 5y - 28 = 0$ **31. (a)** $y - \frac{2}{3} = \frac{3}{5}(x + 3)$; **(b)** $y = \frac{3}{5}x + \frac{37}{15}$; **(c)** $9x - 15y + 37 = 0$
33. (a) $y - \frac{5}{7} = -\frac{7}{3}(x - \frac{2}{3})$; **(b)** $y = -\frac{7}{3}x + \frac{143}{63}$; **(c)** $147x + 63y - 143 = 0$ **35. (a)** $y - 2 = -\frac{1}{2}(x - 5)$; **(b)** $y = -\frac{1}{2}x + \frac{9}{2}$; **(c)** $x + 2y - 9 = 0$
37. (a) $2x + 3y - 8 = 0$; **(b)** $76y - 42x - 127 = 0$

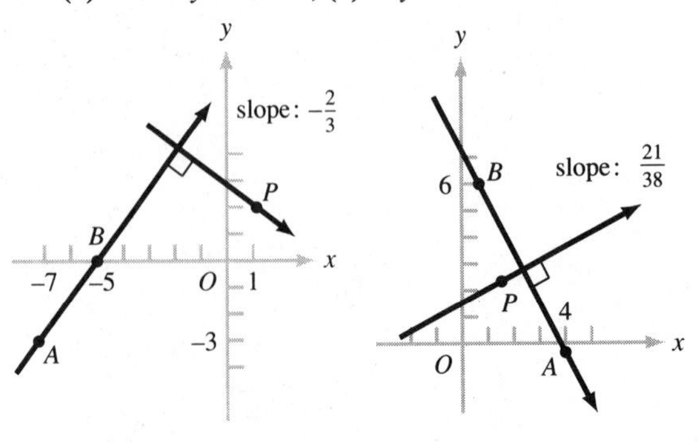

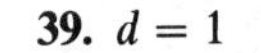
39. $d = 1$

41. $y = 0.2x + 22N$

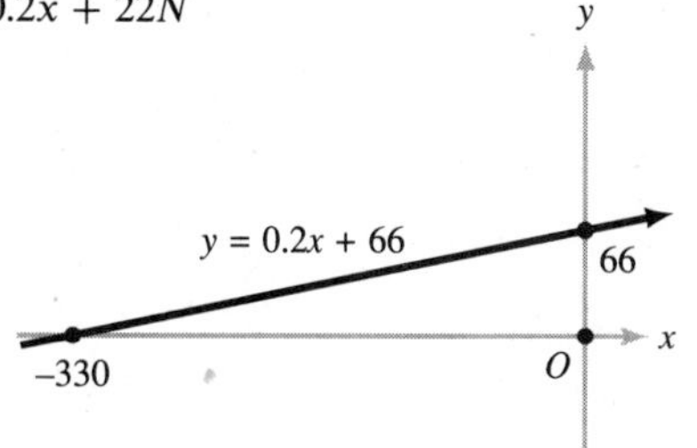

43. $y = -0.75x + 7$. Free of pollution in 9 years and 4 months

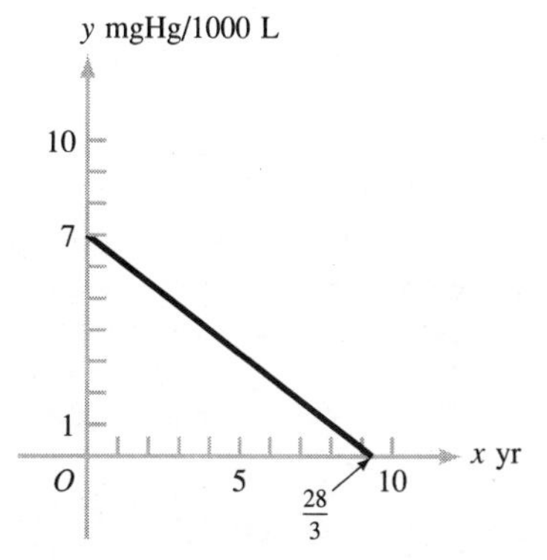

45. **(a)** $F = \frac{9}{5}C + 32$; **(b)** $F = \frac{9}{5}K - 459.67$
47. $V = -1800t + 20{,}000$; \$1800 **49.** 10.9 million

Problem Set 3.3, page 175

1. **(a)** 6; **(b)** -4; **(c)** $b^2 + 3b - 4$; **(d)** $a^2 + 2ab + b^2 - 3a - 3b - 4$ **3.** **(a)** 3; **(b)** $-\sqrt[3]{129}$; **(c)** $\sqrt[3]{x^3 + 3x^2 + 3x - 3}$; **(d)** $-\sqrt[3]{x^3 + 4}$
5. **(a)** 2; **(b)** $\frac{1}{12}$; **(c)** -4 **7.** **(a)** $x \neq 0$; **(b)** $\mathbb{R}$; **(c)** $x \geq 0$; **(d)** $[-3, 3]$; **(e)** $(-\infty, -2) \cup (-2, 2) \cup (2, \infty)$
9. 4 **11.** $2x + h$ **13.** $\dfrac{-1}{x\sqrt{x+h} + (x+h)\sqrt{x}}$

15. **17.**

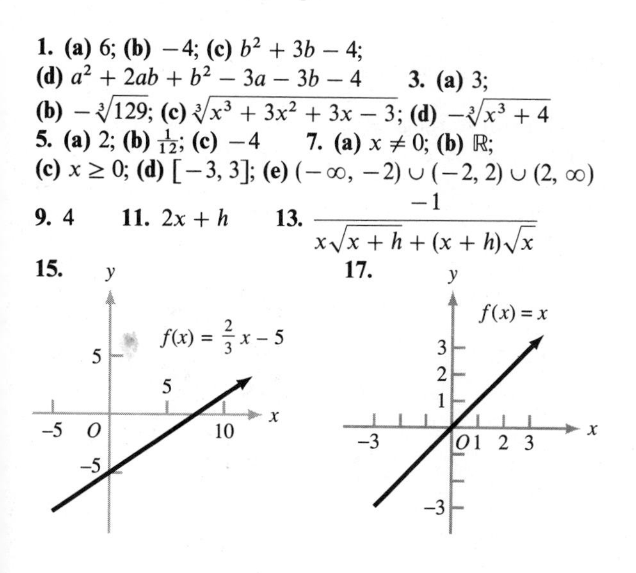

19.

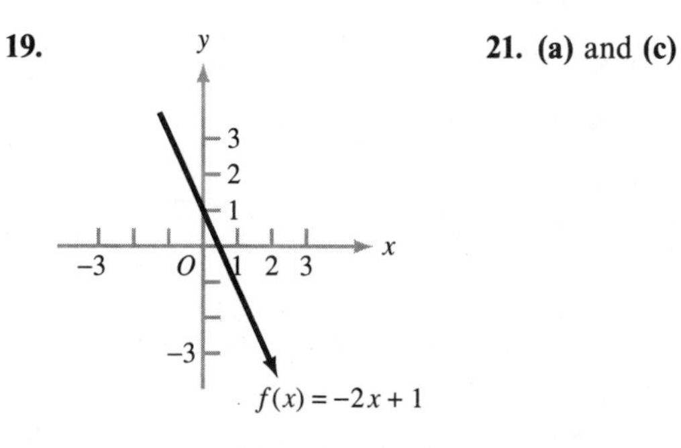

21. **(a)** and **(c)**

23. $f(0) = 32$ or $0 \overset{f}{\mapsto} 32$, $f(15) = 59$ or $15 \overset{f}{\mapsto} 59$, $f(-10) = 14$ or $-10 \overset{f}{\mapsto} 14$, $f(55) = 131$ or $55 \overset{f}{\mapsto} 131$
25. $f(b) - f(a) + f(c) - f(b) = f(c) - f(a)$
27. $T = -\frac{1}{250}h + 65$; when $h = 30{,}000$ ft, $T = -55°\text{F}$

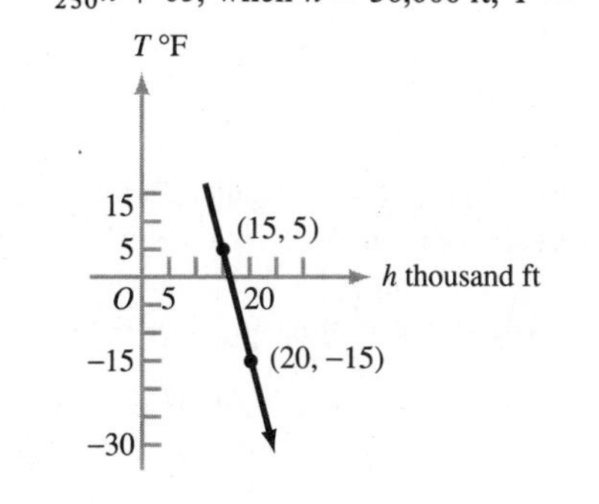

29. **(a)** $(x^2/16)(\pi + 8)$; **(b)** $(x/2)(\pi + 4)$; **(c)** $\dfrac{P^2(\pi + 8)}{4(\pi + 4)^2}$
31. **(a)** $h = \frac{22}{7}l + 63$; **(b)** ≈ 167 cm
33. $V(x) = 4x^3 - 48x^2 + 140x$ in^3
35. $V = \dfrac{\pi r^2}{3}(17.1 \times 10^{-6} - 2r)$

Problem Set 3.4, page 186

1. **3.**

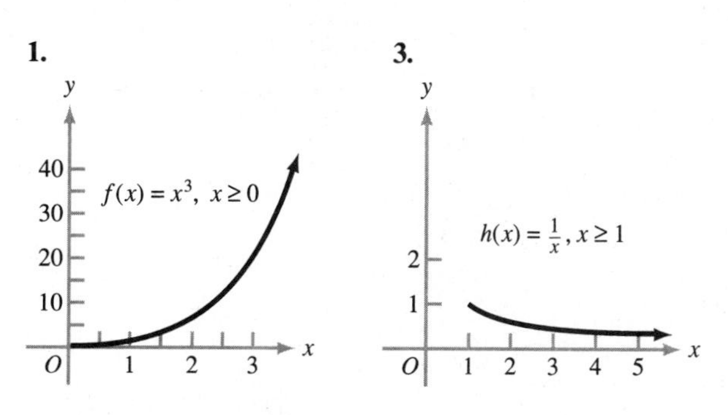

5. **7.**

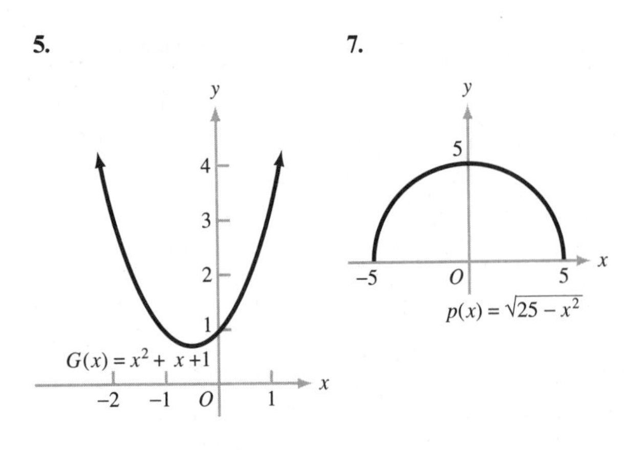

9. **(a)** Domain $\mathbb{R}$, range $(-\infty, 2]$, increasing on $(-\infty, -2]$ and $[0, 2]$, decreasing on $[-2, 0]$ and $[2, \infty)$, even; **(b)** domain $[-5, 5]$, range $[-3, 3]$, increasing on $[-1, 1]$ and $[3, 5]$, decreasing on $[-5, -1]$ and $[1, 3]$, neither; **(c)** domain $\left[-\frac{3\pi}{2}, \frac{3\pi}{2}\right]$, range $[-1, 1]$, increasing on $\left[-\frac{3\pi}{2}, -\frac{\pi}{2}\right]$ and $\left[\frac{\pi}{2}, \frac{3\pi}{2}\right]$, decreasing on $\left[-\frac{\pi}{2}, \frac{\pi}{2}\right]$, odd; **(d)** domain $\mathbb{R}$, range $[-2, 1]$, constant on $(-\infty, 0]$ and $[\pi, \infty)$, increasing on $[0, \pi]$, neither

11. Even, symmetric about y axis

13. Odd, symmetric about origin

15. Even, symmetric about y axis **17.** Neither

19. Odd, symmetric about origin

21. Domain $\mathbb{R}$, range $\mathbb{R}$, increasing on $\mathbb{R}$, neither even nor odd

23. Domain $\mathbb{R}$, range $\{5\}$, constant on $\mathbb{R}$, even, symmetric about y axis

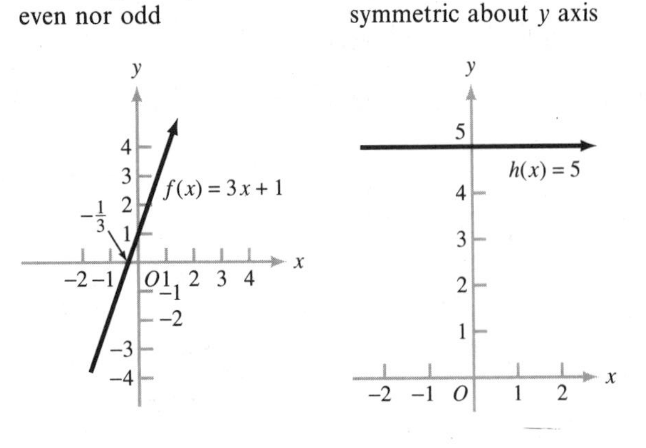

25. Domain $\mathbb{R}$, range $(-\infty, 2]$, increasing on $(-\infty, 0]$, decreasing on $[0, \infty)$, even, symmetric about y axis

27. Domain $\mathbb{R}$, range $\mathbb{R}$, increasing on $\mathbb{R}$, odd, symmetric about origin

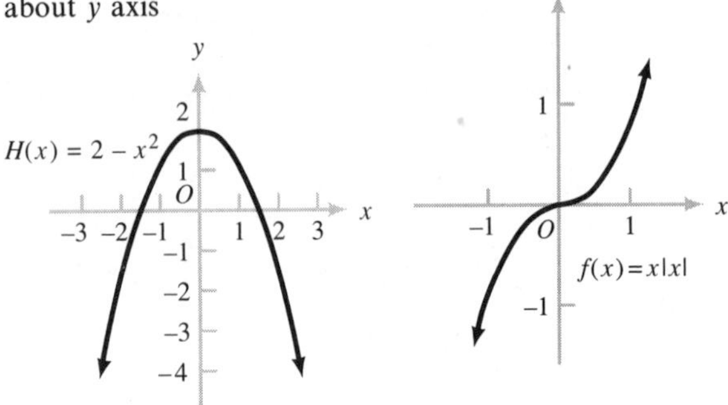

29. Domain $[1, \infty)$, range $[1, \infty)$, increasing on $[1, \infty)$, neither even nor odd

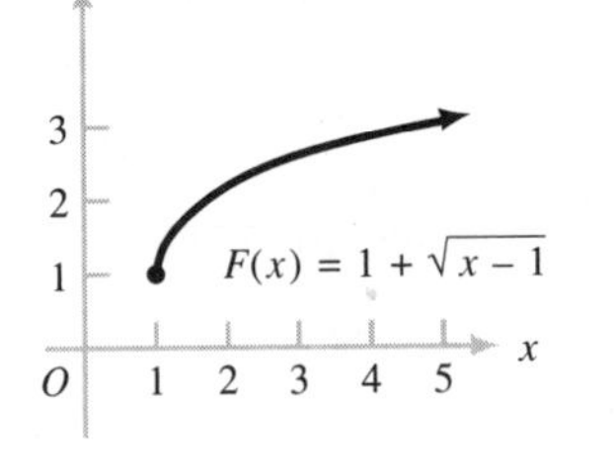

31. Domain $(-\infty, 0)$ and $(0, \infty)$, range $\{-1, 1\}$, constant on $(-\infty, 0)$ and on $(0, \infty)$, odd, symmetric about origin

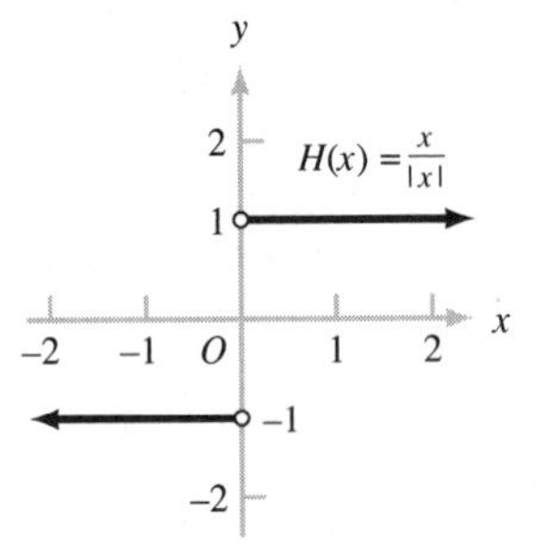

33. Domain $\mathbb{R}$, range $\mathbb{R}$, increasing on $\mathbb{R}$, odd, symmetric about origin

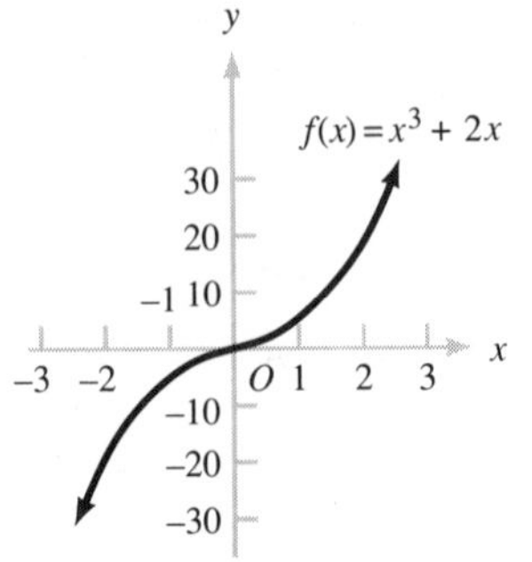

35.

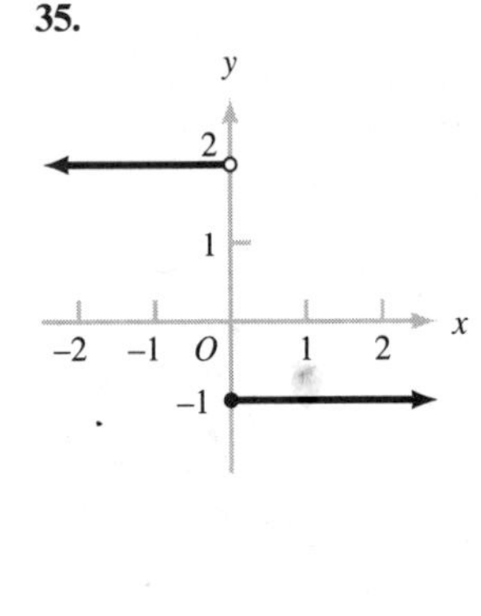

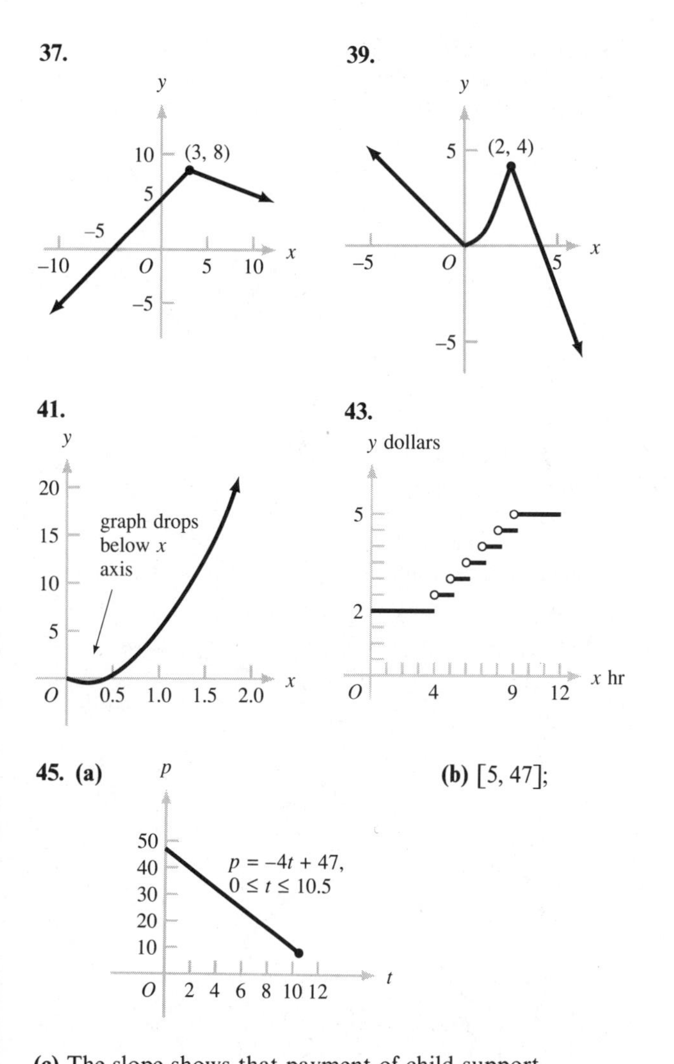

(b) [5, 47];

(c) The slope shows that payment of child support declines by 4% each year. The p intercept shows the initial child support payment is made in only 47% of cases; **(d)** 19% **47.** $A(x) = (2x - x^2)^2,\ 0 < x < 1$

49.

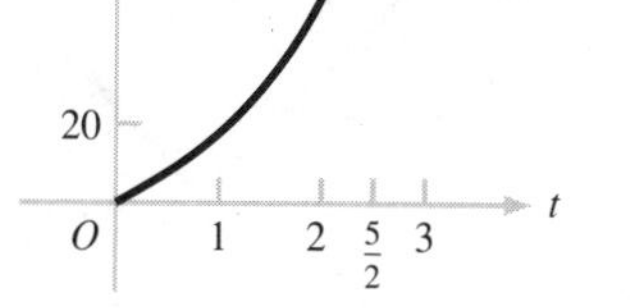

51. Domain: [0, 2.5], range: [0, 100]

Problem Set 3.5, page 196

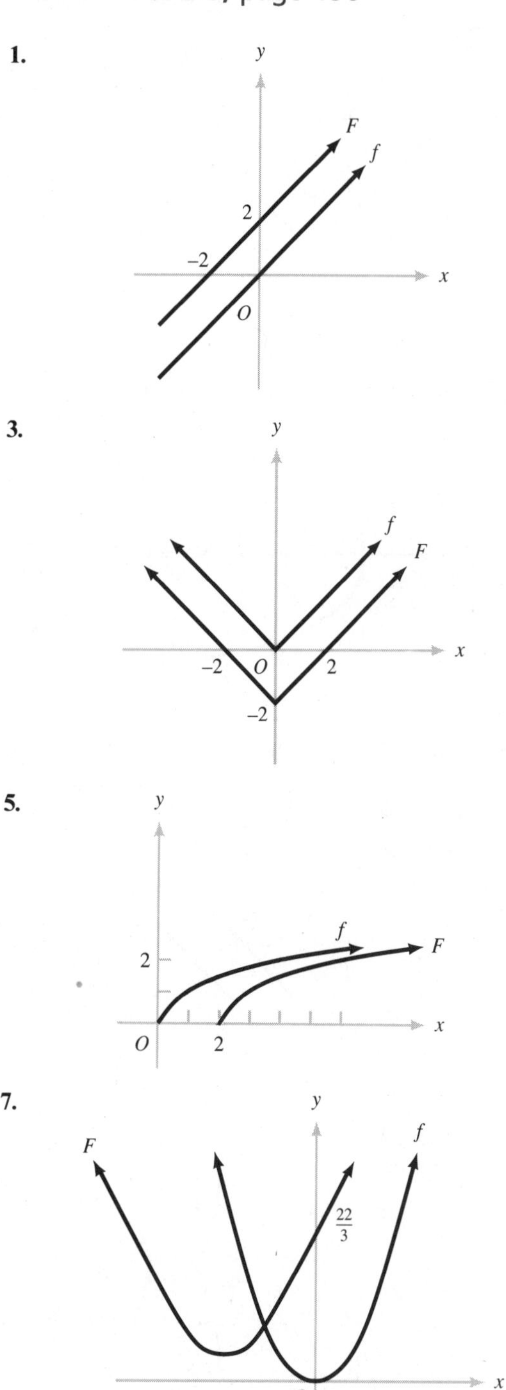

9.

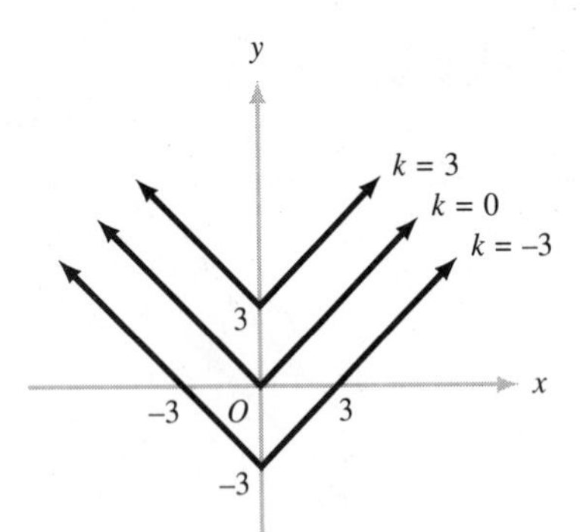

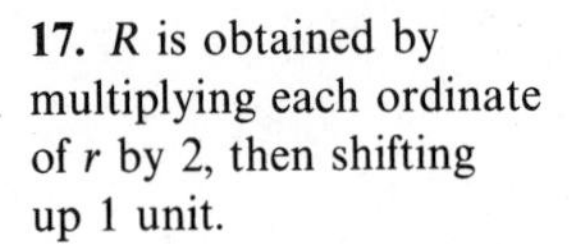

17. R is obtained by multiplying each ordinate of r by 2, then shifting up 1 unit.

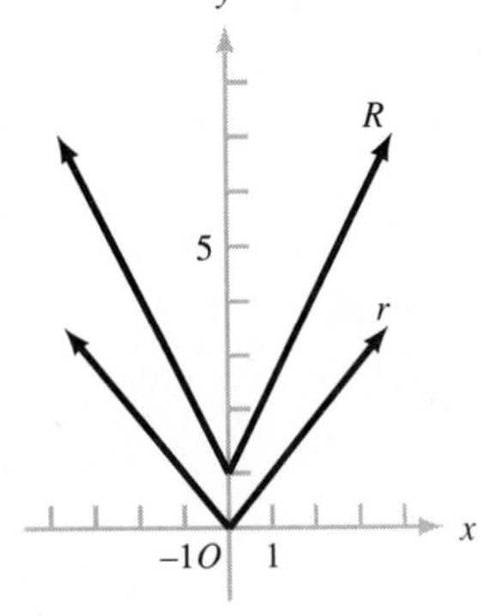

11.

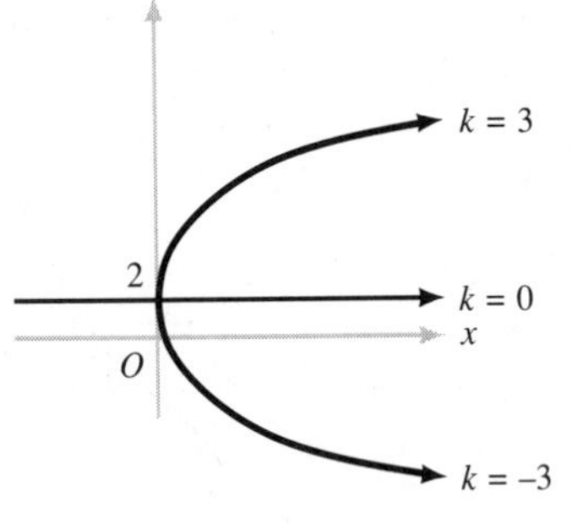

19. T is obtained by shifting t by 2 units to the right, multiplying resulting ordinates by 2, then shifting up 4 units.

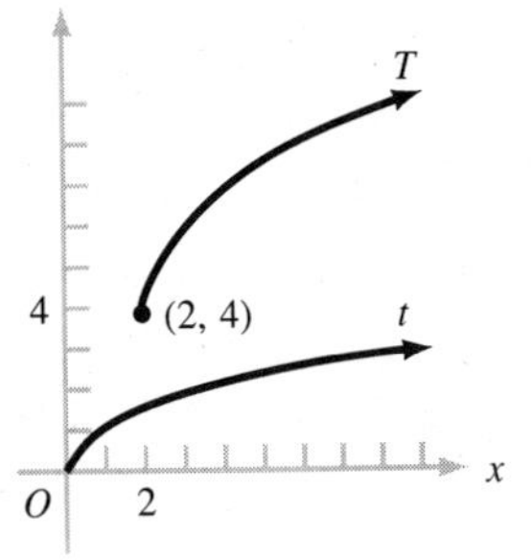

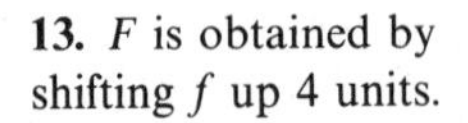

13. F is obtained by shifting f up 4 units.

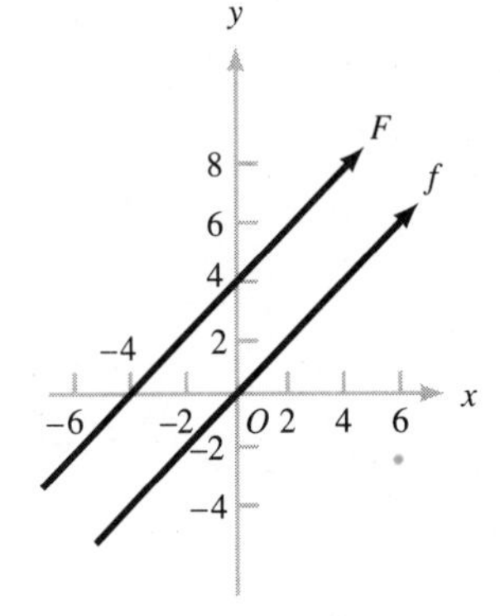

21.

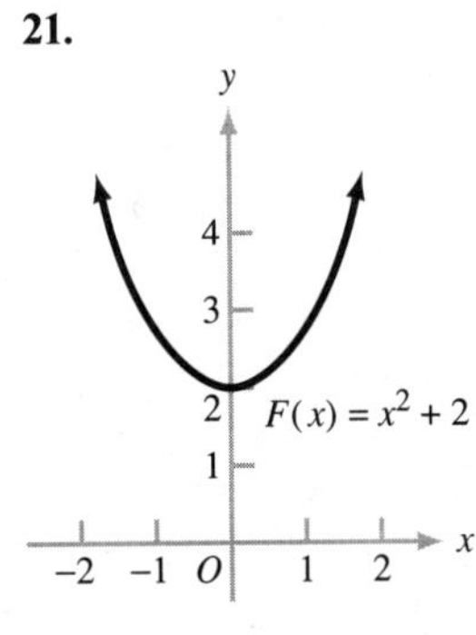

23.

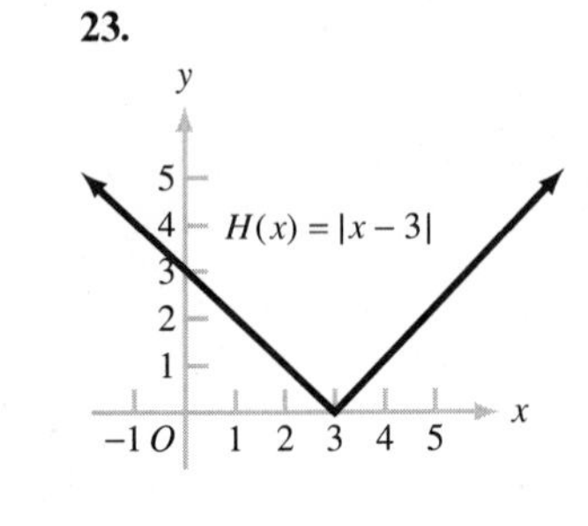

15. P is obtained by reflecting p about the x axis, then shifting up 1 unit.

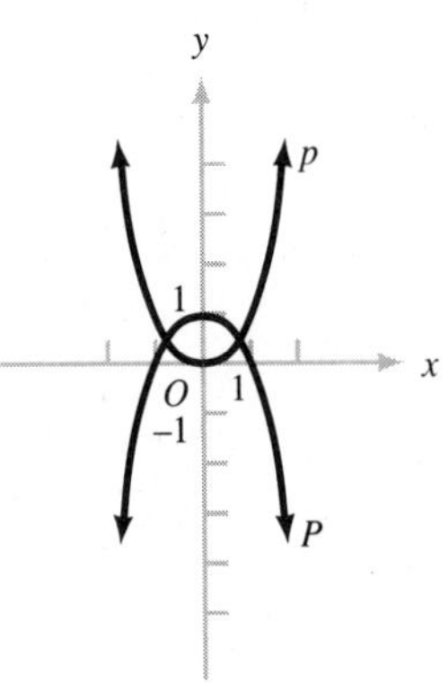

25.

27.

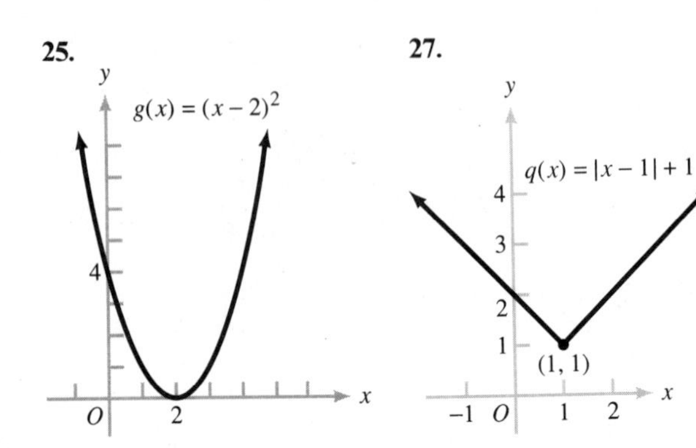

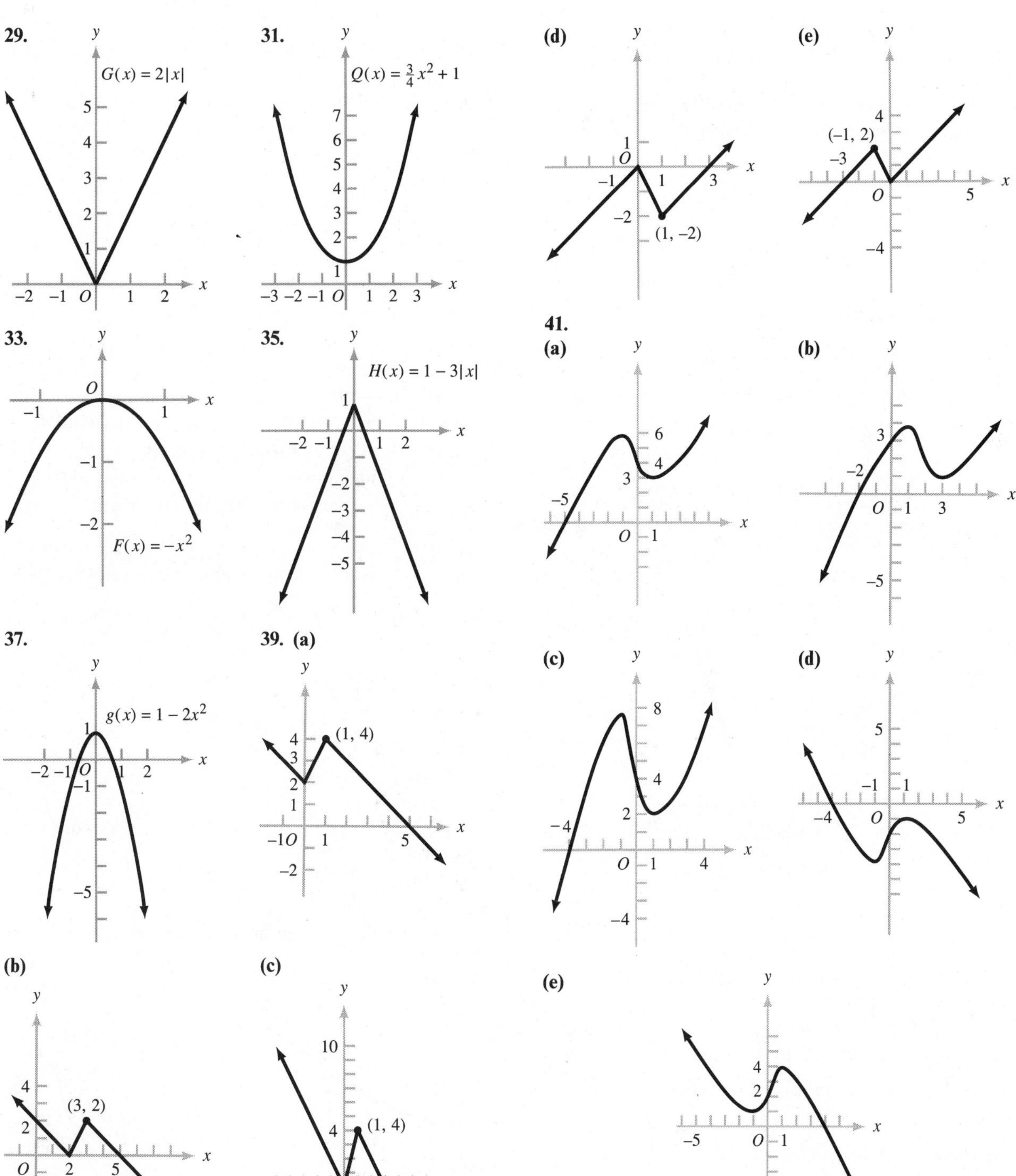
29.
$G(x) = 2|x|$
31.
$Q(x) = \frac{3}{4}x^2 + 1$
33.
$F(x) = -x^2$
35.
$H(x) = 1 - 3|x|$
37.
$g(x) = 1 - 2x^2$
39. (a)
(1, 4)
(b)
(3, 2)
(c)
(1, 4)
(d)
(1, −2)
(e)
(−1, 2)
41.
(a)
(b)
(c)
(d)
(e)

43. (x, y) belongs to graph of F if and only if $y = f(-x)$; that is, $(-x, y)$ belongs to graph of f
45. (a) No; **(b)** yes
47. The graph of F is obtained from the graph of f by stretching, flattening, and/or reflecting about the y axis depending on the value of c.

Problem Set 3.6, page 204

1. (a) $7x - 3$; **(b)** $3x + 7$; **(c)** $10x^2 - 21x - 10$; **(d)** $\dfrac{5x+2}{2x-5}$; **(e)** all reals except $\frac{5}{2}$ **3. (a)** $x^2 + 4$; **(b)** $x^2 - 4$; **(c)** $4x^2$; **(d)** $x^2/4$; **(e)** $\mathbb{R}$ **5. (a)** $x^2 + 2x - 4$; **(b)** $-x^2 + 2x - 6$; **(c)** $2x^3 - 5x^2 + 2x - 5$; **(d)** $\dfrac{2x-5}{x^2+1}$; **(e)** $\mathbb{R}$ **7. (a)** $\dfrac{8x+1}{2x-1}$; **(b)** 1 if $x \neq \frac{1}{2}$; **(c)** $\dfrac{15x^2+5x}{4x^2-4x+1}$; **(d)** $\dfrac{5x}{3x+1}$; **(e)** all reals except $\frac{1}{2}$ and $\frac{1}{3}$
9.

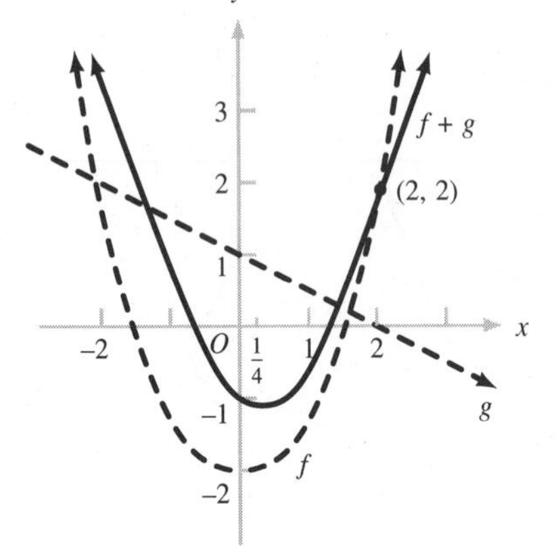

11. (a) 17; **(b)** 3; **(c)** 6.9929; **(d)** 29.8064
13. (a) $3x + 3$; **(b)** $3x + 1$; **(c)** $9x$; domain $\mathbb{R}$ in (a), (b), and (c)
15. (a) x, domain $[0, \infty)$; **(b)** $|x|$, domain $\mathbb{R}$; **(c)** x^4, domain $\mathbb{R}$ **17. (a)** $\dfrac{x}{\sqrt{x-1}}$, domain $(1, \infty)$; **(b)** $\sqrt{x + \dfrac{1}{x} - 1}$, domain $(0, \infty)$ **(c)** $\dfrac{x^4 + 3x^2 + 1}{x^3 + x}$, domain $(-\infty, 0) \cup (0, \infty)$
19. (a) $\dfrac{1}{4x-9}$; domain: all reals except $\frac{3}{2}$ and $\frac{9}{4}$; **(b)** $\dfrac{11-6x}{2x-3}$; domain: all reals except $\frac{3}{2}$; **(c)** $\dfrac{2x-3}{11-6x}$; domain: all reals except $\frac{3}{2}$ and $\frac{11}{6}$
21. (a) $h \circ g$; **(b)** $g \circ h$; **(c)** $h \circ f$; **(d)** $f \circ g$
23. $f(x) = x^{-7}$, $g(x) = 2x^2 - 5x + 1$
25. $f(x) = x^5$, $g(x) = \dfrac{1+x^2}{1-x^2}$
27. $f(x) = \sqrt{x}$, $g(x) = \dfrac{x+1}{x-1}$
29. $f(x) = |x|/x$, $g(x) = x + 1$ **31. (a)** x; **(b)** x; **(c)** $\frac{1}{2}x^2 - 3x + \frac{15}{2}$; **(d)** $\frac{1}{2}x^2 - 3x + \frac{15}{2}$; **(e)** $2x^2 + x + 3$
33. (a) $R(x) = 10x$; **(b)** $P(x) = 10x - 50{,}000 - 10{,}000\sqrt[3]{x+1}$; **(c)** \$56,550
35. (a) $f(x) = \sqrt{x^2 + 8100}$; $g(t) = 50t$; **(b)** $10\sqrt{25t^2 + 81}$; **(c)** $y = f(x)$ and $x = g(t)$, so $y = f[g(t)] = (f \circ g)(t)$
37. $\sqrt{3}p^2/36$ = area of equilateral triangle with perimeter p **39.** $k = \frac{3}{2}$
41. If $f: x \to ax + b$ and $g: x \to cx + d$, then $f \circ g: x \to acx + (ad + b)$ **45.** 1.73205080757
47. $f(p) = \dfrac{1}{2}\left(p + \dfrac{c}{p}\right) = \dfrac{1}{2}(2p)$, $\dfrac{c}{p} = p$, $c = p^2$, $p = \sqrt{c}$

Problem Set 3.7, page 213

1. $f[g(x)] = f[(x-1)/2] = 2[(x-1)/2] + 1 = x$ and $g[f(x)] = g(2x+1) = [(2x+1) - 1]/2 = x$
3. For $x \neq 0$, $f[g(x)] = f(1/x) = 1/(1/x) = x$ and, likewise, $g[f(x)] = x$
5. $f[g(x)] = f(x^3 - 8) = \sqrt[3]{(x^3 - 8) + 8} = x$ and $g[f(x)] = g(\sqrt[3]{x+8}) = (\sqrt[3]{x+8})^3 - 8 = x$
7. (a) Invertible; **(b)** not invertible; **(c)** not invertible
9.

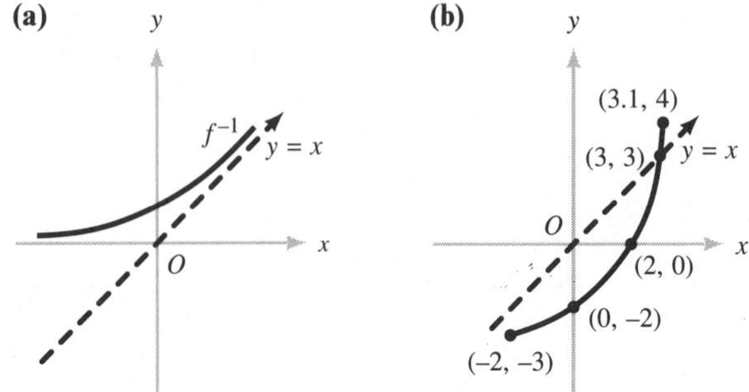

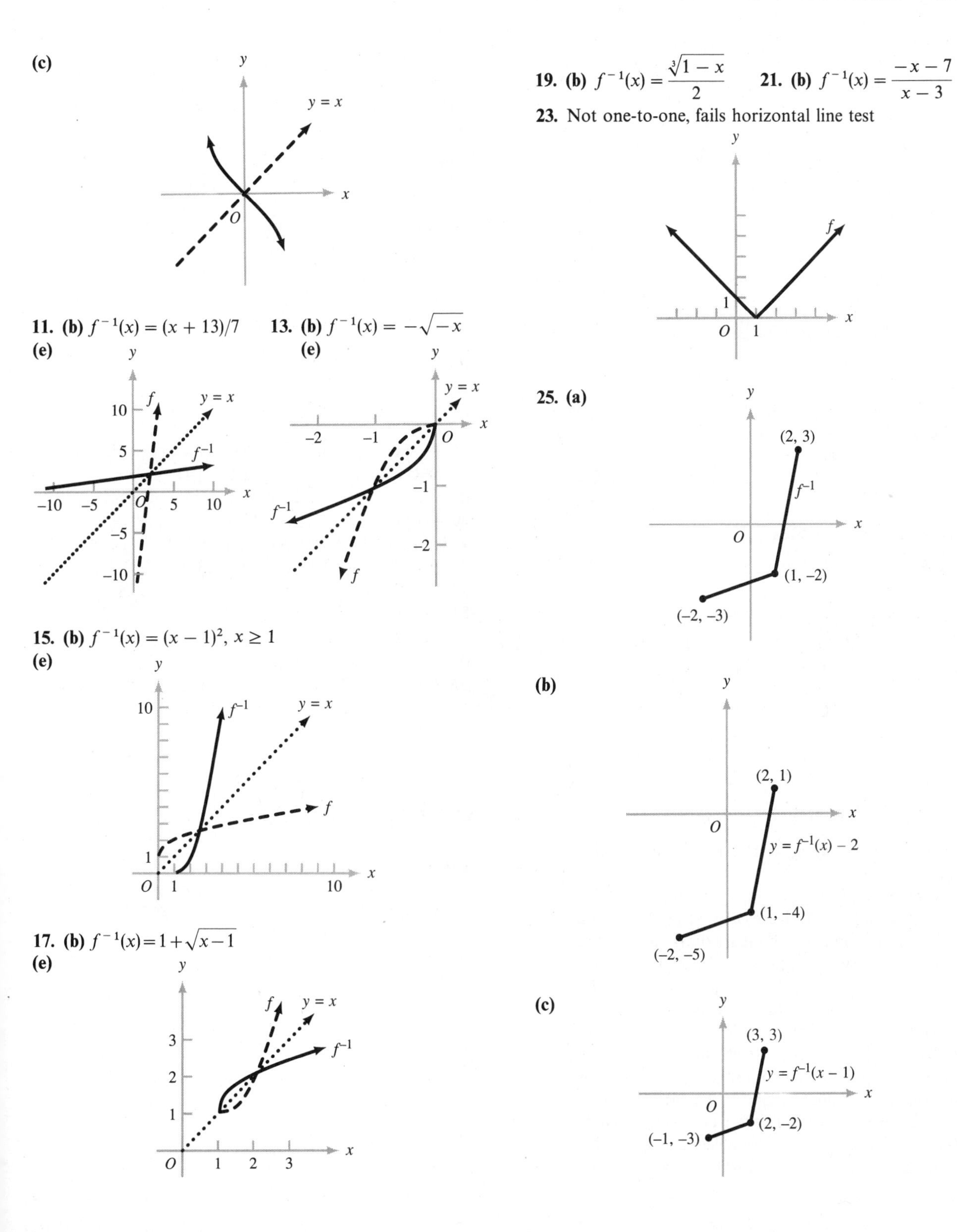

11. (b) $f^{-1}(x) = (x + 13)/7$
(e)

13. (b) $f^{-1}(x) = -\sqrt{-x}$
(e)

15. (b) $f^{-1}(x) = (x - 1)^2,\ x \geq 1$
(e)

17. (b) $f^{-1}(x) = 1 + \sqrt{x - 1}$
(e)

19. (b) $f^{-1}(x) = \dfrac{\sqrt[3]{1 - x}}{2}$ **21. (b)** $f^{-1}(x) = \dfrac{-x - 7}{x - 3}$

23. Not one-to-one, fails horizontal line test

25. (a)

(b)

(c)

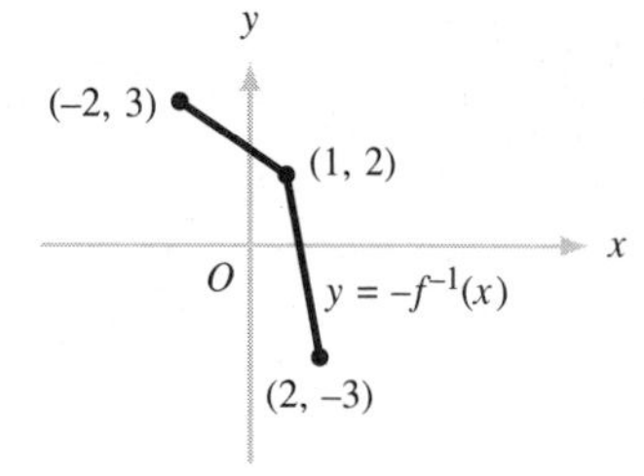

27. $f^{-1}(x) = \dfrac{-dx + b}{cx - a}$, $x \neq a/c$

29. $f(x) = x$, and $f(x) = -x + b$ **31.** $d = -a$

33. The effect is to interchange the x and y axes

35. $f^{-1}(x) = \dfrac{-B + \sqrt{B^2 - 4AC + 4Ax}}{2A}$

37. **(a)** $f(p) = (12{,}500 - 4p)/10$, $g(q) = (12{,}500 - 10q)/4$

39. $C = f^{-1}(t)$

Review Problem Set

CHAPTER 3, page 216

1. **(a)** 5; **(b)** $\frac{4}{3}$; **(c)** $(\frac{5}{2}, 3)$; **(d)** $-4x + 3y + 1 = 0$; **(e)** $(x - \frac{5}{2})^2 + (y - 3)^2 = \frac{25}{4}$ **3.** **(a)** $\sqrt{1381}/6$; **(b)** $\frac{34}{15}$; **(c)** $(-\frac{3}{4}, -\frac{13}{6})$; **(d)** $-34x + 15y + 7 = 0$; **(e)** $(x + \frac{3}{4})^2 + (y + \frac{13}{6})^2 = \frac{1381}{144}$ **5.** 30.35 **7.** **(a)** $y - 2 = -\frac{3}{5}(x - 5)$; **(b)** $y - \frac{1}{2} = \frac{3}{2}(x + \frac{2}{3})$ **9.** **(a)** $y + 2 = \frac{7}{3}(x - 1)$; **(b)** $y = \frac{7}{3}x - \frac{13}{3}$; **(c)** $7x - 3y - 13 = 0$ **11.** $y - b = -(a/b)(x - a)$ **13.** **(a)** $3x^2 - 12$; **(b)** $6xk + 3k^2$; **(c)** $75x^2 - 180x + 104$; **(d)** $\dfrac{19 + 6k}{4 + k}$; **(e)** 12; **(f)** $\sqrt{3x^2 - 4}$; **(g)** $\dfrac{-1}{x(x + k)}$; **(h)** $4 + k$ **15.** **(a)** $(-\infty, 1) \cup (1, \infty)$; **(b)** $(-2, 2)$; **(c)** $[-1, \infty)$; **(d)** $(-\infty, 0)$ **17.** **(a)** Domain $\mathbb{R}$, range $[-1, 1]$, constant on $(-\infty, -2]$, decreasing on $[-2, 0]$, increasing on $[0, 2]$, constant on $[2, \infty)$, even, symmetric about y axis; **(b)** domain $\mathbb{R}$, range $[-3, \infty)$, decreasing on $(-\infty, -3]$, increasing on $[-3, -\frac{2}{3}]$, decreasing on $[-\frac{2}{3}, 2]$, constant on $[2, 4]$, increasing on $[4, \infty)$, neither even nor odd **19.** $f(74) = 99.26$, $f(75) = 99.63$, $f(76) = 100$ **21.** Domain $[2, \infty)$, range $[0, \infty)$, increasing, neither even nor odd **23.** Domain $\mathbb{R}$, range $\mathbb{R}$, increasing on $\mathbb{R}$, odd, symmetric about origin **25.** **(a)** Shift up 1 unit; **(b)** shift down 2 units; **(c)** shift 1 unit to the right; **(d)** shift 2 units to the left; **(e)** reflect across x axis **27.** **(a)** Shift f up 3 units; **(b)** shift g down 5 units; **(c)** reflect h about x axis, then shift up 1 unit; **(d)** shift k 1 unit to the right, then shift up 2 units; **(e)** shift q 2 units to the left, then take half of each ordinate; **(f)** shift r 1 unit to the left, take half of each ordinate, reflect about x axis, and shift up 1 unit **29.** **(a)** $\dfrac{2x}{x^2 - 1}$; **(b)** $\dfrac{2}{x^2 - 1}$; **(c)** $\dfrac{1}{x^2 - 1}$ **(d)** $\dfrac{x + 1}{x - 1}$; **31.** **(a)** $x^4 + \sqrt{x + 1}$; **(b)** $x^4 - \sqrt{x + 1}$; **(c)** $(x^4)(\sqrt{x + 1})$; **(d)** $\dfrac{x^4}{\sqrt{x + 1}}$; **(e)** $(x + 1)^2$ **33.** **(a)** $2x$; **(b)** 2; **(c)** $2x$; **(d)** 2 **35.** **(a)** $g \circ h$; **(b)** $g \circ f$; **(c)** $h \circ g$; **(d)** $f \circ f$; **(e)** $f \circ h$; **(f)** $h \circ h$ **37.** **(a)** $f(x) = x^{-3}$, $g(x) = 4x^3 - 2x + 5$; **(b)** $f(x) = \sqrt[3]{x}$, $g(x) = \dfrac{4 + x^3}{4 - x^3}$; **(c)** $f(x) = \dfrac{2x^2 + x}{\sqrt{x}}$, $g(x) = x^2 + 1$ **39.** $f^{-1}(x) = 5x - 25$ **41.** $f^{-1}(x) = \frac{1}{4}(x + 1)^2$ **43.** **(a)** Not invertible; **(b)** invertible **45.** **(a)** $f^{-1}(x) = 1/(1 - x)$; **(b)** $g^{-1}(x) = (3 - \sqrt{1 + 4x})/2$ **47.** ≈ 14.5 horsepower **49.** **(a)** $F(p) = p[f(p)]$; **(b)** there is a selling price at which no one will buy the product. **51.** **(a)** $f(p) = -1600p + 180{,}000$; **(b)** \$62.50; **(c)** \$112.50; **(d)** $p = f^{-1}(q) = -\frac{1}{1600}q + 112.50$ **53.** $y = 20x + 138{,}000$, \$146,000 **55.** **(a)** $f(t) = 16\pi(9t + 30\sqrt{t} + 25)$, $0 \leq t \leq 4$ cm^2; **(b)** f is increasing on its domain. **57.** **(a)** $f(t) = 62{,}500\, \pi t^2$ m^2; **(b)** f is increasing; **(c)** $t = f^{-1}(A) = \frac{1}{250}\sqrt{A/\pi}$

Chapter 3 Test, page 221

1. **(a)** -3; **(b)** $(1, -1)$, **(c)** $y = -3x + 2$; **(d)** $4\sqrt{10}$; **(e)** $(x - 1)^2 + (y + 1)^2 = 40$ **2.** $3x + 4y - 36 = 0$ **3.** **(a)** $\frac{1}{2}$; **(b)** $2a + h + 1$ **4.** **(a)** $(0, \infty)$; **(b)** $(-\infty, -1] \cup [1, \infty)$; **(c)** $(-\infty, 0)$ **5.** **(a)**

y
6
f
4
2
x
−2 −1 O 1 2

(b) domain; $[-1, 2]$, range $[4, 6]$;
(c) increasing on $[-1, 2]$, decreasing on $[-1, 1]$;
(d) f is not one-to-one.
6. **(a)** f is even, symmetric about the y axis;
(b) g is odd, symmetric about the origin;
(c) h is neither even nor odd. **7.** **(a)** $x^2 + 3x - 7$;
(b) $3x^3 - 7x^2$; **(c)** $(3x - 7)/x^2$; **(d)** $3x^2 - 7$;
(e) $(x + 7)/3$
8. **(a)** **(b)**

(c)

(d)

9. **(a)** $f(K) = \frac{9}{5}K - 459.4 = F$;
(b) assume $f(A) = f(B)$, show $A = B$;
(c) $f^{-1}(F) = \frac{5}{9}F + \frac{2297}{9} = K$
10. **(a)** All reals except $-\frac{2}{3}$;
(b) assume $f(a) = f(b)$ and show that $a = b$;
(c) $f^{-1}(x) = \dfrac{-(2x + 1)}{3x - 2}$

CHAPTER 4

Problem Set 4.1, page 230

1.

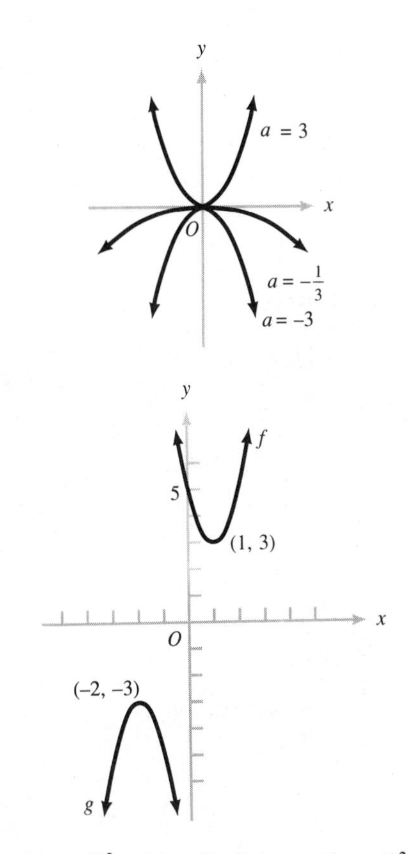

5. $f(x) = 1(x - 2)^2 - 5$ **7.** $f(x) = -1(x + 1)^2 + 9$
9. $f(x) = 2(x - 1)^2 - 1$ **11.** $f(x) = -4(x - 1)^2 - 1$
13. Vertex $(2, 27)$, y intercept 15, x intercepts -1, 5, opens downward, domain $\mathbb{R}$, range $(-\infty, 27]$

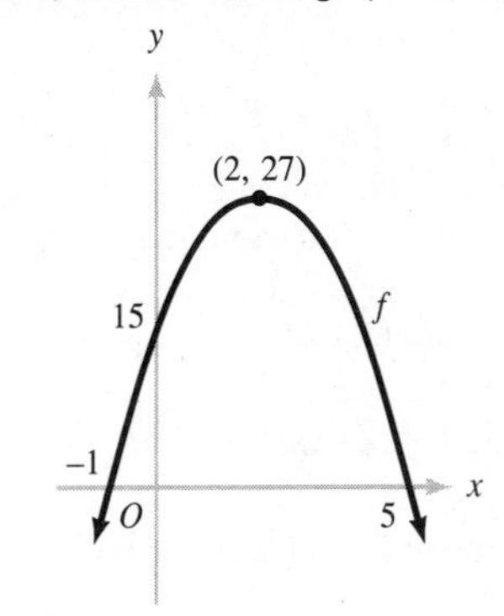

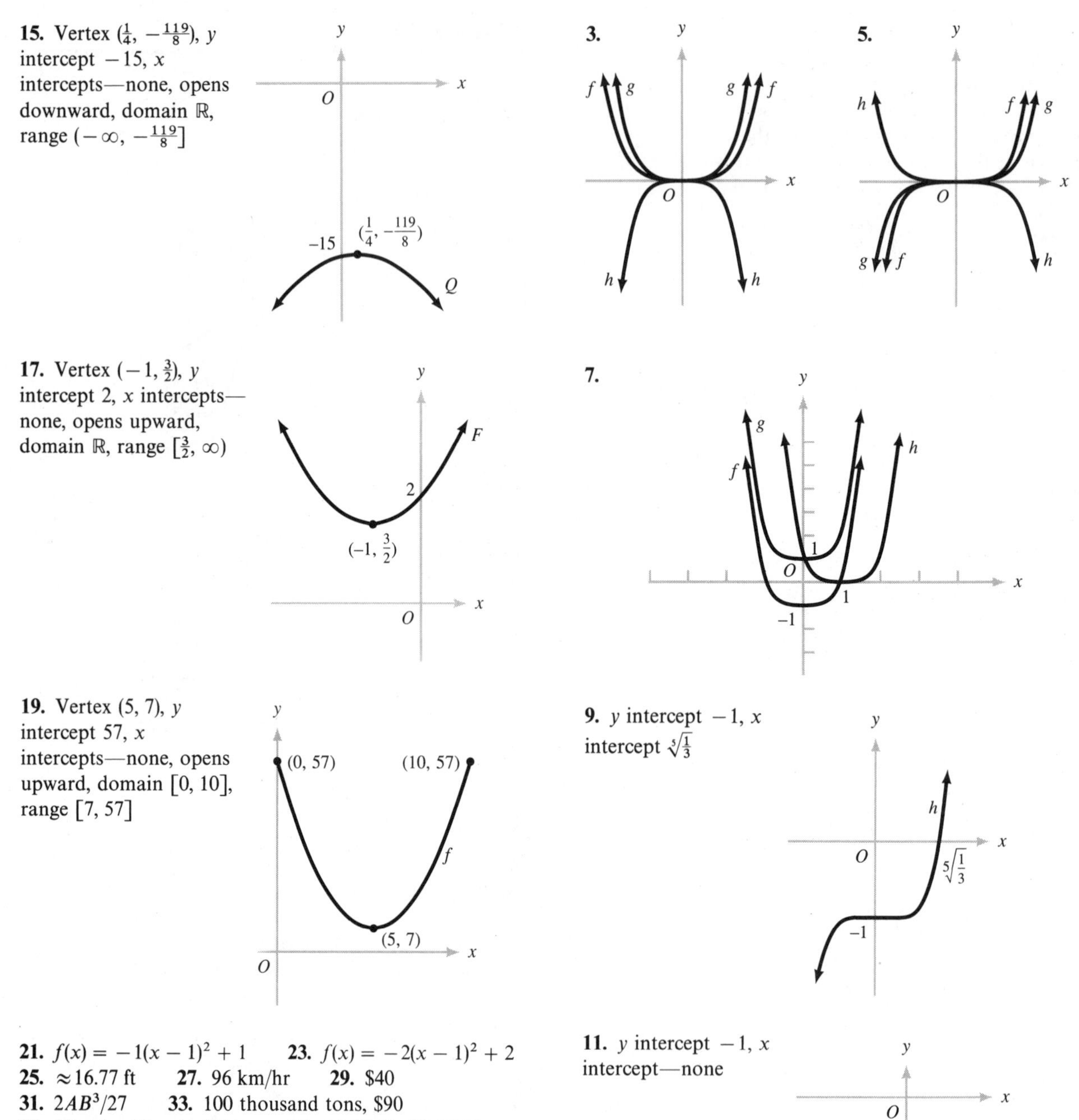

15. Vertex $(\frac{1}{4}, -\frac{119}{8})$, y intercept -15, x intercepts—none, opens downward, domain $\mathbb{R}$, range $(-\infty, -\frac{119}{8}]$

17. Vertex $(-1, \frac{3}{2})$, y intercept 2, x intercepts—none, opens upward, domain $\mathbb{R}$, range $[\frac{3}{2}, \infty)$

19. Vertex (5, 7), y intercept 57, x intercepts—none, opens upward, domain [0, 10], range [7, 57]

21. $f(x) = -1(x-1)^2 + 1$ **23.** $f(x) = -2(x-1)^2 + 2$
25. ≈ 16.77 ft **27.** 96 km/hr **29.** \$40
31. $2AB^3/27$ **33.** 100 thousand tons, \$90
35. 10 m by 20 m by 20 m **37.** **(a)** \$3.20; **(b)** \$3.40
39. 7.5 cm **41.** $(3, \sqrt{6})$

Problem Set 4.2, page 238

1. **(a)** Polynomial; **(b)** polynomial; **(c)** not; **(d)** polynomial; **(e)** not; **(f)** not

3.

5.

7.

9. y intercept -1, x intercept $\sqrt[5]{\frac{1}{3}}$

11. y intercept -1, x intercept—none

13. y intercept $-\frac{7}{8}$, x intercept $\frac{1}{2}$

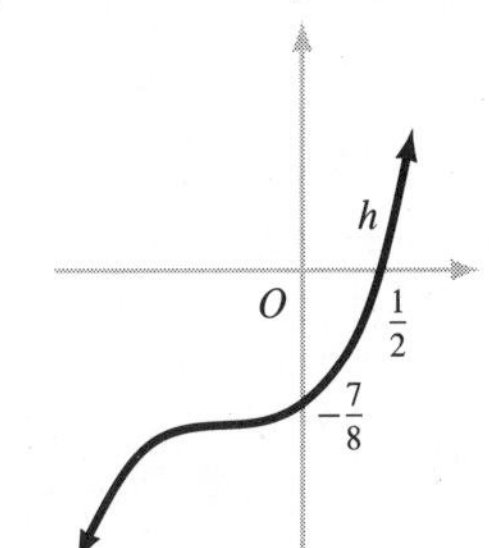

15. y intercept 2593, x intercept—none

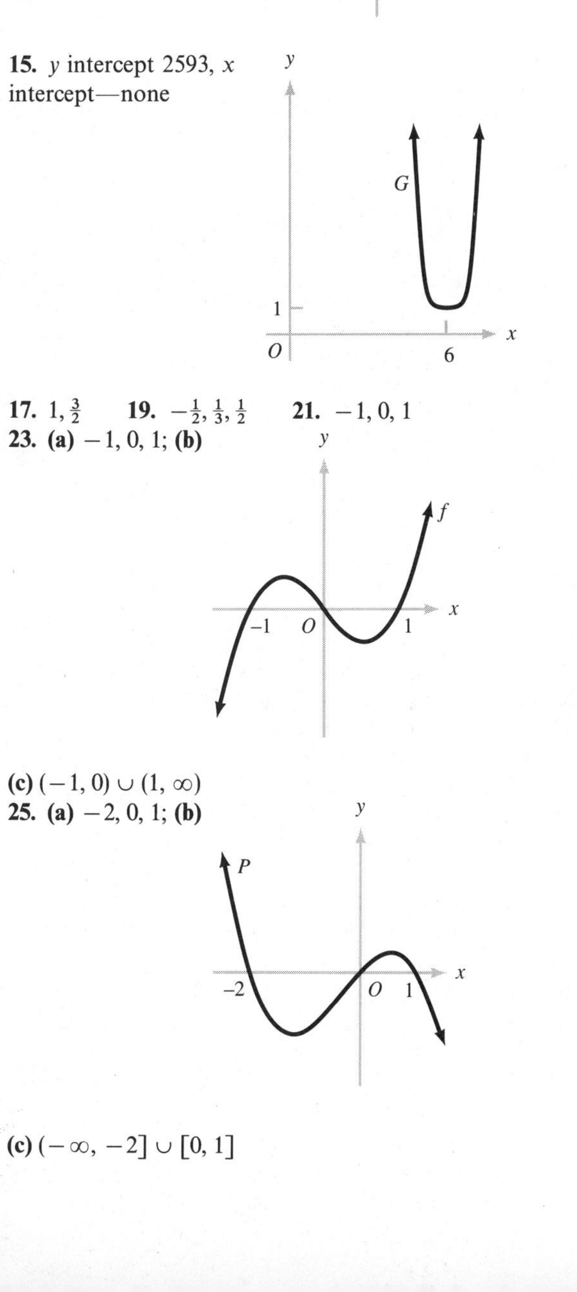

17. $1, \frac{3}{2}$ **19.** $-\frac{1}{2}, \frac{1}{3}, \frac{1}{2}$ **21.** $-1, 0, 1$

23. **(a)** $-1, 0, 1$; **(b)**

(c) $(-1, 0) \cup (1, \infty)$

25. **(a)** $-2, 0, 1$; **(b)**

(c) $(-\infty, -2] \cup [0, 1]$

27. x intercepts: $-4, 0, 3$; $(-\infty, -4)$ and $(0, 3)$

29. x intercepts: $-1, 0, 1$; $(-1, 0)$ and $(0, 1)$

31. x intercepts: $-1, 0, 2$; $[-1, 0]$ and $[2, \infty)$

33. V is maximum for $x \approx 1.7$ in.

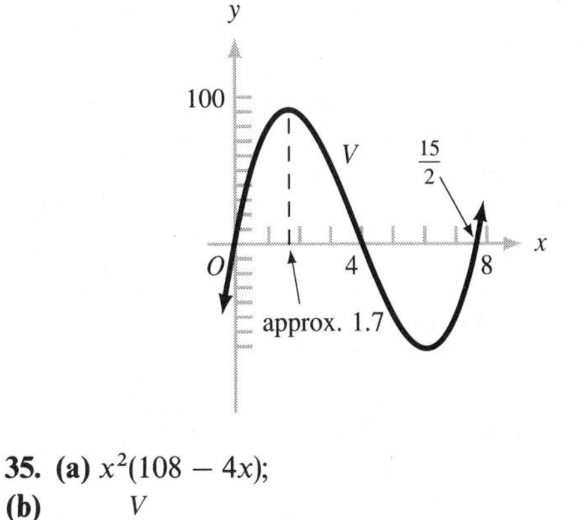

35. **(a)** $x^2(108 - 4x)$;
(b)

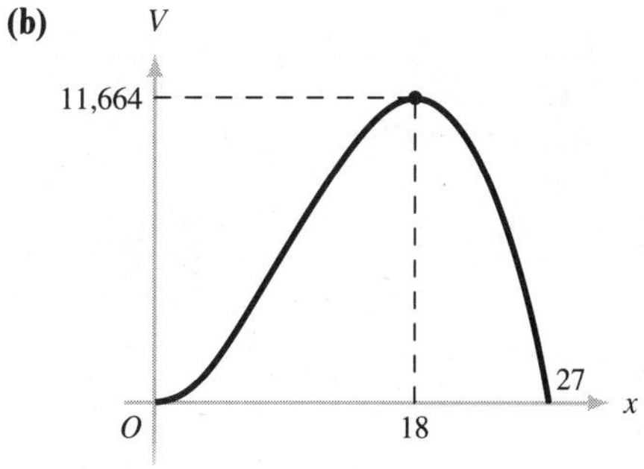

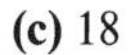
(c) 18

37. **(a)** 40; **(b)**

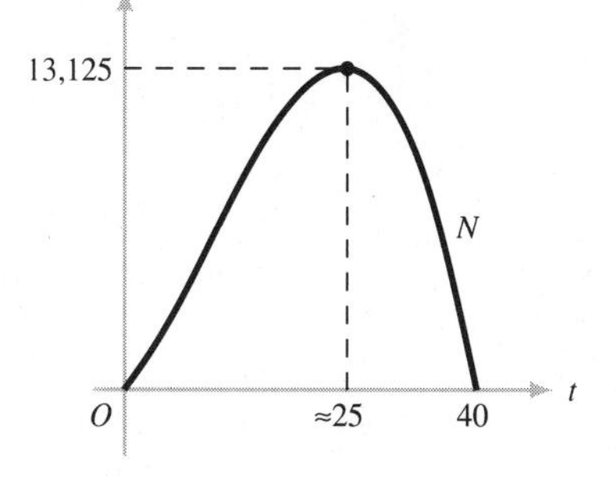

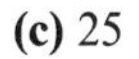
(c) 25

Problem Set 4.3, page 249

1. Q: $x + 8$, R: 30 **3.** Q: $x^2 + x + 1$, R: 0

5. Q: $3x^2 + 6x - 5$, R: -1

7. Q: $2x^3 - 3x^2 + 3x - 4$, R: -12

9. Q: $x + 3$, R: $-x - 13$

11. Q: $x^2 - x + 1$, R: $-x^2 - 1$

13. Q: $(x/2) + (1/2)$, R: $3/2$
15. Q: $(x/3) + (1/9)$, R: $-(5/9)x + (8/9)$
17. Q: $2t + 3$, R: -10
19. **(a)** $5x^2 + 23x + 91 + [366/(x - 4)]$;
(b) $5 + \dfrac{4x^2 - 28x - 16}{x^3 - 2x^2 - 8x}$
21. **(a)** and **(b)** Q: $5x^2 + 4x - 2$, R: -16
23. Q: $3x^2 + 4x + 7$, R: 18
25. Q: $x^4 - 2x^3 - x^2 + 2x - 3$, R: -10
27. Q: $-16x^2 - 20x - 8$, $R = 3$ **29.** 0 **31.** -456
33. 4 **35.** 7.0672 **37.** Yes **39.** No **41.** No
45. 0 **47.** 15 **49.** All positive integers n
51. ≈ 8.9 lb/in^2 **53.** **(a)** ≈ 0.000073; **(b)** ≈ 0.0000037;
(c) ≈ 0.0018

Problem Set 4.4, page 260

1. 2 **3.** 3 **5.** 3 **7.** **(a)** 2; **(b)** 1 **9.** **(a)** 3; **(b)** 2
11. **(a)** 2; **(b)** 3 **13.** Positive: 1, negative: 2 or 0
15. Positive: 3 or 1, negative: 0
17. Positive: 3 or 1, negative: 1
19. Positive: 0; negative: 2 or 0 **21.** $-1, 4$
23. $-3, 4$ **25.** $-2, 2$ **27.** $-3, 5$ **29.** 1, 3, 5
31. $-\frac{1}{2}$ **33.** 2, 2 **35.** $-\frac{1}{2}, -\frac{1}{2}$ **37.** 0, 3
39. 3 **41.** $-\sqrt{3}, -\frac{3}{4}, \frac{1}{5}, 1, \sqrt{3}$
43. **(a)** $f(x) = (x - 6)(x + 1)(x - 1)$;
(b)

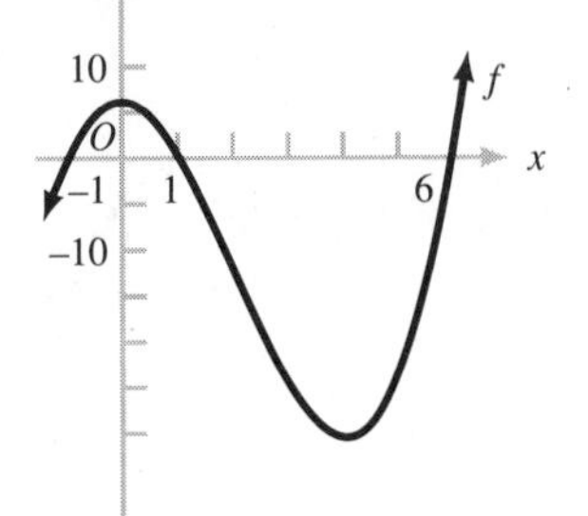

45. **(a)** $Q(x) = (x - 2)(x - 1)(2x - 1)(x + 2)$;
(b)

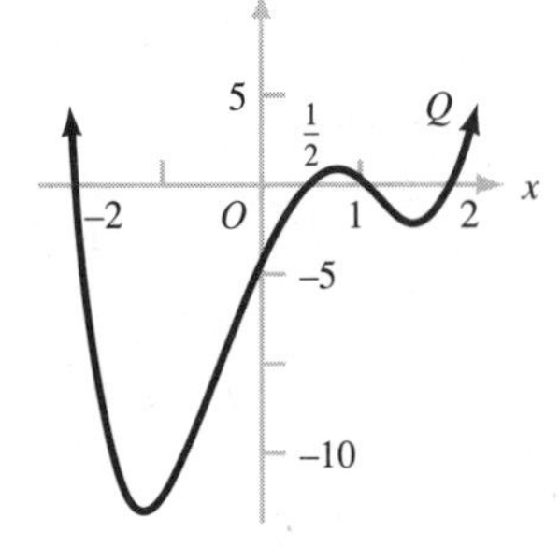

47. Width 5 in, length 10 in, depth 7 in
49. **(a)** $V = \pi\left(12r^2 - \dfrac{r^3}{3}\right)$; **(b)** 6 m **51.** 9
53. Zeros of $f(-x)$ are negatives of zeros of $f(x)$
55. Coefficients of odd powers change signs
57. Use the fact that $(x - p)(x - q) =$
$x^2 + (-p - q)x + pq$

Problem Set 4.5, page 266

1. $f(1) < 0$, $f(2) > 0$ **3.** $f(1.5) < 0$, $f(1.6) > 0$
5. $h(-2) < 0$, $h(-1) > 0$ **7.** Zero in $[1.5, 1.75]$
9. 1.51
11. Zeros in $[-2, -1.75]$, $[0.25, 0.50]$, and $[1.5, 1.75]$
13. -1.88 **15.** 1.532088886 **17.** 1.29 **19.** 2.42
21. 10 **23.** -0.682327804 **25.** 1.679729452
27. 10.93 m
29. **(b)** Possible solutions: 98.04 m and 163.21 m

Problem Set 4.6, page 274

1. $z^3 - (6 + i)z^2 + (9 + 6i)z - 9i$
3. $z^3 + (-1 + 2i)z^2 - (1 + 2i)z + 1$
5. $z^4 + 4z^3 + 5z^2 + 4z + 4$
7. $z^4 - 3z^3 + 3z^2 - 3z + 2$
9. $z^6 - 4z^5 + 16z^4 - 26z^3 + 25z^2 - 22z + 10$
11. $z^5 - 3z^4 + z^3 - 3z^2$ **13.** $\pm\left(\dfrac{\sqrt{2}}{2} + \dfrac{\sqrt{2}}{2}i\right)$
15. $\pm\left(\dfrac{3\sqrt{2}}{2} - \dfrac{\sqrt{2}}{2}i\right)$
17. $\pm[\sqrt{(2 + \sqrt{3})/2} + i\sqrt{(2 - \sqrt{3})/2}]$ **19.** $1 \pm i$
21. $(-i \pm \sqrt{3})/2$ **23.** 2, $1 + i$
25. **(a)** $1, -1, i, -i$; **(b)** $(z - 1)(z + 1)(z - i)(z + i)$;
(c) each zero has multiplicity 1.
27. **(a)** $-2, -2, 2i, -2i$; **(b)** $(z + 2)^2(z + 2i)(z - 2i)$;
(c) -2 has multiplicity 2; others 1.
29. **(a)** $-\frac{1}{2}, -\frac{1}{2}, -\frac{1}{2}, -1, 3$; **(b)** $(2z + 1)^3(z + 1)(z - 3)$;
(c) $-\frac{1}{2}$ has multiplicity 3; others 1.
31. **(a)** $1, i, -i$; **(b)** $(z - 1)(z + i)(z - i)$;
(c) each zero has multiplicity 1.
33. **(a)** $3, 1 + i, 1 - i$; **(b)** $(z - 3)(z - 1 - i)(z - 1 + i)$;
(c) each zero has multiplicity 1.
35. **(a)** $2, 2, (-1 + i\sqrt{3})/2, (-1 - i\sqrt{3})/2$;
(b) $\frac{1}{4}(z - 2)^2(2z + 1 - \sqrt{3}i)(2z + 1 - \sqrt{3}i)$;
(c) 2 is a zero of multiplicity 2; others 1.
37. $az^2 + bz + c = a(z - z_1)(z - z_2) =$
$az^2 + a(-z_1 - z_2)z + az_1z_2$
39. Complex zeros come in *pairs*.
41. 1, and $(-1 + \sqrt{3}i)/2$

Problem Set 4.7, page 282

1. $(-\infty, 0) \cup (0, \infty)$
3. $(-\infty, -1) \cup (-1, \frac{3}{2}) \cup (\frac{3}{2}, \infty)$
5. $(-\infty, -5) \cup (-5, -3) \cup (-3, \infty)$

7. Horizontal asymptote x axis, vertical asymptote y axis

9. Horizontal asymptote x axis, vertical asymptote y axis

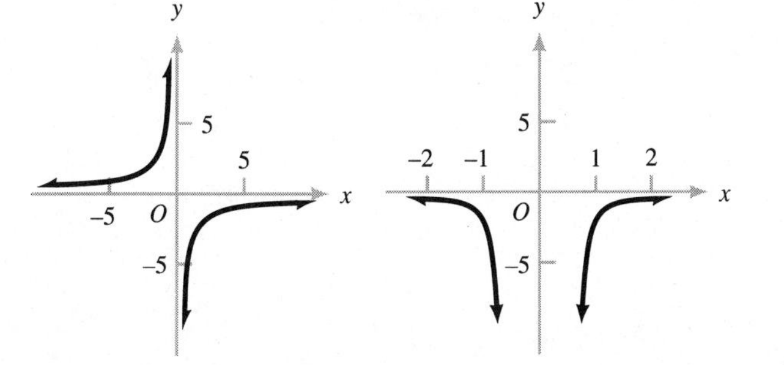

11. Horizontal asymptote $y = 1$, vertical asymptote y axis

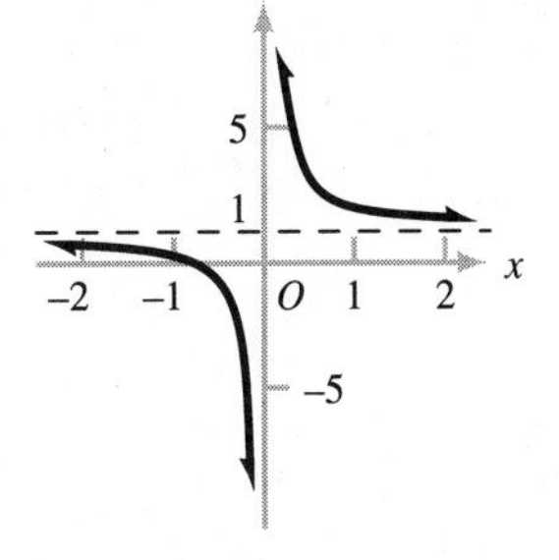

13. Horizontal asymptote $y = -4$, vertical asymptote y axis

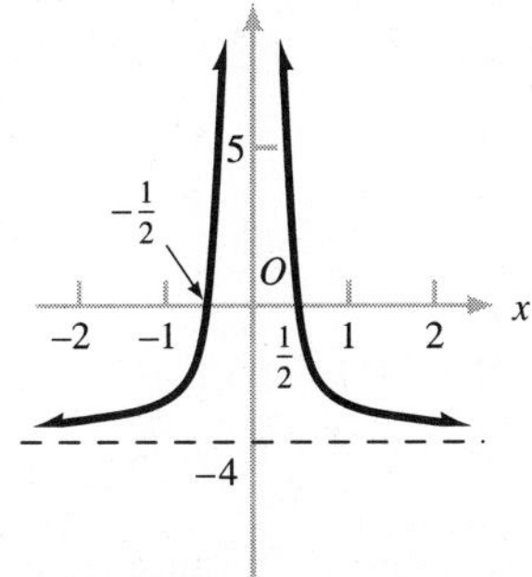

15. Horizontal asymptote x axis, vertical asymptote $x = -1$

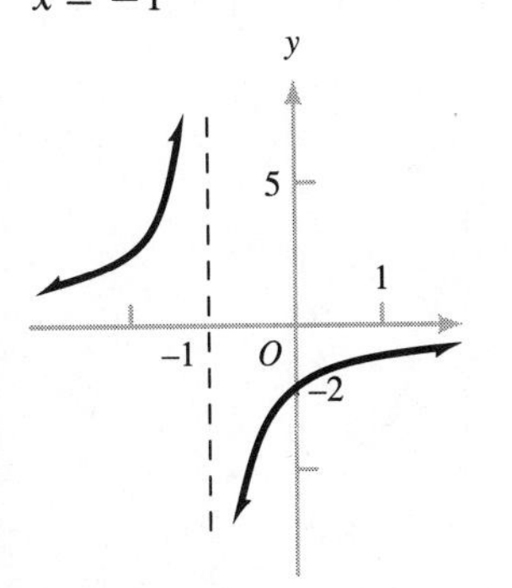

17. Horizontal asymptote $y = 6$, vertical asymptote $x = 2$

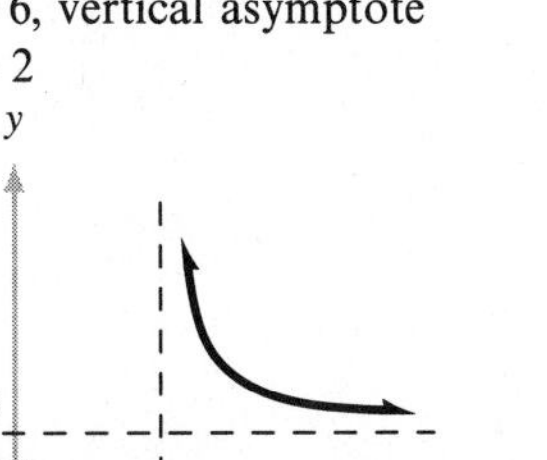

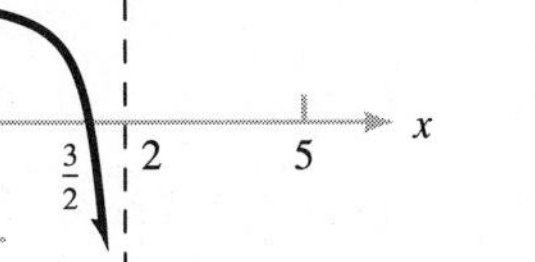

19.

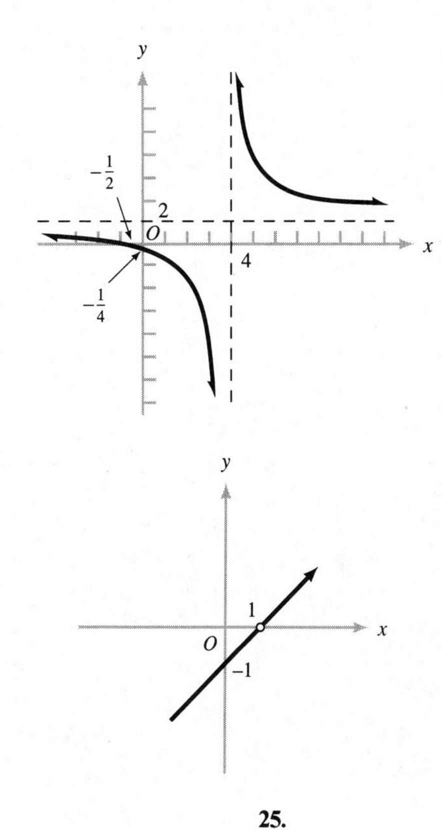

21.

23.

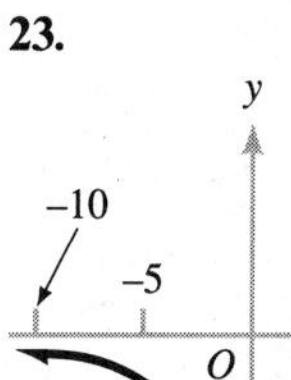

25.

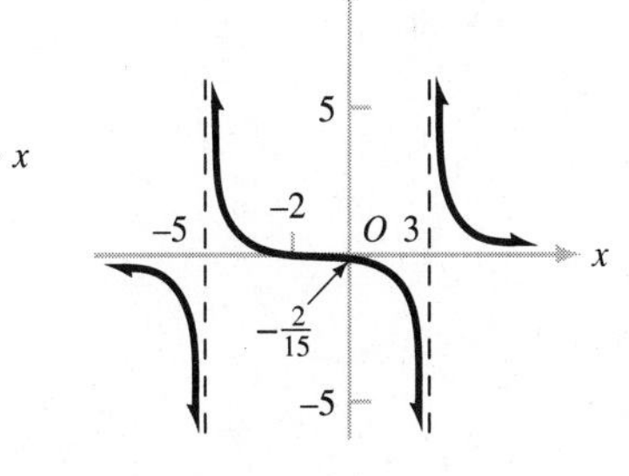

27.

29.

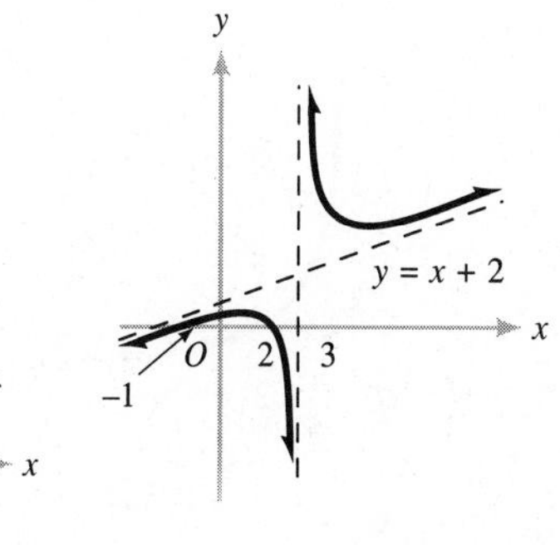

31.

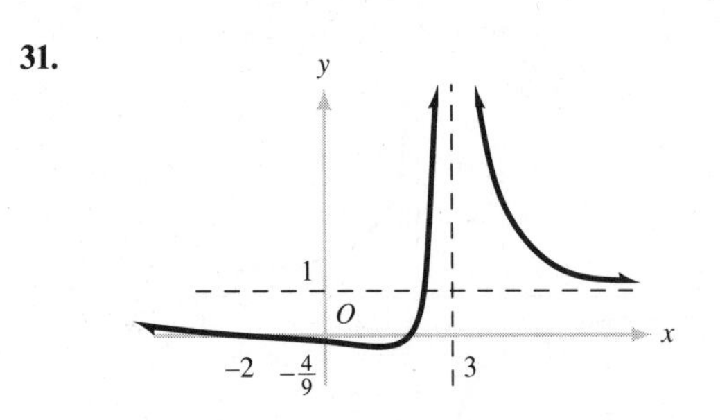

33.

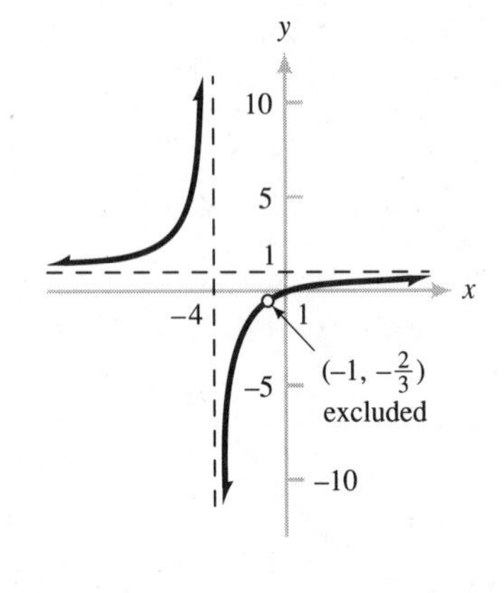

35. **37.**

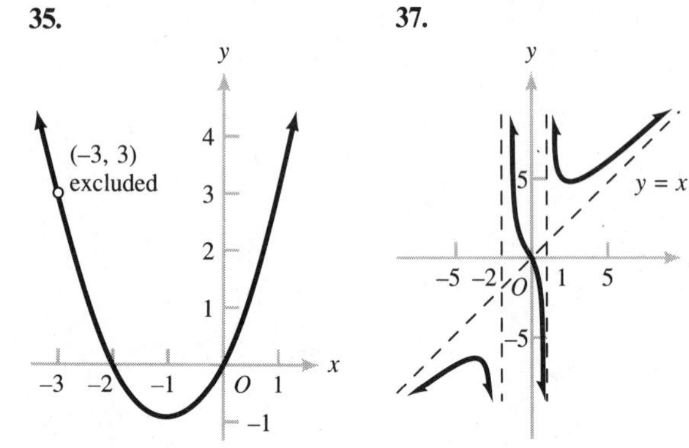

39. (1, 3)

41. (a)

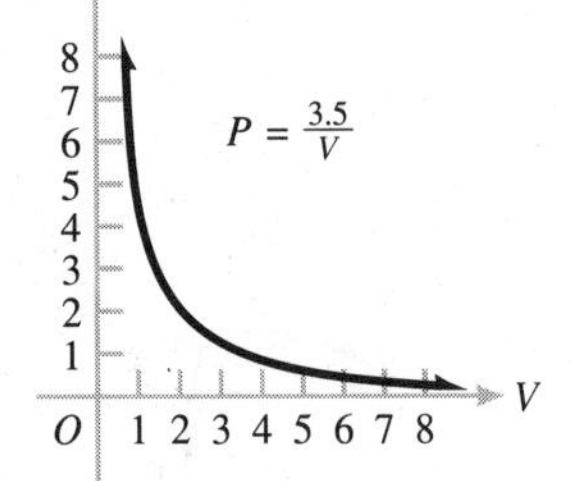

(b) $(0, \infty)$; **(c)** decreases

43. (a) **45.**

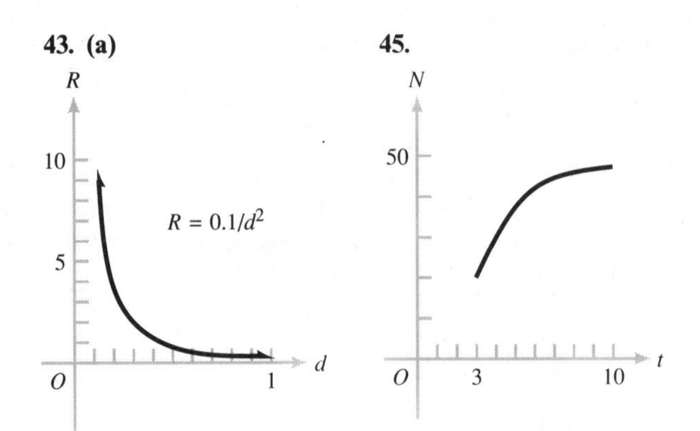

(b) $(0, \infty)$; **(c)** R increases when d decreases

47. (e)

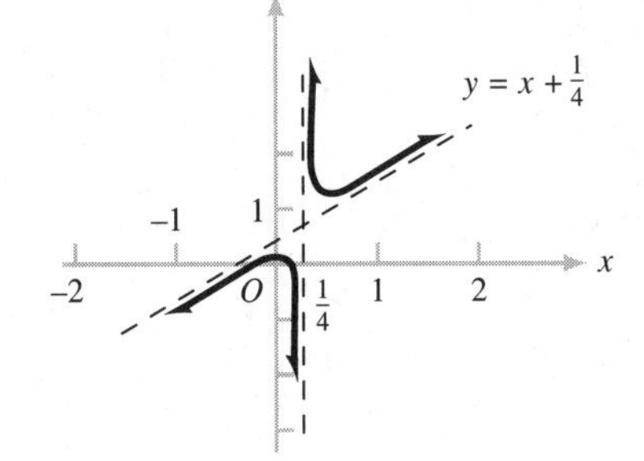

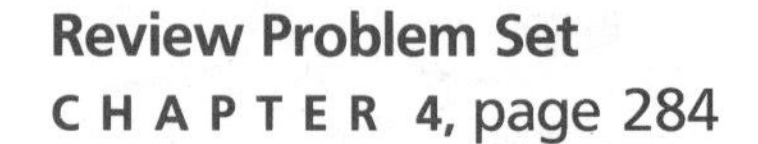

Review Problem Set
CHAPTER 4, page 284

1. Domain $\mathbb{R}$, range $[-\frac{1}{4}, \infty)$, x intercepts 1 and 2, y intercept 2, vertex $(\frac{3}{2}, -\frac{1}{4})$, opens upward
3. Domain $\mathbb{R}$, range $(-\infty, 0]$, x intercept 5, y intercept -25, vertex $(5, 0)$, opens downward
5. $(-\infty, 1]$ and $[2, \infty)$ **7.** $\frac{21}{2}, \frac{21}{2}$ **9.** 15°F
13. (a) $-\frac{6}{5}, -1, \frac{3}{2}$; **(b)**

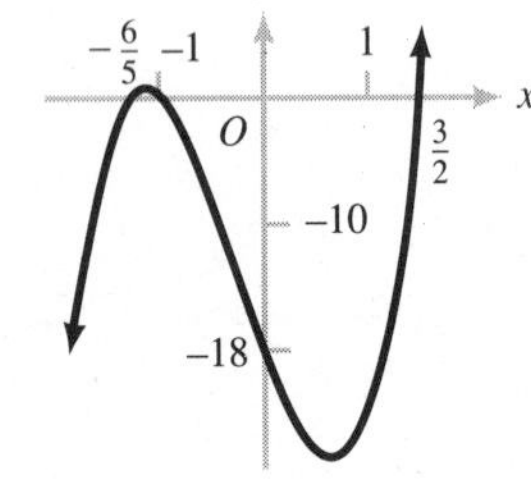

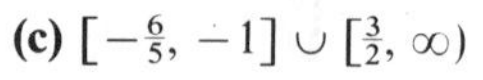

(c) $[-\frac{6}{5}, -1] \cup [\frac{3}{2}, \infty)$

15. **(a)** $-3, 2$; **(b)**

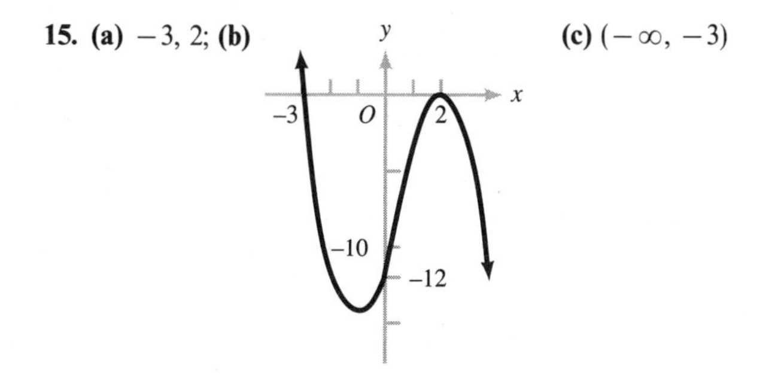

(c) $(-\infty, -3)$

17. Q: $x + 2$, R: -4 **19.** Q: x^2, R: $8x^2 - 32$
21. $4x^3 - 3x^2 + 3x - 5 + 8/(x + 1)$
23. Q: $x^2 - 4x + 11$, R: -27
25. Q: $x^4 - x^3 + x^2 - x + 6$, R: -19 **27.** ≈ 1637.75
29. **(a)** and **(b)** -27 **31.** **(a)** and **(b)** -7 **33.** No
35. 2 **37.** **(a)** 1, 2; positive 1, negative 2 or 0: $[-3, 1]$;
(b) 2, 0; positive 2 or 0, negative 0; $[0, 2]$
39. $-1, 2, 7$ **41.** $-1, -\frac{1}{2}, 1, \frac{3}{2}$
43. $(x + 2)(x + 1)(x - 1)$ **45.** 0.71
47. -0.197429337 **49.** **(a)** $-3, 1, \dfrac{-3 \pm i\sqrt{11}}{2}$;
(b) $(z - 1)(z + 3)\left(z + \dfrac{3}{2} - \dfrac{i\sqrt{11}}{2}\right)\left(z + \dfrac{3}{2} + \dfrac{i\sqrt{11}}{2}\right)$;
(c) each zero has multiplicity 1
51. **(a)** $-2, 2, 3 \pm \sqrt{2}$;
(b) $(z - 2)(z + 2)(z - 3 + \sqrt{2})(z - 3 - \sqrt{2})$;
(c) each zero has multiplicity 1
53. $z^5 - 6z^4 + 13z^3 - 14z^2 + 12z - 8$
55. Domain $(-\infty, 1) \cup (1, \infty)$, asymptotes $x = 1$, $y = 2$, y intercept 0, x intercept 0
57. Domain $(-\infty, -2) \cup (-2, 0) \cup (0, \infty)$, asymptotes $x = 0$, $y = 1$, no intercepts
59. Domain $(-\infty, 0) \cup (0, 3) \cup (3, \infty)$, asymptotes $x = 0$, $x = 3$, $y = 1$, no intercepts
61. Domain $(-\infty, 2) \cup (2, \infty)$, asumptotes $x = 2$, $y = x$, y intercept $\frac{3}{2}$, x intercepts $-1, 3$
63. 30 thousand trophies
65.

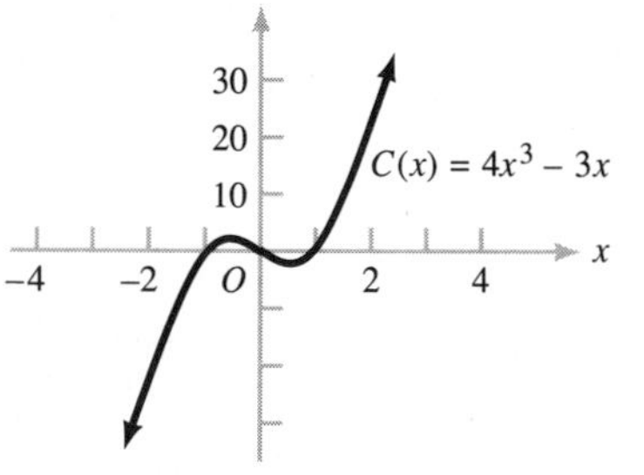

67. 8 cm **69.** ≈ 1.44 m
71. **(a)**

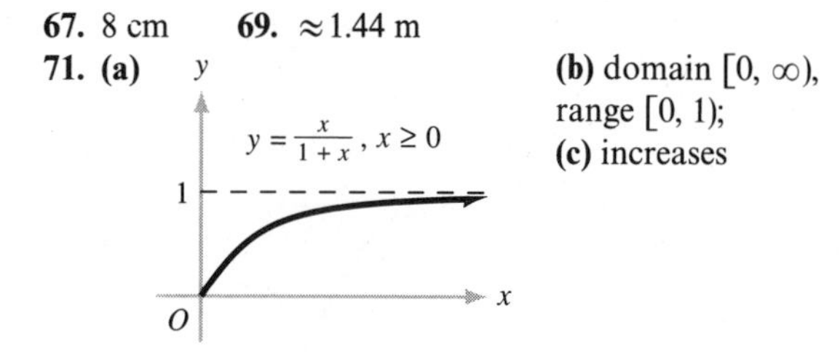

(b) domain $[0, \infty)$, range $[0, 1)$;
(c) increases

Chapter 4 Test, page 288

1. **(a)** $3(x + 2)^2 + 38$; **(b)** $(-2, 38)$;
(c)
(d) domain: $\mathbb{R}$, range: $[38, \infty)$

2. **(a)**

(b)

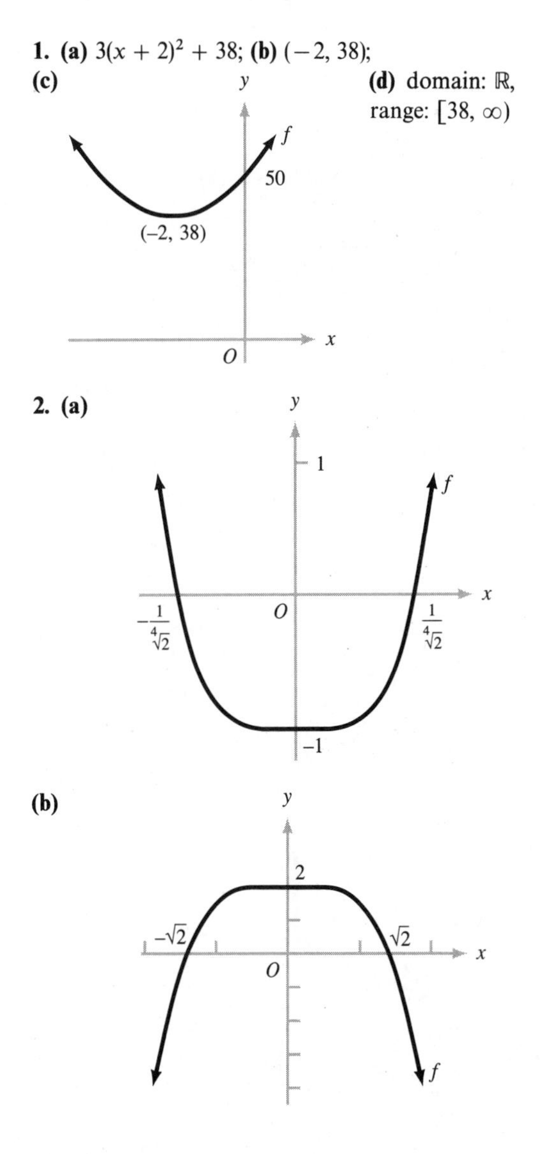

3. **(a)** $q(x) = 4x^4 + 9x^3 + 34x^2 + 204x + 874$;
(b) $f(-2) = 492$, $r(x) = 5244$
4. **(a)** $-1, 2, 3$; **(b)**

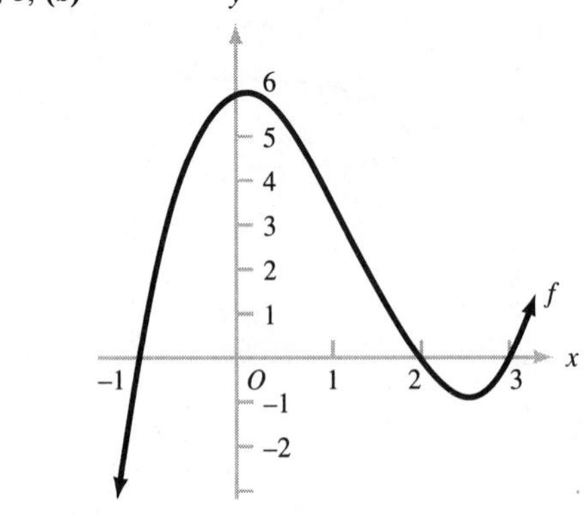

(c) $(-\infty, -1) \cup (2, 3)$
5. **(a)** $f(1) = -3 < 0$; $f(2) = 11 > 0$; **(b)** $[1.25, 1.375]$;
(c) 1.3 **6.** 1.33058740

7. **(a)**

Zero	Multiplicity
$-\frac{5}{2}$	1
$-\frac{1}{3}$	4
0	1
$\frac{5}{2}$	2

(b) $x^3 - 4x^2 + x + 26$

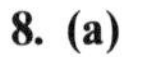
8. **(a)**

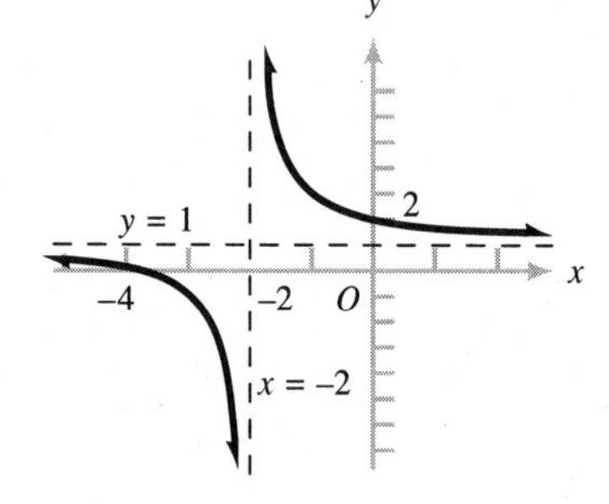

(b)

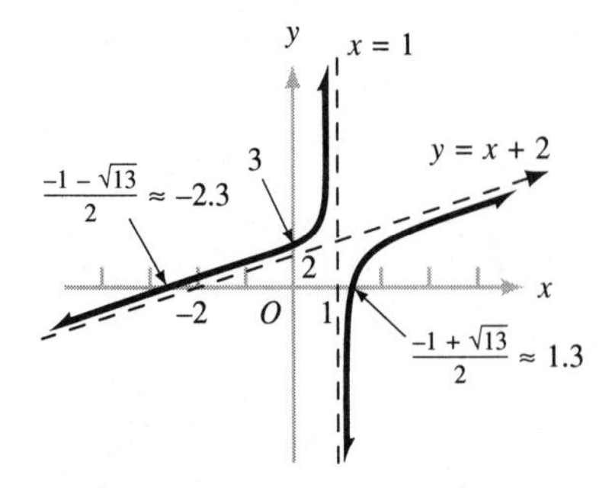

9. 312.5 m^2 **10.** $n = f(t) = \dfrac{t(t+1)}{2t+1}$

CHAPTER 5

Problem Set 5.1, page 299

1. **3.**

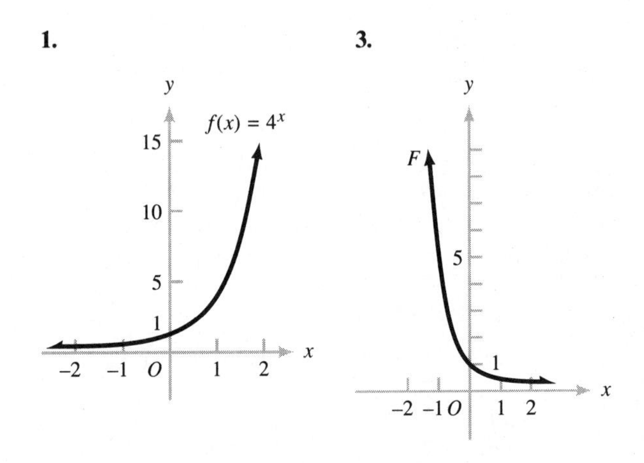

5. **(a)** 2.665144142; **(b)** 0.3752142273; **(c)** 8.824977830;
(d) 0.0119935487; **(e)** 1.63252692; **(f)** 36.462159607;
(g) 0.2927940320
7. **(a)** $a^x a^y = 38.80960175$, $a^{x+y} = 38.80960175$;
(b) $a^{x+y} = 147.3210081$, $a^x a^y = 147.3210081$;
(c) $(a^x)^y = 0.01937829357$, $a^{xy} = 0.01937829357$
9. Domain $\mathbb{R}$, range $(1, \infty)$, increasing, asymptote $y = 1$
11. Domain $\mathbb{R}$, range $(-1, \infty)$, increasing, asymptote $y = -1$

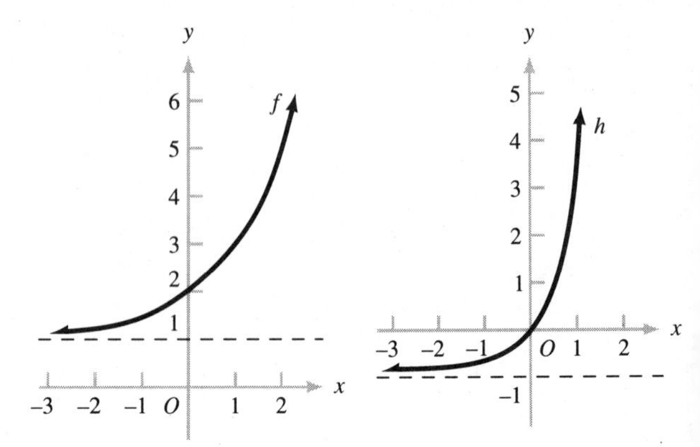

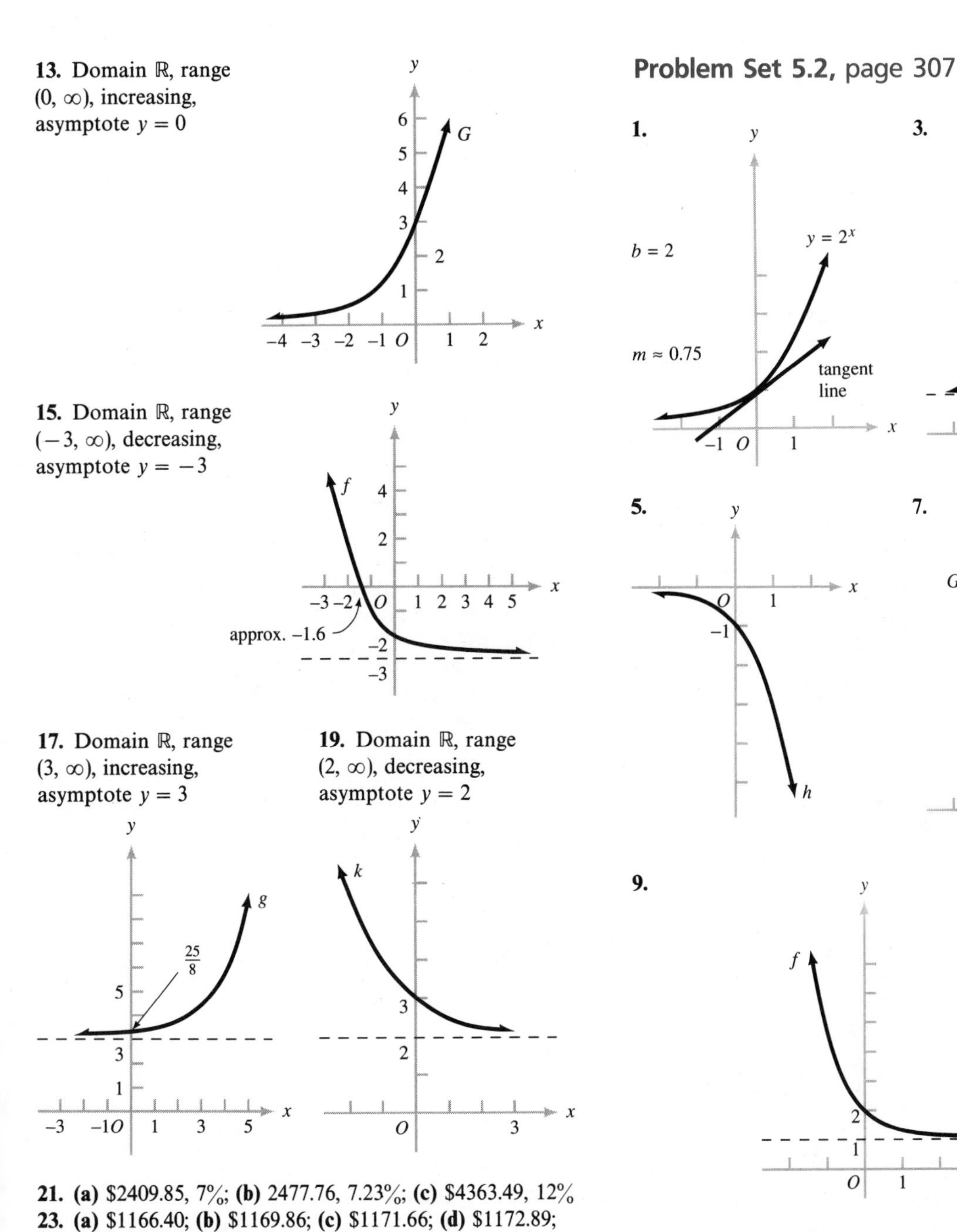

13. Domain $\mathbb{R}$, range $(0, \infty)$, increasing, asymptote $y = 0$

15. Domain $\mathbb{R}$, range $(-3, \infty)$, decreasing, asymptote $y = -3$

17. Domain $\mathbb{R}$, range $(3, \infty)$, increasing, asymptote $y = 3$

19. Domain $\mathbb{R}$, range $(2, \infty)$, decreasing, asymptote $y = 2$

21. **(a)** \$2409.85, 7%; **(b)** 2477.76, 7.23%; **(c)** \$4363.49, 12%
23. **(a)** \$1166.40; **(b)** \$1169.86; **(c)** \$1171.66; **(d)** \$1172.89; **(e)** \$1173.37; **(f)** \$1173.49; **(g)** \$1173.51 **25.** \$549.19
27. \$96.09 **29.** **(a)** \$331.02; **(b)** \$1921.72
31. **(a)** \$818.35; **(b)** \$12,591.67; **(c)** \$75,303.00
33. \$485,989.52 **35.** \$98,496.62
37. First plan pays \$99,587.14, second plan pays \$167,772.16

Problem Set 5.2, page 307

1. $b = 2$, $m \approx 0.75$

3.

5.

7.

9.

11. **(a)** $e^x e^y = 23.24905230$, $e^{x+y} = 23.24905230$; **(b)** $e^{x+y} = 0.404329687$, $e^{xy} = 0.404329687$ **13.** 0, 2
15. **(a)** Discrepancy in third decimal place; **(b)** discrepancy in third decimal place; **(c)** discrepancy in first decimal place; **(d)** discrepancy in first decimal place

17. **19.** **21.**

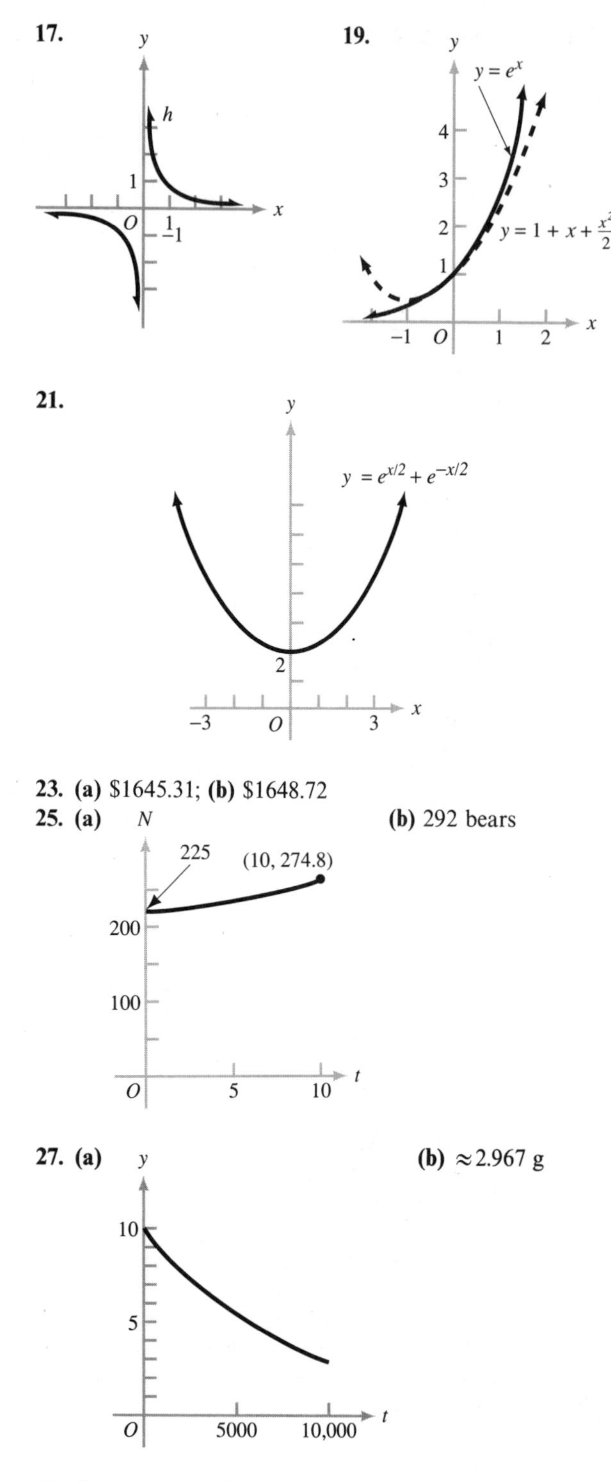

23. **(a)** \$1645.31; **(b)** \$1648.72

25. **(a)** (graph) **(b)** 292 bears

27. **(a)** (graph) **(b)** ≈ 2.967 g

29. **(a)** 2,000,000; **(b)** 3,644,238

31. **(a)** 2000; **(b)** $\approx 8{,}900{,}000$; **(c)**

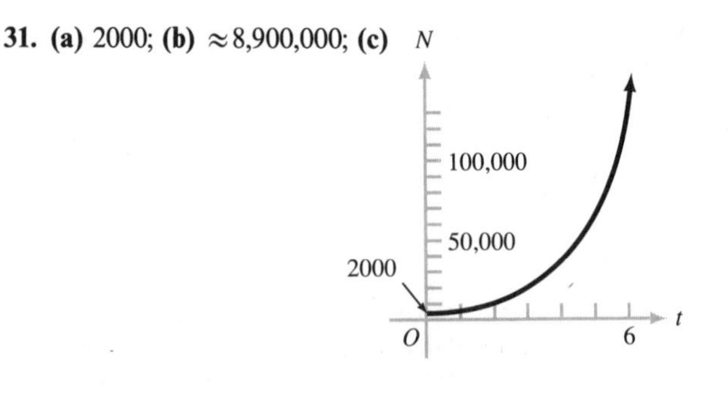

(d) ≈ 4 hr **33.** 0.0005%

Problem Set 5.3, page 316

1. 3 **3.** $\frac{1}{2}$ **5.** 1 **7.** 0 **9.** $-2, 1$ **11.** 1
13. **(a)** 2; **(b)** 3; **(c)** 4; **(d)** 5; **(e)** -2; **(f)** -2; **(g)** 5; **(h)** $-\frac{3}{4}$
15. **(a)** $2^5 = 32$; **(b)** $16^{1/4} = 2$; **(c)** $9^{-1/2} = \frac{1}{3}$; **(d)** $e^1 = e$; **(e)** $(\sqrt{3})^4 = 9$; **(f)** $10^n = 10^n$; **(g)** $x^5 = x^5$
17. **(a)** $\log_8 1 = 0$; **(b)** $\log_{10} 0.0001 = -4$; **(c)** $\log_4 256 = 4$; **(d)** $\log_{27} \frac{1}{3} = -\frac{1}{3}$; **(e)** $\log_8 4 = \frac{2}{3}$; **(f)** $\log_a y = c$ **19.** 2 **21.** 5 **23.** $\frac{1}{4}$ **25.** $\frac{9}{2}$
27. $\frac{85}{3}$ **29.** $-2, -1$ **31.** $-\frac{5}{3}, 1$ **33.** 2
35. **(a)** 2.31; **(b)** 2.70; **(c)** -0.95; **(d)** 1.87; **(e)** 0.74; **(f)** 0.95
37. 2 **39.** $\log_b x + \log_b(x + 1)$, $x > 0$
41. $2 \log_{10} x + \log_{10}(x + 1)$, $x > 0$
43. $3 \log_3 x + 2 \log_3 y - \log_3 z$
45. $\frac{1}{2} \log_e x + \frac{1}{2} \log_e(x + 3)$ **47.** $\log_3 x^9$ **49.** $\log_5 \frac{2}{3}$
51. $\log_e(x + 1)$ **53.** $\log_3 \dfrac{x + 9}{x + 5}$ **55.** 2 **57.** $\frac{1}{9}$
59. 2 **61.** 67 **63.** $\log_e 5/\log_e 10$
65. $\log_a b = \log_b b/\log_b a = 1/\log_b a$ **67.** ≈ 983 m/s
69. $t = 200 \log_e(y_0/y)$

Problem Set 5.4, page 324

1. **(a)** **(b)**

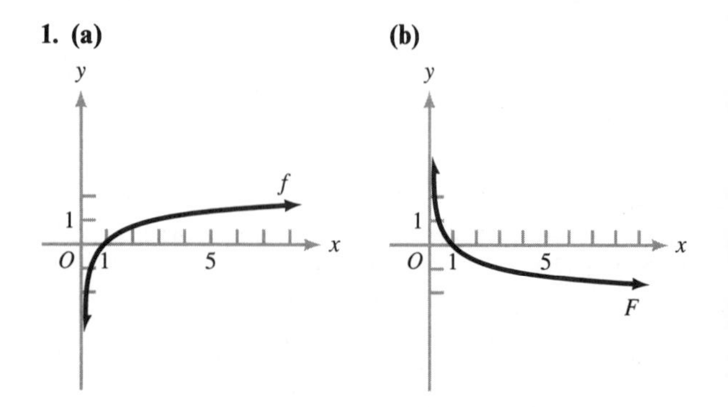

3. **(a)** 0.8043439185; **(b)** 3.090046322; **(c)** -1.453333975; **(d)** 11.48387245; **(e)** -8.182963774

5. **(a)** 8.325063694; **(b)** 0.9947321582; **(c)** -3.208826489; **(d)** 20.41142767; **(e)** -15.56881884
7. **(a)** $\ln e^{\pi} \approx \ln 23.14069264 \approx 3.141592654$; **(b)** $e^{\ln \pi} \approx e^{1.144729886} \approx 3.141592654$
9. **(a)** 4.643856190; **(b)** 0.6309297537; **(c)** 0.4808983469; **(d)** 1.405954306; **(e)** 7.551524220 **11.** **(a)** $(2, \infty)$; **(b)** $(0, \infty)$; **(c)** $(-\infty, 4)$; **(d)** $(-\infty, 0) \cup (0, \infty)$; **(e)** $(0, 1) \cup (1, \infty)$; **(f)** $(-1, 1)$; **(g)** $\mathbb{R}$; **(h)** $\mathbb{R}$
13. **(a)** $f(x) = \log_3 x$; **(b)** $g(x) = -\log_3 x$; **(c)** $h(x) = \log_3(-x + 1)$; **(d)** $F(x) = \log_3(x + 1)$
15. Domain $(0, \infty)$, range $\mathbb{R}$, x intercept 0.5, increasing, asymptote $x = 0$

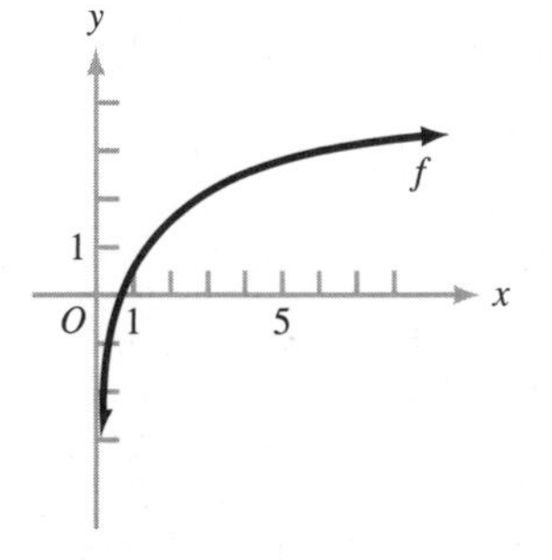

17. Domain $(-\infty, 0) \cup (0, \infty)$, range $\mathbb{R}$, x intercepts -1, 1, increasing on $(-\infty, 0)$, decreasing on $(0, \infty)$, asymptote $x = 0$

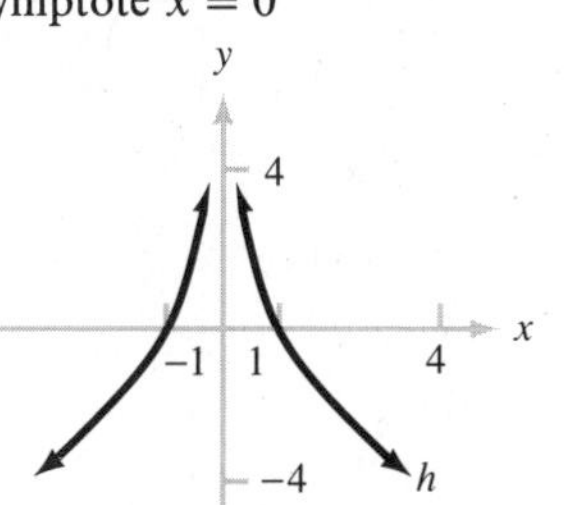

19. Domain $(0, \infty)$, range $\mathbb{R}$, x intercept 1, increasing, asymptote $x = 0$

21. Domain $(1, \infty)$, range $\mathbb{R}$, x intercept 2, increasing, asymptote $x = 1$

23. Domain $(0, \infty)$, range $\mathbb{R}$, x intercept 1, decreasing, asymptote $x = 0$

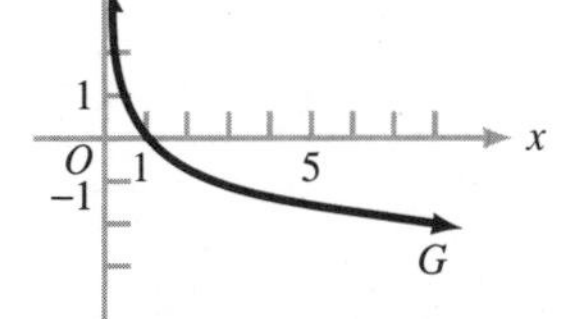

25. Domain $(0, \infty)$, range $\mathbb{R}$, x intercept $1 + e^{-2}$, increasing, asymptote $x = 1$

27. Domain $(0, \infty)$, range $\mathbb{R}$, x intercept 1, decreasing, asymptote $x = 0$

29. **(a)** $e^0 = 1$; **(b)** $\ln x^{-1} = -\ln x$; **(c)** $e^1 = e$; **(d)** $\log_e x = \log x / \log e$ **31.** **(a)** 7; **(b)** x^{-3}; **(c)** $e^6 t^5$; **(d)** $1/x$; **(e)** e^{-4}/p^3; **(f)** $x^2 - 9$; **(g)** $e^{-3}r^2$; **(h)** x^2/y^7
33. $\frac{1}{3}[\ln(x + 2) + \ln(3x - 7) + \ln(4x + 1)]$
35. ln is the inverse of $f(x) = e^x$. Because the domain of f is $\mathbb{R}$, the range of ln is also $\mathbb{R}$.
37.

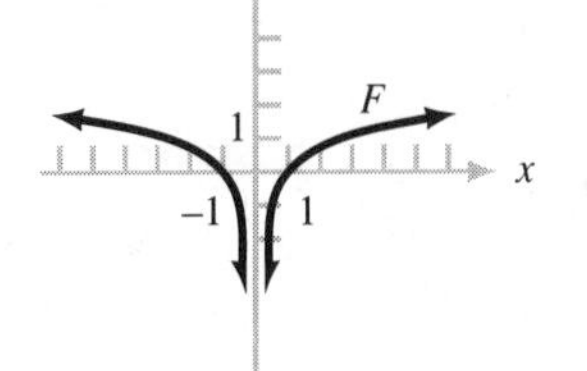

39. 2

41. Tangent line is $y = x - 1$. **43.** $e^{x \ln y} = (e^{\ln y})^x = y^x$
45. True only for $x > 0$
47. **(a)**

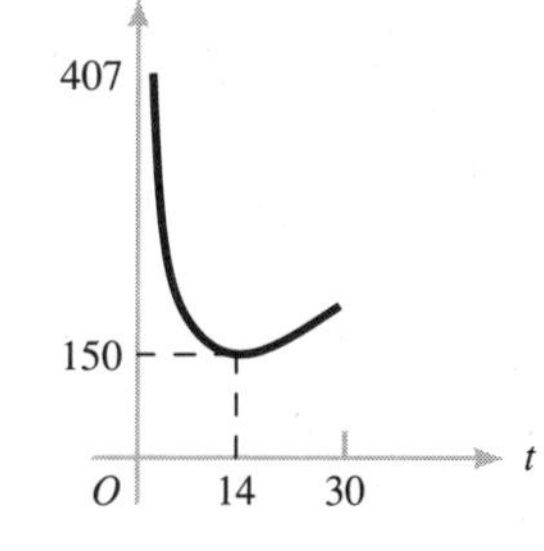

(b) 14 days

Problem Set 5.5, page 335

1. 0.7924812504 **3.** -1.413390105 **5.** $\frac{1}{27}$
7. 0.4785236727 **9.** -0.3420914978

11. $y = \dfrac{\ln(x-2) - \ln 3 - 1}{4}$ **13. (a)** ≈ 601 wk; **(b)** ≈ 607 wk **15. (a)** 7.8; **(b)** 4.2; **(c)** 6.4; **(d)** 7.4 **17.** ≈ 8863 m **19.** ≈ 5545 m **21.** ≈ 113 dB **23.** 10^{12} **25.** $10^{8/5} \approx 3.98$ **27.** $10^{2.1} \approx 126$ **29. (a)** $y = y_0 e^{-0.000418571t}$; **(b)** 0.0418%; **(c)** 0.9917 g **31. (a)** 34.7; **(b)** 693 days **33.** 20.35 yr **35.** 9692 yr **37. (a)** $k = \dfrac{1}{2 \ln 4} \approx 0.361$; **(b)** ≈ 0.354 or 35.4%

Problem Set 5.6, page 342

1. (a) $N = 10^7 e^{kt}$; $k = \ln 1.03$; **(b)** $\approx 18{,}000{,}000$; **(c)** ≈ 23.45 yr **3.** ≈ 9.97 hr **5. (a)** $N = 10e^{(t \ln 2)/20}$; **(b)** ≈ 80 *E. Coli* **7.** $\approx 41.42\%$ **9. (a)** ≈ 700; **(b)** ≈ 421; **(c)** ≈ 2.88 yr **11.** Notice that the inflection point is not dramatic enough to show. **13.** Graph rises more slowly than graph in Problem 11.

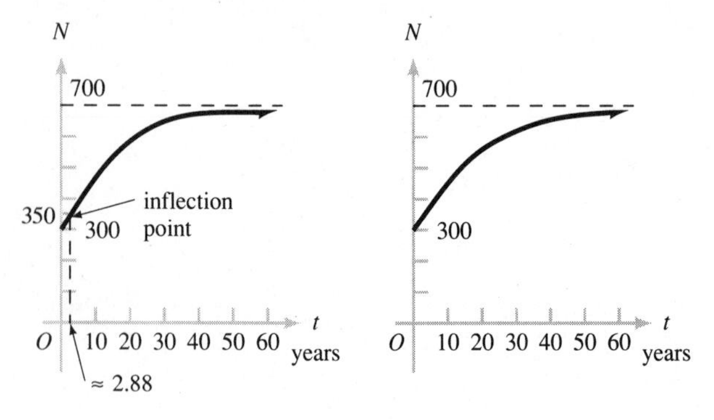

15. Graph rises more slowly than graph in Problem 11, more rapidly than graph in Problem 13.

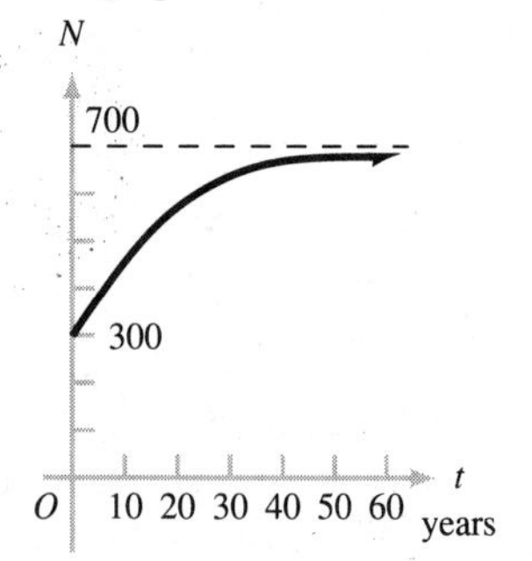

Review Problem Set

CHAPTER 5, page 344

1. (a) 27; **(b)** 1; **(c)** 5; **(d)** $\frac{1}{4}$; **(e)** $\sqrt{3}$; **(f)** $\sqrt{5}/5$; **(g)** 2; **(h)** 0; **(i)** -2; **(j)** $-x$ **3.** Domain $\mathbb{R}$, range $(0, \infty)$, asymptote $y = 0$, increasing **5.** Domain $\mathbb{R}$, range $(0, \infty)$, asymptote $y = 0$, decreasing **7.** Domain $\mathbb{R}$, range $(3, \infty)$, asymptote $y = 3$, decreasing **9.** Domain $\mathbb{R}$, range $(0, \infty)$, asymptote $y = 0$, increasing **11.** Domain $\mathbb{R}$, range $(2, \infty)$, asymptote $y = 2$, increasing **13.** Domain $(1, \infty)$, range $\mathbb{R}$, asymptote $x = 1$, x intercept $1 + (1/e)$, increasing **15.** Domain $(-\infty, 4)$, range $\mathbb{R}$, asymptote $x = 4$, x intercept 3, y intercept log 2, decreasing **17. (a)** $\frac{4}{5}$; **(b)** $\frac{5}{2}$; **(c)** $\frac{1}{7}$; **(d)** 2, $\frac{4}{3}$; **(e)** 2; **(f)** $\pm\sqrt{5}$; **(g)** 0, 2; **(h)** 0 **19. (a)** $\log \pi/\log e \approx 1.145$, $\log \pi - \log e \approx 0.063$; **(b)** $1/\log(\frac{2}{3}) \approx -5.679$, $\log \frac{3}{2} \approx 0.176$; **(c)** $\log_2 3 \approx 1.585$, $\log 8 \approx 0.903$ **21. (a)** $\ln \dfrac{(x-2)(2x-1)}{(x-4)(x-5)}$; **(b)** $\ln \sqrt{\dfrac{x-2}{x^3}}$ **23. (a)** $\frac{1}{2}\log(4-x) - \frac{1}{2}\log(4+x)$; **(b)** $-2\log y + \frac{1}{2}\log(4-y^2) - 4\log 2 - \frac{3}{2}\log(3y+7)$ **25. (a)** 4; **(b)** $\frac{1}{4}$; **(c)** $\frac{1}{2}$; **(d)** 0; **(e)** -3; **(f)** 33; **(g)** -1.4; **(h)** 49; **(i)** $\frac{1}{8}$ **27. (a)** 3.459431619; **(b)** 11.07309365; **(c)** 0.3156023436; **(d)** 0.4306765582 **29. (a)** 1.301029996; **(b)** 6.218875824; **(c)** 0.6538311574; **(d)** 4.773778224 **31. (a)** \$6734.28; **(b)** 11 yr, 8 mo **33.** \$889 **35. (a)** \$702.06; **(b)** \$12,875.42 **37. (a)** \$263.81; **(b)** \$1497.16 **39.** \$5210.39 **41. (a)** 66079.39737; **(b)** 1505.971060; **(c)** $5\sqrt{35}/7$; **(d)** 98.17% **43. (a)** 4255 cells; **(b)** 198 days **45.** 0.9715 g **47.** 123 dB **49.** 5.3 **51.** 12.4 mo **53. (a)** $\approx 75{,}000$; **(b)** $\approx 58{,}000$; **(c)** ≈ 8.1 yr

Chapter 5 Test, page 348

1. (a) Domain $\mathbb{R}$, range $(-1, \infty)$, increasing, horizontal asymptote $y = -1$, no vertical asymptote

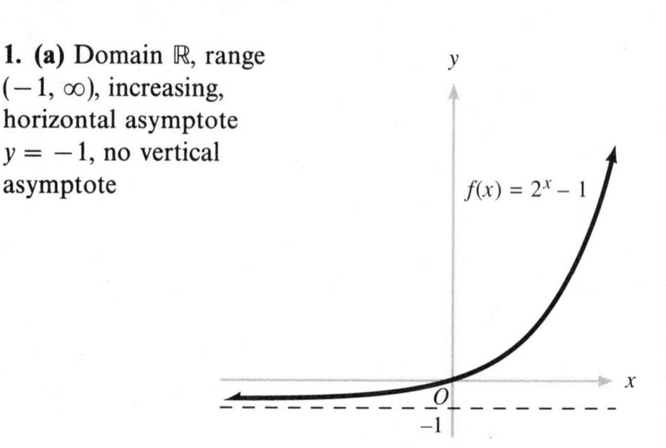

(b) domain $\mathbb{R}$, range $(3, \infty)$, decreasing, horizontal asymptote $y = 3$, no vertical asymptote

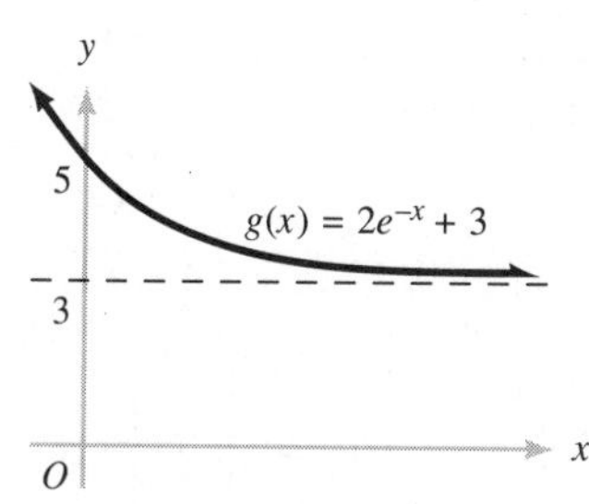

(c) domain $(-1, \infty)$, range $\mathbb{R}$, increasing, vertical asymptote $x = -1$, no horizontal asymptote

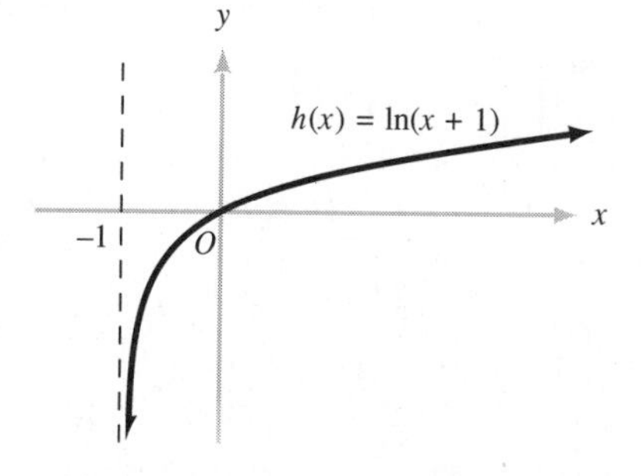

(d) domain $(-\infty, -2) \cup (-2, \infty)$, range $\mathbb{R}$, decreasing on $(-\infty, -2)$ and increasing on $(-2, \infty)$, vertical asymptote $x = -2$, no horizontal asymptote

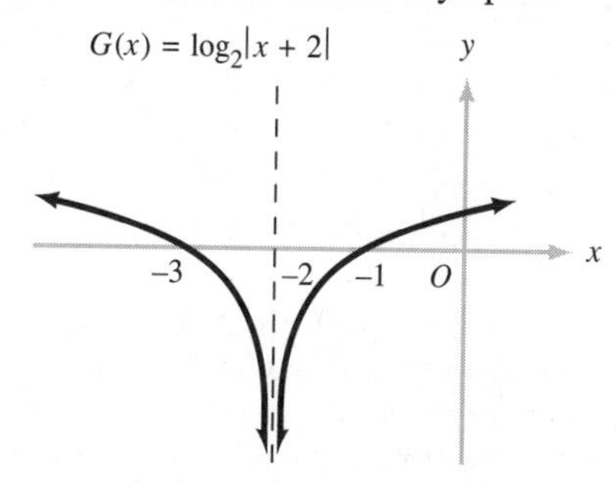

(e) domain $\mathbb{R}$, range $(0, \infty)$, decreasing, horizontal asymptote $y = 0$, no vertical asymptote

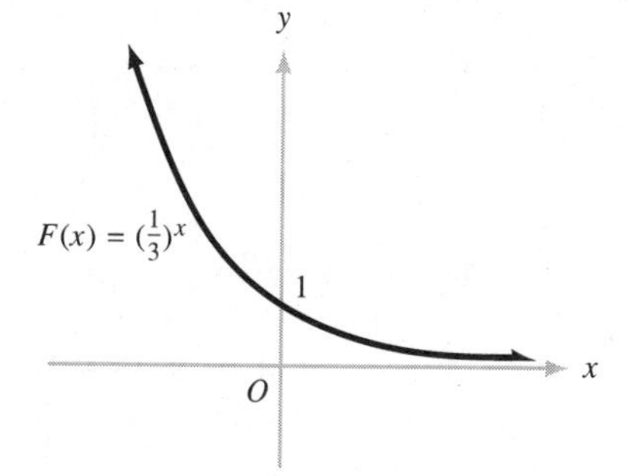

2. $-\frac{4}{3}, -1$ **3.** **(a)** $2x^2 + x$; **(b)** $(5x + 1)^8$; **(c)** x^5/y^3
4. **(a)** $-1, 3$; **(b)** 2; **(c)** $\frac{1}{9}$ **5.** **(a)** $\frac{1}{3}\ln(3x + 4) - \frac{5}{3}\ln x$;
(b) $\frac{3}{4}\log_3 x + \frac{1}{4}\log_3 y$ **6.** 0.73635 **7.** **(a)** 0.05466;
(b) -4.00085; **(c)** 0.68261; **(d)** 3.50133
8. A: \$10,832.87, B: \$10,832.78 **9.** 10^{17} J
10. 78 mg

CHAPTER 6

Problem Set 6.1, page 355

1. Well-determined **3.** Dependent

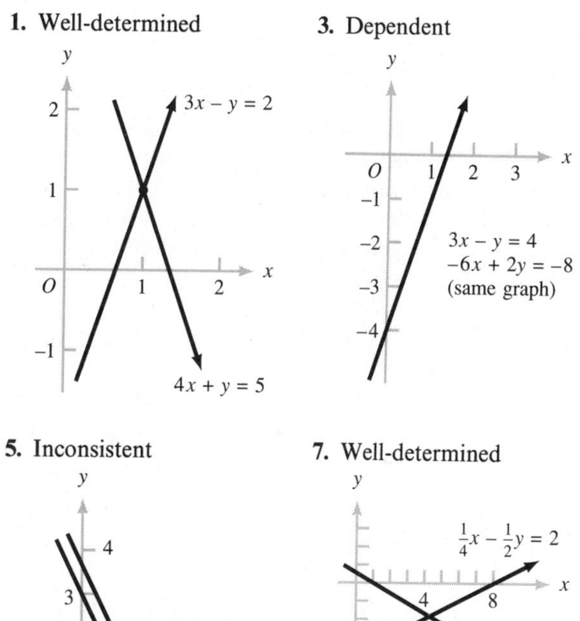

5. Inconsistent **7.** Well-determined

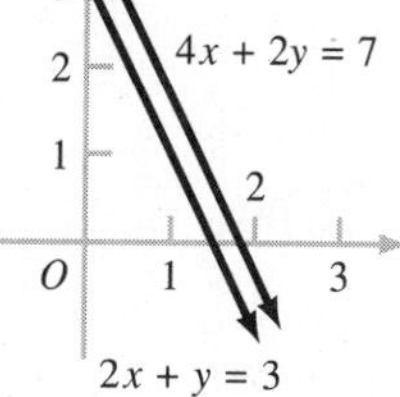

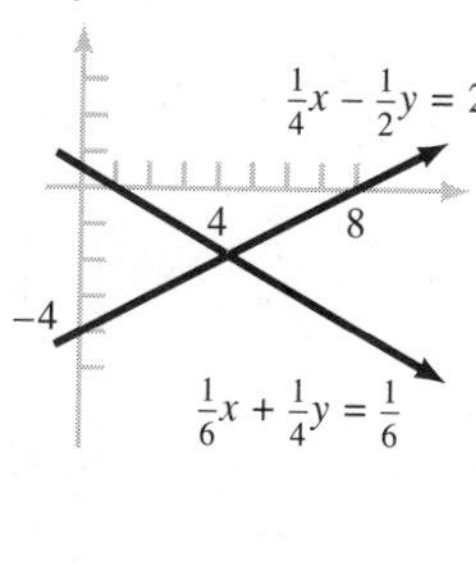

9. (1, 1), all points on the graph of $y = 3x - 4$, no solution
11. (1, 3) **13.** $(4, -1)$ **15.** $(\frac{7}{11}, -\frac{10}{11})$
17. $(1, -1, -2)$ **19.** (3, 4) **21.** (3, 4)
23. (1, 2, 3) **25.** (8, 4, 0) **27.** $(\frac{1}{2}, -1)$
29. **(a)** $x + y = 39$, $380x + 515y = 16{,}575$; **(b)** (26, 13)
31. 12.5 km/hr; 2.5 km/hr
33. 225 adults, 700 children
35. 135 adults, 90 children, 45 senior citizens

37. $x = 50°, y = 40°, z = 90°$ **39.** $p = 30, q = 90$
41. $x = \$8000, y = \$9000$ **43.** $r_1 = 3, r_2 = 4, r_3 = 7$

Problem Set 6.2, page 364

1. (a) $x + 3y = 0, 2x - 4y = 1$;
(b) $2x + 5y + 3z = 1, -3x + 7y + \frac{1}{2}z = \frac{3}{4}, \frac{2}{3}y = -\frac{4}{5}$ **3. (a)** $\left[\begin{array}{cc|c} \frac{3}{4} & -\frac{2}{3} & \frac{1}{7} \\ -1 & 5 & 6 \end{array}\right]$;
(b) $\left[\begin{array}{ccc|c} 40 & 22 & -1 & -17 \\ 0 & 1 & 1 & 0 \\ -13 & 17 & 12 & 5 \end{array}\right]$ **5.** $(\frac{24}{5}, -\frac{4}{5})$ **7.** $(5, 0)$
9. $(2, 3)$ **11.** Inconsistent **13.** $(3, 0)$
15. $(4, -2, 1)$ **17.** $(-\frac{11}{4}, 1, -\frac{1}{4})$ **19.** No solutions
21. $x = \frac{24}{11} - \frac{13}{11}t, y = -\frac{3}{11} + \frac{14}{11}t, z = t$ **23.** $(0, 0, 0)$
25. 9 hr for first survey team, 8 hr for second survey team **27.** 45 of type I, 20 of type II, 15 of type III
29. Coffee is \$2.80/lb, milk is \$0.60/qt, tuna is \$1.20/can
31. 2 metric tons for solvent A, 4 metric tons for solvent B, 6 metric tons for solvent C
33. $A = 2800, B = 2000, C = 3500$
35. Top loop: $E_1 + I_2R_2 = I_1R_1 + E_2$,
bottom loop: $E_2 = I_2R_2 + I_3R_3$,
outer loop: $E_1 = I_1R_1 + I_3R_3$, junction P: $I_1 + I_2 = I_3$,
junction Q: $I_3 = I_1 + I_2$

Problem Set 6.3, page 375

1. (a) $\begin{bmatrix} 8 & 1 \\ 8 & -1 \end{bmatrix}$; **(b)** $\begin{bmatrix} -4 & -7 \\ 2 & 3 \end{bmatrix}$; **(c)** $\begin{bmatrix} -6 & 9 \\ -15 & -3 \end{bmatrix}$;
(d) $\begin{bmatrix} 6 & 17 \\ -9 & -7 \end{bmatrix}$ **3. (a)** $\begin{bmatrix} 1 & 5 \\ 1 & 6 \\ 6 & -1 \\ -3 & 4 \end{bmatrix}$; **(b)** $\begin{bmatrix} 5 & -1 \\ -5 & 4 \\ -2 & 3 \\ -5 & 4 \end{bmatrix}$;
(c) $\begin{bmatrix} -9 & -6 \\ 6 & -15 \\ -6 & -3 \\ 12 & -12 \end{bmatrix}$; **(d)** $\begin{bmatrix} -13 & 0 \\ 12 & -13 \\ 2 & -7 \\ 14 & -12 \end{bmatrix}$
5. (a) $\begin{bmatrix} 4 & -6 & 2 & -1 \\ 0 & 4 & 0 & 6 \\ 4 & -1 & -2 & 4 \end{bmatrix}$; **(b)** $\begin{bmatrix} 0 & 0 & 2 & -5 \\ -6 & 0 & 2 & -4 \\ 4 & 3 & -4 & 4 \end{bmatrix}$;
(c) $\begin{bmatrix} -6 & 9 & -6 & 9 \\ 9 & -6 & -3 & -3 \\ -12 & -3 & 9 & -12 \end{bmatrix}$;
(d) $\begin{bmatrix} -2 & 3 & -6 & 13 \\ 15 & -2 & -5 & 7 \\ -12 & -7 & 11 & -12 \end{bmatrix}$
7. (a) $\begin{bmatrix} 2 & -\frac{2}{3} & \sqrt{2} \\ \frac{17}{6} & \pi + 1 & -2 \\ 4 & \frac{5}{2} & \frac{5}{3} \end{bmatrix}$; **(b)** $\begin{bmatrix} 0 & 1 & -\sqrt{2} \\ -\frac{1}{6} & \pi - 1 & -2 \\ -2 & -\frac{5}{2} & \frac{5}{3} \end{bmatrix}$;
(c) $\begin{bmatrix} -3 & -\frac{1}{2} & 0 \\ -4 & -3\pi & 6 \\ -3 & 0 & -5 \end{bmatrix}$; **(d)** $\begin{bmatrix} -1 & -\frac{13}{6} & 2\sqrt{2} \\ -1 & 2 - 3\pi & 6 \\ 3 & 5 & -5 \end{bmatrix}$
21. $\begin{bmatrix} 14 & -3 \\ 6 & -2 \end{bmatrix}$ **23.** $\begin{bmatrix} 2 & 6 \\ 5 & 10 \end{bmatrix}$ **25.** $\begin{bmatrix} -2 & 9 \\ -3 & 1 \end{bmatrix}$
27. A **29.** $\begin{bmatrix} 2 & -5 & 2 \\ -2 & 0 & 6 \\ 7 & -14 & 1 \end{bmatrix}$
31. $\begin{bmatrix} -2 & 7 & 6 \\ 8 & -2 & -4 \\ 1 & -10 & -8 \end{bmatrix}$ **33.** $\begin{bmatrix} 5 & -10 & 17 \\ 10 & -14 & 30 \\ 8 & -20 & 30 \end{bmatrix}$
35. $\begin{bmatrix} 17 & 36 \\ 8 & 14 \end{bmatrix}$ **37.** $\begin{bmatrix} -6 & 4 \\ -23 & 16 \\ -20 & -2 \end{bmatrix}$ **39.** $\begin{bmatrix} 13 & -6 \\ 7 & -4 \end{bmatrix}$
41. Undefined **45.** $\begin{bmatrix} \frac{4}{5} & \frac{1}{5} \\ \frac{1}{5} & -\frac{1}{5} \end{bmatrix}$ **47.** $\begin{bmatrix} -\frac{5}{34} & \frac{1}{17} \\ \frac{1}{17} & \frac{3}{17} \end{bmatrix}$
49. $\begin{bmatrix} \frac{1}{4} & \frac{1}{12} \\ -\frac{3}{4} & \frac{1}{12} \end{bmatrix}$ **51.** No inverse
53. $\begin{bmatrix} \frac{3}{10} & -\frac{2}{5} & \frac{1}{2} \\ \frac{3}{10} & \frac{3}{5} & -\frac{1}{2} \\ -\frac{1}{10} & \frac{4}{5} & -\frac{1}{2} \end{bmatrix}$ **55.** $\begin{bmatrix} \frac{1}{3} & \frac{1}{3} & 0 \\ 0 & -\frac{3}{10} & \frac{1}{10} \\ -\frac{2}{3} & -\frac{4}{15} & \frac{1}{5} \end{bmatrix}$
57. $\begin{bmatrix} -\frac{3}{2} & \frac{9}{4} & -\frac{5}{2} \\ 1 & -1 & 1 \\ -\frac{1}{2} & \frac{3}{4} & -\frac{1}{2} \end{bmatrix}$ **59.** $(\frac{24}{5}, -\frac{4}{5})$ **61.** $(-\frac{9}{34}, \frac{12}{17})$
63. $(\frac{8}{3}, -\frac{10}{3})$ **65.** $(-1, 2, -3)$ **67.** $(-\frac{11}{4}, 1, -\frac{1}{4})$
69. (a) The entry in the ith row of TX is the number of units of commodity number i used in unit time in the production of all other commodities; **(b)** the entry in the ith row of $X - TX$ is the surplus number of units of commodity number i produced in unit time.
71. (a) $\begin{bmatrix} 2s + 3g = a \\ 3s + 4g = t; \end{bmatrix}$ **(b)** $A^{-1} = \begin{bmatrix} -4 & 3 \\ 3 & -2 \end{bmatrix}$;
(c) $\begin{bmatrix} 2 & 3 \\ 3 & 4 \end{bmatrix}\begin{bmatrix} s \\ g \end{bmatrix} = \begin{bmatrix} a \\ t \end{bmatrix}$; **(d)** $\begin{bmatrix} s \\ g \end{bmatrix} = \begin{bmatrix} -4a + 3t \\ 3a - 2t \end{bmatrix}$

Problem Set 6.4, page 384

1. 5 **3.** 54 **5.** 54 **7.** 4 **9.** $(1, -1)$
11. $(2, 1)$ **13.** $(\frac{5}{4}, -\frac{1}{3})$ **15.** 28 **17.** 38 **19.** 9
21. $-\frac{53}{24}$ **23.** $(1, -1, 2)$ **25.** $(-\frac{10}{11}, \frac{18}{11}, \frac{38}{11})$

27. No solution **29.** -14 **31.** -80 **33.** 22
35. 25 **37.** -130.9347340 **39.** -12 **41.** 0
43. -1 **45.** 3 **47.** -3 **49.** -3
57. $(2 \pm \sqrt{58})/3$

59. $A = ad - A_1 - A_2 - A_3 = \frac{1}{2}ad - \frac{1}{2}cb = \frac{1}{2}\begin{vmatrix} a & b \\ c & d \end{vmatrix}$

61. The triangle formed by O and the head ends of **V** and **W** has half the area of the parallelogram.

Problem Set 6.5, page 390

1. $\dfrac{3}{x-3} - \dfrac{3}{x-2}$ **3.** $\dfrac{\frac{1}{2}}{x+5} + \dfrac{\frac{1}{2}}{x-1}$

5. $\dfrac{\frac{3}{4}}{x} - \dfrac{\frac{9}{8}}{x-2} + \dfrac{\frac{11}{8}}{x+2}$ **7.** $\dfrac{-2}{x} + \dfrac{5}{x-1} - \dfrac{3}{x+1}$

9. $3 + \dfrac{\frac{1}{6}}{x} + \dfrac{\frac{1}{3}}{x+3} + \dfrac{\frac{1}{2}}{x-2}$

11. $\dfrac{-4}{x-3} + \dfrac{2}{x-1} + \dfrac{1}{(x-1)^2}$

13. $1 - \dfrac{1}{x} - \dfrac{1}{x^2} + \dfrac{1}{x-1}$

15. $\dfrac{4}{x-1} - \dfrac{1}{x+2} + \dfrac{3}{(x+2)^2}$

17. $\dfrac{\frac{5}{8}}{x+2} - \dfrac{\frac{13}{24}}{3x-2} + \dfrac{\frac{8}{3}}{(3x-2)^2}$

19. $\dfrac{\frac{9}{2}}{t+1} + \dfrac{-\frac{9}{2}t + \frac{11}{2}}{t^2+1}$ **21.** $x^2 + \dfrac{\frac{1}{9}}{x} - \dfrac{\frac{1}{9}x}{x^2+9}$

23. $\dfrac{3}{t} + \dfrac{-3t+1}{t^2+1}$ **25.** $\dfrac{\frac{1}{4}}{u-1} + \dfrac{\frac{1}{4}}{u+1} - \dfrac{\frac{1}{2}u}{u^2+1}$

27. $A = \frac{3}{2}, B = 0, C = -2, D = -\frac{3}{2}$
29. $A = 1, B = -1, C = -3, D = 3$

Problem Set 6.6, page 395

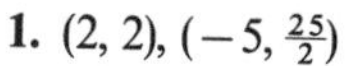
1. $(2, 2), (-5, \frac{25}{2})$

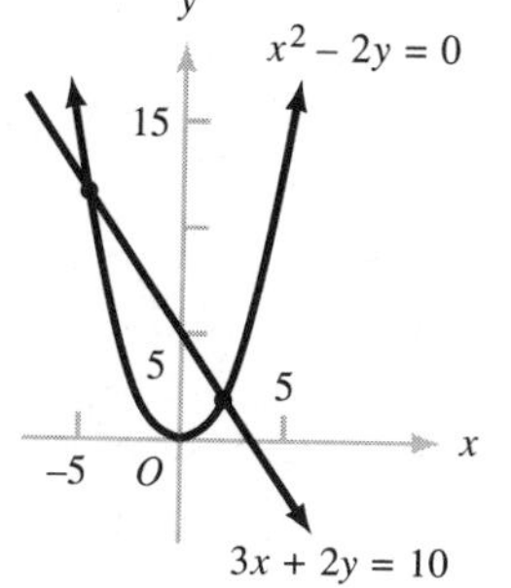

3. $(3, 3), (\frac{3}{4}, -\frac{3}{2})$ **5.** $(0, -2), (\frac{8}{5}, -\frac{6}{5})$

7. $\left(\dfrac{-1+\sqrt{11}}{2}, \sqrt{11}-2\right), \left(\dfrac{-1-\sqrt{11}}{2}, -\sqrt{11}-2\right)$
9. No real solution
11. $(2, 3), (2, -3), (-2, 3), (-2, -3)$
13. $(\sqrt{7}, \frac{1}{2}\sqrt{6}), (-\sqrt{7}, \frac{1}{2}\sqrt{6}), (\sqrt{7}, -\frac{1}{2}\sqrt{6}), (-\sqrt{7}, -\frac{1}{2}\sqrt{6})$
15. $(1, 2), (1, -2), (-1, 2), (-1, -2)$
17. $(3, 7), (-1, -1)$
19. $(2, 2), (-2, -2), (2\sqrt{2}, \sqrt{2}), (-2\sqrt{2}, -\sqrt{2})$
21. $(25, 4)$ **23.** $(16, 2)$
25. $(\frac{5}{2}, -\frac{3}{2}), (-\frac{3}{2}, \frac{5}{2}), (\frac{3}{2}, -\frac{5}{2}), (-\frac{5}{2}, \frac{3}{2})$
27. $(-1, 1 + \log_6 2)$ **29.** $(5, 1)$
31. No solution **33.** $h = 10, b = 6$
35. 12 m by 9 m, or 18 m by 6 m
37. $r = 3$ cm, $l = 450$ cm **39.** 56 or 65
41. $s \approx 2.5$ m, $l \approx 2.2$ m **43.** 4000 units
45. **(a)** (7, 1) is the equilibrium point;
(b) equilibrium price is \$7000.

Problem Set 6.7, page 404

1. **3.**

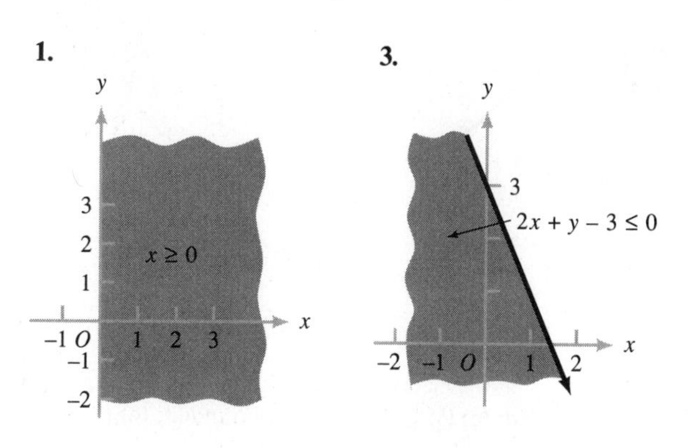

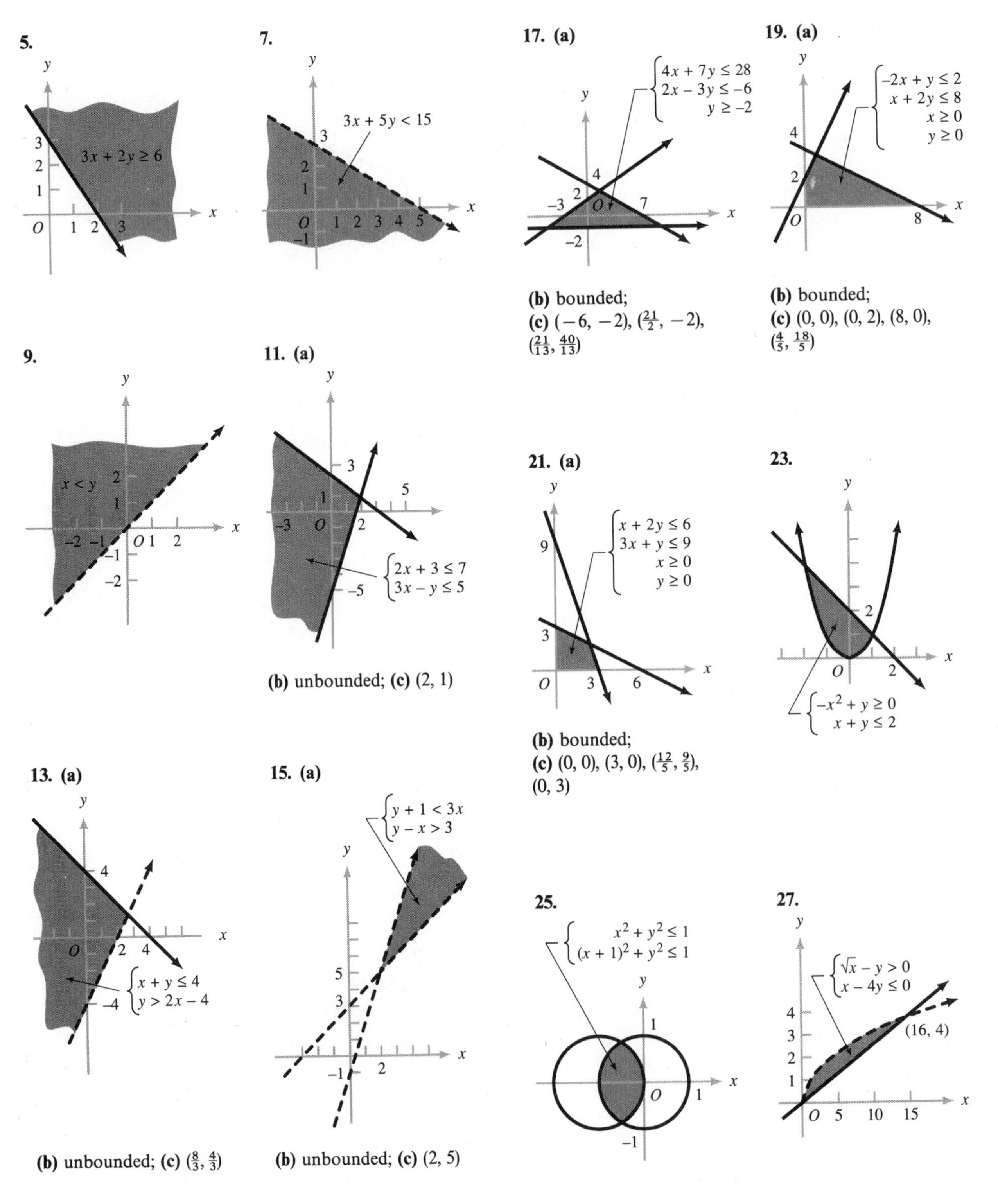
5.
3x + 2y ≥ 6
7.
3x + 5y < 15
17. (a)
4x + 7y ≤ 28
2x − 3y ≤ −6
y ≥ −2
(b) bounded;
(c) (−6, −2), (21/2, −2), (21/13, 40/13)
19. (a)
−2x + y ≤ 2
x + 2y ≤ 8
x ≥ 0
y ≥ 0
(b) bounded;
(c) (0, 0), (0, 2), (8, 0), (4/5, 18/5)
9.
x < y
11. (a)
2x + 3 ≤ 7
3x − y ≤ 5
(b) unbounded; (c) (2, 1)
21. (a)
x + 2y ≤ 6
3x + y ≤ 9
x ≥ 0
y ≥ 0
(b) bounded;
(c) (0, 0), (3, 0), (12/5, 9/5), (0, 3)
23.
−x² + y ≥ 0
x + y ≤ 2
13. (a)
x + y ≤ 4
y > 2x − 4
(b) unbounded; (c) (8/3, 4/3)
15. (a)
y + 1 < 3x
y − x > 3
(b) unbounded; (c) (2, 5)
25.
x² + y² ≤ 1
(x + 1)² + y² ≤ 1
27.
√x − y > 0
x − 4y ≤ 0
(16, 4)

29.

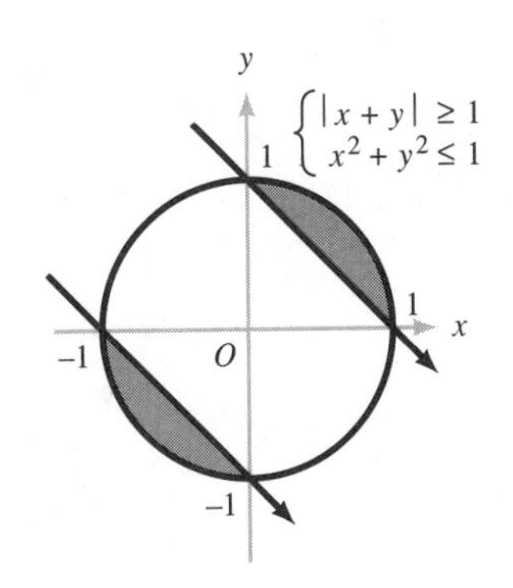

31.

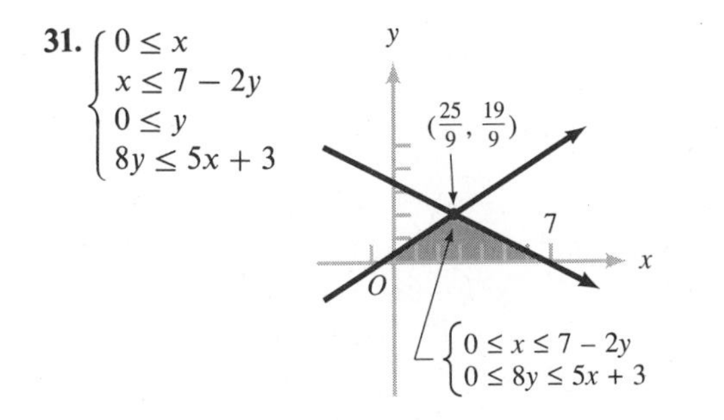

33. max 5, min 2
35. max 25, min 6
37. max 32, min -28
39. max 11, min 5
41. max 30, min -10
43. Vertices are (10, 10), (10, 30), (30, 10)
45. For maximum profit, 50 units of model A, 40 units of model B.
47. 80 units of model A, 0 units of model B.
49. Town should keep center I open 6 days and center II 2 days.
51. $3a = 2b$

Review Problem Set

CHAPTER 6, page 407

1. **(a)** Inconsistent; **(b)** well-determined; **(c)** dependent
3. $(\frac{11}{7}, \frac{8}{7})$ **5.** Inconsistent **7.** (1, 1) **9.** $(3, 4, -6)$
11. (5, 6) **13.** $(1, -3, 2)$ **15.** $(-1, 2, -2)$

17. **(a)** $\begin{bmatrix} -7 & 5 & 6 \\ 16 & -4 & 16 \end{bmatrix}$; **(b)** $\begin{bmatrix} 7 & -4 & 5 \\ -3 & 6 & 4 \end{bmatrix}$;
(c) $\begin{bmatrix} 9 & -\frac{7}{2} & -\frac{3}{2} \\ -\frac{13}{2} & 6 & -4 \end{bmatrix}$; **(d)** $\begin{bmatrix} -5 & 7 & -10 \\ 8 & -6 & -4 \end{bmatrix}$

19. **(a)** $AB = \begin{bmatrix} 0 & 3 \\ -1 & 20 \end{bmatrix}$; **(b)** $BA = \begin{bmatrix} 4 & 1 & 5 \\ 16 & 6 & 6 \\ 15 & 5 & 10 \end{bmatrix}$

21. **(a)** $A^{-1} = \begin{bmatrix} \frac{7}{26} & \frac{2}{13} \\ -\frac{3}{26} & \frac{1}{13} \end{bmatrix}$; **(b)** $AA^{-1} = \begin{bmatrix} 1 & 0 \\ 0 & 1 \end{bmatrix}$;
(c) $A^{-1}A = \begin{bmatrix} 1 & 0 \\ 0 & 1 \end{bmatrix}$

23. $A^{-1} = \begin{bmatrix} \frac{2}{3} & 0 & -\frac{1}{3} \\ \frac{1}{3} & 0 & -\frac{2}{3} \\ -\frac{2}{3} & 1 & \frac{1}{3} \end{bmatrix}$

25. **(a)** $A^{-1} = \begin{bmatrix} \frac{3}{17} & \frac{2}{17} \\ \frac{4}{17} & -\frac{3}{17} \end{bmatrix}$, (3, 1);
(b) $A^{-1} = \dfrac{1}{37}\begin{bmatrix} 3 & -1 & 6 \\ 15 & -5 & -7 \\ -2 & 13 & -4 \end{bmatrix}$, $(-1, 3, 4)$

27. **(a)** $\frac{11}{3}$; **(b)** 133 **29.** **(a)** 0; **(b)** 0 **31.** **(a)** 2; **(b)** 16; **(c)** -2 **33.** $(\frac{26}{19}, \frac{29}{19})$ **35.** $(3, -3, 5)$
37. **(a)** 16 square units; **(b)** 51 square units

39. $\dfrac{1}{x-1} + \dfrac{1}{x+1}$

41. $\dfrac{3}{x} - \dfrac{1}{x+3} + \dfrac{2}{x-1}$ **43.** $\dfrac{4}{x} - \dfrac{3x}{x^2+1}$

45. (1, 2) and $(-1, -2)$ **47.** No real solution

49. $\left(\dfrac{-1+\sqrt{11}}{2}, \sqrt{11}-2\right)$ and $\left(\dfrac{-1-\sqrt{11}}{2}, -\sqrt{11}-2\right)$

51. **(a)**

(b) Unbounded; **(c)** $(-\frac{1}{4}, \frac{3}{4})$

53. **(a)**

(b) Bounded; **(c)** (2, 0), $(-\frac{2}{3}, 0)$, $(\frac{2}{5}, \frac{16}{5})$

55.

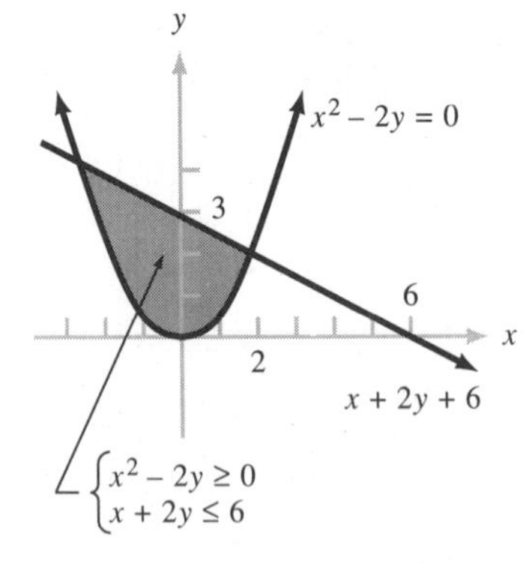

57. Maximum is 60, minimum is 36
59. 20 kW, 40 kW **61.** 18, 48 **63.** $(\frac{14}{5}, \frac{6}{5})$
65. \$13,000 at 8.5%, \$27,000 at 11.2%
67. Byron's is \$6; Jason's is \$4; Adrian's is \$2
69. Sides are 7 and 24 **71.** 75 m by 100 m
73. (0, 20), (15, 20), (0, 40), (15, 25)
75. 400 units of A, 800 units of B

Chapter 6 Test, page 411

1. **(a)** Inconsistent; **(b)** well-determined; **(c)** dependent
2. **(a)** (159, −186); **(b)** (159, −186);
(c) $\begin{bmatrix} 13 & 11 \\ 7 & 6 \end{bmatrix}\begin{bmatrix} x \\ y \end{bmatrix} = \begin{bmatrix} 21 \\ -3 \end{bmatrix}$; **(d)** (159, −186)
3. **(a)** $\begin{bmatrix} 6 & 16 \\ 26 & 0 \\ -16 & 2 \end{bmatrix}$; **(b)** $\begin{bmatrix} -10 & -10 \\ -8 & 14 \\ 28 & -11 \end{bmatrix}$;
(c) $\begin{bmatrix} 28 & 14 \\ -28 & -7 \end{bmatrix}$ **4.** **(a)** −75; **(b)** $\frac{7}{3}, -\frac{2}{5}$ **5.** 6
6. $\dfrac{x^2 - x - 21}{(2x-1)(x^2+4)} = \dfrac{-5}{2x-1} + \dfrac{3x+1}{x^2+4}$
7. $\left(\sqrt{\dfrac{\sqrt{65}-1}{8}}, \dfrac{\sqrt{65}-1}{8}\right), \left(-\sqrt{\dfrac{\sqrt{65}-1}{8}}, \dfrac{\sqrt{65}-1}{8}\right)$
8. **(a)**

(b) bounded; **(c)** vertices (0, 0), (0, 4), $(\frac{7}{4}, 0)$, (1, 3);
(d) minimum 0 at (0, 0), maximum 20 at (0, 4)
9. $l = 500$ m, $w = 400$ m
10. 20,000 units of bubble gum and 40,000 units of candy bars

CHAPTER 7

Problem Set 7.1, page 416

17. $n(n+1)/2$

Problem Set 7.2, page 424

1. 362,880 **3.** 7 **5.** 35 **7.** 1 **9.** 20
11. (1, 7, 21, 35, 35, 21, 7, 1)
13. $a^4 + 12a^3x + 54a^2x^2 + 108ax^3 + 81x^4$
15. $x^5 - 5x^4y + 10x^3y^2 - 10x^2y^3 + 5xy^4 - y^5$
17. $c^6 + 12c^5 + 60c^4 + 160c^3 + 240c^2 + 192c + 64$
19. $1 - 7c^3 + 21c^6 - 35c^9 + 35c^{12} - 21c^{15} + 7c^{18} - c^{21}$
21. $x^3 - 6x^2\sqrt{xy} + 15x^2y - 20xy\sqrt{xy} + 15xy^2 - 6y^2\sqrt{xy} + y^3$
23. $\frac{1}{64}x^6 - \frac{3}{8}x^5y + \frac{15}{4}x^4y^2 - 20x^3y^3 + 60x^2y^4 - 96xy^5 + 64y^6$ **25.** $10s^3t^2$ **27.** $\frac{105}{2}x^{12}y^{12}$
29. $5xy^4$ **31.** $3360c^6d^4$

Problem Set 7.3, page 428

1. $\frac{1}{2}, \frac{1}{3}, \frac{1}{4}, \frac{1}{5}, \frac{1}{6}$ **3.** −1, 2, −3, 4, −5 **5.** $\frac{1}{3}, \frac{3}{5}, \frac{5}{7}, \frac{7}{9}, \frac{9}{11}$
7. $\frac{1}{3}, \frac{1}{9}, \frac{1}{27}, \frac{1}{81}, \frac{1}{243}$ **9.** 4, 9, 16, 25, 36 **11.** 33
13. 100 **15.** 6.4 **17.** $m + 8r$ **19.** $-11\sqrt{5}$
21. $\frac{1}{64}$ **23.** 162 **25.** −0.009 **27.** $2x^{15}$
29. 0.000003 **31.** 2, −6, 18, −54, 162
33. 5, −3, 13, −19, 45 **35.** $1, \frac{1}{2}, \frac{1}{6}, \frac{1}{24}, \frac{1}{120}$
39. 1, 1, 2, 3, 5, 8, 13, 21, 34, 55 **41.** **(c)** ≈ 0.024%
43. **(a)** $a_1 = 300{,}000$ and $r = 1.05$; **(b)** 382,884

Problem Set 7.4, page 437

1. $1 + 4 + 7 + 10 = 22$ **3.** $\frac{13}{4} + \frac{7}{2} + \frac{15}{4} + 4 + \frac{17}{4} = \frac{75}{4}$
5. $1 + \frac{1}{2} + \frac{4}{7} + \frac{4}{5} + \frac{16}{13} + 2 = \frac{5553}{910}$ **7.** 2460
9. $2^{25} + 22$ or 33,554,454 **11.** $\frac{1}{9}[1 - (\frac{1}{10})^{100}]$
13. 55 **15.** 624 **17.** −390 **19.** $-35x$
21. 44,444 **23.** $-\frac{1}{2}[1 - (\frac{1}{3})^{10}]$

25. $(5^{12}/24)[1-(\frac{1}{5})^{18}]$ **27.** 0.333333 **29.** 3
31. $\frac{5}{7}$ **33.** $\frac{25}{8}$ **35.** **(a)** $\frac{4}{9}$; **(b)** $\frac{80}{11}$; **(c)** $\frac{58}{2475}$
37. False, $0.\bar{9}=1$ **39.** $\dfrac{a^n+(-1)^{n+1}b^n}{a^{n-2}b(a+b)}$
43. Problem 1: $\sum_{k=1}^{n}(2k-1)=n^2$;
Problem 3: $\sum_{k=1}^{n}k^2=\frac{1}{6}n(n+1)(2n+1)$;
Problem 6: $\sum_{k=1}^{n}k^3=[n(n+1)/2]^2$ **45.** \$2,440,000
47. **(b)** $T_n=n(n+1)/2$; **(c)** 55 **49.** ≈ 36.2 cm
51. $768-2^{10-n}$ **53.** 768 cm
55. The right idea: The bicycles will meet after they have traveled 10 mi, so the fly will have traveled for 1 hr or 15 mi.

Problem Set 7.5, page 445

1. 8 **3.** 676 **5.** 32
7. *ab, ac, ad, ae, ba, bc, bd, be, ca, cb, cd, ce, da, db, dc, de, ea, eb, ec, ed*; 20 **9.** 24 **11.** 9! or 362,880
13. 20 **15.** 990 **17.** ${}_{10}P_5=30{,}240$
19. $7!=5040$ **21.** 40,320 **23.** ${}_{10}P_3=720$
25. $\{a, b, c\}$, $\{a, b, d\}$, $\{a. c. d\}$, $\{b, c, d\}$; 4
27. 1 **29.** 10 **31.** 20 **33.** 10 **35.** 1
37. ${}_8C_2=28$ **39.** ${}_{10}C_4=210$
41. **(a)** Let c correspond to a corner that is not on the border. If there are b ways to get to c via the corner at b and d ways to get to c via the corner at d, then there are $b+d$ ways to get to c. **(b)** The addition rules for the Pascal triangle are the same as for Figure 1(b).
43. 12 **45.** 43,243,200 **47.** 3780 **49.** 3360

Problem Set 7.6, page 452

1. $\frac{1}{2}$ **3.** $\frac{4}{7}$ **5.** $\frac{1}{6}$ **7.** $(13\cdot 48)/{}_{52}C_5=1/4165$
9. $4/{}_{52}C_5=1/649{,}740$ **11.** $\frac{5}{36}$ **13.** $\frac{11}{36}$
15. $({}_{10}C_3)({}_3C_2)/{}_{13}C_5=40/143$ **17.** $\frac{1}{3}$ **19.** $\frac{1}{2}$
21. **(a)** $\frac{5}{11}$; **(b)** 5 to 6 **23.** $-\frac{100}{37}\approx -2.70$
25. \$1,947,792 **29.** **(b)** Np_k

Problem Set 7.7, page 461

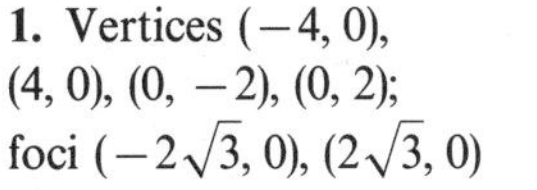
1. Vertices $(-4, 0)$, $(4, 0)$, $(0, -2)$, $(0, 2)$; foci $(-2\sqrt{3}, 0)$, $(2\sqrt{3}, 0)$

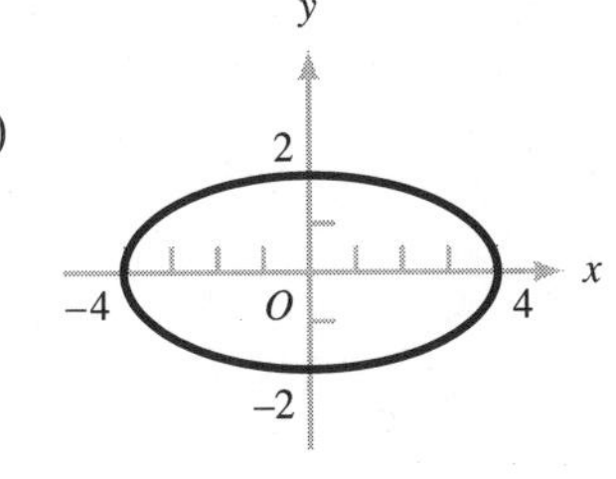

3. Vertices $(-2, 0)$, $(2, 0)$, $(0, -4)$, $(0, 4)$; foci $(0, -2\sqrt{3})$, $(0, 2\sqrt{3})$

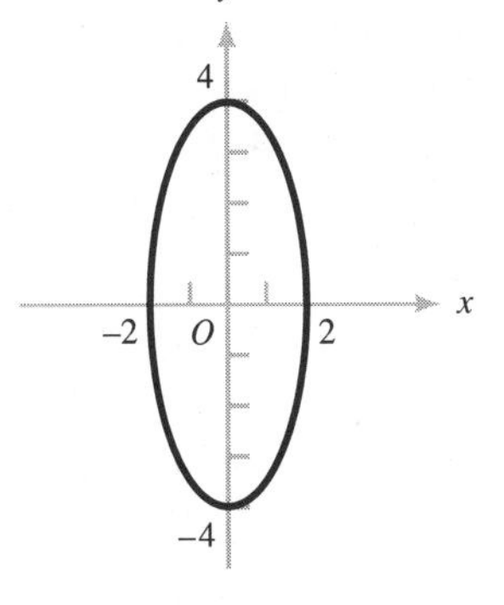

5. Vertices $(-4, 0)$, $(4, 0)$, $(0, -1)$, $(0, 1)$; foci $(-\sqrt{15}, 0)$, $(\sqrt{15}, 0)$

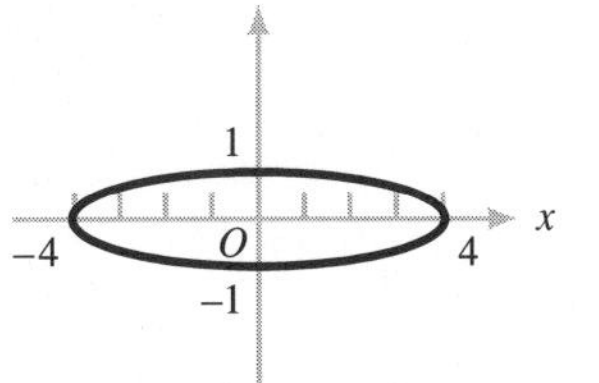

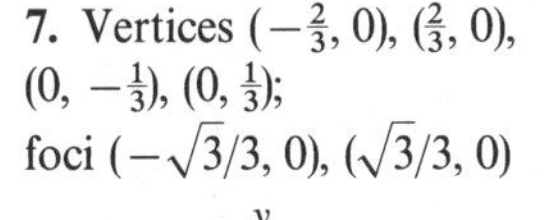
7. Vertices $(-\frac{2}{3}, 0)$, $(\frac{2}{3}, 0)$, $(0, -\frac{1}{3})$, $(0, \frac{1}{3})$; foci $(-\sqrt{3}/3, 0)$, $(\sqrt{3}/3, 0)$

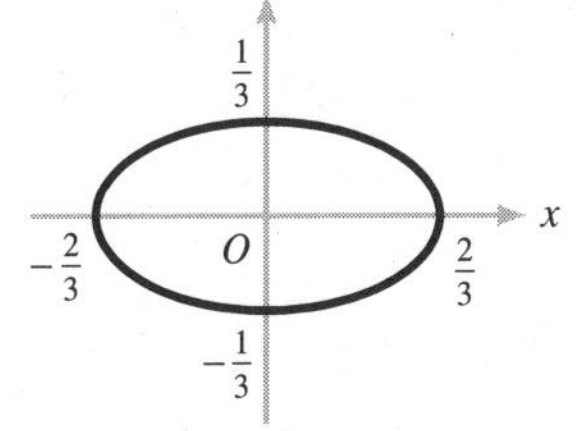

9. $\dfrac{x^2}{25}+\dfrac{y^2}{9}=1$

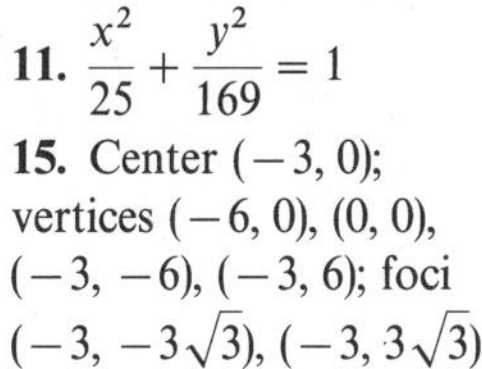
11. $\dfrac{x^2}{25}+\dfrac{y^2}{169}=1$

13. Center $(1, -2)$; vertices $(-2, -2)$, $(4, -2)$, $(1, -4)$, $(1, 0)$; foci $(1-\sqrt{5}, -2)$, $(1+\sqrt{5}, 2)$

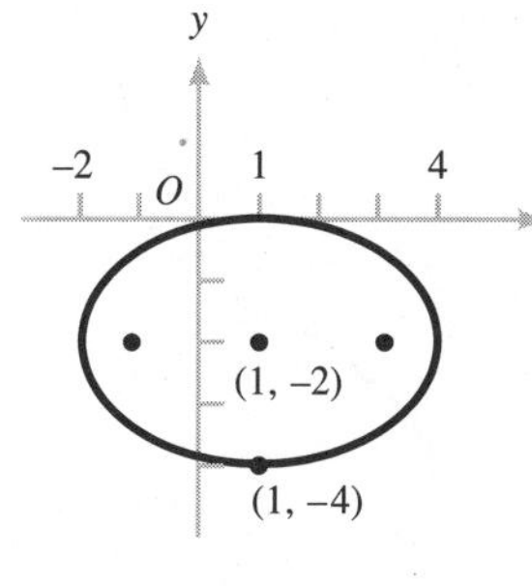

15. Center $(-3, 0)$; vertices $(-6, 0)$, $(0, 0)$, $(-3, -6)$, $(-3, 6)$; foci $(-3, -3\sqrt{3})$, $(-3, 3\sqrt{3})$

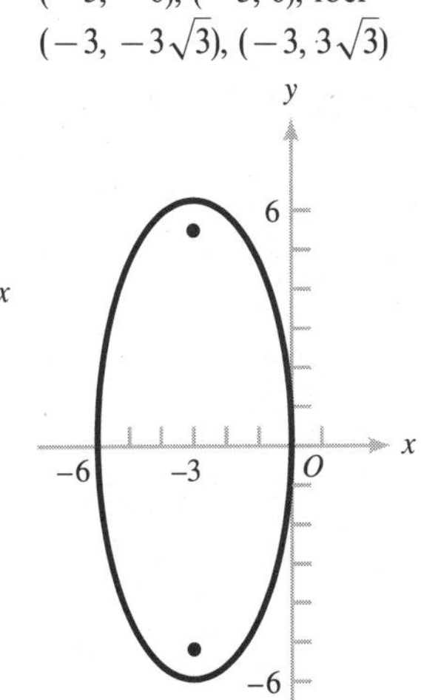

17. Center $(-3, 0)$; vertices $(-3-\sqrt{2}, 0)$, $(-3+\sqrt{2}, 0)$, $(-3, -1)$, $(-3, 1)$; foci $(-4, 0)$, $(-2, 0)$

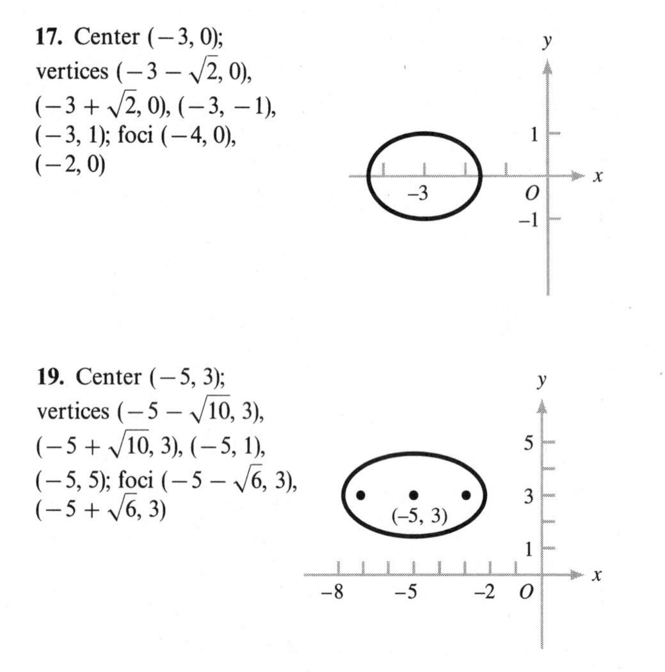

19. Center $(-5, 3)$; vertices $(-5-\sqrt{10}, 3)$, $(-5+\sqrt{10}, 3)$, $(-5, 1)$, $(-5, 5)$; foci $(-5-\sqrt{6}, 3)$, $(-5+\sqrt{6}, 3)$

21. **(a)** If q is the length of the focal chord, then $(c, q/2)$ belongs to the ellipse $c = \sqrt{a^2 - b^2}$; substitute into the equation of the ellipse and solve for q; **(b)** $\frac{9}{2}$ units

23. Length $20 + 3\sqrt{39}$ m, $2c = 3\sqrt{39}$ m

25. $2c = 4\sqrt{217}$ m **27.** $8\sqrt{901} \approx 240$ m

29. 3,107,000 mi

Problem Set 7.8, page 468

1. $V = (0, 0)$, $F = (1, 0)$, D: $x = -1$, F.C. $= 4$

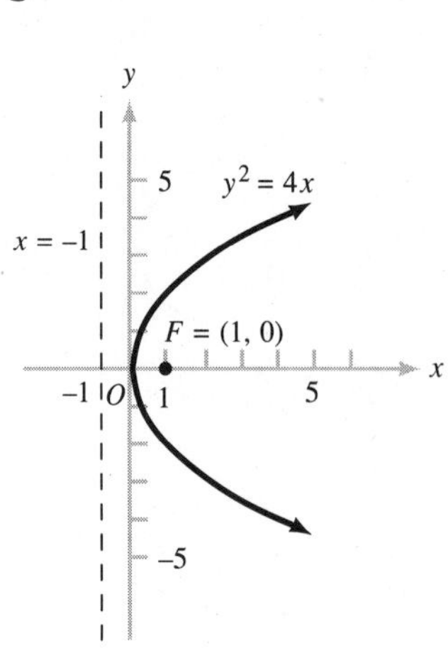

3. $V = (0, 0)$, $F = (0, \frac{1}{4})$, D: $y = -\frac{1}{4}$, F.C. $= 1$

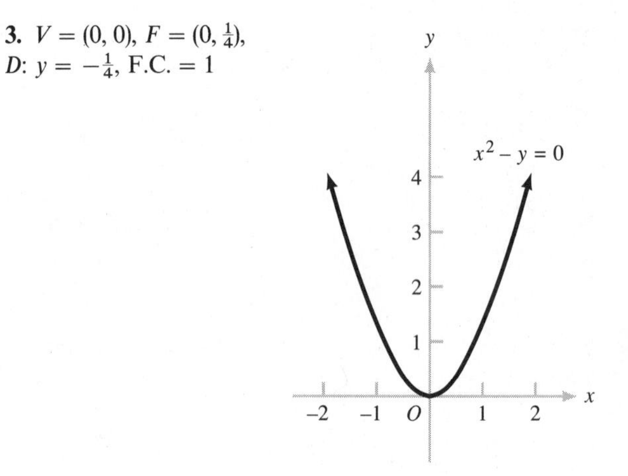

5. $V = (0, 0)$, $F = (0, -\frac{9}{4})$, D: $y = \frac{9}{4}$, F.C. $= 9$

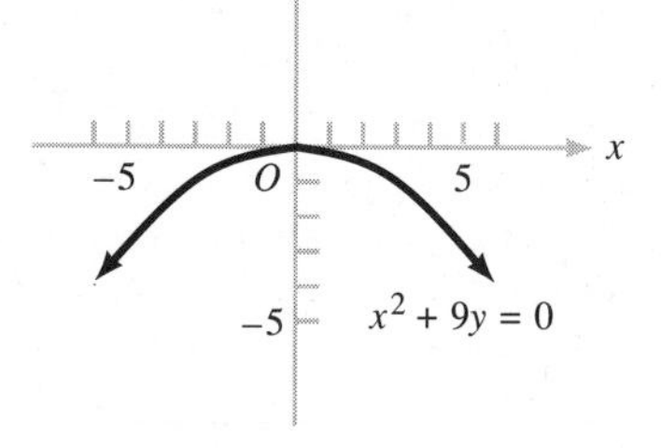

7. $y = \frac{1}{12}x^2$

9. $V = (-3, 2)$, $F = (-1, 2)$, D: $x = -5$, F.C. $= 8$

11. $V = (4, -7)$, $F = (4, -4)$, D: $y = -10$, F.C. $= 12$

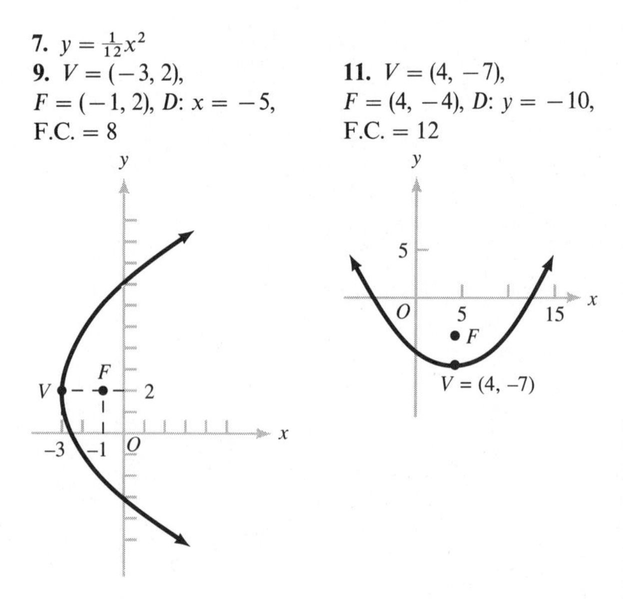

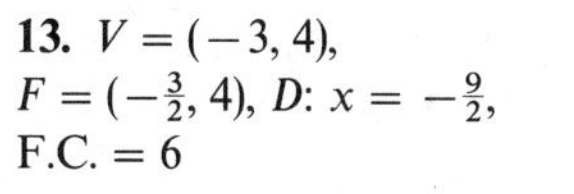

13. $V = (-3, 4)$, $F = (-\frac{3}{2}, 4)$, D: $x = -\frac{9}{2}$, F.C. $= 6$

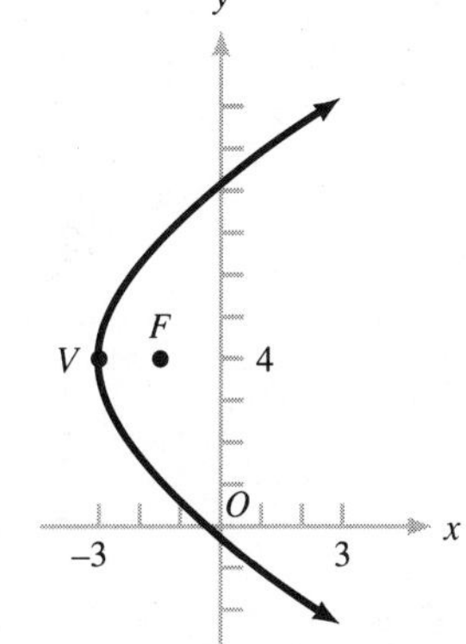

15. $V = (3, -1)$, $F = (3, 1)$, D: $y = -3$, F.C. $= 8$

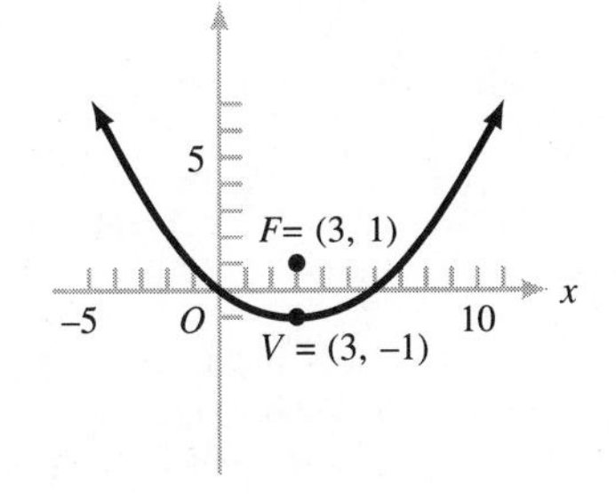

17. $x - 5 = -\frac{1}{4}(y - 2)^2$ **19.** $x + 6 = \frac{1}{32}(y + 5)^2$
21. $x + \frac{1}{2} = \frac{1}{8}(y + 1)^2$ **23.** $x^2 = 2560y$
25. 58, 28, 10, 4, 10, 28, and 58 m
27. Equation of the parabola is $y = x^2/4p$, and $(a/2, b)$ is a point on the parabola. Substitute $x = a/2$, $y = b$ into the equation and solve for p to get $p = a^2/(16b)$.
29. $\frac{8}{3}$ cm **31.** $(x - 15)^2 = -\frac{9}{2}(y - 50)$

Problem Set 7.9, page 476

1. Vertices $(3, 0)$, $(-3, 0)$; foci $(\sqrt{13}, 0)$, $(-\sqrt{13}, 0)$; asymptotes $y = \pm\frac{2}{3}x$

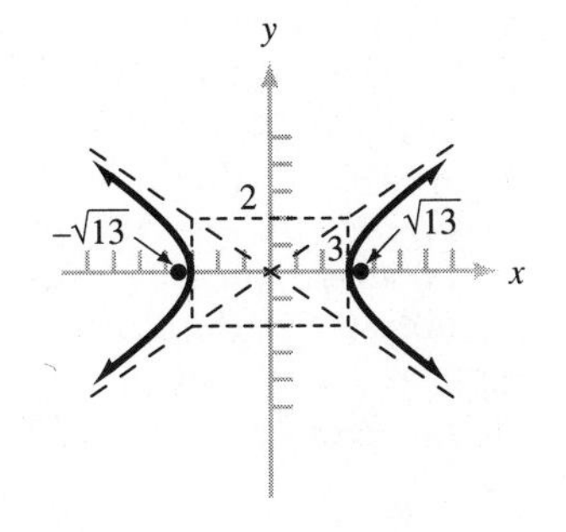

3. Vertices $(0, 4)$, $(0, -4)$; foci $(0, -2\sqrt{5})$, $(0, 2\sqrt{5})$; asymptotes $y = \pm 2x$

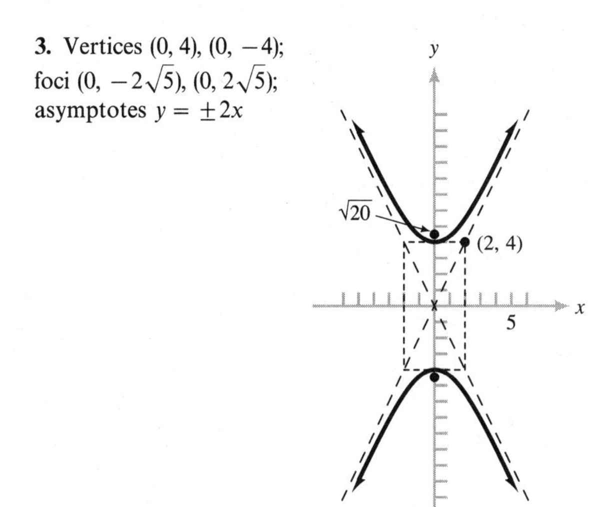

5. Vertices $(4, 0)$, $(-4, 0)$; foci $(2\sqrt{5}, 0)$, $(-2\sqrt{5}, 0)$; asymptotes $y = \pm\frac{1}{2}x$

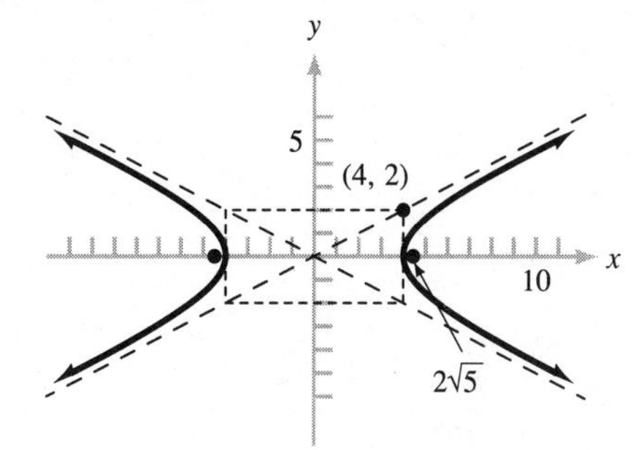

7. Vertices $(0, -\sqrt{10})$, $(0, \sqrt{10})$; foci $(0, -\sqrt{46})$, $(0, \sqrt{46})$; asymptotes $y = \pm\dfrac{\sqrt{10}}{6}x$

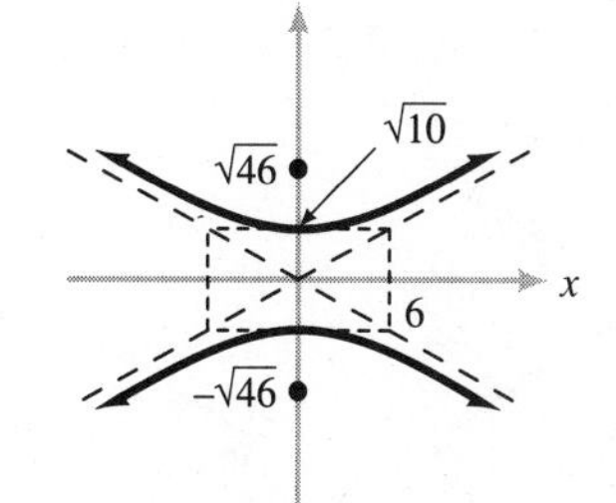

9. $\dfrac{x^2}{16} - \dfrac{y^2}{20} = 1$ **11.** $\dfrac{x^2}{16} - \dfrac{y^2}{25} = 1$

13. Center $(1, -2)$; vertices $(4, -2), (-2, -2)$; foci $(1 + \sqrt{13}, -2), (1 - \sqrt{13}, -2)$; asymptotes $3y - 2x + 8 = 0, 3y + 2x + 4 = 0$

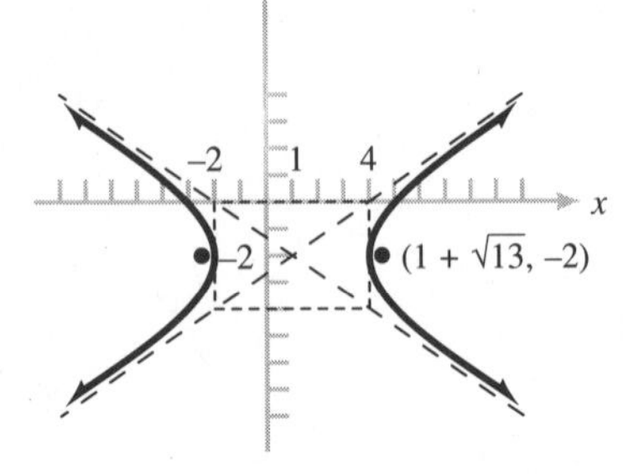

15. Center $(-2, -1)$; vertices $(-2, 3), (-2, -5)$; foci $(-2, -1 - \sqrt{41}), (-2, -1 + \sqrt{41})$; asymptotes $5y - 4x - 3 = 0, 5y + 4x + 13 = 0$

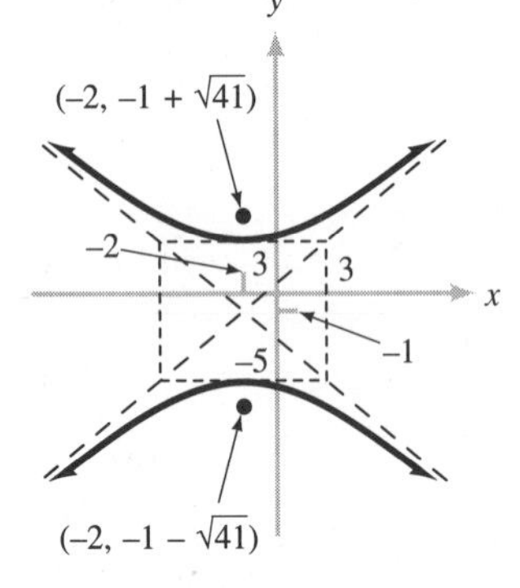

17. Center $(2, -1)$; vertices $(4, -1), (0, -1)$; foci $(2 - \sqrt{5}, -1), (2 + \sqrt{5}, -1)$; asymptotes $2y - x + 4 = 0, 2y + x = 0$

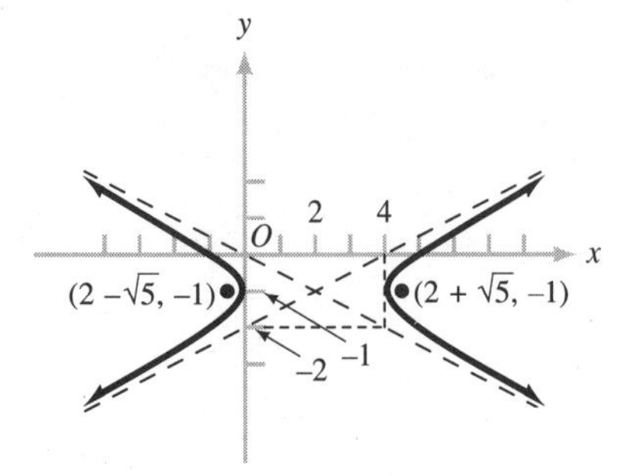

19. Center $(-4, -2)$; vertices $(-4, 1), (-4, -5)$; foci $(-4, -2 - \sqrt{34}), (-4, -2 + \sqrt{34})$; asymptotes $5y - 3x - 2 = 0, 5y + 3x + 22 = 0$

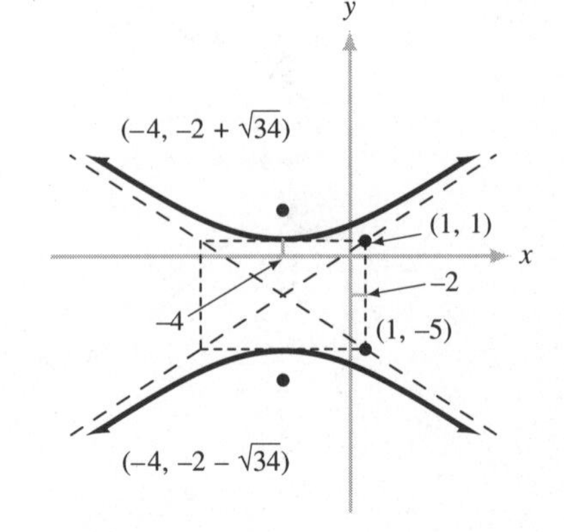

21. **(a)** $\dfrac{(x-4)^2}{1} - \dfrac{(y+1)^2}{8} = 1$; **(b)** $\dfrac{(x+2)^2}{4} - \dfrac{(y-3)^2}{\frac{9}{4}} = 1$; **(c)** $\dfrac{(y-3)^2}{25} - \dfrac{(x-2)^2}{11} = 1$

23. $x^2 - y^2 = a^2$

25. The hyperbola begins to look more and more like the pair of intersecting lines.

27. Points on right-hand branch of hyperbola $\dfrac{b^2x^2}{h^2s^2} - \dfrac{b^2y^2}{h^2(b^2 - s^2)} = 1$ (except at vertex, where bullet will pass)

29. $\dfrac{x^2}{1600} - \dfrac{y^2}{900} = 1, x > 0, y > 0$

33. $F_1 = (-a\sqrt{1 + m^2}, 0)$

35. P moves along the hyperbola $\dfrac{x^2}{a^2} - \dfrac{y^2}{a^2k} = 1$

Review Problem Set

CHAPTER 7, page 479

5. $32 + 80x + 80x^2 + 40x^3 + 10x^4 + x^5$

7. $2187x^{14} - 10{,}206x^{12}y^2 + 20{,}412x^{10}y^4 - 22{,}680x^8y^6 + 15{,}120x^6y^8 - 6048x^4y^{10} + 1344x^2y^{12} - 128y^{14}$

9. **(a)** $5376x^6y^3$; **(b)** $61{,}236x^5y^5$ **11.** $-\frac{1}{2}, \frac{1}{3}, -\frac{1}{4}, \frac{1}{5}, -\frac{1}{6}$

13. $0, \frac{1}{2}, 0, \frac{1}{8}, 0, \frac{1}{48}$ **15.** $\frac{4567}{90}$ **17.** 110 **19.** $\frac{7}{4}$

21. **(a)** $4n - 2$; **(b)** 450 **23.** **(a)** $(10 - n)/3$; **(b)** $\frac{44}{3}$

25. **(a)** $48(2)^{n-1}$; **(b)** 12,240 **27.** **(a)** $\frac{3}{4}(4)^{n-1}$; **(b)** $(4^{10} - 1)/4$ **29.** 9 **31.** $(7 + \sqrt{35})/2$

33. **(a)** $\frac{83}{99}$; **(b)** $\frac{91}{99}$ **35.** **(a)** 360; **(b)** 110; **(c)** 120; **(d)** 84;

(e) 1 **37.** \$50.50 **39.** \$3000 **41.** 1200 cm
43. $\frac{1}{6}n(n+1)(2n+1)$ **45.** 12 **47.** 120 **49.** 325
51. 66 **53.** ${}_{25}P_{20} \approx 1.2926 \times 10^{23}$ **55.** 27,720
57. $\frac{1}{3}$ **59.** $\frac{3}{4}$ **61.** $\frac{1}{16}$ **63.** **(a)** 1 to 35; **(b)** $-\frac{1}{6}$
65. Ellipse; center (0, 0); vertices $(\pm 2\sqrt{2}, 0)$, $(0, \pm 2\sqrt{3})$; foci $(0, \pm 2)$
67. Parabola; vertex (0, 0); focus (1, 0); directrix $x = -1$
69. Hyperbola; center (0, 0); vertices $(\pm 6\sqrt{2}, 0)$; foci $(\pm 4\sqrt{5}, 0)$; asymptotes $y = \pm\frac{1}{3}x$
71. Ellipse; center $(-1, 1)$; vertices $(4, 1)$, $(-6, 1)$, $(-1, 4)$, $(-1, -2)$; foci $(-5, 1)$, $(3, 1)$
73. Ellipse; center $(-4, 6)$; vertices $(0, 6)$, $(-8, 6)$, $(-4, 0)$, $(-4, 12)$; foci $(-4, 6 + 2\sqrt{5})$, $(-4, 6 - 2\sqrt{5})$
75. Parabola; vertex (0, 1); focus (0, 0); directrix $y = 2$
77. Hyperbola; center $(-2, 3)$; vertices $(2, 3)$, $(-6, 3)$; foci $(-2 + 2\sqrt{5}, 3)$, $(-2 - 2\sqrt{5}, 3)$; asymptotes $y - 3 = \pm\frac{1}{2}(x + 2)$
79. Parabola; vertex $(0, -1)$; focus (0, 1); directrix $y = -3$
81. Hyperbola; center (1, 3); vertices $(1, \frac{7}{2})$, $(1, \frac{5}{2})$; foci $\left(1, 3 + \frac{\sqrt{5}}{2}\right)$, $\left(1, 3 - \frac{\sqrt{5}}{2}\right)$; asymptotes $y - 3 = \pm\frac{1}{2}(x - 1)$
83. Circle; center $(2, -1)$; radius 3
85. $\frac{(x-1)^2}{16} + \frac{(y-1)^2}{25} = 1$ **87.** $x = -(y + 6)^2$
89. $\frac{x^2}{\frac{3}{4}} - \frac{y^2}{3} = 1$ **91.** $y = \frac{4H}{L^2}x^2$
93. From left to center: 60 m, $\frac{386}{9}$ m, $\frac{260}{9}$ m, 18 m, $\frac{92}{9}$ m, $\frac{50}{9}$ m, 20 m; from center to right: 20 m, $\frac{50}{9}$ m, $\frac{92}{9}$ m, 18 m, $\frac{260}{9}$ m, $\frac{386}{9}$ m, 60 m
95. $\frac{x^2}{5476} + \frac{y^2}{1764} = 1$ **97.** $\frac{x^2}{52{,}441} + \frac{y^2}{38{,}025} = 1$

Chapter 7 Test, page 483

2. $243x^5 - 810x^4y^2 + 1080x^3y^4 - 720x^2y^6 + 240xy^8 - 32y^{10}$ **3.** **(a)** $15{,}120a^3b^4$; **(b)** $489{,}888x^4y^5$
4. $\frac{1}{6}, -\frac{1}{13}, \frac{1}{22}, -\frac{1}{33}, \frac{1}{46}$ **5.** $\frac{3}{64}$; $\frac{1533}{64}$
6. **(a)** $\frac{4567}{90}$; **(b)** 2 **7.** **(a)** 84; **(b)** 60 **8.** \$16,500
9. $1000(1.15)^8 \approx 3059$ **10.** ${}_8P_3 = 336$
11. $({}_7C_3)({}_5C_2) = 350$ **12.** **(a)** $\frac{1}{1000}$; **(b)** 1 to 999; **(c)** -0.25

13. Vertex (0, 0); focus (1, 0); parabola

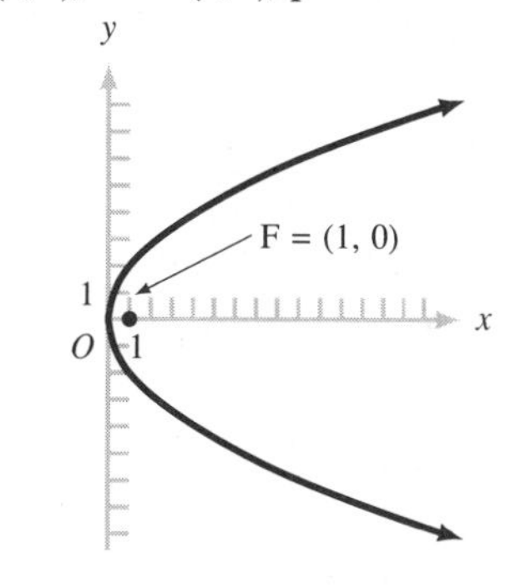

14. Vertex $(-5, -2)$; focus $(-\frac{39}{8}, -2)$; parabola

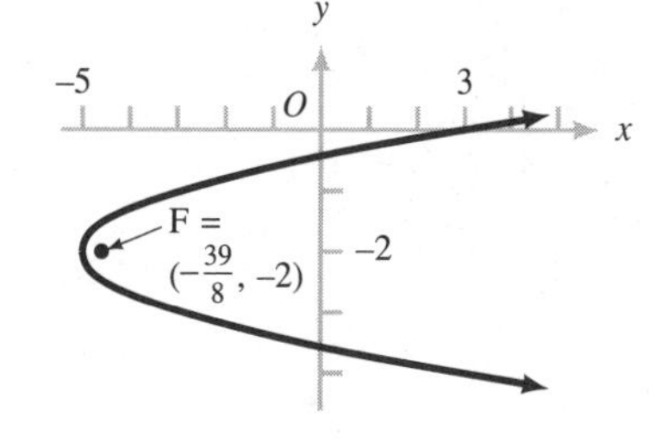

15. Center (0, 0); vertices $(\pm 2\sqrt{3}/3, 0)$, $(0, \pm\sqrt{2})$; foci $(0, \pm\sqrt{6}/3)$; ellipse

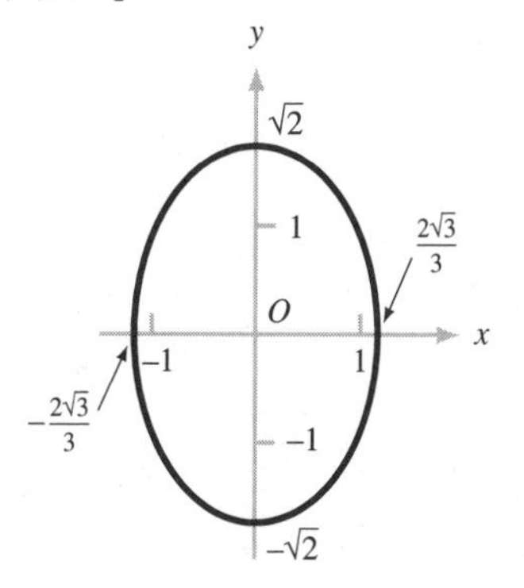

16. Center $(1, -2)$; vertices $(-1, -2)$, $(3, -2)$, $(1, -10)$, $(1, 6)$; foci $(1, -2 \pm 2\sqrt{15})$; ellipse

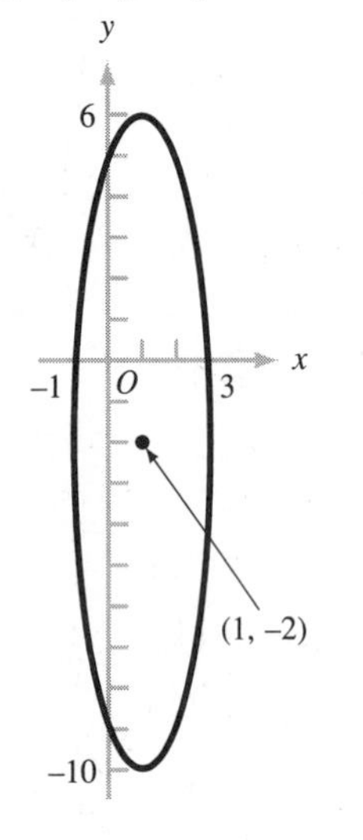

17. Center $(0, 0)$; vertices $(0, \pm 4)$; foci $(0, \pm 5)$; asymptotes $y = \pm\frac{4}{3}x$; hyperbola

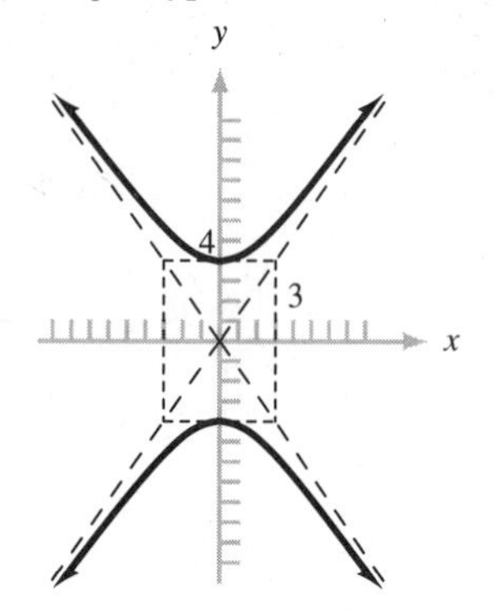

18. Center $(1, 1)$; vertices $(-1, 1)$, $(3, 1)$; foci $(1 \pm \sqrt{5}, 1)$; asymptotes $y - 1 = \pm\frac{1}{2}(x - 1)$; hyperbola

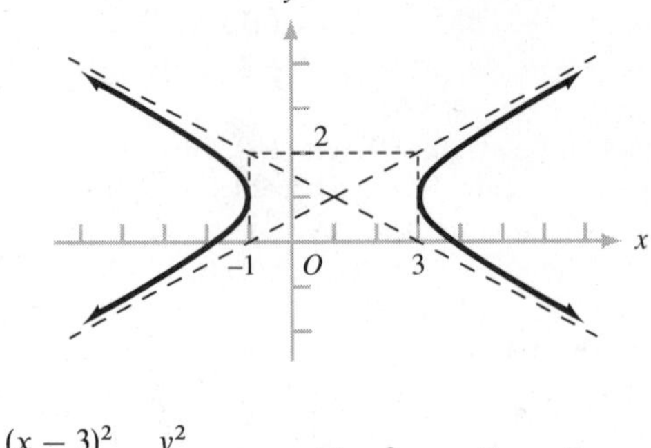

19. $\dfrac{(x-3)^2}{9} + \dfrac{y^2}{25} = 1$ **20.** $y^2 = -4(x - 1)$

21. $\dfrac{x^2}{9} - \dfrac{y^2}{16} = 1$ **22.** Hyperbola

Photo Credits

Cover John Giannicchi/Science Source Photo Photo Researchers

CHAPTER 1

p. 1 Opener © NASA; p. 11 © Blake J. Discher; p. 16 © Culver Pictures, Inc.; p. 18 © The Bettmann Archive; p. 22 Courtesy of Coin World; p. 24 © Charles Kennard/Stock, Boston, Inc.; p. 47 © Allen Rokach/Animals, Animals; p. 57 Courtesy of National Optical Astronomy Observatories; p. 64 © The Bettmann Archive; p. 65 © The Bettmann Archive; p. 69 Courtesy of U.S. Postal Service; p. 72 © Peter Menzel/Stock, Boston, Inc.

CHAPTER 2

p. 77 Opener © Catherine Ursillo/Photo Researchers; p. 89 © Harriet Gans/The Image Works; p. 92 © T.D. Lovering/Stock, Boston, Inc.; p. 93 (left) © Lionel Delevingne/Stock, Boston, Inc.; (right) © Mary Evans Picture Library/Photo Researchers; p. 94 © George Bellerose/Stock, Boston, Inc.; p. 95 (top left) © Peter Menzel/Photo Researchers; (bottom left) © Donald Dietz/Stock, Boston, Inc.; (right) © Focus on Sports; p. 103 © Margot Granitsas/Photo Researchers; p. 105 (left) © Michael Hayman/Stock, Boston, Inc.; (right) © Gary Walts/The Image Works; p. 106 © Fritz Henle/Photo Researchers; p. 111 © Owen Franken/Stock, Boston, Inc.; p. 122 (left) © Jim Fossett/The Image Works; (right) © Will and Deni McIntyre/Photo Researchers; p. 141 © Jim Pickerell/Stock, Boston, Inc.; p. 142 (top) Courtesy of Panasonic Company; (bottom) © UPI/Bettmann Newsphotos; p. 143 © Culver Pictures, Inc.

CHAPTER 3

p. 147 Opener © Ira Kirschenbaum/Stock, Boston, Inc.; p. 147 © The Bettmann Archive; p. 155 Courtesy of U.S. Navy; p. 163 © Eric Neurath/Stock, Boston, Inc.; p. 166 (top) © Robert Houser/Photo Researchers; (bottom) Courtesy of GTE; p. 167 © Marty Snyderman; p. 173 © Custom Medical Stock Photo; p. 174 © Blake J. Discher; p. 178 Courtesy NASA; p. 179 © Blake J. Discher; p. 190 © Blake J. Discher; p. 206 © Ellis Herwig/Stock, Boston, Inc.

CHAPTER 4

p. 223 Opener © Spencer Grant/Stock, Boston, Inc.; p. 230 © George Bellerose/Stock, Boston, Inc.; p. 233 © Jeff Albertson/Stock, Boston, Inc.; p. 238 © Blake J. Discher; p. 250 © Phaneuf/Gurdziel/The Picture Cube; p. 269 © The Bettmann Archive/Hulton; p. 276 © Blake J. Discher; p. 287 © Carl Hanninen/Photo/Nats

CHAPTER 5

p. 291 Opener © Teri Leigh Stratford/Photo Researchers; p. 292 © Blake J. Discher; p. 298 © Sam C. Pierson, Jr./Photo Researchers; p. 307 © UPI/The Bettmann Archive; p. 308 Courtesy of Missouri Division of Tourism; p. 309 © Leonard Lee Rue III/Photo Researchers; p. 329 © Culver Pictures, Inc.; p. 330 Courtesy of AT&T Archives; p. 331 © UPI/Bettmann Newsphotos; p. 333 © Reuters/The Bettmann Archive;

p. 334 © UPI/The Bettmann Archive; p. 336 © Nita Winter/The Image Works; p. 338 © The Bettmann Archive; p. 343 © Michael J. Okoniewski/The Image Works; p. 346 © UPI/Bettmann Newsphotos; p. 347 © Julie O'Neil/Stock, Boston, Inc.

CHAPTER 6

p. 349 Opener © Joel Gordon; p. 357 © Alan Carey/The Image Works; p. 364 © Jeffrey Dunn Studios; p. 366 © The Bettmann Archive; p. 391 © Blake J. Discher; p. 406 © Mark Antman/The Image Works

CHAPTER 7

p. 413 Opener © Joel Gordon; p. 413 © Jim Pickerel/FPG International; p. 421 © The Bettmann Archive; p. 429 © Estate of Harold E. Edgerton/Courtesy of Palm Press, Inc.; p. 451 © Joel Gordon; p. 452 © Paul Silverman/Fundamental Photographs; p. 462 © The Church of Jesus Christ of Latter Day Saints; p. 463 © J. Waring Stinchcomb/FPG International; p. 464 (left) © Fundamental Photographs; (center) © Jeff Albertson/Stock, Boston, Inc.; (right) © Owen Franken/Stock, Boston, Inc.; p. 468 © Nita Winter/The Image Works; p. 470 © Fundamental Photographs; p. 470 (left) © Marley Cooling Tower Co.; (right) © University of California Regents; p. 480 © The Bettmann Archive; p. 482 © UPI/The Bettmann Archive

Index of Applications

(Problem numbers are enclosed in parentheses.)

Calculus

Chemistry

Computer Science

Earth Science

Economics

Education

Energy, Ecology, and Environment

Engineering

Everyday Life

Psychology

Sociology

Sports, Games, and Entertainment

Transportation

Index